Monographs on
Pathology of Laboratory Animals

Sponsored by the
International Life Sciences Institute

W0050834

The following volumes have appeared so far

Endocrine System
1983. 346 figures. XV, 366 pages. ISBN 3-540-11677-X

Respiratory System
1985. 279 figures. XV, 240 pages. ISBN 3-540-13521-9

Digestive System
1985. 352 figures. XVIII, 386 pages. ISBN 3-540-15815-4

Urinary System
1986. 362 figures. XVIII, 405 pages. ISBN 3-540-16591-6

Genital System
1987. 340 figures. XVII, 304 pages. ISBN 3-540-17604-7

Nervous System
1988. 242 figures. XVI, 233 pages. ISBN 3-540-19416-9

Integument and Mammary Glands
1989. 468 figures. XI, 347 pages. ISBN 3-540-51025-7

Hemopoietic System
1990. 351 figures. XVIII, 336 pages. ISBN 3-540-52212-3

Cardiovascular and Musculoskeletal Systems
1991. 390 figures. XVII, 312 pages. ISBN 3-540-53876-3

Eye and Ear
1991. 141 figures. XIII, 170 pages. ISBN 3-540-54044-X

Nonhuman Primates I
1993. 235 figures. XIII, 221 pages. ISBN 3-540-56465-9
                                              0-944398-15-4

Nonhuman Primates II
1993. 264 figures. XVI, 248 pages. ISBN 3-540-56527-2
                                              0-944398-16-2

2nd editions available

Endocrine System
1996. 521 figures. XVIII, 521 pages. ISBN 3-540-59477-9
                                              0-944398-64-2

Respiratory System
1996. 382 figures. XV, 357 pages. ISBN 3-540-60383-2
                                              0-944398-69-3

2nd edition to follow

Urinary System
1997. 360 figures. Approx. 450 pages. ISBN 3-540-61847-3
                                              0-944398-76-6

T.C. Jones   J.A. Popp   U. Mohr (Eds.)

Digestive System

Second Edition

Completely Revised and Updated
with 394 Figures and 30 Tables

Springer

Thomas Carlyle Jones, D.V.M., D.Sc.
Professor of Comparative Pathology
Emeritus, Harvard Medical School
ILSI Research Foundation
1126 Sixteenth Street. N.W., Washington, DC 20036, USA

James A. Popp, D.V.M., Ph.D.
International Director of Toxicology
Vice President, Toxicology U.S.
Sanofi Winthrop
Nine Great Valley Parkway
Malvern, PA 19355, USA

Ulrich Mohr, M.D.
Professor of Experimental Pathology
Medizinische Hochschule Hannover
Institut für Experimentelle Pathologie
Konstanty-Gutschow-Strasse 8
30625 Hannover, Germany

Distribution rights for North America, Canada, and Mexico by
International Life Sciences Institute (ILSI)
1126 Sixteenth Street NW, Washington, DC 20036, USA

2nd Edition

ISBN 978-0-944398-75-3   ISBN 978-3-662-25996-2 (eBook)
DOI 10.1007/978-3-662-25996-2

1st Edition

Library of Congress Cataloging-in-Publication Data. Digestive system/T.C. Jones,
J.A. Popp, U. Mohr, (eds.). – 2nd ed.   p.   cm. – (Monographs on pathology of laboratory
animals) Includes bibliographical references and index. 1. Laboratory animals – Diseases.
2. Veterinary gastroenterology. 3. Digestive organs – Diseases. 4. Rodents – Diseases.
5. Rodents as laboratory animals. I. Jones, Thomas Carlyle. II. Popp, James A.,
1945– . III. Mohr, U. (Ulrich) IV. Series. SF996.5.D54 1996 616.3 – dc20 96-
27162

Cover design: E. Kirchner, Springer-Verlag

Typesetting: Best-set Typesetter Ltd., Hong Kong

SPIN: 10521008   25/3134/SPS – 5 4 3 2 1 0 – Printed on acid-free paper

Foreword to the Second Edition

The International Life Sciences Institute (ILSI) is a nonprofit, worldwide foundation established in 1978 to advance the understanding of scientific issues relating to nutrition, food safety, toxicology, risk assessment, and the environment. By bringing together scientists from academia, government, industry, and the public sector, ILSI seeks a balanced approach to solving problems of common concern for the well-being of the general public. This volume is the third of the Second Edition of *Monographs on Pathology of Laboratory Animals*. The series is designed to facilitate communication among those involved in the safety testing of foods, drugs, and chemicals. The complete set covers cardiovascular/musculoskeletal, digestive, endocrine, genital, hemopoietic, nervous, respiratory and urinary systems, eye and ear, integument and mammary glands, and nonhuman primates. The series is intended for use by pathologists, toxicologists, and others concerned with evaluating toxicity and carcinogenicity studies. ILSI is committed to supporting programs to harmonize toxicologic testing, to advance a more uniform interpretation of bioassay results worldwide, to promote a common understanding of lesion classifications, and to encourage wide discussion of these topics among scientists. Scientific understanding and cooperation will be improved worldwide through the series and this international project.

ILSI accomplishes its work through its branches and institutes. ILSI's branches currently include Argentina, Australasia, Brazil, Europe, Japan, Korea, Mexico, North America, Southeast Asia, and Thailand, and a focal point in China. The ILSI Health and Environmental Sciences Institute focuses on global environmental issues. ILSI Research Foundation includes the ILSI Allergy and Immunology Institute, ILSI Human Nutrition Institute, ILSI Pathology and Toxicology Institute, and ILSI Risk Science Institute.

Alex Malaspina
President
International Life Sciences Institute

Preface to the Second Edition

During the years that have intervened since the third volume of the International Life Sciences Institute (ILSI) Monographs on Pathology of Laboratory Animals, Digestive System, was published, new information of interest to pathologists has developed at a rather remarkable pace. Standardized nomenclature has been proposed and gained significant acceptance during the period since the first edition and is being utilized on an international basis. This has resulted in improvement in communication of pathologic data to regulatory agencies and in scientific publications worldwide. This monograph series and others sponsored by ILSI have had significant effects on improved communications and the international acceptance of standardized nomenclature.

In this second edition, new formats have been used in some instances where more appropriate for the subjects to be covered. The format introduced in the first edition remains useful as it emphasizes the necessity to recognize the morphologic features of pathologic lesions to identify them precisely. This identification is an essential first step toward development of new insights into pathogenetic mechanisms and their use in decisions eventually applicable to public health.

New information is included in this edition on the nature and variability of preneoplastic lesions in the liver of laboratory rodents. Data on the accompanying changes in enzyme activity in affected liver cells are expanded, and additional information on spongiosis hepatis in the rat and its relation to spongiotic pericytoma is a further feature. In a few instances, research on a pathologic entity has been limited but its recognition remains important. This justifies inclusion of such entities in the new edition.

We are very grateful to the dedicated scientists from lands all around the world who have contributed to this volume. The authors are named in the list of contributors to this volume, in the table of contents, and at the heading of each individual manuscript. The members of the editorial board are listed in the title pages. They are particularly to be thanked for their efforts in identifying authors and subject areas and for the scientific review of individual manuscripts. The editors are especially grateful for the steadfast support of Dr. Alex Malaspina, and to members of the ILSI staff and others who have helped in so many ways. We particularly wish to mention Ms. Sharon Weiss, the Associate Director as well as Ms. Frances DeLuca, Executive Assistant, ILSI Pathology and Toxicology Institute and ILSI Research Foundation.

We are grateful to Prof. Dr. Dietrich Goetze, Prof. Dr. Thomas Thiekötter, Ms. Barbara Montenbruck, Dr. Agnes Heinz and others on the staff of Springer-Verlag for the quality of the finished book.

August 1996

T.C. Jones
J.A. Popp
U. Mohr

Contents

Contributors

Goro Asano, M.D.
Professor, Second Department of Pathology, Nippon Medical School,
Tokyo, Japan

Peter Bannasch, Prof.Dr.med.
Head, Division of Cell Pathology, Deutsches Krebsforschungszentrum,
Heidelberg, Germany

Stephen W. Barthold, D.V.M., Ph.D.
Professor of Comparative Medicine, Yale University School
of Medicine, New Haven, Connecticut, USA

Hendrick G. Bedigian, Ph.D.
Director of Quality Control Laboratories, Senior Staff Scientist,
The Jackson Laboratory, Bar Harbor, Maine, USA

Josep A. Bombi, M.D.
Associate Professor, Department of Anatomic Pathology, Hospital
Clinic, University of Barcelona Medical School, Barcelona, Spain

Gary A. Boorman, D.V.M., M.S., Ph.D.
Chief, Pathology Branch, National Institute of Environmental Health
Sciences, Research Triangle Park, North Carolina, USA

Paul N. Brooks, BSc, MIBiol, CBiol
Saffron Walden, Essex, England

Gary T. Burger, D.V.M.
R.J. Reynolds Tobacco Company, Winston-Salem, North Carolina,
USA

Antonio Cardesa, M.D.
Professor and Chairman, Department of Anatomic Pathology,
Hospital Clinic, University of Barcelona Medical School, Barcelona,
Spain

Russell C. Cattley, V.M.D., Ph.D.
Scientist, Chemical Industry Institute of Toxicology, Research Triangle
Park, North Carolina, USA

Clare G. Collier, Ph.D., M.I.Biol.
AEA Technology, Didcot, Oxfordshire, England

Linden E. Craig, D.V.M., Diplomate A.C.V.P.
Post-Doctoral Fellow, Division of Comparative Medicine, Johns
Hopkins University School of Medicine, Baltimore, Maryland, USA

John M. Cullen, V.M.D., Ph.D.
Department of Microbiology, Pathology and Parasitology,
College of Veterinary Medicine, North Carolina State University,
Raleigh, North Carolina, USA

Bhalchandra A. Diwan, Ph.D.
Senior Scientist, SAIC Frederick, NCI/FCRDC, Frederick, Maryland,
USA

Natalia V. Engelhardt
Doctor of Biology, Cancer Research Center, Moscow, Russian
Federation

Scot L. Eustis, D.V.M., Ph.D.
SmithKline Beecham Pharmaceuticals, King of Prussia, Pennsylvania,
USA

Raymond Everett, D.V.M., Ph.D.
Sanofi Winthrop Research Division, Malvern, Pennsylvania, USA

Pedro Luis Fernandez, M.D.
Assistant Professor, Department of Anatomic Pathology, Hospital
Clinic, University of Barcelona Medical School, Barcelona, Spain

James G. Fox, D.V.M.
Professor and Director, Division of Comparative Medicine,
Massachusetts Institute of Technology, Cambridge, Massachusetts,
USA

Craig L. Franklin, D.V.M., Ph.D.
Department of Veterinary Pathobiology, University of Missouri
Columbia, Missouri, USA

Charles H. Frith, D.V.M., Ph.D.
Consultant, Toxicology Pathology Associates, Little Rock, Arkansas,
USA

Shoji Fukushima, M.D., D.M.S.
Professor, Department of Pathology, Osaka City University Medical
School, Osaka, Japan

John E. Greenlee, M.D.
Chief, Neurology Service, Veteran Affairs Medical Center, Professor
and Vice Chairman, Department of Neurology, The University
of Utah School of Medicine, Salt Lake City, Utah, USA

Hans Jörg Hacker, Dr.rer.nat.
Division of Cell Pathology, Deutsches Krebsforschungszentrum,
Heidelberg, Germany

James E. Heath, D.V.M.
Senior Pathologist, Southern Research Institute, Birmingham,
Alabama, USA

David E. Hinton, Ph.D.
Department of Medicine, School of Veterinary Medicine, Anatomy, Physiology & Cell Biology, University of California/Davis, Davis, California, USA

Masao Hirose, M.D.
Associate Professor, Nagoya City University Medical School, Nagoya, Japan

Carel F. Hollander, M.D., Ph.D.
Senior Director, Centre de Recherche, Laboratoires Merck Sharp & Dohme, Riom, France

Robert O. Jacoby, D.V.M., Ph.D.
Professor and Chairman, Section of Comparative Medicine, Yale University School of Medicine, New Haven, Connecticut, USA

Richard Kociba, D.V.M., Ph.D.
Toxicology Research Laboratory, Dow Chemical Company, Midland, Michigan, USA

Yoichi Konishi, M.D.
Professor, Department of Oncological Pathology, Cancer Center, Nara Medical University, Nara, Japan

Norio Matsukura, M.D.
Assistant Professor, First Department of Surgery, Nippon Medical School, Tokyo, Japan

Paul M. Newberne, DVM, Ph.D.
Professor of Pathology, Mallory Institute of Pathology, Boston University School of Medicine, Boston, Massachusetts, USA

Akiyoshi Nishikawa, M.D., Ph.D.
Section Chief, Division of Pathology, National Institute of Health Sciences, Tokyo, Japan

Maria Yolanda Ovelar, M.D.
Assistant Professor, Department of Anatomic Pathology, Hospital Clinico, University of Valladolid Medical School, Valladolid, Spain

Manuel Pera, M.D.
Consultant Surgeon, Department of Surgery, Hospital Clinic, University of Barcelona Medical School, Barcelona, Spain

James A. Popp, D.V.M., Ph.D.
International Director of Toxicology, Vice President, Toxicology U.S., Sanofi Winthrop, Malvern, Pennsylvania, USA

Jerold E. Rehg, D.V.M.
Director, Comparative Medicine Division, St. Jude Children's Research Hospital, Memphis, Tennessee, USA

Jerry M. Rice, Ph.D.
Chief, Laboratory of Comparative Carcinogenesis, National Cancer
Institute, Frederick, Maryland, USA

Lela K. Riley, Ph.D.
Associate Professor, Department of Veterinary Pathobiology,
University of Missouri, Columbia, Missouri, USA

Francis J.C. Roe, DM(Oxon), DSc(London), FRCPath, FATS
London, England

Adrianne E. Rogers, M.D.
Department of Pathology and Laboratory Medicine, Boston
University Medical Center, Boston, Massachusetts, USA

Boris H. Ruebner, M.D.
Department of Medical Pathology, University of California/Davis,
Davis, California, USA

Dante G. Scarpelli, M.D., Ph.D.
Pathology Department, Northwestern University Medical School,
Chicago, Illinois, USA

Robert C. Sills, D.V.M., Ph.D.
Veterinary Pathologist, National Institute of Environmental Health
Sciences, Research Triangle Park, North Carolina, USA

A.J. Spencer, BVMS, MRCVS, Ph.D., FRCPath
Sanofi Winthrop, Alnwick, Northumberland, England

John D. Strandberg, D.V.M., Ph.D.
Associate Professor and Director, Division of Comparative Medicine,
Johns Hopkins University School of Medicine, Baltimore, Maryland,
USA

János Sugar, M.D., D.MsC.
Budapest, Hungary

John P. Sundberg, D.V.M., Ph.D.
Head of Pathology, The Jackson Laboratory, Bar Harbor, Maine,
USA

Zoltán Szentirmay, M.D.
National Oncological Institute, Research Institute of Oncopathology,
Budapest, Hungary

Michihito Takahashi, M.D., Ph.D.
Director, Division of Pathology, National Institute of Health Sciences,
Tokyo, Japan

James W. Townsend, Ph.D.
Director of Computer Operations for Anatomic Pathology,
Department of Pathology, University of Arkansas for Medical
Sciences, Little Rock, Arkansas, USA

Vladimir S. Turusov, M.D.
Professor, Cancer Research Center, Russian Academy of Medical
Science, Moscow, Russian Federation

Matthew J. van Zwieten, D.V.M., Ph.D.
Department of Safety Assessment, Merck Research Labs, West Point,
Pennsylvania, USA

Hideki Wanibuchi, M.D.
Assistant Professor, Osaka City University Medical School, Osaka,
Japan

Jerrold M. Ward, D.V.M., Ph.D.
National Cancer Institute, Frederick, Maryland, USA

Klaus Wayss, Dr.rer.nat.
Division of Cell Biology, Deutsches Krebsforschungszentrum,
Heidelberg, Germany

Heide Zerban, Dr.rer.nat.
Division of Cell Pathology, Deutsches Krebsforschungszentrum
Heidelberg, Germany

The Liver

Foci of Altered Hepatocytes, Rat

Peter Bannasch, Heide Zerban, and Hans J. Hacker

Synonyms. Hyperplastic foci, phenotypically altered foci, enzyme-altered foci, preneoplastic foci

Gross Appearance

The foci of altered hepatocytes are usually not visible with the naked eye, but they can occasionally be recognized as small, white spots on the liver surface.

Microscopic Features

Various types of focal lesions composed of phenotypically altered hepatocytes have been observed prior to the appearance of hepatic adenomas (neoplastic nodules) and carcinomas (Bannasch 1968; Squire and Levitt 1975; Schauer and Kunze 1976; ILAR, NCR 1980; Bannasch and Zerban 1990; Pitot 1990). These foci of altered hepatocytes (FAH) are considered to be preneoplastic lesions which indicate an early response to carcinogenic agents; they may consequently be useful in evaluating carcinogenesis bioassays (Bannasch 1986; Montesano et al. 1986; Rinde et al. 1986; US National Institute of Environmental Health Sciences 1989; Ito et al. 1992). The classification shown in Table 1 is based on characteristic tinctorial changes of the hepatocytes, which are due to quantitative alterations in certain macromolecules or cytoplasmic organelles (i.e., glycogen, endoplasmic reticulum, ribosomes, mitochondria, peroxisomes) and can be readily detected in paraffin sections stained with hematoxylin and eosin (H&E). There is increasing evidence for differences in carcinogenic potential between types of foci (Bannasch 1968; Rabes et al. 1972; Emmelot and Scherer 1980;

Bannasch et al. 1985; Enzmann and Bannasch 1987; Weber et al. 1988a; Harada et al. 1989a; Zerban et al. 1989, 1994; Kraupp-Grasl et al. 1990; Weber and Bannasch 1994a–c; Marsman and Popp 1994; Metzger et al. 1995). Careful subclassification of foci of altered hepatocytes is important for improving the predictive value of histopathological data from carcinogenesis bioassays and for unravelling the process of hepatocarcinogenesis (Bannasch and Zerban 1992, 1994; Goodman et al. 1994).

Clear Cell Areas

In the context of the focal liver lesions, the term "area" was recommended for designating lesions approximately as large as or larger than a hepatic lobule (Squire and Levitt 1975). A slightly different definition has been given by Bannasch et al. (1982). According to this definition, clear cell areas (Fig. 1) are not sharply demarcated and occupy large portions of the liver parenchyma which are predominantly localized in the first and second zone of the functional liver acinus. The clear cells are enlarged and store more glycogen than normal hepatocytes. The glycogen can be demonstrated by the periodic acid-Schiff (PAS) reaction in alcohol-fixed material, but is water soluble and is eluted during the usual tissue preparations. Histochemical investigations of a number of enzymes of carbohydrate metabolism, such as glycogen synthase, glycogen phosphorylase, or glucose-6-phosphate dehydrogenase, revealed no significant changes in the activity of these enzymes in the clear cell areas that appear early after administration of N-nitrosomorpholine (Hacker et al. 1982). The same holds true for the activity of acid or alkaline nucleases (Taper and Bannasch 1976).

Table 1. Classification of foci of altered hepatocytes (FAH) in the rat according to cytomorphological and simple cytochemical criteria (from Bannasch and Zerban 1990)

Type of focus	Glycogen	Acidophilia	Basophilia
Clear cell focus	+++	o	o
Acidophilic cell focus	++	+++ Reticular	o
Amphophilic cell focus	–	+++ Granular	+ Scattered
Intermediate cell focus	++	o/++ Reticular	++ Patchy
Tigroid basophilic cell focus	–/o	o/+	++/+++ Tigroid
Mixed cell focus	+++→–[a]	+++→–[a]	o→+++[a]
Vacuolated cell focus	++→–[a] Abundant fat	++→–[a]	o→++[a]
Homogeneous basophilic cell focus	–	–	++/+++ Diffuse

o, normal; +, slightly increased; ++, increased; +++, strongly increased; –, strongly reduced.
[a] Depending on the cell type within the mixed cell population.

Clear and Acidophilic Cell Foci (Figs. 1–3)

In comparison to the clear cell areas, the foci are smaller but much more prominent (Fig. 1). They are usually well demarcated from the surrounding liver tissue, although the plates composed of the altered hepatocytes merge imperceptibly with those of the adjacent normal parenchyma. Sometimes the foci are localized within areas of clear cells. In this case, gradual transitions between foci and areas may considerably hamper classification. The clear cell foci consist of an abnormal cell population (Fig. 2) that stores glycogen in excessive amounts (glycogenosis) (Bannasch 1968). Klimek et al. (1984) have demonstrated by microdissection and biochemical microanalysis that the glycogenotic foci contain on average 100% more glycogen than normal hepatocytes. The clear hepatocytes are polyhedral and markedly enlarged. Their cytoplasm appears almost empty in sections stained with H&E. The basophilic bodies, which correspond to the rough endoplasmic reticulum, are displaced toward peripheral or paranuclear regions of the cell. The nuclei of the clear cells may be small and dense, but frequently their volume is considerably in-

creased, and they have less condensed chromatin and prominent nucleoli (Romen et al. 1972; Abmayr et al. 1983).

In addition to or in place of the clear cells, many glycogen storage foci contain acidophilic cells (Fig. 1d) which are especially voluminous (Bannasch 1968). Foci exclusively or predominantly composed of this cell type have been called acidophilic cell foci. The alternative designation of this type of focus as "eosinophilic" (Squire and Levitt 1975; ILAR, NCR 1980) restricts its definition to tissue sections stained with H&E, whereas the term "acidophilic" (and its counterpart "basophilic") indicates a biologic property that is independent of a specific staining procedure. Glycogen storage foci consisting of both clear and acidophilic cells should be classified as combined clear/acidophilic cell foci rather than mixed cell foci, which contain additional basophilic cells and represent a later stage of hepatocarcinogenesis (Bannasch and Zerban 1992). The characteristic ground glass appearance of the cytoplasm of the acidophilic cells is due to a proliferation of the smooth endoplasmic reticulum, as described for the first time by Porter and Bruni (1959). Typically, the nuclei of the acidophilic cells are enlarged and reveal basically the same texture as the large nuclei of the clear cells (Romen et al. 1972). Mitotic figures are rare in both clear and acidophilic cells and incorporation of [3H]thymidine into nuclear DNA is only slightly elevated over the normal level (Zerban et al. 1989, 1994). In contrast to the clear cell areas, a decrease or an increase of the activity of various enzymes is a characteristic histochemical feature of the foci which store excess glycogen (Fig. 3). Examples of enzymes with a decreased or increased activity in such foci are listed in Table 2.

However, enzyme histochemical patterns of the foci may be rather heterogeneous and appear to be influenced by many factors, such as the nutritional state of the animals, the circadian rhythm in the metabolism of hepatocytes, the method of induction of the foci, the localization of the foci within the liver lobule, and the time of investigation after the beginning of treatment with the respective carcinogen (Bannasch et al. 1980; Emmelot and Scherer 1980; Farber 1980; Pitot and Sirica 1980; Williams 1980; Moore and Kitagawa 1986; Pitot 1990). In this context, it should be mentioned that, in livers of animals that die spontaneously, the glycogen is rapidly broken down by autolytic processes and the clear and acidophilic

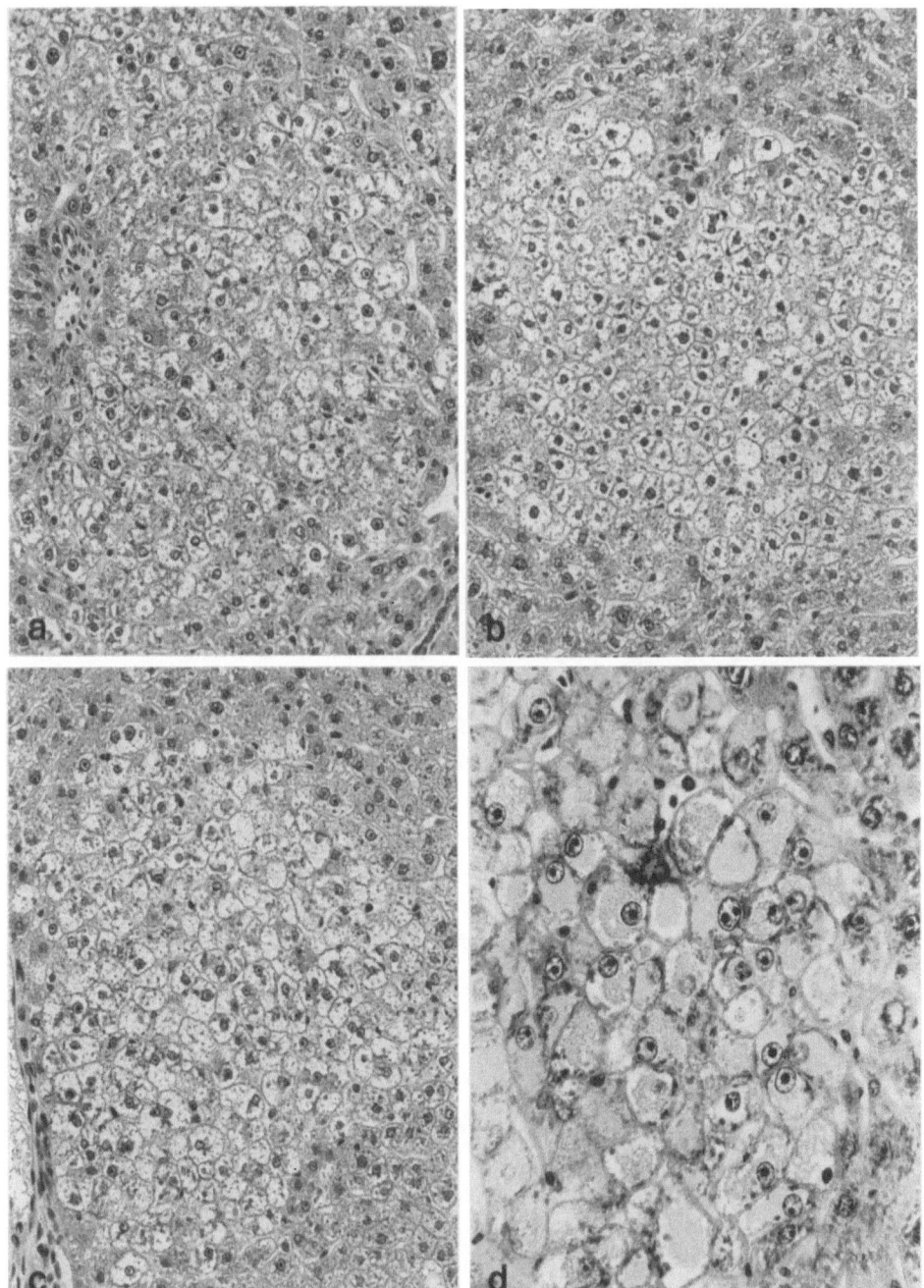

Fig. 1a–d. Early focal lesions induced in rat liver by *N*-nitrosomorpholine. **a** Clear cell area, not sharply demarcated. H&E, ×190. **b** Clear cell focus, well demarcated, composed of enlarged hepatocytes with dense nuclei. H&E, ×200. **c** Clear cell focus, well demarcated, composed of considerably enlarged hepatocytes. H&E, ×220. **d** Portion of a focus composed of very large acidophilic hepatocytes. H&E, ×350

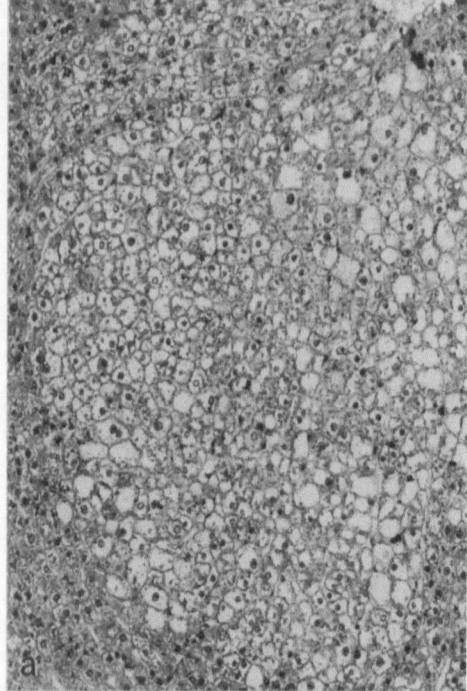

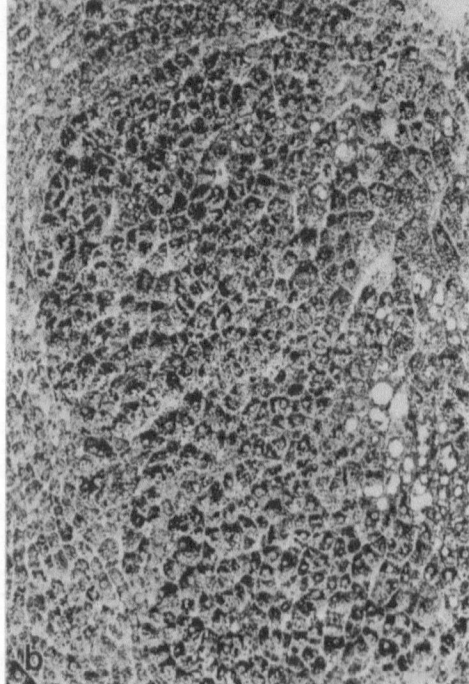

Fig. 2a,b. Glycogen storage focus induced in rat liver by *N*-nitrosomorpholine. **a** Focus composed of clear, acidophilic, and some vacuolated (fat-storing) cells. H&E, ×150. **b** Serial section demonstrating glycogen. Periodic acid-Schiff. ×150

Table 2. Selected cytochemical markers of foci of altered hepatocytes (FAH) in rats (from Bannasch and Zerban 1992)

Cytochemical marker	Type of focus	Reference
Increased content and/or activity		
Glycogen	CCF, ACF	Bannasch and Müller 1964
Glucose-6-phosphate dehydrogenase	Various types	Hacker et al. 1982
γ-Glutamyltransferase	Various types	Kalengayi and Desmet 1975
Glutathione *S*-transferase, placental form	n.s.	Sato et al. 1984
Uridine-diphosphate-glucuronyltransferase	n.s.	Fischer et al. 1983
Epoxide hydrolase	n.s.	Kuhlmann et al. 1981
	n.s.	Enomoto et al. 1981
Glucose-6-phosphatase	EHF	Friedrich-Freksa et al. 1969
Glycerol-3-phosphate dehydrogenase, mitochondrial form	EHF	Enzmann et al. 1989
Pyruvate kinase	EHF	Reinacher et al. 1986
Decreased content and/or activity		
Glycogen	BCF	Grundmann and Sieburg 1962
	TCF	Bannasch et al. 1985
	APF	Weber et al. 1988a
Glycogen phosphorylase	Various types	Hacker et al. 1982
Glucose-6-phosphatase	Various types	Gössner and Friedrich-Freksa 1964
Pyruvate kinase	n.s.	Reinacher et al. 1986
	BCF, MCF	Klimek and Bannasch 1990
Adenylate cyclase	Various types	Ehemann et al. 1986
Adenosine triphosphatase	Various types	Schauer and Kunze 1968
Tryptophane oxygenase	n.s.	Moore et al. 1986b
Iron in siderotic liver	Various types	Williams et al. 1976

CCF, clear cell focus; ACF, acidophilic cell focus; EHF, enzymatically hyperactive focus; BCF, basophilic cell focus; TCF, tigroid cell focus; APF, amphophilic cell focus; MCF, mixed cell focus; n.s., not specified.

storage cells can then no longer be detected. Disappearance of clear and acidophilic cell foci may also be due to glycogen reduction by prolonged starvation, cachexia, and additional neoplastic or infectious diseases (Bannasch and Zerban 1992).

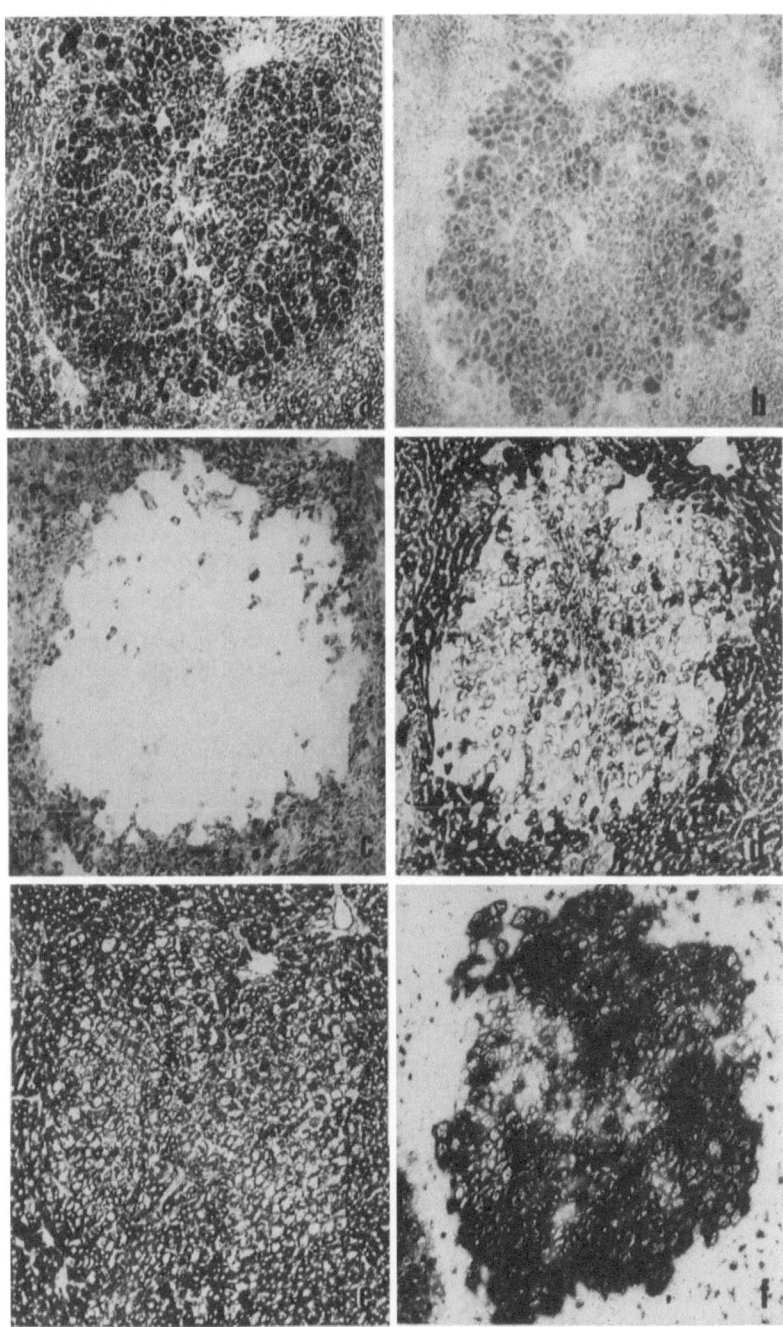

Fig. 3a–f. Serial sections through a glycogen storage focus induced in rat liver by *N*-nitrosomorpholine. **a** Periodic acid-Schiff (PAS) reaction. **b** Glycogen synthetase activity. **c** Glycogen phosphorylase activity. **d** Glucose-6-phosphatase activity. **e** Glyceraldehyde-3-phosphate dehydrogenase activity. **f** Glucose-6-phosphate dehydrogenase activity. (From Bannasch et al. 1984) ×60

Amphophilic Cell Foci

Amphophilic cell foci (Figs. 4a,b) have been described in rats treated with N-nitrosomorpholine and the adrenal hormone dehydroepiandrosterone (Weber et al. 1988a), and with dehydroepiandrosterone alone (Metzger et al. 1995), which has been shown to induce peroxisomal proliferation (Frenkel et al. 1990) and hepatocellular carcinomas (Rao et al. 1992; Hayashi et al. 1994; Metzger et al. 1995). The foci are composed of large cells with a homogeneous, granular acidophilic cytoplasm with faint, randomly scattered or nearly homogeneous basophilia (Fig. 4a). An increase in the activity of the mitochondrial enzyme succinate dehydrogenase and the peroxisomal enzyme catalase, as demonstrated histochemically in frozen sections, suggested that granular cytoplasmic acidophilia is due mainly to a multiplication of mitochondria and peroxisomes (Weber et al. 1988a; see also "Ultrastructure"). Other enzymes which have shown increased activity in amphophilic foci are glucose-6-phosphate dehydrogenase, glyceraldehyde-3-phosphate dehydro-

genase, and acid phosphatase, while glucose-6-phosphatase may be either increased or decreased, and both γ-glutamyltransferase and glutathione S-transferase, placental form (GST-P) are totally lacking (Bannasch et al. 1989). There is evidence for a progression from amphophilic foci to amphophilic/tigroid cell adenomas and highly differentiated hepatocellular carcinomas (Weber et al. 1988a, b; Metzger et al. 1995). Harada et al. (1989a) reported the induction of morphologically similar focal lesions (which they called "atypical" eosinophilic foci) in rats by the peroxisomal proliferator 1-amino-2,4-dibromoanthraquinone. The results of a combined morphological and stereological analysis of a 2-year carcinogenicity study suggested that these focal lesions were also involved in the development of hepatocellular carcinomas.

Vacuolated Cell Foci

Synonym. Fat storage foci

Vacuolated cell foci may be mistaken for clear cell foci in H&E-stained tissue sections (Bannasch and

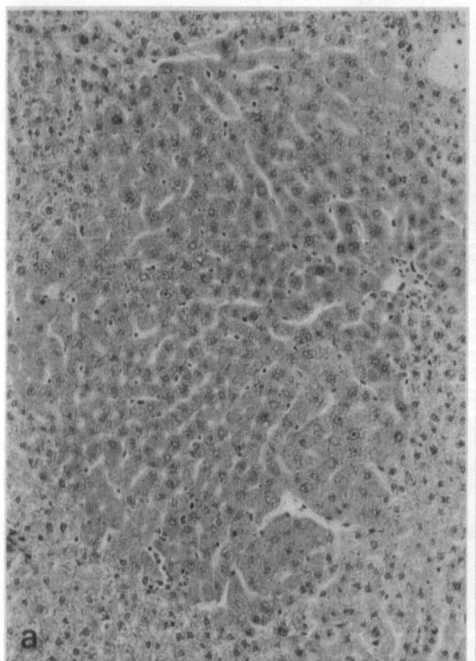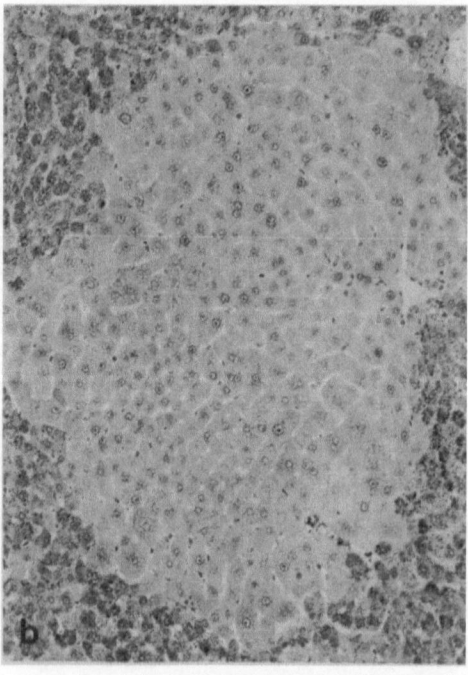

Fig. 4a,b. Serial sections through amphophilic cell focus induced in rat liver by N-nitrosomorpholine. **a** A well-demarcated focus composed of enlarged hepatocytes with a homogeneous granular acidophilic cytoplasm with faint ran-domly scattered basophilia. H&E, ×130. **b** Loss of glycogen from amphophilic cells compared to glycogen-rich surrounding parenchyma. Periodic acid-Schiff, ×130

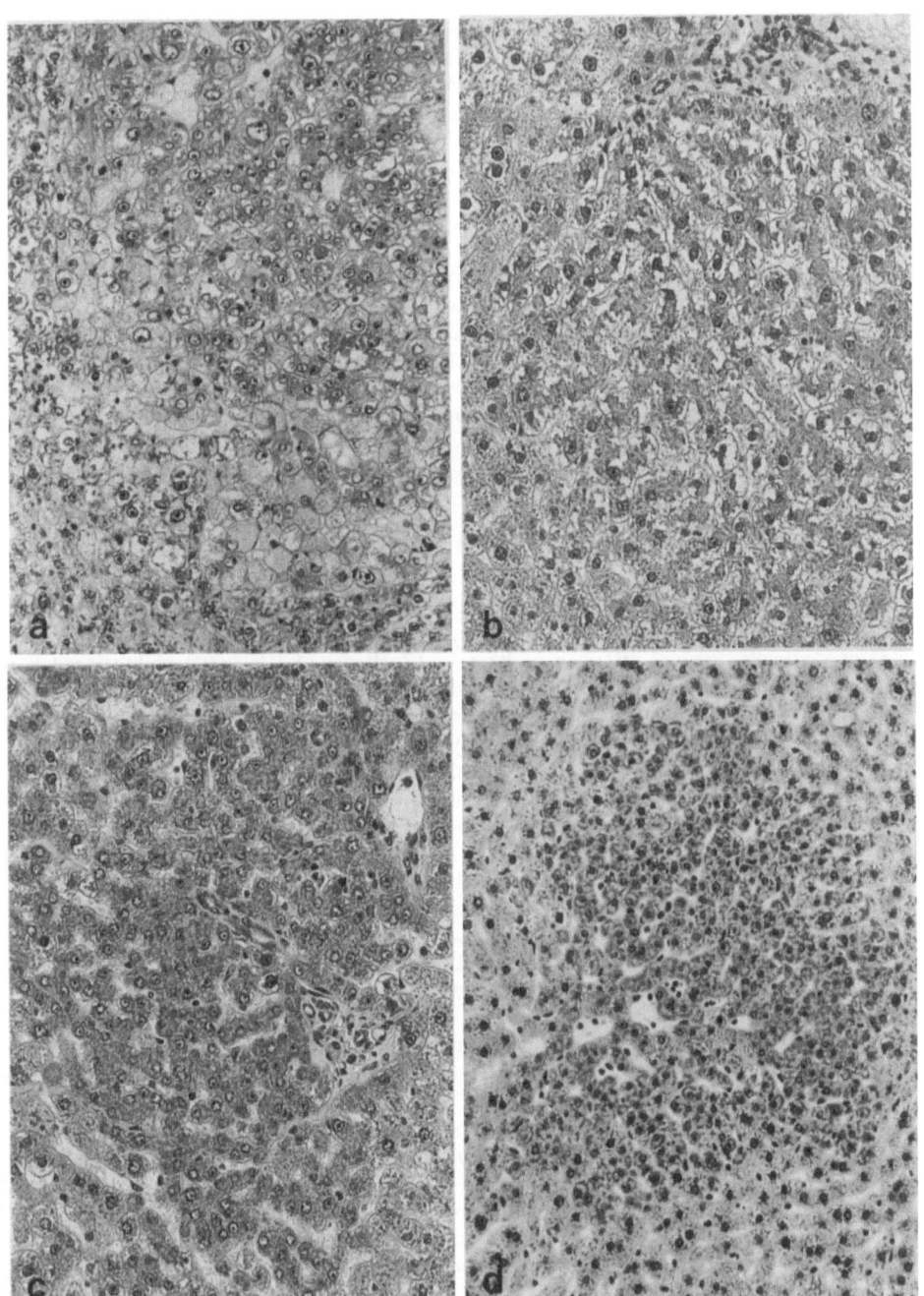

Fig. 5a–d. Late focal lesions induced in rat liver by **a–c** *N*-nitrosomorpholine or **d** aflatoxin. **a** Mixed cell focus composed of clear, acidophilic, basophilic, and intermediate cell types. H&E, ×220. **b** Intermediate cell focus. H&E, ×220. **c** Basophilic focus composed of cells with a homogeneous cyto- plasmic basophilia and large "vesicular" nuclei. H&E, ×200. **d** Tigroid cell focus composed of cells with a nonhomogeneous (tigroid) cytoplasmic basophilia (see also Fig. 14, inset) and dense nuclei. H&E, ×180

Zerban 1990). However, the cytoplasmic vacuo-lization is predominantly a consequence of fat accumulation (leading to round or oval vacuoles) in this case, though some glycogen may be found in addition to fat. In the course of hepatocarcinogenesis, fat storage foci usually emerge later than glycogen storage foci. There is circumstantial evidence based on morphometric investigations that the fat storage foci may de-velop from glycogen storage foci and progress to mixed cell foci and neoplastic liver lesions (Bannasch et al. 1972). The enzyme histochemical pattern of fat storage foci closely resembles that of mixed cell foci (Fig. 5a) and adenomas, which frequently contain vacuolated fat-storing cells in addition to several other cell types.

Intermediate Cell Foci

Various types of intermediate cells have been de-scribed in foci of altered hepatocytes (Bannasch et al. 1980). The cytoplasm of these cells is partly clear or vacuolated (storing glycogen or fat, re-spectively) and partly glycogen poor and basophilic (rich in ribosomes). The basophilic material mixed with acidophilic components may appear either homogeneously or in a striped ("tigroid") pattern, as described in some detail below (see "Ultrastructure"). Another type of in-termediate cell can be detected with the light mi-croscope only by close inspection (Bannasch 1976). The cytoplasm of these cells is very similar to that of clear cells. However, the clear cytoplas-mic regions are interwoven by a loose meshwork that is predominantly acidophilic but contains some basophilic spots, which correspond to pock-ets of the rough endoplasmic reticulum in large cisternae of smooth endoplasmic reticulum. Foci predominantly made up of one type of these inter-mediate cells (Fig. 5b) have been defined as inter-mediate cell foci (Bannasch and Zerban 1990). Recent studies on the [3H]thymidine incorpora-tion of these lesions revealed that their prolifera-tive activity corresponds to that of clear and acidophilic cell foci rather than that of mixed or basophilic cell foci (Zerban et al. 1994).

Mixed Cell Foci

Mixed cell foci (Figs. 5a,b, 6a–f; Squire and Levitt 1975; ILAR, NRC 1980) may have very heteroge-neous cytology, but two main variants of this type of focus can be distinguished (Bannasch and Zerban 1990): (1) foci composed of relatively small cells which are characterized by a clear, vacuolated, intermediate or strongly basophilic cytoplasm and (2) foci consisting of large cells which show an acidophilic, intermediate, vacuolated or slightly basophilic cytoplasm. All possible combinations between these two main types of mixed cell foci may occur. Goodman et al. (1994) have recently proposed that the term "mixed cell focus" be omitted and the predomi-nant cell type be used for the designation of this lesion. Although this proposal is reasonable, we prefer to keep the category of mixed cell foci, which has proved to be helpful under many ex-perimental conditions but should only be applied when basophilic cell types indicating more ad-vanced stages of neoplastic development are present in addition to clear, acidophilic, and inter-mediate forms (Bannasch and Zerban 1992, 1994). [3H]Thymidine incorporation in mixed cell foci according to this definition is strongly increased (Zerban et al. 1989, 1994), and their mitotic rate is high. The enzyme histochemical pattern of the mixed cell foci (Figs. 6a–f) differs in some respects from that of the clear cell foci (Hacker et al. 1982). Glucose-6-phosphatase activity may be normal within the clear cells of the mixed foci; glucose-6-phosphate dehydrogenase activity is usually greatly increased (Klimek et al. 1984). Activity of the glycolytic enzyme glyceraldehyde-3-phos-phate dehydrogenase is also higher in many of the mixed cell foci (Hacker et al. 1982).

Homogeneous Basophilic Cell Foci

Homogeneous basophilic cell foci (Fig. 5c) have been well known as "basophilic cell foci" and widely accepted as prestages of hepatocellular car-cinomas for many years (Grundmann and Sieburg 1962; Daoust and Calamai 1971; Bannasch 1975, and references therein), but have been called more precisely "homogeneous basophilic cell foci" by Goodman et al. (1994). The foci are made up of a homogeneous population of more or less basophilic cells which may be arranged in somewhat irregular plates and occasionally show polymorphic cells with marked nuclear atypia. Mitotic figures are frequent. The cyto-plasm of the cells is poor in glycogen and diffusely basophilic. The basophilia may be very intense

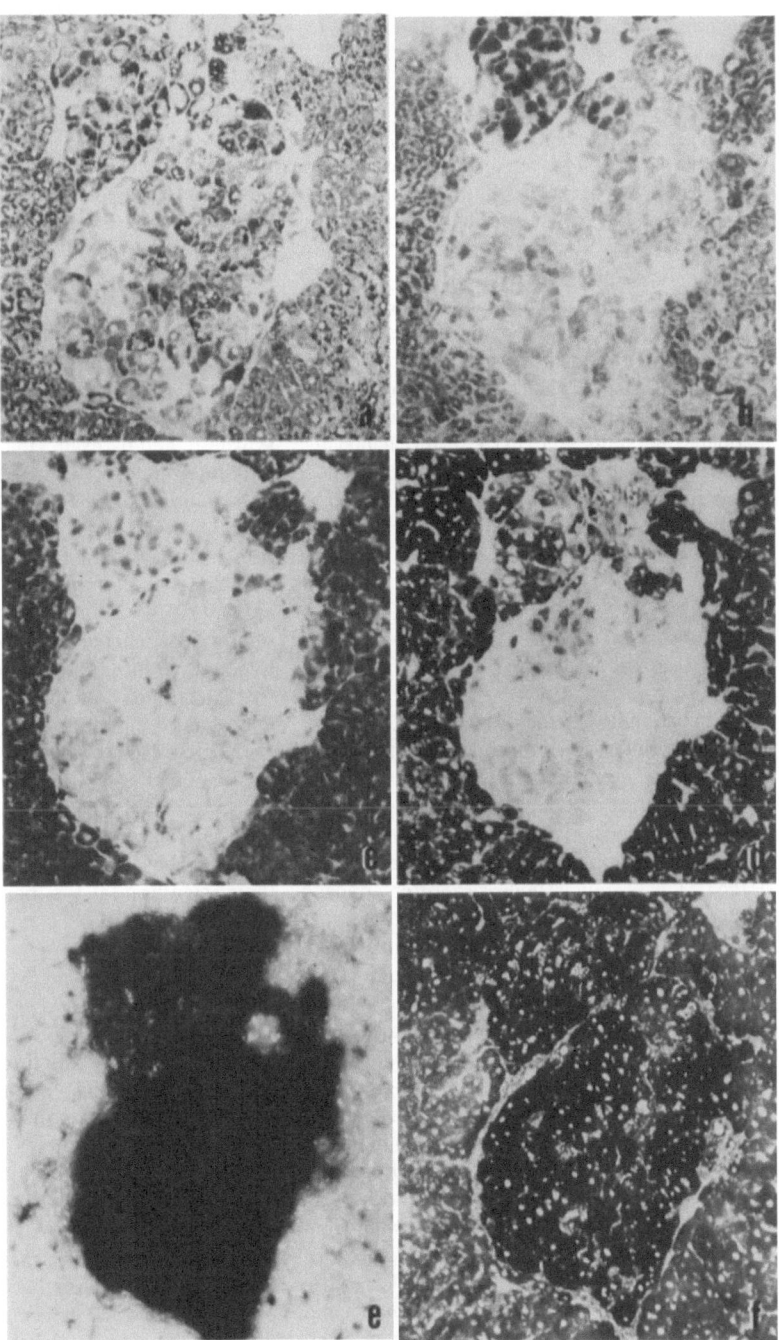

Fig. 6a–f. Serial sections through a mixed cell focus induced in rat liver by *N*-nitrosomorpholine. **a** Periodic acid-Schiff (PAS) reaction. **b** Glycogen synthetase activity. **c** Glycogen phosphorylase activity. **d** Glucose-6-phosphatase activity. **e** Glucose-6-phosphate dehydrogenase activity. **f** Glyceraldehyde-3-phosphate dehydrogenase activity. (From Bannasch et al. 1984) ×72

("hyperbasophilia" of Daoust and Calamai 1971) due to an increase in RNA ("chromatogenesis"), as suggested by Opie half a century ago (Opie 1946). However, the basophilia of the glycogen-poor cells does not always exceed that seen in normal hepatocytes (Taper and Bannasch 1979). The replacement of the basophilic bodies present in normal hepatocytes by a diffuse basophilia is a more general feature than hyperbasophilia. For historical reasons, the significance of the basophilic cell foci in hepatocarcinogenesis has been overestimated. Foci composed exclusively of cells which exhibit a pronounced diffuse basophilia are actually much rarer than intermediate and mixed cell foci, which, in addition to basophilic elements, contain clear and acidophilic components that store glycogen. Moreover, foci with a tigroid pattern of cytoplasmic basophilia and amphophilic cell foci have been separated more recently from the homogeneous basophilic cell foci.

Tigroid Basophilic Cell Foci

Tigroid basophilic cell foci (Fig. 5d) are made up of a distinct cell population resembling that of the basophilic foci, but have some important differences. Instead of a homogeneous basophilia, the cells contain large and abundant basophilic bodies on a clear or acidophilic background (Bannasch et al. 1985). The basophilic bodies are often arranged in long bands with a striped ("tigroid") pattern in paranuclear or peripheral regions of the cytoplasm. In this respect, the cells of the tigroid cell foci have some similarity to a certain type of intermediate cell frequently occurring in mixed cell foci. However, the cells of the distinct foci composed predominantly or exclusively of tigroid cells are usually smaller than their counterparts in the mixed cell foci. The nuclei of cells within tigroid cell foci may be small and dense, but frequently they are large and transparent. The histochemical pattern of tigroid cell foci differs significantly from that of other types of foci. For example, glucose-6-phosphatase and adenosine triphosphatase activity often are normal or only slightly reduced, and γ-glutamyltransferase and GST-P are almost always absent (Bannasch et al. 1989). Thus, a number of enzymatic changes which have been widely used as "markers" for preneoplastic hepatic foci in rats cannot detect tigroid cell foci. Transitions of tigroid cell foci into focal

lesions and adenomas in which cells with a diffusely basophilic cytoplasm appear next to typical tigroid cells have been observed after long lag periods (Bannasch et al. 1985; Weber and Bannasch 1994a, b). Tigroid cell foci apparently develop mainly after low-dose treatment with hepatocarcinogenic agents.

Enzymatically Hyperactive Foci

An enzymatically hyperactive focus which, as a rule, has only increased activities of various enzymes (mitochondrial glycerol-3-phosphate dehydrogenase, glucose-6-phosphatase, glycogen synthase, and pyruvate kinase) was characterized in detail in rats treated with N-nitrosomorpholine (Enzmann et al. 1989). This type of focus has also been observed in rats after combined administration of the peroxisomal proliferator ciprofibrate and the dietary antioxidant vitamin E (Glauert et al. 1990). In contrast to the other types of foci of altered hepatocytes, enzymatically hyperactive foci are not detectable after staining with H&E; however, they can progress to hepatic adenomas and, perhaps, also carcinomas (Enzmann et al. 1989). Glycogen phosphorylase hyperactive foci may occur in aged, untreated rats (Enzmann et al. 1992b) and after exposure to different carcinogenic agents (Seelman-Eggebert et al. 1987; Hacker et al. 1992; Ober et al. 1994), but their significance for hepatocarcinogenesis is unclear.

Ultrastructure

Changes in the fine structure of the foci of altered hepatocytes involve mainly glycogen, endoplasmic reticulum, ribosomes, peroxisomes, and mitochondria. In *clear cells* (Fig. 7), the glycogen is predominantly localized in the cytoplasmic matrix in the form of α- or β-particles, but it may also be enclosed in large autophagic vacuoles (Bannasch 1968; Bannasch et al. 1980). In the latter case, the glycogen becomes finely granular, probably as a consequence of lysosomal degradation. The rough endoplasmic reticulum is for the most part pushed toward peripheral or paranuclear regions of the cell and may be severely reduced per unit volume of the cytoplasm. In spite of this displacement and relative reduction of the granular reticulum, the fine structure

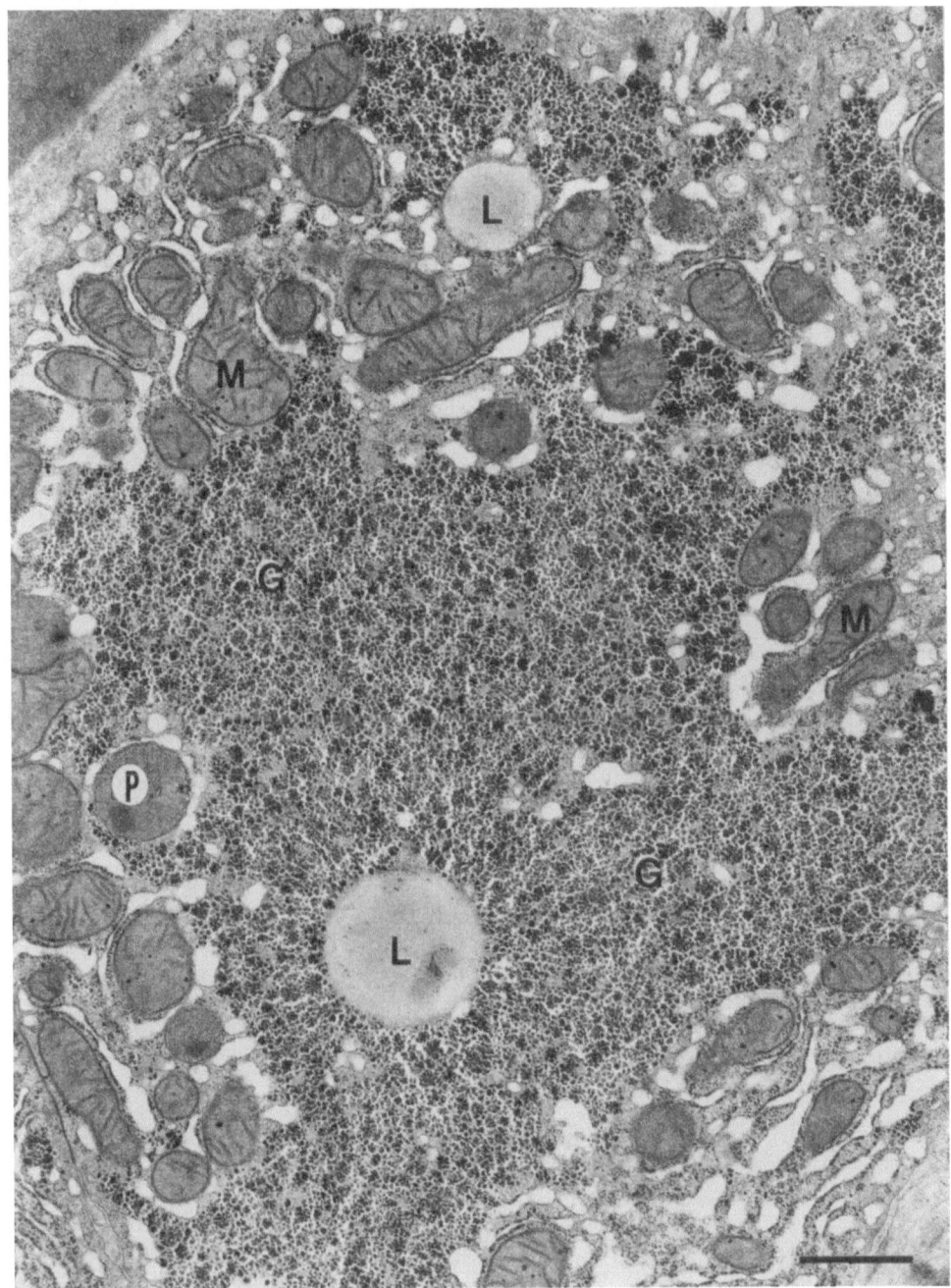

Fig. 7. Clear glycogen storage cell induced in rat liver by *N*- nitrosomorpholine. Abundant glycogen (*G*), lipid (*L*), mitochondria (*M*), and peroxisomes (*P*). TEM, lead citrate, ×18 500

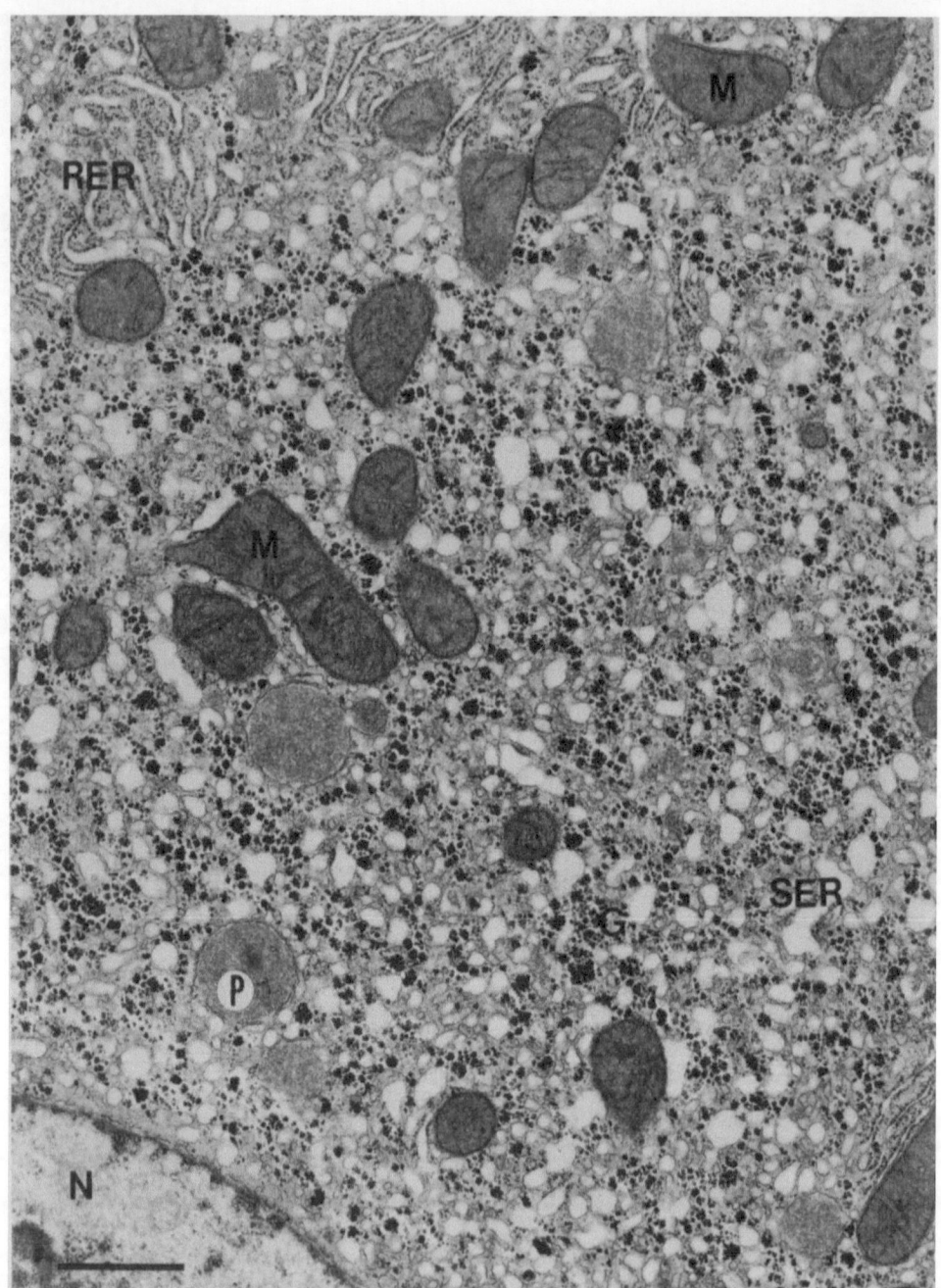

Fig. 8. Acidophilic glycogen storage cell induced in rat liver by *N*-nitrosomorpholine. Glycogen (*G*), abundant smooth endoplasmic reticulum (*SER*), rough endoplasmic reticulum (*RER*), mitochondria (*M*), peroxisomes (*P*), and nucleus (*N*). TEM, lead citrate, ×20000

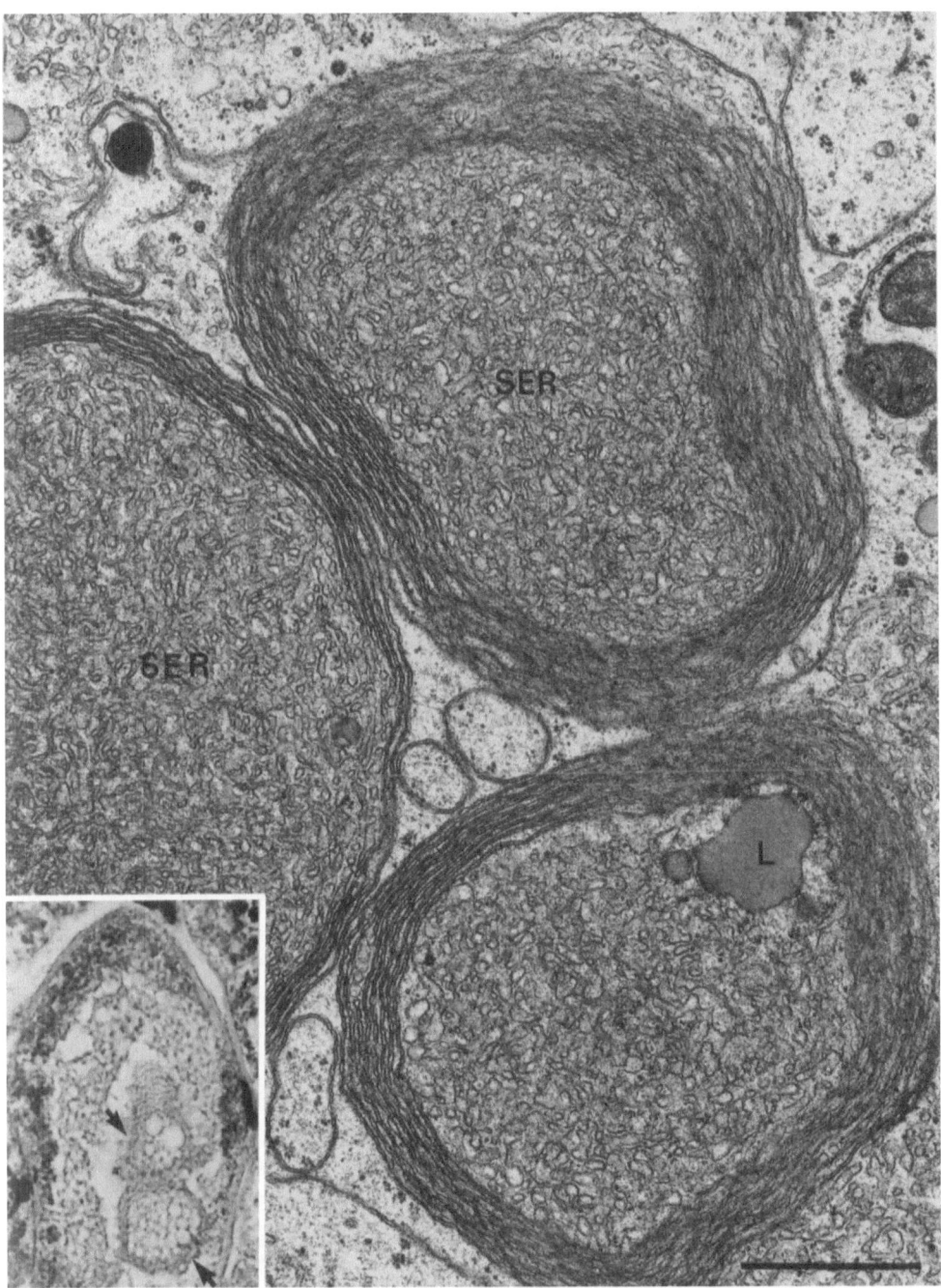

Fig. 9. Portion of acidophilic cell induced in rat liver by *N*-nitrosomorpholine. Abundant smooth endoplasmic reticulum (*SER*) arranged in concentric lamellar formations surrounding a highly condensed network of smooth membranes and an occasional lipid droplet (*L*). TEM, lead citrate, ×30000. *Inset,* acidophilic cell with whorl-shaped figures (*arrows*) in addition to an acidophilic network as seen under the light microscope. H&E, ×1200

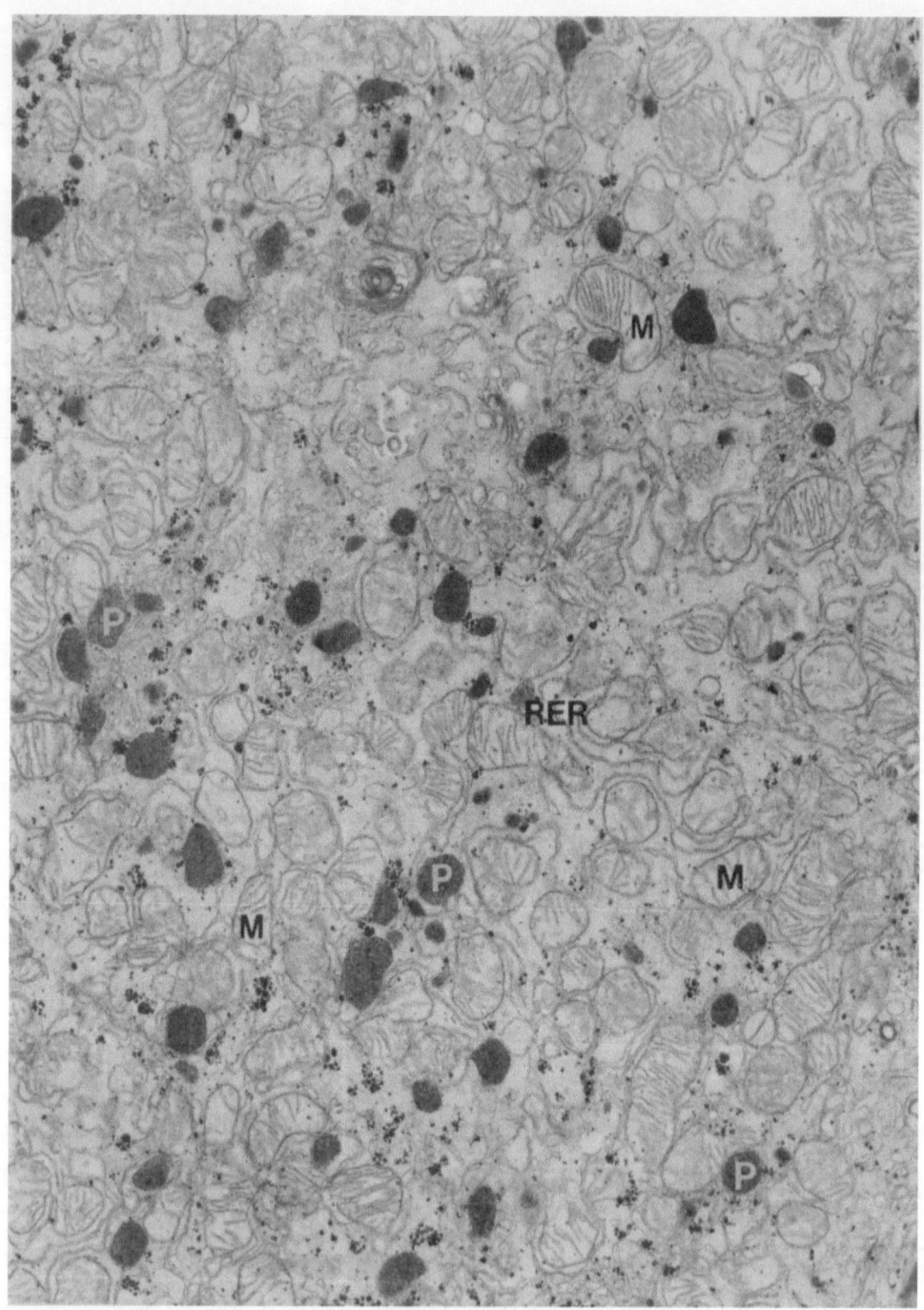

Fig. 10. Portion of amphophilic cell induced in rat liver by dehydroepiandrosterone. Abundant mitochondria (*M*) and peroxisomes (*P*) stained by the diaminobenzidine (DAB) method, and small stacks or single cristae of the rough endoplasmic reticulum (*RER*). TEM, lead citrate, ×12800

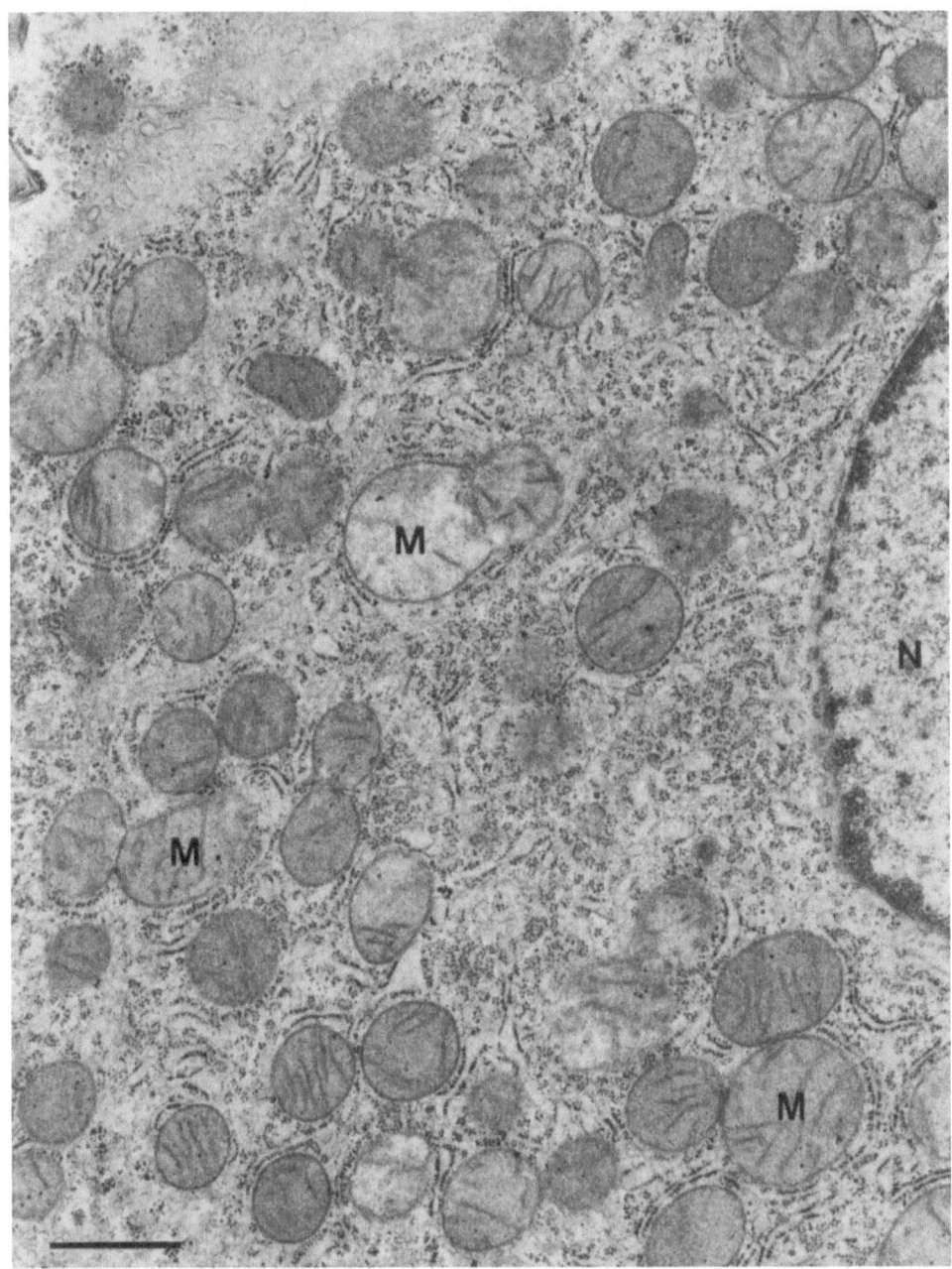

Fig. 11. Basophilic cell induced in rat liver by *N*-nitrosomorpholine. Abundant free and membrane-housed ribosomes, mitochondria (*M*), and nucleus (*N*). TEM, lead citrate, ×22500

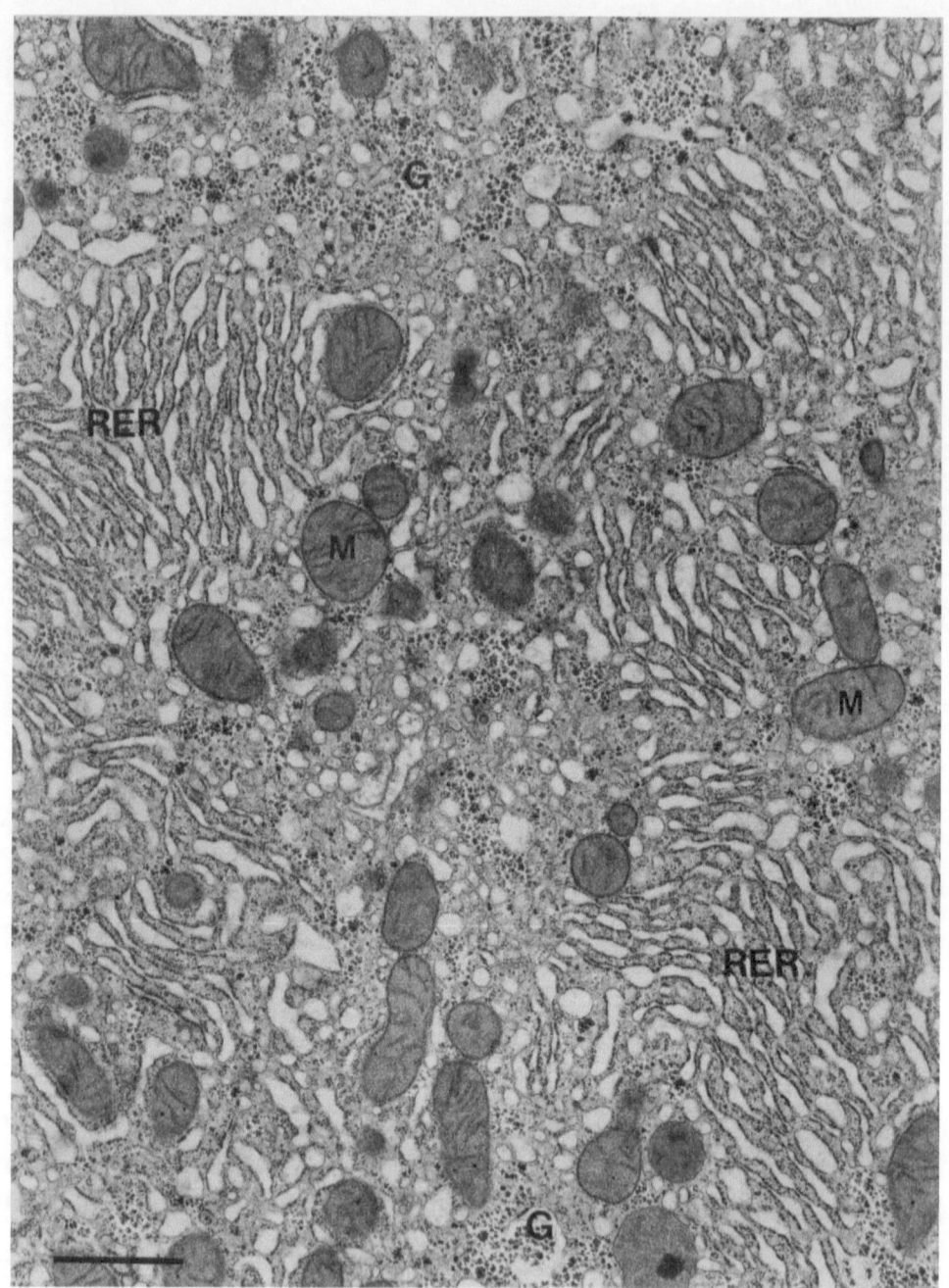

Fig. 12. Intermediate cell induced in rat liver by *N*-nitrosomorpholine. Abundant rough endoplasmic reticulum (*RER*) distributed between remnants of glycogen (*G*) and smooth endoplasmic reticulum. Mitochondria (*M*). TEM, lead citrate, ×21 500

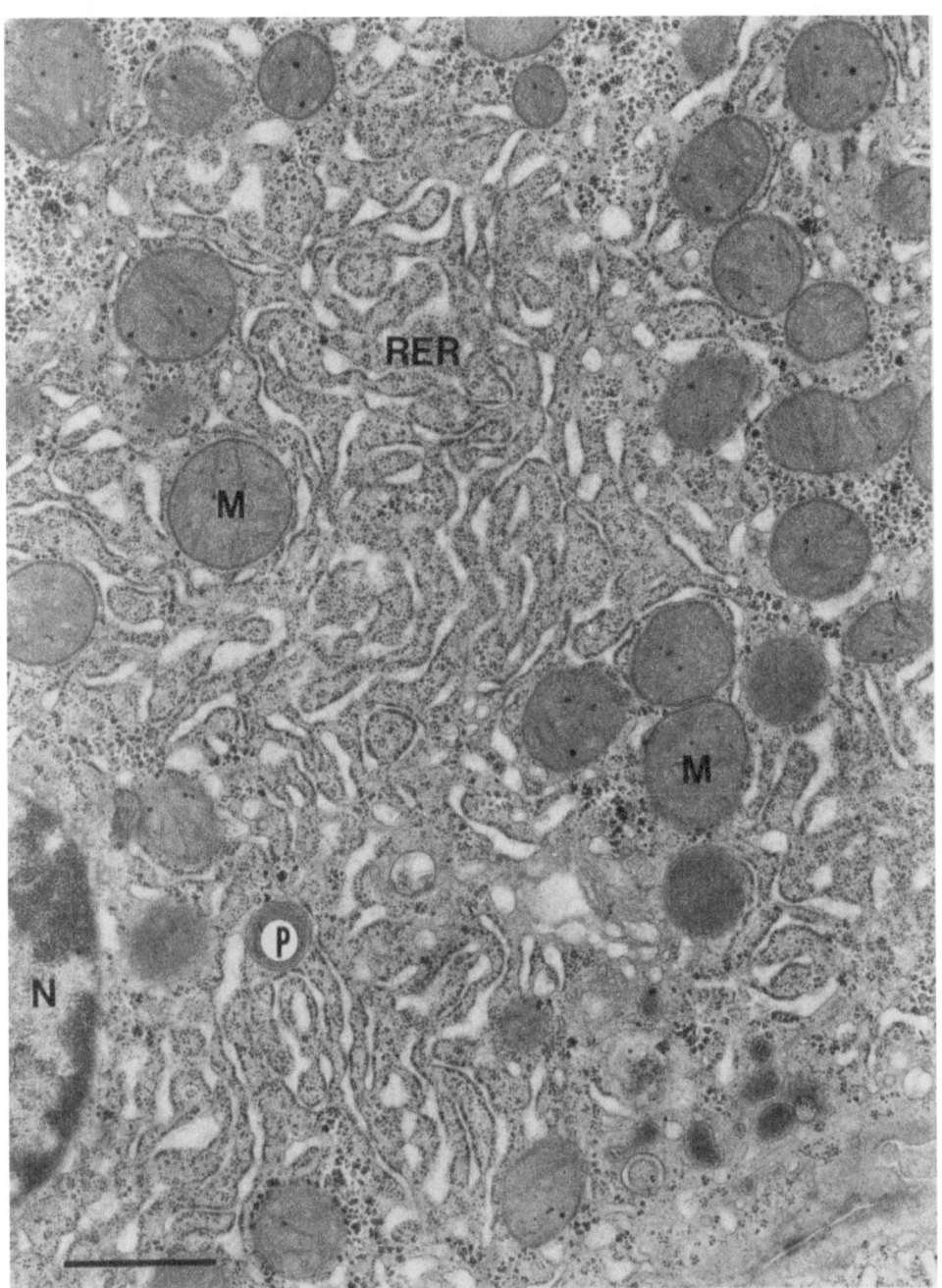

Fig. 13. Intermediate cell induced in rat liver by *N*-nitrosomorpholine. Abundant rough endoplasmic reticulum (*RER*) arranged in a complex network which is at some places smooth and closely associated with glycogen particles. Mitochondria (*M*), peroxisomes (*P*), and nucleus (*N*). TEM, lead citrate, ×25000

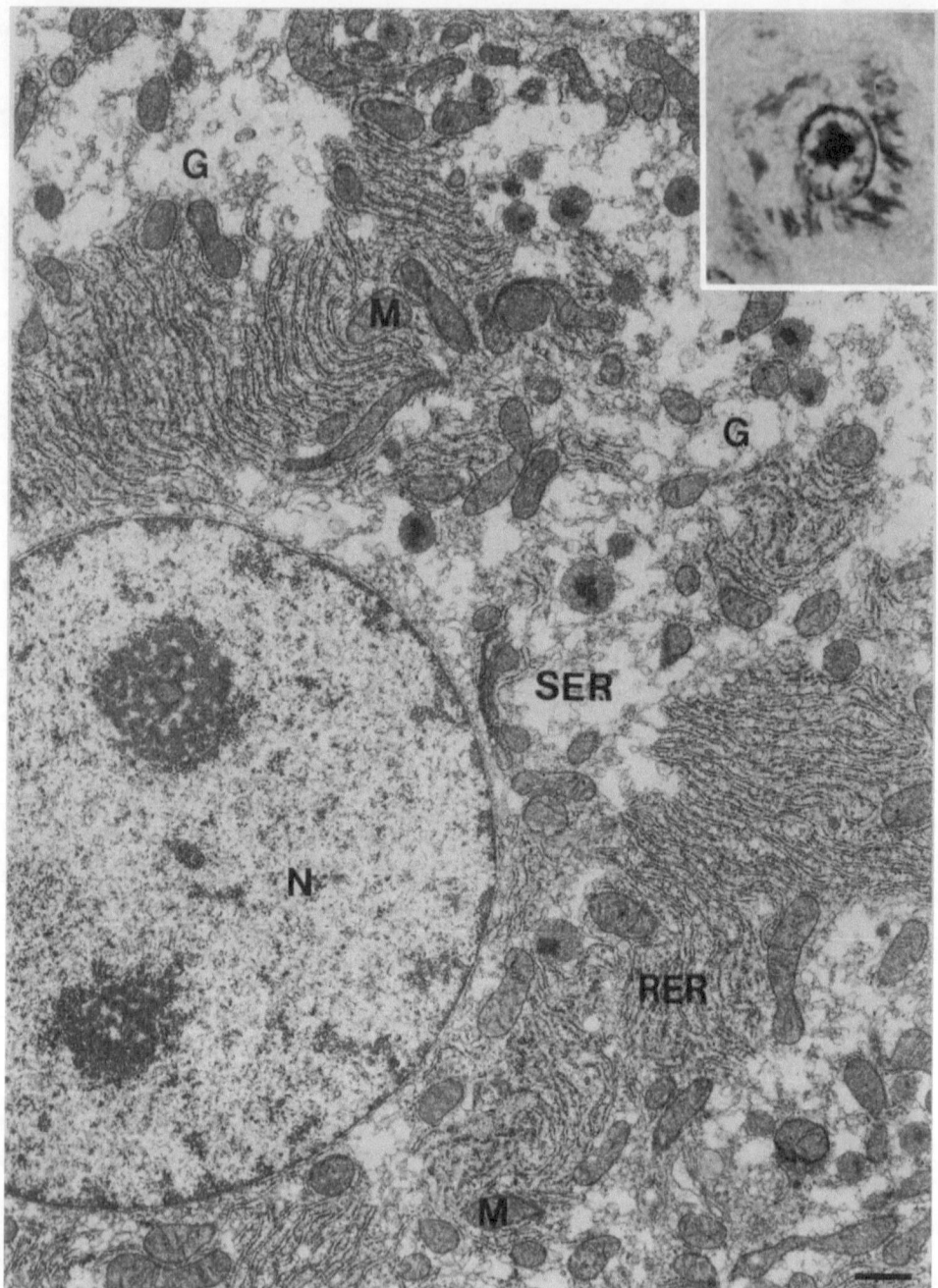

Fig. 14. Tigroid cell induced in rat liver by aflatoxin. Abundant rough endoplasmic reticulum (*RER*) arranged in highly ordered stacks which are frequently connected with membranes of the smooth endoplasmic reticulum (*SER*). Cloudy glycogen zones (*G*), mitochondria (*M*), and nucleus (*N*). TEM, uranyl acetate, ×10500. *Inset*, light microscopic appearance of tigroid cell. Note stripy arrangement of basophilic bodies corresponding to the RER. H&E, ×1400

of this organelle remains almost invariably unchanged.

Acidophilic Cells

Acidophilic (ground glass) cells (Fig. 8) are distinguished by a considerable hypertrophy of the smooth endoplasmic reticulum, which is often combined with an excessive accumulation of glycogen (Bannasch 1968; Bruni 1973; Bannasch et al. 1980; Jack et al. 1990b). There is usually a close relationship between the proliferated smooth membranes and the glycogen particles. Although the smooth membranes maintain a typical arrangement in most cases, they may form unusual concentric lamellar complexes with or without glycogen (Fig. 9; Steiner et al. 1964; Stenger 1966; Bannasch 1968; Feldman et al. 1981; Jack et al. 1990b). These well-known structures, often called fingerprints, appear to be only a morphological variant of hypertrophy of the smooth endoplasmic reticulum. By light microscopy, they can be seen as whorl-like acidophilic figures (Fig. 9, inset; Altmann and Osterland 1961).

Amphophilic Cells

The *amphophilic cells* (Fig. 10) are characterized by abundant mitochondria and often also peroxisomes (Metzger et al. 1995), which correspond to the histochemical finding of an increased activity of mitochondrial and peroxisomal enzymes (Weber et al. 1988a) and to the granular acidophilia of the cytoplasm seen under the light microscope. In addition, small stacks or single cisternae of the rough endoplasmic reticulum represent the basophilic components of the cytoplasm in amphophilic cells. In contrast to the acidophilic cells, amphophilic cells are nearly free of glycogen.

Homogeneous Basophilic Cells

Homogeneous basophilic cells (Fig. 11) also contain little or no glycogen, fat, and smooth membranes of the endoplasmic reticulum (Bannasch 1968; Karasaki 1969; Hirota and Williams 1982); the number of peroxisomes is variable (Bannasch 1968). The ultrastructural equivalent of the cytoplasmic basophilia observed with the light microscope is an unusual abundance of free or membrane-bound ribosomes.

Intermediate Cell Types

Diverse *intermediate cell types* (Figs. 12, 13) have been described in detail (Bannasch 1968, 1976; Bannasch et al. 1980). They may contain unusual formations of endoplasmic reticulum, which are characterized by a combination of smooth and rough components of this organelle; the rough parts form pockets which are poor in glycogen and rich in ribosomes. Many intermediate cells contain fat, which may not only form droplets within the cytoplasmic matrix, but also occur as liposomes in the cisternae of the endoplasmic reticulum.

Tigroid Basophilic Cells

The *tigroid basophilic cells* (Fig. 14), which may also represent an intermediate cell type, have a very characteristic ultrastructure (Bannasch et al. 1985). The fine structural equivalents of the intense basophilic bodies seen under the light microscope are highly ordered stacks of cisternae of rough endoplasmic reticulum with a typical pattern of membrane-bound ribosomes. They are frequently connected with smooth components of the endoplasmic reticulum, which form the usual network and are closely associated with α- or β-glycogen particles. Sometimes the glycogen zones are free of smooth endoplasmic reticulum. In addition to glycogen particles, large osmiophilic bodies are often observed in the cytoplasm. These bodies contain either an amorphous osmiophilic material or have a curvilinear pattern.

Alterations of the plasma membrane, especially abnormalities at the bile canalicular pole of the hepatocytes (e.g., blebbing, reduction, and/or elongation of microvilli), have been observed in the persistent foci produced by the Solt-Farber procedure (Ogawa et al. 1979) and by cycles of feeding N-2-fluorenylacetamide (Hirota and Williams 1982).

Differential Diagnosis

Differentiation between acidophilic glycogen storage foci and amphophilic cell foci may be difficult in H&E-stained tissue sections, but it is easy after additional demonstration of the glycogen by the PAS reaction, since acidophilic cell foci are strongly PAS positive, while amphophilic cell foci are largely or completely PAS negative (Bannasch and Zerban 1990, 1994; Goodman et al. 1994). A less simple task is the discrimination between

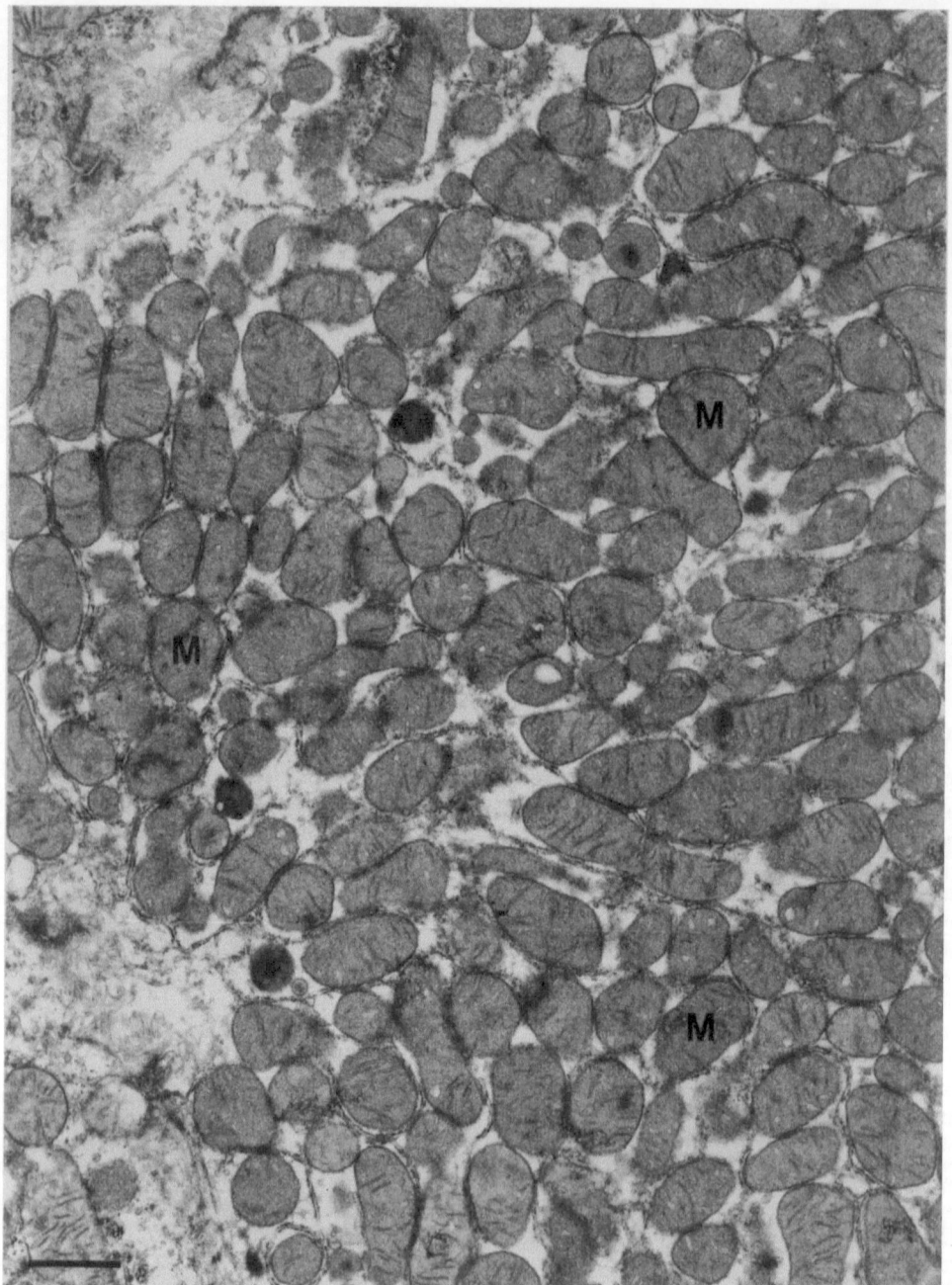

Fig. 15. Granular acidophilic cell induced in rat liver by *N*-nitrosomorpholine. The cytoplasm is crowded with mitochondria (*M*) and resembles that of an oncocyte. TEM, lead citrate, ×15 500

amphophilic cells and granular acidophilic cells resembling oncocytes, which have been rarely described in rat liver (Reznik-Schüller and Gregg 1983). The cytoplasm of these "oncocytes" is rich in mitochondria, but not in peroxisomes (Fig. 15). Foci composed of these cells might also represent a preneoplastic cell population (Metzger et al. 1995).

The separation of amphophilic cell foci (which by definition show both acidophilic and basophilic cytoplasmic components) from homogeneous basophilic cell foci may also be difficult, particularly since both types of focus are poor in glycogen. In this case, other histochemical parameters appear to be the most reliable criteria for differential diagnosis. Whereas amphophilic cell foci are always negative for γ-glutamyltransferase activity and do not react with antibodies to GST-P, the majority of the diffusely basophilic cell foci are positive for γ-glutamyltransferase and react with antibodies to GST-P (Bannasch et al. 1989). It remains to be clarified whether the focal lesions which were induced by peroxisomal proliferators and were described as weakly or homogeneously basophilic (Kraupp-Grasl et al. 1990, 1991; Marsman and Popp 1994) should be grouped with amphophilic rather than basophilic cell foci. The differential diagnosis of basophilic cell foci is further complicated by the fact that enhanced cytoplasmic basophilia due to an increase in ribosomes and accompanied by a reduction in glycogen may develop under various pathologic conditions, particularly in so-called megalocytosis (Theodossiou et al. 1971; Taper and Bannasch 1979). Enzyme histochemical markers may help to distinguish these changes from preneoplastic basophilic cell populations (Taper and Bannasch 1979).

Evaluation of the significance of foci of altered hepatocytes in hepatocarcinogenesis may be seriously hampered by reversion-linked phenotypic instability (Bannasch 1986). This type of instability has been observed mainly after withdrawal of chemicals that produce foci of altered hepatocytes under conditions of pronounced hepatotoxicity, such as repeated administration of high, sublethal doses of hepatocarcinogens or the Solt-Farber schedule. The reversibility of biochemical and morphological changes in focal lesions has been described in terms of phenotypic "maturation" (Kitagawa 1971), "reversion" (Williams and Watanabe 1978), "remodeling" (Solt et al. 1977; Farber and Sarma 1987), "redifferentiation" (Tatematsu et al. 1983), and phenotypic "instabil-

ity" (Moore et al. 1983). These various terms mirror the uncertainty about the biologic significance of this phenomenon. The loss of phenotypic properties in the altered cell populations is apparently not brought about by cell death and compensatory proliferation of normal hepatocytes (Kitagawa and Sugano 1973; Enomoto and Farber 1982; Tatematsu et al. 1983). Some observations indicate that remodeled lesions may reappear after long lag periods (Watanabe and Williams 1978; Tatematsu et al. 1983). Recently, it was shown in stop experiments with high doses of N-nitrosomorpholine that there was no reversion of focal lesions to a normal parenchyma, but a replacement of the majority of mixed cell foci to the less advanced clear and acidophilic (glycogenotic) cell foci after withdrawal of the carcinogen (Weber and Bannasch 1994b). When administration of the same dose was continued, a high incidence of hepatocellular carcinomas was observed (Weber and Bannasch 1994c). A similar phenotypic instability has been reported by Marsman and Popp (1994), who observed a replacement of homogeneous basophilic foci by less altered phenotypes, particularly clear cell populations, when they switched a diet containing the strong peroxisomal proliferator Wy-14,643 to basal diet.

The problems resulting from reversion-linked phenotypic instability for the evaluation of foci of altered hepatocytes in carcinogenicity testing should not be overestimated, however, since this phenomenon does not seem to play an important role at the dose levels usually administered in long-term carcinogenesis bioassays (Bannasch and Zerban 1992). The results of several stereological studies on the development of foci of altered hepatocytes after long-term or limited exposure of rats to hepatocarcinogens (with and without additional administration of phenobarbital) indicated that all or at least the vast majority of foci of altered hepatocytes persisted, showing signs of progression rather than regression until neoplasms appeared or the study was terminated (Scherer and Emmelot 1975; Moore et al. 1982; Goldsworthy et al. 1984; Enzmann and Bannasch 1987; Hendrich et al. 1987; Soffritti and McConnell 1988; Zerban et al. 1989; Harada et al. 1989b). Whenever foci of altered hepatocytes of dubious significance occur, stop experiments are recommended, which permit discrimination between reversible and persistent lesions (Bannasch et al. 1982; Tatematsu et al. 1988b).

Biologic Features

Experimental Induction

Persistent foci of altered hepatocytes have been produced in rat liver by a variety of chemical hepatocarcinogens (Bannasch and Zerban 1992; Hasegawa and Ito 1994) and by radiation (Ober et al. 1994), and it appears that they represent an obligatory precursor of hepatic tumors, no matter what the inducing oncogenic agent was. However, genetic factors are most probably also involved. Thus a high incidence of foci of altered hepatocytes have been described in a mutant rat strain that has hereditary hepatitis and in which hepatic neoplasia develops in long-living animals (Sawaki et al. 1990). Hully et al. (1994) observed foci of altered hepatocytes, the phenotype of which was similar to those found in chemical hepatocarcinogenesis in transgenic rats carrying the simian virus 40 (SV40) T antigen gene and developing a high incidence of hepatocellular neoplasms. The occurrence and age-dependent increase in the incidence of foci of altered hepatocytes have also been reported in untreated rats of several strains (Burek 1978; Goodman et al. 1994; Ogawa et al. 1981; Ward 1981; Schulte-Hermann et al. 1983; Popp et al. 1985; Mitaka and Tsukada 1987; Harada et al. 1989b). Burek (1978) reported that, in the WAG/Rij strain and its outcrosses, as many as 35%–85% of the aged rats reveal foci of altered hepatocytes. A high spontaneous incidence of such foci has also been observed in Fischer 344 rats, in which the percentage of animals with foci sometimes approaches or even exceeds 90% at 2 years (Ward 1981; Harada et al. 1989b). In addition to genetic factors, the possible contamination of food or environment has to be taken into account in the explanation of these findings.

Location and Clonal Development

The foci are usually located in peripheral or intermediate parts of the liver lobule (Bannasch 1968; Maguire and Rabes 1989), but after continuous administration of low doses of carcinogens they may start to develop in centrilobular regions (Bannasch et al. 1974). In spite of the pronounced phenotypic heterogeneity in the cellular phenotype of foci of altered hepatocytes, observations in different laboratories suggest a clonal origin

for the focal lesions (Scherer and Hoffmann 1971; Rabes et al. 1982; Tsuji et al. 1988; Weinberg and Iannaccone 1988). The results of stereological studies do not exclude this possibility for the earliest foci, but they argue in favor of simultaneous alterations of many hepatocytes in larger cell populations rather than repeated clonal selection in the progressive development of phenotypic heterogeneity during hepatocarcinogenesis (Enzmann and Bannasch 1987).

Cell Proliferation and Cell Death (Apoptosis)

Cell proliferation plays an important part in different stages of the development of the focal lesions (Rabes 1988). However, the biochemical and morphological phenotypes of glycogenotic (clear and acidophilic), amphophilic, and tigroid cell foci by no means correspond to that of proliferating liver parenchyma, as observed, for example, after partial hepatectomy. Only the mixed and diffusely basophilic cell foci, poor in glycogen, show some phenotypic similarity to the regenerating parenchyma and are at the same time characterized by rapid cell proliferation (Bannasch and Zerban 1992; Zerban et al. 1994). An enhancing effect of previous partial hepatectomy has been demonstrated for the induction of foci of altered hepatocytes (Scherer et al. 1972; Rabes 1983). Tanaka et al. (1986) did not observe a similar effect when partial hepatectomy was performed after exposure to a carcinogen; however, other authors (some of whom used more complex experimental models) did find an enhancing effect (Cayama et al. 1978; Pound and McGuire 1978; Ishikawa et al. 1980; Columbano et al. 1981). A significantly higher incidence of foci of altered hepatocytes was also recorded in the rapidly proliferating liver of neonatal rats as compared to adult rats treated with various carcinogens (Peraino et al. 1981; Decloitre et al. 1990; Hasegawa et al. 1991; Mathur et al. 1992). When proliferation of rat hepatocytes was synchronized by hydroxyurea after partial hepatectomy, they had the highest risk of being initiated when they traversed the early S phase of the cell cycle (Rabes et al. 1986; Kaufmann et al. 1987). Unlike compensatory cell proliferation induced by partial hepatectomy or necrogenic doses of carbon tetrachloride, the cell proliferation induced by mitogens such as lead nitrate and cyproterone acetate did not result in an increased number of foci, despite the fact that the extent of cell proliferation at the time of administration of carcinogen was similar with the two types of proliferative stimulus (Columbano et al. 1987, 1990).

It has been known for a long time that foci of altered hepatocytes are characterized by enhanced cell proliferation, which increases with time (Schauer and Kunze 1968; Rabes 1988) and is correlated with increasing conformity of the expression of different marker enzymes (Pugh and Goldfarb 1978; Baba et al. 1989; Ito et al. 1989). However, autoradiographic studies on the different types of focal lesions induced in the stop model and defined by cytomorphological criteria revealed that the incorporation of [3H]thymidine is only slightly enhanced in the clear and acidophilic glycogen storage foci, but increases gradually with increasing appearance of basophilic cells in mixed and basophilic cell foci and in adenomas and carcinomas (Zerban et al. 1989, 1994). There was no indication of any inhibition of cell proliferation in extrafocal hepatic tissue, as described by some authors under other experimental conditions (Rotstein et al. 1986; Tatematsu et al. 1988a). Tsuda et al. (1995) have reported that the increase in cell proliferation from preneoplastic hepatic foci to hepatocellular adenomas and carcinomas is associated with a progressive decrease in the expression of connexin 32, a major liver gap junction protein. In addition to cell proliferation, cell death (apoptosis) is frequently increased in foci of altered hepatocytes and particularly in hepatocellular neoplasms (Bursch et al. 1984; Columbano et al. 1984; Schulte-Hermann et al. 1990; Zerban et al. 1994). Whereas some authors feel that apoptosis plays a major role in counterbalancing cell replication (Bursch et al. 1984; Schulte-Hermann et al. 1990), others emphasize that cell death occurs more frequently in the course of hepatocarcinogenesis the further neoplastic development advances (Columbano et al. 1984; Zerban et al. 1994).

Nuclear Morphology and Ploidy

A great variety of changes in nuclear morphology and ploidy of clear, acidophilic, and basophilic preneoplastic cell populations was described in detail earlier in rat and mouse liver, but these were considered to be facultative rather than obligatory events in neoplastic transformation of the hepatocytes (Bannasch and Müller 1964; Romen et al. 1972, 1973; Bannasch 1975; Abmayr et al. 1983, and literature therein). In the last few years, controversial results have been published on the distribution of ploidy during hepatocarcinogenesis in rodents. Whereas some authors emphasized the emergence of diploid cell populations early on in hepatocarcinogenesis (Schwarze et al. 1986; Deleener et al. 1987; Styles et al. 1987; Haesen et al. 1988; Saeter et al. 1988; Sargent et al. 1989), others found variable ploidy distribution in foci of altered hepatocytes (Mori et al. 1982; Digernes 1983; Sarafoff et al. 1986; Danielsen et al. 1988; Pitot et al. 1989; Jack et al. 1990a; Wang et al. 1990; Sudilovsky and Hei 1991). Gil and coworkers (1988) reported that foci of clear cells, mixed cells, and large basophilic cells induced in rats with N-nitrosomorpholine or aflatoxin B1 had a ploidy distribution similar to that of extrafocal parenchyma, while foci consisting of small hyperbasophilic cells, which correspond to tigroid cell foci, were predominantly diploid. In accordance with earlier interpretations by Bannasch (1975), Danielsen et al. (1988) and Wang et al. (1990) maintained that changes in ploidy distribution are not essential in hepatocarcinogenesis. The appearance of aneuploidy in foci of altered hepatocytes has, however, been regarded by several authors as a risk factor that increases the probability of neoplastic progression of preneoplastic focal lesions (Mori et al. 1982; Sarafoff et al. 1986; Wang et al. 1990; Sudilovsky and Hei 1991). Jack et al. (1990a) found a striking reduction in binucleate cells but an increase in mononucleate tetraploid cells in foci of altered hepatocytes, suggesting an altered mitotic mechanism.

Oncogenes, Tumor Suppressor Genes, and Growth Factors

There is scanty and controversial information on the role of oncogenes and tumor suppressor genes in the development and progression of preneoplastic hepatic foci in the rat. According to Beer et al. (1986), altered hepatocytes derived from focal lesions defined by an increased activity of γ-glutamyltransferase did not differ from normal hepatocytes in their expression of c-myc and c-H-ras protooncogenes. In line with these observations, Embleton and Butler (1988) using an immunohistochemical approach did not observe an increase in the c-H-*ras* oncoprotein in preneoplastic and neoplastic liver lesions induced in rats by four cycles of treatment with acetylaminofluorene, with and without additional administration of single doses of diethylnitrosamine and carbon tetrachloride. Galand et al. (1988) and Alexandre et al. (1990) found elevated levels of the c-H-*ras* and c-*myc* oncoprotein, respectively, by a similar approach in preneoplastic lesions induced in rats by a single dose of diethylnitrosamine administered after partial hepatectomy. Stumpf (1992) failed to detect an overexpression of c-H-*ras* mRNA by in situ hybridization with antisense mRNA in various types of foci of altered hepatocytes induced in rats by limited exposure (stop model) to *N*-nitrosomorpholine. Using the same animal model and technique of in situ hybridization, this author was able to demonstrate an overexpression of c-*myc* in clear and acidophilic cell foci excessively storing glycogen, and particularly in mixed and basophilic cell foci (Stumpf 1992). Sarafoff and Rabes (1991) had previously reported a similar finding in hepatocellular adenomas. Preneoplastic hepatic foci induced in rats by the Solt-Farber procedure and identified by their immunohistochemically demonstrated expression of GST-P showed a partial overexpression of c-*myc* (Simile et al. 1994) and c-*jun* (Suzuki et al. 1995) oncoproteins. An overexpression of c-*raf* was also found in different types of preneoplastic hepatic foci (Stumpf 1992) and in hepatocellular adenomas and carcinomas (Beer et al. 1988). Evidence indicating one or more p53 gene mutations in foci of altered hepatocytes induced by diethylnitrosamine has been provided by immunohistochemical studies with antibodies directed against mutant forms of the p53 protein (Smith et al. 1991).

Immunohistochemical approaches revealed an increased expression of transforming growth factor-α in preneoplastic hepatic foci positive for GST-P and progressing to hepatocellular neoplasms, which were induced in a two-stage protocol by different mutagenic chemicals and phenobarbital (Kaufmann et al. 1992; Dragan et al. 1995). However, in different experimental models of hepatocarcinogenesis, Perez-Tomas et al. (1992) were unable to identify tumor growth factor-α-immunoreactive cells in preneoplastic hepatic lesions. Immunoreaction was only observed in some cells of hepatocellular neoplasms. In rats treated with the peroxisomal proliferator Wy-14,643 with and without prior administration of diethylnitrosamine, only eosinophilic cell foci, and not "homogeneous basophilic cell foci" (which may correspond to amphophilic cell foci as defined in this chapter), overexpressed transforming growth factor-α. These results indicate that overexpression of this growth factor is not a reliable marker of tumor progression as proposed by Dragan et al. (1995) under all experimental conditions.

Phenotypic Heterogeneity and Instability

The significance of the phenotypic heterogeneity in foci of altered hepatocytes is controversial. Several authors described phenotypic stability of individual foci of altered hepatocytes induced in the liver by a single dose of a carcinogen in newborn or partially hepatectomized adult rats followed by phenobarbital (Peraino et al. 1984, 1988; Goldsworthy and Pitot 1985). These authors concluded that the phenotypic heterogeneity of foci of altered hepatocytes indicated random initiating events, resulting in diverse phenotypes, each of which might represent a specific set of cellular changes; they considered that foci of altered hepatocytes that appear early do not evolve through progressively more deviated forms to hepatic neoplasms. This interpretation is at variance with the conception of a progression-linked phenotypic instability which mirrors different stages in one or several cell lineages leading to hepatic tumors (Fig. 16; Bannasch et al. 1989; Bannasch and Zerban 1994). A number of results, which have been reviewed repeatedly, indicate that the predominant cell lineage related to hepatocarcinogenesis in rats is characterized by an ordered sequence of metabolic and morphological changes leading from glycogenotic foci via various intermediate cell populations to glycogen-poor hepatocellular carcinomas (Bannasch 1968; Bannasch et al. 1980, 1984; Hirota and Yokoyama 1985; Steinberg et al. 1991; Weber and Bannasch 1994a–c). Establishment of this sequence permits separation of the carcinogenic process into different stages on the basis of biologic rather than operational criteria. The associated metabolic aberrations have been analyzed to some extent by means of cytochemical and microbiochemical methods.

Metabolic Aberrations

Hepatocellular glycogenosis is usually associated with a disturbance in phosphorylytic glycogen breakdown (Hacker et al. 1982) which is not due to a loss of the phosphorylase protein (Seelmann-Eggebert et al. 1987) but is apparently the consequence of alterations in superordinate regulatory mechanisms, such as dysfunction of signal transduction, as demonstrated by a reduction in

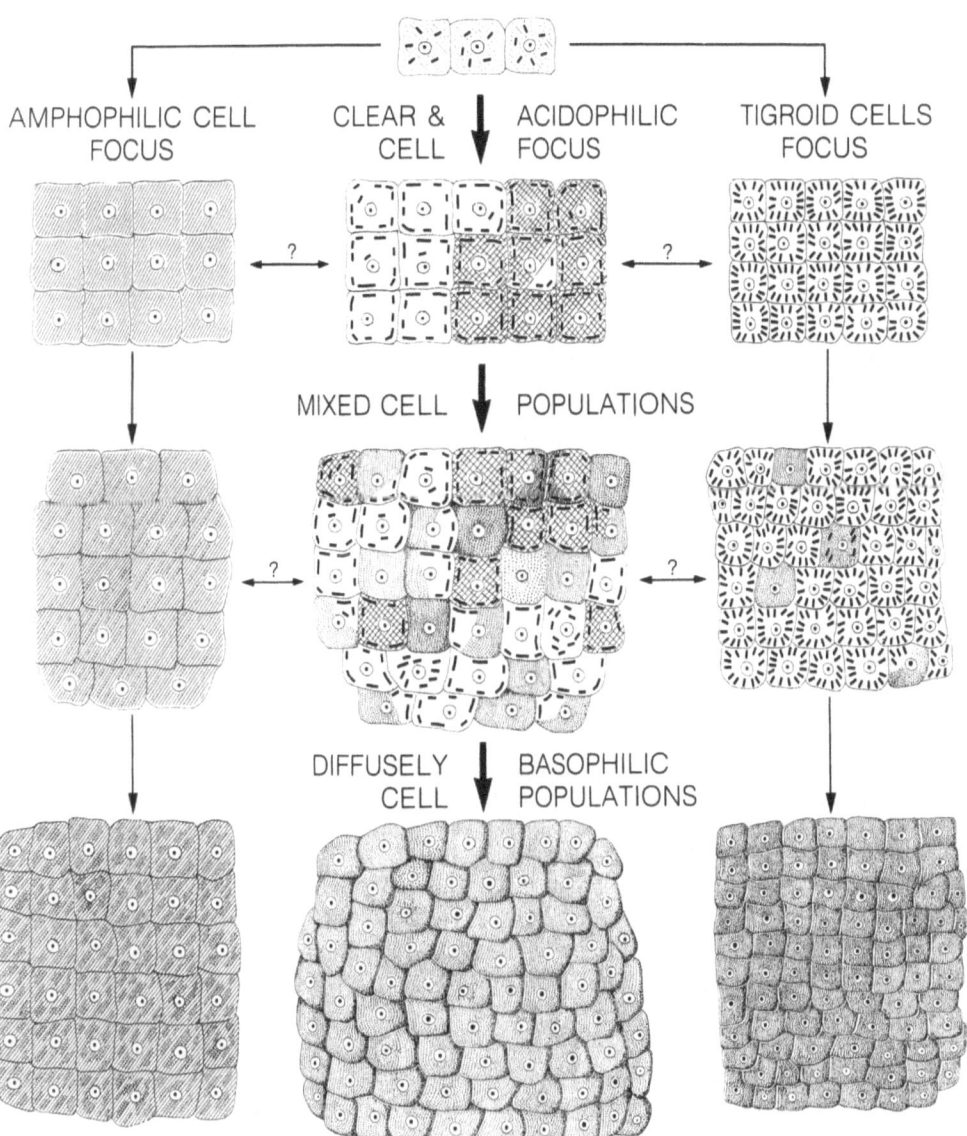

Fig. 16. Sequential cellular changes during the development of hepatocellular neoplasms in rat liver. In addition to the predominant sequence (*center*), two alternative sequences have been established (*left* and *right*), which may either occur independently or represent phenotypic modulation of the main sequence. (From Bannasch and Zerban 1994)

the activity of adenylate cyclase (Ehemann et al. 1986). Moreover, many glycogenotic foci have reduced activity of the microsomal enzyme glucose-6-phosphatase (Friedrich-Freksa et al. 1969) and of the lysosomal α-glucosidase (Klimek and Bannasch 1989) and a reduced content of glucose transporter protein GLUT2 (Grobholz et al. 1993). These alterations might act in concert with the disturbance in phosphorylytic glycogen breakdown to intensify the excessive storage of glycogen. An increase in the concentration of the central metabolite of carbohydrate metabolism, glucose-6-phosphate, was measured in homogenates of livers treated with N-nitrosomorpholine for 7 weeks (Enzmann et al. 1988). Of particular interest are the increases in the content or activity of key enzymes in the pentose phosphate pathway, glucose-6-phosphate dehydrogenase (Hacker et al. 1982; Klimek et al. 1984; Greaves et al. 1986; Moore et al. 1986a; Gerbracht et al. 1993), and the glycolytic enzyme pyruvate kinase in glycogen storage foci (Klimek and Bannasch 1990). These findings indicate the beginning of a metabolic shift in glycogenotic hepatocytes toward alternative metabolic pathways. Microbiochemical studies (Klimek et al. 1984) have shown that there is a gradual increase in glucose-6-phosphate dehydrogenase activity from small to large glycogenotic foci to mixed and basophilic cell populations, which prevail in adenomas and carcinomas. The increase in enzyme activity is accompanied by increasing levels of the enzyme protein (Moore et al. 1986a) and an overexpression of the mRNA coding for glucose-6-phosphate dehydrogenase, as demonstrated by in situ hybridization with antisense mRNA (Stumpf and Bannasch 1994). Baba et al. (1989) showed, by autoradiographic studies in foci of altered hepatocytes negative and positive for glucose-6-phosphate dehydrogenase, that the increased expression of this enzyme is closely related to cell proliferation.

When glycogen storage foci give rise to mixed or basophilic cell foci, adenomas, and carcinomas, additional metabolic changes occur. The glycogen initially stored in excess is reduced (Bannasch and Müller 1964; Klimek et al. 1984); the activities of glyceraldehyde-3-phosphate dehydrogenase (Hacker et al. 1982) and α-glucosidase (Klimek and Bannasch 1989) usually increase, while the content and activity of pyruvate kinase decreases (Fischer et al. 1987a; Klimek et al. 1988; Klimek and Bannasch 1990). Microbiochemical studies

have shown that these changes are accompanied by a decrease in glucokinase and an increase in hexokinase activity, which is not an early (Fischer et al. 1987b) but a late event (Klimek and Bannasch 1993) in hepatocarcinogenesis. Those foci of altered hepatocytes described as tigroid cell foci (Bannasch et al. 1985), amphophilic cell foci (Weber et al. 1988a), and enzymatically hyperactive foci (Enzmann et al. 1989) are apparently not preceded by pronounced hepatocellular glycogenosis, but they share a number of enzymatic changes with the other types of foci.

Observations that suggest an ordered pattern of metabolic changes during hepatocarcinogenesis have also been described for a number of enzymes involved in drug metabolism, such as various cytochrome P-450 isoenzymes and epoxide hydrolase (Buchmann et al. 1985; Schwarz et al. 1989). Histochemical investigations of the activity of several dehydrogenases in focal hepatic lesions induced in rats with N-ethyl-N-hydroxyethylnitrosamine revealed an increase in reduced nicotinamide adenine dinucleotide phosphate (NADPH)-generating potential in foci of altered hepatocytes and particularly in hepatocellular adenomas (Moore et al. 1986c). Glucose-6-phosphate dehydrogenase, malic enzyme, and isocitrate dehydrogenase all showed enhanced activity, while the activities of succinate dehydrogenase and β-hydroxybutyrate dehydrogenase were reduced. These alterations in enzyme activity again indicate an adaptive metabolic shift in hepatocarcinogenesis (Moore et al. 1986c); the increased levels of the enzymes responsible for generation of NADPH possibly result in increased drug detoxification (Farber and Sarma 1987; Sato 1988; Gerbracht et al. 1993) or biosynthetic potential, such as cholesterogenesis and DNA synthesis (Ledda-Columbano et al. 1985; Gerbracht et al. 1993).

Phenotypic Modulation

Different types of phenotypic modulation in foci of altered hepatocytes have been described for two classes of chemicals: (1) microsomal enzyme inducers and (2) peroxisome inducers. The microsomal enzyme inducer studied most extensively is phenobarbital. Additional administration of phenobarbital to rats previously exposed to hepatocarcinogens leads to a predominance of acidophilic cells (exhibiting marked proliferation of smooth endoplasmic reticulum) in early foci of altered hepatocytes (Ward and Ohshima 1985; Evans et al. 1986; Jack et al. 1990a) and to a more rapid shift toward basophilia, which characterizes later stages of hepatocarcinogenesis (Moore et al. 1983; Ito et al. 1984; Schulte-Hermann et al. 1990). In addition, a more pronounced expression of a number of enzymatic changes, such as increased activity of γ-glutamyltransferase and decreased activity of adenosine triphosphatase, was observed in foci of altered hepatocytes under these experimental conditions (Pitot et al. 1978; Watanabe and Williams 1978; Moore et al. 1983; Ito et al. 1984; Schulte-Hermann et al. 1986; Maruyama et al. 1990; Tsuda et al. 1992). A reduction in single-cell necrosis (apoptosis) by additional treatment with phenobarbital was also noted in foci of altered hepatocytes in rats previously exposed to N-nitrosomorpholine (Bursch et al. 1984, 1990; Schulte-Hermann et al. 1990).

In contrast, several peroxisome inducers, such as nafenopin, clofibrate, and ciprofibrate, inhibit the expression of certain enzymes such as γ-glutamyltransferase and GST-P in foci of altered hepatocytes (Numoto et al. 1984; Stäubli et al. 1984; Furukawa et al. 1985; Glauert et al. 1986; Préat et al. 1986; Hosokawa et al. 1989; Gerbracht et al. 1990; Maruyama et al. 1990; Tsuda et al. 1992). Gerbracht et al. (1990) emphasized that there was no increase in apoptosis within foci of altered hepatocytes under these conditions, which might have been an alternative explanation for the reduction in the number and size of enzyme-altered foci. After additional administration of clofibrate to rats pretreated with N-nitrosodiethylamine, Hosokawa and colleagues (1989) found that foci of altered hepatocytes both positive and negative for GST could be identified morphologically. The total number of foci of altered hepatocytes, both positive and negative for GST, was higher than in rats treated with N-nitrosodiethylamine alone, indicating an enhancing rather than a reducing effect on the development of foci of altered hepatocytes. A higher incidence of hepatocellular carcinomas was also seen after additional treatment with clofibrate. An enhancing effect of nafenopin on rat hepatocarcinogenesis was described by Kraupp-Grasl et al. (1990, 1991). In addition, these authors observed a variable expression of enzymes of the peroxisomal β-oxidation, ranging

from reduced to increased contents in preneoplastic hepatic foci that were negative for γ-glutamyltranspeptidase and showed low levels or absence of several GST isoenzymes (Grasl-Kraupp et al. 1993a, b). A general reduction in the expression of peroxisomal enzymes was described in preneoplastic and neoplastic lesions induced in rat liver by the peroxisomal proliferator clofibrate (Yokoyama et al. 1992). The same authors reported a diametrically opposed expression of peroxisomal enzymes and GST-P in focal liver lesions induced by the Solt-Farber protocol followed by clofibrate (Yokoyama et al. 1993).

The heterogeneity and instability of the phenotypic cellular changes that characterize foci of altered hepatocytes have serious implications for quantitative assessment, which have been discussed in detail in a recent review (Bannasch and Zerban 1992).

Comparison with Other Species

Clear and acidophilic glycogen storage foci, mixed cell foci, basophilic cell foci, and enzyme-altered foci preceding the development of hepatic tumors have also been described in a number of other species, including primates (Bannasch and Zerban 1992). Induction of foci of altered hepatocytes by chemicals has been particularly observed in mice (Ward 1984; Vesselinovitch et al. 1985), hamsters (Stenbäck et al. 1986; Thamavit et al. 1987), and fish (Couch and Courtney 1987; Hinton et al. 1988). A high incidence of foci of altered hepatocytes was also found in woodchucks infected with the woodchuck hepatitis virus (Abe et al. 1988; Toshkov et al. 1990; Bannasch et al. 1995), which is an oncogenic DNA virus closely related to the human hepatitis B virus; a high incidence was also found in transgenic mice that are prone to the development of hepatic tumors (Kim et al. 1991; Bannasch and Zerban 1992; Toshkov et al. 1994). Enzmann et al. (1992a, 1995) reported that foci of altered hepatocytes may also be induced in embryonal turkey liver by injection of N-nitrosomorpholine, urethane, or diethylnitrosamine into fertilized turkey eggs. This may be a promising approach for the replacement of classical animal experiments in carcinogenesis bioassays. Although it appears that the sequential cellular changes during hepatocarcinogenesis are very similar in all spe-

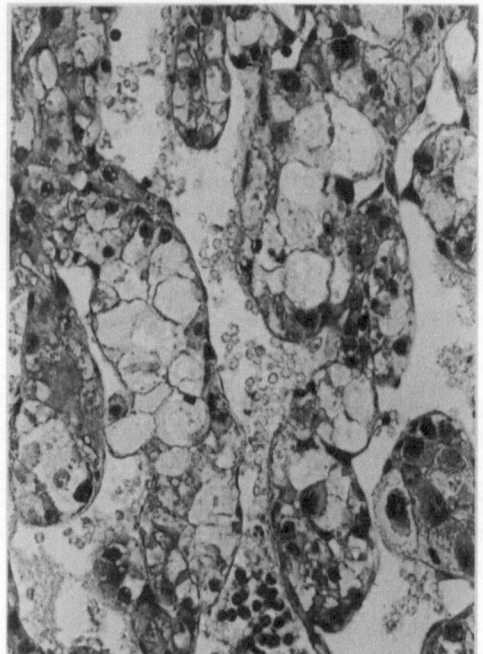

Fig. 17. Clear cell carcinoma induced in rat liver by *N*-nitrosomorpholine. Clear cells predominated also in lung metastases of this tumor. H&E, ×300

cies investigated, some interesting differences have been reported, for instance in mice (see p. 38). Of particular interest are findings suggesting that the sequence of cellular changes in man and experimental animals is, in principle, identical. Foci of clear or acidophilic (ground class) cells storing glycogen in excess have been found in livers of patients suffering from liver cirrhosis and bearing multicentric hepatocellular carcinomas (Bannasch et al. 1992; Altmann 1994). The predominance of glycogen-rich cells in many adenomas and in clear cell carcinomas (Fig. 17) of the liver is well known (Wu et al. 1983). However, the most challenging aspect of human pathology in this context is the appearance of hepatic tumors increasingly reported in patients suffering from inborn hepatic glycogenosis, mostly that of the von Gierke type (Bannasch et al. 1984; Bianchi 1993). Liver tumors were detected by scintigrams, ultrasonograms, biopsies, or autopsies in many patients with this genetically fixed disease. Whereas the tumors were relatively rare in the first decade of life, they developed in most patients who had passed through adolescence.

Histologically, both adenomas and carcinomas were diagnosed, and in some cases the clinical course of the disease suggested transformation from adenomas into carcinomas. These clinical observations support the hypothesis (Bannasch 1968; Bannasch et al. 1984) that the molecular changes underlying the aberrations in carbohydrate metabolism that emerge during hepatocarcinogenesis are causally related to neoplastic transformation of the hepatocytes.

References

Abe K, Kurata T, Shikata T, Tennant BC (1988) Enzyme-altered liver cell foci in woodchucks infected with woodchuck hepatitis virus. Jpn J Cancer Res 79:466–472

Abmayr W, Deml E, Oesterle D, Gössner W (1983) Nuclear morphology in preneoplastic lesions of rat liver. Anal Quant Cytol 5:275–284

Alexandre K, Jacobovitz D, Galand P (1990) Immunohistochemical detection of *c-myc* and c-*erbA* products in diethylnitrosamine-induced preneoplastic and neoplastic liver lesions. Carcinogenesis 11:1189–1194

Altmann H (1994) Hepatic neoformations. Pathol Res Pract 190:513–577

Altmann HW, Osterland U (1961) Über cytoplasmatische Wirbelbildungen in den Leberzellen der Ratte bei chronischer Thioacetamidvergiftung. Beitr Pathol Anat 124:1–18

Baba M, Yamamoto R, Iishi H, Tatsuta M, Wada A (1989) Role of glucose-6-phosphate dehydrogenase on enhanced proliferation of pre-neoplastic and neoplastic cells in rat liver induced by N-nitrosomorpholine. Int J Cancer 43:892–895

Bannasch P (1968) The cytoplasm of hepatocytes during carcinogenesis electron and light microscopical investigations of the nitrosomorpholine-intoxicated rat liver. In: Rentchnick P, Herfarth C, Senn HJ (eds) Recent results in cancer research, vol 19. Springer, Berlin Heidelberg New York, pp 1–100

Bannasch P (1975) Die Cytologie der Hepatocarcinogenese. In: Grundmann E (ed) Handbuch der allgemeinen Pathologie, vol VI/VII. Springer, Berlin Heidelberg New York, pp 123–276

Bannasch P (1976) Cytology and cytogenesis of neoplastic (hyperplastic) hepatic nodules. Cancer Res 36:2555–2562

Bannasch P (1986) Preneoplastic lesions as end points in carcinogenicity testing. Carcinogenesis 7:849–852

Bannasch P, Müller HA (1964) Lichtmikroskopische Untersuchungen über die Wirkung von N-Nitrosomorpholin auf die Leber von Ratte und Maus. Arzneimittelforschung 14:805–814

Bannasch P, Zerban H (1990) Tumours of the liver. In: Turusov V, Mohr U (eds) Pathology of tumours in laboratory animals, vol I: tumours of the rat. International Agency for Cancer, Lyon, pp 199–240

Bannasch P, Zerban H (1992) Predictive value of hepatic preneoplastic lesions as indicators of carcinogenic response. In: Vainio H, Magee P, McGregor D, McMichael AJ (eds)

Mechanisms of carcinogenesis in risk identification. International Agency for Research on Cancer, Lyon, pp 381–419

Bannasch P, Zerban H (1994) Preneoplastic and neoplastic lesions of the liver of the rat. In: Bannasch P, Gössner W (eds) Pathology of neoplasia and preneoplasia in rodents. EULEP color atlas. Schattauer, Stuttgart, pp 18–63

Bannasch P, Papenburg J, Ross W (1972) Cytomorphologic and morphometric investigations of hepatocarcinogenesis. I. Reversible and irreversible cytoplasmic alterations of hepatocytes in nitrosomopholine-intoxicated rats. Z Krebsforsch 77:108–133

Bannasch P, Hesse J, Angerer H (1974) Hepatocellular glycogenosis and the genesis of so-called hyperplastic liver nodules in thioacetamide intoxicated rats. Virchows Arch [B] Cell Pathol 7:29–50

Bannasch P, Mayer D, Hacker HJ (1980) Hepatocellular glycogenosis and hepatocarcinogenesis. Biochim Biophys Acta 605:217–245

Bannasch P, Moore MA, Klimek F, Zerban H (1982) Biological markers of preneoplastic foci and neoplastic nodules in rodent liver. Toxicol Pathol 10:19–34

Bannasch P, Hacker HJ, Klimek F, Mayer D (1984) Hepatocellular glycogenosis and related pattern of enzymatic changes during hepatocarcinogenesis. Adv Enzyme Regul 22:97–121

Bannasch P, Benner U, Enzmann H, Hacker HJ (1985) Tigroid cell foci and neoplastic nodules in the liver of rats treated with a single dose of aflatoxin B1. Carcinogenesis 6:1641–1648

Bannasch P, Enzmann H, Klimek F, Weber E, Zerban H (1989) Significance of sequential cellular changes inside and outside foci of altered hepatocytes during hepatocarcinogenesis. Toxicol Pathol 17:617–628

Bannasch P, Jahn UR, Zerban H (1992) Preneoplastic lesions as early indicators of neoplastic development. In: Bannasch P (ed) Cancer diagnosis, early detection. Springer, Berlin Heidelberg New York, pp 178–190

Bannasch P, Imani NK, Hacker HJ, Radaeva S, Mrozek M, Zillmann U, Kopp-Schneider A, Haberkorn U, Elgas M, Tolle T, Roggendorf M, Toskov I (1995) Synergistic hepatocarcinogenic effect of hepadnaviral infection and dietary aflatoxin B₁ in woodchucks. Cancer Res 55:3318–3330

Beer DG, Schwarz M, Sawada N, Pitot HC (1986) Expression of H-ras, c-myc protooncogenes in isolated gamma-glutamyl transpeptidase-positive rat hepatocytes and in hepatocellular carcinomas induced by diethylnitrosamine. Cancer Res 46:2435–2441

Beer DG, Neveu MJ, Paul DL, Rapp UR, Pitot HC (1988) Expression of the c-raf protooncogene, gamma-glutamyl transpeptidase, and gap junction protein in rat liver neoplasms. Cancer Res 48:1610–1617

Bianchi L (1993) Glycogen storage disease I and hepatocellular tumours. Eur J Pediatr 152 [Suppl 1]:S63–S70

Bruni C (1973) Distinctive cells similar to fetal hepatocytes associated with liver carcinogenesis by diethylnitrosamine. Electron microscopic study. J Natl Cancer Inst 50:1513–1528

Buchmann A, Kuhlmann W, Schwarz M, Kunz W, Wolf CR, Moll E, Friedberg T, Oesch F (1985) Regulation and expression of four cytochrome P-450 isoenzymes, NADPH-cytochrome P-450 reductase, the glutathione transferase B, C, and microsomal epoxide hydrolase in preneoplastic and neoplastic lesions in rat liver. Carcinogenesis 6:513–521

Burek JD (1978) Pathology of aging rats. A morphological, experimental study of the age-associated lesions in aging BN/B, WA6/Rij and (WAxBN)F1 rats. CRC, West Palm Beach, pp 58–68

Bursch W, Lauer B, Timmermann-Trosiener I, Barthel G, Schuppler J, Schulte-Hermann R (1984) Controlled cell death (apoptosis) of normal and putative preneoplastic cells in rat liver following withdrawal of tumor promoters. Carcinogenesis 5:453–458

Bursch W, Paffe S, Putz B, Barthel G, Schulte-Hermann R (1990) Determination of the length of the histological stages of apoptosis in normal liver and in altered hepatic foci of rats. Carcinogenesis 11:847–853

Cayama E, Tsuda H, Sarma DS, Farber E (1978) Initiation of chemical carcinogenesis requires cell proliferation. Nature 275:60–62

Columbano A, Rajalakshmi S, Sarma DS (1981) Requirement of cell proliferation for the initiation of liver carcinogenesis as assayed by three different procedures. Cancer Res 41:2079–2083

Columbano A, Ledda-Columbano GM, Rao PM, Rajalakshmi S, Sarma DS (1984) Occurrence of cell death (apoptosis) in preneoplastic and neoplastic liver cells: a sequential study. Am J Pathol 116:441–446

Columbano A, Ledda-Columbano GM, Lee G, Rajalakshmi S, Sarma DS (1987) Inability of mitogen-induced liver hyperplasia to support the induction of enzyme-altered islands induced by liver carcinogens. Cancer Res 47:5557–5559

Columbano A, Ledda-Columbano GM, Ennas MG, Curto M, Chelo A, Pani P (1990) Cell proliferation, promotion of rat liver carcinogenesis: different effect of hepatic regeneration and mitogen induced hyperplasia on the development of enzyme-altered foci. Carcinogenesis 11:771–776 .

Couch JA, Courtney LA (1987) N-nitrosodiethylamine-induced hepatocarcinogenesis in estuarine sheephead minnow (Cyprinodon variegatus): neoplasms and related lesions compared with mammalian lesions. J Natl Cancer Inst 79:297–321

Danielsen HE, Steen HB, Lindmo T, Reith A (1988) Ploidy distribution in experimental liver carcinogenesis in mice. Carcinogenesis 9:59–63

Daoust R, Calamai R (1971) Hyperbasophilic foci as sites of neoplastic transformation in hepatic parenchyma. Cancer Res 31:1290–1296

Decloitre F, Lafarge-Frayssinet C, Barroso M, Lechner MC, Oudelhkim M, Frayssinet C (1990) Effect of rat developmental stage at initiation on the expression of biochemical markers during liver tumor promotion. Tumour Biol 11:295–305

Deleener A, Castelain P, Preat V, de Gerlache J, Alexandre H, Kirsch-Volders M (1987) Changes in nucleolar transcriptional activity and nuclear DNA content during the first steps of rat hepatocarcinogenesis. Carcinogenesis 8:195–201

Digernes V (1983) Chemical liver carcinogenesis: monitoring of the process by flow cytometric DNA measurements. Environ Health Perspect 50:195–200

Dragan Y, Teeguarden J, Campbell H, Hsia S, Pitot H (1995) The quantitation of altered hepatic foci during multistage hepatocarcinogenesis in the rat: transforming growth factor α expression as a marker for the stage of progression. Cancer Lett 93:73–83

Ehemann V, Mayer D, Hacker HJ, Bannasch P (1986) Loss of adenylate cyclase activity in preneoplastic and neoplastic lesions induced in rat liver by N-nitrosomorpholine. Carcinogenesis 7:567-573

Embleton MJ, Butler PC (1988) Reactivity of monoclonal antibodies to oncoproteins with normal rat liver, carcinogen-induced tumours, and premalignant liver lesions. Br J Cancer 57:48-53

Emmelot P, Scherer E (1980) The first relevant cell stage in rat liver carcinogenesis: a quantitative approach. Biochim Biophys Acta 605:247-304

Enomoto K, Farber E (1982) Kinetics of phenotypic maturation of remodeling of hyperplastic nodules during liver carcinogenesis. Cancer Res 42:2330-2335

Enomoto K, Ying TS, Griffin MJ, Farber E (1981) Immunohistochemical study of epoxide hydrolase during experimental liver carcinogenesis. Cancer Res 41:3281-3287

Enzmann H, Bannasch P (1987) Potential significance of phenotypic heterogeneity of focal lesions at different stages in hepatocarcinogenesis. Carcinogenesis 8:1607-1612

Enzmann H, Dettler T, Ohlhauser D, Bannasch P (1988) Elevation of glucose-6-phosphate in early stages of hepatocarcinogenesis induced in rats by N-nitrosomorpholine. Horm Metabol Res 20:128-129

Enzmann H, Ohlhauser D, Enzmann H, Dettler T, Benner U, Hacker HJ, Bannasch P (1989) Unusual histochemical pattern in preneoplastic hepatic foci characterized by hyperactivity of several enzymes. Virchows Arch B Cell Pathol Incl Mol Pathol 57:99-108

Enzmann H, Kahner G, Watta-Gebert B, Löser E (1992a) Foci of altered hepatocytes induced in embryonal turkey liver. Carcinogenesis 13:943-946

Enzmann H, Zerban H, Löser E, Bannasch P (1992b) Glycogen phosphorylase hyperactive foci of altered hepatocytes in aged rats. Virchows Arch [B] Cell Pathol 62:3-8

Enzmann H, Kühlem C, Kaliner G, Löser E, Bannasch P (1995) Rapid induction of preneoplastic liver foci in embryonal turkey liver by diethylnitrosamine. Toxicol Pathol 23:560-569

Evans JG, Collins MA, Savage SA, Lake BG, Butler WH (1986) The histology and development of hepatic nodules in C3H/He mice following chronic administration of phenobarbitone. Carcinogenesis 7:627-631

Farber E (1980) The sequential analysis of liver cancer induction. Biochim Biophys Acta 605:149-166

Farber E, Sarma DS (1987) Hepatocarcinogenesis: a dynamic cellular perspective. Lab Invest 56:4-22

Feldman D, Swarm RL, Becker J (1981) Ultrastructural study of rat liver and liver neoplasms after long-term treatment with phenobarbital. Cancer Res 41:2151-2162

Fischer G, Ullrich D, Katz N, Bock KW, Schauer A (1983) Immunohistochemical and biochemical detection of uridinediphosphate-glucuronyl-transferase (UDP-GT) activity in putative preneoplastic liver foci. Virchows Arch [B] Cell Pathol Mol Pathol 42:193-200

Fischer G, Domingo M, Lodder D, Katz N, Reinacher M, Eigenbrodt E (1987a) Immunohistochemical demonstration of decreased L-pyruvate kinase in enzyme altered rat liver lesions produced by different carcinogens. Virchows Arch B Cell Pathol Mol Pathol 53:359-364

Fischer G, Ruschenburg J, Eigenbrodt E, Katz N (1987b) Decrease in glucokinase, glucose-6-phosphatase and increase in hexokinase in putative preneoplastic lesions of rat liver. J Cancer Res Clin Oncol 113:430-436

Frenkel RA, Slaughter CA, Orth K, Moomaw CR, Hicks SH, Snyder JM, Bennet M, Prough RA (1990) Peroxisome proliferation and induction of enzymes in mouse and rat liver by dehydroepiandrosterone feeding. J Steroid Biochem 35:333-342

Friedrich-Freksa H, Papadopulu G, Gössner W (1969) Histochemical investigations of carcinogenesis in rat liver after time-limited application of diethylnitrosamine. Z Krebsforsch 72:240-253

Furukawa K, Numoto S, Furuya K, Furukawa NT, Williams GM (1985) Effects of the hepatocarcinogen nafenopin, a peroxisome-proliferator, on the activities of rat liver glutathione/requiring enzymes and catalase in comparison to the action of phenobarbital. Cancer Res 45:5011-5019

Galand P, Jacobovitz D, Alexandre K (1988) Immunohistochemical detection of c-Ha-ras oncogene p21 product in pre-neoplastic and neoplastic lesions during hepatocarcinogenesis in rats. Int J Cancer 41:155-161

Gerbracht U, Bursch W, Kraus P, Putz B, Reinacher M, Timmermann-Trosiener I, Schulte-Hermann R (1990) Effects of hypolipidemic drugs nafenopin and clofibrate on phenotypic expression and cell death (apoptosis) in altered foci of rat liver. Carcinogenesis 11:617-624

Gerbracht U, Eigenbrodt E, Simile MM, Pascale RM, Gaspa L, Daino L, Seddaiu MA, De Miglio MR, Nufris A, Feo F (1993) Effect of S-adenosyl-L-methionine on the development of preneoplastic foci and the activity of some carbohydrate metabolizing enzymes in the liver, during experimental hepatocarcinogenesis. Anticancer Res 13:1965-1972

Gil R, Callaghan R, Boix J, Pellin A, Llombart-Bosch A (1988) Morphometric, cytophotometric nuclear analysis of altered hepatocyte foci induced by N-nitrosomorpholine (NNM), aflatoxin B1 (AFB1) in liver of Wistar rats. Virchows Arch [B] Cell Pathol Mol Pathol 54:341-349

Glauert HP, Schwarz M, Pitot HC (1986) The phenotypic stability of altered hepatic foci: effect of the short-term withdrawal of phenobarbital and of long-term feeding of purified diets after the withdrawal of phenobarbital. Carcinogenesis 7:117-121

Glauert HP, Beaty MM, Clark TD, Greenwell WS, Tatum V, Chen LC, Borges T, Clark TL, Srinivasan SR, Chow CK (1990) Effect of dietary vitamin E on the development of altered hepatic foci and hepatic tumors induced by the peroxisome proliferator ciprofibrate. J Cancer Res Clin Oncol 116:351-356

Goldsworthy TL, Pitot HC (1985) The quantitative analysis, stability of histochemical markers of altered hepatic foci in rat liver following initiation by diethylnitrosamine administration, promotion with phenobarbital. Carcinogenesis 6:1261-1269

Goldsworthy TL, Campbell HA, Pitot HC (1984) The natural history, dose-response characteristics of enzyme-altered foci in rat liver following phenobarbital and diethylnitrosamine administration. Carcinogenesis 5:67-71

Goodman DG, Maronpot RR, Newberne PM, Popp JA, Squire RA (1994) Proliferative and selected other lesions in the liver of rats. In: Streett CS, Burek JD, Hardisty JF, Garner FM, Leininger JR, Pletscher JM, Moch RW (eds) Guides for toxicologic pathology. STP/ARP/AFIP, Washington, pp G1-5, 1-24

Gössner W, Friedrich-Freksa H (1964) Histochemische Untersuchungen über die Glucose-6-Phosphatase in der

Rattenleber während der Kanzerisierung durch Nitrosamine. Z Naturforsch 19b:862–863

Grasl-Kraupp B, Huber W, Just W, Gibson G, Schulte-Hermann R (1993a) Enhancement of peroxisomal enzymes, cytochrome P-452 and DNA synthesis in putative preneoplastic foci of rat liver treated with the peroxisome proliferator nafenopin. Carcinogenesis 14:1007–1012

Grasl-Kraupp B, Waldhör T, Huber W, Schulte-Hermann R (1993b) Glutathione S-transferase isoenzyme patterns in different subtypes of enzyme-altered rat liver foci treated with the peroxisome proliferator nafenopin or with phenobarbital. Carcinogenesis 14:2407–2412

Greaves P, Irisarri E, Monroe AM (1986) Hepatic foci of cellular and enzymatic alteration and nodules in rats treated with clofibrate or diethylnitrosamine followed by phenobarbital: their rate of onset and their reversibility. J Natl Cancer Inst 76:475–484

Grobholz R, Hacker HJ, Thorens B, Bannasch P (1993) Reduction in the expression of glucose transporter protein GLUT 2 in preneoplastic and neoplastic hepatic lesions and reexpression of GLUT 1 in late stages of hepatocarcinogenesis. Cancer Res 53:4204–4211

Grundmann E, Sieburg H (1962) Die Histogenese, Cytogenese des Lebercarcinoms der Ratte durch Diäthylnitrosamin im lichtmikrokopischen Bild. Beitr Pathol Anat 126:57–90

Hacker HJ, Moore MA, Mayer D, Bannasch P (1982) Correlative histochemistry of some enzymes of carbohydrate metabolism in preneoplastic and neoplastic lesions in the rat liver. Carcinogenesis 3:1265–1272

Hacker HJ, Steinberg P, Toshkov I, Oesch F, Bannasch P (1992) Persistence of the cholangiocellular and hepatocellular lesions observed in rats fed a choline-deficient/DL-ethionine-supplemented diet. Carcinogenesis 13:271–276

Haesen S, Derijcke T, Deleener A, Castelain P, Alexandre H, Preat V, Kirsch-Volders M (1988) The influence of phenobarbital and butyrated hydroxytoluene on the ploidy rate in rat hepatocarcinogenesis. Carcinogenesis 9:1755–1761

Harada T, Maronpot RR, Morris RW, Boorman GA (1989a) Observations on altered hepatocellular foci in National Toxicology Program two-year carcinogenicity studies in rats. Toxicol Pathol 17:690–706

Harada T, Maronpot RR, Morris RW, Stitzel KA, Boorman GA (1989b) Morphological and stereological characterization of hepatic foci of cellular alteration in control Fischer 344 rats. Toxicol Pathol 17:579–593

Hasegawa R, Ito N (1994) Hepatocarcinogenesis in the rat. In: Waalkes MP, Ward JM (eds) Carcinogenesis. Raven, New York, pp 39–65

Hasegawa R, Takahashi S, Imaida K, Yamaguchi S, Shirai T, Ito N (1991) Age-dependent induction of preneoplastic liver cell foci by 2-acetylaminofluorene, phenobarbital and acetaminophen in F344 rats initially treated with diethylnitrosamine. Jpn J Cancer Res (Gann) 82:293–297

Hayashi F, Tamura H, Yamada J, Kasai H, Suga T (1994) Characteristics of the hepatocarcinogenesis caused by dehydroepiandrosterone, a peroxisome proliferator, in male F-344 rats. Carcinogenesis 15:2215–2219

Hendrich S, Campbell HA, Pitot HC (1987) Quantitative stereological evaluation of four histochemical markers of altered foci in multistage hepatocarcinogenesis in the rat. Carcinogenesis 8:1245–1250

Hinton DE, Couch JA, Teh SJ, Courtney LA (1988) Cytological changes during progression of neoplasia in selected fish species. Aquatic Toxicol 11:77–112

Hirota N, Williams GM (1982) Ultrastructural abnormalities in carcinogen-induced hepatocellular altered foci identified by resistance to iron accumulation. Cancer Res 42:2298–2309

Hirota N, Yokoyama T (1985) Comparative study of abnormality in glycogen storing capacity, other histochemical phenotypic changes in carcinogen-induced hepatocellular preneoplastic lesions in rats. Acta Pathol Jpn 35:1163–1179

Hosokawa S, Tatematsu M, Aoki T, Nakanowatari J, Igarashi T, Ito N (1989) Modulation of diethylnitrosamine-initiated placental glutathione S-transferase positive preneoplastic and neoplastic lesions by clofibrate, a hepatic peroxisome proliferator. Carcinogenesis 10:2237–2241

Hully JR, Su Y, Lohse JK, Griep AE, Sattler CA, Haas MJ, Dragan Y, Peterson J, Neveu M, Pitot HC (1994) Transgenic hepatocarcinogenesis in the rat. Am J Pathol 145:384–397

ILAR, NCR (1980) Histological typing of liver tumors of the rat. J Natl Cancer Inst 64:179–206

Ishikawa T, Takayama S, Kitagawa T (1980) Correlation between time of partial hepatectomy after a single treatment with diethylnitrosamine and induction of adenosine triphosphatase deficient islands in rat liver. Cancer Res 40:4261–4264

Ito N, Moore MA, Bannasch P (1984) Modification of the development of N-nitrosomorpholine-induced hepatic lesions by 2-acetylaminofluorene, phenobarbital and 4,4′-diaminodiphenylmethane: a sequential histological and histochemical analysis. Carcinogenesis 5:335–342

Ito N, Tatematsu M, Hasegawa R, Tsuda H (1989) Medium-term bioassay system for detection of carcinogens and modifiers of hepatocarcinogenesis utilizing the GST-P positive liver cell focus as an endpoint marker. Toxicol Pathol 17:630–641

Ito N, Shirai T, Hasegawa R (1992) Medium-term bioassays for carcinogens. In: Vainio H, Magee PN, McGregor DB, McMichael AJ (eds) Mechanisms of carcinogenesis in risk identification. IARC scientific publications, International Agency for Research on Cancer, Lyon, pp 353–388

Jack EM, Bentley P, Bieri F, Muakkassah-Kelly SF, Stäubli W, Suter J, Waechter F, Cruz-Orive LM (1990a) Increase in hepatocyte and nuclear volume and decrease in the population of binucleated cells in preneoplastic foci of rat liver: a stereological study using the nucleator methods. Hepatology 11:286–297

Jack EM, Stäubli W, Waechter F, Bentley P, Suter J, Bieri F, Muakkassah-Kelly SF, Cruz-Orive LM (1990b) Ultrastructural changes in chemically induced preneoplastic focal lesions in the rat liver: a stereological study. Carcinogenesis 11:1531–1538

Kalengayi MM, Desmet VJ (1975) Sequential histological and histochemical study of the rat liver during aflatoxin B1-induced carcinogenesis. Cancer Res 35:2845–2852

Karasaki S (1969) The fine structure of proliferating cells in preneoplastic rat liver during azo-dye carcinogenesis. J Cell Biol 40:322–335

Kaufmann WK, Rahija RJ, MacKenzie SA, Kaufman DG (1987) Cell cycle-dependent initiation of hepatocarcinogenesis in rats by (±)7r-8t-dihydroxy-9t,10t-epoxy-7,8,9,10-tetrahydrobenzo(a) pyrene. Cancer Res 47:3771–3775

Kaufmann WK, Zhang Y, Kaufman DG (1992) Association between expression of transforming growth factor-alpha and progression of hepatocellular foci to neoplasms. Carcinogenesis 13:1481–1483

Kim CM, Koike K, Saito I, Miyamura T, Jay G (1991) HBx gene of hepatitis B virus induces liver cancer in transgenic mice. Nature 351:317–320

Kitagawa T (1971) Histochemical analysis of hyperplastic lesions, hepatomas of the liver of rats fed 2-fluorenylacetamide. Gann 62:207–216

Kitagawa T, Sugano H (1973) Combined enzyme histochemical and radioautographic studies on areas of hyperplasia in the liver of rats fed N-2-fluorenylacetamide. Cancer Res 33:2993–3001

Klimek F, Bannasch P (1989) Biochemical microanalysis of alpha glucosidase activity in preneoplastic and neoplastic hepatic lesions induced in rats by N-nitrosomorpholine. Virchows Arch [B] Cell Pathol Mol Pathol 57:245–250

Klimek F, Bannasch P (1990) Biochemical microanalysis of pyruvate kinase activity in preneoplastic and neoplastic liver lesions induced in rats with N-nitrosomorpholine. Carcinogenesis 11:1377–1380

Klimek F, Bannasch P (1993) Isoenzyme shift from glucokinase to hexokinase is not an early but a late event in hepatocarcinogenesis. Carcinogenesis 14:1857–1861

Klimek F, Mayer D, Bannasch P (1984) Biochemical microanalysis of glycogen content and glucose-6-phosphate dehydrogenase activity in focal lesions of the rat liver induced by N-nitrosomorpholine. Carcinogenesis 5:265–268

Klimek F, Moore MA, Schneider E, Bannasch P (1988) Histochemical and microbiochemical demonstration of reduced pyruvate kinase activity in thioacetamide-induced neoplastic nodules of rat liver. Histochemistry 90:37–42

Kraupp-Grasl B, Huber W, Putz B, Gerbracht U, Schulte-Hermann R (1990) Tumor promotion by the peroxisome proliferator nafenopin involving a specific subtype of altered foci in rat liver. Cancer Res 50:3701–3708

Kraupp-Grasl B, Huber W, Taper H, Schulte-Hermann R (1991) Increased susceptibility of aged rats to hepatocarcinogenesis by the peroxisome proliferator nafenopin and the possible involvement of altered liver foci occurring spontaneously. Cancer Res 51:666–671

Kuhlmann WD, Krischan R, Kunz W, Guenthner TM, Oesch F (1981) Focal elevation of liver microsomal epoxide hydrolase in early preneoplastic stages and its behaviour in the further course of hepatocarcinogenesis. Biochem Biophys Res Commun 98:417–423

Ledda-Columbano GM, Columbano A, Dessi S, Coni P, Chiodino C, Pani P (1985) Enhancement of cholesterol synthesis and pentose phosphate pathway activity in proliferating hepatocyte nodules. Carcinogenesis 6:1371–1373

Maguire S, Rabes HM (1989) Intralobular distribution of preneoplastic foci in rat liver after a single dose of N-methyl-N-nitrosourea (MNU) following partial hepatectomy. Carcinogenesis 10:871–874

Marsman DS, Popp JA (1994) Biological potential of basophilic hepatocellular foci and hepatic adenoma induced by the peroxisome proliferator, Wy-14,643. Carcinogenesis 15:111–117

Maruyama H, Tanaka T, Williams GM (1990) Effects of the peroxisome proliferator di(2-ethylhexyl)phthalate on enzymes in rat liver and on carcinogen-induced liver altered

foci in comparison to the promoter phenobarbital. Toxicol Pathol 18:257–267

Mathur M, Rizvi TA, Nayak NC (1992) Aflatoxin B1 induced hepatocarcinogenesis in neonatal rats. Indian J Exp Biol 30:165–168

Metzger C, Mayer D, Hoffmann H, Bocker T, Hobe G, Benner A, Bannasch P (1995) Sequential appearance and ultrastructure of amphophilic cell foci, adenomas and carcinomas in the liver of male and female rats treated with dehydroepiandrosterone. Toxicol Pathol 23:591–605

Miller RT, Cattley RC, Marsman DS, Lyght O, Popp JA (1995) TGFα differentially expressed in liver foci induced by diethylnitrosamine initiation and peroxisome proliferator promotion. Carcinogenesis 16:77–82

Mitaka T, Tsukada H (1987) Sexual differences in the histochemical characteristics of "altered cell foci" in the liver of aged Fischer 344 rats. Jpn J Cancer Res 78:785–790

Montesano R, Bartsch H, Vaino H, Wilboum, J, Yamasaki H (eds) (1986) Long-term and short-term assays for carcinogens: a critical appraisal. IARC, Lyon (IARC scientifique publications, no 83)

Moore MA, Kitagawa T (1986) Hepatocarcinogenesis in the rat: the effect of the promoters and carcinogens in vivo and in vitro. Int Rev Cytol 101:125–173

Moore MA, Mayer D, Bannasch P (1982) The dose-dependence, sequential appearance of putative pre neoplastic populations induced in the rat liver by stop experiments with N-nitrosomorpholine. Carcinogenesis 3:1429–1436

Moore MA, Hacker HJ, Bannasch P (1983) Phenotypic instability in focal and nodular lesions induced in a short term system in the rat liver. Carcinogenesis 4:595–603

Moore MA, Nakamura T, Shirai T, Ichihara A, Ito N (1986a) Immunohistochemical demonstration of increased glucose-6-phosphate dehydrogenase in preneoplastic and neoplastic lesions induced by propylnitrosamine in F344 rats and Syrian hamsters. Jpn J Cancer Res (Gann) 77:131–138

Moore MA, Nakamura T, Thamavit W, Ichihara A, Ito N (1986b) Immunohistochemically demonstrated suppressed expression of tryptophan oxygenase, a marker for liver differentiation, within putative preneoplastic rat liver lesions. Carcinogenesis 7:1393–1396

Moore MA, Tsuda H, Ito N (1986c) Dehydrogenase histochemistry of N-ethyl-N-hydroxyethylnitrosamine-induced focal liver lesions in the rat – increase in NADPH-generating capacity. Carcinogenesis 7:339–342

Mori H, Tanaka T, Sugie S, Takahashi M, Williams G (1982) DNA content of liver cell nuclei of N-2-fluorenylacetamide-induced altered foci and neoplasms in rats and human hyperplastic foci. J Natl Cancer Inst 69:1277–1282

Numoto S, Furukawa K, Furuya K, Williams GM (1984) Effects of the hepatocarcinogenic peroxisome-proliferating hypolipidemic agents clofibrate and nafenopin on the rat liver cell membrane enzymes gamma-glutamyl transpeptidase and alkaline phosphatase and on the early stages of liver carcinogenesis. Carcinogenesis 5:1603–1611

Ober S, Zerban H, Spiethoff A, Wegener K, Schwarz M, Bannasch P (1994) Preneoplastic foci of altered hepatocytes induced in rats by irradiation with a-particles of Thorotrast and neutrons. Cancer Lett 83:81–88

Ogawa K, Medline A, Farber E (1979) Sequential analysis of hepatic carcinogenesis. A comparative study of the ultrastructure of preneoplastic, malignant, prenatal, postnatal, and regenerating liver. Lab Invest 41:22–35

Ogawa K, Onoe T, Takeuchi M (1981) Spontaneous occurrence of gamma-glutamyl transpeptidase-positive hepatocytic foci in 105-week-old Wistar and 72-week-old Fischer 34 male rats. J Natl Cancer Inst 67:407–412

Opie EL (1946) Mobilisation of basophile substance (ribonucleic acid) in the cytoplasm of liver cells with the production of tumors by butter yellow. J Exp Med 84:91–106

Peraino C, Staffeldt EF, Ludeman VA (1981) Early appearance of histochemically altered hepatocyte foci and liver tumors in female rats treated with carcinogens one day after birth. Carcinogenesis 2:463–465

Peraino C, Staffeldt EF, Carnes BA, Ludeman VA, Blomquist JA, Vesselinovitch SD (1984) Characterization of histochemically detectable altered hepatocyte foci and their relationship to hepatic tumorigenesis in rats treated once with diethylnitrosamine or benzo(a)pyrene within one day after birth. Cancer Res 44:3340–3347

Peraino C, Carnes BA, Stevens FJ, Staffeldt EF, Russel JJ, Prapoulenis A, Blomquist JA, Vesselinovitch SD, Maronpot RR (1988) Comparative developmental and phenotypic properties of altered hepatocyte foci and hepatic tumors in rats. Cancer Res 48:4171–4178

Perez-Tomas R, Mayol Y, Cullere X, Diaz-Ruiz C, Domingo J (1992) Transforming growth factor-alpha expression in rat experimental hepatocarcinogenesis. Histol Histopathol 7:757–762

Pitot HC (1990) Altered hepatic foci: their role in murine hepatocarcinogenesis. Annu Rev Pharmacol Toxicol 30:465–500

Pitot HC, Sirica AE (1980) The stages of initiation and promotion in hepatocarcinogenesis. Biochim Biophys Acta 605:191–215

Pitot HC, Barsness L, Goldsworthy TL, Kitagawa T (1978) Biochemical characterization of stages of hepatocarcinogenesis after a single dose of diethylnitrosamine. Nature 271:456–458

Pitot HC, Campbell HA, Maronpot RR, Bawa N, Rizvi TA, Xu YH, Sargent L, Dragan Y, Pyron M (1989) Critical parameters in the quantitation of the stages of initiation, promotion, and progression in one model of multistage hepatocarcinogenesis in the rat. Toxicol Pathol 17:594–612

Popp JA, Scortichini BH, Garvey LK (1985) Quantitative evaluation of hepatic foci of cellular alteration occurring spontaneously in Fischer 344 rats. Fund Appl Toxicol 5:314–319

Porter KR, Bruni C (1959) An electron microscope study of the early effects of 3'-methyl-DAB on rat liver cells. Cancer Res 19:997–1009

Pound AW, McGuire LJ (1978) Repeated partial hepatectomy as a promoting stimulus for carcinogenic response of liver to nitrosamines in rats. Br J Cancer 37:585–594

Préat V, De Gerlache D, Lans M, Taper H, Roberfroid M (1986) Influence of the nature and the dose of the initiator on the development of premalignant and malignant lesions in rat hepatocarcinogenesis. Teratogenesis Carcinog Mutagen 6:165–172

Pugh TD, Goldfarb S (1978) Quantitative histochemical and autoradiographic studies of hepatocarcinogenesis in rats fed 2-acetylaminofluorene followed by phenobarbital. Cancer Res 38:4450–4457

Rabes HM (1983) Development, growth of early preneoplastic lesions induced in the liver by chemical carcinogens. J Cancer Res Clin Oncol 106:85–92

Rabes HM (1988) Cell proliferation and hepatocarcinogenesis. In: Roberfroid MB, Préat V (eds) Experimental hepatocarcinogenesis. Plenum, New York, pp 121–132

Rabes HM, Scholze P, Jantsch B (1972) Growth kinetics of diethylnitrosamine-induced enzyme-deficient preneoplastic liver cell populations in vivo and in vitro. Cancer Res 32:2577–2586

Rabes HM, Bücher T, Hartmann A, Linke I, Dünnwald M (1982) Clonal growth of carcinogen-induced enzyme-deficient preneoplastic cell populations in mouse liver. Cancer Res 42:3220–3227

Rabes HM, Müller L, Hartmann A, Kerler R, Schuster C (1986) Cell cycle-dependent initiation of adenosine triphosphatase-deficient populations in adult rat liver by a single dose of N-methyl-N-nitrosourea. Cancer Res 46:645–650

Rao MS, Subbarao V, Yeldani AV, Reddy JK (1992) Hepatocarcinogenicity of dehydroepiandrosterone in the rat. Cancer Res 52:2977–2979

Reinacher M, Eigenbrodt E, Gerbracht U, Zeuk G, Timmermann-Trosiener I, Bentley P, Waechter F, Schulte-Hermann R (1986) Pyruvate kinase isoenzymes in altered foci, carcinoma of rat liver. Carcinogenesis 7:1351–1357

Reznik-Schüller HM, Gregg M (1983) Sequential morphologic changes during metapyrilene-induced hepatocellular carcinogenesis in rats. J Natl Cancer Inst 71:1021–1031

Rinde E, Hill R, Chiu A, Haberman B (1986) Proliferative hepatocellular lesions of the rat: review and future use in risk assessment. US Environmental Protection Agency, Washington DC, pp 1–22

Romen W, Ross W, Bannasch P (1972) Cytomorphologic and morphometric studies on hepatocarcinogenesis. II. Reversibility of cell-nucleus changes in nitrosomopholine poisoned rat liver. Z Krebsforsch 77:134–140

Romen W, Bannasch P, Aziz A, Reuss W (1973) Karyokinesis and nuclear morphology during hepatocarcinogenesis. II. The fine structure of the nuclei in hepatocytes and hepatoma cells of the nitrosomorpholine-intoxicated rat liver. Virchows Arch Cell Pathol 13:267–296

Rotstein J, Sarma DS, Farber E (1986) Sequential alterations in growth control and cell dynamics of rat hepatocytes in early precancerous steps in hepatocarcinogenesis. Cancer Res 46:2377–2385

Saeter G, Schwarze PE, Nesland JM, Juul N, Pettersen EO, Seglen PO (1988) The polyploidizing growth pattern of normal rat liver is replaced by divisional and diploid growth in hepatocellular nodules and carcinomas. Carcinogenesis 9:939–945

Saraoff M, Rabes HM (1991) Demonstration of temporal and topical c-myc expression in regenerating rat liver and in neoplastic nodules by in situ hybridization. J Cancer Res Clin Oncol 117:S37–S37

Saraoff M, Rabes HM, Dormer P (1986) Correlations between ploidy and initiation probability determined by DNA cytophotometry in individual altered hepatic foci. Carcinogenesis 7:1191–1196

Sargent L, Xu YH, Sattler GL, Meisner L, Pitot HC (1989) Ploidy and karyotype of hepatocytes isolated from enzyme-altered foci in two different protocols of multistage hepatocarcinogenesis in the rat. Carcinogenesis 10:387–391

Sato K (1988) Glutathione S-transferases and hepatocarcinogenesis. Jpn J Cancer Res (Gann) 79:556–572

Sato K, Kitahara A, Satoh K, Ishikawa T, Tatematsu M, Ito N (1984) The placental form of glutathione S-transferase as a new marker protein for preneoplasia in rat chemical hepatocarcinogenesis. Gann 75:199–202

Sawaki M, Enomoto K, Takahashi H, Nakajima Y, Mori M (1990) Phenotype of preneoplastic and neoplastic liver lesions during spontaneous liver carcinogenesis of LEC rats. Carcinogenesis 11:1857–1861

Schauer A, Kunze E (1968) Enzyme histochemical and radioautographic studies during cancerization of the rat liver with diethylnitrosamine. Z Krebsforsch 70:252–266

Schauer A, Kunze E (1976) Tumours of the liver. In: Turusov V (ed) Pathology of tumours in laboratory animals, vol I. Tumours of the rat, part II. International Agency for Cancer Research, Lyon, pp 41–72

Scherer E, Emmelot P (1975) Foci of altered liver cells induced by a single dose of diethylnitrosamine and partial hepatectomy: their contribution to hepatocarcinogenesis in the rat. Eur J Cancer 11:145–154

Scherer E, Hoffmann M (1971) Probable clonal genesis of cellular islands induced in rat liver by diethylnitrosamine. Eur J Cancer 7:369–371

Scherer E, Hoffmann M, Emmelot P, Friedrich-Freksa M (1972) Quantitative study on foci of altered liver cells induced in the rat by a single dose of diethylnitrosamine and partial hepatectomy. J Natl Cancer Inst 49:93–106

Schulte-Hermann R, Timmermann-Trosiener I, Schuppler J (1983) Promotion of spontaneous preneoplastic cells in rat liver as a possible explanation of tumor production by nonmutagenic compounds. Cancer Res 43:839–844

Schulte-Hermann R, Roome N, Timmermann-Trosiener I, Schuppler J (1984) Immunocytochemical demonstration of a phenobarbital-inducible cytochrome P450 in putative preneoplastic foci of rat liver. Carcinogenesis 5:143–153

Schulte-Hermann R, Timmermann-Trosiener I, Schuppler J (1986) Facilitated expression of adaptive responses to phenobarbital in putative pre-stages of liver cancer. Carcinogenesis 7:1651–1655

Schulte-Hermann R, Timmermann-Trosiener I, Barthel G, Bursch W (1990) DNA synthesis, apoptosis, and phenotypic expression as determinants of growth of altered foci in rat liver during phenobarbital promotion. Cancer Res 50:5127–5135

Schwarz M, Buchmann A, Schulte M, Pearson D, Kunz W (1989) Heterogeneity of enzyme-altered foci in rat liver. Toxicol Lett 49:297–317

Schwarze PE, Petterson EO, Tolleshaug H, Seglen PO (1986) Isolation of carcinogen-induced diploid rat hepatocytes by centrifugal elutriation. Cancer Res 46:4732–4737

Seelmann-Eggebert G, Mayer D, Mecke D, Bannasch P (1987) Expression and regulation of glycogen phosphorylase in preneoplastic and neoplastic hepatic lesions in rats. Virchows Arch [B] Cell Pathol Mol Pathol 53:44–51

Simile MM, Pascale R, De Miglio MR, Nufris A, Daino L, Seddaiu MA, Gaspa L, Feo F (1994) Correlation between S-adenosyl-L-methionine content and production of c-myc, c-Ha-ras, and c-Ki-ras mRNA transcripts in the early stages of rat liver carcinogenesis. Cancer Lett 79:9–16

Smith MC, Yeleswarapu L, Locker J, Lombardi B (1991) Expression of p53 mutand proteins in diethylnitrosamine-induced foci of enzyme-altered hepatocytes in male Fischer-344 rats. Carcinogenesis 12:1137–1141

Soffritti M, McConnell EE (1988) Liver foci formation during aflatoxin B1 carcinogenesis in the rat. Ann N Y Acad Sci 534:531–540

Solt DB, Medline A, Farber E (1977) Rapid emergence of carcinogen-induced hyperplastic lesions in a new model for the sequential analysis of liver carcinogenesis. Am J Pathol 88:595–618

Squire RA, Levitt MH (1975) Report of a workshop on classification of specific hepatocellular lesions in rats. Cancer Res 35:3214–3223

Stäubli W, Bentley P, Bieri F, Fröhlich E, Waechter F (1984) Inhibitory effect of nafenopin upon the development of diethylnitrosamine-induced enzyme-altered foci within the rat liver. Carcinogenesis 5:41–46

Steinberg P, Hacker HJ, Dienes HP, Oesch F, Bannasch P (1991) Enzyme histochemical and immunohistochemical characterization of oval and parenchymal cells proliferating in livers of rats fed a choline-deficient/DL-ethionine-supplemented diet. Carcinogenesis 12:225–231

Steiner JW, Miyai K, Phillips MJ (1964) Electron microscopy of membrane-particle arrays in liver cells of ethionine-intoxicated rat. Am J Pathol 44:169–213

Stenbäck F, Mori H, Furuya K, Williams GM (1986) Pathogenesis of dimethylnitrosamine-induced hepatocellular cancer in hamster liver and lack of enhancement by phenobarbital. J Natl Cancer Inst 76:327–333

Stenger RJ (1966) Concentric lamellar formations in hepatic parenchymal cells of carbon tetrachloride-treated rats. J Ultrastruct Res 14:240–253

Stumpf H (1992) Untersuchungen zu Struktur- und Expressionsveränderungen zellulärer Onkogene und des Gens der Glukose-6-phosphatdehydrogenase bei der Hepatokarzinogenese der Ratte. Inaugural-dissertation, University of Gießen

Stumpf H, Bannasch P (1994) Overexpression of glucose-6-phosphate dehydrogenase in rat hepatic preneoplasia and neoplasia. Int J Oncol 5:1255–1260

Styles JA, Kelly M, Elcombe CR (1987) A cytological comparison between regeneration, hyperplasia and early neoplasia in the rat liver. Carcinogenesis 8:391–399

Sudilovsky O, Hei TK (1991) Aneuploidy and progression in promoted preneoplastic foci during chemical hepatocarcinogenesis in the rat. Cancer Lett 56:131–135

Suzuki S, Satoh K, Nakano H, Hatayama I, Sato K, Tsuchida S (1995) Lack of correlated expression between the glutathione S-transferase P-form and the oncogene products c-Jun and c-Fos in rat tissues and preneoplastic hepatic foci. Carcinogenesis 16:567–571

Tanaka T, Mori H, Hirota N, Furuya K, Williams GM (1986) Effect of DNA synthesis on induction of preneoplastic and neoplastic lesions in rat liver by a single dose of methylazoxymethanol acetate. Chem Biol Interact 58:13–27

Taper HS, Bannasch P (1976) Histochemical correlation between glycogen, nucleic acids and nucleases in preneoplastic and neoplastic lesions of rat liver after short-term administration of N-nitrosomorpholine. Z Krebsforsch 87:53–65

Taper HS, Bannasch P (1979) Histochemical differences between so-called megalocytosis and neoplastic liver lesions induced by N-nitrosomorpholine. Eur J Cancer 15:189–196

Tatematsu M, Nagamine Y, Farber E (1983) Redifferentiation as a basis for remodeling of carcinogen-induced hepatocyte nodules to normal appearing liver. Cancer Res 43:5049–5058

Tatematsu M, Aoki T, Kagawa M, Mera Y, Ito N (1988a) Reciprocal relationship between development of glutathione S-transferase positive liver foci and proliferation of surrounding hepatocytes in rats. Carcinogenesis 9:221–225

Tatematsu M, Mera Y, Inoue T, Satoh K, Sato K, Ito N (1988b) Stable phenotypic expression of glutathione S-transferase placental type and unstable phenotypic expression of gamma-glutamyltransferase in rat liver preneoplastic and neoplastic lesions. Carcinogenesis 9:215–220

Thamavit W, Ngamying M, Boonpucknavig V, Boonpucknavig S. Moore MA (1987) Enhancement of DEN-induced hepatocellular nodule development by Opisthorchis viverrini infection in Syrian golden hamsters. Carcinogenesis 8:1351–1353

Theodossiou A, Bannasch P, Reuss R (1971) Glycogen and endoplasmic reticulum of the liver cell after high doses of the carcinogen N-nitrosomorpholin. Virchows Arch [B] Cell Pathol 7:126–146

Toshkov I, Hacker HJ, Roggendorf M, Bannasch P (1990) Phenotypic patterns of preneoplastic and neoplastic lesions in woodchucks infected with woodchuck hepatitis virus. J Cancer Res Clin Oncol 116:581–590

Toshkov I, Chisari F, Bannasch P (1994) Hepatic preneoplasia in hepatitis B virus transgenic mice. Hepatology 20:1162–1172

Tsuda H, Ozaki K, Uwagawa S, Yamaguchi S, Hakoi K, Aoki T, Kato T, Sato K, Ito N (1992) Effects of modifying agents on conformity of enzyme phenotype and proliferative potential in focal preneoplastic and neoplastic liver cell lesions in rats. Jpn J Cancer Res 83:1154–1165

Tsuda H, Asamoto M, Baba H, Iwahori Y, Matsumoto K, Iwase T, Nishida Y, Nagao S, Hakoi K, Yamaguchi S, Ozaki K, Yamasaki H (1995) Cell proliferation and advancement of hepatocarcinogenesis in the rat are associated with a decrease in connexin 32 expression. Carcinogenesis 16:101–105

Tsuji S, Ogawa K, Takasaka H, Sonoda T, Mori M (1988) Clonal origin of gamma-glutamyl transpeptidase-positive hepatic lesions induced by initiation-promotion in ornithine carbamoyltransferase mosaic mice. Jpn J Cancer Res 79:148–151

US National Institute of Environmental Health Sciences (1989) Significance of foci of cellular alteration in the rat liver. A symposium. Toxicol Pathol 17:557–735

Vesselinovitch SD, Hacker HJ, Bannasch P (1985) Histochemical characterization of focal hepatic lesions induced by single diethylnitrosamine treatment in infant mice. Cancer Res 45:2774–2780

Wang JH, Hinrichsen LI, Whitacre CM, Cechner RL, Sudilovsky O (1990) Nuclear DNA content of altered hepatic foci in a rat liver carcinogenesis model. Cancer Res 50:7571–7576

Ward JM (1981) Morphology of foci of altered hepatocytes and naturally-occurring hepatocellular tumours in F344 rats. Virchows Arch Pathol Anat 390:339–345

Ward JM (1984) Morphology of potential preneoplastic hepatocyte lesions and liver tumors in mice and a comparison with other species. In: Popp JA (ed) Mouse liver neoplasia: current perspective. Hemisphere Publishing Corporation, Washington, pp 1–26

Ward JM, Ohshima M (1985) Evidence for lack of promotion of the growth of the common naturally occurring basophilic focal hepatocellular proliferative lesions in aged F344/NCr rats by phenobarbital. Carcinogenesis 6:1255–1259

Watanabe K, Williams GM (1978) Enhancement of rat hepatocellular-altered foci by the liver tumor promoter phenobarbital: evidence that foci are precursors of neoplasms and that the promoter acts on carcinogen-induced lesions. J Natl Cancer Inst 61:1311–1314

Weber E, Bannasch P (1994a) Dose- and time-dependence of the cellular phenotype in rat hepatic preneoplasia and neoplasia induced by single oral exposure to N-nitrosomorpholine. Carcinogenesis 15:1219–1226

Weber E, Bannasch P (1994b) Dose- and time-dependence of the cellular phenotype in rat hepatic preneoplasia and neoplasia induced in stop experiments by oral exposure to N-nitrosomorpholine. Carcinogenesis 15:1227–1234

Weber E, Bannasch P (1994c) Dose- and time-dependence of the cellular phenotype in rat hepatic preneoplasia and neoplasia induced by continuous oral exposure to N-nitrosomorpholine. Carcinogenesis 15:1235–1242

Weber E, Moore MA, Bannasch P (1988a) Enzyme histochemical and morphological phenotype of amphophilic foci and amphophilic/tigroid cell adenomas in rat liver after combined treatment with dehydroepiandrosterone and N-nitrosomorpholine. Carcinogenesis 9:1049–1054

Weber E, Moore MA, Bannasch P (1988b) Phenotypic modulation of hepatocarcinogenesis and reduction in N-nitrosomorpholine-induced hemangiosarcoma and adrenal lesion development in Sprague-Dawley rats by dehydroepiandrosterone. Carcinogenesis 9:1191–1195

Weinberg WC, Iannaccone PM (1988) Clonality of preneoplastic liver lesions: histological analysis in chimeric rats. J Cell Sci 89:423–431

Williams GM (1980) The pathogenesis of rat liver cancer caused by chemical carcinogens. Biochim Biophys Acta 605:167–189

Williams GM, Watanabe K (1978) Quantitative kinetics of development of N-2-fluorenylacetamide-induced, altered-(hyperplastic)-hepatocellular foci resistant to iron accumulation and of their reversion or persistence following removal of carcinogen. J Natl Cancer Inst 61:113–121

Williams GM, Klaiber M, Parker SE, Farber E (1976) Nature of early appearing, carcinogen-induced liver lesions to iron accumulation. J Natl Cancer Inst 57:157–165

Wu PC, Lai CL, Lam KC, Lok AS, Lin HJ (1983) Clear cell carcinoma of liver. An ultrastructural study. Cancer 52:504–507

Yokoyama Y, Tsuchida S, Hatayama I, Sato K (1993) Lack of peroxisomal enzyme inducibility in rat hepatic preneoplastic lesions induced by mutagenic carcinogens: contrasted expression of glutathione S-transferase P form and enoyl CoA hydratase. Carcinogenesis 14:393–398

Yokoyama Y, Tsuchida S, Hatayama I, Satoh K, Narita T, Rao MS, Reddy JK, Yamada J, Suga T, Sato K (1992) Loss of peroxisomal enzyme in preneoplastic and neoplastic lesions induced by peroxisome proliferator in rat livers. Carcinogenesis 13:265–269

Zerban H, Rabes HM, Bannasch P (1989) Sequential changes in growth kinetics and cellular phenotype during hepatocarcinogenesis. J Cancer Res Clin Oncol 115:329–334

Zerban H, Radig S, Kopp-Schneider A, Bannasch P (1994) Cell proliferation and cell death (apoptosis) in hepatic preneoplasia and neoplasia are closely related to phenotypic cellular diversity and instability. Carcinogenesis 15:2467–2473

Foci of Altered Hepatocytes, Mouse

Boris H. Ruebner, Peter Bannasch, David E. Hinton, John M. Cullen, and Jerrold M. Ward

Synonyms. Hepatocellular foci, hyperplastic foci, preneoplastic foci, enzyme-altered foci, phenotypically altered foci

Gross Appearance

Foci of altered hepatocytes in the mouse liver are usually invisible with the naked eye. However, they may occasionally be recognizable grossly on careful examination, as small, whitish spots, 1–2 mm in diameter, on the liver surface.

Microscopic Features

Foci or islands of hepatocellular alteration in mice were clearly described by Bannasch and Müller (1964) and later by Frith and Ward (1980, 1988). These lesions are the result of alterations in the tinctorial qualities and textural appearance of the cytoplasm of hepatocytes. Only one or two or, according to some authors (Maronpot et al. 1987), up to four adjacent lobules or acini may be involved. The overall architecture therefore is not disturbed, and there is little or no compression of adjacent hepatic parenchyma (Frith and Ward 1980, 1988; Ward 1984). In many studies, these foci were not thought to have a specific lobular or acinar localization. However, in some they were considered to be centrilobular, as, for instance, in the case of lesions induced by aldrin and dieldrin (Reuber 1975).

Hepatic cords in the foci may not be contiguous with those of the adjacent normal liver and may differ in architectural pattern from those of the adjacent liver. The cells in the foci may be either larger or smaller than adjacent hepatocytes, and on the basis of cytoplasmic characteristics, they have been classified acidophilic, basophilic, clear cell, vacuolated mixed, or amphophilic (see below). The nuclear alterations in the hepatocytic foci in mice have been studied relatively less. However, most of the cells in the foci appear to be diploid and have an increased rate of DNA synthesis (Fig. 18; Vesselinovitch et al. 1985; Ward et al. 1990; Siglin et al. 1991). Nuclear dysplasia has been described very occasionally, particularly in

transgenic mice (Takagi et al. 1993). Apoptosis has been studied predominantly in foci of rat liver. The tinctorial properties of the cytoplasm of component hepatocytes in hematoxylin and eosin (H&E)-stained sections have been employed as the principal differentiating feature for the classification of foci in mouse liver. The different types of foci are described below.

Basophilic cell foci are composed of hepatocytes which are distinctly more basophilic and usually somewhat smaller than the adjacent normal hepatocytes (Fig. 19). Their cytoplasm is usually finely granular. However, in some studies, basophilic foci with hepatocytes larger than normal have been reported (Fig. 20; Ward et al. 1983). Foci apparently invading hepatic veins were described by Goldfarb in mice treated with diethylnitrosamine (DEN) (Koen et al. 1983a) and were considered to be microcarcinomas. Most of these are predominantly basophilic (Fig. 21). Similar foci have been observed in transgenic mice. Because of the absence of dysplasia or a trabecular pattern, some pathologists consider this to be pseudoinvasion and do not believe these foci to be true carcinomas.

In *acidophilic (eosinophilic) cell foci*, the hepatocytes are generally larger than adjacent normal hepatocytes. The cytoplasm of the hepatocytes in these foci is distinctly acidophilic and may have a ground glass or lacy appearance (Fig. 22). Hepatocytes with unusually abundant eosinophilic cytoplasm may be seen, particularly in foci of transgenic mice. Such foci often also have dysplastic nuclei (Fig. 23).

Clear cell foci stand out from the normal adjacent hepatocytes by their lack of or decrease in cytoplasmic staining. The cytoplasm of the cells composing these foci has been shown to store excessive amounts of glycogen (Fig. 24; Ruebner et al. 1984a). Glycogen storage foci composed of both clear and acidophilic cells are not infrequent and have been classified in rats as combined clear cell/acidophilic cell foci (see the chapter on "Foci of Altered Hepatocytes, Rat," this volume).

Vacuolated cell foci may be mistaken for clear cell foci. However, they differ from them in that distinct vacuoles are present in the cytoplasm, which are either small or large, in which case the nuclei may be displaced towards the periphery of the cell. These cells probably contain lipid droplets (Frith and Ward 1980).

Mixed cell foci differ from clear cell and combined clear cell foci in that there is not excessive glycogen. They are composed of two or more of the cell types described in the preceding paragraphs in varying proportions and are quite common. They should be identified by their constituent cells, as for example in "mixed basophilic and vacuolated foci." They may also be identified by their predominant cell type.

Amphophilic cell foci have been described in rats as composed of hepatocytes with a homogeneous cytoplasm which is both acidophilic and basophilic (Fig. 25; Weber et al. 1988; Goodman et al. 1994). Such foci are identified by their overall architecture and may be difficult to differentiate from surrounding uninvolved liver. Histochemistry may be helpful in their identification (see below). Foci in mice which are not easily classified into one of the subtypes above generally fit into this category.

Other unusual types of foci have been observed in certain studies. Foci with a stroma composed of prominent sinusoidal lining were seen in mice fed tetrachlorvinphas (Fig. 26; Ward et al. 1979) and in mice infected with *Helicobacter hepaticus* (J.M. Ward, unpublished).

Hepatic foci in transgenic mice also frequently have unusual morphologies and may be composed of hepatocytes with pleomorphic nuclei or unusually large eosinophilic hepatocytes (Fig. 21; Takagi et al. 1993; Tamano et al. 1994).

A focus apparently invading a hepatic vein is shown in Fig. 24. Foci with a trabecular plate pattern have been observed (Fig. 27) and may represent very early carcinomas.

Histochemistry

A large number of metabolic alterations have been identified in mouse liver foci. These include quantitative alterations in chemical constituents, such as glycogen, which may be either increased or decreased when compared with adjacent hepatic tissue. Vesselinovitch et al. (1985), for instance, found that hepatic foci of animals injected at 15 days with DEN generally contained decreased amounts of glycogen. However, some foci contained increased glycogen. After induced siderosis, foci are generally iron deficient (Lipsky et al. 1979; Williams et al. 1979). Enzyme levels in

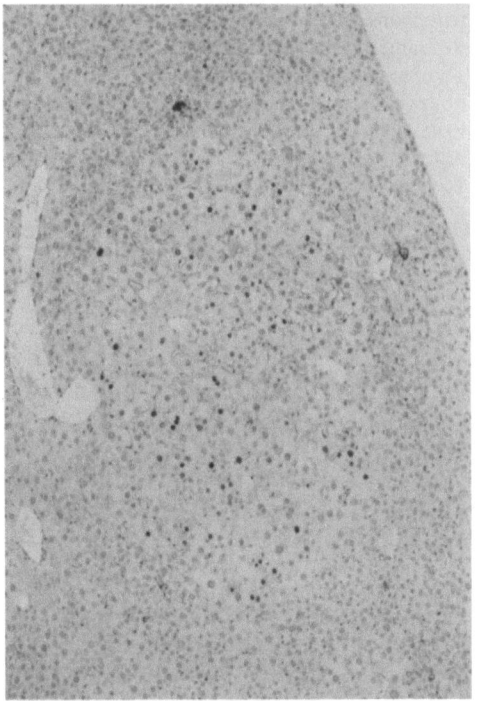

Fig. 18. Focus induced by diethylnitrosamine (DEN), liver, mouse. Increased DNA synthesis is indicated by nuclear staining for proliferating cell nuclear antigen (PCNA). Peroxidase immunohistochemistry, ×100

the foci have been studied both by enzyme and immunohistochemical methods. When studied by such methods, enzymes have been observed to be either increased or decreased; in addition, there may be a change in cellular localization of the enzyme within the hepatocyte, as for instance with γ-glutamyl transpeptidase (Table 3).

Some general statements about these chemical compounds and enzymatic reactions can be made. Spontaneous hepatocytic foci often differ from those induced by various chemicals in their morphology. In sections stained by H&E, they are more often basophilic than acidophilic (Ward 1984). Histochemically, they generally do not stain for γ-glutamyl transpeptidase (Kyriazis and Vesselinovitch 1973; Williams et al. 1980). The histochemical reactions in the foci may vary in different species such as the mouse and the rat. DEN, for instance, generally produces foci with excessive glycogen in rats, but produced foci with decreased glycogen in mice (Vesselinovitch et al. 1985). Most likely there are also differences in histochemical reactions between foci produced by the same chemical in different strains of mice (Ruebner et al. 1984a). Differences in dosage and other experimental details may also influence these reactions (Bannasch et al., p. 3, this volume). Different promoting agents, for instance, produce different foci even if the initiator is identical (Hanigan et al. 1993; Ward et al. 1983; Becker 1984, 1985).

Histochemical reactions of altered foci in the mouse liver have been studied particularly by Vesselinovitch et al. (1985) and by Lipsky et al.

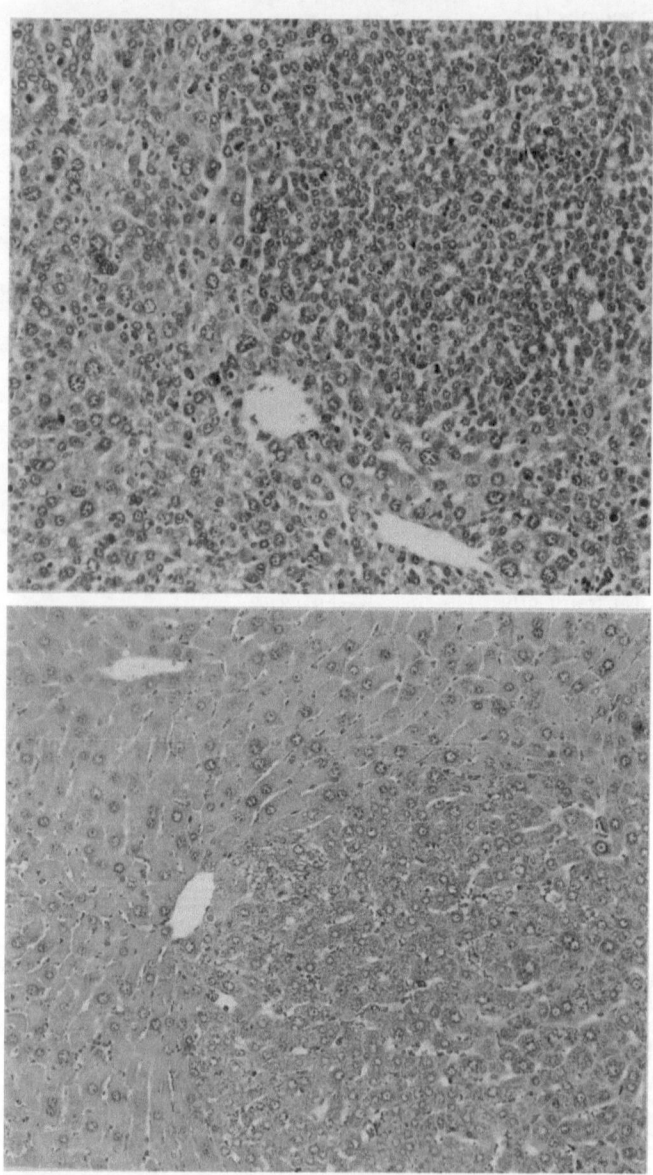

Fig. 19. (*above*) Basophilic cell focus, liver, mouse. The hepatocytes composing this focus are somewhat smaller than the surrounding hepatocytes. The hepatic plates are arranged in a circular pattern. H&E, ×200

Fig. 20. (*below*) Basophilic cell focus, liver, mouse, containing hepatocytes somewhat larger than surrounding hepatocytes induced by diethylnitrosamine (DEN) and di-(2-ethylhexyl) phthalate. H&E, ×200

(1984). None of the histochemical reactions identified so far will demonstrate every focus. Glutathione *S*-transferase, placental form, which approximates this ideal in rats, falls considerably short of this and identifies only a proportion of foci in mice. γ-Glutamyl transpeptidase is often quite useful in mice (Table 3). However, glucose-6-phosphatase may be more useful as an enzyme marker in some systems. A representative result in mice injected with DEN at the age of 15 days is included in Table 3. Table 4 (Lipsky et al. 1984)

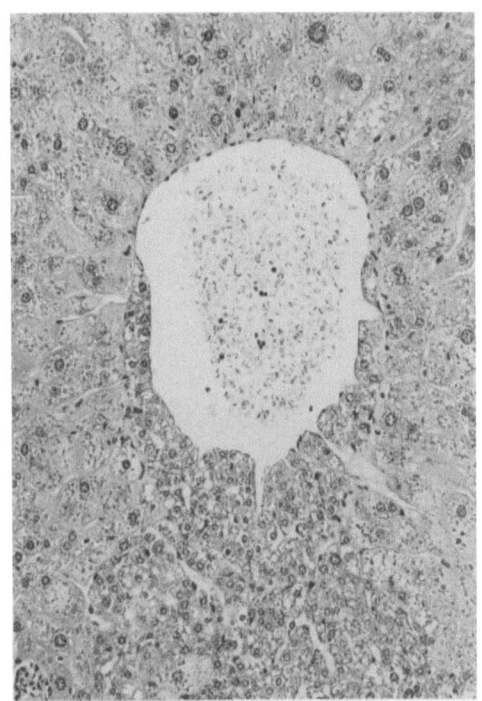

Fig. 21. Focus, liver, mouse, composed predominantly of basophilic cells which appears to have invaded a vein. H&E, ×200

Table 4. Distribution of hepatocellular foci by enzymatic phenotype[a]

Enzymatic phenotype	Percentage
1. GGT[b]	77.7
2. G-6-pase	8.7
3. G-6-PDH	4.9
4. GGT + G-6-pase[b]	3.8
5. GGT + G-6-PDH[b]	4.0
6. GGT + G-6-pase + G-6-PDH[b]	1.2
7. G-6-pase + G6PDH	0

GGT, gamma-glutamyltranspeptidase; G6pase, glucose-6-phosphatase; G6PDH, glucose-6-phosphate dehydrogenase.
[a] All sections sampled from ten mice after 36 weeks of saffrole treatment.
[b] Total foci marked by GGT = 86.4%.

Table 3. Percentage of foci showing increase, decrease, or no change in the level of investigated histochemical markers

Histochemical markers	12 weeks		18 weeks		24 weeks	
	Males (*n* = 36)[a]	Females (*n* = 16)[a]	Males (*n* = 12)[a]	Females (*n* = 16)[a]	Males (*n* = 36)[a]	Females (*n* = 15)[a]
TB	100↑	100↑	100↑↑	100↑	100↑↑	100↑↑
PAS	82↓	57↓	83↓↓	100↓	97↓↓↓	100↓
SYN	100 NC	93 NC	100 NC	83 NC	95 NC	100 NC
		7↓		17↓	2.5↑	
					2.5↑	
PHO	100 NC	100 NC	100 NC	100 NC	47 NC	100 NC
					39↓	
					14↑	
G6Pase	100↓	100↓	100↓↓	100↓	100↓↓	100↓↓
ATPase						
Cytoplasmic	22↓	31↓	63↓	100↓	84↓	100↓
Membrane	100 NC	100 NC	7↑	100 NC	65↑	7↑
G6PDH	84↑	94↑	94↑	88↑	95↑↑	94↑
GAPDH	56↑	87↑	83↑	59↑	68↑↑	87↑
GGT	100 NC	100 NC	100 NC	100 NC	100 NC	100 NC
AcPase	11↓	18↓	75↓	19↓	78↓	37↓

Reproduced by permission from Vesselinovitch et al. (1985).
↑, increase; ↓, decrease; NC, no change; TB, toluidine blue; PAS, periodic acid-Schiff; PHO, glycogen phosphorylase; SYN, glycogen synthetase; G6Pase, glucose-6-phosphatase; G6PDH, glucose-6-phosphate dehydrogenase; GAPDH, glyceraldehyde-3-phosphate dehydrogenase; ATPase, adenosine triphosphatase; G-GT, G-glutamyltranspeptidase; AcPase, acid phosphatase.
[a] Number of foci examined.

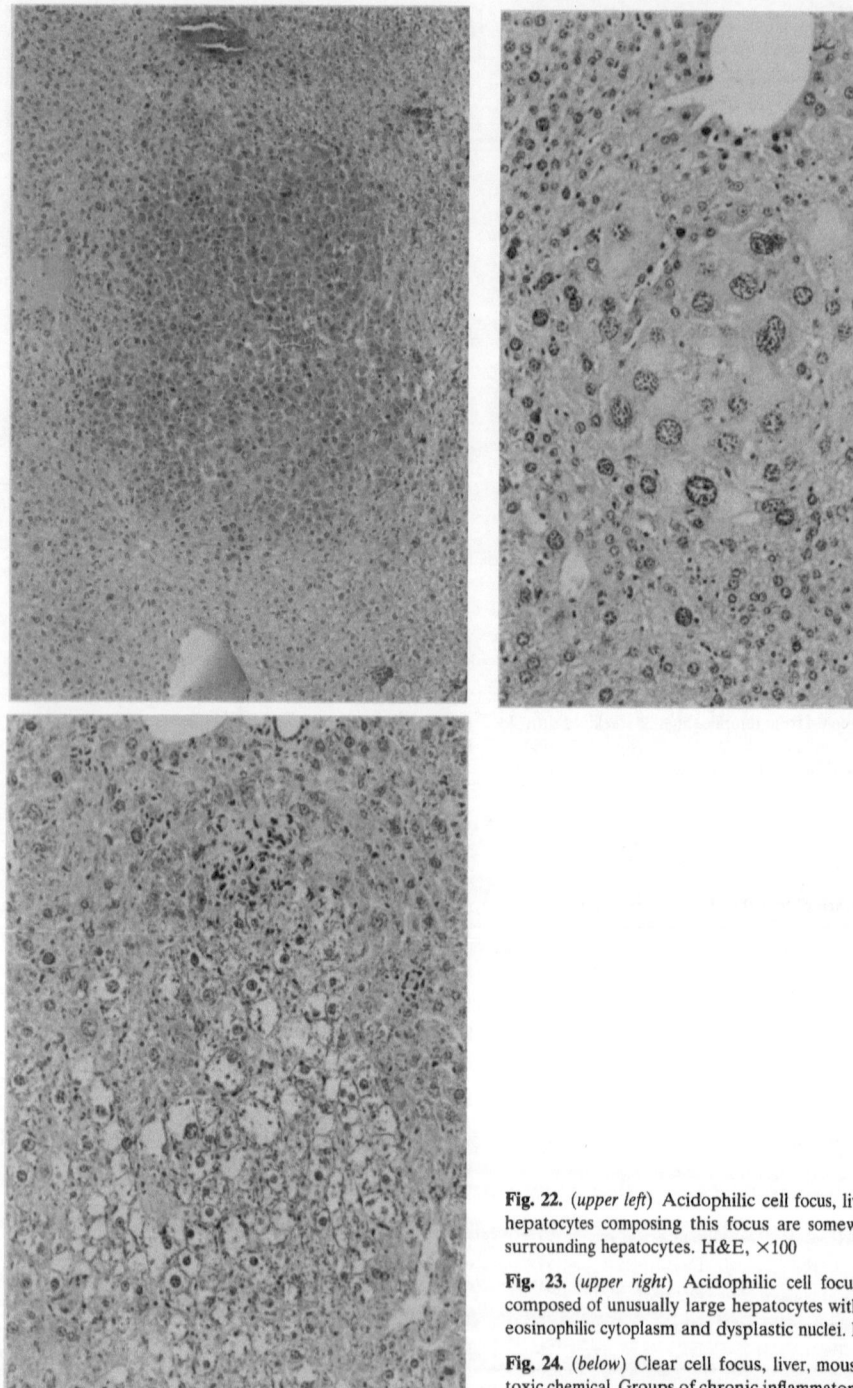

Fig. 22. (*upper left*) Acidophilic cell focus, liver, mouse. The hepatocytes composing this focus are somewhat larger than surrounding hepatocytes. H&E, ×100

Fig. 23. (*upper right*) Acidophilic cell focus, liver, mouse, composed of unusually large hepatocytes with abundant pale eosinophilic cytoplasm and dysplastic nuclei. H&E, ×200

Fig. 24. (*below*) Clear cell focus, liver, mouse, induced by a toxic chemical. Groups of chronic inflammatory cells indicating areas of necrosis are also seen. H&E, ×200

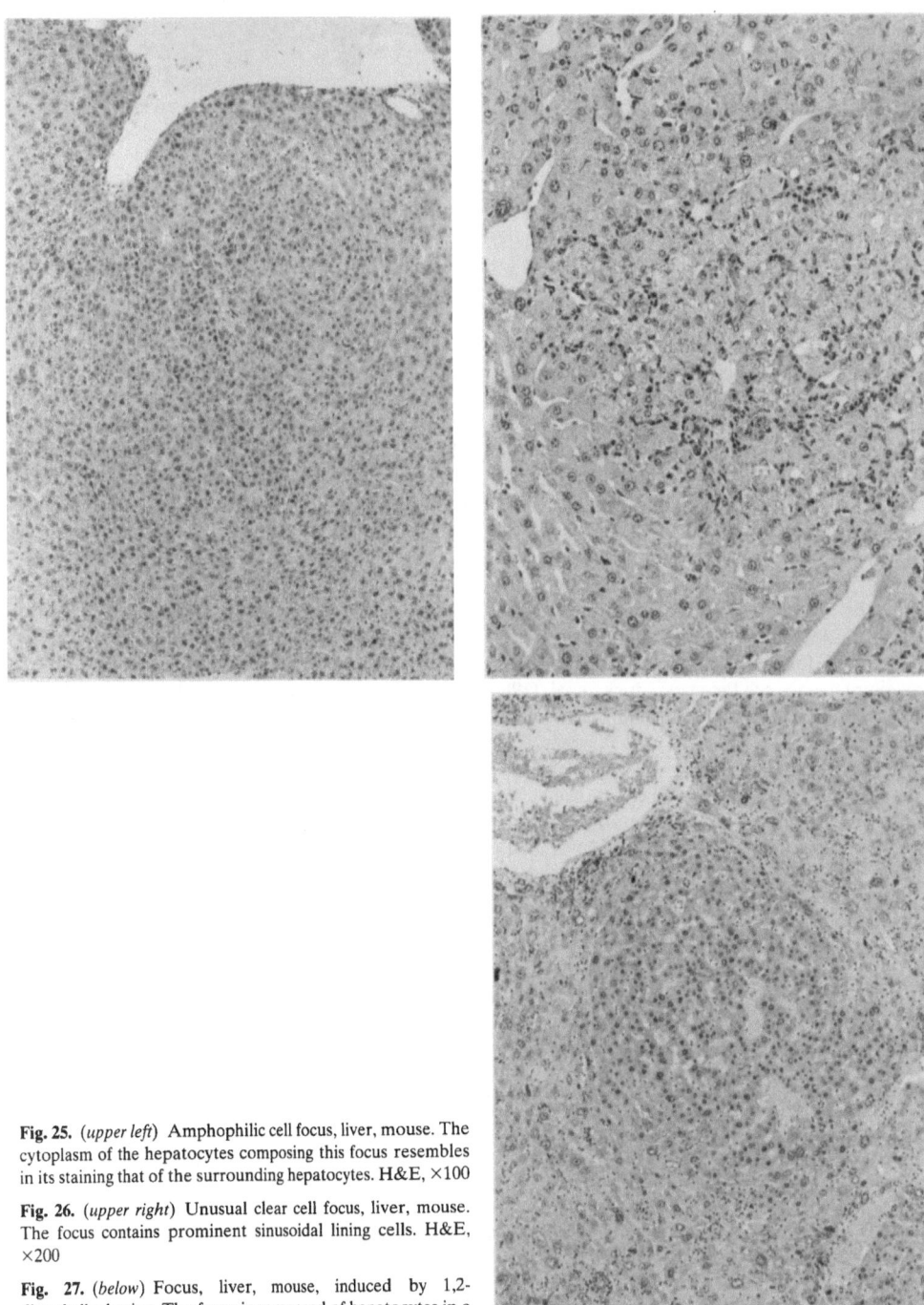

Fig. 25. (*upper left*) Amphophilic cell focus, liver, mouse. The cytoplasm of the hepatocytes composing this focus resembles in its staining that of the surrounding hepatocytes. H&E, ×100

Fig. 26. (*upper right*) Unusual clear cell focus, liver, mouse. The focus contains prominent sinusoidal lining cells. H&E, ×200

Fig. 27. (*below*) Focus, liver, mouse, induced by 1,2-dimethylhydrazine. The focus is composed of hepatocytes in a trabecular pattern with wide-branching hepatic plates. H&E, ×100

identifies the enzymatic phenotype of hepatocellular foci induced by safrole, particularly with respect to the multiplicity of enzyme defects seen in various foci 36 weeks after initiation of safrole feeding. Because of the lack of a universal histochemical reaction, foci visible in H&E sections at this time still appear to be the single most useful indicator of hepatic foci in mice.

Stereology

A quantitative approach to observations on foci is becoming essential in virtually all types of investigation (Pugh et al. 1983). The simplest approach would be to compare the total surface area of the foci produced by various chemicals with that of controls. However, such a method is rarely adequate because it does not take into account the number or size of lesions. These parameters are important, since it is generally assumed that the number of lesions is a function of the initiator, while their volume correlates with the effectiveness of the promoter. Various quantitative methods suitable for foci have been described. Among the simpler ones are those of Lipsky et al. (1984) and Enzmann et al. (1987). The results of one study are shown in Table 5. New "design-based" procedures (Gunderson and Jensen 1987; Gunderson et al. 1988) may find enhanced usage, since assessment of three-dimensional properties of foci is desirable. Software for computational biology including applications is available (Bolender 1993).

Ultrastructure

Although there are several ultrastructural studies of adenomas, relatively few studies of foci have been performed in mice (Lipsky et al. 1981).

Differential Diagnosis

Acidophilic, clear cell, vacuolated, and basophilic cell foci are usually easily identifiable by routine light microscopy. However, sometimes lesions are seen in H&E sections which are suggestive, but not diagnostic of foci. A battery of histochemical reactions usually establishes clearly whether these lesions are indeed foci or not. Performance of histochemical reactions is thus particularly valuable in newly established experimental protocols. The histochemical reactions selected in a particular experimental protocol will depend on a review of the literature concerning the specific compounds and strains to be studied as well as on reports of studies on related compounds in similar or identical strains of mice. Such preliminary studies will suggest the spectrum of lesions to be expected in the experimental system to be studied. It is important to apply the classification of foci outlined above, because different histologic types of foci are observed after different carcinogens and in different strains. Moreover, different types of foci are probably associated with different enzymatic alterations, different molecular events, and probably also varying risks of progression to carcinoma.

Table 5. Stereology of hepatocellular foci (from Lipsky et al. 1984)

Weeks on protocol	Vv	Nv	Vq
16 S	–	–	–
16 S + 36 cont.	0.0043 ± 0.0009	2.21 ± 0.416	0.013 ± 0.009
16 S + 52 cont.	0.0044 ± 0.0011	0.212 ± 0.054	0.062 ± 0.021
36 S	0.011 ± 0.0004	1.077 ± 0.196	0.0013 ± 0.0002
36 S + 36 cont.	0.031 ± 0.0024	0.113 ± 0.017	0.025 ± 0.022
52 S	0.012 ± 0.003	0.312 ± 0.081	0.136 ± 0.055
Control	0	0	0

The results are expressed as mean values ± standard error of the mean.
Analysis based on seven to ten sections (6–7 μm) from five to seven animals stained for γ-glutamyltranspeptidase.
S, safrole feeding; cont., control diet feeding; Vv, volume of lesions in unit volume of structure, a ratio of volume of lesions to test or reference volume (mm^3/mm^3); Nv, number of lesions per unit test volume (no./mm^3); Vq, mean volume of lesions (mm^3).

Combined routine histologic and histochemical studies may establish that there are histochemically altered foci which appear histologically unremarkable in sections stained by H&E. Foci are generally easily distinguished from hepatocellular adenomas. Adenomas, while sharing many cytologic features with foci, differ histologically and in their distinct gross appearance. Hepatocellular carcinomas, on the other hand, have well-defined cytologic and histologic characteristics.

Hepatocarcinomatous "foci within foci" have apparently not been described in foci of mice, but foci of this type within mouse hepatocellular adenomas have been reported (Jang et al. 1992). Foci with a trabecular plate pattern which may represent early carcinomas have also been observed (see above).

The distinction of foci from areas of altered hepatocytes is somewhat problematic and arbitrary. Areas were originally defined as lesions larger than a hepatic lobule (Squire and Levitt 1975). If such lesions are invisible grossly and occupy only up to four lobules, they are probably best classified as foci (see above). If they are visible grossly and have histologic evidence of compression of adjacent parenchyma, they are probably best classified as adenomas.

Biologic Features

Pathogenesis

The process of hepatocarcinogenesis has been studied intensively in many animal species, both for its intrinsic interest as a contribution to understanding of the neoplastic process as well as for its practical usefulness in the assessment of the risks posed by environmental carcinogens to humans. In most of these studies, rodents, particularly rats, have been employed because of their convenience (Bannasch et al., p. 3 this volume). In this review, we will concentrate on foci of altered hepatocytes in mice (Maronpot et al. 1987; Pitot 1990). Mice have for a long time been employed in the investigation of hepatic carcinogenesis and in the testing of chemical compounds for carcinogenicity.

It has generally become accepted that carcinogenesis in the liver is a multistep process. Several morphological lesions have been considered to represent the successive steps in this pathway. Apart from the identification of individual putative preneoplastic cells in the liver, which are difficult to study (Moore et al. 1987), foci have generally been considered to be the earliest or at least an early putative premalignant lesion in the liver. Friederich-Freksa and coworkers (Gössner and Friedrich-Freksa 1964) and Bannasch and Müller appear to have been the first to identify these lesions in rats and mice (Bannasch and Müller 1964). Later in the carcinogenic process, a small proportion of foci develop into nodules which in experimental models at first have the characteristics of benign lesions or adenomas; still later, typical hepatocellular carcinomas are seen, apparently derived from some of the adenomas. The exact interrelationship of these various morphological lesions is still debated, and it is likely that individual foci or islands of altered hepatic cells may take one of several pathways, including regression and progression. Regression appears to affect a proportion of foci when the carcinogenic stimulus is removed (Lipsky et al. 1984). Other foci appear to progress, predominantly to adenomas. However, in some cases there appears to be a progression from foci immediately to hepatocellular carcinoma (Ward 1984).

Etiologic Factors and Mechanisms

Hepatocellular neoplasms have been observed in virtually all strains of mice under a great variety of conditions. Both infant (Vesselinovitch et al. 1985) and aging (Ward et al. 1988) mice appear to be particularly susceptible to chemically induced hepatic carcinogenesis. Most of the hepatic neoplasms in the mouse have been associated with or preceded by foci. The spontaneous incidence of tumors and foci in different strains of mice is very variable, and some strains have a considerably higher incidence than others. Chemicals which have increased the incidence of foci and tumors above background levels include not only chemical carcinogens which may be found in the environment, but also hormonal, dietary, and genetic factors (Ward 1984; Pitot 1990). Chemicals which are known to be genotoxic are considered tumor initiators, while chemicals such as phenobarbital, sex steroids, bile acids, selenium, choline deficiency, unsaturated fat, dietary tryptophan, and sucrose, which are not mutagenic and, after initiation, still increase the incidence and size of foci in the mouse liver, are considered to be tumor promoters. Two stage carcinogenesis studies have shown that differences in spontaneous and carcinogen-induced tumor susceptibility in different strains may be the result of different sensitivity to promotion (Della Porta et al. 1987).

Certain strains of transgenic mice, such as those expressing transforming growth factor-α (Takagi et al. 1993), simian virus 40 (SV40) T antigens (Cullen et al. 1993; Tamano et al. 1994), the HB S gene (Pasquinelli et al. 1992), or ZZ α-1-antitrypsin (Geller et al. 1994), develop a high incidence of foci, adenomas, and hepatocellular carcinomas. These lesions may be composed of clear, acidophilic, or basophilic cells, often with dysplasia, and generally resemble the well-recognized typical foci long associated with chemical carcinogenesis. They may, however, also have unusual features, particularly hypertrophy and pleomorphism of nuclei, both within (Takagi et al. 1993) and outside (Cullen et al. 1993) the foci.

Among recent studies which have shed new light on the mechanisms of hepatocellular neoplasia are the studies on transgenic mice described above, as well as studies of growth factors, growth factor receptors, oncogenes, and tumor suppressor genes. Specific mutations in codon 61 of the H-*ras* gene in tumor-susceptible mouse strains, such as B6C3F1, were demonstrated in approximately one third of altered hepatic foci or small adenomas by Buchmann et al. (1991). These mutations appear to be characteristic of many spontaneous and induced hepatic tumors in tumor-susceptible mice (Devereux et al. 1993). However, differences in oncogene spectra have been described in tumors induced by different carcinogens in different strains (Anderson et al. 1992). Most of the lesions studied were actually adenomas. In a different transgenic model, c-*myc* and a secretable analogue of epidermal growth factor appear to cooperate in carcinogenesis (Paul 1993). However, in hepatitis B virus transgenic mice with hepatocellular neoplasms, no alterations in oncogenes or tumor suppressor genes have so far been identified (Pasquinelli et al. 1992).

Progression and Regression of Foci

In several experimental systems, basophilic foci appear to represent the earliest stage of neoplastic development. Foci generally have higher rates of cell proliferation than adjacent normal hepatocytes (Fig. 18; Ward et al. 1990). Rabes et

al. (1982), Tsuji et al. (1988), and others have demonstrated that foci are clonal in origin. While several observers have demonstrated a quantitative relationship between the dose of the carcinogen and the number of foci in rats, this has rarely been done in mice (Klaunig et al. 1988; Vesselinovitch et al. 1985). Morphometric data indicate that, after cessation of safrole feeding, a high proportion of tinctorially and enzyme-histochemically altered foci regressed, but that the surviving foci increased in size, so that the total volume of the liver involved by foci actually increased (Table 5; Lipsky et al. 1984). Some tumors and foci also apparently regressed after cessation of administration of hexachlorocyclohexane (Ito et al. 1976). We have pointed out previously, however, that most hepatic neoplastic lesions in mice, whether spontaneous or induced, do not characteristically regress and/or disappear (Ruebner et al. 1984b). This may be the result of the strong promoting action inherent in the liver of certain inbred strains of mice by genetic or other endogenous factors. Limited evidence has been presented that some hepatic foci in mice may transform directly into carcinomas without the formation of adenomas (Koen et al. 1983a, b; Ward 1984). Pugh and Goldfarb (1978) and others (Dragan and Pitot 1992) observed that, in rats, foci with different morphological and biochemical characteristics also differ in their susceptibility to promotion by chemicals. For instance, foci which are altered with respect to several enzymes are more likely to progress than foci altered with respect to only one or two enzymes.

Comparison with Other Species

Foci have been observed in many species, particularly in rats (see p. 3, this volume), but also in ducks, hamsters, woodchucks, turkeys, fish, and even primates (Bannasch and Zerban 1992). Histologically, these lesions are essentially similar in different species. However, certain subtypes, such as tigroid foci and oncocytic foci, observed in rats have not so far been described in mice. In addition, the spectrum of histochemical reactions differs in different species. The best example of this is probably the relative rarity of γ-glutamyl-transpeptidase- and glutathone S-transferase (placental form)-positive foci in mice, although these foci are frequent in rats. Dispute continues about the identity of preneoplastic hepatic lesions in human patients. Classic foci, as described in this review, have rarely been observed in humans. However, focal hepatocytic dysplasia (Anthony et al. 1973), macroregenerative nodules, and atypical adenomas or adenomatous hyperplasia in cirrhosis (Nakanuma et al. 1990; Matsuno et al. 1990; Sakamoto et al. 1991) may be important precursors. The high spontaneous incidence of hepatic neoplasms in some strains of mice has led some investigators to question the significance of hepatocarcinogenesis studies in mice and specifically the results of carcinogenesis bioassays. It seems likely that some strains of mice are genetically predisposed to the development of neoplastic hepatocellular lesions and may be considered to be genetically initiated or subject to spontaneous tumor progression. However, if experiments are carefully controlled and if good quantitative morphometric methods are used, important results can clearly be obtained from studies of foci, hepatocellular adenomas, and carcinomas in the mouse.

References

Anderson M, Stanley L, Devereux T, Reynolds S, Maronpot R (1992) Oncogenes in mouse liver tumors. Prog Clin Biol Res 376:187–201

Anthony PP, Vogel CL, Barker LF (1973) Liver cell dysplasia: a premalignant condition. J Clin Pathol 26:217–223

Bannasch P, Müller HA (1964) Lichtmikroskopische Untersuchungen Über die Wirkung von N-Nitrosomorpholin auf die Leber von Ratte und Maus. Arzneimittelforschung 14:805–814

Bannasch P, Zerban H (1992) Predictive value of hepatic preneoplastic lesions as indicators of carcinogenic response. In: Vainio H, Magee PN, McGregor DB, McMichael AJ (eds) Mechanisms of carcinogenesis risk identification. IARC Scientific Publications, Lyon

Becker FF (1984) The direct and indirect effects of promoters may depend upon the nature of the initiated cell. In: Fujiki H, Hecker E, Moore RE, Sugimura T, Weinstein IB (eds) Cellular interactions by environmental tumor promoters. VNU Science, Tokyo, pp 349–359

Becker FF (1985) Tumor phenotype and susceptibility to progression as an expression of subpopulations of initiated murine cells. Cancer Res 45:768–773

Bolender RP (1993) Current methods in quantitative morphology. QM 2000 Version 2.0. Lecture notes and software for computational biology. University of Washington, Seattle

Buchmann A, Bauer-Hofmann R, Mahr J, Drinkwater NR, Schwartz M (1991) Mutational activation of the C-Ha-ras gene in liver tumors of different rodent strains. Correlation with susceptibility to hepatocarcinogenesis. Proc Natl Acad Sci USA 88:911–915

Cullen JM, Sandgren EP, Brinster RL, Maronpot RR (1993) Histologic characterization of hepatic carcinogenesis in

transgenic mice expressing SV40 T antigens. Vet Pathol 30:111–118

Della Porta GD, Dragani TA, Manenti G (1987) Two-stage liver carcinogenesis in the mouse. Toxicol Pathol 15:229–233

Devereux TR, Foley JF, Maronpot RR, Kari F, Anderson MW (1993) Ras proto-oncogene activation in liver and lung tumors from B6C3F1 mice exposed chronically to methylene chloride. Carcinogenesis 14:795–801

Dragan YP, Pitot HC (1992) The role of the stages of initiation and promotion in phenotype diversity during hepatocarcinogenesis in the rat. Carcinogenesis 13:739–750

Enzmann H, Edler L, Bannasch P (1987) Simple elementary method for the quantification of focal liver lesions induced by carcinogens. Carcinogenesis 8:231–235

Frith CH, Ward JM (1980) A morphologic classification of proliferative and neoplastic hepatic lesions in mice. J Environ Pathol Toxicol 3:329–351

Frith CH, Ward JM (1988) Color atlas of neoplastic and non-neoplastic lesions in aging mice. Elsevier, Amsterdam, p 109

Geller SA, Nichols WS, Kim S, Tolmachoff T, Lee S, Dycaico MJ, Felts KA, Sorge JA (1994) Hepatocarcinogenesis is the sequel to hepatitis in Z2 alpha-1-antitrypsin transgenic mice: Histopathological and DNA ploidy studies. Hepatology 19:389–397

Goodman DG, Maronpot PR, Newberne PM, Popp JA, Squire RA (1994) Proliferative and selected other lesions of the liver in rats. In: Streett CS, Burek JD, Hardisty JF, Garner FM, Leininger JR, Pletscher JM, Moch RW (eds) Guides for toxicologic pathology. STP/ARP/AFIP, Washington, pp GI-5, 1–24

Gössner VW, Friedrich-Freksa H (1964) Histochemische Untersuchungen über die glucose-6-phosphatase in der Rattenleber wahrend der Kanzerisierung durch Nitrosamine. Z. Naturforsch 19:862–864

Gundersen HJ, Jensen EB (1987) The efficiency of systematic sampling in stereology and its prediction. J Microsc 147:229–263

Gundersen HJ, Bagger P, Bendtsen TF, Evans SM, Korbo L, Marcussen N, Moller A, Nielsen K, Nyengaard JR, Pakkenberg B (1988) The new stereological tools: disector, fractionater, nucleator and point sampled intercepts and their use in pathological research and diagnosis. APMIS 96:857–881

Hanigan MH, Winkler WL, Drinkwater NR (1993) Induction of three histochemically distinct populations of hepatic foci in C57BL/67 mice. Carcinogenesis 14:1035–1040

Ito N, Hamanouchi M, Sugihara S, Shirai T, Tsuda H (1976) Reversibility and irreversibility of liver tumors in mice induced by the alpha-isomer of 1,2,3,4,5,6 hexachlorocyclohexane. Cancer Res 36:2227–2234

Jang JJ, Weghorst CM, Henneman JR, Devor DE, Ward JM (1992) Progressive atypia in spontaneous and N-nitrosodiethylamine induced hepatocellular adenomas of C3H/HeNCr mice. Carcinogenesis 13:1541–1547

Klaunig JE, Pereira MA, Ruch RJ, Weghorst CM (1988) Dose-response relationship of diethylnitrosamine-initiated tumors in neonatal balb/c mice: effect of phenobarbital promotion. Toxicol Pathol 16:381–385

Koen H, Pugh TD, Nychka D, Goldfarb S (1983a) Presence of alpha-fetoprotein-positive cells in hepatocellular foci and microcarcinomas induced by single injections of diethylnitrosamine in infant mice. Cancer Res 43:702–708

Koen H, Pugh, TD, Goldfarb S (1983b) Hepatocarcinogenesis in the mouse. Combined morphologic-stereologic studies. Am J Pathol 112:89–100

Kyriazis AP, Vesselinovitch SD (1973) Transplantability and biological behavior of mouse liver tumors induced by ethylnitrosourea. Cancer Res 33:332–338

Lipsky MM, Hinton DE, Goldblatt PJ, Klaunig JE, Trump BF (1979) Iron negative foci and nodules in safrole-exposed mouse liver made siderotic by iron-dextran injection. Pathol Res Pract 164:178–185

Lipsky MM, Hinton DE, Klaunig JE, Trump BF (1981) Biology of hepatocellular neoplasia in the mouse. III. Electron microscopy of safrole-induced hepatocellular adenomas and hepatocellular carcinomas. J Natl Cancer Inst 67:393–405

Lipsky MM, Tanner DC, Hinton DE, Trump BF (1984) Reversibility, persistence, and progression of safrole-induced mouse liver lesions following cessation of exposure. In: Popp JA (ed) Mouse liver neoplasia: current perspectives. Hemisphere, Washington, pp 161–177

Maronpot RR, Haseman JK, Boorman GA, Eustis SE, Rao GN, Huff JE (1987) Liver lesions in B6C3F1 mice: the National Toxicology Program, experience and position. Arch Toxicol Suppl 10:10–26

Matsuno Y, Hirohashi S, Furuya S, Sakamoto M, Mukai K, Shimosato Y (1990) Heterogeneity of proliferative activity in nodule-in-nodule lesions of small hepatocellular carcinoma. Jpn J Cancer Res 81:1137–1140

Moore MA, Nakagawa K, Satoh K, Ishikawa T, Sato K (1987) Single GST-P positive liver cells – putative initiated hepatocytes. Carcinogenesis 8:483–486

Moore MR, Drinkwater NR, Miller EC, Miller JA, Pitot HC (1981) Quantitative analysis of the time dependent development of glucose-6-phosphatase deficient foci in the livers of mice treated neonatally with diethylnitrosamine. Cancer Res 41:1585–1593

Nakanuma Y, Terada T, Terasaki S, Ueda K, and others (1990) Atypical adenomatous hyperplasia in liver cirrhosis: lowgrade hepatocellular carcinoma or borderline lesions? Histopathology 17:27–35

Pasquinelli C, Bhavani K, Chisari FV (1992) Multiple oncogenes and tumor suppressor genes are structurally and functionally intact during hepatocarcinogenesis in hepatitis B virus transgenic mice. Cancer Research 52:2823–2829

Paul D (1993) Hepatocarcinogenesis in transgenic mice. Joint conference of the European Association for Cancer Research and Abteilung für experimentelle Krebsforschung, Heidelberg

Pitot HC (1990) Altered hepatic foci: their role in murine hepatocarcinogenesis. Annu Rev Pharmacol Toxicol 30:465–500

Pugh TD, Goldfarb S (1978) Quantitative histochemical and autoradiographic studies of hepatocarcinogenesis in rats fed 2-acetylaminofluorene followed by phenobarbitol. Cancer Res 38:4450–4457

Pugh TD, King JH, Koen H, Nychka D, Chover J, Wahba G, He Y, Goldfarb S (1983) Reliable stereological method for estimating the number of microscopic hepatocellular foci from their transections. Cancer Res 43:1261–1268

Rabes HM, Bucher T, Hartmann A, Linke I, Dunnwald M (1982) Clonal growth of carcinogen-induced enzyme-deficient preneoplastic cell populations in mouse liver. Cancer Res 42:3220–3227

Reuber MD (1975) Histogenesis of hyperplasia and carcinomas of the liver arising around central veins in mice ingesting chlorinated hydrocarbons. Pathol Microbiol 43: 287–298

Ruebner BH, Gershwin ME, French SW, Meierhenry E, Dunn P, Hsieh LS (1984a) Mouse hepatic neoplasia: differences among strains and carcinogens. In: Popp JA (ed) Mouse liver neoplasia: current perspectives. Hemisphere, Washington, pp 115–143

Ruebner BH, Gershwin ME, Meierhenry EF, Hsieh LS, Dunn PL (1984b) Irreversibility of liver tumors in C3H mice. J Natl Cancer Inst 73:493–498

Sakamoto M, Hirohashi S, Shimosato Y (1991) Early stages of multistep hepatocarcinogenesis: adenomatous hyperplasia and early hepatocellular carcinoma. Hum Pathol 22:172–178

Siglin JC, Weghorst CM, Klaunig JE (1991) Role of hepatocyte proliferation in α-hexachlorocyclohexane and phenobarbital tumor promotion in B6C3F1 mice. Prog Clin Biol Res 369:407–416

Squire RA, Levitt MN (1975) Report of a workshop on classification of specific hepatocellular lesions in rats. Cancer Res 35:3214–3223

Takagi H, Sharp R, Takayama H, Anver MR, Ward JM, Merlino G (1993) Collaboration between growth factors and diverse chemical carcinogens in hepatocarcinogenesis of transforming growth factor alpha transgenic mice. Cancer Research 53:4329–4336

Tamano S, Merlino GT, Ward JM (1994) Rapid development of hepatic tumors in transforming growth factor alpha (TGF-α) transgenic mice associated with increased cell proliferation in precancerous hepatocellular lesions initiated by N-nitrosodiethylamine and promoted by phenobarbital. Carcinogenesis 15:1791–1798

Tsuji S, Ogawa K, Takasaka H, Sonoda T, Mori M (1988) Clonal origin of gamma-glutamyl transpeptidase-positive hepatic lesions induced by initiation-promotion of ornithine carbamoyltransferase mosaic mice. Jpn J Cancer Res 79:148–151

Vesselinovitch SD, Hacker HJ, Bannasch P (1985) Histochemical characterization of focal hepatic lesions induced by single diethylnitrosamine treatment in infant mice. Cancer Res 45:2774–2780

Ward JM (1984) Morphology of potential preneoplastic hepatocyte lesions and liver tumors in mice and a comparison with other species. In: Popp JA (ed) Mouse liver neoplasia. Current perspectives. Hemisphere, Washington, pp 1–26

Ward JM, Bernal E, Buratto B, Goodman DG, Strandberg JD, Schueler R (1979) Histopathology of neoplastic and nonneoplastic hepatic lesions in mice fed diets containing tetrachlorvinphos. J Natl Cancer Inst 63:111–118

Ward JM, Rice JM, Creasia D, Lynch P, Riggs C (1983) Dissimilar patterns of promotion by di (2-ethylhexyl) phthalate and phenobarbital of hepatocellular neoplasia initiated by diethylnitrosamine in B6C3F1 mice. Carcinogenesis 4:1021–1029

Ward JM, Lynch P, Riggs C (1988) Rapid development of hepatocellular neoplasms in aging male C3H/HeNCr mice given phenobarbital. Cancer Lett 39:9–18

Ward JM, Diwan BA, Lubet RA, Henneman JR, Devor DE (1990) Liver tumor promoters and other mouse liver carcinogens. In: Stevenson DE, McClain R, Popp JA, Slaga TJ,

Ward JM, Pitot HC (eds) Mouse liver carcinogenesis: mechanisms and species comparisons. Wiley-Liss, New York, pp 85–108

Weber E, Moore MA, Bannasch P (1988) Enzyme histochemical and morphological phenotype of amphophilic foci and amphophilic/tigroid cell adenomas in rat liver after combined treatment with dehydroepiandrosterone and N-nitrosomorpholine. Carcinogenesis 9:1049–1054

Williams GM, Hirota N, Rice JM (1979) The resistance of spontaneous mouse hepatocellular neoplasms to iron accumulation during rapid iron loading by parenteral administration and their transplantability. Am J Pathol 94:65–74

Williams GM, Oamori T, Katayama S, Rice JM (1980) Alteration by phenobarbital of membrane-associated enzymes including gamma glutamyl transpeptidase in mouse liver neoplasms. Carcinogenesis 1:813–818

Hepatocellular Adenoma, Liver, Rat

Paul N. Brooks and Francis J.C. Roe

Synonyms. Benign liver cell tumor, liver parenchymal cell adenoma, hepatoma, neoplastic nodule

Gross Appearance

Hepatocellular adenomas vary in size and multiplicity. Most are nonfatal and are discovered incidentally when animals are killed or die for other reasons. If a liver tumor is located close to the ventral body wall, it may be detectable by palpation in a nonobese animal. However, this is not a reliable way to detect liver tumors in living animals. Moreover, since heavy palpation may cause a tumor to bleed, this method of detecting liver tumors in living animals is not recommended. If an adenoma arises near the liver surface, it may be noticed at necropsy even if it has a mean diameter of less than 1 mm. Otherwise the presence of small adenomas may not be suspected until tissues are trimmed after fixation or until sections are examined under a microscope. Even the presence of a very large tumor that replaces a whole lobe of the liver may not be suspected until necropsy unless it gives rise to abdominal distension. In general, tumors discovered at necropsy in young animals tend to be smaller and more spherical than those found in older animals. Large tumors are often molded by the shape of the liver lobe in which they arise.

Tumors may be solitary or multiple with an increasing tendency to multiplicity with age. Where there are multiple liver tumors, they may be of the same or different histologic type, degree of malignancy, size, and appearance. The lesions may be the same color as the surrounding liver or may be darker or lighter, depending on the relative degrees of congestion and steatosis in the tumor and in the surrounding liver tissue.

Occasionally, liver cell adenomas become pedunculated, with the risk that the pedicle will become twisted. Infarction can also occur in nonpedunculated tumors located at the edges of the lobes. Such tumors tend to be red or pale, depending on how long before death the infarction occurred. Another rare event is that an infarcted tumor may lose all contact with the liver and end up as a free body floating around the peritoneal cavity. A similar discrete lesion resulting from herniation of liver parenchyma through an esophageal hiatus is described on p. 167.

Although the vast majority of liver cell adenomas are without obvious effect on health, occasionally even a relatively small lesion of this kind may cause death from intraperitoneal hemorrhage.

Size, or for that matter any other macroscopically observable characteristic, is not a reliable indicator of malignancy and cannot, therefore, be of specific value in the diagnosis of hepatocellular adenoma.

Microscopic Features

Small adenomas, greater in diameter than one liver lobule, are generally spherical and well circumscribed, but progressive proliferation of the lesion may result in a nodular neoplasm with an irregular boundary (Fig. 28). During this process,

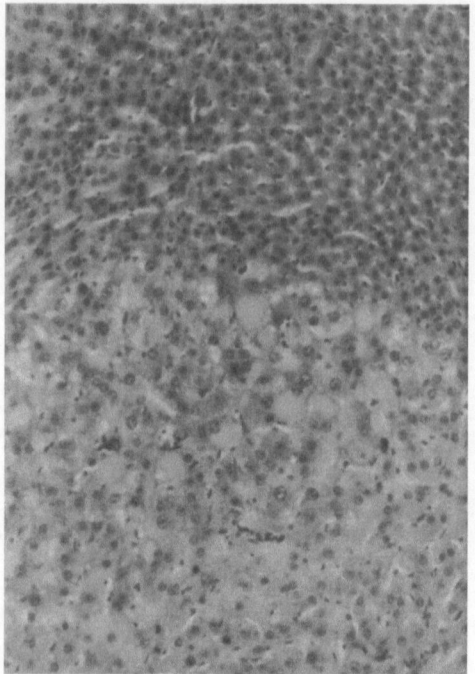

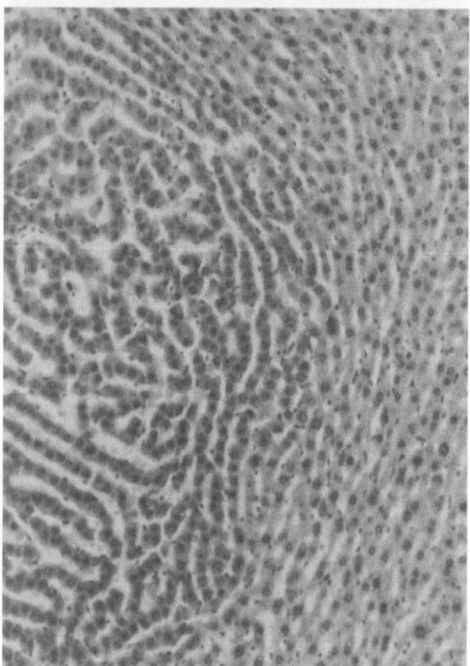

the normal hepatic architecture is lost, although central veins and portal tracts are not necessarily always absent from the edges of the lesion (Fig. 29). Structures resembling central veins can be produced within adenomas, and portal tracts can become engulfed during parenchymal proliferation and remain within the lesion (Fig. 30). However, the architecture within the adenoma is always atypical in that the normal topography is not maintained and consequently there is no lobular arrangement.

The cytology of hepatocellular adenomas varies considerably. They commonly consist of cells with eosinophilic, basophilic, clear, or vacuolated cytoplasm or various combinations of the different cell types, sometimes with islands of one cell type within a lesion consisting mainly of another cell type. Adenoma cells commonly have slightly enlarged clear nuclei with prominent nucleoli, the heterogeneity of cells increasing as the neoplasm becomes more developed. The trabecular arrangement of cords, which are usually one to two cells thick, is maintained with the formation of discontinuous plates at the boundary, adenoma cords being perpendicular or oblique to those in the normal parenchyma. Adjacent normal liver cords are often compressed, and there is a sharp demarcation of the neoplastic lesion from the surrounding parenchyma (Figs. 28, 29). Sinusoids within the adenoma have a variable width, depending upon whether or not the neoplastic cells are enlarged, and any increase in mitotic rate of neoplastic cells may not be obvious compared with the surrounding parenchyma. Invasive growth of hepatocellular adenomas is not observed.

In some instances, changes which some pathologists regard as regressive may occur within hepatocellular adenomas. These are more commonly cystic or fatty changes. Neither of these types of changes necessarily constitutes evidence of regres-

◄

Fig. 28. (*above*) Hepatocellular adenoma, rat. Note the boundary of the neoplasm with adjacent normal liver parenchyma (*above*). In general, neoplastic cells are larger than normal and have more eosinophilic cytoplasom. There are no clearly defined sinusoids, and slight compression of adjacent normal liver cords is evident. There is no invasive growth. H&E, ×430

Fig. 29. (*below*) Hepatocellular adenoma, rat. The cytology of neoplastic cells is very similar to that of normal hepatocytes except that the former have slightly more basophilic cytoplasm. Cords one cell thick form clear sinusoids. There is no evidence of invasive growth. H&E, ×430

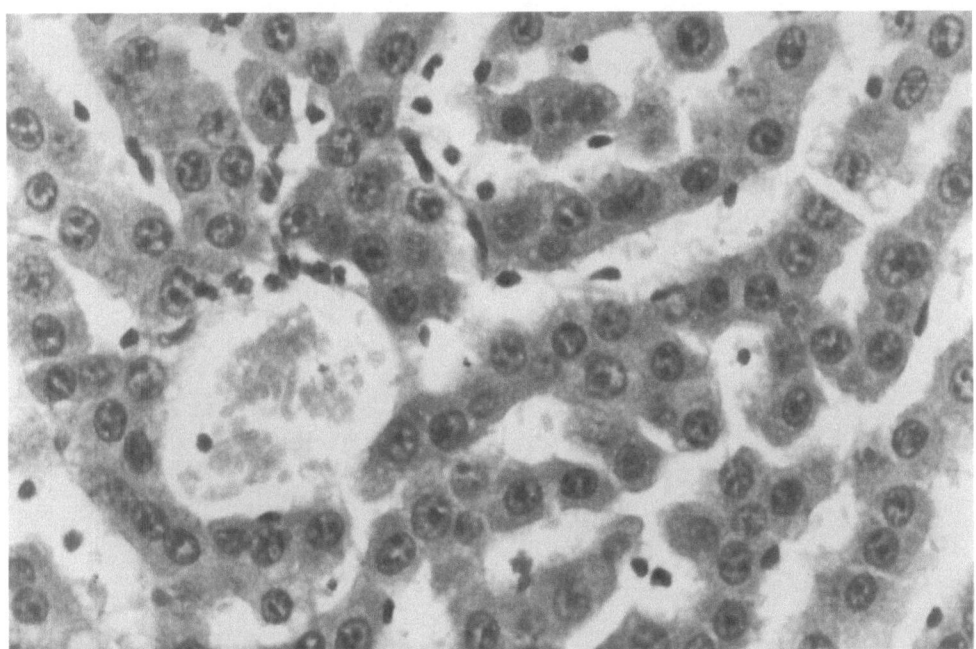

Fig. 30. A higher magnification of the same lesion as in Fig. 29, illustrating the formation of a structure resembling that of a central vein. There were no portal tracts within the neoplasm. H&E, ×1720

sion per se, although spontaneous regression undoubtedly can occur. Cystic and fatty changes can lead to considerable distortion of the microscopic appearance of adenomas and make diagnosis difficult (Figs. 31–33).

Ultrastructure

Shoji et al. (1994) undertook electron microscopic examination of 2-acetylaminofluorene-induced hepatocellular adenomas in male rats. The authors concluded that the ultrastructure of early-stage hepatocellular adenomas is almost the same as that of the normal liver with fenestration of endothelial cells and a discontinuous basement membrane. At later stages of development, endothelial sieve plates remained well preserved, but the basement membrane became thicker and continuous. Vitamin A-lipid droplets were observed with decreasing frequency in the stellate cells of the space of Disse as hepatocellular adenomas became more extensive.

The ultrastructure of 2-actylaminofluorene-induced liver lesions in male rats revealed the nuclear pore density from preneoplastic and neoplastic lesions to be higher than that of normal hepatocytes. However, the area of nuclear pores from preneoplastic lesions, or altered foci, did not differ from normal hepatocytes, whereas the area of pores from adenoma cells was found be increased (Sugie et al. 1994). The authors suggest that the number and the area of nuclear pores are related to the progress of tumorigenesis.

Differential Diagnosis

Criteria for the diagnosis of benign hepatocellular tumor (adenoma) have been subject to considerable debate and disagreement. Classification schemes have been proposed for the rat (Squire and Levitt 1975; Stewart et al. 1980) and for the mouse (Frith and Ward 1980). In theory, the adenoma lies between "hyperplastic lesion" and "hepatocellular carcinoma." In practice, there are few clear-cut criteria for distinguishing between adenoma and either hyperplasia or adenocarcinoma. The situation has been further complicated by the identification of "foci and areas of hepatocellular alteration," since there is no clear demarcation line between some forms of

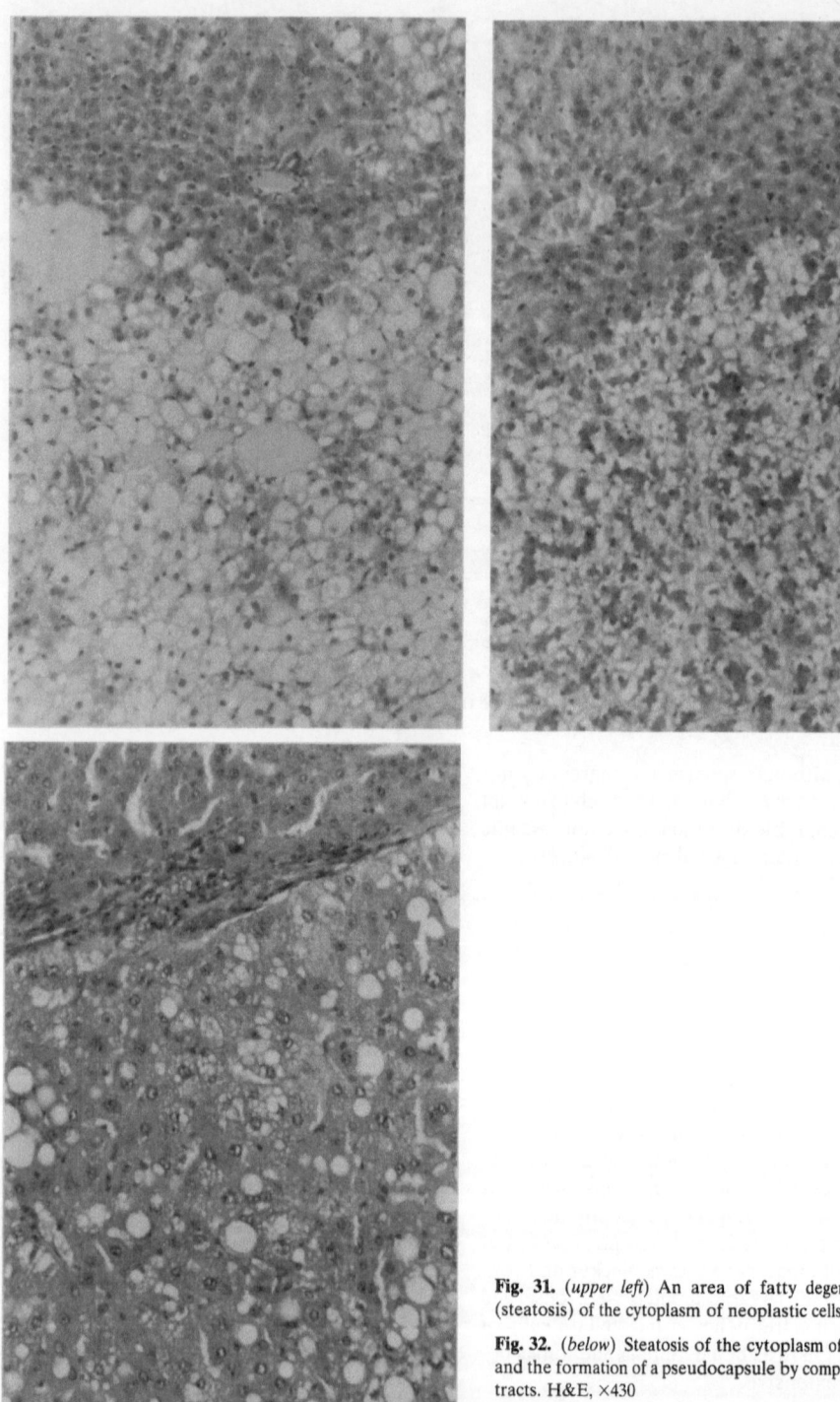

Fig. 31. (*upper left*) An area of fatty degenerative change (steatosis) of the cytoplasm of neoplastic cells. H&E, ×430

Fig. 32. (*below*) Steatosis of the cytoplasm of neoplastic cells and the formation of a pseudocapsule by compression of portal tracts. H&E, ×430

Fig. 33. (*upper right*) Vacuolated neoplastic cells associated with cytoplasmic accumulation of glycogen. H&E, ×172

hepatocellular alteration and hyperplasia or even adenoma. A few pathologists faced with these difficulties have decided to regard all the above lesions as hepatocellular carcinomas. To do this, however, is to waste data, to exaggerate the incidence of malignant neoplasia, and, in fact, to risk getting both false-positive and false-negative results in carcinogenicity tests.

It is better practice, therefore, to attempt to distinguish between the various kinds of lesions in the best way possible, knowing that the borderlines between them are indistinct. In the interpretation of chronic toxicity and carcinogenicity tests, the precise diagnosis of each proliferative liver lesion is often of less importance than the avoidance of drift in the use of criteria, such that comparisons of the incidence of lesions in treated and control groups of animals are biased.

For a lesion to be regarded as hyperplastic or neoplastic, there has to be evidence of proliferative changes (e.g., an expansive lesion with higher mitotic activity than in the surrounding liver). Groups of hepatocytes demonstrating cytoplasmic storage, tinctorial, degenerative, or enzymatic changes in the absence of evidence of cellular proliferation would lead to a diagnosis of focus of hepatocellular alteration.

If a lesion shows expansion with compression of surrounding liver tissue, evidence of cellular proliferation, and loss of lobular structure, it must be regarded as neoplastic. Preservation of lobular structure with the presence of portal triads within a lesion suggests it to be hyperplastic rather than neoplastic. For proliferative, expansive lesions that are less than one lobule in size, there is no sure way of distinguishing between hyperplasia and neoplasia.

In animals in which there has been previous liver injury, the whole organ or whole lobes of it may consist of regenerative nodules separated by bands of scar tissue. The term *regenerative nodular hyperplasia* is applicable to this condition, although it is recognized that it may coexist with hepatocellular neoplasia.

At the other end of the spectrum, there is no difficulty in identifying a lesion as a hepatocellular carcinoma if it has clearly invaded through some definable barrier, e.g., through the liver capsule or into a walled blood vessel, or if it has metastasized to the lungs or other tissues. However, the interpretation of lesser signs of possible malignancy, such as irregular extension into surrounding liver tissue without penetration of a definable barrier,

leads to disagreements between pathologists. It is particularly in this situation that great attention should be given to cytologic changes suggestive of malignancy (see p. 57) and to variations from normal hepatocellular arrangement.

Because of the difficulties involved in distinguishing between hyperplasia, benign neoplasia, and malignant neoplasia, there is often a case for blind rereading of sets of liver sections. This may not result in more accurate diagnosis, but does serve to eliminate intergroup biases.

Lesions of endothelial and bile duct origin also must be considered during diagnosis, but these lesions are usually more of a problem in the differential diagnosis of hepatocellular carcinoma than of adenoma.

Biologic Features

Hepatocellular adenomas generally do not result in the death of animals, although expansive growth may ultimately lead to functional anomalies. Of greater significance is the possibility of progression of hepatocellular adenoma to carcinoma. The earliest indication of localized functional disturbances in response to hepatic carcinogens is focal hepatocellular alteration (Williams 1980), and it is possible that such foci of proliferating cells are clones that developed from scattered isolated liver cells (see p. 24).

Foci and areas of hepatocellular alteration have a variety of enzymatic and other changes that facilitate their demonstration by histochemical techniques. These changes include the following: loss of adenosine triphosphatase (Hirota and Williams 1979a) and glucose-6-phosphate activity (Hirota and Williams 1979a), accumulation of glycogen (Williams et al. 1976; Williams and Watanabe 1978), increased γ-glutamyl transpeptidase activity (Hirota and Williams 1979a), resistance to iron accumulation (Hirota and Williams 1979a), and hyperbasophilic staining with toluidine blue (Hirota and Williams 1979b). According to Williams and Yamamoto 1972 and Williams et al. (1976), nodules (hepatocellular adenomas) occur only after the appearance of foci in the process of carcinogenesis, and this view is supported by the work of Reznik-Schüller and Gregg (1983) (see also p. 3). More recently, Williams et al. (1993) quantified neoplastic conversion among hepatocellular lesions in the rat following the administration of diethylnitrosamine. The authors demonstrated that enzymatic changes can reveal the presence of groups of altered hepatocytes which were otherwise undetectable by conventional histologic evaluation and identified a precursor role for such foci in the formation of hepatocellular neoplasms, although a high level of induction of foci was required for the evolution of neoplasms.

Many of these indicators of functional abnormalities are also shared by hepatocellular adenomas, although they do not usually exhibit α-fetoprotein secretion (Kroes et al. 1972; Tchipysheva et al. 1977; Kuhlmann 1978), which is a feature of some hepatocellular carcinomas.

Both adenomas and foci and areas of hepatocellular alteration are resistant to the cytotoxic effects of the inducing carcinogen and those of other toxic carcinogens requiring metabolic activation. Reduction of cytochrome P450 within the lesions is probably the basis for this diminished metabolic capability (Gravela et al. 1975).

Since some foci of hepatocellular alteration are able to persist after removal of the chemical stimulus, these must be regarded as irreversible (Schauer and Kunze 1976; Williams 1980). In the rat, foci of areas of clear, eosinophilic, or basophilic hepatocellular alteration have been demonstrated to be stages in the development of neoplasia (Bannasch 1968), and this is supported to some extent by the cytologic similarity between lesions of hepatocellular alteration and hepatocellular adenoma.

Adenomas, as might be expected, have been found to persist and increase in size after removal of carcinogen (Hirota and Williams 1979b), but many that persist for 1 year after removal of carcinogen undergo extensive cystic change. This cystic change may not necessarily be a regressive alteration, but closely resembles the lesion referred to as spongiosis hepatis by Bannasch et al. (1981) (see also p. 104), which probably results from the extracellular accumulation of acid mucopolysaccharides and/or proteins such as collagen. Such accumulation could result from an overproduction or impaired degradation of these substances within the neoplastic lesion.

In transplantation studies, hepatocellular adenomas persisted for months at the transplantation site, but did not exhibit continued growth, although cell division was maintained (Williams et al. 1977). This would suggest that the rodent hepatocellular adenomas require the environment of the intact liver for progressive growth (Williams

1980), an observation supported by the more recent findings of Marsman and Popp (1994) who examined the biologic potential of foci of alteration and hepatocellular tumors following the administration of a peroxisome proliferator to male rats. The authors reported a significant morphological continuity between basophilic foci and hepatocellular tumors. Although such lesions are the likely precursors of malignant hepatocellular lesions, they have little ability to maintain autonomous growth.

Much of the work undertaken in elucidating the morphological identification and behavior of proliferative parenchymal lesions of the rodent liver has been based on experiments employing known carcinogens. Although the morphology and presumably also the behavior and progression of spontaneous and induced neoplastic lesions is generally similar, it is quite possible that carcinogens, particularly if there is continuous exposure to them, to some extent modify the natural progression and behavior of such lesions, thereby providing a somewhat different basis for classification.

References

Bannasch P (1968) The cytoplasm of hepatocytes during carcinogenesis. II. The ascinus peripheral cytotoxic pattern. A. Light microscopy. B. Electron microscopy. In: Rentchnick P, Herfarth C, Senn HJ (eds) Recent results in cancer research, vol 19. Springer, Berlin Heidelberg New York, pp 18–48

Bannasch P, Bloch M, Zerban H (1981) Spongiosis hepatis. Specific changes of the perisinusoidal liver cells induced in rats by N-nitrosomorpholine. Lab Invest 44:252–264

Frith CH, Ward JM (1980) A morphologic classification of proliferative and neoplastic hepatic lesions in mice. J Environ Pathol Toxicol 3:329–351

Gravela E, Feo F, Canuto RA, Garcea R, Gabriel L (1975) Functional and structural alterations of liver ergastoplasmic membranes during DL-ethionine hepatocarcinogenesis. Cancer Res 35:3041–3047

Hirota N, Williams GM (1979a) The sensitivity and heterogeneity of histochemical markers for altered foci involved in liver carcinogenesis. Am J Pathol 95:317–328

Hirota N, Williams GM (1979b) Persistence and growth of rat liver neoplastic nodules following cessation of carcinogen exposure. J Natl Cancer Inst NCI 63:1257–1265

Kroes R, Williams GM, Weisburger JH (1972) Early appearance of serum alpha-fetoprotein during hepatocarcinogenesis as a function of age of rats and extent of treatment with 3′-methyl-4-dimethylaminoazobenzene. Cancer Res 32:1526–1532

Kuhlmann WD (1978) Localization of alpha 1-fetoprotein and DNA-synthesis in liver cell populations during experimental hepatocarcinogenesis in rats. Int J Cancer 21:368–380

Marsman DS, Popp JA (1994) Biological potential of basophilic hepatocellular foci and hepatic adenoma induced by the peroxisome proliferator, Wy-14,643. Carcinogenesis 15:111–117

Reznik-Schüller HM, Gregg M (1983) Sequential morphologic changes during methapyrilene-induced hepatocellular carcinogenesis in rats. J Natl Cancer Inst 71:1021–1031

Schauer A, Kunze E (1976) Tumours of the liver. In: Turusov VS (ed) Pathology of tumours in laboratory animals, vol 1. Tumours of the rat, part 2. IARC, Lyon, pp 41–72 (IARC scientific publications no 5)

Shoji Y, Kaneda K, Wake K, Mishima Y (1994) Light and electron microscopic analysis of liver sinusoids during hepatocarcinogenesis with 2-acetylaminofluorene in rats. Jpn J Cancer Res 85:491–498

Squire RA, Levitt MH (1975) Report of a workshop on classification of specific hepatocellular lesions in rats. Cancer Res 35:3214–3223

Stewart HL, Williams GM, Keysser CH, Lombard LS, Montali RJ (1980) Histologic typing of liver tumors of the rat. J Natl Cancer Inst 64:177–206

Sugie S, Yoshimi N, Tanaka T, Mori H, Williams GM (1994) Alterations of nuclear pores in preneoplastic and neoplastic rat liver lesions induced by 2-acetylaminofluorene. Carcinogenesis 15:95–98

Tchipysheva TA, Guelstein VI, Bannikov GA (1977) Alpha-fetoprotein-containing cells in the early stages of liver carcinogenesis induced by 3'-methyl-4-dimethyl-aminoazobenzene and 2-acetylaminofluorene. Int J Cancer 20:388–393

Williams GM (1980) The pathogenesis of rat liver cancer caused by chemical carcinogens. Biochim Biophys Acta 605:167–189

Williams GM, Watanabe K (1978) Quantitative kinetics of development of N-2-fluorenylacetamide-induced, altered (hyperplastic) hepatocellular foci resistant to iron accumulation and of their reversion or persistence following removal of carcinogen. J Natl Cancer Inst 61:113–121

Williams GM, Yamamoto RS (1972) Absence of stainable iron from preneoplastic and neoplastic lesions in rat liver with 8-hydroxyquinoline-induced siderosis. J Natl Cancer Inst 49:685–692

Williams GM, Klaiber M, Parker SE, Farber E (1976) Nature of early appearing carcinogen-induced liver lesions resistant to iron accumulation. J Natl Cancer Inst 57:157–165

Williams GM, Klaiber M, Farber E (1977) Differences in growth of transplants of liver, liver hyperplastic nodules, and hepatocellular carcinomas in the mammary fat pad. Am J Pathol 89:379–390

Williams GM, Gebhardt R, Sirma H, Stenbäck F (1993) Non-linearity of neoplastic conversion induced in the rat liver by low exposures to diethylnitrosamine. Carcinogenesis 14:2149–2156

Hepatocellular Carcinoma, Liver, Rat

James A. Popp and Russell C. Cattley

Synonyms. Hepatic cell carcinoma, hepatoma, malignant hepatoma, hepatocarcinoma, liver cell carcinoma, hepatocellular carcinoma (trabecular, adenocarcinoma, poorly differentiated, well differentiated, moderately differentiated)

Gross Appearance

Hepatocellular carcinomas are usually visible grossly and range in size from several millimeters to several centimeters (Farber 1976; Greenblatt and Lijinsky 1972; Goodman et al. 1994; Fig. 34). They are roughly spherical, although multilobulation may distort the shape (Fig. 35), and appear as single or multiple, light-tan to dark-red lesions in the liver. Rats chronically exposed to potent hepatocarcinogens frequently have multiple hepatocellular carcinomas alone or in combination with hepatocellular adenomas (Fig. 34). The liver of such an animal may be almost totally replaced by numerous, variably sized neoplastic and preneoplastic lesions.

In general, the smaller lesions tend to have a uniform texture and grayish-white color, while larger lesions frequently have a mottled appearance due to necrosis and hemorrhage (Fig. 35). The dark color, usually due to hemorrhage, is frequently found toward the center of the lesion, while the edge of the lesion is typically lighter than the surrounding liver tissue. Hepatocellular carcinomas are invariably softer than normal liver tissue. As lesions enlarge and become necrotic, they typically become even more soft and friable. Rupture of such a lesion occasionally results in hemoperitoneum (Goodall and Butler 1969). These tumors do not have a visible capsule (Fig. 35) and, although several morphological variants

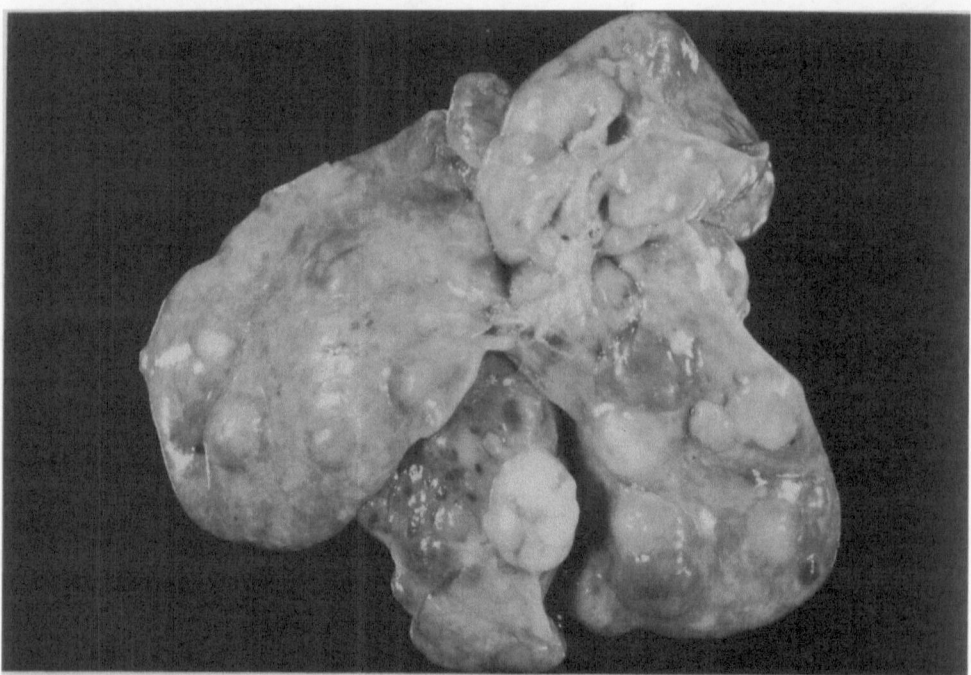

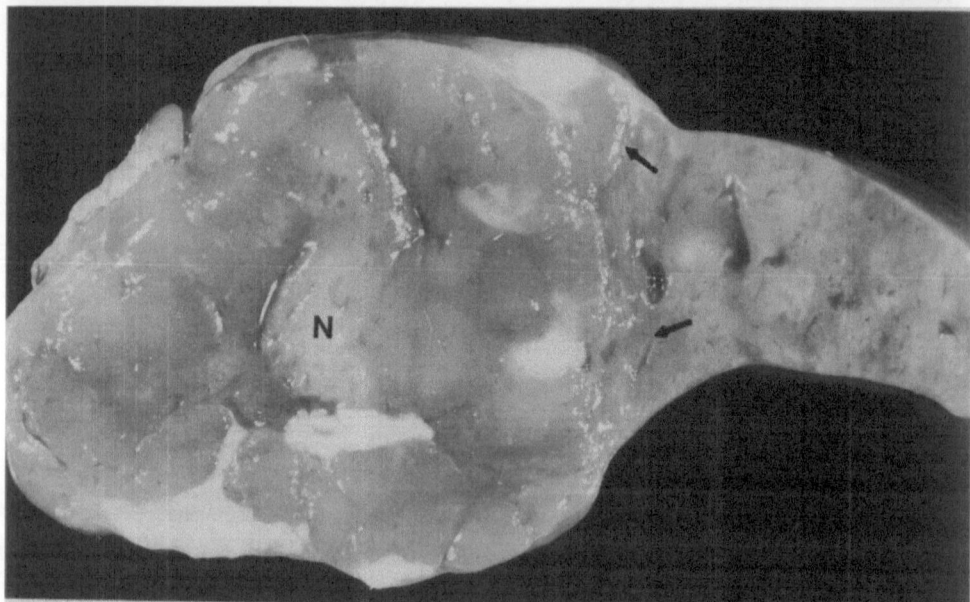

Fig. 34. (*above*) Hepatocellular carcinoma, Fischer 344 rat. This liver contains numerous hepatocellular carcinomas and adenomas which bulge from the capsular surfaces. ×5.2

Fig. 35. (*below*) Hepatocellular carcinoma, Fischer 344 rat. On cross-section, the multilobulated appearance can be seen as well as an area of necrosis in the center of the lesion. No evidence of encapsulation is evident at the edge of the lesion (*arrows*). *N*, necrosis. ×24

can be distinguished by light microscopy, they cannot be identified grossly.

Microscopic Features

The histologic features of hepatocellular carcinomas have been described in varying detail in numerous publications (Institute of Laboratory Animal Resources 1980; Ito et al. 1969; Mikol et al. 1983; Squire and Levitt 1975; Stewart and Snell 1959; Maronpot et al. 1986; Goodman et al. 1994). All hepatocellular carcinomas lack true encapsulation, although a pseudocapsule may be found along the edge of some lesions. This is probably the result of the accumulation of normal delicate supporting structures which condense at the edge of the lesion as normal adjacent hepatocytes are lost due to compression.

Except in rare instances, hepatocellular carcinomas are composed of cells readily recognizable as hepatocytes. The large, spherical basophilic nucleus is generally surrounded by relatively abundant eosinophilic cytoplasm, although the nucleus to cytoplasm ratio is generally greater in the neoplasm than in normal liver (Fig. 36). The arrangement of the neoplastic hepatocytes is abnormal, with variations in this arrangement forming the basis for the several histologic subtypes to be discussed. Hepatocellular carcinomas occasionally have a small number of bile ducts that may be present singly or in small clusters and, although not generally associated with fibrosis, lie adjacent to the neoplastic hepatocytes. There is some debate as to whether these bile ducts are part of the neoplastic response or are simply normal bile ducts, which are remnants of preexisting portal triads that have been enveloped in the expanding neoplasm as it encroaches on adjacent liver tissue. These neoplasms grow by progressive extension into adjacent tissue, sometimes resulting in the invasion of vessels (Greenblatt and Lijinsky 1972; Ito et al. 1969). Carcinomas on the surface of the liver appear to be retained by the liver capsule, which is frequently thickened.

Several histologic types of hepatocellular carcinomas are recognizable in the rat, and numerous synonyms have been used to describe them (Insti-

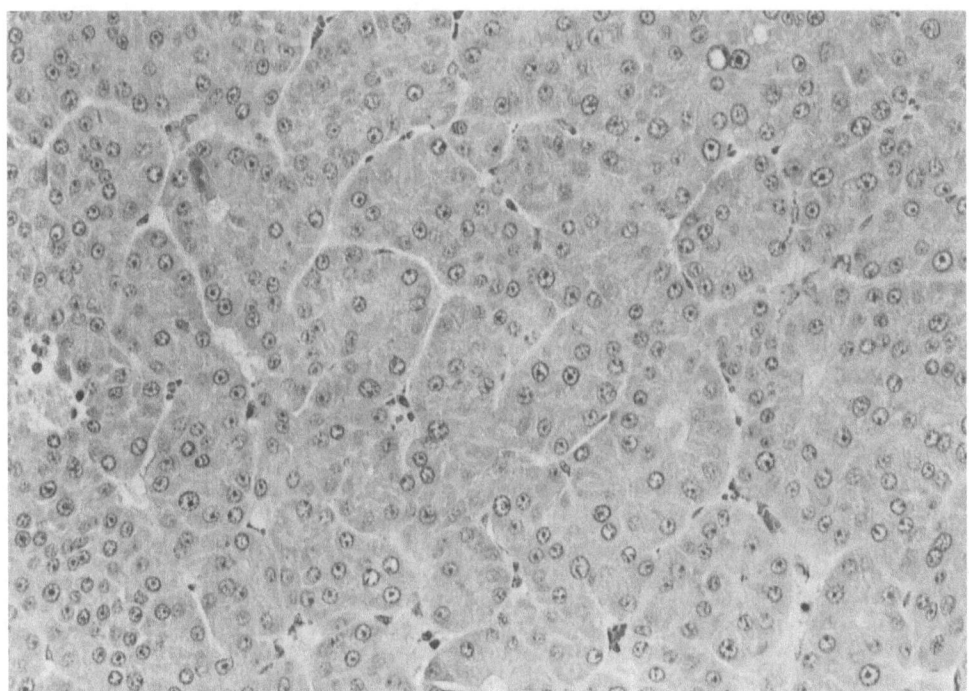

Fig. 36. Hepatocellular carcinoma, trabecular pattern, Fischer 344 rat. The trabeculae are composed of many layers of neoplastic hepatocytes. The cytoplasm is eosinophilic and the nuclei are slightly variable in size with large nucleoli. H&E, ×535

tute of Laboratory Animal Resources 1980; Stewart and Snell 1959). The value of subclassification of these neoplasms for developing an understanding of the lesion or predicting its biologic potential has not been demonstrated.

Most hepatocellular carcinomas have a trabecular pattern and are well differentiated (Fig. 36). In these lesions, the neoplastic hepatocytes form multicellular, thick trabeculae, which are arranged in an irregular pattern throughout the lesion. These trabeculae frequently branch or intersect to form a quilt-like pattern. Curiously, the most obvious trabecular pattern is frequently in the center of the lesion. The center of trabeculae, more than six to eight cells in thickness, usually undergoes necrosis, evidenced by individual necrotic hepatocytes. Loss of necrotic cells from the centers of trabeculae may result in structures with empty spaces that resemble glandular acini (so-called pseudoglandular pattern). Individual cells frequently have a remarkable resemblance to normal hepatocytes with extensive cytoplasm which can be either lightly basophilic (Bannasch 1976; Ito et al. 1969; Stewart and Snell 1959; Ward 1981, 1983) or eosinophilic (Mikol et al. 1983). However, in some neoplasms, the cytoplasm will be reduced and more basophilic, leading to the diagnostic categories of "moderately differentiated" and "poorly differentiated" used by some pathologists (Fig. 37).

A glandular pattern, identified in up to 40% of hepatocellular carcinomas (Goodall and Butler 1969), has led some pathologists to subclassify this type as an adenocarcinoma (Fig. 38). In this lesion, the pattern results from a variable number of neoplastic hepatocytes forming a single layer around a central clear space, which probably represents a greatly dilated bile canaliculus, although the space does not contain evidence of blocked biliary excretion. Individual hepatocytes are typically well differentiated, similar to those of well-differentiated trabecular carcinoma. The glandular pattern rarely involves more than 50% of the neoplasm and is interspersed with areas that have either trabecular or anaplastic architecture.

Anaplastic hepatocellular carcinoma is the least common form. This type appears as a sheet of cells with no obvious organization. The cells vary in size and shape and contain scant, frequently basophilic cytoplasm and relatively large hyperchromatic nuclei.

A distinct type of carcinoma that is very rarely observed in rat liver contains some areas com-

posed of neoplastic hepatocytes and other areas composed of neoplastic biliary epithelial cells forming bile ducts (Mikol et al. 1983). Such lesions are referred to as hepatocholangiocellular carcinomas, reflecting their mixed epithelial components (Goodman et al. 1994). The neoplastic hepatocytes are usually well differentiated, with equally well differentiated bile ducts interspersed among them. The structures resembling bile ducts are composed of cuboidal or sometimes low-columnar epithelium surrounding a narrow lumen. Individual ductular cells have smaller and more pale-staining nuclei than the hepatocytic nuclei, and their cytoplasm is clearly, but lightly, basophilic. Occasionally, hepatocytic cells with large hyperchromatic nuclei and eosinophilic cytoplasm can be found adjacent to a bile duct cell with smaller nuclei and basophilic cytoplasm. Both of these cell types form the wall of the ductular structure. Fibrosis is generally not found in this lesion in contrast to cholangiocarcinomas, in which it is extensive.

In contrast to benign neoplastic and hyperplastic lesions, cells of hepatocellular carcinomas are generally pleomorphic with atypical nuclei characterized by abnormal chromatin patterns (Williams 1980). Increased heterogeneity of both cell area and nuclear area has been demonstrated by morphometric analysis of hepatocellular carcinomas, in contrast to non-malignant lesions (Broxup et al. 1988). The cells within hepatocellular carcinomas also have various cytologic characteristics. As demonstrated by special stains, the cytoplasm of some tumor cells may contain glycogen or lipid (Mikol et al. 1983). Several different types of pigment have been described and characterized by special stains (Stewart and Snell 1959). Hyaline bodies (Stewart and Snell 1959) and Mallory bodies (Borenfreund and Bendich 1978) have been described within the cytoplasm of hepatocellular carcinomas. Histochemical stains such as γ-glutamyltranspeptidase (γ-GT) are widely used to identify foci and nodules in the livers of rats given hepatocarcinogens; however, the resultant neoplastic cells do not uniformly stain with γ-GT (Ohshima et al. 1984; Rao et al. 1982; Ward 1983). In some cases, the neoplasms induced by a single carcinogen have variable responses when stained for γ-GT activity (Ohshima et al. 1984).

Hepatocellular carcinomas are generally believed to arise from preexisting adenomas or persistent hyperplastic nodular lesions, which in turn are believed to arise from smaller foci of cellular al-

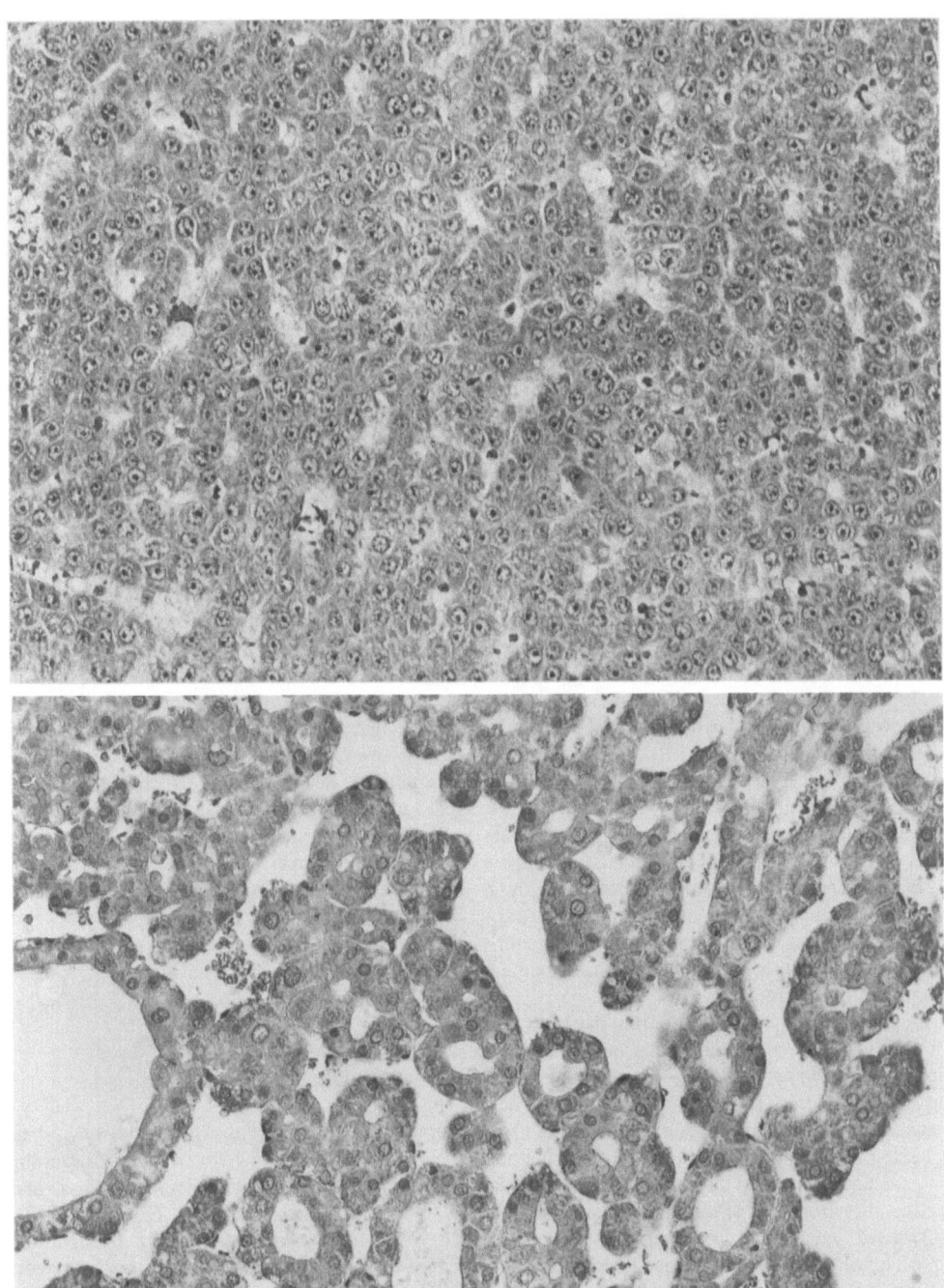

Fig. 37. (*above*) Hepatocellular carcinoma, trabecular, Fischer 344 rat, composed of small hepatocytes with a large nuclear to cytoplasmic ratio and basophilic cytoplasm. H&E, ×535

Fig. 38. (*below*) Hepatocellular carcinoma, adenocarcinoma, Fischer 344 rat. Large areas of this neoplasm have a glandular pattern in which a single row of neoplastic hepatocytes surround an empty lumen. H&E, ×535

teration (Bannasch 1976; Farber 1974, 1976, 1980; see p. 3). These foci and neoplasms have a monoclonal origin, with each lesion arising from one cell (Weinberg et al. 1987; Weinberg and Iannccone 1988). Within a small number of adenomas or persistent hyperplastic nodules, a hepatocellular carcinoma can be observed. These early carcinomas usually have a trabecular pattern and are composed of moderately well differentiated hepatocytes. Although hepatocellular carcinomas undoubtedly arise within preexisting hepatocellular adenomas, some evidence suggests that they may also arise directly from foci of altered hepatocytes (Williams 1980).

Ultrastructure

The ultrastructural characteristics of hepatocellular carcinomas have been described (Bannasch et al. 1980; Flaks et al. 1983; Merkow et al. 1969; Ogawa et al. 1979, 1984; Reznik-Schüller and Gregg 1983). The most striking feature is the variability of the cells, even within a single tumor (Flaks et al. 1983; Merkow et al. 1969; Ogawa et al. 1979), in contrast to the uniformity among cells composing hepatocellular adenomas. The nuclear membrane of neoplastic hepatocytes may have a corrugated appearance. As a result, pockets of cytoplasm may appear to be within the nucleus when a nuclear indentation is cut in cross-section. The cytoplasm of the neoplastic cells appears to be less well organized than the cytoplasm of normal hepatocytes. For example, instead of forming parallel stacks, rough endoplasmic reticulum generally consists of single, irregular cisternae wrapped around other organelles.

The smooth endoplasmic reticulum has been described as either reduced (Flaks et al. 1983) or increased (Merkow et al. 1969; Ogawa et al. 1979) in neoplastic cells. These differences may be determined by the chemical used to induce the neoplasms. Mitochondria are unevenly distributed, are variable in size, and have cristae that are either reduced or increased in length and number. The increased number of mitochondria found in methapyrilene-induced hepatocellular carcinomas (Reznik-Schüller and Gregg 1983) is not usually a feature of neoplastic hepatocytes induced by other chemicals. Annulate lamaellae, myelin figures, cytoplasmic inclusions, and an increase in microfilaments have also been described in neoplastic cells.

Ultrastructural changes of the cell surface have also been described (Ogawa et al. 1979, 1984). Deep invaginations of the cell membrane into the cytoplasm have been noted. The bile canaliculi are frequently dilated, irregular in shape, and lined by microvilli of variable length. This irregular canalicular pattern is particularly well demonstrated by scanning electron microscopy (Ogawa et al. 1984).

The vascular spaces within hepatocellular carcinomas are usually lined by nonfenestrated endothelium (Ogawa et al. 1984), in contrast to the fenestrated endothelium of normal liver tissue. Occasionally, basement membranes have been observed in the space of Disse (Ogawa et al. 1979) of carcinomatous liver, while they are not found under the sinusoidal endothelium of normal liver tissue. These observations indicate that the vascular lining within hepatocellular carcinomas tends to be more characteristic of normal veins rather than the normal sinusoidal lining found in the liver. The ultrastructural characteristics of trabecular versus glandular lesions have been reported (Flaks et al. 1983; Ogawa et al. 1984). While differences exist in the relationship of one neoplastic cell to another in these two lesions, the intracellular characteristics are remarkably similar irrespective of the cell arrangement.

In summary, hepatocellular carcinomas have altered ultrastructural characteristics that vary from cell to cell within an individual neoplasm as well as between neoplasms. No unique ultrastructural characteristics distinguish malignant neoplastic cells from normal or benign hepatocytes. For these reasons, electron microscopy is not particularly useful in the differential diagnosis of these neoplasms.

Differential Diagnosis

Hepatocellular carcinomas must be distinguished from other primary liver tumors, including hemangiosarcomas, Kupffer cell sarcomas, and cholangiocarcinomas (see pp. 63, 86, & 101). These lesions are relatively easily distinguished with the light microscope, but not grossly. Grossly, the hemangiosarcoma usually has a mottled appearance which may distinguish it, unless the hepatocellular carcinoma has undergone necrosis and hemorrhage. Kupffer's cell sarcomas generally do not reach the size of the largest hepatocellular carcinomas and tend to have less distinct borders. Cholangiocarcinomas are usually firmer due to their extensive fibrous connective tissue stroma. Microscopically, hepatocellular carcinomas generally have distinct cell borders and regular patterns of cell arrangement. However, it may be difficult to distinguish anaplastic types from very cellular hemangiosarcomas. The pleomorphic cell types of Kupffer's cell sarcomas usually distinguish this lesion, but the rare anaplastic carcinoma may be difficult to differentiate. Immunofluorescence microscopy can be used to distinguish hepatocellular carcinomas from neoplasms of mesenchymal origin (Bannasch et al. 1980). With this technique, normal and neoplastic hepatocytes stain with antibodies to nonfilament prekeratin, and normal and neoplastic mesenchymal cells of the liver stain with antibody to vimentin.

Well-differentiated hepatocellular carcinomas are easily distinguished from metastatic neoplasms, while more anaplastic hepatocellular lesions must be differentiated from metastatic neoplasms of endocrine origin. Most other metastatic carcinomas induce a substantial desmoplastic response.

Biologic Features

In chronic bioassay studies, hepatocellular carcinomas may be encountered in 0%–100% of the animals, depending on the strain of rat and the chemical studied. These lesions also occur spontaneously in male and female rats of inbred and outbred stains. Cumulative incidence rates for untreated F344 and Sprague-Dawley rats are summarized in Table 6.

The frequency of metastases is extremely variable, ranging from 0% to 100% of the animals with the tumors (Goodall and Butler 1969; Lijinsky et al. 1980; Mabuchi 1979; Mikol et al. 1983; Ohshima et al. 1984; Reuber et al. 1972). In general, metastasis of hepatocellular carcinoma in the rat is considered to be an uncommon finding (Goodman et al. 1994). Metastases may depend, in part, on the specific carcinogen and its dose. However, it should be remembered that a high incidence of animals with metastasis does not necessarily indicate an equally high incidence of neoplasms that metastasize, since an individual animal may have multiple neoplasms. Metastatic lesions are not necessarily observed grossly (Greenblatt and

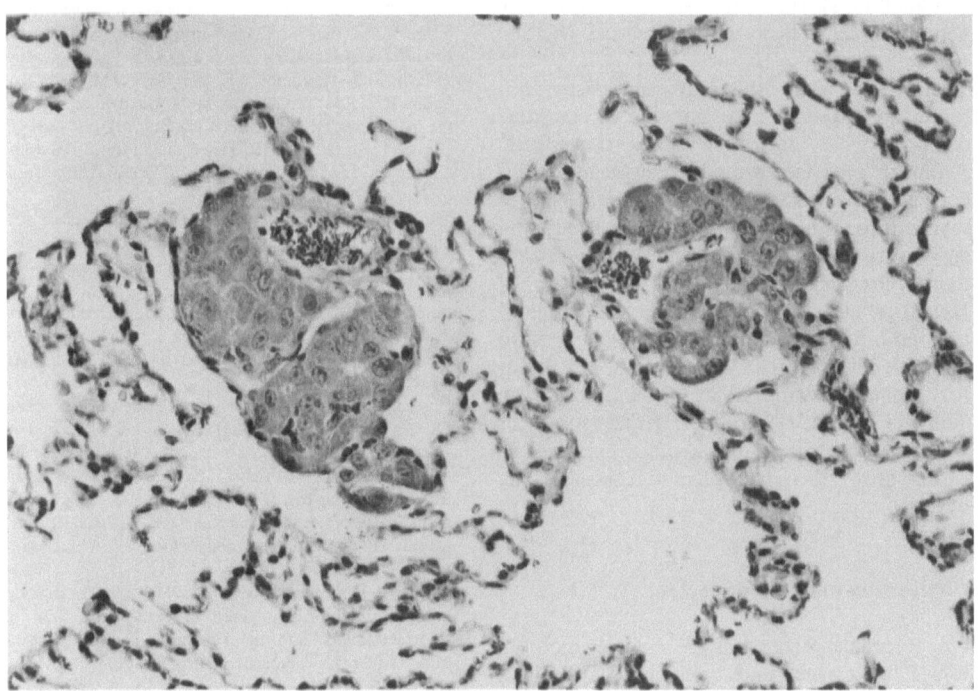

Fig. 39. Hepatocellular carcinoma, in the lung, Fischer 344 rat. Tumor emboli form metastatic sites as they grow through vessel walls. H&E, ×535

Table 6. Spontaneous incidence of hepatocellular carcinomas in aging rats

	Number examined	Rats with hepatocellular carcinoma (%)[b]	Reference
F344			
Male	1928	1.0	Haseman et al. 1990
Female	1979	0.2	
Crl:CD[a]			
Male	1258	2.6	Lang 1992
Female	1263	0.4	
Crl:CD			
Male	585	2.4	McMartin et al. 1992
Female	585	0.2	
Crl:CD			
Male	1340	0.4	Chandra et al. 1992
Female	1329	0.1	

[a] CD rats are derived from the Sprague-Dawley strain.
[b] Based on rats surviving up to 25–26 months of age.

Lijinsky 1972) and are most commonly found in the lung and peritoneum (Hamm 1983). In the lung, neoplastic cells apparently become lodged in small capillaries and grow until the vessel wall is distended (Fig. 39). Metastasis has also been observed in the lymph nodes, pancreas, omentum, testes, kidney, spleen, adrenal gland, and stomach (Farber 1976; Ito et al. 1969; Stewart and Snell 1959). The metastatic route is generally hematogenous rather than lymphatic (Becker 1978).

There was no relationship between metastatic potential and growth rate, tumor size, chromosome composition, or other functional characteristics. Although metastasis would be expected to decrease longevity, such a relationship was not observed in one study; however, in this study, the survival time was linked to the grade of malignancy (Greenblatt and Lijinsky 1972).

Hepatocellular carcinomas are readily transplantable into syngeneic animals and less so into noninbred lines (Farber 1976; Institute of Laboratory Animal Resources 1980; Ohshima et al. 1984; Williams et al. 1977). In contrast, hepatocellular adenomas and hyperplastic hepatocellular lesions rarely grow when transplanted conventionally even into syngeneic hosts.

Comparison with Other Species

Hepatocellular carcinomas are similar in morphological appearance and biologic behavior in a number of species, including rat, mouse, and humans (Ward 1984).

References

Bannasch P (1976) Cytology and cytogenesis of neoplastic (hyperplastic) hepatic nodules. Cancer Res 36:2555–2562

Bannasch P, Zerban H, Schmid E, Franke WW (1980) Liver tumors distinguished by immunofluorescence microscopy with antibodies to proteins of intermediate-sized filaments. Proc Natl Acad Sci USA 77:4948–4952

Becker FF (1978) Patterns of spontaneous metastasis of transplantable hepatocellular carcinomas. Cancer Res 38:163–167

Borenfreund H, Bendich A (1978) In vitro demonstration of Mallory body formation in liver cells from rats fed diethylnitrosamine. Lab Invest 38:295–303

Broxup BR, Valli VEO, Losos GL, Percy DH, Farber E, McMillan I (1988) Morphometric evaluation of hepatocellular proliferative lesions in the rat liver. Toxicol Pathol 16:401–417

Chandra M, Riley MGI, Johnson DE (1992) Spontaneous neoplasms in aged Sprague-Dawley rats. Arch Toxicol 66:496–502

Farber E (1974) Pathogenesis of liver cancer. Arch Pathol 98:145–148

Farber E (1976) The pathology of experimental liver cell cancer. In: Cameron HM (ed) Liver cell cancer. Elsevier, New York, pp 243–277

Farber E (1980) The sequential analysis of liver cancer induction. Biochim Biophys Acta 605:149–166

Flaks B, Trevan MT, Flaks A (1983) An electron microscopy study of hepatocellular changes in the rat during chronic treatment with acetamide. Parenchyma, foci and neoplasms. Carcinogenesis 4:1117–1125

Goodall CM, Butler WH (1969) Aflatoxin carcinogenesis: inhibition of liver cancer induction in hypophysectomized rats. Int J Cancer 4:422–429

Goodman DG, Maronpot RR, Newberne PM, Popp JA, Squire RA (1994) Proliferative and selected other lesions in the liver in rats, GI-5. In: Streett CS, Burek JD, Hardisty JF, Garner FM, Leininger JR, Pletscher JM, Moch RW (eds) Guides for toxicologic pathology. STP/ARP/AFIP, Washington, pp GI-5, 1–24

Greenblatt M, Lijinsky W (1972) Nitrosamine studies: neoplasms of liver and genital mesothelium in nitrosopyrrolidine-treated MRC rats. J Natl Cancer Inst 48:1687–1969

Hamm TE Jr (1983) The occurrence of neoplasms in long-term in vivo studies. In: Milman HA, Sell S (eds) Application of biological markers to carcinogen testing. Plenum, New York, pp 9–23

Haseman JK, Arnold J, Eustis SL (1990) Tumor incidences in Fischer 344 rats: NTP historical data. In: Boorman GA, Eustis, SL, Elwell MR, Montgomery CA Jr, MacKenzie WF (eds) Pathology of the Fischer rat, reference and atlas. Chapter 35. Academic, San Diego, pp 555–564

Institute of Laboratory Animal Resources, National Research Council (1980) Histologic typing of liver tumors of the rat. J Natl Cancer Inst 64:177–206

Ito N, Hiasa Y, Konishi Y, Marugami M (1969) The development of carcinoma in liver of rats treated with m-toluylenediamine and the synergistic and antagonistic effects with other chemicals. Cancer Res 29:1137–1145

Lang P (1992) Spontaneous neoplastic lesions and selected non-neoplastic lesions in the Crl:CD®BR rat. Charles River Laboratories, February, 1992

Lijinsky W, Reuber MD, Manning WB (1980) Potential carcinogenicity of nitrosodiethanolamine in rats. Nature 288:589–590

Mabuchi M (1979) Sequential hepatic changes during sterigmatocystin-induced carcinogenesis in the rat. Jpn J Exp Med 49:365–372

McMartin DN, Sahota PS, Gunson DE, Hsu HH, Spaet RH (1992) Neoplasms and related proliferative lesions in control Sprague-Dawley rats from carcinogenicity studies. Historical data and diagnostic considerations. Toxicol Pathol 20:212–225

Maronpot RR, Montgomery CA Jr, Boorman GA, McConnell EE (1986) National toxicology program nomenclature for hepatoproliferative lesions of rats. Toxicol Pathol 14:263–273

Merkow LP, Epstein SM, Farber E, Pardo M, Bartus B (1969) Cellular analysis of liver carcinogenesis. III. Comparison of the ultrastructure of hyperplastic liver nodules and hepatocellular carcinomas induced in rat liver by 2-fluorenylacetamide. J Natl Cancer Inst 43:33–63

Mikol YB, Hoover KL, Creasia D, Poirier LA (1983) Hepatocarcinogenesis in rats fed methyl-deficient, amino acid-defined diets. Carcinogenesis 4:1619–1629

Ogawa K, Medline A, Farber E (1979) Sequential analysis of hepatic carcinogenesis. Lab Invest 41:22 (35 DIG/NEO)

Ogawa H, Itoshima T, Ukida M, Ito T, Kiyotoshi S, Kitadai M, Hattory S, Mizutani S, Kita K, Tanaka R, Robe K, Nagashima H, Kobayashi T, Hamaya K (1984) Scanning electron microscopy of experimentally induced hepatocellular carcinoma. Scan Electron Microsc 41:359–368

Ohshima M, Ward JM, Brennan LM, Creasia DA (1984) A sequential study of methapyrilene hydrochloride-induced liver carcinogenesis in male F344 rats. J Natl Cancer Inst 72:759–768

Rao MS, Lalwani ND, Scarpelli DG, Reddy JK (1982) The absence of gamma-glutamyl transpeptidase activity in putative preneoplastic lesions and in hepatocellular carcinomas induced in rats by the hypolipidemic peroxisome proliferator Wy-14,643. Carcinogenesis 3:1231–1233

Reuber MD, Stromberg K, Glover EL (1972) Influence of age and sex on hepatic lesions induced by chemical carcinogens: ingestion of 3'-methyl-4-dimethylaminoazabenzene by Buffalo strain rats. J Natl Cancer Inst 48:675–683

Reznik-Schüller HM, Gregg M (1983) Sequential morphologic changes during methapyrilene, induced hepatocellular carcinogenesis in rats. J Natl Cancer Inst 71:1021–1031

Squire RA, Levitt MH (1975) Report of a workshop on classification of specific hepatocellular lesions in rats. Cancer Res 35:3214–3223

Stewart HL, Snell KC (1959) The histopathology of experimental tumors of the liver of the rat. In: Homberger F (ed) Pathophysiology of cancer, 2nd edn Hoeber, New York, pp 85–126

Ward JM (1981) Morphology of foci of altered hepatocytes and naturally-occurring hepatocellular tumors in F344 rats. Virchows Arch (Pathol Anat) 390:339–345

Ward JM (1983) Increased susceptibility of livers of aged F344/NCr rats to the effects of phenobarbital on the incidence, morphology, and histochemistry of hepatocellular foci and neoplasms. J Natl Cancer Inst 71:815–823

Ward JM (1984) Morphology of potential preneoplastic hepatocyte lesions and liver tumors in mice and a comparison with other species. In: Popp JA (ed) Current perspectives in mouse liver neoplasia. Hemisphere, New York, pp 1–26

Weinberg WC, Iannaccone PM (1988) Clonality of preneoplastic liver lesions: histological analysis in chimeric rats. J Cell Sci 89:423–431

Weinberg WC, Berkwits L, Iannaccone PM (1987) The clonal nature of carcinogen-induced altered foci of γ-glutamyl transpeptidase expression in rat liver. Carcinogenesis 8:545–570

Williams GM (1980) The pathogenesis of rat liver cancer caused by chemical carcinogens. Biochim Biophys Acta 605:167–189

Williams GM, Klaiber M, Farber E (1977) Differences in growth of transplants of liver, liver hyperplastic nodules, and hepatocellular carcinomas in the mammary fat pad. Am J Pathol 89:379–390

Cholangiofibroma and Cholangiocarcinoma, Liver, Rat

Peter Bannasch and Heide Zerban

Synonyms. Nodules of cholangiofibrosis, cholangiocellular carcinoma, cholangiolar adenocarcinoma bile duct carcinoma, malignant cholangioma, adenocarcinoma

Gross Appearance

Both cholangiofibroma and cholangiocarcinoma appear macroscopically as firm nodules frequently

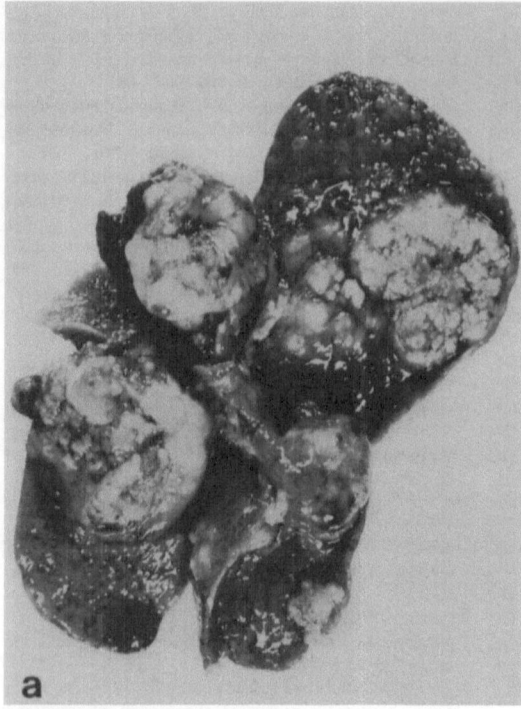

Fig. 40a,b. Cholangiofibromas induced in rat with *N*-nitrosomorpholine. **a** Macroscopic appearance, ×1.5. **b** Overview at low magnification, ×4

distributed in the liver in a multinodular fashion (Fig. 40). The tumor tissue, which usually has a grayish-white color, may also show yellow areas. The macroscopic picture may become very complex and colorful when the cholangiocellular tumors are combined with hepatocellular carcinomas or malignant mesenchymal tumors, such as angiosarcomas.

Microscopic Features

Cholangiofibromas and cholangiocarcinomas are collectively classified as carcinomas by many authors (Stewart et al. 1980; Maronpot et al. 1991; Goodman et al. 1994). Based on detailed studies on the morphogenesis and biologic behavior of these tumors, Bannasch et al. (1985; Bannasch and Massner 1976) have proposed that the cholangiofibroma be classified as a separate pathomorphological entity which is only potentially malignant and may progress to cholangiocarcinomas after long lag periods. This classification has widely been used (Ohshima et al. 1984; Moore

et al. 1986; Evans et al. 1989; Bannasch and Zerban 1990, 1994; Steinberg et al. 1991; Hacker et al. 1992; Elmore and Sirica 1993), but it has not yet been accepted unanimously (Maronpot et al. 1991; Goodman et al. 1994). Whereas autochthonous cholangiofibromas may reach a considerable size without leading to distant metastases (Bannasch and Massner 1976; Bannasch et al. 1985; Evans et al. 1989), the metastatic potential of cholangiocarcinomas has been established beyond doubt (Bannasch and Massner 1976; Maronpot et al. 1991; Elmore and Sirica 1993).

Cholangiofibroma

The cholangiofibroma (Figs. 40–42) is composed of atypical ductules and large amounts of collagen-rich connective tissue (Bannasch and Massner 1976; Bannasch et al. 1985). As a rule, the epithelium of the neoplastic ductular structures is composed of one cell layer (Fig. 41), which contains many goblet cells storing and secreting abundant

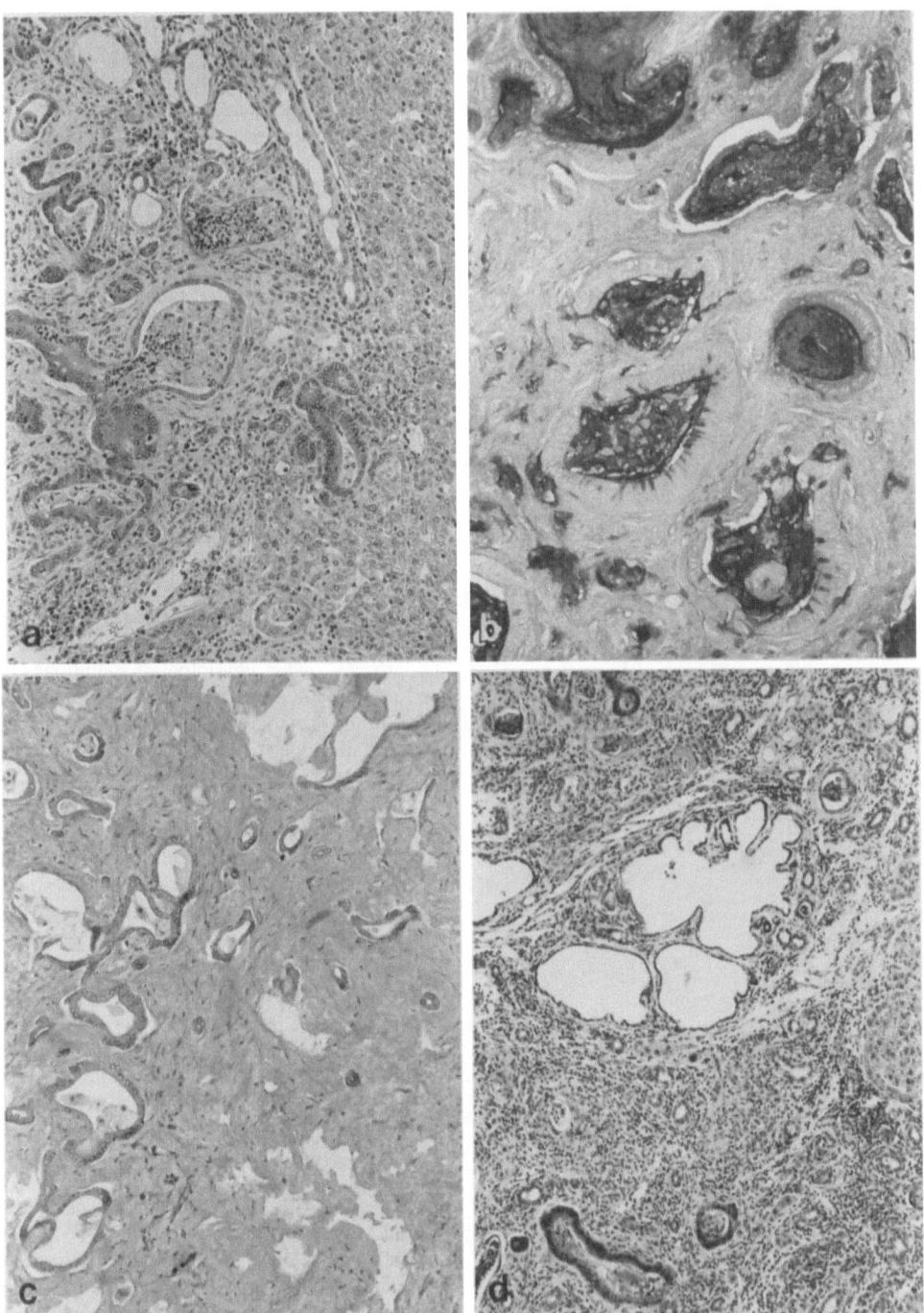

Fig. 41a–d. Cholangiofibromas induced in rat with *N*-nitrosomorpholine. **a** Periphery of cholangiofibroma exhibiting atypical ductules and a considerable proliferation of mesenchymal cells. H&E, ×140. **b** Accumulation of mucus (*black*) in goblet cells and duct lumina. Periodic acid-Schiff reaction, ×140. **c** Center of cholangiofibroma with advanced sclerosis and partial degeneration of epithelia of the neoplastic ductules. H&E, ×140. **d** Cystic changes within a cholangiofibroma. H&E, ×90

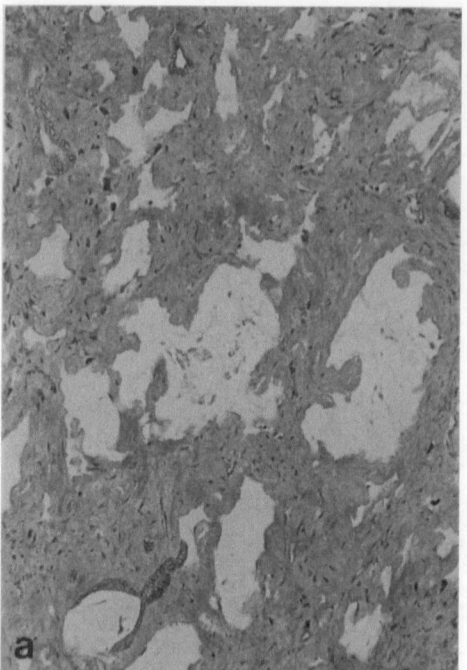

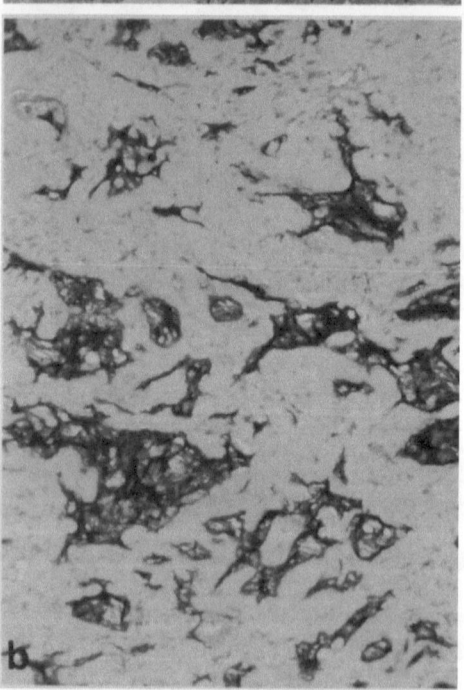

Fig. 42a,b. Central parts of a cholangiofibroma induced in rat with *N*-nitrosomporhpoline. **a** Pronounced sclerosis and nearly complete loss of ductular epithelia H&E, ×140. **b** Mucous material surrounded by connective tissue rich in collagenic fibers. Periodic acid-Schiff reaction, ×140

mucous substances (Fig. 41b). The remaining epithelia are usually intensely basophilic and may appear rather atypical. Mitotic figures (which are frequently pathologically altered), but also necrotic cells, are often present within the epithelium of the glandular structures. The lumina of the ductules are filled with mucous substances which may be mixed with shedded, lipid-laden, or necrotic epithelial cells and polymorphonuclear leukocytes (Fig. 41a). The mucous substances contain both acid and neutral mucopolysaccharides, as demonstrated histochemically by staining with alcian blue and by the Hale or periodic acid-Schiff reaction (Figs. 41b, 42b). Glycogen particles can be seen within the ductular epithelium. In addition, the majority of the epithelial tumor cells are very rich in intermediate filaments of the cytokeratin type; these have been identified by immunofluorescence microscopy (Fig. 43a,b) using antibodies produced against the constitutive proteins of these filaments (Bannasch et al. 1980, 1981b). More specifically, the expression of cytokeratin 19, which is typical of bile duct and ductular (cholangiolar) cells but not of hepatocytes under physiological conditions, has recently been demonstrated in the ductular components of cholangiofibromas, whereas preneoplastic and neoplastic hepatocellular lesions were not detected by antibodies to this type of cytokeratin (Steinberg et al. 1991).

In central parts of the cholangiofibromas, the lining epithelia of the atypical ductules may, in part or even totally, degenerate and disappear so that the mucus is surrounded by connective tissue only (Figs 41, 42a,b, 43b). Rarely, cysts lined by a flat epithelium and free of mucus may be found within cholangiofibromas (Fig. 41d).

The connective tissue of the cholangiofibroma consists predominantly of fibroblasts and more or less abundant collagenic fibers (Figs. 41, 42). In contrast to the epithelial cells, the fibroblasts contain intermediate filaments of the vimentin type (Fig. 43c), which are characteristic of both normal and neoplastic mesenchymal cells (Bannasch et al. 1980, 1981b). The collagen fibers are closely associated with glycosaminoglycans, as shown by positive staining with alcian blue. Whereas fibroblasts prevail over fibers in the periphery of the tumors, central parts are frequently characterized by an advanced sclerosis. This is especially true for large tumors. Occasionally, small groups of hepatocytes with normal appearance may be enclosed in the cholangiofibromas. The adjacent liver paren-

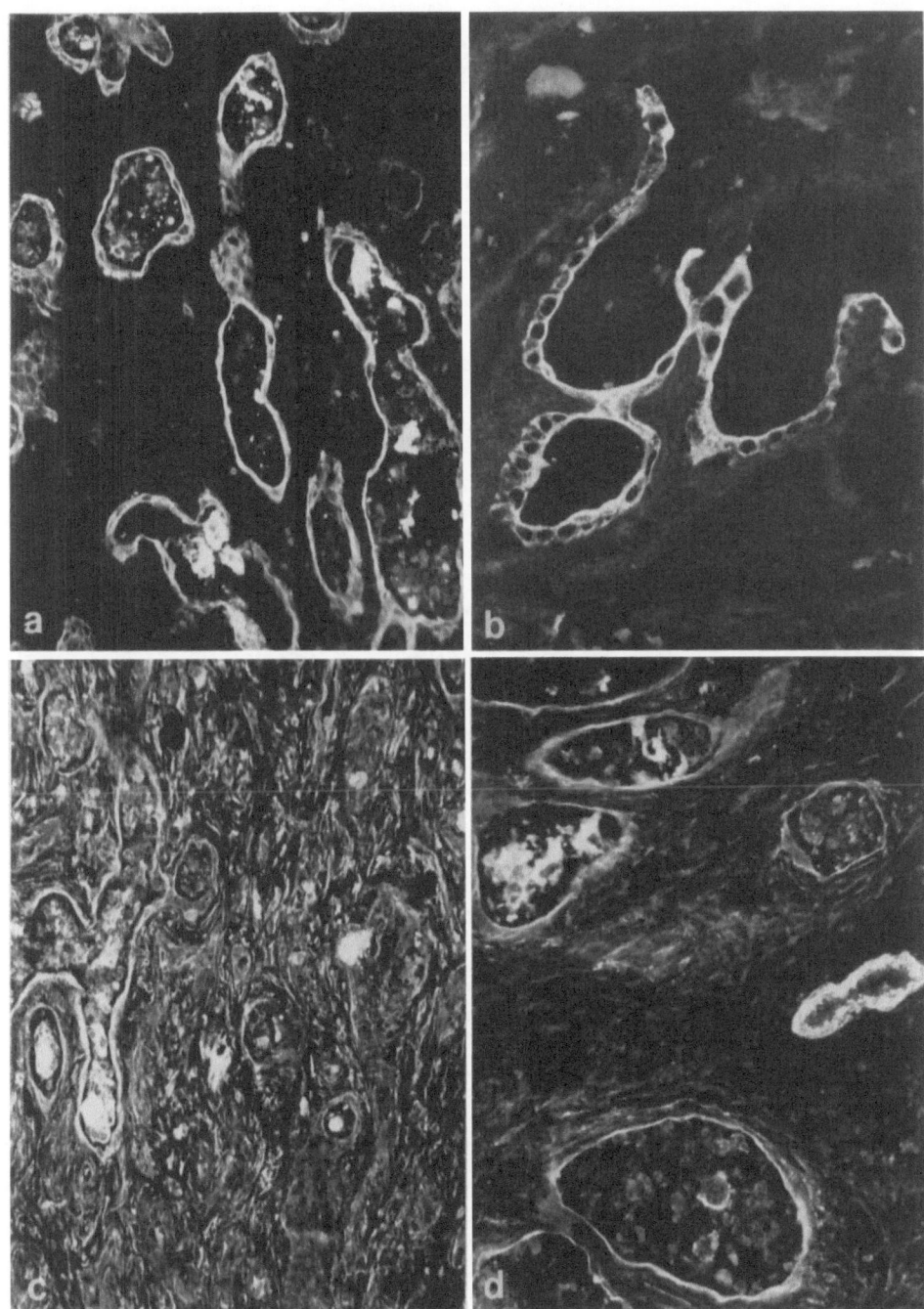

Fig. 43a–d. Cytoskeletal components by fluorescent antibodies to cytoskeletal proteins in cholangiofibromas induced in rat with *N*-nitrosomorpholine. **a** Strong reaction for cytokeratins in epithelial but not in mesenchymal tumor components, ×225. **b** Partial loss of cytokeratin-positive epithelia in neoplastic ductules, ×290. **c** Strong reaction for vimentin in mesenchymal but not in epithelial cells, ×225. **d** Strong reaction for actin in apical part of ductular epithelia and in wall of a small vessel (*middle right*), ×290

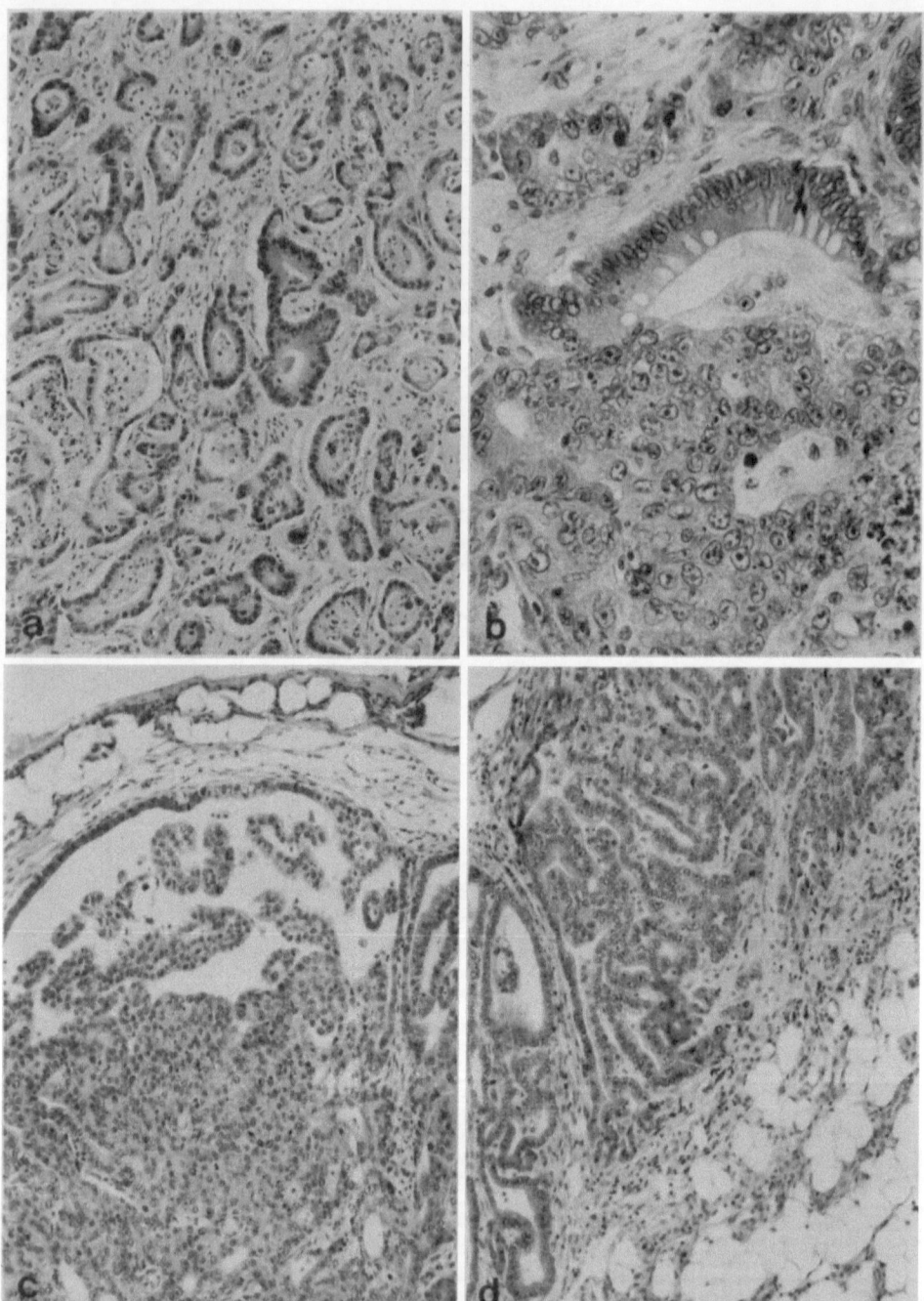

Fig. 44a–d. Cholangiocarcinomas induced in rat with *N*-nitrosomorpholine. **a** Ductular formations of cholangiocarcinoma. H&E, ×140. **b** Transition from mucus-producing, ductule-containing goblet cells into a highly undifferentiated tumor formation free of mucus. H&E, ×250. **c** Papillary tumor formation invading the liver capsule. H&E, ×140. **d** Metastatic growth of cholangiocarcinoma in liver capsule and adjacent extrahepatic connective tissue. H&E, ×140

chyma is often compressed due to the expansive growth of the tumors. In addition, single ductules of the tumors are sometimes localized within the neighboring parenchyma (Fig. 41a). Large numbers of fibroblasts within peripheral tumor formations may mimic invasive growth. However, in our experience with more than 100 cholangiofibromas induced in rats by various chemical carcinogens (N-nitrosomorpholine, thioacetamide, 2-acetylaminofluorene, choline-deficient/D-ethionine-supplemented diet), this tumor type does not metastasize (Bannasch and Zerban 1990, 1994; Steinberg et al. 1991). Although it therefore appears to be reasonable to classify the cholangiofibroma as a benign tumor, a number of observations suggest that it is potentially malignant (Bannasch and Massner 1976; Maronpot et al. 1991).

Cholangiocarcinoma

The majority of cholangiocellular tumors hitherto described, seem to be related to cholangiofibromas rather than to cholangiocarcinomas. Metastasizing cholangiocarcinomas, which apparently develop from cholangiofibromas, have only rarely been distinguished (Nakano 1974; Bannasch and Massner 1976; Ohshima et al. 1984). The tumors are glandular, solid, or papillary in structure (Fig. 44). The tumor cells are polygonal and contain large nuclei with prominent nucleoli. Their cytoplasm is usually intensely basophilic, but may also be clear due to the eluation of accumulated glycogen. Sometimes, transitions occur from mucus-producing ductular tumor components resembling those in mucous cholangiofibromas to frank malignant formations which are poor in, or totally free of mucous material (Fig. 44b).

In contrast to cholangiofibromas, cholangiocarcinomas undergo clear-cut invasive growth into intrahepatic blood and lymph vessels and into the connective tissue surrounding the liver (Fig. 44c,d; Bannasch and Massner 1976; Maronpot et al. 1991; Elmore and Sirica 1993).

Ultrastructure

The fine structure of the cholangiofibroma has been described in detail (Bannasch and Massner 1977; Bannasch et al. 1985). It is very similar, if not identical, to the ultrastructure of tumors classified by others as cholangiocarcinomas (David 1962; Onoe and Fuse 1966; Reddy et al. 1977), adenocarcinomas (Svoboda 1964; Ma and Webber 1966; Elmore and Sirica 1993), or "nodules" of cholangiofibrosis (Terao and Nakano 1974). The epithelial component of the cholangiofibroma is characterized by mucus-producing tubules which are surrounded by a basal lamina and have an apical brush border (Fig. 45). The microvilli of the brush border contain axial bundles of actin filaments (Figs. 43d, 45a) and are associated with a thick glycocalix (Fig. 45a). The basal lamina may be split into several layers (Fig. 45c). Neighboring ductular epithelial cells are connected by elaborate interdigitations which form many desmosomal junctions. The cytoplasm of the epithelial tumor cells is rich in free ribosomes and/or granular endoplasmic reticulum (Fig. 46a). Large bundles of intermediate filaments (Fig. 46b) correspond to the cytokeratins, as demonstrated by immunofluorescence microscopy (Fig. 43a,b).

The goblet cells seen by light microscopy (Fig. 41b) exhibit abundant mucous substances (Fig. 47a–c) and prominent Golgi complexes which frequently contain mucous secretary droplets. Cells storing glycogen in excess may also be found in the lining epithelium of the neoplastic ductules (Fig. 47a,d). Like other ductular epithelia, they rest on a basal lamina (Fig. 47d) and have an apical brush border or microvilli. The glycogen is stored mainly within the cytoplasmic matrix, but it may also be enclosed in large autophagic vacuoles. Unlike the predominantly α-particle glycogen depots in hepatocytes or hepatoma cells, large aggregates of single β-particles are typical of the accumulated glycogen in cholangiocellular tumors (Fig. 47c,d). It is noteworthy that cells which excessively store glycogen are sometimes located next to goblet cells which store mucous substances (Fig. 47c). Lipid bodies with a lamellar substructure can be sporadically detected in the cytoplasm. Microbodies (peroxisomes) are completely absent. The nuclei of the neoplastic ductular epithelium frequently have indentations of the nuclear envelope and prominent nucleoli with normal fine structures.

The mesenchymal component of the cholangiofibroma consists mainly of fibroblasts, collagenic fibers, and capillaries. Both the fibroblasts and the endothelial cells of the capillaries contain bundles of intermediate filaments of the vimentin type. Mast cells are loosely distributed within the tumor tissue. Some authors have described

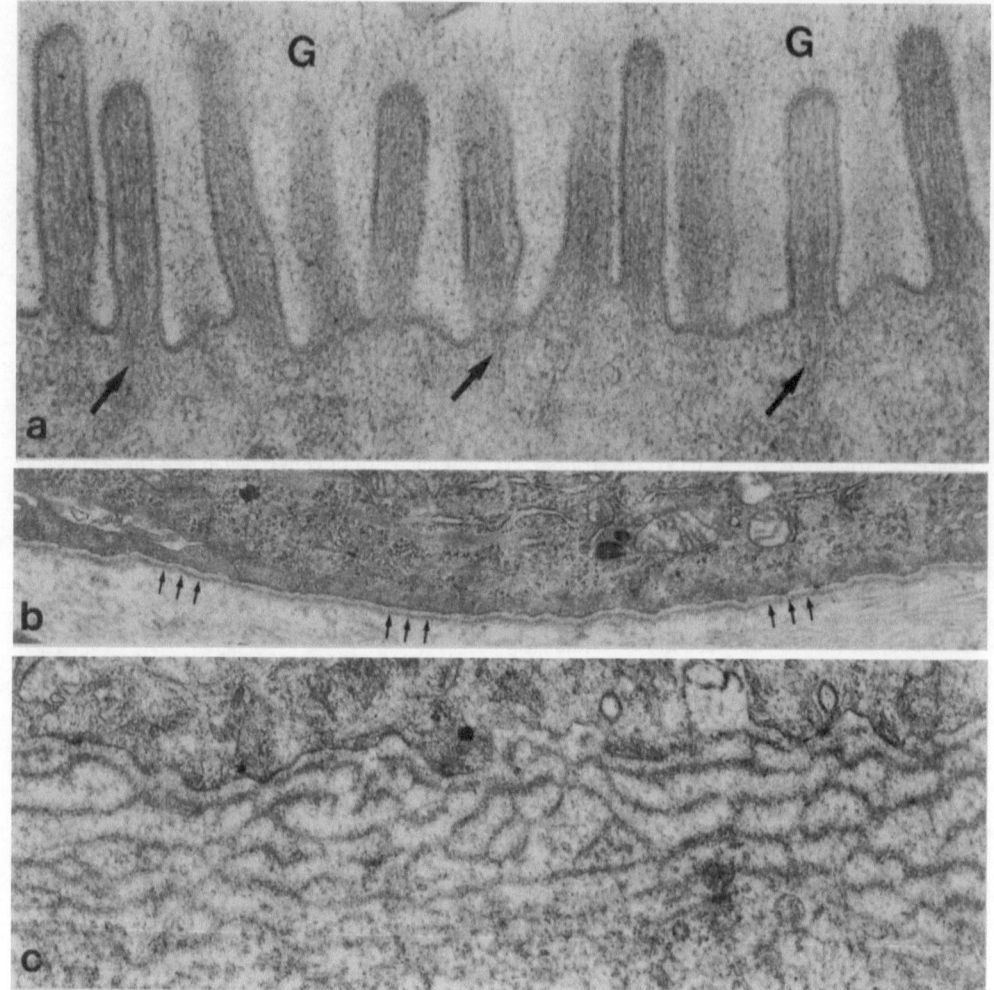

Fig. 45a–c. Apical and basal parts of ductular epithelia in cholangiofibromas induced in rat by *N*-nitrosomorpholine. **a** Apical brush border showing axial bundles of microfilaments with in microvilli (*arrows*) and close association with a marked glycocalix (*G*), ×80 000. **b** Lining of the neoplastic epithelium by a well-developed basement membrane (*arrows*), TEM, ×11 500. **c** Splitting of basement membrane, TEM, ×25 500

endocrine-like cells as a rare component of cholangiocellular liver tumors in rats (Terao and Nakano 1974; Reddy et al. 1977; Elmore and Sirica 1993).

Differential Diagnosis

The differential diagnosis of both cholangiofibroma and cholangiocarcinoma may be difficult. For cholangiofibroma, it is imperative to distinguish between cholangiofibrosis on the one hand and cholangiocarcinoma on the other. *Cholangiofibrosis* ("adenofibrosis") is frequently observed after treatment with chemical carcinogens (Opie 1944; Stewart and Snell 1959; Bannasch 1975). Whereas cholangiofibroma undergoes an expansive growth with compression of the adjacent parenchyma (Bannasch and Massner 1976), the surrounding hepatic tissue is retracted from all directions towards cholangiofibrosis (Fig. 48). This indicates that the process is one of contraction rather than proliferation and expansion (Stewart et al. 1980; Bannasch and Zerban 1990).

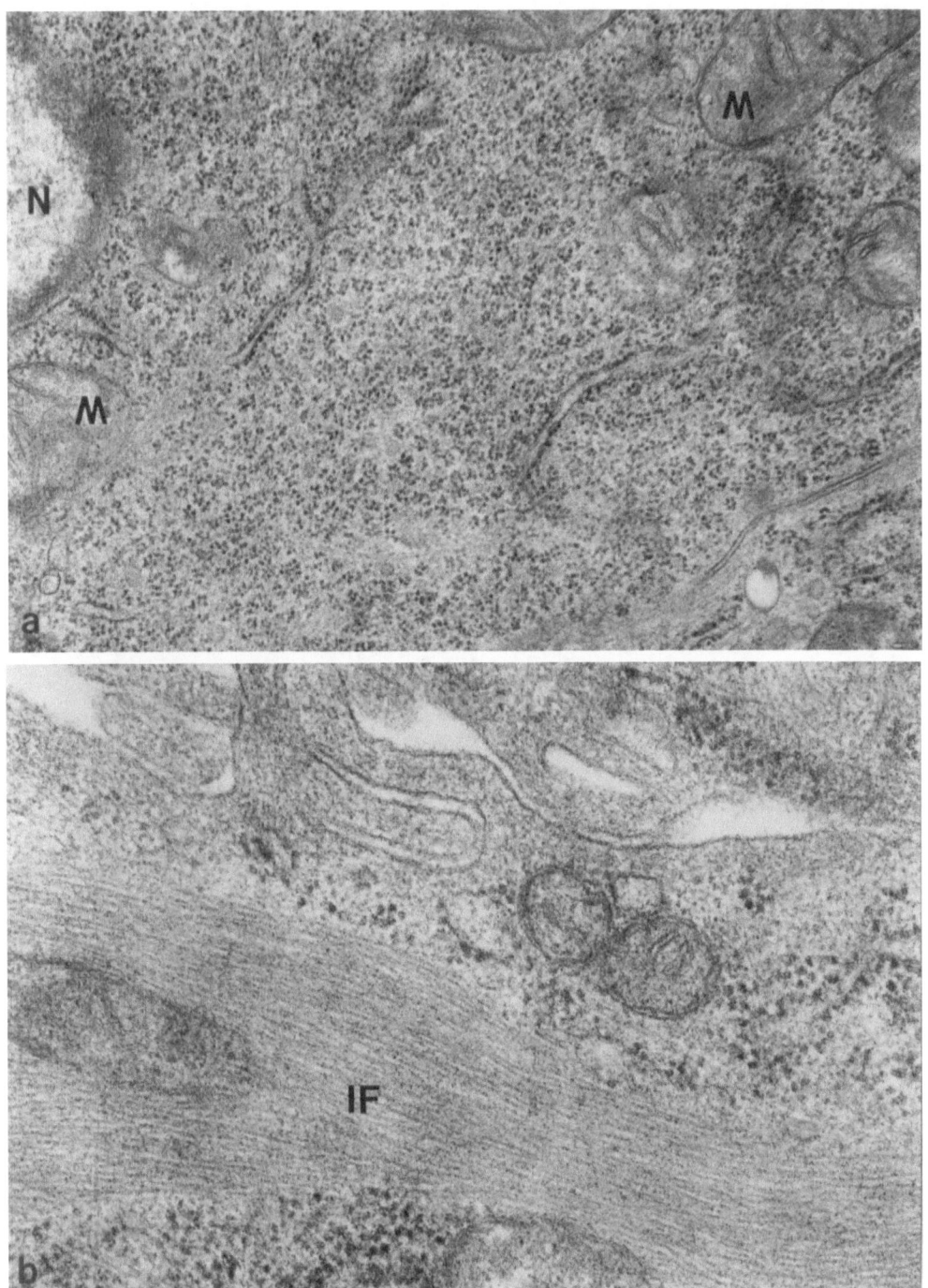

Fig. 46a,b. Portions of the cytoplasm of ductular epithelia in cholangiofibromas induced in rat by *N*-nitrosomorpholine. **a** Abundant free ribosomes, some profiles of rough endoplasmic reticulum, and parts of mitochondria (*M*) and nucleus (*N*). TEM, ×36500. **b** Bundle of intermediate filaments (*IF*). TEM, ×99500

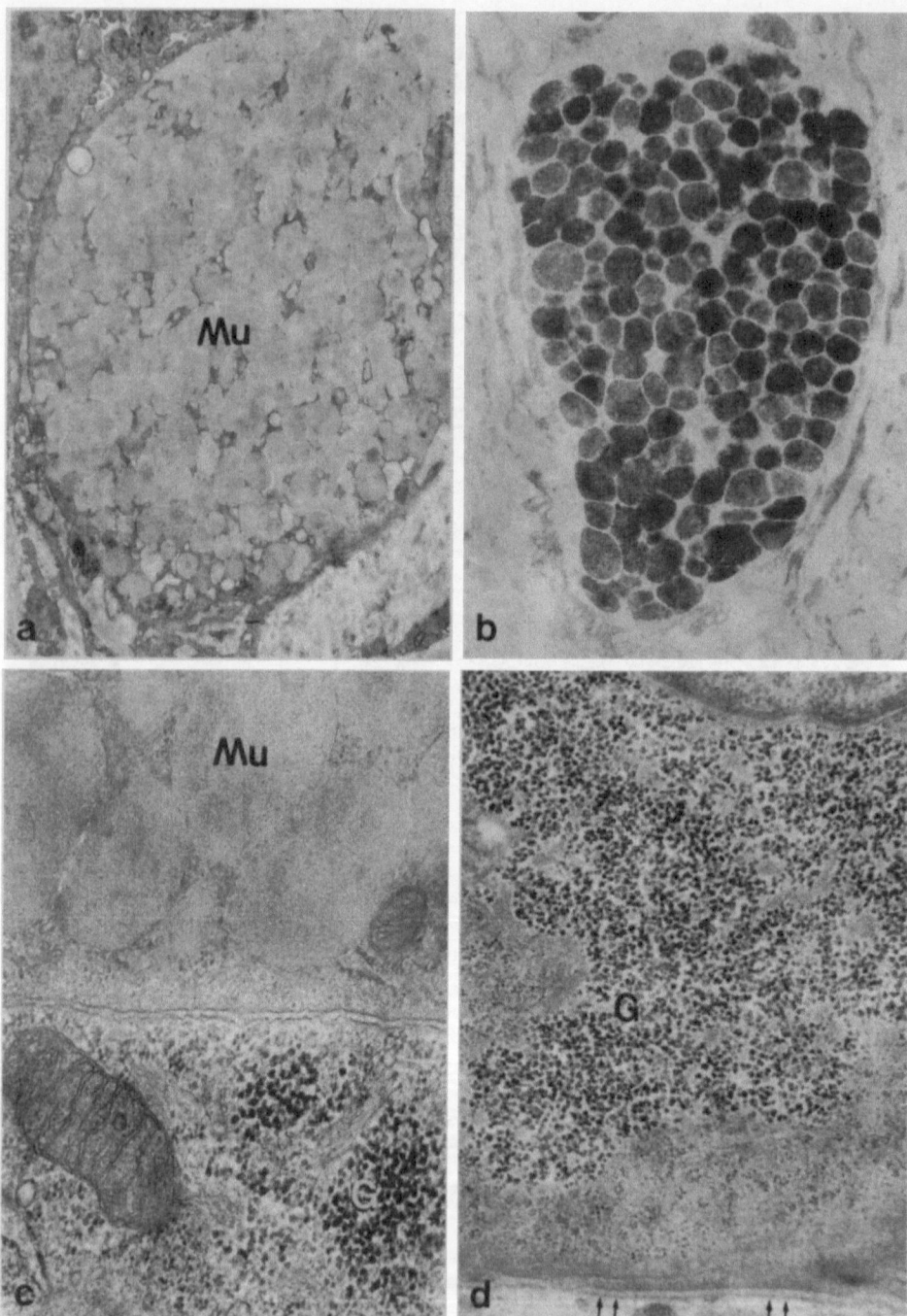

Fig. 47a–b. Storage of polysaccharides in ductular epithelia of cholangiofibromas induced in rat with *N*-nitrosomorpholine. **a** Goblet cell storing and secreting mucus (*Mu*), ×3500. **b** Selective staining of mucus in goblet cell with silver proteinate. TEM, ×10000. **c** Close association of goblet cell storing mucus (*Mu*) and neoplastic ductular cell storing beta particles of glycogen (*G*). TEM, ×43500. **d** Neoplastic ductular cell excessively storing glycogen in the form of beta particles (*G*). Note basement membrane (*arrows*). TEM, ×30000

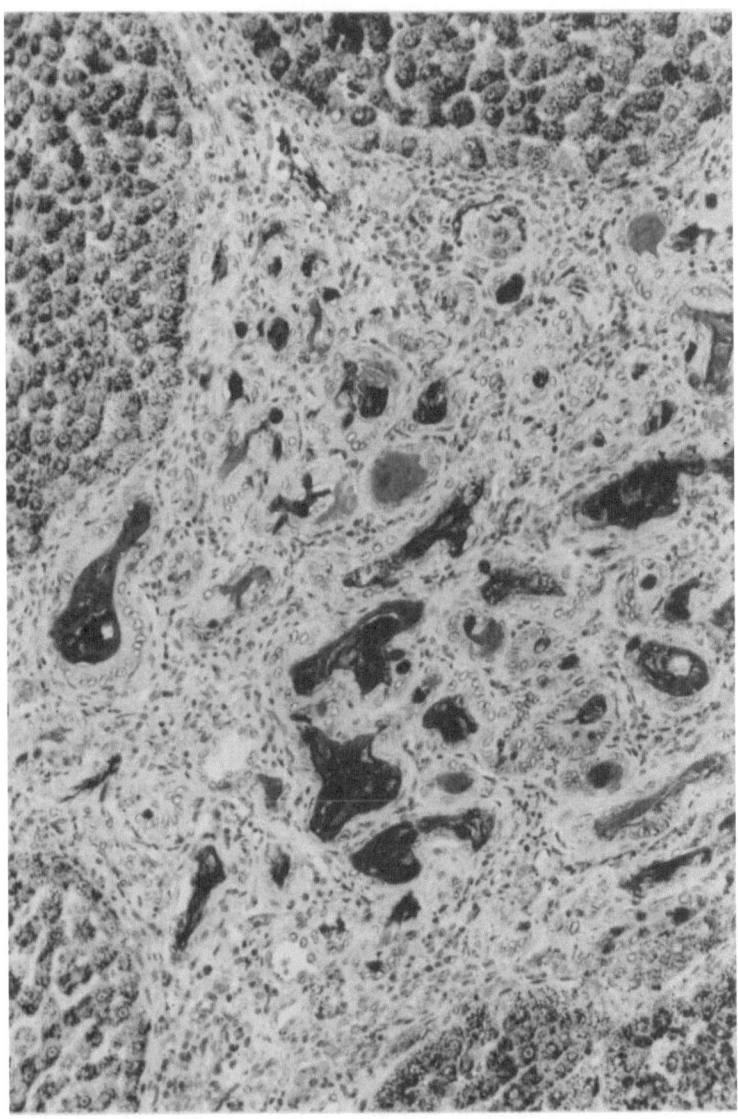

Fig. 48. Cholangiofibrosis induced in rat liver with *N*-nitrosomorpholine. Note abundant mucus production (cholangiolar mucopolysaccharidosis) in ductular component and contraction rather than expansion of the lesion. (From Bannasch and Reiss 1971.) Periodic acid-Schiff reaction, ×230

Cholangiofibrosis has been considered as a non-neoplastic lesion (Opie 1944; Firminger and Mulay 1952; Kinosita 1955; Firminger 1955; Farber 1956; Bannasch and Reiss 1971; Chou and Gibson 1972; Terao and Nakano 1974), the epithelial component of which may very occasionally undergo degeneration and involution, leaving a scar-like alteration (Fig. 49a,b; Evans et al. 1989; Hacker et al. 1992).

As will be discussed in some detail, it has been shown in different experimental models, however, that *cholangiofibrosis* usually is a preneoplastic lesion which may progress to cystic cholangioma (Fig. 50a,b; Bannasch and Reiss 1971; Dominis and Damjanov 1977), cholangiofibroma (Bannasch and Massner 1976; Ito et al. 1984; Ohshima et al. 1984; Evans et al. 1989; Steinberg et al. 1991; Hacker et al. 1992), and eventually also

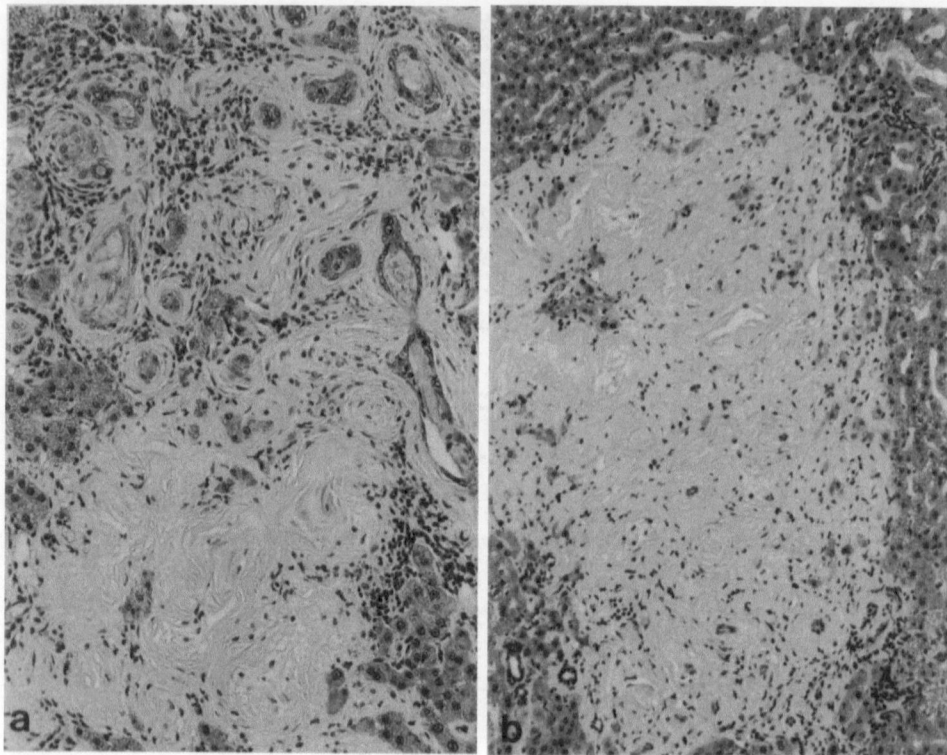

Fig. 49a,b. Involuted cholangiofibrotic lesions induced in rat liver by a choline-deficient/ethionine-supplemented diet. **a** Note remaining cholangiolar components in the upper and complete loss of ductules in lower part of the lesion. H&E, ×170. **b** Nearly complete loss of ductular components throughout the lesion. H&E, ×130

cholangiocarcinoma (Opie 1944; Nakano 1974; Bannasch and Massner 1976; Reddy et al. 1977; Praet and Roels 1984; Ohshima et al. 1984; Peterson 1990; Elmore and Sirica 1993). In line with the interpretation of cholangiofibrosis as a preneoplastic lesion, Ohshima et al. (1984) failed to produce tumors by transplantation of mucous cholangiofibrosis (induced in rats with methapyrilene) into the fat pad of syngeneic animals (observation period, 6 months). However, Maronpot et al. (1991) questioned the preneoplastic nature of cholangiofibrosis and considered this lesion to be essentially malignant or premalignant, since they did not observe qualitative differences between furan-induced mild cholangiofibrosis and neoplasms diagnosed as cholangiocarcinomas, some of which were transplantable into syngeneic rats and metastasized after eight serial passages in the recipients. It is evident that these findings support the concept of a precursor relationship between cholangiofibrosis and cholangiocarcinoma

(Opie 1944), but the view of Maronpot et al. (1991) does not take into consideration the following factors:

1. Even extended cholangiofibrosis may only replace lost liver parenchyma without any indication of expansion beyond the confines of the liver lobe (Fig. 50a; Bannasch and Reiss 1971; Bannasch 1975; Peterson 1990).
2. The ductular component of individual cholangiofibrotic lesions may sometimes undergo involution (Fig. 49a,b; Evans et al. 1989; Hacker et al. 1992).
3. Progression of cholangiofibrosis frequently leads to cystic cholangiomas (Fig. 50b), which represent a benign end-stage lesion (Bannasch and Reiss 1971; Bannasch 1975; Dominis and Damjanov 1977; Brooks and Roe 1985), while only a second neoplastic derivative of cholangiofibrosis, the cholangiofibroma (Figs. 40, 43), is potentially malignant and may eventually

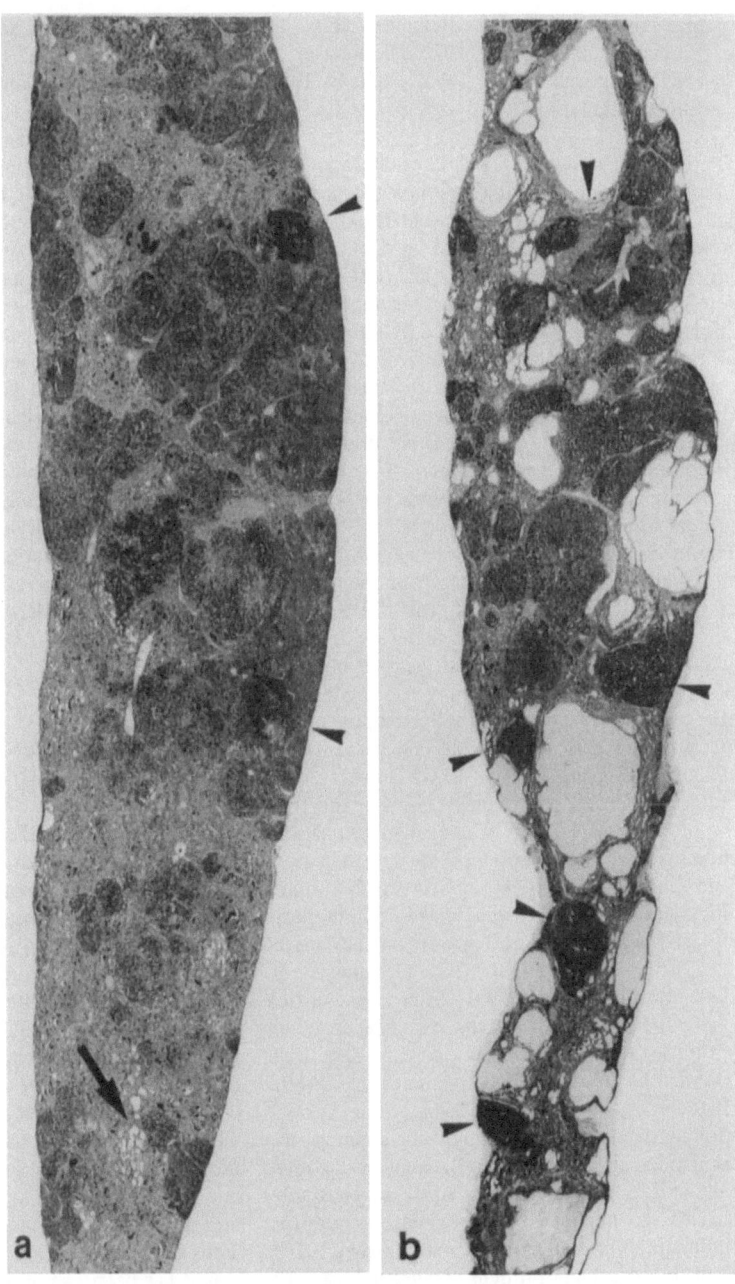

Fig. 50. a Cholangiofibrosis and **b** polycystic disease of the liver induced in rat with *N*-nitrosomorpholine. **a** Note replacement of large portions of the liver parenchyma by cholangiofibrosis without remarkable changes in the confines of the liver lobe. In places, formation of cysts (*arrow*) and glycogen storage foci in the remaining liver parenchyma (*ar-* *rowhead*) has begun. **b** Multiple cystic cholangiomas and some remaining cholangiofibrotic areas. Several strongly Periodic acid-Schiff (PAS)-positive foci excessively storing glycogen in the liver parenchyma (*arrowhead*). (From Bannasch and Reiss 1971) PAS reaction, ×16

give rise to cholangiocarcinomas (Fig. 44; Bannasch and Massner 1976; Ohshima et al. 1984; Bannasch et al. 1985; Peterson 1990; Elmore and Sirica 1993).

Distinguishing between cholangiofibrosis and cystic cholangioma is not a major diagnostic problem, since the cysts of benign cholangiomas are usually free of mucus and are lined by a single layer of isomorphic flat or cuboidal cells. The differential diagnosis between cholangiofibrosis and cholangiofibroma is complicated by the fact that all possible transitions between these two lesions exist. However, in contrast to cholangiofibrosis, the cholangiofibroma does not only undergo expansion instead of contraction, but usually also produces much more mucus and particularly collagenic fibers, which prevail in central parts of these tumors (Bannasch et al. 1985).

The distinction between cholangiofibroma and cholangiocarcinoma relies mainly on the absence of metastases in cholangiofibromas (Bannasch and Massner 1976; Bannasch and Zerban 1990). This biologic behavior contrasts with the observation that cholangiofibromas may exhibit considerable cellular atypia of its epithelial component and often mimic invasive growth (Bannasch et al. 1985; Bannasch and Zerban 1994). However, only metastases prove malignancy (Fig. 44), and thus the proposed separation of cholangiofibroma and cholangiocarcinoma appears to be indicated. Since in a number of studies in which cholangiocarcinomas were diagnosed extrahepatic metastases were not found or were rare (Reddy et al. 1977; Praet and Roels 1984; Ohshima et al. 1984; Maronpot et al. 1991; Elmore and Sirica 1993), the justification of the classification of all of these tumors as carcinomas is debatable (Bannasch et al. 1985; Evans et al. 1989). Histochemically, the transition from benign cholangiofibroma to cholangiocarcinoma is accompanied by a reduction in the number of goblet cells and by progressive loss of the mucous substances accumulated in great excess in the cholangiofibroma (Bannasch and Massner 1976). Thus the reduction of mucus appears to be a valid criterion for differential diagnosis. Concomitantly with the loss of mucus, the amount of collagen fibers is also reduced in cholangiocarcinomas as compared with cholangiofibromas.

A hitherto unresolved problem is the differential diagnosis of cholangiocellular tumors and hepatocellular carcinomas with a glandular (ad-enoid) pattern (Butler and Jones 1978; Jones and Butler 1978; Ruan et al. 1989). More than 40 years ago, Firminger and Mulay (1952) proposed that the differential diagnosis be based on the observation that mucus is present in cholangiocellular but not in hepatocellular tumors, and this has subsequently been supported by others (Bannasch and Massner 1976; Reddy et al. 1977; Weinbren 1984; Ruan et al. 1989). However, in advanced cholangiocarcinomas, which may have lost their mucus totally, this is no longer a convincing criterion (Bannasch et al. 1985). It is, therefore, sometimes impossible to discriminate clearly between cholangiocellular and hepatocellular carcinomas on purely morphological grounds. Antibodies to the different types of cytokeratin now available may not solve this problem, since the adenoid structures in hepatocellular carcinomas seem to develop from hepatocyte-derived cells which undergo transdifferentiation (metaplasia) and hence acquire the cytokeratin pattern of cholangiocellular tumors (Ruan et al. 1989; Bannasch and Zerban 1990).

Biologic Features

Natural History

The sequence of cellular changes leading to cholangiocellular tumors in rats has been investigated in detail (Bannasch 1975; Bannasch and Massner 1976). In particular, these tumors developed after application of high doses of carcinogens (e.g., N-nitrosomorpholine), which produced pronounced necrotic alterations of the liver parenchyma (Bannasch 1980). Under these experimental conditions, the changes in the bile duct epithelia (Fig. 48) have been shown to start with cholangiolar (oval) cell proliferation (Fig. 51a,b), which may progress to cholangiofibrotic areas, benign cystic cholangiomas, cholangiofibromas, and eventually cholangiocarcinomas (Bannasch and Reiss 1971; Bannasch and Massner 1976). Thus a morphogenetic sequence with four distinct stages has been established (Fig. 48). Similar sequential cellular changes leading to cholangiocellular neoplasms were observed in rats treated with the banned antihistaminic drug methapyrilene (Ohshima et al. 1984), the plant product coumarin (Evans et al. 1989), the mycotoxin phomopsin (Peterson 1990), the industrial chemical furan (Maronpot et al. 1991; Elmore

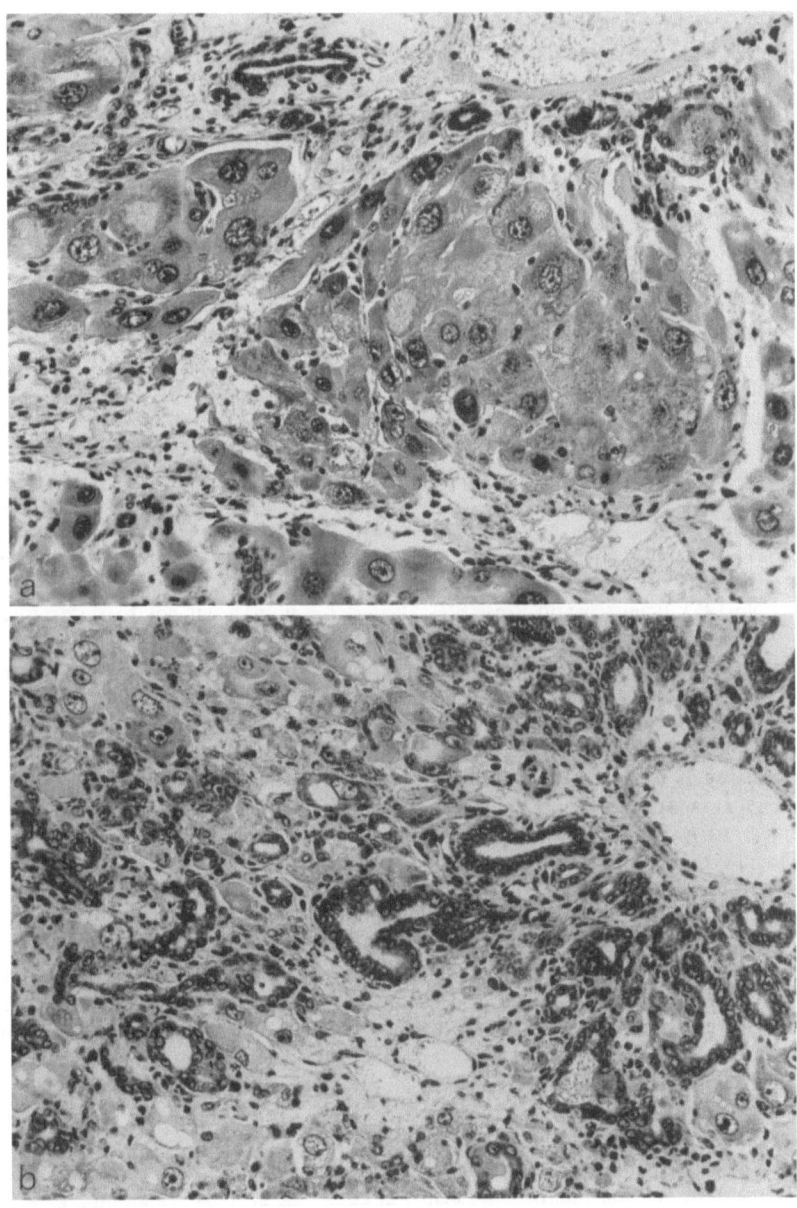

Fig. 51a,b. Early lesions in cholangiocarcinogenesis induced in rat liver with *N*-nitrosomorpholine. **a** Severe toxic necrosis and parenchymal loss, in addition to "megalocytosis" of remaining hepatocytes. H&E, ×250. **b** Pronounced cholangiolar (oval) cell proliferation leading to replacement of lost parenchyma by ductular formations. Cresyl violet, ×250. (From Bannasch and Reiss 1971)

and Sirica 1993), and a choline-deficient/ethionine-supplemented diet (Steinberg et al. 1991; Hacker et al. 1992). The same holds true for hamsters treated with propylnitrosamine (Moore et al. 1986; Thamavit et al. 1988) or dimethylnitrosamine (Thamavit et al. 1987), with and without additional infestation by the liver fluke *Opisthorchis viverrini*, which significantly increases the incidence of cholangiocarcinomas. An additional increase in the incidence of bile ductules, mucus-producing goblet cells, and cholangiocarcinomas was likewise observed after incomplete bile duct obstruction in hamsters treated with diisopropanolnitrosamine (Kinami et al. 1990).

The origin of oval cells and their significance for cellular differentiation and carcinogenesis in the liver has long been under discussion (Farber 1956; Inaoka 1967; Bannasch et al. 1985; Evarts et al. 1987; Tsao and Grisham 1987; Sell and Dunsford 1989; Steinberg et al. 1991, 1994a; Hacker et al. 1992; Sirica 1992; Pack et al. 1993; Radaeva and Steinberg 1995). In recent years, the concept that the oval cell might represent (or be closely related to) a liver stem cell has been emphasized by a number of authors (Sell 1990; Aterman 1992; Sirica 1992; Radaeva and Steinberg 1995). However, there is no definite proof for the suggested stem cell character of oval cells (Pack et al. 1993; Steinberg et al. 1994a). All electron microscopy, autoradiographic, and histochemical investigations point to the cholangioles as the site of origin of the oval cells (Grisham and Hartroft 1961; Rubin 1964; Bannasch and Reiss 1971; Bannasch et al. 1985; Iwai et al. 1988; Plenat et al. 1988; Sell and Dunsford 1989; Steinberg et al. 1991; Lenzi et al. 1992; Sirica 1992). The demonstration of glucose-6-phosphatase activity in oval cells (Ogawa et al. 1974; Plenat et al. 1988) does not substantiate the frequently assumed transition to hepatocytes, since electron microscopy studies have clearly shown that this enzyme is also present in cells of normal cholangioles (Benner et al. 1979) as well as in ductular components of cholangiofibrotic lesions and cholangiofibromas (Bannasch et al. 1981a). Using the expression of α-fetoprotein-positive oval cells as a marker at the ultrastructural level, Iwai et al. (1988) also provided evidence for a different origin of α-fetoprotein-positive cells in hepatocellular lesions and α–fetoprotein-positive oval cells. Makino and coworkers (1988) likewise concluded from studies of the three-dimensional arrangement of ductular structures formed by oval cells (as demonstrated by scanning electron microscopy) that oval cells have characteristics more similar to those of biliary epithelium than of hepatocytes and that oval cells have no relation to the development of nodular hepatic lesions induced in rats by 2-acetylaminofluorene. Most convincing are recent results of immunohistochemical studies on the expression of cytokeratin 19 (which is characteristic of normal bile duct and ductular epithelia, but not of hepatocytes) in preneoplastic and neoplastic liver lesions induced in rats by a choline-deficient/ethionine-supplemented diet. Whereas oval cells,

cholangiofibroses, cystic cholangiomas, and cholangiofibromas expressed cytokeratin 19, parenchymal cells, foci of altered hepatocytes, and hepatocellular adenomas did not (Steinberg et al. 1991; Hacker et al. 1992). These observations were confirmed for oval cells, ductular formations, and early hepatocyte nodules produced by the Solt-Farber procedure (Anilkumar et al. 1995). A positive reaction with antibodies to cytokeratin 19 has also been observed in preneoplastic and neoplastic cholangiocellular lesions induced in rats by furan (Elmore and Sirica 1993). Similar findings had previously been reported by Carthew et al. (1989), who used a less well defined polyclonal antibody to cytokeratins. In contrast to other authors who described the development of hepatocellular carcinomas from isolated oval cells (Tsao and Grisham 1987; Braun et al. 1989), Steinberg et al. (1994a) have shown in a similar experimental approach that oval cell lines give rise to cholangiocellular and undifferentiated carcinomas, but not to differentiated hepatocellular carcinomas.

All these findings strongly support the sequence of cellular changes during cholangiocarcinogenesis previously described (Bannasch and Massner 1976; Bannasch et al. 1985).

The early proliferation of cholangiolar cells seems to be a reactive response to severe hepatic necrosis rather than a neoplastic process (Popper et al. 1957; Bannasch and Reiss 1971; Terao and Nakano 1974; Bannasch et al. 1985; Tatematsu et al. 1985). Many of the proliferated cholangiolar cells may undergo necrotic changes and disappear later on (Hutterer et al. 1961; Grisham and Porta 1964; Rubin 1964; Bannasch and Massner 1976; Tatematsu et al. 1984, 1985). However, persisting complexes developing from the cholangiolar cell reaction (which is regularly accompanied by a proliferation of fibroblasts) proceed to a preneoplastic mucous cholangiofibrosis (Bannasch and Reiss 1971; Terao and Nakano 1974; Ohshima et al. 1984; Tatematsu et al. 1985; Evans et al. 1989; Peterson 1990; Hacker et al. 1992; Tarsetti et al. 1993). The term "adenofibrosis," used by some authors (Stewart et al. 1980) for this lesion, does not appear to be justified any longer, since it is based on the assumption that the oval cells are derived from hepatocytes.

Independent of further action of the carcinogen, mucous cholangiofibrosis persists for weeks and months. After long lag periods, it may give rise to two types of benign cholangiocellular tumors: cystic cholangioma and cholangiofibroma

(Bannasch and Reiss 1971; Terao and Nakano 1974; Bannasch and Massner 1976; Dominis and Damjanov 1977; Ohshima et al. 1984; Evans et al. 1989; Peterson 1990; Steinberg et al. 1991; Hacker et al. 1992; Elmore and Sirica 1993). It is not yet clear which factors determine the development of one or the other tumor type, but it may be speculated that the degree of cellular injury by the carcinogen is different in both cases. This suggestion would be compatible with the observation that cholangiofibromas appear especially after high doses of hepatocarcinogens, while cystic cholangiomas develop after lower doses (Bannasch 1980).

The progression of cholangiofibroma to cholangiocarcinoma marks the fourth stage in the morphogenesis of cholangiocellular tumors (Bannasch and Massner 1976). It is a rare event which, as a rule, takes place only after very long lag periods. It is accompanied by a reduction in the number of goblet cells and in the production of mucus. Interestingly, in line with these observations, the only cholangiocellular tumor exhibiting extrahepatic metastases detected by Elmore and Sirica (1993) in rats treated with furan was characterized by "a more native biliary rather than interstitial-type of differentiation." In serially passaged transplants of furan-induced cholangiocellular tumors in syngeneic rats, these authors generally found a lower percentage of goblet cells than in the original primary tumors. This interesting observation is in contrast to results reported by Maronpot et al. (1991), who did not recognize differences between primary tumors and transplants in the same animal model. More detailed investigations on the correlation between the morphology and the biologic behavior of these tumors should be conducted in order to clarify this discrepancy (Maronpot et al. 1991; Elmore and Sirica 1993), but the results of both studies are consistent with the concept that the mucous cholangiofibroma is potentially malignant, leading to metastasizing cholangiocarcinomas (Bannasch and Massner 1976; Bannasch et al. 1985).

It should be mentioned, however, that not only short-term administration of high doses, but also long-term application of low doses of hepatocarcinogens can lead to bile duct proliferation, benign cystic cholangiomas, and perhaps also cholangiocarcinomas without cholangiofibrosis as an early stage (Bannasch 1980). The morphogenesis of these cholangiocellular tumors is poorly understood.

Pathogenesis

Cholangiofibromas or cholangiocarcinomas rarely occur "spontaneously" (Butler 1971), but usually are the consequence of severe liver damage by hepatocarcinogens (Opie 1944; Firminger and Mulay 1952; Bannasch and Massner 1976; Reddy et al. 1977; Ohshima et al. 1984; Evans et al. 1989; Peterson 1990; Steinberg et al. 1991; Maronpot et al. 1991; Elmore and Sirica 1993). In the previous paragraph, we stated that early cholangiolar cell proliferation most probably develops as a reaction to severe hepatic necrosis. However, an increased expression of the tumor suppressor gene p53 has been demonstrated immunohistochemically in oval cells appearing early during hepatocarcinogenesis induced by N-nitrosomorpholine or 2-acetylaminofluorene (Wirnitzer et al. 1995). A most characteristic phenomenon that may help to understand the pathogenesis of cholangiocellular tumors is the transient excessive accumulation of mucous substances which contain both acid and neutral mucopolysaccharides. The early accumulation of mucus in cholangiofibrotic lesions, which has been called "cholangiolar mucopolysaccharidosis" by Bannasch and Reiss (1971) and "intestinal metaplasia" by Terao and Nakano (1974), appears to be specific for a ductular proliferation induced by hepatocarcinogens (Bannasch and Reiss 1971; Chou and Gibson 1972; Malvaldi et al. 1984). This phenomenon might be analogous to the storage of different polysaccharides or lipids as observed early during the development of a number of tumor types (Bannasch et al. 1985; Moore et al. 1986; Steinberg et al. 1991; Hacker et al. 1992). The cellular thesaurismoses usually persist until the tumors start to grow rapidly. The storage of metabolites appears to indicate an early carcinogen-induced aberration of carbohydrate metabolism. Many interesting similarities, but also some important differences in changes of the activity or content of enzymes of the carbohydrate metabolism during neoplastic transformation of cholangiolar (oval) and parenchymal cells have been demonstrated (Steinberg et al. 1991, 1994b). More detailed investigations on the neoplastic transformation of hepatocytes suggest that the later reduction of the polysaccharides, initially produced in excess, might be due to a shift of carbohydrate metabolism by an adaptation of cellular enzymes gradually activating alternative metabolic pathways, such as the pentose phosphate pathway or glycolysis (Bannasch et al. 1984).

Comparison with Other Species

In accordance with the observation in rats, cholangiofibromas have been distinguished from cholangiocarcinomas in hamsters (Moore et al. 1986). Cholangiocarcinomas have been described in mice (Reuber 1967), hamsters (Della Porta et al. 1959; Tomatis et al. 1964; Reuber 1968; Reznik and Mohr 1977; Moore et al. 1982; Thamavit et al. 1987, 1988; Kinami et al. 1990), fish (Stanton 1965; Pliss and Khudoley 1975; Scarpelli 1975; Simon et al. 1980; Couch and Courtney 1987; Stehr and Myers 1990), and other species (Butler 1971; Schmähl et al. 1978), including human patients (Weinbren 1984; Sugihara and Kojiro 1987). An accumulation of mucous substances, as observed during the development of cholangiofibromas and cholangiocarcinomas in the rat, also appears to characterize the pathogenesis of these tumors in other species. This has been reported in hamsters by Reuber (1968), Reznik and Mohr (1977), and Moore et al. (1986) and in fish by Simon et al. (1980). According to Chou and Gibson (1970), an accumulation of neutral and acid mucopolysaccharides may also play an important role in the development of human cholangiocellular tumors.

References

Anilkumar TV, Golding M, Edwards RJ, Lalani EN, Sarraf CE, Alison MR (1995) The resistant hepatocyte model of carcinogenesis in the rat: the apparent independent development of oval cell proliferation and early nodules. Carcinogenesis 16:845–853

Aterman K (1992) The stem cells of the liver – a selective review. J Cancer Res Clin Oncol 118:87–115

Bannasch P (1975) Die Cytologie der Hepatocarcinogenese. In: Grundmann E (ed) Geschwülste/Tumors 111. Springer, Berlin Heidelberg New York, pp 123–276 (Handbuch der Allgemeinen Pathologie, vol 6/7)

Bannasch P (1980) Dose-dependence of early cellular changes during liver carcinogenesis. Arch Toxicol [Suppl] 3:111–128

Bannasch P, Massner B (1976) Histogenese und Cytogenese von Cholangiofibromen und Cholangiocarcinomen bei Nitrosomorpholin-vergifteten Ratten. Z Krebsforsch 87:239–255

Bannasch P, Massner B (1977) Die Feinstruktur des Nitrosomorpholin-induzierten Cholangiofibroms der Ratte. Virchows Arch [B] Cell Pathol 24:295–315

Bannasch P, Reiss W (1971) Histogenese und Cytogenese cholangiocellulärer Tumoren bei Nitrosomorpholin-vergifteten Ratten. Zugleich ein Beitrag zur Morphogenese der Cystenleber. Z Krebsforsch 76:193–215

Bannasch P, Zerban H (1990) Tumours of the liver. In: Turusov VS, Mohr U (eds) Pathology of tumours in labora-

tory animals, vol I: tumours of the rat, 2nd edn. IARC, Lyon, pp 199–240

Bannasch P, Zerban H (1994) Preneoplastic and neoplastic lesions of the liver of the rat. In: Bannasch P, Gössner W (eds) Pathology of neoplasia and preneoplasia in rodents. EULEP Color Atlas. Schattauer, Stuttgart, pp 18–63

Bannasch P, Zerban H, Schmid E, Franke WW (1980) Liver tumors distinguished by immunofluorescence microscopy with antibodies to proteins of intermediate-sized filaments. Proc Natl Acad Sci USA 77:4948–4952

Bannasch P, Benner U, Hacker HJ, Klimek F, Mayer D, Moore M, Zerban H (1981a) Cytochemical and biochemical microanalysis of carcinogenesis. Histochem J 13:799–820

Bannasch P, Zerban H, Schmid E, Franke WW (1981b) Characterization of cytoskeletal components in epithelial and mesenchymal liver tumors by electron and immunofluorescence microscopy. Virchows Arch (Cell Pathol) 36:139-158

Bannasch P, Hacker HJ, Klimek F, Mayer D (1984) Hepatocellular glycogenosis and related pattern of enzymatic changes during hepatocarcinogenesis. Adv Enzyme Regul 22:97–121

Bannasch P, Benner U, Zerban H (1985) Cholangiofibroma and cholangiocarcinoma, liver, rat. In: Jones TC, Mohr U, Hunt RD (eds) Digestive system. Monographs on pathology of laboratory animals. Springer, Berlin Heidelberg New York, pp 52–65

Benner U, Hacker HJ, Bannasch P (1979) Electron microscopical demonstration of glucose-6-phosphatase in native cryostat sections fixed with glutaraldehyde through semipermeable membranes. Histochemistry 65:41–47

Braun L, Mikumo R, Fausto N (1989) Production of hepatocellulara caracinoma by oval cells: cell cycle expression of c-myc and p53 at different stages of oval cell proliferation. Cancer Res 49:1554–1561

Brooks PN, Roe FJC (1985) Cholangioma, liver, rat. In: Jones TC, Mohr U, Hunt RD (eds) Digestive system. Monographs on pathology of laboratory animals. Springer, Berlin Heidelberg New York, pp 66–69

Butler WH (1971) Pathology of liver cancer in experimental animals. In: Liver cancer. IARC, Lyon, pp 30–41 (IARC scientific publication no 1)

Butler WH, Jones G (1978) Ultrastructure of hepatic neoplasia. In: Newberne PM, Butler WH (eds) Rat hepatic neoplasia. MIT Press, Cambridge, pp 144–179

Carthew P, Edwards RE, Hill RJ, Evans JG (1989) Cytokeratin expression in cells of the rodent bile duct developing under normal and pathological conditions. Br J Exp Pathol 70:717–725

Chou ST, Gibson JB (1970) The histochemistry of biliary mucins and the changes caused by infestation by Clonorchis sinensis. J Pathol 101:185–197

Chou ST, Gibson JB (1972) A comparative histochemical study of rat livers in alpha-naphthyl-iso-thiocyanate (ANIT) and DL-ethionine intoxication. J Pathol 108:73–83

Couch JA, Courtney LA (1987) N-nitrosodiethylamine-induced hepatocarcinogenesis in estuarine sheepshead minnow (Cyprinodon variegatus): neoplasms and related lesions compared with mammalian lesions. J Natl Cancer Inst 79:297–321

David H (1962) Die submikroskopische Struktur des DMAB-induzierten cholangio-cellularen Lebercarcinoms der Ratte. Z Krebsforsch 65:130–138

Della Porta G, Shubik P, Scortecci V (1959) The action of N-2-fluorenylacetamide in the Syrian golden hamster. J Natl Cancer Inst 22:463–487

Dominis M, Damjanov I (1977) Cystic cholangiofibrosis of the liver. Arch Geschwulstforsch 47:661–669

Elmore LW, Sirica AE (1993) "Intestinal-type" of adenocarcinoma preferentially induced in right/caudate liver lobes of rats treated with furan. Cancer Res 53:254–259

Evans JG, Appleby EC, Lake BG, Conning DM (1989) Studies on the induction of cholangiofibrosis by coumarin in the rat. Toxicology 55:207–224

Evarts RP, Nagy P, Marsden E, Thorgeirsson SS (1987) A precursor-product relationship exists between oval cells and hepatocytes in rat liver. Carcinogenesis 8:1737–1740

Farber E (1956) Similarities in the sequence of early histological changes induced in the liver of the rat by ethionine, 2-acetylaminofluorene and 3'-methyl-4- dimethylaminoazobenzene. Cancer Res 16:142–148

Firminger HI (1955) Histopathology of carcinogenesis and tumors of the liver in rats. J Natl Cancer Inst 15:1427–1442

Firminger HI, Mulay AS (1952) Histochemical and morphologic differentiation of induced tumors of the liver in rats. JNCI 13:19–33

Goodman DG, Maronpot RR, Newberne PM, Popp JA, Squire RA (1994) Proliferative and selected other lesions in the liver of rats. In: Streett CS, Burek JD, Hardisty JF, Garner FM, Leininger JR, Pletscher JM, Moch RW (eds) Guides for toxicologic pathology. STP/ARP/AFIP, Washington DC, pp G1–5, 1–24

Grisham JW, Hartroft WS (1961) Morphologic identification by electron microscopy of "oval" cells in experimental hepatic degeneration. Lab Invest 10:317–332

Grisham JW, Porta EA (1964) Origin and fate of proliferated hepatic ductal cells in the rat: electron microscopic and autoradiographic studies. Exp Mol Pathol 3:242–261

Hacker HJ, Steinberg P, Toshkov I, Oesch F, Bannasch P (1992) Persistence of the cholangiocellular lesions observed in rats fed a choline-deficient/DL-ethionine supplemented diet. Carcinogenesis 13:271–276

Hutterer F, Rubin E, Singer EJ, Popper H (1961) Quantitative relation of cell proliferation and fibrogenesis in the liver. Cancer Res 21:206–215

Inaoka Y (1967) Significance of the so-called oval cell proliferation during azo-dye hepatocarcinogenesis. Gann 58:355–366

Ito N, Moore MA, Bannasch P (1984) Modification of the development of N-nitrosomorpholine-induced hepatic lesions by 2-acetylaminofluorene, phenobarbital and 4,4'-diaminodiphenylmethane: a sequential histological and histochemical analysis. Carcinogenesis 5:335–342

Iwai M, Kashiwadni M, Takino T, Ibata Y (1988) Demonstration by light and ultrastructural immunoperoxidase study of a-fetoprotein-positive non-hepatoma cells and hepatoma during 3'-methyl-4-dimethylaminoazobenzene hepatocarcinogenesis. Virchows Arch (Cell Pathol) 55:117–123

Jones G, Butler WH (1978) Light microscopy of rat hepatic neoplasia. In: Newberne PM, Butler WH (eds) Rat hepatic neoplasia. MIT Press, Cambridge, pp 114–140

Kinami Y, Ashida Y, Seto K, Takashima S, Kita I (1990) Influence of incomplete bile duct obstruction on the occurrence of cholangiocarcinoma induced by diisopropanolnitrosamine in hamsters. Oncology 47:170–176

Kinosita R (1955) Some recent findings concerning hepatomas induced with p-dimethylaminoazobenzene. J Natl Cancer Inst 15:1443–1445

Lenzi R, Liu MH, Tarsetti F, Slott PA, Alpini G, Zhai WR, Paronetto F, Lenzen R, Tavoloni N (1992) Histogenesis of bile duct-like cells proliferating during ethionine hepatocarcinogenesis. Evidence for a biliary nature of oval cells. Lab Invest 66:390–402

Ma MH, Webber AJ (1966) Fine structure of liver of tumors induced in the rat by 3'-methyl-4-dimethylaminoazobenzene. Cancer Res 26:935–946

Makino Y, Yamamoto K, Tsuji T (1988) Three-dimensional arrangement of ductular structures formed by oval cells during hepatocarcinogenesis. Acta Med Okayama 42:143–150

Malvaldi G, Chieli E, Saviozzi M (1984) Biliary cirrhosis and tumors induced by chronic administration of thiobenzamide to rats. Arch Toxicol 55:34–38

Maronpot RR, Giles HD, Dykes DJ, Irwin RD (1991) Furan-induced hepatic cholangiocarcinomas in Fischer 344 rats. Toxicol Pathol 19:561–570

Moore MA, Fukushima S, Iciara A, Sato K, Ito N (1986) Intestinal metaplasia and altered enzyme expression in propylnitrosamine-induced syrian hamster cholangiocellular and gallbladder lesions. Virchows Arch (Cell Pathol) 51:29–38

Moore MR, Pitot HC, Miller EC, Miller JA (1982) Cholangiocellular carcinomas induced in Syrian golden hamsters administered aflatoxin B1 in large doses. J Natl Cancer Inst 68:271–278

Nakano M (1974) Cholangiofibrosis induced in short-term feeding of 3-methyl-4-dimethylaminoazobenzene. Kanzo 15:292–300

Ogawa K, Minase T, Onhoe T (1974) Demonstration of glucose 6-phosphatase activity in the oval cells of rat liver and the significance of the oval cells in azo dye carcinogenesis. Cancer Res 34:3379–3386

Ohshima M, Ward JM, Brennan LM, Creasia DA (1984) Sequential study of methapyrilene hydrochloride-induced liver carcinogenesis in male F344 rats. J Natl Cancer Inst 72:759–768

Onoe T, Fuse Y (1966) Electron microscopic study on azo-dye carcinogenesis. Tumor Res 1:143–173

Opie EL (1944) The pathogenesis of tumors of the liver produced by butter yellow. J Exp Med 80:231–246

Pack R, Heck R, Dienes HP, Oesch F, Steinberg P (1993) Isolation, biochemical characterization, long-term culture, and phenotype modulation of oval cells from carcinogen-fed rats. Exp Cell Res 204:198–209

Peterson JE (1990) Biliary hyperplasia and carcinogenesis in chronic liver damage induced in rats by phomopsin. Pathology 22:213–222

Plenat F, Braun L, Fausto N (1988) Demonstration of glucose-6-phosphatase and peroxisomal catalase activity by ultrastructural cytochemistry in oval cells from livers of carcinogen-treated rats. Am J Pathol 130:91–102

Pliss GB, Khudoley VV (1975) Tumor induction by carcinogenic agents in aquarium fish. J Natl Cancer Inst 55:129–136

Popper H, Kent G, Stein R (1957) Ductular cell reaction in the liver in hepatic injury. J Mt Sinai Hosp 24:551–556

Praet MM, Roels HJ (1984) Histogenesis of cholangiomas and cholangiocarcinomas in thioacetamide fed rats. Exp Pathol 26:3–14

Radaeva S, Steinberg P (1995) Phenotype and differentiation patterns of the oval cells lines OC/CDE 6 and OC/CDE 22 derived from the livers of caracinogen-treated rats. Cancer Res 55:1028–1038

Reddy KP, Buschmann RJ, Chomet B (1977) Cholangiocarcinomas induced by feeding 3'-methyl-4-dimethylaminoazobenzene to rats. Am J Pathol 87:189–204

Reuber MD (1967) Poorly differentiated cholangiocarcinomas occurring "spontaneously" in C3H and C3HxY hybrid mice. J Natl Cancer Inst 38:901–907

Reuber MD (1968) Histogenesis of cholangiofibrosis and well-differentiated cholangiocarcinoma in Syrian hamster given 2-acetamidofluorene or 2-diacetamidofluorene. Gann 59:239–246

Reznik G, Mohr U (1977) Colangiomas y colangiocarcinomas en el hamster europeo tras tratamiento con di-iso-propanol-nitrosamina. Pathol No Extraordinario 10 (11):171–204

Ruan Y, Hacker HJ, Zerban H, Bannasch P (1989) Ultrastructural and immunocytochemical characterization of the cellular phenotype in primary adenoid liver tumours of the rat. Pathol Res Pract 184:223–233

Rubin E (1964) The origin and fate of proliferated bile ductular cells. Exp Mol Pathol 3:279–286

Scarpelli DG (1975) Neoplasia in poikilotherms. In: Becker FF (ed) Cancer. A comprehensive treatise, vol 4. Plenum, New York, pp 375–410

Schmähl D, Habs M, Ivankovic S (1978) Carcinogenesis of N-nitrosodiethylamine (DENA) in chickens and domestic cats. Int J Cancer 22:552–557

Sell S (1990) Is there a liver stem cell? Cancer Res 50:3811–3815

Sell S, Dunsford HA (1989) Evidence for the stem cell origin of hepatocellular carcinoma and cholangiocarcinoma. Am J Pathol 134:1347–1363

Simon K, Tajti L, Lapis K (1980) Hepatic lesions in Lebistes reticulatus induced by N-nitrosodiethylamine. In: Walker EA, Castegnaro M, Griciute L, Borzsonyi M (eds) N-Nitrosocompounds: analysis, formation and occurrence. IARC, Lyon, pp 717–729 (IARC scientific publications no 31)

Sirica AE (1992) (ed) The role of cell types in hepatocarcinogenesis. CRC, Boca Raton

Stanton MF (1965) Diethylnitrosamine-induced hepatic degeneration and neoplasia in the aquarium fish, Brachydanio rerio. J Natl Cancer Inst 34:117–130

Stehr CM, Myers MS (1990) The ultrastructure and histology of cholangiocellular carcinomas in English sole (Parophrys vetulus) from Puget Sound, Washington. Toxicol Pathol 18:362–372

Steinberg P, Hacker HJ, Dienes HP, Oesch F, Bannasch P (1991) Enzyme histochemical and immunohistochemical characterization of oval and parenchymal cells proliferating in livers of rats fed a choline-deficient/DL-ethionine-supplemented diet. Carcinogenesis 12:225–231

Steinberg P, Steinbrecher R, Radaeva S, Schirmacher P, Dienes HP, Oesch F, Bannasch P (1994a) Oval cell lines OC/CDE 6 and OC/CDE 22 give rise to cholangio-cellular and undifferentiated carcinomas after transformation. Lab Invest 71:700–709

Steinberg P, Weiss G, Eigenbrodt E, Oesch F (1994b) Expression of L-and M_2-pyruvate kinases in proliferating oval cells and cholangiocellular lesions developing in the livers of rats fed a methyl-deficient diet. Carcinogenesis 15:125–127

Stewart HL, Snell KC (1959) The histopathology of experimental tumors of the liver of the rat. In: Hamburger F, Fischman WH (eds) The physiopathology of cancer. Hoeber, New York, pp 85–126

Stewart HL, Williams G, Keysser CH, Lombard LS, Montali RJ (1980) Histologic typing of liver tumors of the rat. J Natl Cancer Inct 64:179–206

Sugihara S, Kojiro M (1987) Pathology of cholangiocarcinoma. In: Okuda K, Ishak KG (eds) Neoplasms of the liver. Springer, Berlin Heidelberg New York, pp 143–158

Svoboda DJ (1964) Fine structure of hepatomas induced in rats with p-dimethylaminoazobenzene. J Natl Cancer Inst 33:315–338

Tarsetti F, Lenzi R, Salvi R, Schuler E, Rijhsinghani K, Lenzen R, Tavoloni N (1993) Liver carcinogenesis associated with feeding of ethionine in a choline-free diet: evidence against a role of oval cells in the emergence of hepatocellular carcinoma. Hepatology 18:596–603

Tatematsu M, Ho RH, Kaku T, Ekem JK, Farber E (1984) Studies on the proliferation and fate of oval cells in the liver of rats treated with 2-acetylaminofluorene and partial hepatectomy. Am J Pathol 114:418–430

Tatematsu M, Kaku T, Medline A, Farber E (1985) Interstitial metaplasia as a common option of oval cells in relation to cholangiofibrosis in liver of rats exposed to 2-acetylaminofluorene. Lab Invest 52:354–362

Terao K, Nakano M (1974) Cholangiofibrosis induced by short-term feeding of 3'-methyl-4-(dimethylamino)azobenzene: an electron microscopic observation. Gann 65:249–260

Thamavit W, Kongkanuntn R, Tiwawech D, Moore MA (1987) Level of Opisthorchis infestation and carcinogen dose-dependence of cholangiocarcinoma induction in Syrian golden hamsters. Virchows Arch (Cell Pathol) 54:52–58

Thamavit W, Moore MA, Hiasa Y, Ito N (1988) Enhancement of DHPN induced hepatocellular, cholangiocellular and pancreatic carcinogenesis by Opisthorchis viverrini infestation in Syrian golden hamsters. Carcinogenesis 9:1095–1098

Tomatis I, Magee PN, Shubik P (1964) Induction of liver tumors in the Syrian golden hamster by feeding dimethylnitrosamine. J Natl Cancer Inst 33:341–345

Tsao M-S, Grisham JW (1987) Hepatocarcinomas, cholangiocarcinomas, and hepatoblastomas produced by chemically transformed cultured rat liver epithelial cells. Am J Pathol 127:168–181

Weinbren K (1984) Precancerous states in the liver. In: Carter RL (ed) Precancerous states. Oxford Medical, London, pp 254–277

Wirnitzer U, Enzmann H, Rosenbruch M, Bomhard EM (1995) Accumulation of p53 protein in chemically induced oval cells during early stages of rodent hepatocarcinogenesis. Carcinogenesis 16:687–701

Cholangioma, Liver, Rat

Paul N. Brooks and Francis J.C. Roe

Synonyms. Bile duct adenoma, biliary adenoma, bile duct cystadenoma, biliary cystadenoma

Gross Appearance

The macroscopic appearance of cholangioma is variable. Usually the neoplasms are observed as raised, grayish-white areas of firm consistency. However, in the more cystic forms, cholangiomas have a spongy texture and, in the presence of appreciable associated vascular proliferation, the neoplasm is dark-red in color. Multilocular cystic forms have a more irregular surface than those in which there is no significant cyst formation.
Macroscopic differentiation between hepatocellular neoplasms and those of bile duct origin can be difficult and in some cases impossible.

Microscopic Features

The cholangioma is a generally uniform, well-circumscribed neoplasm composed of glandular acini of uniform size. The acini are lined by a single layer of cuboidal cells with somewhat basophilic cytoplasm and a round or oval nucleus, which is occasionally vesicular and contains one or two conspicuous nucleoli. The nucleus is usually located at or toward the base of the cell. Simple cholangiomas have a sparse vascular stroma and rarely contain evidence of mitotic activity. Cholangiomas can become large enough to cause considerable distortion, but there is never invasive growth. The glandular acini may vary in size and shape, and the lining epithelium, which can range from columnar to flattened, is occasionally multilayered (Figs. 52, 53).
Cystic cholangiomas are characteristically composed of dilated glandular acini lined by flattened, almost atrophic epithelium. The stroma in cystic forms is less well vascularized and contains more fibrous tissue and collagen than in the simple cholangioma. Papillary structures are occasionally observed projecting into the lumen of cystic acini, and clumps of liver cells are commonly seen between the cysts (Fig. 54).

In hemangiomatous cholangioma, the acini are morphologically similar to those in simple cholangioma, but the stroma contains cystic, blood-filled spaces lined by endothelium (Fig. 55).

Differential Diagnosis

The morphological forms of cholangioma must be distinguished from other proliferative lesions of bile duct, hepatocellular, and endothelial origin. With the exception of some forms of hepatocellular carcinoma, the morphology of hepatocellular and endothelial neoplasms is quite distinct and rarely confounds the differential diagnosis of cholangioma.
Occasionally, acinar structures are formed within hepatocellular carcinomas, and these can appear very similar in morphology to the glandular acini of cholangiomas. Importantly, the morphology of such hepatocellular carcinomas is variable. The more solid areas of carcinoma, frequently demonstrating evidence of invasion, will clarify diagnosis. This distinction is more of a problem in the differential diagnosis of cholangiocarcinoma than cholangioma.
Other proliferative lesions of bile duct origin are non-neoplastic bile duct proliferation (hyperplasia), cholangiofibrosis, and cholangiocarcinoma. Simple non-neoplastic bile duct proliferation is a variable lesion which occurs spontaneously in aging rats and as a result of exposure to hepatotoxins, including carcinogens. Cells proliferate in strands, with the progressive formation of a lumen giving the appearance of bile ducts. Proliferation extends into the liver parenchyma and may be associated with fibroblastic proliferation, which can lead to cirrhosis. Bile duct proliferation, of this type, unlike cholangioma, is multifocal and usually widespread throughout the affected liver. In more advanced lesions, the fibrous proliferation is much greater than that usually observed in the stroma of bile duct neoplasms, and cystic hyperplasia is also occasionally observed.
Cholangiofibrosis, regarded by many as preneoplastic lesion, has the characteristic appear-

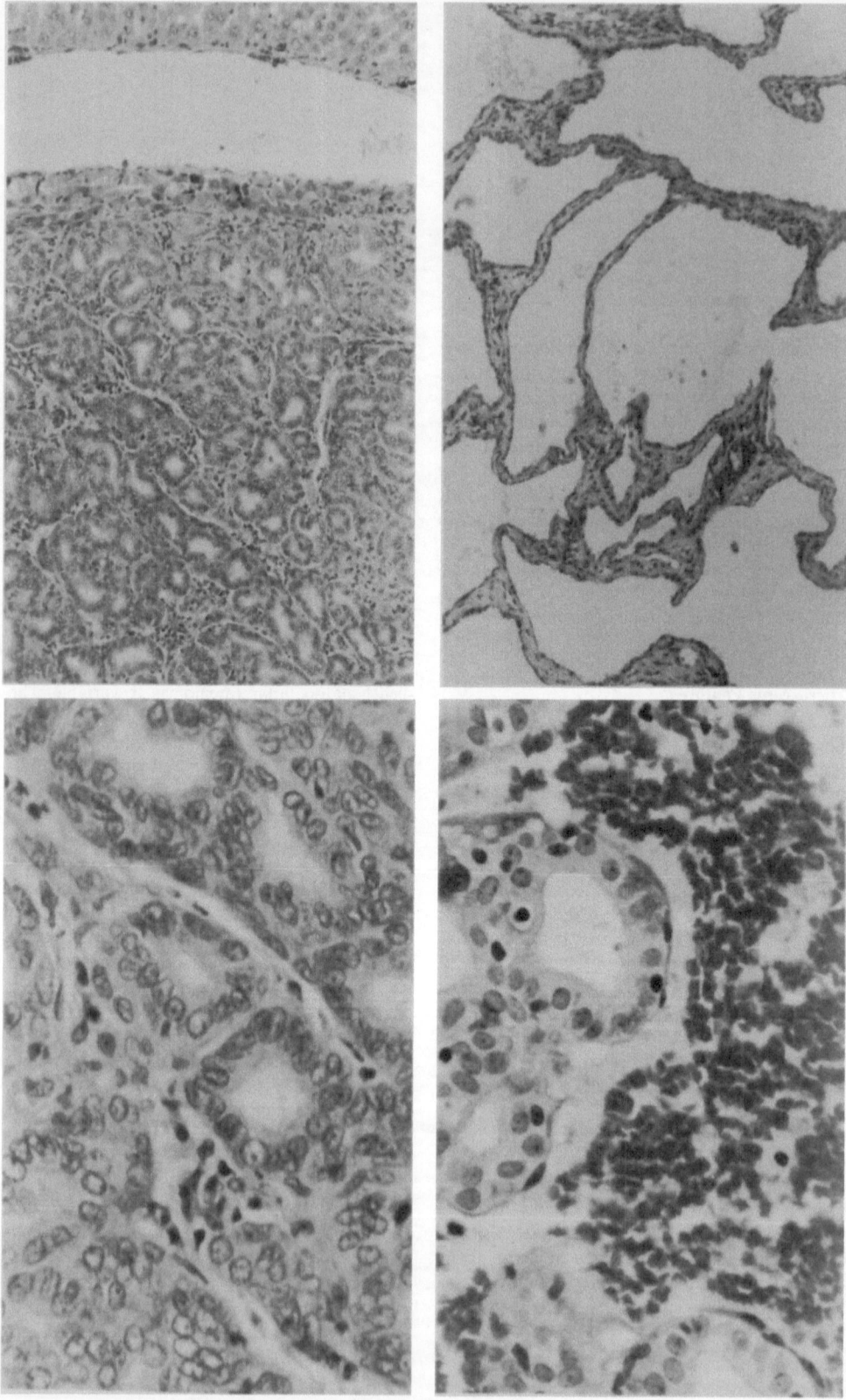

ance of clumps of glandular structures surrounded by dense connective tissue. The glandular proliferation is from bile ducts, and the lesion can be single or multiple, but usually has an initial periportal distribution. The proliferation of connective tissue may be great enough to result in atrophy of the glandular elements. There is a marked production of mucus by the acini in cholangiofibrosis, which is not a characteristic of the cholangioma (see p. 83).

In contrast to cholangiomas, cholangiocarcinomas are made up of acini lined by atypical epithelium, the cells of which frequently contain mucus. Cholangiocarcinomas have evidence of invasive growth, which is never observed in cholangiomas.

Biologic Features

An early change in the process of chemically induced carcinogenesis in the rat liver is the proliferation of bile ducts (Farber 1963; Schaffner and Popper 1961; Bannasch and Reiss 1971). This proliferation is considered to be a reparative lesion rather than a direct cellular response to carcinogenic agents (Bannasch and Reiss 1971), and the initial bile duct proliferation does not represent preneoplasia (Schauer and Kunze 1976). However, the long-term administration of carcinogenic compounds can result in the formation of adenomatous hyperplasia, which can progress to cholangioma (Schauer and Kunze 1976).

Of greater significance in the development of cholangiomatous neoplasms is cholangiofibrosis, which is regarded as irreversible (Farber 1963) and preneoplastic (Bannasch and Reiss 1971), although cholangiomas do not always pass through cholangiofibrosis (Schauer and Kunze 1976). Bannasch and Reiss (1971) demonstrated the progression of cholangiofibrosis to cholangioma following the administration of N-nitrosomorpholine.

Susceptibility to cholangioma formation has been demonstrated in partially hepatectomized rats following the administration of 2-acetylaminofluorene (Richards et al. 1982).

It is probable that cystic cholangiomas result from the accumulation of secretion following partial or complete obstruction of the biliary drainage; this is supported by the appearance of the flattened epithelium of the cystic acini. Such an obstruction could result either from internal epithelial proliferation or from compression resulting from the external proliferation of connective tissue (Schauer and Kunze 1976). Apart from the cystic appearance of the acini, this form of cholangioma is morphologically similar to simple cholangioma and presumably behaves in much the same way. Cystic transformation of simple cholangioma seems to be a likely means by which cystic cholangioma can be formed, although the possibility that some cystic forms arise de novo cannot be excluded, and the appearance of islands of hepatocytes trapped between the cystic acini in some instances is support for a primarily proliferative origin rather than a consequence of cystic transformation of a preexisting simple cholangioma.

Where hemangiomatous areas are present within a cholangioma, there is no indication of a primary proliferation of endothelial cells. Simple endothelium-lined, blood-filled spaces surround the glandular acini.

There is no evidence to suggest that, as a general rule, either the simple or cystic forms of cholangioma progress to cholangiocarcinoma.

Comparison with Other Species

Cholangiomas are observed rarely in humans and are usually only encountered as an incidental finding at autopsy. In the cat and dog, simple and cystic forms of neoplasm exist with much the same morphology as that described for the rat. In particular, multicellular cystic forms demonstrate the same pressure-related epithelial flattening, with occasional papillomatous outgrowths and collagenous septal stroma. Cholangiomas have been induced in the hamster following bile duct ligation after dimethylnitrosamine (DMN) initiation (Thamavit et al. 1993). Hamsters with liver fluke

◄

Fig. 52. (*upper left*) Liver, rat. A well-circumscribed simple cholangioma illustrating the acinar structure of the neoplasm and the delicate connective tissue capsule. No invasive growth is evident. H&E, ×430

Fig. 53. (*lower left*) Cytologic detail of the same neoplasm as in Fig. 52 illustrating the single-celled cuboidal epithelial lining of the acini and sparse connective tissue stroma. H&E, ×1720

Fig. 54. (*upper right*) Cystic cholangioma with prominent connective tissue stroma. H&E, ×430

Fig. 55. (*lower right*) Hemangiomatous cholangioma demonstrating cystic, endothelial-lined spaces in the stroma. H&E, ×1720

infestation also develop bile duct neoplasms after administration of DMN.

References

Bannasch P, Reiss W (1971) Histogenese and Cytogenese cholangiocellularer Tumoren bei Nitrosomorpholin-vergifteten Ratten. Zugleich ein Beitrag zur Morphogenese der Cystenleber. Z Krebsforsch 76:193–215

Farber E (1963) Ethionine carcinogenesis. Adv Cancer Res 7:383–474

Richards WL, Pitot HC, Potter VR (1982) Hepatic cyst formation in male rats partially hepatectomized as weanlings during short-term feeding of 2-acetylaminofluorene. Carcinogenesis 3:1355–1359

Schaffner F, Popper H (1961) Electron microscopic studies of normal and proliferated bile ductules. Am J Pathol 38:393–410

Schauer A, Kunze E (1976) Tumours of the liver, In: Turusov VS (ed) Pathology of tumours in laboratory animals, vol 1: tumours of the rat, part 2. IARC, Lyon, pp 41–72 (IARC scientific publications no 6)

Thamavit W, Pairojkul C, Tiwawech D, Itoh M, Shirai T, Ito N (1993) Promotion of cholangiocarcinogenesis in the hamster liver by bile duct ligation after dimethylnitrosamine initiation. Carcinogenesis 14:2415–2417

Hemangiosarcoma, Liver, Rat

James A. Popp and Russell C. Cattley

Synonyms. Angiosarcoma, malignant hemangio-endothelioma

Gross Appearance

The gross appearance is variable. Lesions may be barely visible or over 1 cm in diameter. They frequently bulge above the surface of the liver capsule, but are sometimes deeply embedded in the liver parenchyma and not visible on the capsular surface. The lesion usually lacks a capsule and has poorly defined borders. Perhaps the greatest variability occurs in the color, usually red, reddish brown, or black with a mottled appearance, while the more cellular and less vascular areas may be white to light tan. Hemangiosarcomas are typically soft and pliable and ooze blood or blood-tinged fluid when sectioned. Cysts are common due either to greatly dilated vascular spaces or to areas of necrosis which subsequently fill with blood (Feron et al. 1981). Rupture of cysts or friable vascular walls results in hemoperitoneum or hemothorax from metastatic lesions in over 75% of rats dying of hemangiosarcoma (Ward et al. 1975). Hemangiosarcomas are frequently multicentric within the liver. Metastatic sites in other tissues have a gross appearance similar to the primary lesion in the liver.

Microscopic Features

Early developing and small hemangiosarcomas consist of small areas (1 mm in diameter) in which the sinusoids are lined by numerous neoplastic endothelial cells that may be multilayered (Fig. 56). The sinusoids are frequently dilated with underlying atrophied hepatocytes. However, with some carcinogens there may be an associated neoplastic response of hepatocytes with development of foci, adenomas, and carcinomas (Feron et al. 1981; Lijinsky and Reuber 1984). The neoplastic endothelial cells vary in both size and shape, with individual large cells occasionally observed. Mitotic figures may be relatively numerous even in the smaller lesions.

As the hemangiosarcoma enlarges, hepatocytes are no longer found within the developing tumor. Large sheets of neoplastic cells with little evidence of vascular formation are seen. However, in all or nearly all lesions, some vessels can be identified even though they may occupy a small percentage of the lesion. The vascular channels may be large, dilated cysts or thin, capillary-sized openings in the neoplastic tissue. Individual cells of the larger lesions tend to be more pleomorphic, assuming either a polyhedral or spindle shape in different areas of a single lesion (Fig. 57). Large cells with large single nuclei as well as abnormal mitotic figures and numerous mitotic cells are observed; multinucleated giant cells are rare. Necrotic areas are invariably found in large lesions (more than 1 cm in diameter) and frequently contain hemorrhage.

The border of all hemangiosarcomas, irrespective of size, is indistinct. Neoplastic cells may merge with adjacent normal tissue or may actively invade the sinusoids of adjacent tissue. Encapsulation is never observed, although a thin fibrous zone may be found along some edges associated with compression.

Ultrastructure

Limited information is available on the ultrastructural features of hepatic hemangiosarcomas in rats. The large nuclei have an irregular outline, but are not indented. Some cells have numerous mitochondria and a well-developed rough endoplasmic reticulum, which frequently surrounds small mitochondria (Shinohara et al. 1977; Hadjiolov and Markow 1973; Ward et al. 1975). Bannasch et al. (1980) reported aggregates of intermediate filaments in the cytoplasm of hemangiosarcoma cells, which are composed of vimentin but not prekeratin or actin, as determined by immunofluorescence. The presence of so-called vimentin storage cells containing large, fibrillar, vimentin-positive inclusions is distinctive of hemangiosarcomas (Bannasch et al. 1981). In contrast, the filaments noted in neoplastic epithelial cells of hepatic origin are stained by labeled antibody to prekeratin.

Desmosomes have not been observed in hemangiosarcoma cells (Bannasch et al. 1980), although tight junctions have been described (Hadjiolov and Markow 1973). Basement membranes are found partially or entirely surrounding many neoplastic endothelial cells (Hadjiolov and Markow 1973).

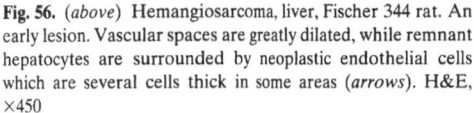

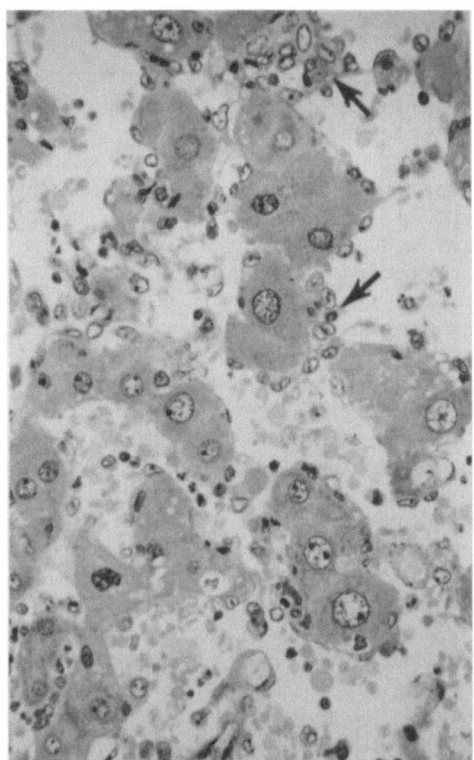

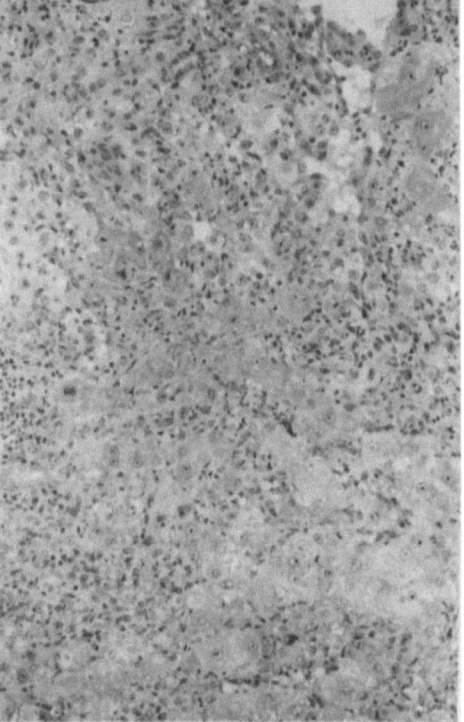

Fig. 56. (*above*) Hemangiosarcoma, liver, Fischer 344 rat. An early lesion. Vascular spaces are greatly dilated, while remnant hepatocytes are surrounded by neoplastic endothelial cells which are several cells thick in some areas (*arrows*). H&E, ×450

Fig. 57. (*below*) Hemangiosarcoma, liver, Fischer 344 rat. Small neoplastic cells are pleomorphic and invade adjacent tissue. Note the small clusters of hepatocytes trapped within the neoplasm. H&E, ×160

Differential Diagnosis

The presence of vascular spaces lined by obviously neoplastic cells is a significant aid in diagnosis. Nevertheless, primary hemangiosarcoma of the rat liver must be differentiated from anaplastic hepatocellular carcinomas, Kupffer's cell sarcomas, and other primary sarcomas of the liver such as fibrosarcoma. Even the most anaplastic hepatocellular carcinoma differentiates toward hepatocytes, which aids in distinguishing the neoplasm from a hemangiosarcoma. Kupffer's cell sarcoma is probably the lesion most similar to hemangiosarcoma. Kupffer's cell sarcomas arise from Kupffer's cells normally found in the sinusoidal wall and may closely resemble hemangiosarcomas which arise from endothelial cells of the sinusoidal wall. The two neoplasms are frequently considered synonymous (Institute of Laboratory Animal Resources, National Research Council 1980), but attempts should be made to distinguish them. Although morphological features noted light microscopically are helpful, the use of both electron microscopy and immunologic markers may be necessary to make the distinction between the more anaplastic hemangiosarcoma and Kupffer's cell sarcoma (see the chapter on "Kupffer's Cell Sarcoma," this volume). Following one study of vinyl chloride-induced hemangiosarcoma, it was concluded that ultrastructural findings provided no evidence to doubt an endothelial cell origin (Spit et al. 1981). Primary hemangiosarcomas must be distinguished from metastatic hemangiosarcomas from other sites.

Biologic Features

Hemangiosarcoma of the liver tends to invade locally by first extending along hepatic cords. As it becomes larger and more cellular, extension across hepatic cords and other normal structures is noted. Primary hepatic hemangiosarcomas metastasize to other organs, primarily the lung. Although they are induced by various chemical carcinogens, hemangiosarcomas are rarely spontaneous neoplasms. In large studies of spontaneously occurring neoplasms comprising 3907 F344 (Haseman et al. 1990) and 6360 Sprague-Dawley male and female rats (Chandra et al. 1992; Lang 1992; McMartin et al. 1992), only two of 3177 female Sprague-Dawleys (0.06%) had primary hepatic hemangiosarcomas. In terms of chemically induced hemangiosarcomas, one study (using vinyl chloride) demonstrated an increased susceptibility of female rats as compared to males, as well as greater susceptibility with increasing age (32 weeks or older) of starting exposure (Groth et al. 1981).

Comparison with Other Species

The histology of primary hepatic hemangiosarcomas is similar in most species, including humans (Popper et al. 1977; Wayss et al. 1979). The development and various histologic stages of the lesion appear to be equivalent in humans and animals when the lesions are caused by exposure to vinyl chloride, inorganic arsenicals, and thorium dioxide (Thorotrast) (Popper et al. 1977, 1981).

References

Bannasch P, Zerban H, Schmid E, Franke WW (1980) Liver tumors distinguished by immunofluorescence microscopy with antibodies to proteins of intermediate-sized filaments. Proc Natl Acad Sci USA 77:4948–4952

Bannasch P, Zerban H, Schmid E, Franke WW (1981) Characterization of cytoskeletal components in epithelial and mesenchymal liver tumors by electron and immunofluorescence microscopy. Virchows Arch (Cell Pathol) 36:139–158

Chandra M, Riley MGI, Johnson DE (1992) Spontaneous neoplasms in aged Sprague-Dawley rats. Arch Toxicol 66:496–502

Feron VJ, Hendriksen CFM, Speek AJ, Til HP, Spit BJ (1981) Lifespan oral toxicity study of vinyl chloride in rats. Food Cosmet Toxicol 19:317–333

Groth DH, Coate WB, Ulland BM, Hornung RW (1981) Effects of aging on the induction of angiosarcoma. Environ Health Perspect 41:53–57

Hadjiolov D, Markow D (1973) Fine structure of hemangioendothelial sarcomas in the rat liver induced with n-nitrosodimethylamine. Arch Geschwulstforsch 42:120–126

Haseman JK, Arnold J, Eustis SL (1990) Tumor incidences in Fischer 344 rats: NTP historical data. In: Boorman GA, Eustis SL, Elwell MR, Montgomery CA Jr, MacKenzie WF (eds) Pathology of the Fischer rat, reference and atlas, chap 35. Academic, San Diego, pp 555–564

Institute of Laboratory Animal Resources, National Research Council (1980) Histologic typing of liver tumors of the rat. J Natl Cancer Inst 64:179–206

Lang P (1992) Spontaneous neoplastic lesions and selected non-neoplastic lesions in the Crl:CD®BR rat. Charles River Laboratories, Feb 1992

Lijinsky W, Reuber MD (1984) Carcinogenesis in rats by nitrosodimethylamine and other nitrosomethylalkylamines at low doses. Cancer Lett 22:83–88

McMartin DN, Sahota PS, Gunson DE, Hsu HH, Spaet RH (1992) Neoplasms and related proliferative lesions in control Sprague-Dawley rats from carcinogenicity studies. Historical data and diagnostic considerations. Toxicol Pathol 20:212–225

Popper H, Selikoff IJ, Maltoni C, Squire RA, Thomas LB (1977) Comparison of neoplastic hepatic lesions in man and experimental animals. In: Hiatt HH, Watson JD, Winsten JA (eds) Origins of human cancer, vol 4, book C, Cold Spring Harbor Conference on Cell Proliferation. Cold Spring Harbor Laboratory, Cold Spring Harbor, pp 1359–1382

Popper H, Maltoni C, Selikoff IJ (1981) Vinyl chloride-induced hepatic lesions in man and rodents. A comparison. Liver 1:7–20

Shinohara Y, Ogiso T, Hananouchi M, Nakanishi K, Yoshimura T, Ito N (1977) Effect of various factors on the induction of liver tumors in animals by quinoline. Gann 68:785–796

Spit BJ, Feron VJ, Hendriksen CFM (1981) Ultrastructure of hepatic angiosarcoma in rats induced by vinyl chloride. Exp Mol Pathol 35:277–284

Ward JM, Sontag JM, Weisburger EK, Brown CA (1975) Effect of lifetime exposure to aflatoxin B_1 in rats. J Natl Cancer Inst 55:107–113

Wayss K, Bannasch P, Mattern J, Volm M (1979) Vascular liver tumors induced in Mastomys (Praomys) natalensis by single or twofold administration of dimethylnitrosamine. J Natl Cancer Inst 62:1199–1207

Hemangioma, Liver, Rat

Paul N. Brooks and Francis J.C. Roe

Synonym. Hemangioendothelioma

Gross Appearance

Hemangiomas appear macroscopically as dark, raised foci or small nodular areas and are generally soft in texture. They occasionally occur in association with hepatocellular neoplasms and tend to hemorrhage easily, which may result in fatal intra-abdominal bleeding.

Microscopic Features

The diagnosis of hemangioma is made upon evidence of a primary proliferation of endothelial cells, which is initially apparent in association with cords of liver cells, but subsequently hepatocytes are excluded from the neoplasm except for clumps of liver cells trapped during expansive growth. Eventually, these trapped hepatocytes become atrophic and degenerate and are replaced by fibrous tissue. In other areas, a definite connective tissue stroma is evident in association with the proliferating endothelial cells.

Variability is an obvious feature in the micromorphology of hemangiomas of the liver, even within individual tumors. Frequently, cords of liver cells lined by neoplastic endothelium give rise to pseudoglandular, papillary, and cystic structures. These areas contrast with those of more solid appearance containing none or only a very few hepatocytes, but characterized by a prominent connective tissue stroma. The endothelial lining is usually only one cell thick, but locally the depth can be greater.

Sinusoids within the neoplasm develop to a variable degree and become filled with blood. In some instances, cystic blood-filled spaces are a prominent feature.

Neoplastic endothelial cells are larger than normal, but are occasionally oval or round rather than flattened. The nucleus is generally large in proportion to cell size, with only scant cytoplasm in most cases. Benign neoplastic endothelial cells are easily recognizable as endothelial in origin, in contrast to the malignant variants, which have a more diverse morphology. Mitotic figures are only rarely observed within the neoplasm and there is never invasive growth, although such lesions are not always clearly delineated from surrounding tissue by any distinct capsule (Figs. 58, 59).

Differential Diagnosis

The differential diagnosis of hemangioma can be quite difficult when such neoplasms occur in the liver. In particular, hemangioma must be distinguished from morphological forms of hepatocellular proliferation, those of hepatocellular alteration, and hemangiosarcoma. Usually, the characteristic epithelial appearance of cholangioma does not lead to confusion in the diagnosis of hemangioma.

Within hepatocellular adenomas and carcinomas, there are occasionally hemangiomatous areas in which sinusoidal dilatation is conspicuous. The associated endothelial cells are, however, quite normal and without neoplastic features.

Foci or areas of telangiectasis, peliosis hepatis, and spongiosis hepatis can each be confused with early neoplastic lesions of vascular origin (see pp. 104, 154). In areas of telangiectasis, the vessels undergo dilatation without proliferation of the endothelium or hepatocytes, and both elements are morphologically normal except for dilatation of sinusoids. Peliosis hepatis is characterized by blood-filled cystic spaces bounded by hepatocytes and having no endothelium, and in spongiosis hepatis, a lesion described by Bannasch et al. (1981) (see p. 104) in which there is an extracellular accumulation of mucopolysaccharides and/or protein, no proliferation of endothelial cells is seen, although there may be proliferation of surrounding hepatocytes.

The morphological features of the malignant endothelial cells observed in hemangiosarcomas are usually sufficient to delineate this from hemangioma. In general, malignant endothelial

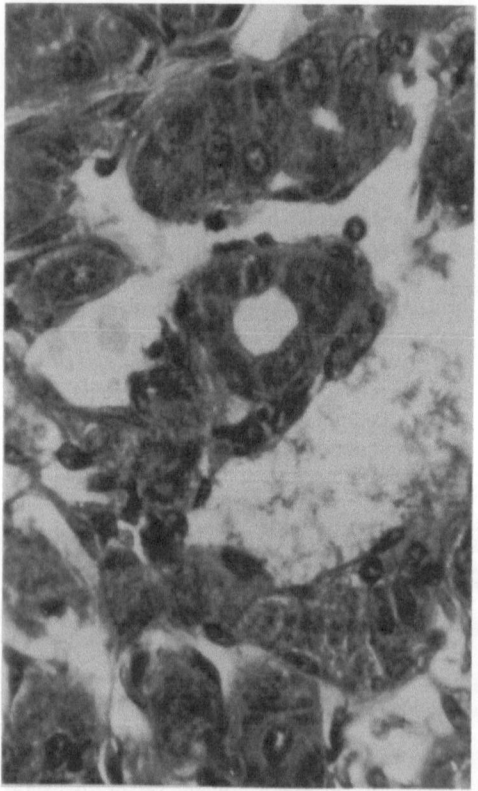

Fig. 58. Hemangioma, liver, rat. Cords of hepatocytes are lined by plump, but otherwise normal, endothelial cells. H&E, ×1720

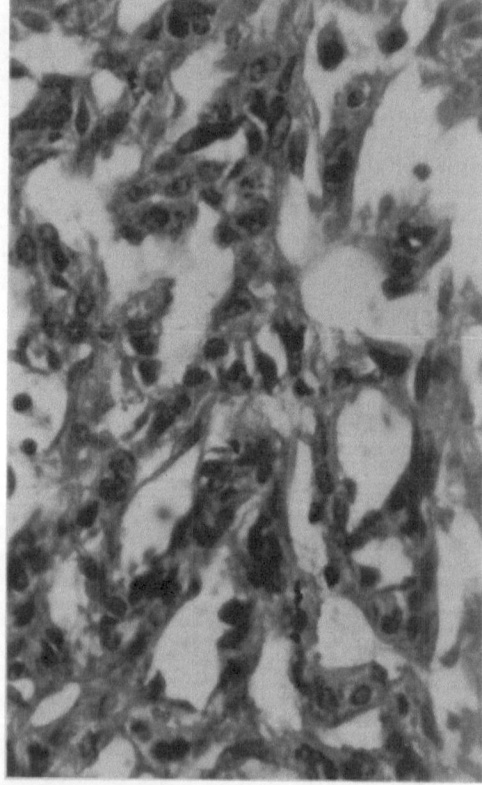

Fig. 59. Hemangioma, liver, rat. Vascular spaces lined by endothelium; hepatocytes are sparse, and those present are atrophic. H&E, ×1720

cells vary widely in size and shape and line irregular vascular spaces often to a depth of several cells. The cells have considerable nuclear polymorphism and a relatively high mitotic rate, and they invade adjacent tissues.

Biologic Features

The relatively low frequency with which hemangioma occurs spontaneously in the rat liver makes comment on the behavior of such tumors difficult. Hemangiomas have, however, been induced experimentally by the administration of nitrosamines, nitrosamides, or vinyl chloride, but there remains confusion as to whether or not hepatic hemangiomas progress to malignant forms in the rat.

Comparison with Other Species

Hemangioma of the liver is rare in most species. The neoplasm does occur in mice (Frith and Ward 1980), and there are suggestions that in the mouse the development of hemangiosarcoma is a progression from hemangioma. In humans, cavernous hemangiomas of the liver are not uncommon.

References

Bannasch P, Zerban H (1994) Preneoplastic and neoplastic lesions of the rat liver. In: Bannasch P, Gössner W (eds) Pathology of neoplasia and preneoplasia in rodents. EULEP Color Atlas, Schattauer, Stuttgart, pp 18–63

Bannasch P, Bloch M, Zerban H (1981) Spongiosis hepatis. Specific changes of the perisinusoidal liver cells induced in rats by N-nitrosomorpholine. Lab Invest 44:252–264

Frith CH, Ward JM (1980) A morphologic classification of proliferative and neoplastic hepatic lesions in mice. J Environ Pathol Toxicol 3:329–351

Hepatoblastoma, Mouse

Vladimir S. Turusov, Bhalchandra A. Diwan, Natalia V. Engelhardt, and Jerry M. Rice

Synonyms. Malignant embryonal tumor, unusual tumor with rosette formation, poorly differentiated cholangiocarcinoma, poorly differentiated hepatocellular carcinoma

Gross Appearance

Macroscopically, hepatoblastomas are grayish or brownish nodules (up to 1.5 cm in diameter) often associated with hemorrhagic cysts, vascular lakes, or areas of necrosis. Hepatoblastoma may have the appearance of a large hemorrhagic cyst with a thick, grayish or whitish wall. It is usually associated with hepatocellular tumors. No site specificity within liver lobes has been noted.

Microscopic Features

Hepatoblastomas are almost invariably observed within hepatocellular proliferative lesions, including preneoplastic hepatocellular foci, hepatocellular adenomas, and carcinomas (Turusov et al. 1973b; Nonoyama et al. 1988; Diwan et al. 1989, 1992). Some small hepatoblastomas lack this association and may represent hepatoblastoma in situ or intrahepatic metastases, especially in livers with large hepatoblastomas. In contrast to eosinophilia of hepatocellular adenomas and carcinomas in hematoxylin and eosin (H&E) preparations, hepatoblastomas are intensely basophilic, consisting of mixtures of dense cellular areas, spongiform

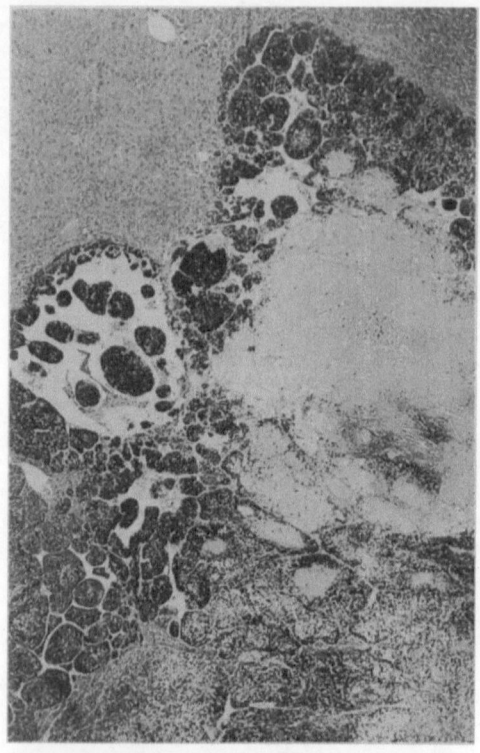

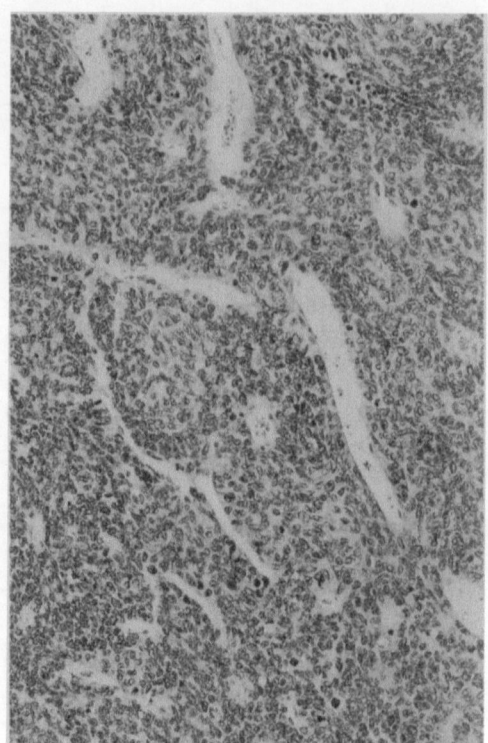

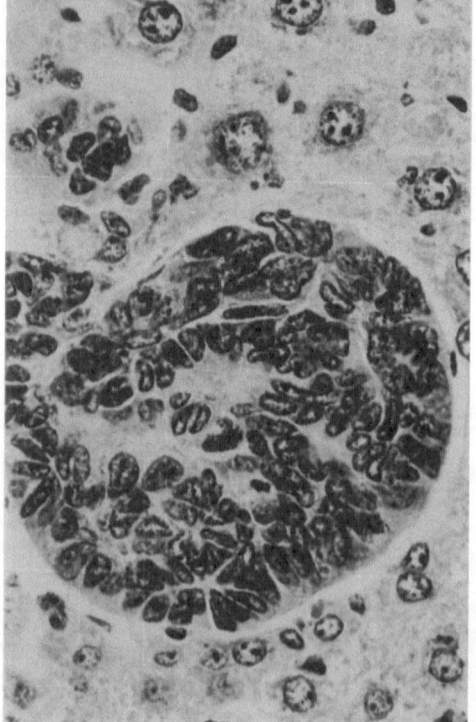

Fig. 60. (*upper left*) Liver, CF-1 mouse. General view of hepatoblastoma. Note the cellular areas separated by cystic areas filled with necrotic debris. H&E, ×32

Fig. 61. (*upper right*) Liver, CF1-1 mouse. Hepatoblastoma with neoplastic cells arranged in sheets and pseudorosettes. H&E, ×200

Fig. 62. (*below*) Liver, mouse. Early hepatoblastoma sharply demarcated from the surrounding hepatic tissue. At the center, tumor cells form a rosette. H&E, ×500

►

Fig. 63. (*upper left*) Liver, mouse. Organoid structures in hepatoblastoma separated either by cystic spaces or by dark elongated cells. H&E, ×80

Fig. 64. (*upper right*) **a** Lumina in the center of structures are lined by columnar to cuboidal epithelium and some contain blood (*top right*). **b** Cells within the structures form rosettes. H&E, ×200

Fig. 65. (*lower left*) Liver, mouse. Infiltrative growth of hepatoblastoma. H&E, ×200

Fig. 66. (*lower right*) Squamous metaplasia in hepatoblastoma. Hepatoblastoma induced by initiation/promotion. Areas of squamous metaplasia (*arrows*). H&E, ×200

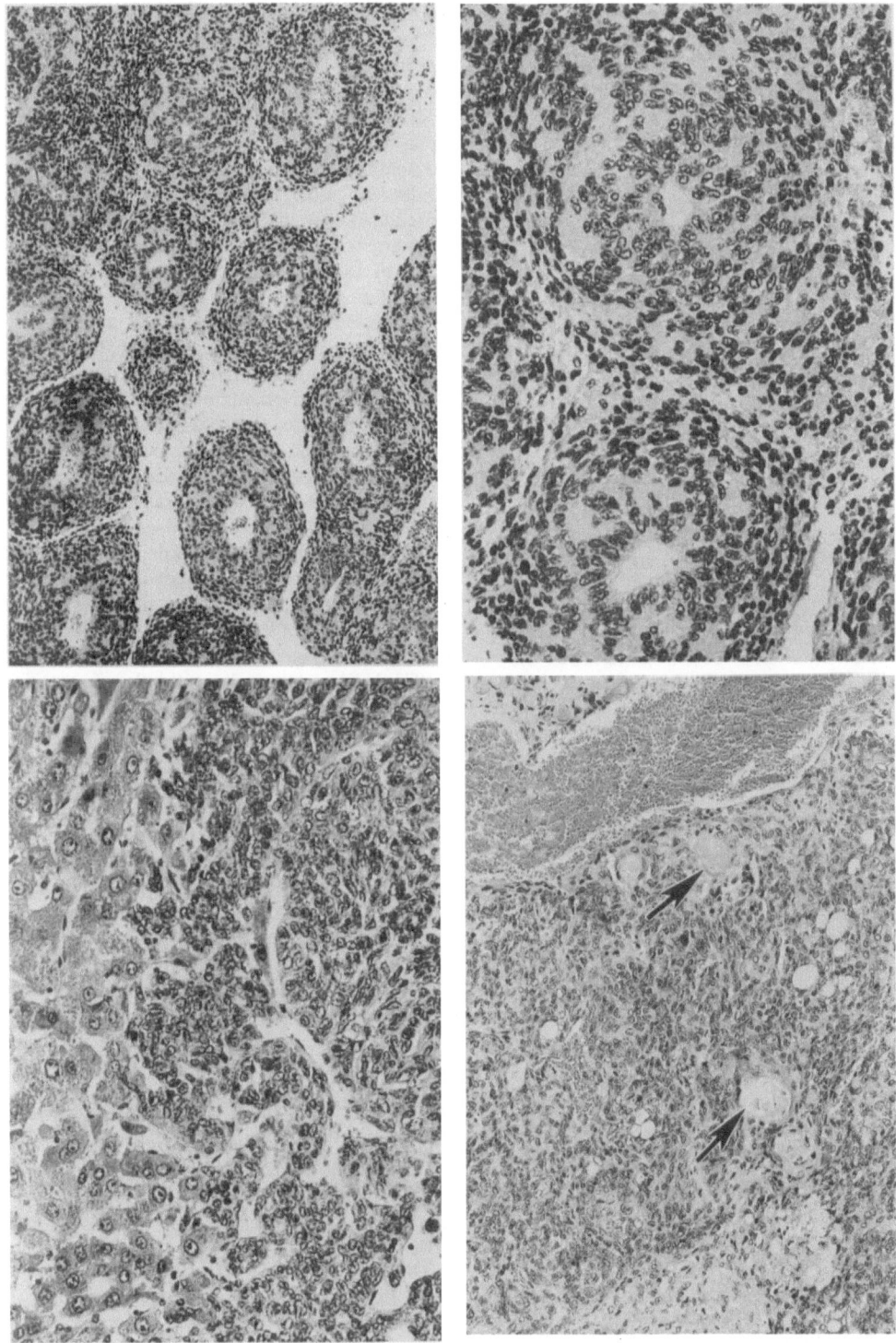

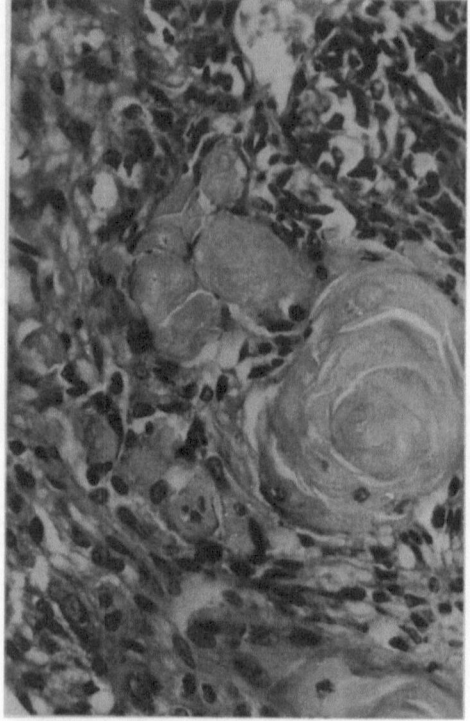

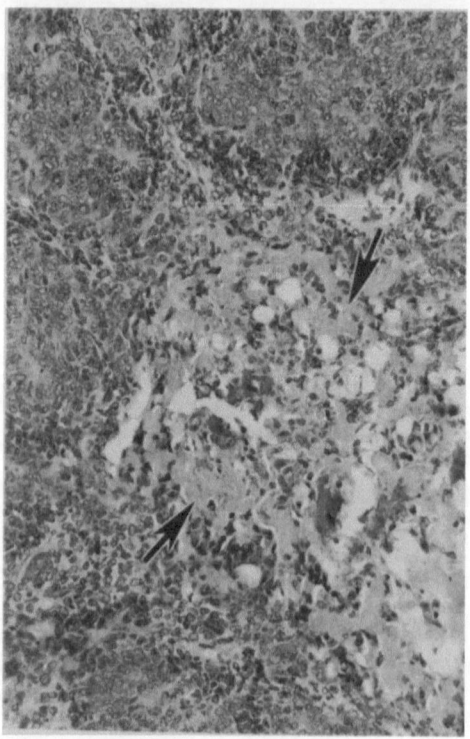

areas, and large vascular lakes or cystic spaces filled with necrotic debris, blood, or blood-tinged proteinous liquid (Fig. 60). Cellular areas are composed of several morphological variants of neoplastic cells. Elongated (spindle-shaped) neoplastic cells with hyperchromatic nuclei and eosinophilic cytoplasm predominate. These tumor cells are arranged in the form of rosettes, pseudorosettes, ribbons, or solid cords of cells. Rosette (pseudorosette) formation is a typical feature of advanced tumor, but can be seen in very early lesions (Figs. 61, 62). Another frequent variant consists of small islands of cells with round to oval nuclei and scant basophilic cytoplasm. These cells may be arranged in compact sheets or small islands of cells (Nonoyama et al. 1988; Marceau et al. 1989; Diwan et al. 1989, 1992). In many cases, hepatoblastomas consist of peculiar organoid structures (Figs. 63, 64b) made up of vascular channels surrounded by several layers of tumor cells arranged either radially or concentrically. The vascular channels are lined with flattened endothelium and often contain blood elements. These structures are separated from one another by a few layers of dark, elongated cells located at the periphery of each complex (Turusov et al. 1973b). Some tumors have tubules of atrophied hepatic cords with associated hepatic stroma and areas of adenofibrosis. Although most hepatoblastomas are sharply demarcated from surrounding neoplastic hepatocytes, we have seen several cases of infiltrative growth of hepatoblastoma cells into the surrounding parenchyma (Fig. 65). Multiple small (early) lesions without distinct demarcation from surrounding parenchyma are occasionally seen within large, proliferative hepatocellular lesions. Mitoses are abundant, even in early lesions. Squamous metaplasia with keratin formation is occasionally seen (Figs. 66, 67). Osteoid or cartilage formation, a very common observation in human hepatoblastoma, has also been reported in some mouse hepatoblastomas (Figs. 68, 69b; Turusov et al. 1973b; Nonoyama et al. 1988), but occurred in only one of the Frederick Cancer Research and Development Center series (Diwan et al. 1989,

◀

Fig. 67. (*above*) Transplanted spontaneous hepatoblastoma with keratin pearl formation. H&E, ×500

Fig. 68. (*below*) Liver, mouse. Osteoid formation in hepatoblastoma. Stromal osseous metaplasia in hepatoblastoma induced by initiation/promotion. H&E, ×200

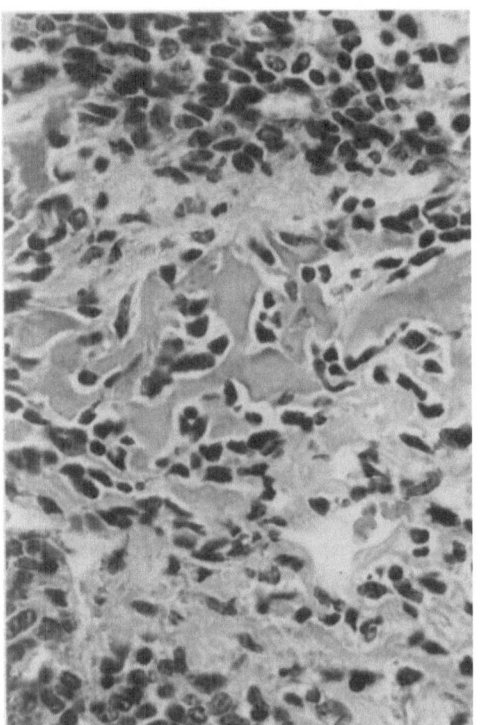

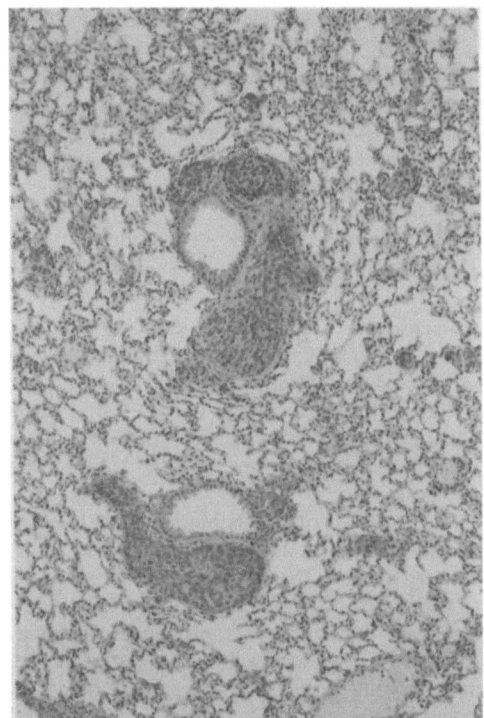

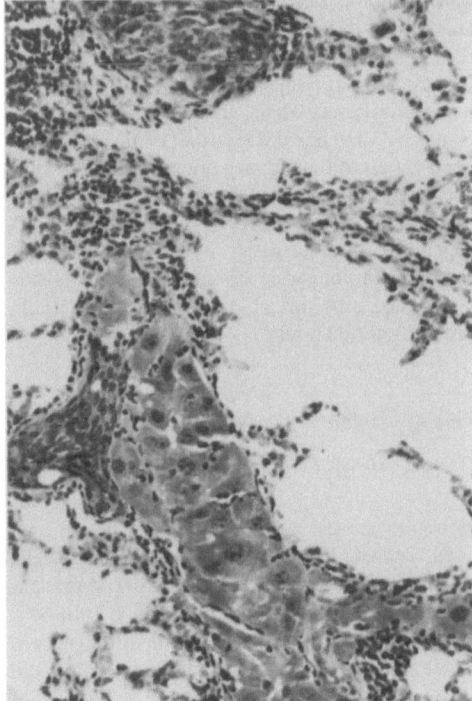

Fig. 69. (*upper left*) Osteoid formation within parenchyma of hepatoblastoma in CF-1 mouse treated with dichlorodiphenyltrichloroethane (DDT). ×500

Fig. 70. (*upper right*) Pulmonary metastasis of hepatoblastoma. Multiple metastases of hepatoblastoma induced by initiation/promotion. H&E, ×200

Fig. 71. (*below*) Metastases of hepatoblastoma and of liver cell carcinoma in CF-1 mouse treated with dichlorodiphenyltrichloroethane (DDT). H&E, ×500

1992). Vascular channels may be so abundant as to give the impression of vascular tumor.

There is a high incidence of metastases to the lung (25%–37%; Figs. 70, 71b). Hepatoblastomas are transplantable (Kharkavskaya et al. 1990). The histologic appearance of the transplanted tumors is similar to that of primary tumors.

Differential Diagnosis

Spontaneous liver tumors exhibiting certain morphological characteristics of human hepatoblastoma were first reported by Deringer (1970) in (YBR × AKR)FL mice carrying the lethal yellow gene (AY). She described these tumors as unusual liver tumors with rosette formations.

There is great confusion in the literature between diagnosis of hepatoblastomas and poorly differentiated cholangiocarcinomas and cholangiomas (Reuber 1967; Vlahakis and Heston 1971). Cholangiolar tumors resembling hepatoblastomas have been shown to arise from periportal areas in mice carrying the viable yellow allele at the agouti locus (C3H-Avy fB). Some of the lesions reported as cholangiocarcinomas (Reuber 1967) or cholangiomas (Vlahakis and Heston 1971) were reexamined by Turusov et al. (1973b) and were regarded to be similar to those diagnosed as hepatoblastomas. The structures regarded as glands by Vlahakis and Heston (1971) are actually lined by endothelium and are therefore vascular channels, commonly seen in mouse hepatoblastoma. Furthermore, ducts, tubules, or acini, which are characteristics of cholangiomas and cholangiocarcinomas, are not seen in hepatoblastomas, and this absence should aid in the differential diagnosis.

Biologic Features

Natural History

Hepatoblastomas occur spontaneously, but have been reported very rarely, in several strains of mice (Jones and Butler 1975; Turusov and Takayama 1979; Maronpot et al. 1987; Turusov et al. 1991). There are considerable differences in incidence between various mice strains. In one study, a low incidence was found in C57BL and CBA mice (0.5%) and a much higher incidence in

their hybrids, (CBA × C57BL)Fl mice (5.0%) (Turusov et al. 1991), thus indicating the role of genetic factors.

Hepatoblastomas have been induced by chemical treatment in some strains of mice. A few tumors resembling human hepatoblastomas are described by Rice (1973) in C3H mice exposed transplacentally to N-nitrosoethylurea. Turusov et al. (1973a) have found similar tumors in untreated and dichlorodiphenyltrichloroethane (DDT)-treated CF1 mice and were the first to tentatively call them hepatoblastomas. Hepatoblastoma induction by DDT showed a clear-cut dose response: 0.9%, 1.4%, 3.9%, 3.1%, and 7.1% in mice fed 0, 2, 10, 50, and 250 ppm DDT, respectively (Turusov et al. 1973a). Hepatoblastomas have been recently reported to occur in BALB/C and B6C3F1 mice after chronic exposure to 2-acetylaminofluorene (Nonoyama et al. 1986, 1988) and in B6C3F1 male mice initiated with N-nitrosodiethylamine (NDEA) and promoted by either phenobarbital or di-(2-ethylhexyl) phthalate (Ward et al. 1983). The incidence of these tumors reported by the above investigators, however, did not exceed 6%–7% even in lifetime studies. This failure to induce or increase the incidence of hepatoblastomas is probably related to the strains of mice used as well as the nature of the carcinogens employed and the experimental protocol followed.

In the experimental studies reported from the Frederick Cancer Research and Development Center (Diwan et al. 1989, 1992), no hepatoblastomas were observed in untreated D2B6Fl male mice during the first 60 weeks of life. The

Table 7. Features of hepatoblastoma in D2B6F1 mice

Feature	n
Mice with hepatoblastoma	63
Mice with pulmonary metastases	13
Total hepatoblastomas	173
Located within eosinophilic foci	2
Located within eosinophilic lesions[a]	131
Mice with large vascular spaces	95
Prominent pseudorosettes	48
Organoid structure	43
Squamous metaplasia	3
Bone or cartilage formation	1

Tumors were initiated with N-nitrosodiethylamine (NDEA) and promoted with phenobarbital or other promoters.
[a] No hepatoblastomas developed within or adjacent to basophilic hepatocellular neoplasms or foci of cellular alteration.

carcinogen, NDEA, at initiating doses used in our studies (90mg/kg body weight) has produced a very low incidence (3.3%) of hepatoblastomas in these mice. Recently, Yamate et al. (1990) have reported spontaneous occurrence of hepatoblastomas in 6% of B6D2F1 (reciprocal F1 hybrid) male mice in a lifetime (127 weeks) study. In contrast, we have seen a very high incidence (70%–85%) of hepatoblastomas in D2B6F1 mice initiated with NDEA and promoted with phenobarbital, while a low incidence (10%) of such tumors occurred in B6D2F1 mice fed phenobarbital for 53 weeks following NDEA initiation. Thus D2B6F1 mice are remarkably more susceptible than the reciprocal F1 hybrid males to the development of these malignant embryonal tumors. Obviously both genetic and epigenetic factors play an important role in the development of mouse hepatoblastomas (see Table 7).

Pathogenesis

The definitive histogenesis of hepatoblastomas remains uncertain. In the case of those tumors that arise in large numbers in the course of our tumor promotion studies, the preponderance of evidence is that they generally evolve from preneoplastic or neoplastic hepatocytes in foci or hepatocellular tumors (Diwan et al. 1992). Consistent with this pathway are the following observations:

1. Most of the hepatoblastomas were found within hepatocellular lesions, including hepatocellular foci, while none were found in bile duct lesions induced or promoted by Aroclor (Ar-1254) in our earlier study (Diwan et al. 1992). Whether foci of hepatoblastoma that are not surrounded by neoplastic or preneoplastic hepatocytes represent intrahepatic metastases of another hepatoblastoma or overgrowth of a preexistent hepatocellular lesion by a primary hepatoblastoma originating there has not been established.
2. No hepatoblastomas were found in basophilic hepatocellular proliferative lesions induced by the carcinogen alone in our study. On the other hand, hepatoblastomas were almost always found in eosinophilic hepatocellular neoplasms promoted by phenobarbital or phenobarbital-type promoters (Table 8).
3. In several cases, dark, basophilic, poorly differentiated hepatoblastoma cells appeared to

evolve from hepatocellular adenoma or carcinoma cells that were keratin negative (Fig. 72).

Thus hepatoblastomas in D2B6F1 mice appear to evolve as a progression of the hepatocyte lineage, known to be susceptible to promotion, from foci to adenomas to hepatocellular carcinomas and, as a result of further loss of differentiation or dedifferentiation of neoplastic hepatocytes, to hepatoblastomas.

Immunohistochemical studies to characterize mouse hepatoblastomas and help define their histogenesis have been undertaken. Thus far these studies have provided more questions than answers regarding the cellular origins of these tumors. There are some remarkable similarities and differences between mouse and human hepatoblastomas. Similarities include the overall histologic structure comprising a thick fibrous capsule, poorly differentiated cells arranged in sheets or rosettes, vascular channels, areas of fibrosis, and hemorrhagic and necrotic areas. Osteoid

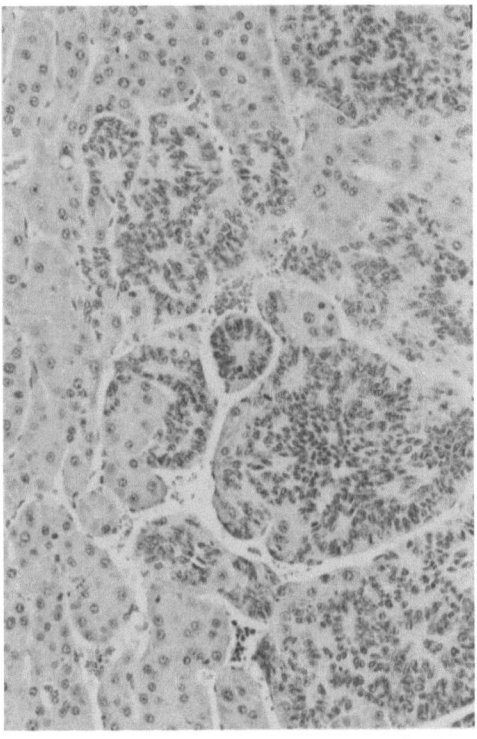

Fig. 72. Liver, mouse. Hepatoblastoma cells arising within neoplastic hepatocytic trabeculae after *N*-nitrosodiethylamine (NDEA) initiation and phenobarbital promotion. H&E, ×100

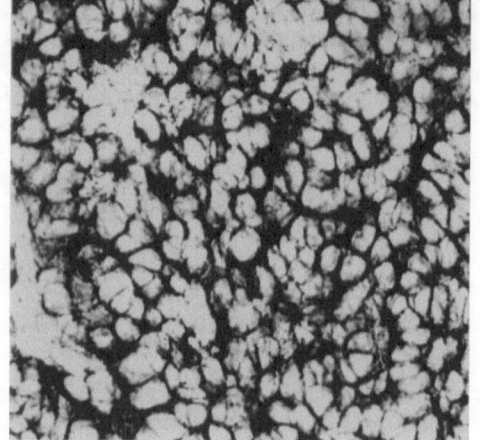

Fig. 73a,b. Biliary epithelial and oval cell marker G7 in sections of **a** normal mouse liver and **b** hepatoblastoma II. Indirect immunoperoxidase technique, ×420

metaplasia, a very common observation in human hepatoblastomas, is also known to occur in mouse hepatoblastomas (Turusov et al. 1973b, 1991; Nonoyama et al. 1986), but very rarely.

Detailed immunohistochemistry has been carried out on several generations of two transplantable mouse hepatoblastomas (Table 9). Hepatoblastoma I, which occurred in an untreated aged (CBA × C57BL/6)Fl male, was transplanted subcutaneously and has been maintained for more than 30 passage generations (Kharkovskaya et al. 1990). Hepatoblastoma II was induced in a D2B6F1 mouse by NDEA initiation followed by phenobarbital promotion and is currently in its eighth subcutaneous passage generation. Histologically, both tumors are similar and resemble primary hepatoblastomas. In addition to α-fetoprotein (AFP), two other hepatocyte markers and two markers common to mouse oval and biliary epithelial cells were evaluated in hepatoblastoma I and II (Table 10; N.V. Engelhardt et al., in preparation).

Table 9. Comparative phenotypes of mouse liver cell lineages, oval cells, and transplantable hepatoblastomas (from N.V. Engelhardt et al., in preparation)

Cell lineage/ population	AFP	AgB10	AgA1	AgA6	AgG7
Hepatocytes	+	+	+	+/−	−
Biliary epithelial cells	−	−	+	+	+
Oval cells	+/−	−	n/d	+	+
Hepatoblastoma I	−	−	n/d	−	+
Hepatoblastoma II	−	−	−	−/+	+

n/d, not determined.

Table 8. Cumulative incidence and multiplicity of hepatic tumors in various groups of D2B6F1 male mice

Treatment groups	Mice with tumors (n)	Total at risk (n)	Hepatocellular adenoma		Hepatocellular carcinoma		Incidence (%) of lung metastasis	Hepatoblastoma		Incidence (%) of lung metastasis
			Incidence (%)	Mean per tumor-bearing animal (n)	Incidence (%)	Mean per tumor-bearing animal (n)		Incidence (%)	Mean per tumor-bearing animal (n)	
NDEA	12	30	40	3.4	−	−	−	3.3	1.0	0
NDEA + PB	29	29[a]	100	16.1[a]	95	7.9	3.5	73	2.7	29.0
NDEA + Ar-1254	24	24[a]	100	10.8[a]	100	15.4[b]	29.2	33	1.5	−
NDEA + DDT	22	22[a]	82	5.1	82	5.1	−	27.3	1.0	16.6
NDEA + TCPOBOP	30	30[a]	100	11.3	100	11.3	40.0	90	3.2	37.0
Vehicle + TCPOBOP	30	30[a]	100	3.8	100	3.8	13.3	70	2.1	19.0

Modified from Diwan et al. (1992).
TCPOBOP, 1,4-bis[2-(3,5-dichloropyridyloxy)]benzene; NDEA, N-nitrosodiethylamine; DDT, dichlorodiphenyltrichloroethane; PB, phenobarbital; Ar-1254, aroclor.
[a] Significantly different ($p < 0.05$) from the group treated with NDEA only.
[b] Significantly different ($p < 0.05$) from the group treated with NDEA + PB.

Table 10. Differentiation antigens of mouse liver cells studied in transplantable hepatoblastomas I and II

Antigen	Localization in sections of fetal liver	Localization in sections of adult liver	Specific reagent	Immunizing material	References
AFP	Cytoplasm of hepatoblasts and fetal hepatocytes	Absent	Polyclonal monospecific Ab	AFP	Engelhardt et al. 1971
Antigen of bile canaliculi (B10)	Apical surfaces of hepatocytes beginning from GD 16	Bile canaliculi	mAb B10	"Ghosts" of mouse liver cells	Kuprina et al. 1990
A1	Apical surfaces of some hepatoblasts beginning from GD 12	Bile canaliculi and cytoplasm of hepatocytes Luminal surface of bile ducts	mAb A1	Yolk sac cells of 10-day-old embryo	Medvinsky et al., in preparation
A6	Cytoplasm of hepatoblasts on GD 12–15 Plasma membranes of developing biliary epithelial cells beginning from GD 16 Erythroblasts beginning from GD 11.5	Plasma membranes of epithelial cells of all bile ducts and ductules	mAb A6	Oval cell-enriched fraction from the liver of mice treated with Dipin and partial hepatectomy	Engelhardt et al. 1990, 1993
G7	Plasma membranes of developing biliary epithelial cells beginning from GD 16 Glisson's capsule	Plasma membranes of epithelial cells of all bile ducts and ductules Sinusoidal surfaces of hepatocytes Glisson's capsule	mAb G7	Oval cell-enriched fraction from the liver of mice treated with Dipin and partial hepatectomy	Engelhardt et al. 1990

GD, gestational day; AFP, α-fetoprotein; mAb, monoclonal antibody; DIPIN, 1,4-bis(diaziridiny Lphosphiny Lidyne)-piperazine.

The antigenic profile of hepatoblastomas was stable during several generations. AFP and hepatocyte markers were never demonstrated in sections of either tumor. However, trace amounts of AFP mRNA were found in hepatoblastoma II by northern dot hybridization (N. Lazarevitz, personal communication). One marker shared by oval and biliary epithelial cells (G7) was present in most tumor cells in both hepatoblastomas (Fig. 73a,b). The second marker of similar specificity, A6, is absent in sections of hepatoblastoma I, but occasionally observed in hepatoblastona II in a few elongated cells lining small lumens. Antigen G7 in mouse liver is found both in binary epithelial cells and in Glisson's capsule (Table 10); moreover, it is not liver specific and is present in different epithelia, including foregut during early ontogenesis (Engelhardt et al. 1990). Thus the antigenic profile of hepatoblastoma I and II differed from that of hepatocytes and biliary epithelial and oval cells (Table 9); further, hepatoblastoma II was found to be negative for cytokeratin 19 and γ-glutamyltranspeptidase, which are markers of biliary epithelial cells.

Comparison with Other Species

The most notable difference between species is the consistent occurrence of this lesion in older rats and mice, while in humans it is primarily a disease of childhood. Histologically, mouse hepatoblastomas are similar to those described in the rat (Institute of Laboratory Animal Resources 1980). Unlike human or mouse hepatoblastomas, however, these tumors have not been reported to arise spontaneously in the rat. This primitive type of neoplasm is most likely to occur in rats subjected to potent carcinogenic regimens (Tsao and Grisham 1987).

Ishak and Glunz (1967) have identified two types of cells, the fetal type and the embryonal type, within epithelial components of human hepatoblastomas. The fetal type cells have large, round, darkly stained nuclei, while the embryonal-type cells are less well differentiated and form rows or rosettes. This "small cell" variant of human hepatoblastoma closely resembles the predominant component of murine hepatoblastona. Thus mouse tumors are much more dedifferentiated than those in humans and correspond to the least differentiated variant of human hepatoblastoma (Turusov et al. 1973b, 1991). Nonoyama et al. (1988) have also described some ultrastructural differences between human and mouse hepatoblastomas. Unlike hepatoblastoma cells in human lesions, mouse hepatoblastoma cells lack crystalloid inclusions in mitochondria and are characterized by scarcity of smooth endoplasmic reticulum, Golgi apparatus, and glycogen granules.

Finally, most human hepatoblastomas are immunoreactive for AFP, while murine hepatoblastoma cells stain consistently negative for this fetal protein. However, only highly differentiated (fetal-type) human hepatoblastomas are AFP positive, while AFP is sometimes absent in the embryonal type and regularly absent in the anaplastic type (Schmidt et al. 1985). According to Abenoza et al. (1987), AFP-positive human hepatoblastomas are vimentin negative, while anaplastic AFP-negative hepatoblastomas are vimentin positive, thus showing the capacity of anaplastic hepatoblastoma to have mesenchymal differentiation patterns. Mouse hepatoblastoma can be induced by a number of carcinogens. It is of interest that a higher risk of hepatoblastoma is observed in children whose mothers were exposed before or during pregnancy to dyes, pigments, mineral oils, coal products, or metals (Buckley et al. 1989).

Acknowledgement. This study was partially supported by a grant from the Russian Foundation for Fundamental Research N 93-04-21793 and by the National Cancer Institute contract N01-CO-74102.

References

Abenoza P, Manivel JC, Wick MR, Hagen K, Dehner LP (1987) Hepatoblastoma: an immunohistochemical and ultrastructural study. Hum Pathol 18:1025–1035

Buckley JD, Sather H, Ruccione K, Rogers PC, and others (1989) A case-control study of risk factors for hepatoblastoma. A report from the children's cancer study group. Cancer 64:1169–1176

Deringer MK (1970) Influence of the lethal yellow (A) gene on development of reticular neoplasms. J Natl Cancer Inst 45:1205–1210

Diwan BA, Ward JM, Rice JM (1989) Promotion of malignant "embryonal" liver tumors by phenobarbital: increased incidence and shortened latency of hepatoblastomas in (DBA/2

× C57BL/6)F1 mice initiated with N-nitrosodiethylamine. Carcinogenesis 10:1345–1348

Diwan BA, Ward JM, Rice JM (1992) Origin and pathology of hepatoblastoma in mice. In: Sirica AE (ed) The role of cell types in hepatocarcinogenesis. CRC, Boca Raton, pp 71–87

Engelhardt NV, Goussev AI, Shipova LJ, Abelev GI (1971) Immunofluorescent study of α-foetoprotein (alpha-fp) in liver tumours. I. Technique of (alpha-fp) localization in tissue sections. Int J Cancer 7:198–206

Engelhardt NV, Factor VM, Yazova AK, Poltoranina VS, Baranov VN, Lazareva MN (1990) Common antigens of mouse oval and biliary epithelial cells. Expression on newly formed hepatocytes. Differentiation 45:29–37

Engelhardt NV, Factor VM, Medvinsky AL, Baranov VN, Lazareva MN, Poltoranina VS (1993) Common antigen of oval and biliary epithelial cells (A6) is a differentiation marker of epithelial and erythroid cell lineages in early development of the mouse. Differentiation 55:19–26

Institute of Laboratory Animal Resources, National Academy of Sciences (1980) Histological typing of liver tumors of the rat. J Natl Cancer Inst 64:179–206

Ishak KG, Glunz PR (1967) Hepatoblastoma and hepatocarcinoma in infancy and childhood. Report of 47 cases. Cancer 20:396–422

Jones G, Butler YM (1975) Morphology of spontaneous and induced neoplasia. In: Butler WE (ed) Mice hepatic neoplasia. Elsevier, Amsterdam, pp 21–57

Kharkovskaya NA, Svinolupova SI, Khrustalev SA, Engelhardt NV, Kondalenko VF, Poltoranina VS, Turusov VS (1990) Transplantable mouse hepatoblastoma: histologic, ultrastructural and immunohistochemical study. Exp Pathol 40:283–289

Kuprina NI, Baranov VN, Yazova AK, Rudinskaya TD, Escribano M, Cordier J, Gleiberman AS, Goussev AI (1990) The antigen of bile canaliculi of the mouse hepatocyte: identification and ultrastructural localization. Histochemistry 94:179–186

Marceau N, Blouin MJ, Germain L, Noel M (1989) Role of different epithelial cell types in liver ontogenesis, regeneration and neoplasia. In Vitro Cell Dev Biol 25:336–341

Maronpot RR, Haseman JK, Boorman GA, Eustis SE, Rao GN, Huff JE (1987) Liver lesions in B6C3F1 mice: the National Toxicology Program, experience and position. Arch Toxicol [Suppl] 10:10–26

Nonoyama T, Reznik G, Bucci TJ, Fullerton F (1986) Hepatoblastoma with squamous differentiation in a B6C3F1 mouse. Vet Pathol 23:619–622

Nonoyama T, Fullerton F, Reznik G, Bucci TJ, Ward JM (1988) Mouse hepatoblastomas: a histologic, ultrastructural, and immunohistochemical study. Vet Pathol 25:286–296

Reuber MD (1967) Poorly differentiated cholangiocarcinomas occurring spontaneously in C3H and C3H Y hybrid mice. J Natl Cancer Inst 38:901–907

Rice JM (1973) The biological behavior of transplacentally induced tumors in mice. In: Davis W, Mohr U, Tomatis L (eds) Transplacental carcinogenesis. IARC, Lyon, pp 71–83 (IARC scientific publication no 4)

Schmidt D, Harms D, Lang W (1985) Primary malignant hepatic tumours in childhood. Virchows Arch [A] Pathol Anat Histopathol 407:387–405

Tsao MS, Grisham JW (1987) Hepatocarcinomas, cholangiocarcinomas, and hepatoblastomas produced by chemically transformed cultured rat liver epithelial cells. A light- and electronmicroscopic analysis. Am J Pathol 127:168–181

Turusov VS, Takayama S (1979) Tumors of the liver. In: Turusov VS (ed) Pathology of tumors in laboratory animals. IARC, Lyon, pp 193–232 (IARC scientific publication no 23)

Turusov VS, Day HE, Tomatis L, Gati E, Charles RT (1973a) Tumors in CF-1 mice exposed for six consecutive generations to DDT. J Natl Cancer Inst 51:983–997

Turusov VS, Deringer MK, Dunn TB, Stewart HL (1973b) Malignant mouse-liver tumors resembling human hepatoblastomas. J Natl Cancer Inst 51:1689–1695

Turusov VS, Kharkovskaya NA, Kondalenko VF, Poltoranina VS, Svinolupova SI, Khrustalev SA, Engelhardt NV (1991) Mouse hepatoblastoma: comparative aspects. Arkh Patol S3:38–43

Vlahakis G, Heston WE (1971) Spontaneous cholangiomas in strain C3H-AvyfB mice and in their hybrids. J Natl Cancer Inst 46:677–683

Ward JM, Rice JM, Creasia D, Lynch P, Riggs C (1983) Dissimilar patterns of promotion by di(2-ethylhexyl)phthalate and phenobarbital of hepatocellular neoplasia initiated by diethylnitrosamine in B6C3F1 mice. Carcinogenesis 4:1021–1029

Yamate J, Tajima M, Kudow S, Sannai S (1990) Background pathology in BDF1 mice allowed to live out their life-span. Lab Anim 24:332–340

Kupffer's Cell Sarcoma, Liver, Rat

James A. Popp

Synonym. Hepatic reticulum cell sarcoma

Gross Appearance

Kupffer's cell sarcoma usually consists of numerous small nodules (1–5 mm) randomly distributed throughout the liver (Chopra et al. 1979; Ford and Becker 1982). The lesions usually are irregular in shape and frequently have indistinct borders. On cut surface, the color is a homogeneous gray–white, although the tissue may be discolored by necrosis and hemorrhage in large lesions.

Microscopic Features

Neoplastic cells may form nodules or irregular sheets which arise in and infiltrate along the sinusoids throughout the lobule (Fig. 74). In the latter case, isolated hepatocytes or remnants of the lobular architecture may be evident (Chopra et al. 1979; Ford and Becker 1982). Individual neoplastic cells are usually round to oval with indistinct cytoplasmic borders. The cytoplasm is abundant, is frequently vacuolated, and may contain phagocytized necrotic debris or red blood cells. The nucleus is typically oval, pale, and often indented (Fig. 75). In some lesions, the cells are pleomorphic and assume a spindle shape with limited cytoplasm. Multinucleated giant cells have been observed but are not common.

Ultrastructure

Ultrastructural features are reminiscent of normal Kupffer's cells or macrophages (Chopra et al. 1979; Gillman and Hallowes 1972; Ford and Becker 1982). The cell membrane is frequently ruffled with short processes. Numerous vacuoles of phagocytized material and dense bodies are commonly present in the cytoplasm. Few mitochondria are observed, while the rough and smooth endoplasmic reticulum and Golgi complexes are well developed. A large number of free ribosomes are also noted.

Differential Diagnosis

Due to the rare occurrence of this neoplasm and therefore limited histologic descriptions in the literature, Kupffer's cell sarcoma is often difficult to diagnose. In a recent classification system of rat liver neoplasms (Institute of Laboratory Animal Resources, National Research Council 1980), Kupffer's cell sarcoma was not defined as a separate entity, but was considered synonymous with

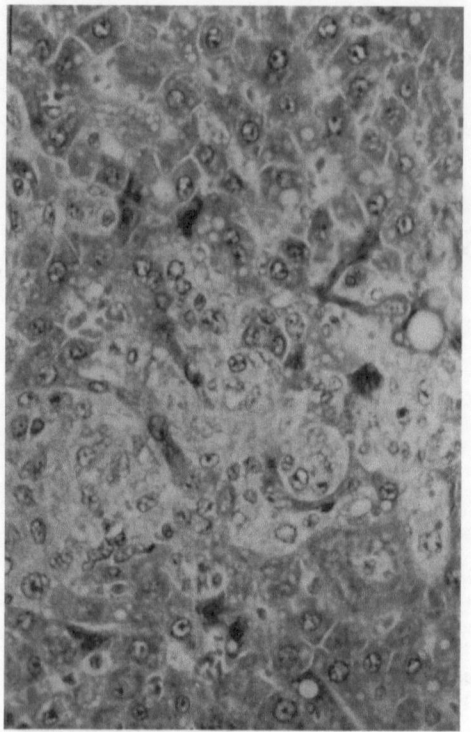

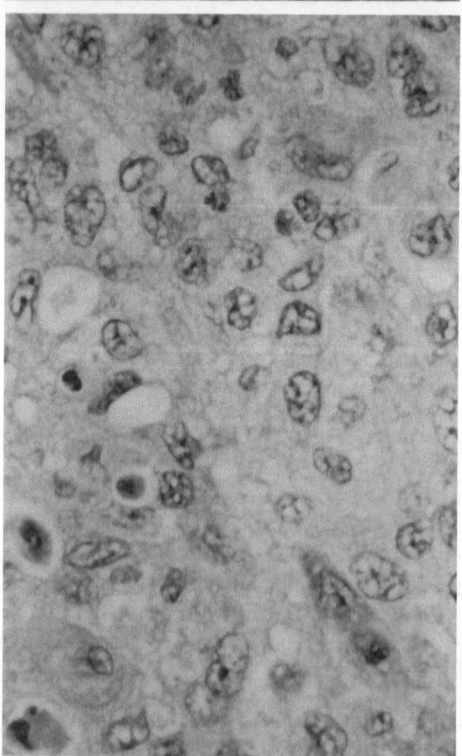

hemangiosarcoma. Later descriptions of Kupffer's cell sarcomas in the literature (Chopra et al. 1979; Ford and Becker 1982; Gillman and Hallowes 1972) have included not only light microscopy characteristics, but supporting ultrastructural, histochemical, and immunologic marker evaluation as well. The neoplastic cells had neither B nor T lymphocyte membrane markers, but did have demonstrable Fc receptors (Ford and Becker 1982). As demonstrated by histochemical staining, a large percentage of the cells had nonspecific esterase compatible with a Kupffer's cell origin (Gillman and Hallowes 1972; Ford and Becker 1982). Such information has clearly shown that at least the trypan blue-induced neoplasms are of macrophage and not endothelial origin. As additional information is obtained on the characteristics of normal Kupffer's cells (Wisse and Knook 1977; Knook and Wisse 1982), the use of histochemical and immunologic markers will become increasingly useful as adjuncts in distinguishing Kupffer's cell sarcomas.

Although recent reports (Chopra et al. 1979; Ford and Becker 1982) describe the trypan blue-induced neoplasms as Kupffer's cell sarcomas, other investigators have classified them as reticuloendothelial (Gillman and Hallowes 1972) or malignant lymphoreticular tumors (Gillman et al. 1973). Ford and Becker (1982) discuss the macrophage origin of the neoplasm and clearly distinguish it from lymphoma. Reticuloendothelial neoplasms have been described by Laqueur et al. (1963) in the livers of rats fed cycad nuts, although the cell of origin was not determined.

Ward et al. (1981) described a malignant fibrous histiocytoma which apparently arises in the liver in some cases and in extrahepatic sites in others. The cytologic appearance of individual cells in fibrous malignant histiocytoma is comparable to those of Kupffer cell sarcoma. However, as the name implies, the malignant fibrous histiocytoma usually has a collagen component not described in trypan blue-induced Kupffer cell sarcomas. The

◄

Fig. 74. (*above*) Kupffer's cell sarcoma, liver, Fischer-344 rat. The neoplasm has extended into the sinusoids of the adjacent liver. H&E, × 320

Fig. 75. (*below*) Higher magnification of Kupffer's cell sarcoma cells. Note the large, pale-staining, irregular nuclei and vacuolated cytoplasm. H&E. × 660

cell of origin of the malignant fibrous histiocytoma has not been determined.

Squire et al. (1981) described a large series of histiocytic sarcomas in Sprague-Dawley rats. The neoplasm is commonly found in the liver, but is also frequently found in many other sites. The lesion varies from a granulomatous to sarcomatous pattern, with the cytologic characteristics suggesting histiocytic origin. Stellate necrosis is often seen, a feature lacking in Kupffer's cell sarcoma. Although some members of a review panel have suggested a Kupffer's cell origin, identification of the actual cell of origin remains uncertain.

The lack of vascular spaces and larger cell size aid in distinguishing Kupffer's cell sarcoma from hemangiosarcomas when examined under the light microscope. Additional use of electron microscopy, histochemistry, and evaluation of immunologic markers should clarify those cases in which the cell of origin is in doubt.

Biologic Features

Little is known about the biologic features of the Kupffer's cell sarcoma. Emboli of neoplastic cells have been reported in vessels of the lung (Chopra et al. 1979). Trypan blue-induced neoplasms have been transplanted into the peritoneum of inbred rats (Ford and Becker 1982; Gillman and Hallowes 1972; Gillman et al. 1973). These transplanted tumors, which infiltrated the liver, spleen, and mesenteric lymph nodes, were morphologically similar to the primary neoplasms.

Comparison with Other Species

Although similar neoplasms have been described in the livers of mice, little attempt has been made to compare directly the lesion in rats and mice. Gillman et al. (1969, 1973) suggest that the trypan blue-induced kupffer's cell sarcoma is a model for

Hodgkin's disease of man. However, more detailed evaluation of the neoplasm (Ford and Becker 1982) suggests that the trypan blue-induced tumor may not be a good model for this disease. Recently, Wegener et al. (1983) reported Kupffer's cell sarcomas as occurring in the livers of rats given thorium dioxide (Thorotrast), although this neoplasm has not been described in humans receiving the drug.

References

Chopra P, Manga A, Nayak NC (1979) Kupffer cell sarcoma in rats after exposure to small doses of dimethylnitrosamine and N-2-acetylaminofluorene during hepatic regeneration. JNCI 62:1089–1095

Ford RJ, Becker FF (1982) The characterization of trypan blue-induced tumors in Wistar rats. Am J Pathol 106:326–331

Gillman T, Hallowes RC (1972) Ultrastructural and histochemical observations on a transplantable reticuloendothelial tumor in rats. Cancer Res 32:2383–2392

Gillman T, Kinns M, Cross RM (1969) Hodgkin's disease: a possible experimental model in rats. Lancet 2:1421–1422

Gillman T, Kinns AM, Hallowes RC, Lloyd JB (1973) Malignant lymphoreticular tumors induced by trypan blue and transplanted in inbred rats. JNCI 50:1179–1193

Institute of Laboratory Animal Resources, National Research Council (1980) Histologic typing of liver tumors of the rat. JNCI 64:179–206

Knook DL, Wisse E (eds) (1982) Sinusoidal liver cells. Elsevier Biomedical, New York

Laqueur GL, Mickelsen O, Whiting MG, Kurland LT (1963) Carcinogenic properties of nuts from Cycas circinalis L. indigenous to Guam. JNCI 31:919–951

Squire RA, Brinkhous KM, Peiper SC, Firminger HI, Mann RB, Strandberg JD (1981) Histiocytic sarcoma with a granuloma-like component occurring in a large colony of Sprague-Dawley rats. Am J Pathol 105:21–30

Ward JM, Kulwich BA, Reznik G, Berman JJ (1981) Malignant fibrous histiocytoma. An unusual neoplasm of soft-tissue origin in the rat that is different from the human counterpart. Arch Pathol Lab Med 105:313–316

Wegener K, Hasenohrl K, Wesch H (1983) Recent results of the German Thorotrast study-pathoanatomical changes in animal experiments and comparison to human thorotrastosis. Health Phys 44 [Supply 1]:307–316

Wisse E, Knook DL (1977) Kupffer cells and other liver sinusoidal cells. Elsevier Biomedical, New York

Spongiosis Hepatis and Spongiotic Pericytoma, Rat

Peter Bannasch and Heide Zerban

Synonym. Cystic degeneration

Gross Appearance

As a rule, the lesions of spongiosis are not visible with the naked eye. However, advanced lesions of this type may look like cysts at the macroscopic level.

Microscopic Features

In conventional hematoxylin and eosin (H&E)-stained tissue sections, the spongiotic lesions are characterized (Bannasch et al. 1981) by cystic multilocular formations filled with a finely granular or flocculent acidophilic material (Figs. 76, 77). Recent immunohistochemical and autoradiographic investigations (see "Histochemistry," p. 104) provided convincing evidence for a regular integration of the spongiotic formations into a larger proliferative lesion deriving from the perisinusoidal (Ito) cells, which is regarded as a benign neoplasm called spongiotic pericytoma (Ströbel et al. 1995). Because the complete neoplastic lesion can only be clearly identified after specific immunohistochemical demonstration of the constituent cells by antibodies to desmin, it appears to be appropriate to keep the designation spongiosis hepatis for the alterations visible in conventional histologic sections (Bannasch et al. 1981, 1985). The spongiotic formations often replace large areas of the liver parenchyma or, sometimes, considerable portions of neoplastic hepatic nodules and hepatocellular carcinomas. The surrounding non-neoplastic liver parenchyma may appear normal, but frequently it is composed of clear, acidophilic, basophilic, or mixed cell populations, as described elsewhere in this volume elsewhere (see "Foci of Altered Hepatocytes," p. 3). Neither the normal or putative preneoplastic parenchyma nor the neoplastic tissue appear to be compressed by the spongiosis.

Closer inspection of the spongiotic lesions reveals that the cavities of the multilocular formations are separated by narrow walls in which, at various places, small fibroblastic cells can be identified.

These septa are never lined by epithelia or endothelia. Sometimes, however, single hepatocytes (which have often undergone degenerative changes) are enclosed within the spongiotic lesions (Fig. 76a). If the lesions are localized close to the portal tracts, they often encompass bile ducts, arterioles, or capillaries (Fig. 76b). Single mononuclear cells are found rarely inside the cavities of the spongiotic lesions. A connection of the cavities to the blood sinuses is excluded by the observation that India ink, injected into the vena portae, never penetrates into the cavities of this lesion (Bannasch et al. 1981). Although erythrocytes can very occasionally be detected within some cavities in animals with disturbed blood circulation due to the presence of large hepatocellular or cholangiocellular tumors, this is probably a consequence of secondary rupture of walls of the spongiosis and bleeding into the cavities.

Histochemistry

Immunocytochemical approaches reveal that the vast majority of the cells composing spongiotic formations are positive for desmin (Fig. 78a) and vimentin, but negative for α-smooth muscle actin (Ströbel et al. 1995), confirming their previously inferred origin from the perisinusoidal (Ito) cells (see "Ultrastructure," p. 106). Moreover, the immunohistochemical demonstration of desmin and vimentin reveals that the spongiotic formations represent only readily visible components of much more expanded focal aggregates of perisinusoidal cells (Fig. 78b) which show a significantly increased incorporation of [3H]thymidine compared with perisinusoidal cells in extrafocal tissue and in the liver tissue of untreated controls (Ströbel et al. 1995). The septa of the spongiotic lesions are stained green by treatment with Masson-Goldner stain, are weakly periodic acid-Schiff (PAS) positive, and occasionally contain faint red spots after staining with van Gieson solution. The silver impregnation according to Gomori is positive in all segments of the septa (Fig. 79a). Immunocytochemical investigations have shown that the septa contain laminin,

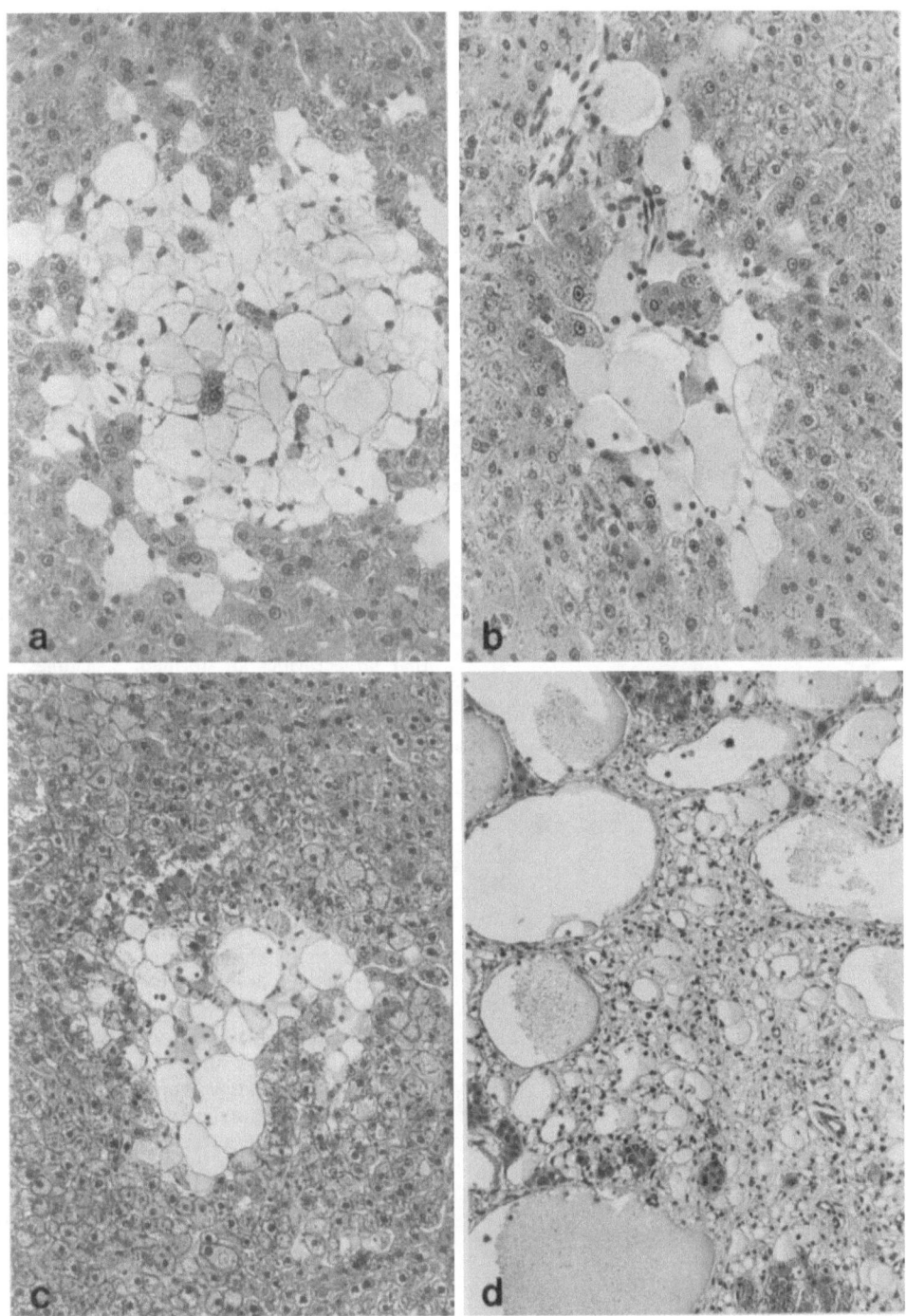

Fig. 76a–d. Spongiotic lesions induced in rat by *N*-nitrosomorpholine. **a** Spongiotic pericytoma surrounded by hepatic tissue of normal appearance. Note inclusion of single hepatocytes within the spongiosis. H&E, ×200. **b** Localization of spongiotic lesion in the neighborhood of a periportal tract. H&E, ×150. **c** Close association of spongiotic lesion with acidophilic cell focus of the liver parenchyma. H&E, ×150. **d** Spongiotic pericytoma with considerably increased cell proliferation. H&E, ×120

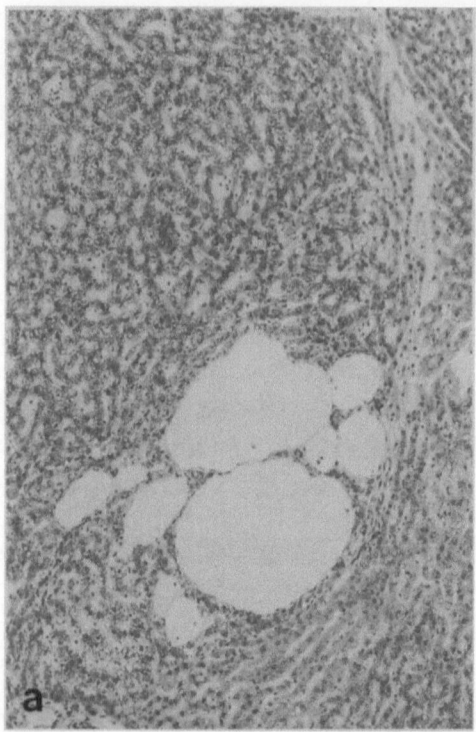

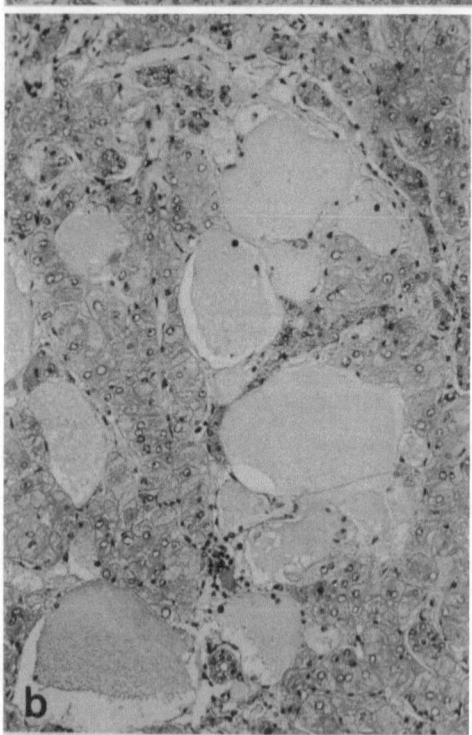

fibronectin, and collagen type III (Fig. 79b) and IV, but not collagen type I (Bannasch et al. 1985). The granular or flocculent acidophilic content of the cavities of the spongiotic lesions stains green with Masson-Goldner stain and is weakly positive with periodic acid-Schiff. Often the acidophilic material is alcianophilic and gives a positive iron-binding reaction, indicating that it contains proteoglycans (Bannasch et al. 1981). This is especially true for small spongiotic lesions which regularly are rich in proteoglycans. However, large lesions are often totally lacking in alcianophilia or iron-binding.

Ultrastructure

Under the electron microscope, the spongiotic lesions have a very characteristic appearance (Bannasch et al. 1981). The walls of the sponge are composed of cells which resemble fibroblasts and possess extremely elongated cytoplasmic processes (Figs. 80, 81). The processes of neighboring cells are in contact with each other in such a way that large cavities are formed. Sometimes, the plasma membranes of neighboring cells are connected by junctions. The luminal surface of the plasma membrane is usually covered by a basement membranous coat approximately 15μm thick. Bundles of collagenic fibers with a regular cross-striated pattern (periodicity, approximately 53 nm) are sometimes enclosed in this coat. The collagenous fibers are encountered especially at places where the cytoplasmic processes closely overlap. The cavities of the spongiosis contain a finely granular or fibrillar material and sometimes also lymphocytes or macrophages.

The fibroblastic cells forming the skeleton of the spongiotic lesions have small, irregular nuclei with much condensed chromatin and small nuclei (Fig. 80). The predominating cytoplasmic organelle is the granular endoplasmic reticulum. Coated vesicles are often found at the cell periphery, and a few mitochondria are irregularly distributed in the cytoplasm. Some cells contain many large, homogenous osmiophilic droplets, thus displaying the typical picture of fat-storing (Ito) cells. In ad-

◄

Fig. 77a,b. Spongiotic lesions induced in rat by *N*-nitrosomorpholine. Localization of spongiotic formations within **a** neoplastic hepatic nodule and **b** highly differentiated hepatocarcinoma. H&E, **a** ×100, **b** ×120

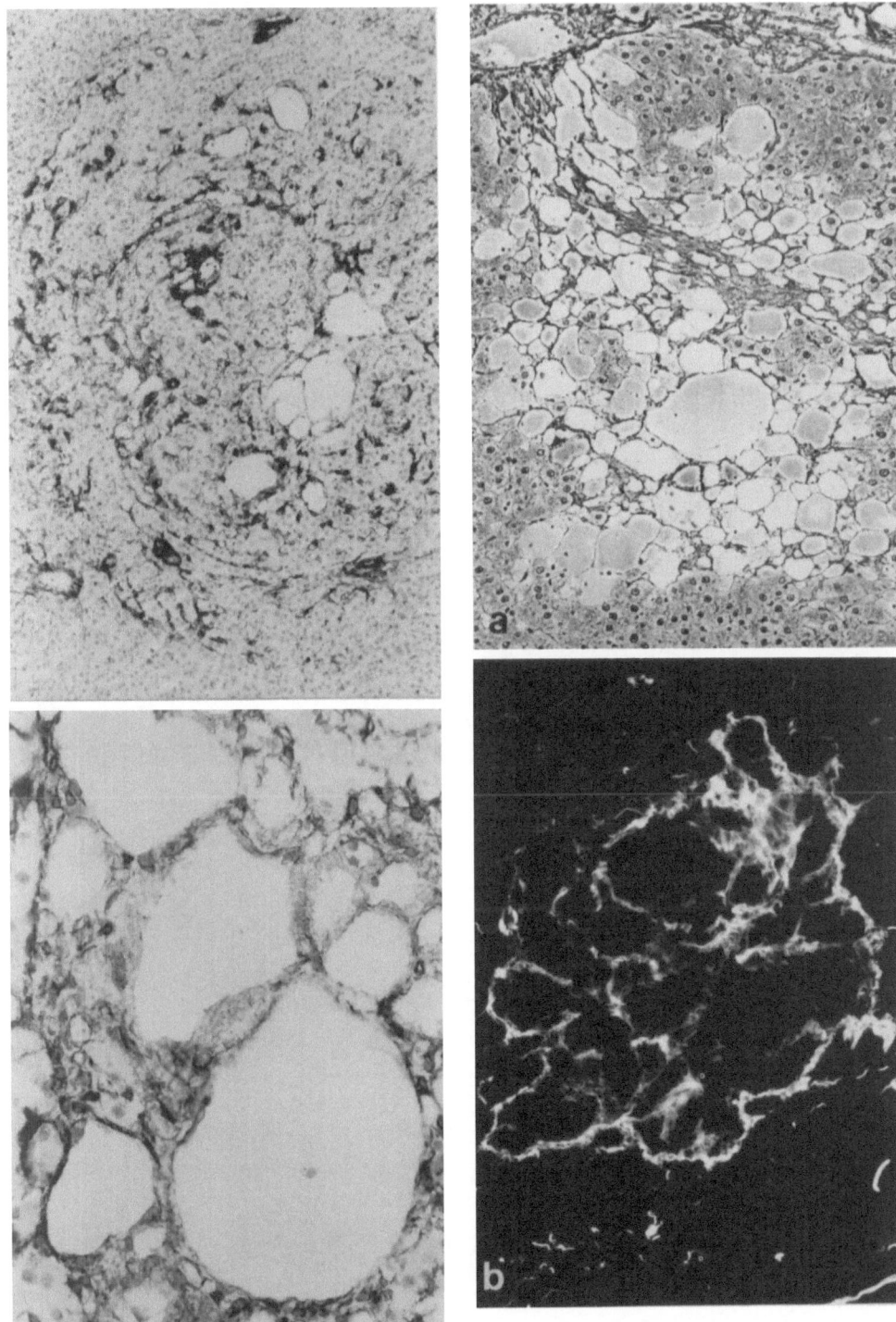

Fig. 78a,b. Spongiotic pericytomas composed of desmin-positive cells. **a** Pericytoma with small spongiotic component. ×75. **b** Pericytoma with pronounced spongiotic formations. ×250

Fig. 79a,b. Demonstration of components of the extracellular matrix of the connective tissue within the septa of spongiotic lesion induced in rat by *N*-nitrosomorpholine. **a** Prominent argyrophilia of the septa as demonstrated by silver impregnation according to Gomori. ×145. **b** Collagen type III within septa as demonstrated by immunofluorescence microscopy using rabbit antibodies to precollagen type III. ×200

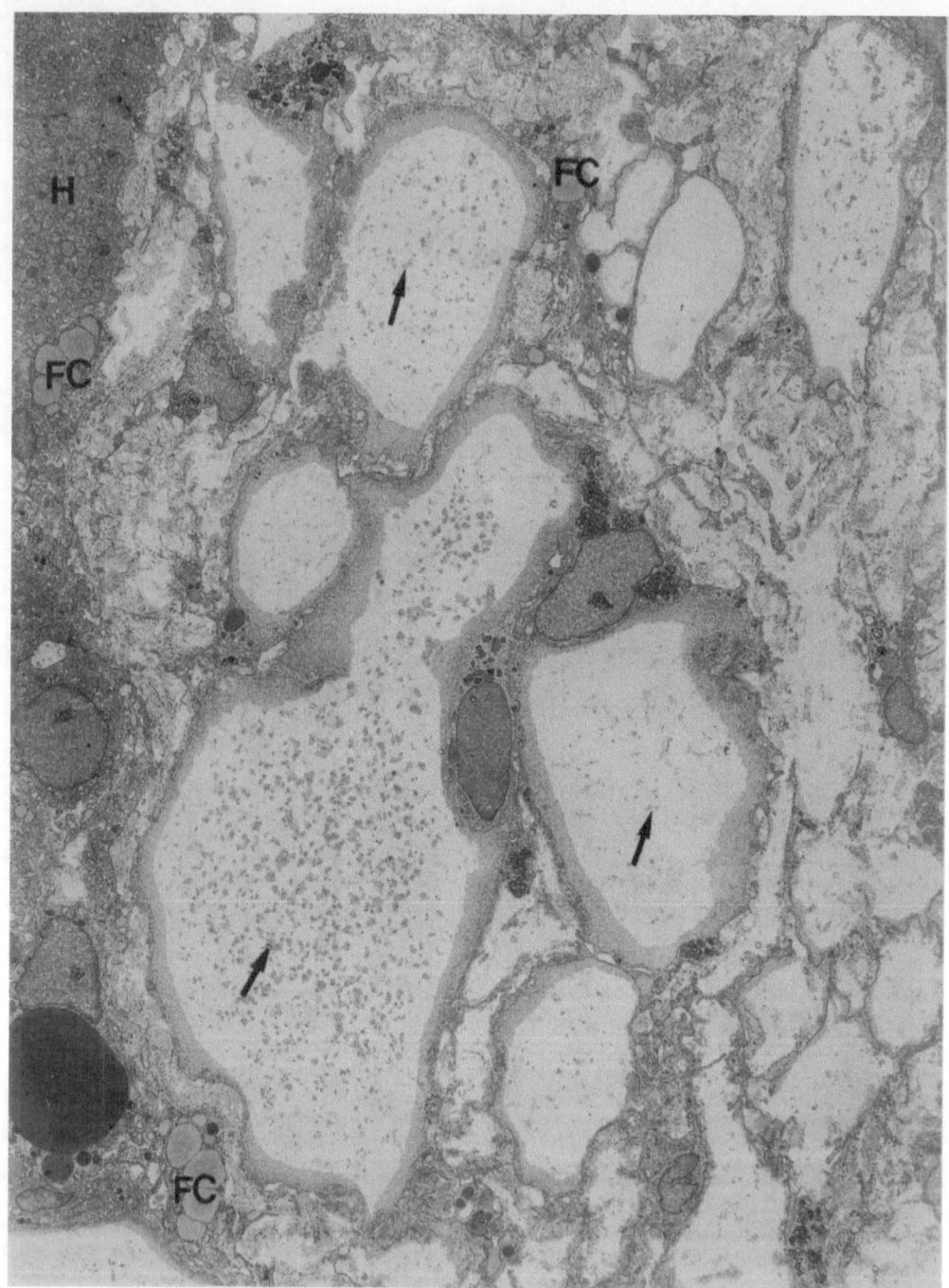

Fig. 80. Spongiotic lesions induced in rat by *N*-nitrosomorpholine. The walls of spongiotic formations are formed by fibroblastic cells and fat-storing cells (*FC*) exhibiting long cytoplasmic processes covered by a basement membranous material. Note the finely flocculent material (*arrows*) within the extracellular cavities. *H*, hepatocytes. TEM, ×2500

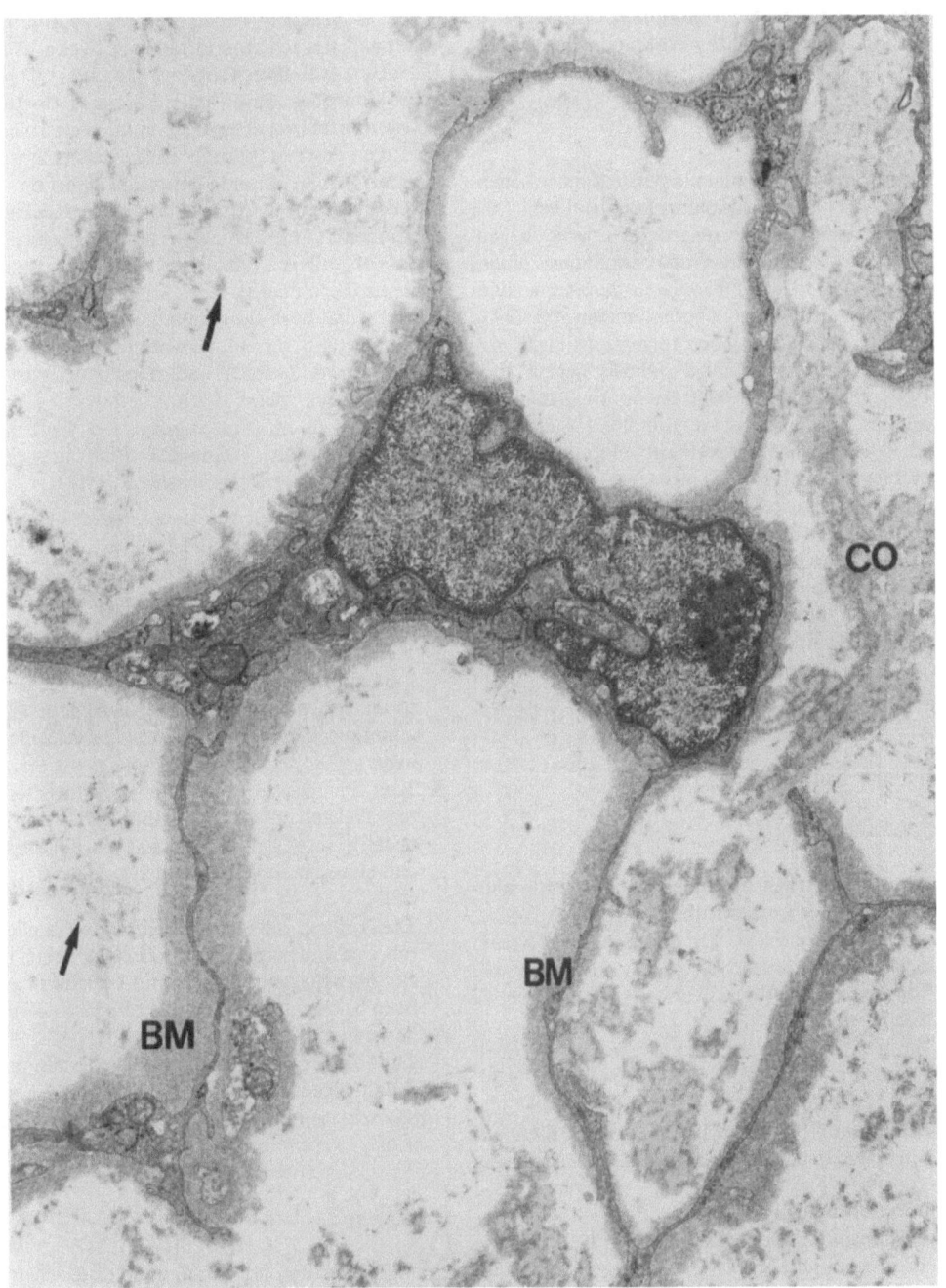

Fig. 81. Fibroblastic cell from spongiotic lesions. Note the long cytoplasmic processes coated with a basement membranous material (*BM*) and associated with collagen fibers (*CO*). Finely flocculent material (*arrows*) is apparent in extracellular cavities. TEM, ×13000

dition, polymorphic osmiophilic inclusions or vacuoles are occasionally observed.

Differential Diagnosis

Prior to the time spongiotic pericytoma was identified as a specific pathomorphological entity, the lesion was sometimes regarded as cystic "degeneration" within hepatocellular carcinomas (Jones and Butler 1978) or neoplastic hepatic nodules (Hirota and Williams 1979; Tatematsu et al. 1980). At first glance, spongiotic formations might also be mistaken for so-called peliosis hepatis (Lee 1983). However, in contrast to the latter, the cavities of spongiosis are not filled with blood, but with a finely flocculent material rich in proteoglycans. It is well known that proliferations of the perisinusoidal cells occur in the course of fibrogenesis (Geerts et al. 1991; Johnson et al. 1992; Ogawa et al. 1986; Ramadori et al. 1990). However, in this case, the perisinusoidal cells transform into myofibroblastic cells and regularly express α-smooth muscle actin, which is usually lacking in neoplastic spongiotic lesions (Ströbel et al. 1995).

Biologic Features

Natural History

From detailed light and electron microscopy studies in rat liver treated with N-nitrosomorpholine, it has been concluded that spongiotic pericytoma is a specific pathomorphological entity originating from the perisinusoidal liver cells (Bannasch et al. 1981).

The perisinusoidal liver cells were first described by Ito and Nemoto (1952) as "fat-storing cells" and analyzed electron microscopically by Yamagishi (1959). The observations of these authors have been confirmed and extended by many reports (Wake 1980). Various terms have been proposed to name the respective cell type, e.g., perisinusoidal liver cell (Wood 1963), lipocyte (Bronfenmajer et al. 1966), or Ito cell (Yamamoto 1975). The "pericytes" described by some authors were interpreted as "empty fat-storing cells" by Ito and Shibasaki (1968). The similarity between empty fat-storing cells and fibroblasts has been stressed repeatedly. The perisinusoidal cells express the intermediate filament protein desmin,

which is generally considered an appropriate marker of this cell type in rat liver (Yokoi et al. 1984; Burt et al. 1986; Tsutsumi et al. 1987). Three main functions have been ascribed to the perisinusoidal liver cells. One of these is a role in lipid metabolism, especially in the metabolism of vitamin A. A second important function of this cell type is the production of extracellular components of the connective tissue, in particular those of collagen, proteoglycans, and matrix proteins (Schäfer et al. 1987; Gressner 1991). Recently, it has been shown that the perisinusoidal cells may also secrete growth factors, such as tumor growth factor-β and hepatocyte growth factor (Michalopoulos 1992).

A number of findings support the view that spongiosis hepatis originates from the perisinusoidal liver cells (Bannasch et al. 1981):

1. In early stages of the development of spongiotic lesions, the altered mesenchymal cells are localized in the perisinusoidal space in contact with hepatocytes on the one side and close to endothelial cells at the other.
2. The vast majority of the cells composing spongiotic pericytoma inside and outside the spongiotic components are positive for desmin, a marker protein for perisinusoidal cells in rat liver.
3. The fine structure of the cells that form the skeleton of the spongiotic formations is usually very similar to that of fibroblasts and sometimes even that of typical fat-storing cells.
4. The cells of the spongiotic lesions are often closely associated with collagen fibers, including collagen types III and IV, which have been shown to be possible products of the perisinusoidal cells.
5. The cavities of the spongiotic lesions are usually filled with proteoglycans, which in the liver are mainly produced by the perisinusoidal cells.

Pathogenesis

Spongiotic pericytoma is rare in untreated control rats, but in old animals the lesion may appear "spontaneously." In rats treated with various hepatocarcinogens, such as N-nitrosomorpholine, dimethylnitrosamine, nitrosopyrrolidine, or 2-acetylaminofluorene, spongiotic lesions develop frequently (Bannasch et al. 1981; Zerban and

Bannasch 1983; Ito et al. 1984). This also holds true for continuous administration of low doses of hepatocarcinogens, while carcinogens with another organotropism, such as the bladder carcinogen butyl-butanolnitrosamine, do not induce spongiosis under the same experimental conditions (Zerban and Bannasch 1983). Ito et al. (1984) have shown that 4,4'-diaminodiphenylmethane considerably enhances the induction of spongiosis hepatis when applied subsequently to N-nitrosomorpholine or 2-acetylaminofluorene. In all events, spongiosis hepatis developed only after very long latent periods. For example, when N-nitrosomorpholine was administered for 7 weeks at a concentration of 12mg per 100ml drinking water, the first spongiotic lesion was observed 33 weeks after cessation of the carcinogen treatment. Later on, the incidence of spongiotic lesions steadily increased until nearly all animals were affected. The increased incidence of spongiosis hepatis was frequently accompanied by an increase in size of the single spongiotic areas. Ströbel et al. (1995) have recently shown that the spongiotic formations are an integral part of larger proliferative lesions which show a slowly progressive behavior independent of an ongoing carcinogenic stimulus and should, hence, be regarded as benign neoplasms.

From the histochemical findings mentioned above, it is evident that the pathogenesis of spongiotic pericytoma is closely associated with an excessive extracellular accumulation of proteoglycans and/or proteins (Bannasch et al. 1981; Ströbel et al. 1995). These alterations of the extracellular compartment might be responsible for the development of the large cavities characteristic of the spongiotic formations. The reason for the excess of proteoglycans and/or proteins might be an overproduction or an impaired degradation of these substances. Interestingly enough, various carcinogens have been shown to induce an intracellular accumulation of polysaccharides, lipids, or proteins, in particular in epithelial cells (Bannasch 1984). These storage phenomena are most probably due to a disturbance of intracellular degradation of the respective macromolecules. One might speculate, therefore, that carcinogens may not only impair degradation of polysaccharides and proteins in the intracellular, but also in the extracellular compartment.

It is reasonable to assume that spongiotic pericytoma is produced by a direct effect of the carcinogen or its metabolites on the perisinusoidal cells. In addition to the significant increase in cell proliferation mentioned above, a much more pronounced cell proliferation may appear in spongiotic lesions after very long lag periods (Bannasch and Zerban 1986; Ströbel et al. 1995). Some findings suggested that spongiosis may progress to a malignant mesenchymal tumor, for which we recently proposed the term "perisinusoidal (Ito) cell sarcoma" (Ströbel et al. 1995). With respect to the pathogenesis of this tumor type, it is of particular interest that our preliminary results suggest that the accumulation of proteoglycans within the spongiotic formations seems to disappear during transformation of spongiosis into pericytoma.

Comparison with Other Species

In recent years, spongiosis hepatis has been repeatedly described in fish exposed to N-nitrosodiethylnitrosamine (Couch and Courtney 1987; Bunton 1990; Couch 1991), to methylazoxymethanolacetate (Hinton et al. 1984), or to the polluted water of the Puget Sound, Washington (Myers et al. 1987). Whether the cystic alterations in fish liver occurring relatively early during the cytotoxic phase of exposure to N-nitrosodiethylamine should also be designated as spongiosis hepatis, is debatable, since they might merely represent parenchymal loss due to toxic hepatocellular necrosis (Laurén et al. 1990; Braunbeck et al. 1992). There is little doubt, however, that the lesions developing in N-nitrosodiethylamine-treated fish after longer latent periods are actually similar to spongiosis hepatis in the rat (Couch and Courtney 1987; Couch 1991). The possible benign neoplastic nature and late progression of these lesions to malignant neoplasms has been discussed in detail (Couch 1991). In line with the interpretation of spongiotic pericytoma in the rat, Couch (1991) considered spongiosis hepatis in fish a histopathologic indicator of exposure to hepatotoxic chemicals. Only in one report of human hepatic adenomas that appeared in users of oral contraceptives has a picture been published with features resembling spongiosis changes within an adenoma (Nime et al. 1979).

References

Ballardini G, Fallani M, Biagini G, Bianchi FB, Pisi E (1988) Desmin and actin in the identification of Ito cells and in monitoring their evolution to myofibroblasts in experimental liver fibrosis. Virchows Arch B Cell Pathol 56:45–49

Ballardini G, Groff P, De Giorgi LB, Schuppan D, Bianchi FB (1994) Ito cell heterogeneity: desmin-negative Ito cells in normal rat liver. Hepatology 19:440–446

Bannasch P (1984) Sequential cellular changes during chemical carcinogenesis. J Cancer Res Clin Oncol 108:11–22

Bannasch P, Zerban H (1986) Pathogenesis of primary liver tumors induced by chemicals. Recent Results Cancer Res 100:1–15

Bannasch P, Bloch M, Zerban H (1981) Spongiosis hepatis. Specific changes of the perisinusoidal liver cells induced in rats by N-nitrosomorpholine. Lab Invest 44:252–264

Bannasch P, Zerban H, Fügel HJ (1985) Spongiosis hepatis, rat. In: Jones TC, Mohr U, Hunt RD (eds) Monographs on pathology of laboratory animals, digestive system. Springer, Berlin Heidelberg New York, pp 116–123

Braunbeck TA, Teh SJ, Lester SM, Hinton DE (1992) Ultrastructural alterations in liver of medaka (Oryzias latipes) exposed to diethylnitrosamine. Toxicol Pathol 20:179–196

Bronfenmajer S, Schaffner F, Popper H (1966) Fat-storing cells (lipocytes) in human liver. Arch Pathol 82:447–453

Bunton TE (1990) Hepatopathology of diethylnitrosamine in the medaka (Oryzias latipes) following short-term exposure. Toxicol Pathol 18:313–323

Burt AD, Robertson JL, Heir J, MacSween RNM (1986) Desmin-containing stellate cells in rat liver: distribution in normal animals and response to experimental acute liver injury. J Pathol 150:29–35

Couch JA (1991) Spongiosis hepatis: chemical induction, pathogenesis and possible neoplastic fate in a teleost fish model. Toxicol Pathol 19:237–250

Couch JA, Courtney LA (1987) N-nitrosodiethylamine-induced hepatocarcinogenesis in estuarine sheepshead minnow (Cyprinodon variegatus): neoplasms and related lesions compared with mammalian lesions. J Natl Cancer Inst 79:297–321

Geerts A, Lazou J-M, De Bleser P, Wisse E (1991) Tissue distribution, quantitation and proliferation kinetics of fat storing cells in carbon tetrachloride-injured rat liver. Hepatology 13:1193–1202

Gressner AM (1991) Proliferation and transformation of cultured liver fat-storing cells (perisinusoidal lipocytes) under conditions of β-D-xyloside abrogation of proteoglyan synthesis. Exp Mol Pathol 55:143–169

Hinton DE, Lantz RC, Hampton JA (1984) Effect of age and exposure to a carcinogen on the structure of the Medaka liver: a morphometric study. Natl Cancer Inst Monogr 65:239–249

Hirota N, Williams GM (1979) Persistence and growth of rat liver neoplastic nodules following cessation of carcinogen exposure. J Natl Cancer Inst 63:1257–1265

Ito T, Nemoto M (1952) Über die Kupfferschen Stemzellen und die "Fettspeicherzellen" ("fat storing cells") in der Blutkapillarenwand der menschlichen Leber. Okajimas Folia Anat Jpn 24:243–258

Ito T, Shibasaki S (1968) Electron microscopic study on the hepatic sinusoidal wall and the fat-storing cells in the normal human liver. Arch Histol Jpn 29:137–192

Ito N, Moore MA, Bannasch P (1984) Modification of the development of N-nitrosomorpholine-induced hepatic lesions by 2-acetylaminofluorene, phenobarbital and 4,4'-diaminodiphenylmethane: a sequential histological and histochemical analysis. Carcinogenesis 5:335–342

Johnson SJ, Hines JE, Burt AD (1992) Immunolocalization of proliferating perisinusoidal cells in rat liver. Histochem J 24:67–72

Jones G, Butler WM (1978) Light microscopy of rat hepatic neoplasia. In: Newberne PM, Butler WH (eds) Rat hepatic neoplasia. MIT Press, Cambridge, pp 114–138

Laurén DJ, Teh SJ, Hinton DE (1990) Cytotoxicity phase of diethylnitrosamine-induced hepatic neoplasia in medaka. Cancer Res 50:5504–5514

Lee KP (1983) Peliosis hepatis-like lesion in aging rats. Vet Pathol 20:410–423

Michalopoulos G (1992) Liver regeneration and growth factors: old puzzles and new perspectives. Lab Invest 67:413–415

Myers MS, Rhodes LD, McCain BB (1987) Pathologic anatomy and patterns of occurrence of hepatic neoplasms, putative preneoplastic lesions, and other idiopathic hepatic conditions in English sole (Parophrys vetulus) from Puget Sound, Washington. J Natl Cancer Inst 78:333–363

Nime F, Pickren JW, Vana J, Aronoff BL, Baker HW, Murphy GP (1979) The histology of liver tumors in oral contraceptive users observed during a national survey by the American College of Surgeons Commission on Cancer. Cancer 44:1481–1489

Ogawa K, Suzuki J-I, Mukai H, Mori M (1986) Sequential changes of extracellular matrix and proliferation of Ito cells with enhanced expression of desmin and actin in focal hepatic injury. Am J Pathol 125:611–619

Ramadori G, Veit T, Schwögler S, Dienes HP, Rieder H, Meyer zum Büschenfelde K-H (1990) Expression of the gene of the α-smooth muscle-actin isoform in rat liver and in rat fat-storing (Ito) cells. Virchows Arch [B] Cell Pathol 59:349–357

Schäfer S, Zerbe O, Gressner AM (1987) The synthesis of proteoglycans in fat storing cells of rat liver. Hepatology 7:680–687

Ströbel P, Mayer F, Zerban H, Bannasch P (1995) Spongiotic pericytoma: a benign neoplasm deriving from the perisinusoidal (Ito) cells in rat liver. Am J Pathol 146:903–913

Tatematsu M, Takano T, Hasegawa R, Imaida K, Nakanowatari J, Ito N (1980) A sequential qualitative study of the reversibility of rat liver hyperplastic nodules in rats exposed to hepatocarcinogens. Gann 71:843–855

Tsutsumi M, Takada A, Takase S (1987) Characterization of desmin-positive rat liver sinusoidal cells. Hepatology 7:277–284

Wake K (1980) Perisinusoidal stellate cells (fat-storing cells, interstitial cells, lipocytes), their related structure in and around the liver sinusoids, and vitamin A storing cells in extrahepatic organs. Int Rev Cytol 66:303–353

Wood RL (1963) Evidence of species differences in the ultrastructure of the hepatic sinusoid. Z Zellforsch Miktosk Anat 58:679–692

Yamagishi M (1959) Electron microscope studies in the fine structure of the sinusoidal wall and fat-storing cells of rabbit liver. Arch Histol Jpn 18:223–261

Yamamoto M (1975) Ultrastructure and function of Ito cell (fat-storing cell) in the liver. Med J Hiroshima Univ 23:245–274

Yokoi Y, Namihisa T, Kuroda H, Komatsu I, Miyazaki A, Watanabe S, Usui K (1984) Immunocytochemical detection

of desmin in fat-storing cells (Ito cells). Hepatology 4:709–714

Zerban H, Bannasch P (1983) Spongiosis hepatis in rats treated with low doses of hepatotropic nitrosamines. Cancer Lett 19:247–252

Focal Carcinoma in Hepatocellular Adenoma, Liver, Mouse

Jerrold M. Ward

Synonyms. Focal carcinoma: hepatocellular carcinoma, focal atypia or dysplasia in hepatocellular adenoma, nodule in nodule, early hepatocellular carcinoma in hepatocellular adenoma; hepatocellular adenoma: focal nodular hyperplasia, nodular hyperplasia; hyperplastic nodule, type A, type 1, or type 2 nodule

Gross Appearance

Hepatocellular adenomas (Figs. 82, 83) in the mouse are seen as sharply demarcated, roughly spherical, nonencapsulated masses that may reach a diameter of 1.5 cm. They usually elevate the hepatic capsule and thus project from the surface of the liver. Some of these masses may have a yellowish tint due to the presence of fat in the cells (Fig. 82); others are somewhat less dark than the adjacent liver (due to the presence of basophilic and clear cells (Fig. 83), and still others may closely resemble the color and pattern of the liver (Fig. 84). The *hepatocellular carcinomas* occasionally found within an adenoma are not usually recognized grossly, although some may be seen at low magnification (Fig. 84). These carcinomas are more apt to be found in adenomas which exceed 1 cm in diameter.

Microscopic Features

Hepatocellular adenomas of mice have a uniform morphological appearance (Frith and Ward 1980). They are made up of cells which resemble normal hepatocytes in some cases; in others, lipid droplets in the cytoplasm expand many of the cells; and in further cases, the cytoplasm of the cells may contain clear areas (glycogen) or may be darker (basophilic) than usual hepatocytes. The trabecular or columnar pattern of plates of hepatic cells is usually maintained in the adenoma; arteries and veins may be seen, but hepatic triads are usually not present. Any of the various histologic types of hepatocellular adenoma may apparently give rise to carcinoma, but more appear to arise in adenomas made up of cells with basophilic cytoplasm (see p. 53, this volume).

Adenomas in which hepatocellular carcinomas are seen initially are usually greater than 1.0 cm in diameter (Andervont and Dunn 1952; Becker 1982; Goldfarb et al. 1980; Frith and Dooley 1976; Frith and Ward 1980; Lipsky et al. 1981; Ward 1980, 1984; Ward and Vlahakis 1978; Ward et al. 1983).

Malignant cells appear in small foci within the adenoma (Fig. 84; Jang et al. 1992). They are usually larger than the cells of the adenoma, are more pleomorphic, have a greater nucleus to cytoplasm ratio, form trabecular plates several cells thick, and usually contain more mitotic figures than the adjacent adenoma. The carcinoma cells may compress or infiltrate into the surrounding adenoma (Fig. 85). The foci usually enlarge and eventually become a major portion of the tumor mass.

Histochemically, carcinoma cells may have a different enzyme profile than cells of adenomas and, thus, the malignant foci may appear enzymatically from the adjacent adenoma cells (Essigmann and Newberne 1981; Ruebner et al. 1982). By using

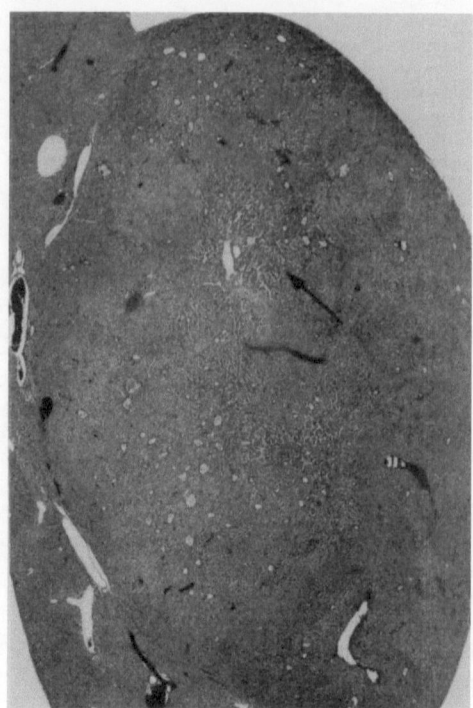

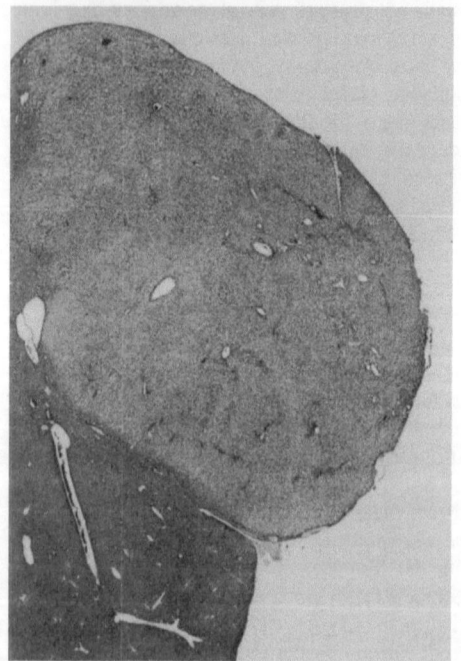

Fig. 82. (*upper left*) Hepatocellular adenoma, liver, mouse. Yellowish color and distinct texture are due to lipid in tumor cells. H&E, ×20

Fig. 83. (*below*) Hepatocellular adenoma, liver, mouse. Color and texture are due to presence of clear and basophilic cells in the adenoma. H&E, ×20

Fig. 84. (*upper right*) Hepatocellular carcinoma (*arrow*) within an adenoma, liver, mouse. Note the color of the adenoma is similar to that of normal liver. H&E, ×11.5

the avidin-biotin-peroxidase complex (ABC) immunocytochemical technique, we have found α-fetoprotein (AFP) in the cells in these foci (Fig. 86) much more frequently than in cells of the adenoma. Although AFP can be found in hepatocyte foci and in adenomas of mice (Koen et al. 1983), the majority of cells in these lesions, in our experience, do not contain AFP.

Ultrastructure

Insofar as we are aware, electron microscopy features of focal carcinomas in adenomas have not been reported.

Differential Diagnosis

The classification of a tumor is usually based on its most malignant morphological area. Thus a diagnosis of hepatocellular carcinoma is warranted when focal carcinoma is found within a hepatocellular adenoma. The diagnostic pathologist should, however, be confident that the lesion is morphologically malignant despite lack of evidence of invasion. One possible interpretation might be that another carcinoma which has arisen nearby may have invaded into the adenoma, and in a cross-section of the adenoma the invasive carcinoma may appear to have arisen from cells of the adenoma. This is a possibility to consider, but could be resolved by studying serial sections of these lesions.

Biologic Features

Spontaneous and induced hepatocellular neoplasms of mice may arise as focal proliferative lesions (Frith and Ward 1980; Koen et al. 1983; Lipsky et al. 1981; Vessclinovitch and Mihailovich 1983; Ward 1984) or as clonal proliferations (Rabes et al. 1982; Williams et al. 1983). These early lesions, considered histologically to be "hepatocellular foci," are believed to progress to

▶

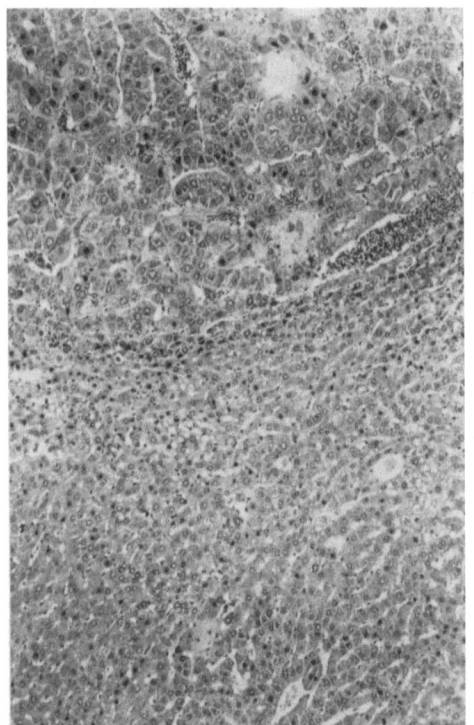

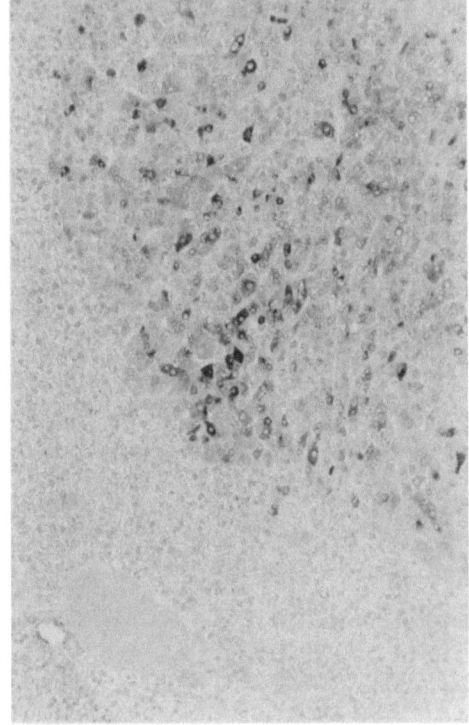

Fig. 85. (*above*) Focus of trabecular carcinoma (*top*) in a large hepatocellular adenoma. H&E, ×80

Fig. 86. (*below*) α-Fetoprotein within cells of a focal carcinoma (*top*) in a hepatocellular adenoma. Avidin-biotin-peroxidase complex (ABC) immunoperoxidase technique and hematoxylin, ×130

adenomas by proliferative growth (Vesselinovitch and Mihailovich 1983).

Hepatocellular carcinomas seen within the large adenomas are made up of atypical or malignant cells which appear morphologically and histochemically distinct (Essigmann and Newberne 1981; Ruebner et al. 1982). They seem to outgrow the benign cells and eventually replace the benign nodule with malignant cells, which may eventually metastasize to lung and other tissues. In mouse liver, small carcinomas that are not within adenomas are almost never seen as spontaneous lesions. These findings suggest that the majority of spontaneous and probably induced carcinomas arise within adenomas. Further research is needed, however, to clarify the histogenesis of hepatocellular carcinoma.

Comparison with Other Species

A similar sequential development of hepatocellular carcinoma within benign liver nodules has been reported in rats (Moore et al. 1982; Williams 1980). The progression and transformation of liver adenomas has been less well studied in other species. Limited evidence exists for such transformation in human benign liver nodules (Stromeyer and Ishak 1981).

References

Andervont HB, Dunn TB (1952) Transplantation of spontaneous and induced hepatomas in inbred mice. JNCI 13:455–503

Becker FF (1982) Morphological classification of mouse liver tumors based on biological characteristics. Cancer Res 42:3918–3923

Essigmann EM, Newberne PM (1981) Enzymatic alterations in mouse hepatic nodules induced by a chlorinated hydrocarbon pesticide. Cancer Res 41:2823–2831

Goldfarb S, Pugh TD, Cripps DJ (1980) Increased alkaline phosphatase activity – a positive histochemical marker for griseofulvin-induced mouse hepatocellular nodules. JNCI 64:1427–1433

Frith CH, Dooley K (1976) Hepatic cytologic and neoplastic changes in mice given benzidine dihydrocholoride. JNCI 56:679–682

Frith CH, Ward JM (1980) A morphologic classification of proliferative and neoplastic hepatic lesions in mice. J Environ Pathol Toxicol 3:329–351

Jang JJ, Weghorst CM, Henneman JR, Devor DE, Ward JM (1992) Progressive atypia in spontaneous and N-nitrosodiethylamine induced hepatocellular adenomas of C3H/HeNG mice. Carcinogenesis 13:1541–1547

Koen H, Pugh TD, Nychka D, Goldfarb S (1983) Presence of alphafetoprotein-positive cells in hepatocellular foci and microcarcinomas induced by single injections of diethylnitrosamine in infant mice. Cancer Res 43:702–708

Lipsky MM, Hinton DE, Klaunig JE, Trump BF (1981) Biology of hepatocellular neoplasia in the mouse. I. Histogenesis of safrole-induced hepatocellular carcinoma. JNCI 67:365–376

Moore MA, Mayer D, Bannasch P (1982) The dose dependence and sequential appearance of putative preneoplastic populations induced in the rat liver by stop experiments with N-nitrosomorpholine. Carcinogenesis 3:1429–1436

Rabes HM, Bucher T, Hartmann A, Linke I, Dunnwald M (1982) Clonal growth of carcinogen-induced enzyme-deficient preneoplastic cell populations in mouse liver. Cancer Res 42:3220–3227

Ruebner BH, Gershwin ME, Meierhenry EF, Dunn P (1982) Enzyme histochemical characteristics of spontaneous and induced hepatocellular neoplasms in mice. Carcinogenesis 3:899–903

Stromeyer FW, Ishak KG (1981) Nodular transformation (nodular "regenerative" hyperplasia) of the liver. A clinicopathologic study of 30 cases. Hum Pathol 12:60–71

Vesselinovitch SD, Mihailovich N (1983) Kinetics of diethylnitrosamine heptocarcinogenesis in the infant mouse. Cancer Res 43:4253–4259

Ward JM (1980) Morphology of hepatocellular neoplasms in B6C3F1 mice. Cancer Lett 9:319–325

Ward JM (1984) Morphology of potential preneoplastic hepatocyte lesions and liver tumors in mice and comparison with other species. In: Popp JA (ed) Current perspectives in mouse liver neoplasia. Hemisphere, Washington DC, pp 1–26

Ward JM, Vlahakis G (1978) Evaluation of hepatocellular neoplasms in mice. JNCI 61:807–811

Ward JM, Rice JM, Creasia D, Lynch P, Riggs C (1983) Dissimilar patterns of promotion by di(2-ethylhexyl)phthalate and phenobarbital of hepatocellular neoplasia initiated by diethylnitrosamine in B6C3F1 mice. Carcinogenesis 4:1021–1029

Williams ED, Wareham KA, Howell S (1983) Direct evidence for the single cell origin of mouse liver cell tumours. Br J Cancer 47:723–726

Williams GM (1980) The pathogenesis of rat liver cancer caused by chemical carcinogens. Biochem Biophys Acta 605:167–189

Hyperplasia, Adenoma, Gallbladder, Hamster

Yoichi Konishi

Synonyms. Hyperplasia: adenomatous hyperplasia; adenoma: papillary adenoma; papilloma; adenomatous polyp; papillary polyp

Gross Appearance

In hamsters experimentally treated with *N*-nitrosobis(2-hydroxypropyl)amine (DHPN) followed by a diet containing 0.5% deoxycholic acid (DCA), the gallbladder is enlarged and its wall thickened. The serosa is smooth and whitish-yellow. The cut surface shows multiple, whitish-gray, soft projections from the mucosa into the lumen (Fig. 87).

Microscopic Features

The gallbladder epithelium of nontreated hamsters consists of a single layer of columnar cells, in contrast to that of hamsters treated with DHPN, which exhibits hyperplasias and adenomas. The hyperplasias are focal (Fig. 88) or diffuse (Fig. 89) thickenings of the mucosa, consisting of villous and spongioid areas (arrows, Fig. 89). The villous area contains long, irregular, and ramifying mucosal folds. In the spongioid area, these long, ramifying folds coalesce to form a reticular structure. Spindle-shaped nuclei predominate in both areas; goblet cells and mitotic figures are scanty. In adenomas, papillary structures predominate; only a few are nonpapillary (Fig. 90). Most adenomas are pedunculated. The tumors have a branching, tree-like configuration with thin, vascular connective tissue stalks covered by a single layer of cuboidal or columnar epithelium. Inflammatory cell infiltration is occasionally found in the interstitial space. Cystic dilatation of the glands is also occasionally seen. The polarity of epithelial cells is preserved with infrequent mitotic figures.

Differential Diagnosis

By definition, hyperplasias and adenomatous tissues always grow into the lumen. The intramural and subserosal lesions, known as adenomyomatous hyperplasia, that are associated with irregular masses of smooth muscle (Edmondson 1967) are not observed in adenomas. Adenomas have also been called papillomas or papillary polyps (Pour et al. 1975) and have been visualized as low, flat elevations on the mucosal surface (Halpert 1977). Based on the lesion's histology, use of the term "papilloma" as synonymous for "adenoma" is confusing, since adenomas contain both papillary and nonpapillary structures. Because the tumors seen in our experiment in hamsters consist predominantly of papillary structures with only a few nonpapillary structures, we feel that "papillary adenoma" is the most appropriate term for the lesion. Papillary adenomas are relatively easy to distinguish from cholesterol polyps, which are yellow, soft, usually pedunculated, and composed of aggregates of foam cells covered by a single layer of columnar epithelium (Christensen and Ishak 1970). Inflammatory polyps have a glandular epithelial proliferation associated with vascular connective tissue stroma and intense chronic inflammatory cell infiltration. In carcinomas, there is loss of polarity of epithelial cells, predominant nuclear irregularity, frequent mitotic figures, and invasion of surrounding tissues.

Biologic Features

Although spontaneous gallbladder tumors in Syrian golden hamsters are extremely rare (Pour et al. 1976, 1979), neoplasms can be induced by repeated subcutaneous injections of DHPN (Pour et al. 1975). We injected hamsters with DHPN at a dose of 500 mg/kg body weight subcutaneously once a week for 5 weeks, followed by a basal diet alone or diets containing 0.05% phenobarbital (PB), 0.5% DCA, or 0.5% lithocholic acid (LCA) for 30 weeks. The results are shown in Table 11. Hyperplasia was seen in 50% of the hamsters receiving the LCA diet only; however, this increased to 89% in hamsters receiving DHPN followed by the LCA diet. The incidence of adenomas in hamsters receiving DHPN followed by DCA or LCA was higher than that of hamsters receiving DHPN followed by a basal diet. The various factors associated with the development of gallbladder

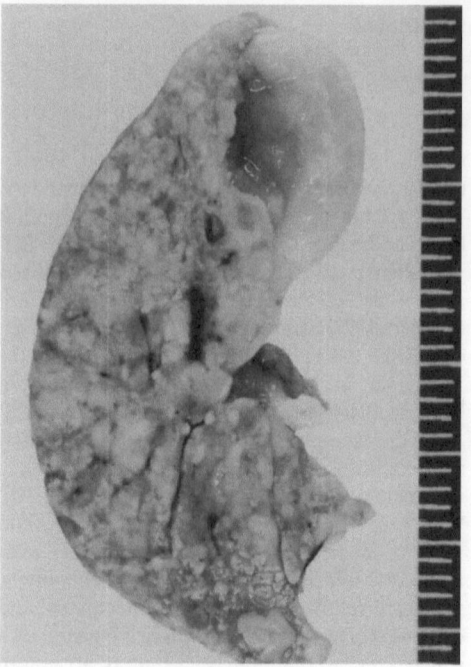

Fig. 87. Hyperplasia and adenoma, gallbladder of a hamster treated with *N*-nitrosobis(2-hydroxypropyl)amine (DHPN) followed by a diet containing 0.5% deoxycholic acid (DCA). Multiple irregular projections from the mucosa into the lumen can be seen. Nodular lesions are numerous in the liver

tumors have been reported (Edmondson 1967). Cholelithiasis, cholecystitis, and bile itself may play some role in tumor development. On the other hand, the carcinogenic acticity of DHPN in the pancreas of hamsters reportedly takes place via the blood stream (Rückert et al. 1981). The present results suggest that secondary bile acids may promote DHPN carcinogenesis in the gallbladder. In this context, further studies are required of DHPN metabolism in relation to bile acid, DCA, and LCA. The fact that no carcinomas were found in the present experiment may only reflect the length of the experimental period. The development of carcinoma from papillary adenoma has not yet been observed in hamsters. Tumors of the gallbladder have also been induced in Syrian golden hamsters (*Mesocricetus auratus*) by the surgical implantation of pellets containing a carcinogen into the gallbladder (Greenblatt 1982). 3-Methylcholanthrene was an effective carcinogen to the gallbladder under these conditions. Instillation of pellets containing chlosterol followed by

feeding of nitrosodimethylamine was also effective. Curiously, surgical implantation of nitrosodiethylamine into the gallbladder alone did not produce tumors.

Comparison with Other Species

The occurrence of adenomas in the gallbladder of humans is relatively rare in comparison with the frequency of carcinomas (Kozuka et al. 1982). Although malignant changes have been reported in adenomas (Christensen and Ishak 1970), the possible role of adenoma as a premalignant lesion in the development of carcinoma remains controversial. Human adenomas are of both papillary and nonpapillary (tubular) types. In hamsters, the papillary adenoma is the more frequent type; in humans, the incidence of both types has been variously reported. Nevertheless, the histology of papillary adenoma in hamsters is quite similar to that in humans; its biologic behavior remains to be elucidated.

Spontaneous proliferative lesions in the gallbladder have been described in a captive colony of 256 fat sand rats (*Psammomys obesus terraesanctea*) (Unger and Adler 1982). Most of these lesions appeared during the second half of their life span and a few more papillomas were recognized in males (15/131 in males versus 6/125 females; chi squared, $0.1 > p > 0.05$. The histologic diagnoses pertaining to the gallbladder among these 256 animals were: normal, 207; diffuse polypoid

Table 11. Incidence of gallbladder lesions in hamsters treated with DHPN followed by diets containing 0.05% PB, 0.5% DCA, or 0.5% LCA

Group	Treatment	Hamsters (*n*)	Hyperplasia (*n*)	Hyperplasia (%)	Adenoma (*n*)	Adenoma (%)
1	Control	10	0	0	0	0
2	0.05% PB	10	0	0	0	0
3	0.5% DCA	10	0	0	1	10
4	0.5% LCA	10	5	50	0	0
5	DHPN	8	4	50	1	13
6	DHPN → PB	10	0	0	1	10
7	DHPN → DCA	9	4	44	5	56
8	DHPN → LCA	9	8	89	6	67

DHPN, N-nitrosobis(2-hydroxypropyl)amine; PB, phenobarbital; DCA, deoxycholic acid; LCA, lithocholic acid.

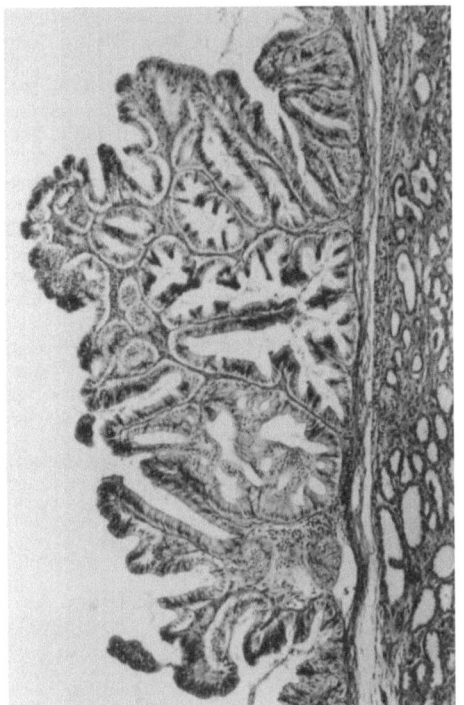

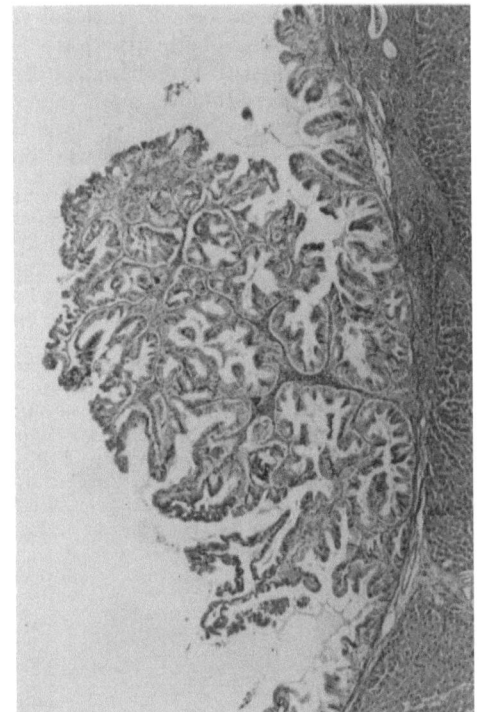

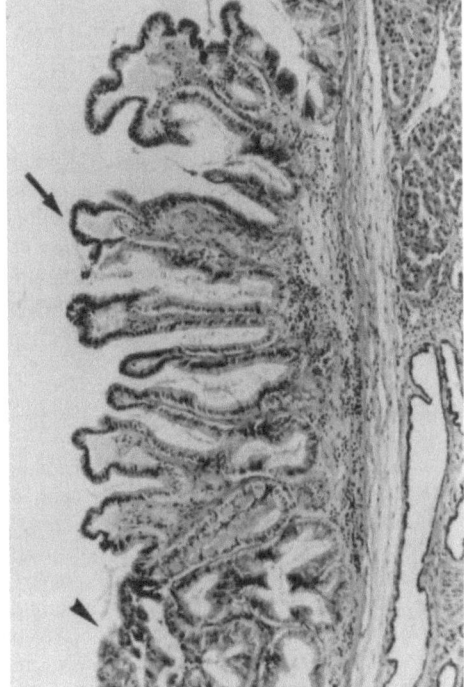

Fig. 88. (*upper left*) Focal hyperplasia, gallbladder of a hamster treated with *N*-nitrosobis(2-hydroxypropyl)amine (DHPN) followed by a diet containing 0.5% deoxycholic acid (DCA). H&E, ×40

Fig. 89. (*below*) Diffuse hyperplasia, gallbladder of a hamster treated with *N*-nitrosobis(2-hydroxypropyl)amine (DHPN) followed by a diet containing 0.5% deoxycholic acid (DCA). Villous (*arrow*) and spongioid areas (*arrowhead*) are present. H&E, ×50

Fig. 90. (*upper right*) Papillary adenoma in the gallbladder of a hamster treated with *N*-nitrosobis(2-hydroxypropyl)amine (DHPN) followed by a diet containing 0.5% deoxycholic acid (DCA). H&E, ×20

hyperplasia, 28; and papilloma, 21. Included were three animals in which both hyperplasia and papilloma were recognized. In addition, one of the papillomas had evidence of invasion into the wall of the gallbladder and a second papilloma was considered to have features of carcinoma in situ.

References

Christensen AH, Ishak KG (1970) Benign tumors and pseudotumors of the gallbladder. Report of 180 cases. Arch Pathol 90:423–432

Edmondson HA (1967) Tumors of the gallbladder and extrahepatic bile ducts. Atlas of tumor pathology, sect VII, fasc 26. Armed Forces Institute of Pathology, Washington DC

Greenblatt M (1982) Tumours of the liver. In: Turosov VS (ed) Pathology of tumours in laboratory animals, vol III. Tumours of the hamster. IARC, Lyon, pp 69–101

Halpert B (1977) Gallbladder and biliary ducts. In: Anderson WAD, Kissane JM (eds) Pathology, 7th edn. Mosby, St Louis, chap 32

Kozuka S, Tsubone N, Yasui A, Hachisuka K (1982) Relation of adenoma to carcinoma in the gallbladder. Cancer 50: 2226–2234

Pour P, Krüger FW, Althoff J, Cardesa A, Mohr U (1975) Effect of beta-oxidized nitrosamines on Syrian hamsters. III. 2,2'-dihydroxy-di-n-propylnitrosamine. JNCI 54: 141–146

Pour P, Kmoch N, Greiser E, Mohr U, Althoff J, Cardesa A (1976) Spontaneous tumors and common diseases in two colonies of Syrian hamsters. I. Incidence and sites. JNCI 56:931–935

Pour P, Althoff J, Salmasi SZ, Stepan K (1979) Spontaneous tumors and common diseases in three types of hamsters. JNCI 63:797–811

Rückert K, Pracht B, Kloppel G (1981) Differences in experimental pancreatic carcinogenesis induced by oral or subcutaneous administration of 2,2'-dihydroxydi-n-propylnitrosamine in duct-ligated hamsters. Cancer Res 41:4715–4719

Ungar H, Adler JH (1982) Naturally occurring polypoid hyperplasia and papilloma in gallbladder of the fat sand rat (Psammomys obesus terraesanctae Thomas). Vet Pathol 19:230–238

Mesothelioma, Peritoneum, Induced by Mineral Fibers, Rat

Paul N. Brooks and Clare G. Collier

Synonyms. Malignant mesothelioma, epithelial mesothelioma, epithelioid mesothelioma, epitheliomatous mesothelioma, fibrous mesothelioma, mesothelial sarcoma, sarcomatoid mesothelioma, sarcomatous mesothelioma, mixed mesothelioma, papillary mesothelioma, vesicular mesothelioma

Gross Appearance

The macroscopic appearance of abdominal mesotheliomas is variable depending upon the extent of organ involvement. Early tumors may be identified as no more than a localized, occasionally nodular thickening of the peritoneal, diaphragmatic, or organ serosal surface, whereas more extensive tumors are typically diffuse, involving almost all abdominal organs with extensive infiltration and adhesions and relatively large masses of neoplastic tissue. Unattached tumor or fibrous

nodules are often reported in the abdominal cavity. Spread beyond the abdominal cavity is rarely observed. Fibrous nodules and early fiber-induced mesotheliomas tend to be discolored with accumulations of injected fibers; otherwise mesotheliomas are white to cream or yellow in color, with a firm consistency. Hemorrhagic ascites frequently accompany mesothelioma development.

Microscopic Features

Mesothelial cells are derived from the mesoderm and form the lining of the serous body cavities and serosal surfaces of organs as well as meninges. The normal mesothelium is comprised of a single, flattened layer of cells on a thin basement membrane (Fig. 91). The mesothelial lining forms a biologic barrier between an organ and adjacent structures, reducing frictional damage during respiration and visceral movement. Mesothelial cells are also very active in the phagocytosis of exogenous particles, with the resultant generation of free radicals.

Mesothelial cells have the potential to differentiate as either mesenchymal stromal cells or epithelial lining cells. This dimorphic potential is also expressed in mesotheliomas which contain a mix of the two morphological cell types, one of which may predominate in individual neoplasms. The intraperitoneal injection of mineral fibers in the rat results in deposition over the surfaces of most of the abdominal organs (Collier et al. 1995). When in excess, the mineral fibers tend to aggregate in clumps, either adherent to the surface of viscera or free in the abdominal cavity, eliciting a foreign body response with granuloma formation (Figs. 92, 93). Long-standing granulomata become fibrotic and sclerosed, effectively isolating mineral fibers from the surrounding environment, and probably also from mesothelial cells. Our own observations concur with those of others (Davis 1979; Hill et al. 1990; Rittinghausen et al. 1991; Fraire et al. 1994) in terms of both the cellular morphology of mesotheliomas and the histologic growth patterns observed. Two fundamental cellular phenotypes, with intermediate forms, are identified as sarcomatoid (Fig. 94) and epithelioid (Fig. 95). Mixtures of these cell types, often with a predominant cell type, grow in at least one of several histologic patterns. Mesotheliomas appear histologically as solid or nodular, papillary, and vesicular neoplastic growths.

In sarcomatoid mesotheliomas, the cellular morphology ranges from fibroblastic cells with oval or rounded spindle nuclei with scant to moderate pale-staining cytoplasm to cells that are more uniform and tightly arranged in bundles in different planes. Epithelioid tumors contain cells that are round, cuboidal, or angular, with large nuclei in an eosinophilic cytoplasm which may be vacuolated or even foamy (Fig. 96). Occasionally epithelioid cells have a flattened appearance and can form papillary and tubular structures (Figs. 97, 98). Nuclei of any cell type can be chromatin dense or sparse and vesicular with variable numbers of nucleoli. Nuclear and cellular pleomorphism are common with variable, but usually intense mitotic activity, often resulting in bizarre forms. Both sarcomatoid and epithelioid phenotypes have a tetraploid DNA profile (Lee et al. 1993).

Mesotheliomas generally have only a sparse stroma, particularly the more densely cellular

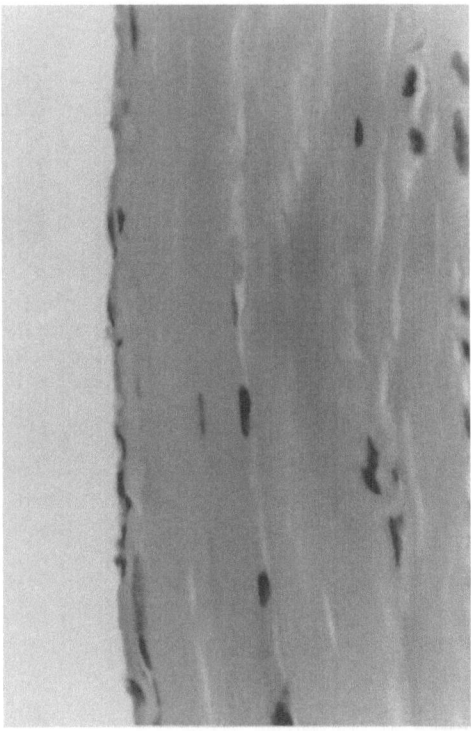

Fig. 91. Mesothelium, peritoneum, rat. A single layer of normal flattened mesothelial cells overlying the diaphragm. H&E, ×400

variants. Sclerotic areas are occasionally observed (Figs. 99, 100), and the mesenchymal origin of mesotheliomas is emphasized in some instances by the presence of cartilaginous and osseous differentiation in sarcomatoid tumor areas (Fig. 101). Areas of hemorrhage and necrosis are commonly observed within the denser areas of tumor and are a frequent consequence of direct invasion of abdominal organs.

The surface of mesotheliomas is often covered by a typical layer of flattened or slightly plump mesothelial cells (Fig. 102), whereas in other areas the surface cells are rounded, in loose contact, and easily fragmented. The tendency for tumor cells to detach is important in intraperitoneal spread and may be one factor in the diffuse organ involvement of mesothelioma. Direct invasion of the liver, spleen, pancreas, mesentery, and serosal surfaces of abdominal organs is frequently observed, but infiltration of blood vessels and lymph channels is less apparent, although clearly evident in some cases. Distant metastases are uncommon.

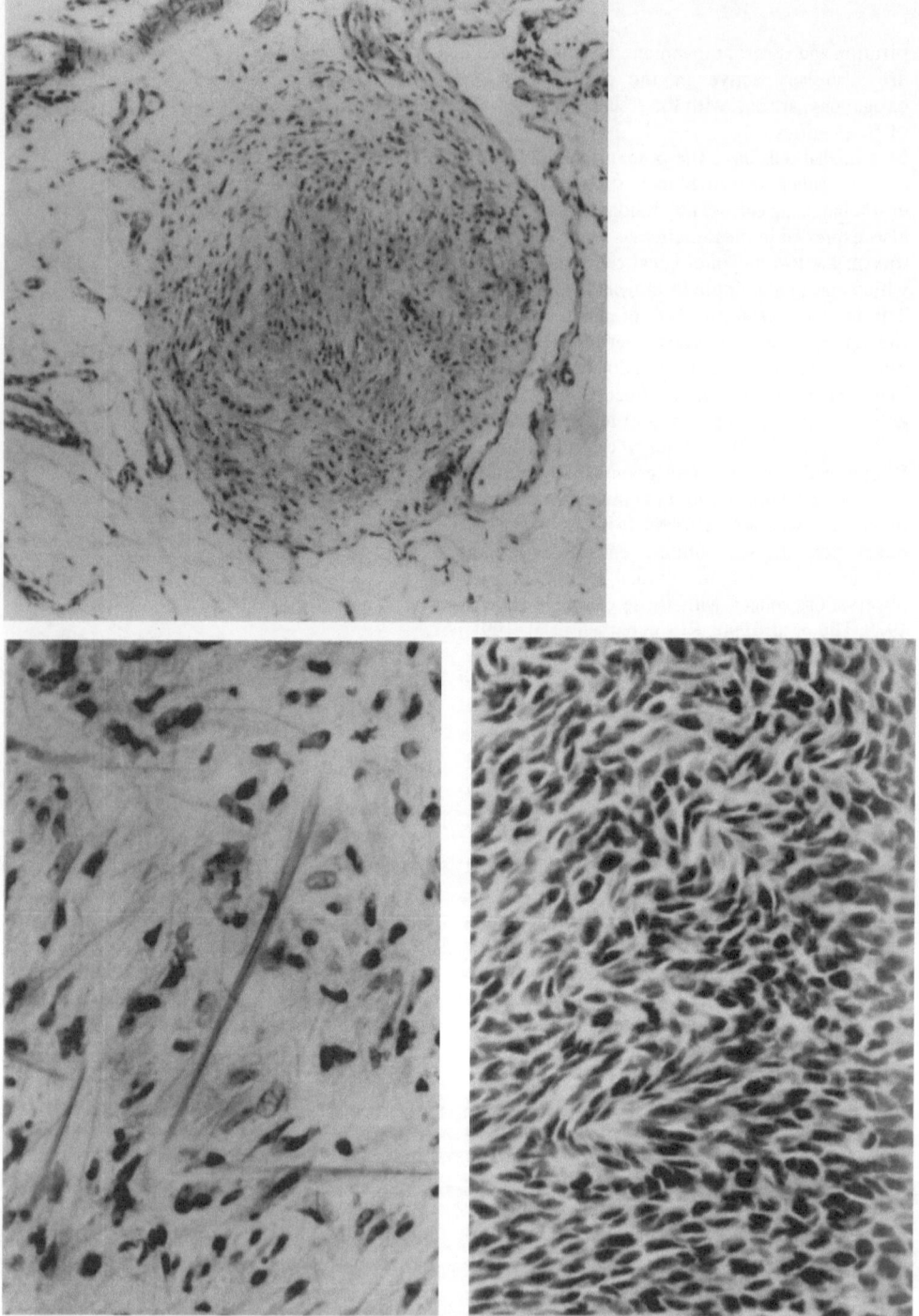

Fig. 92. (*above*) Mesentery, rat. A fibrous nodule induced in the mesentery following the intraperitoneal injection of crocidolite. H&E, ×100

Fig. 93. (*lower left*) Fibrous nodule, mesentery, rat. Higher magnification of Fig. 92, illustrating crocidolite fibers within the lesion. H&E, ×400

Fig. 94. (*lower right*) Sarcomatoid mesothelioma, mesentery, rat. Note spindle cells and "herring bone" growth pattern. H&E, ×400

Edwards et al. (1984) investigated the cellularity and histochemistry of asbestos-related lesions in Wistar and Fischer 344 rats following intrapleural injection. The authors found that crocidolite-induced granulomas showed the presence of lysosomal enzymes and nonspecific esterase in mononuclear cells and giant cells, even 2 years after injection. Mesotheliomas have been demonstrated to secrete hyaluronic acid, and histochemical tests for this in the stromal tissue can provide confirmatory evidence for the diagnosis (Wagner et al. 1962; 1982). Lee et al. (1993) applied immunocytochemical staining techniques to cultured mesothelioma cells. The authors found that sarcomatoid mesotheliomas stained strongly for vimentin and weakly for cytokeratin, and the epithelioid variant stained weakly for vimentin and strongly for cytokeratin. Yang et al. (1988) found that sarcomatoid cells in mixed mesotheliomas demonstrated a moderate to strong vimentin reaction.

There is probably no diagnostic value, at least from the point of view of toxicological histopathology, in attempting to subclassify mesotheliomas as sarcomatoid or epithelioid; most mesotheliomas do, in any event, contain a mixture of both morphological cellular expressions. The phenotypic expression may, to some extent, be dependent upon the location of the cell, since surface cells usually more closely resemble the typical mesothelial epithelial cell, whereas deeper cells tend to have a more spindle cell appearance. There is also probably little to be gained by classifying the growth pattern, since a single mesothelioma may not be exclusively composed of one histologic growth pattern. This local differentiation could be related to cell contact and density.

The microscopic features illustrated are from abdominal mesotheliomas induced in female Wistar rats (Harlan, UK) following the intraperitoneal injection of crocidolite fibers as part of a larger comparative life span toxicity study conducted at AEA Technology UK, according to a protocol designed to follow the proposals of Pott et al. (1990). Rats received a single injection of more than 0.5×10^9 fibers, 35% of which were 10 µm or longer, in saline suspension. Female rats were employed in our studies to minimize the background incidence of spontaneous mesothelioma, as mesothelioma can arise from the tunica vaginalis as a spontaneous entity in male rats.

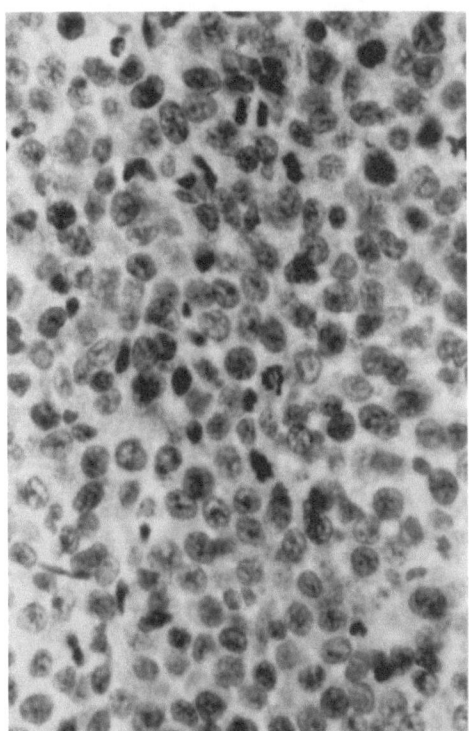

Fig. 95. Epithelioid mesothelioma, mesentery, rat. Note nuclear pleomorphism and frequent mitotic figures. H&E, ×400

Ultrastructure

Davis (1979) examined the ultrastructure of crocidolite-induced mesotheliomas in the rat and detailed the morphological features of the cells found in the different growth patterns. Mesothelioma cells on the surface of tumor nodules were described as loosely arranged rounded, cells with very few desmosomes on the cell surface membrane. Projections, sometimes resembling microvilli, were observed. Within the cells, cytoplasmic lipid droplets were usually present with large quantities of short-lengthened, granulated endoplasmic reticulum. In the deeper layers of the earliest nodules, cells were irregular in shape with a loose network of reticulin fibers. The surface membrane showed only a few processes. Lipid droplets were still commonly found within the cytoplasm, and the granular endoplasmic reticulum was well developed. In the central region of more advanced lesions, and in areas of invasion, the cells were more closely packed and spindle

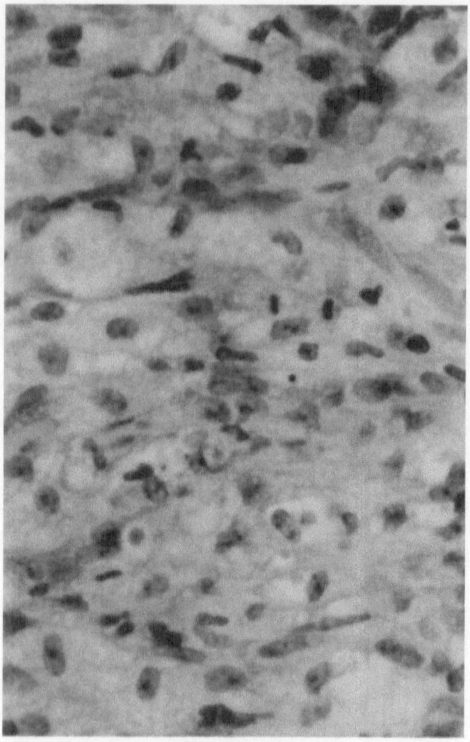

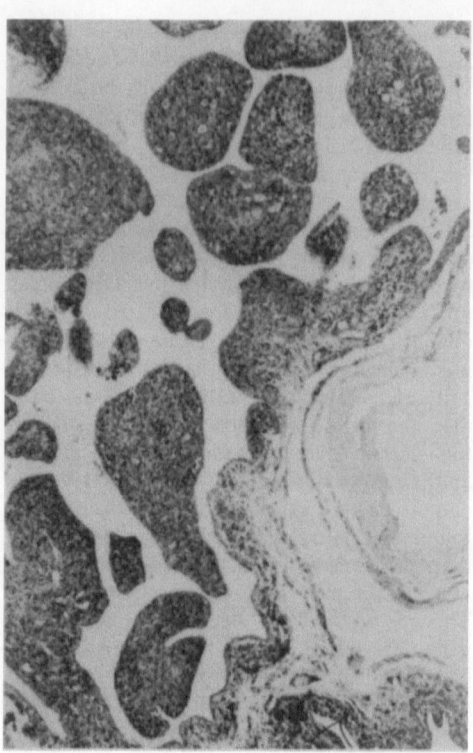

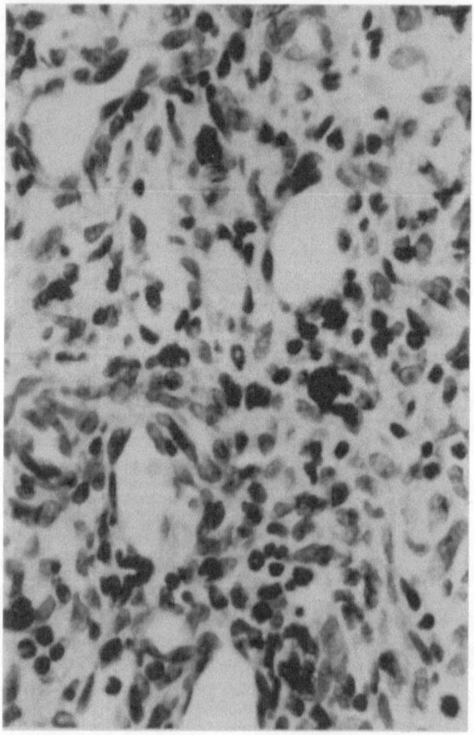

Fig. 96. (*upper left*) Mesothelioma, mesentery, rat. Note the foamy cells. H&E, ×400

Fig. 97. (*upper right*) Mesothelioma, mesentery, rat. Note papillary growth pattern. H&E, ×40

Fig. 98. (*below*) Higher magnification of Fig. 97. Predominantly sarcomatoid cells with the formation of channels that connect with the surface of the tumor mass. H&E, ×400

shaped. However, the cell cytoplasm still contained large amounts of endoplasmic reticulum. Surface processes were much less common at this stage of tumor development. The rounded cells in papillary forms of mesothelioma were usually firmly attached to each other by desmosomes, with the free cell membrane covered with microvilli.

The ultrastructural cellularity of 23 rat mesotheliomas was investigated by Yang et al. (1988), who concluded that mesotheliomas were composed of various types of tumor cells, including primitive mesenchymal cells and fibroblastic, epithelioid, and intermediate cells.

Changes in the ultrastructure of mesothelial cells following exposure to crocidolite fibers was investigated using scanning electron microscopy (SEM) by Lee et al. (1993). These authors found that early changes were characterized by loss of the normal microvillus surface of the mesothelial cells and replacement by an irregular surface, devoid of microvilli, in which partially phagocytosed fibers were observed. SEM examination of the surface of mesothelioma cells revealed a covering layer of loose mesothelial cells, some of which had long microvilli.

Ultrastructural investigations on serially transplanted rat mesotheliomas by Wagner et al. (1982) demonstrated that one cell type may dominate in one transplant generation and not in the following and that a single cell type may ultimately emerge. Brown et al. (1985) investigated the multipotential behavior of cloned mesothelial cells using ultrastructural evaluation of cultured cells. The authors demonstrated that both sarcomatoid and epithelioid mesothelioma phenotypes could be derived from a single mesothelioma cell and that the morphology and growth characteristics of these cells were density dependent.

Differential Diagnosis

Mesotheliomas must be distinguished from abdominal fibrosarcoma, leiomyosarcoma, histiocytoma, liposarcoma, spindle cell sarcoma,

▶

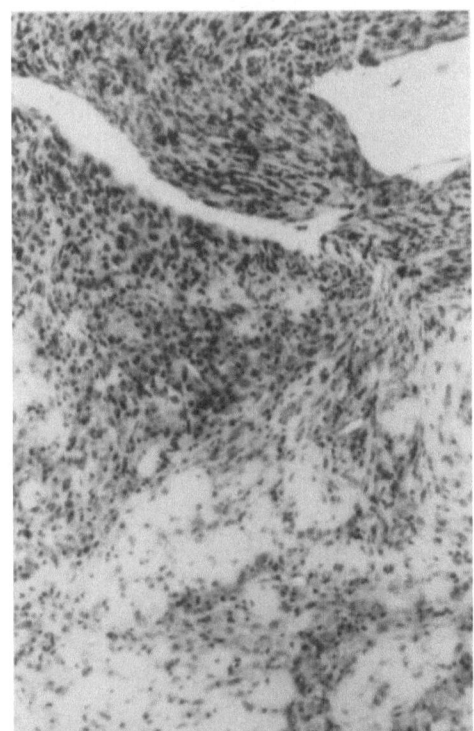

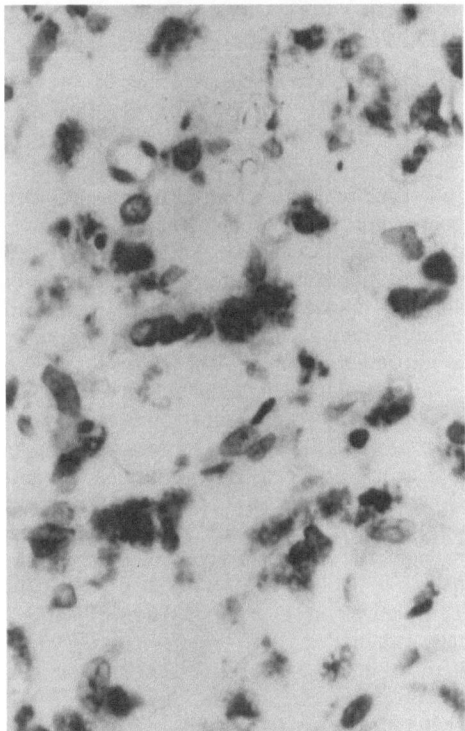

Fig. 99. (*above*) Mesothelioma, mesentery, rat. Areas of sclerosis within a predominantly sarcomatoid mesothelioma. H&E, ×100

Fig. 100. (*below*) Higher magnification of Fig. 99. Intracytoplasmic pigment deposits are also present in macrophages within the sclerotic area, indicating previous hemorrhage. H&E, ×400

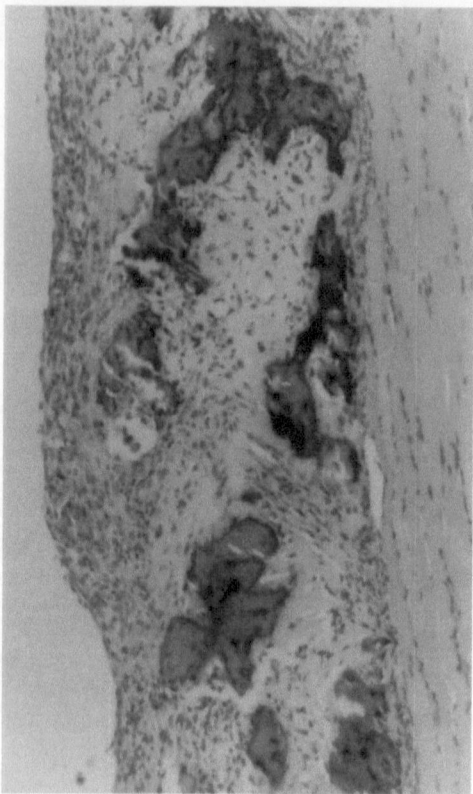

Fig. 101. Mesothelioma, mesentery, rat. Bone formation within an area of sarcomatoid mesothelioma overlying the intestinal musculature. H&E, ×100

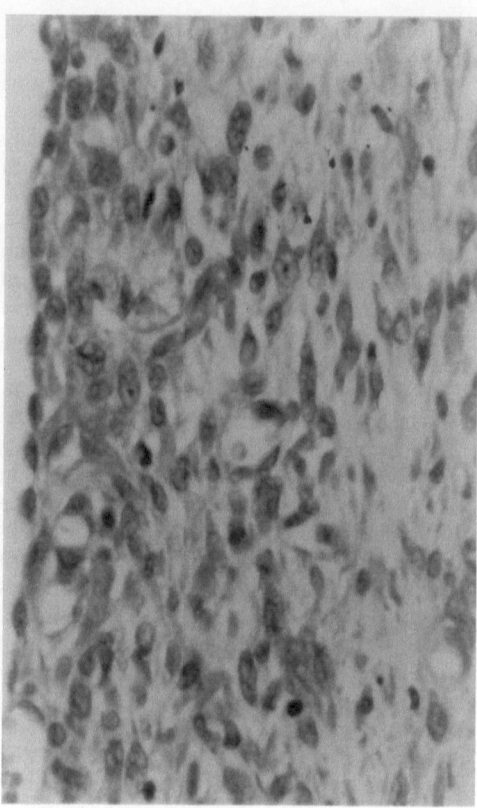

Fig. 102. Mesothelioma, mesentery, rat. The surface of a predominantly sarcomatoid mesothelioma, with an overlying single layer of flattened neoplastic mesothelial cells. H&E, ×400

pancreatic carcinoma, and gastrointestinal carcinoma. For instances involving relatively well differentiated examples of abdominal tumors, identification of the cell of origin is usually reasonably easy. However, when presented with poorly differentiated abdominal tumors, the situation is altogether different. The diffuse nature of peritoneal mesothelioma, in the absence of any obvious primary tumor mass, may be an aid to diagnosis, as is the tendency for mesothelioma surface cells to adopt a more typical appearance. Otherwise, an on balance opinion will result from careful consideration of the presence or absence of diagnostic criteria characteristic of other tumor types.

The relative rarity of abdominal mesotheliomas in rats, other than those related to the tunica vaginalis in males, would not make this tumor the first diagnosis when considering any spontaneous abdominal neoplasm tending towards anaplasia.

However, following intraperitoneal inoculation, and especially the injection of mineral fibers, mesothelioma would have to be an immediate consideration.

Biologic Features

Spontaneous mesothelioma arising from the tunica vaginalis is a relatively common neoplasm of the Fischer 344 male rat and has also been reported for other strains. Spontaneous mesothelioma is rare in female rats, although the neoplasm has been reported in relation to the ovarian bursa (Hall 1990).

Apart from extensive reports of mesothelioma induction by mineral fibers, mesotheliomas have also been produced in rats following exposure to radiation (Hahn and Lundgren 1992; Sanders 1992).

Pathogenesis

In addition to the long-recognized relationship between fiber length and width, it is likely that other factors play an important role in determining the tumorigenic potency of different mineral fibers. Durability has obvious relevance, and attention has recently been given to the chemical and physical properties of the fiber surface.

Fiber dimensions are of importance in the development of mesothelioma. The general consensus is that the carcinogenicity of mineral fibers is related to their length and width, with longer fibers offering the greatest carcinogenic potential. Stanton et al. (1977), Stanton and Layard (1978), Stanton and Wrench (1972), Pott (1978), and Pott and Friedrichs (1972) demonstrated that the development of mesotheliomas as a result of intrapleural or intraperitoneal implantation of asbestos and other mineral fibers was closely related to the number of fibers in any dust preparation that were more than $8\mu m$ in length and less than $0.25\mu m$ in diameter. The type of material appeared to be unimportant. Davis et al. (1986) reported long-term inhalation and injection studies in rats in which a sample of amosite with almost all fibers less than $5\mu m$ in length was compared with a normal amosite dust preparation containing many long fibers. The short-fiber material produced neither fibrosis nor neoplasia, apart from a single mesothelioma in the highest dose injected, while the long fibers were highly pathogenic. A similar study was undertaken on chrysotile (Davis and Jones 1988). Work undertaken by Wagner et al. (1985) and Wagner (1990) demonstrated that an incidence of mesothelioma of almost 100% following the inhalation of erionite in rats was reduced to 0% when short-fiber preparations of the material were used.

The sequence of events leading to mesothelioma formation in male Porton rats following the intrapleural inoculation of erionite was detailed by Hill et al. (1990). Friemann et al. (1990) examined the mesothelial response to different mineral fibers following intraperitoneal injection into female Sprague-Dawley rats after intervals of 8 h to 15 months after injection. The authors observed an initial reparative proliferation of the submesothelial connective tissue leading to focal submesothelial fibrosis. After 15–28 months, atypical mesothelial proliferation was observed. Preneoplastic changes in the process of mesothelioma development have been proposed

by Fraire et al. (1994), who undertook timed kills at intervals ranging from 2 to 430 days after intrapleural inoculation of female Fischer 344 rats with fibrous glass. The authors identified mesothelial hyperplasia and dysplasia among the rats examined and proposed a spectrum of pleural mesothelial histopathologic changes occurring prior to the development of mesothelioma. The sequence of morphological changes induced in peritoneal mesothelial cells was investigated in male Fischer 344 rats following intraperitoneal injection of crocidolite by Lee et al. (1993). The authors looked at mesothelial cell changes by light and electron microscopy and by cytological examination of peritoneal washings. It was observed that the injection of fibers stimulated an acute inflammatory response with rapid phagocytosis of fibers by mesothelial cells and incorporation into the submesothelial tissues. After 7 days, the normal surface of the mesothelium was replaced by a syncytium of proliferating mesothelial cells, with the presence of multifocal mesothelial tumors within the peritoneal cavity 9 months after exposure. Lee et al. (1993) further studied the thermodynamic properties of normal and abnormal mesothelial cells in culture and demonstrated that the adhesive properties of asbestos-stimulated mesothelial cells and mesothelial tumors were lower than those in normal mesothelial tissue.

Although fiber dimensions are clearly important for carcinogenicity, chemical and other physical properties of the fibers, especially those of the fiber surface, may also be significant. The importance of the generation of reactive oxygen metabolites by interaction between the mineral fibers and tissues must be considered, since it has been well recognized that reactive oxygen species can react with a variety of intra- and extracellular components, including DNA (Sahu 1991). Adachi et al. (1994) investigated one aspect of this hypothesis by examining the role of iron existing inside and outside of the crocidolite fiber. The authors looked at the amount of oxidative DNA damage resulting from the incubation of calf thymus DNA with crocidolite and deironized crocidolite and concluded that the induction of oxidative DNA damage could be changed by chemical treatment. The addition of iron, in the form of Fe_2O_3, was observed by the authors to promote the development of mesothelioma which had been initiated by deironized crocidolite. Coffin et al. (1992) investigated the relative potency of different mineral fibers to induce pleural mesotheliomas in Fischer 344 rats. Erinoite was found to be 500–800 times more tumorigenic, and crocidolite 30–60 times more tumorigenic than chrysotile. The authors propose that such differences in potency could be related to differing surface chemical reactivity, resulting in the release of free radicals during the reduction of endogenous ferric iron.

Fiber durability is another important factor in pathogenicity; fibers that readily dissolve will prevent significant accumulation and consequential tissue damage (Pott et al. 1989).

In spite of isolated reports of benign mesotheliomas in experimental animals, usually spontaneous in origin, there is some question as to whether a truly benign variant of mesothelioma exists. All of the fiber-induced peritoneal mesotheliomas examined in our studies, regardless of size, had characteristics of malignancy and were diagnosed as such.

The diffuse nature of mesotheliomas raises interesting questions as to the point of origin and spread of these neoplasms. Many of the tumors encountered experimentally appear to be multifocal. There are two possible explanations for this. First, given the loose surface arrangement of mesothelioma cells in tumor masses, it is likely that cells readily become detached and spread throughout the abdomen, subsequently adhering to tissues at points remote from the origin and commencing fresh tumor growths. Second, growth factors released from transformed cells at one location may diffuse through the peritoneal fluid and stimulate multifocal neoplastic transformation elsewhere within the body cavity. The simultaneous multifocal development of mesothelioma seems otherwise to be a less likely option.

Comparison with Other Species

Both benign and malignant forms of mesothelioma are diagnosed in humans. The benign mesothelioma in humans is a localized pleural growth that is often attached to the pleural surface by a pedicle. The tumor, which can vary greatly in size, always remains confined to the surface of the lung. The micromorphology of this neoplasm is characterized by spindle cells that resemble fibroblasts, leading to the term pleural fibroma by which the tumor is alternatively known. Malignant forms of the disease in humans are morphologically very similar to those observed in experimen-

tal animals. The benign mesothelioma has no relationship to exposure to mineral fibers.

Peritoneal mesotheliomas are extremely rare in humans. When encountered, however, they duplicate the morphology of thoracic mesotheliomas induced following asbestos exposure. The vast majority of mesotheliomas in humans are asbestos related, with a long latent period of 25–45 years. The morphology of human mesotheliomas is as variable as that presented by pleural or intraperitoneal mesotheliomas in the rat, being either mainly epithelioid or spindle celled or of mixed type. Mesotheliomas have also been produced experimentally in the hamster (Smith et al. 1987; Hesterberg et al. 1991).

References

Adachi S, Yoshida S, Kawamura K, Takahashi M, Uchida H, Odagiri Y, Takemoto K (1994) Inductions of oxidative DNA damage and mesothelioma by crocidolite, with special reference to the presence of iron inside and outside of asbestos fiber. Carcinogenesis 15:753–758

Brown DG, Johnson NF, Wagner MM (1985) Multipotential behaviour of cloned rat mesothelium cells with epithelial phenotype. Br J Cancer 51:245–252

Coffin DL, Cook PM, Creason JP (1992) Relative mesothelioma induction in rats by mineral fibers: comparison with residual pulmonary mineral fiber number and epidemiology. Inhal Toxicol 4:273–300

Collier CG, Morris KM, Launder KA, Humphreys JA, Morgan A, Eastes W, Townsend S (1995) The behaviour of glass fibers in the rat following intraperitoneal injection. J Reg Toxicol Pharmacol 20:589

Davis JM (1979) The histopathology and ultrastructure of pleural mesotheliomas produced in the rat by injections of crocidolite asbestos. Br J Exp Pathol 60:642–652

Davis JM, Jones AD (1988) Comparisons of the pathogenicity of long and short fibres of chrysotile asbestos in rats. Br J Exp Pathol 69:717–737

Davis JM, Addison J, Bolton RE, Donaldson K, Jones AD. Smith T (1986) The pathogenicity of long versus short fibre samples of amosite asbestos administered to rats by inhalation and intraperitoneal injection. Br J Exp Pathol 67:415–430

Edwards RE, Wagner MM, Moncrieff CB (1984) Cell population and histochemistry of asbestos related lesions of rat pleural cavity after injection of various inorganic dusts. Br J Ind Med 41:506–513

Fraire AE, Greenburg SD, Spjut HJ, Roggli VL, Dodson RF. Cartwright J, Williams G, Baker S (1994) Effect of fibrous glass on rat pleural mesothelium. Histopathologic observations. Am J Respir Crit Care Med 150:521–527

Friemann J, Müller KM, Pott F (1990) Mesothelial proliferation due to asbestos and man-made fibres. Experimental studies on rat omentum. Pathol Res Pract 186:117–123

Hahn FF, Lundgren DL (1992) Pulmonary neoplasms in rats that inhaled cerium-144 dioxide. Toxicol Pathol 20:169–178

Hall W (1990) Peritoneum, retroperitoneum, mesentry, and abdominal cavity. In: Boorman GA, Eustis SL, Elwell MR (eds) Pathology of the Fischer rat: reference and atlas. Academic, San Diego, pp 63–69

Hesterberg TW, Mast R, McConnell EE, Chevalier J, Bernstein DM, Bun WB, Anderson R (1991) Chronic inhalation toxicity of refractory ceramic fibers in syrian hamsters. In: Brown RC, Hoskins JA, Johnson NF (eds) Mechanisms in fibre carcinogenesis. Plenum, New York, pp 531–538

Hill RJ, Edwards RE, Carthew P (1990) Early changes in the pleural mesothelium following the intrapleural inoculation of the mineral fiber erionite and the subsequent development of mesotheliomas. J Exp Pathol 71:105–118

Lee MM, Green FH, Demetrick DJ, Jiang XX, Schürch S (1993) A study of surface property changes in rat mesothelial cells induced by asbestos using aqueous two-phase polymer solutions. Biochem Biophys Acta 1181:223–232

Pott F (1978) Some aspects on the dosimetry of the carcinogenic potency of asbestos and other fibrous dusts. Staub Reinhalt Luft 38:486–490

Pott F, Friedrichs KH (1972) Tumoren der Ratte nach ip-Injektion faserformiger Staube. Naturwissenschaften 59:318

Pott F, Roller M, Ziem U, Reiffer FJ, Bellmann B, Rosenbruch M, Hath F (1989) Carcinogenicity studies on natural and man-made fibers with the intraperitoneal test in rats. In: Bignon J, Peto J, Saracci R (eds) Non-occupational exposure to mineral fibres. IARC, Lyon, pp 175–179 (IARC scientific publications no 90)

Pott F, Bolme H, Bruch J, Friedberg KD, Rödesisperger K, Woitowitz H-J (1990) Einstufungsvorschlag für Anorganische und Oranische Fasern. Arbeitsmed Soziamed Praventivmed 25:463–466

Rittinghausen S, Ernst H, Muhle H, Fuhst R, Mohr U (1991) Histopathological analysis of tumor types after intraperitoneal injection of mineral fibres in rats. In: Brown RC, Hoskins JA, Johnson NF (eds) Mechanisms in fibre carcinogenesis. Plenum, New York, pp 81–89

Sahu SC (1991) Role of oxygen free radicals in the molecular mechanisms of carcinogenesis: a review. Environ Carcino Ecotox Revs C9:83–112

Sanders CL (1992) Pleural mesothelioma in the rat following exposure to $^{239}PuO_2$. Health Phys 63:695–697

Smith DM, Ortiz LW, Archuleta RF, Johnson NF (1987) Long-term health effects in hamsters and rats exposed chronically to man-made vitreous fibers. Ann Occup Hyg 31:731–754

Stanton MF, Wrench C (1972) Mechanisms of mesothelioma induction with asbestos and fibrous glass. J Natl Cancer Inst 48:797–821

Stanton MF, Laynard M (1978) The carcinogenicity of fibrous materials. In: Gravatt CC, Lafleur PD, Heinrich FJ (eds) Workshop on asbestos: definitions and measurement methods. National Bureau of Standards, Washington DC (NBS special publication no 506)

Stanton MF, Laynard M, Tegeris A, Miller E, May M, Morgan E, Kent E (1977) Carcinogenicity of fibrous glass: pleural response in the rat in relation to fiber dimension. J Natl Cancer Inst 58:587–603

Wagner JC (1990) Significance of the fiber size of erionite. Abstract in National Institute for Occupational Safety and Health. Proceedings of the VIIth international pneumoconioses conference, 23–26 Aug 1988, Pittsburgh, Pennsylvania, USA. Part I. Pittsburgh (PA): US Department of Health and Human Services 158. DHHS (NIOSH) publication no 90–108, part I

Wagner JC, Munday DE, Harrison JS (1962) Histochemical demonstration of hyaluronic acid in pleural mesotheliomas. J Pathol Bacteriol 84:73

Wagner JC, Johnson NF, Brown DG, Wagner MM (1982) Histology and ultrastructure of serially transplanted rat mesotheliomas. Br J Cancer 46:294–299

Wagner JC, Skidmore JW, Hill RJ, Griffiths DM (1985) Erionite exposure and mesotheliomas in rats. Br J Cancer 51:727–730

Yang GH, Tan YS, Liu XZ, Luo SQ (1988) Ultrastructure and immunohistochemical study of mesothelioma induced by asbestos in rat. Hua Hsi I Ko Hsueh Hsueh Pao 19:337–341

Polyploidy, Liver, Rat

Matthew J. van Zwieten and Carel F. Hollander

Synonyms. Increased ploidy level, nuclear hypertrophy, binucleation

Gross Appearance

Polyploidy cannot be observed grossly.

Microscopic Features

Ploidy is the state of the cell nucleus relating to the number of genomes present. An increase in the ploidy of rat hepatocytes from the diploid (2n) state can be inferred histologically by the presence of large nuclei and of binuclear cells. The large nuclei in hepatocytes may be round or oval and are often two or more times the size of normal diploid nuclei (Fig. 103). Occasionally, very large, bizarre, indented, or partially lobulated nuclei may be seen. The chromatin of the large nuclei tends to be arranged in coarse aggregates, and such nuclei are often intensely basophilic. Nucleoli are frequently prominent and multiple. The nuclei of binuclear hepatocytes are usually round and identical in appearance (Fig. 104). They may be equal in size to, or larger than, the normally mononuclear diploid nuclei and often do not differ substantially from them in their staining characteristics.

Ultrastructure

The electron microscopy features of large hepatocellular nuclei or nuclei of binuclear hepatocytes have not been specifically described in rats. Based on electron microscopy studies of rat hepatocytes performed in our institute, it has been observed that the nuclei of binuclear hepatocytes did not differ ultrastructurally from those of mononuclear diploid hepatocytes (A.M. de Leeuw, personal communication). Systematic studies of large nuclei have not been performed.

Differential Diagnosis

Polyploidy should be recognized as a common morphological finding in the normal rat liver. As such, cytologic features and architectural relationships of polyploid hepatocytes are normal. A familiarity with the variation in size and tinctorial properties of polyploid nuclei is essential when assessing nuclear pleomorphism, which may be induced by certain chemical compounds or which may be associated with neoplastic transformation of liver cells. In the latter situations, cytoplasmic changes and alterations in the architecture of liver cell plates will often be evident.

Biologic Features

Polyploidy is defined as the condition in which a cell nucleus has more than twice its normal haploid number of chromosomes. The polyploidy of hepatic nuclei is determined most accurately using isolated hepatocytes. The methods used include counting chromosomes in metaphase "spreads," the determination of the relative amount of DNA by cytofluorometric methods, and measurements of nuclear diameters (Meinders-Groeneveld 1969). The most common ploidy classes found in the rat liver are mononuclear diploid (2n), mononuclear tetraploid (4n), mononuclear octaploid (8n), and their binuclear counterparts (Van Bezooijen et al. 1974). There is evidence that the sequence of the shift to a higher ploidy class is from mononuclear diploid to binuclear diploid to mononuclear tetraploid to binuclear tetraploid, and so on

(Nadal and Zajdela 1966). Moreover, the data indicate that binuclear cells arise by karyokinesis in the absence of cytokinesis, and these subsequently form mononuclear cells of a higher ploidy level through chromosomal fusion during mitotic division (James et al. 1979; Nadal and Zajdela 1966). It is usually stated that polyploidy is an aging phenomenon in the rat liver. There is ample evidence, however, that the greatest shift to an increased ploidy level occurs relatively early in life and that only minor changes are found thereafter. Thus, in the fetal liver and during the first 2 weeks of postnatal life, nearly all (90% or more) hepatocytes are of the mononuclear diploid class, and a rapid shift to higher ploidy classes occurs within the next few weeks, such that, in young adult rats, 50%–70% of hepatocytes are of the mononuclear tetraploid class. However, the actual numbers and time of appearance of cells of the various ploidy classes appear to be strain related, and, within a single strain, some variation in these parameters may be expected. Thus the fraction of cells of a particular ploidy number at a given age should be viewed in terms of its qualitative rather than its quantitative significance. For example, in female WAG/Rij rats (Van Bezooijen et al. 1974), the percentage of mononuclear diploid cells decreases from about 97% at 2 weeks of age to about 36% at 4 weeks of age, whereas the percentage of binuclear diploid cells increases from approximately 2% to 48% at these two ages. Between 4 weeks and 3 months of age, the percentage of both the mononuclear and binuclear diploid cells decreases (to 8% and 17%, respectively), and that of mononuclear tetraploid cells increases from approximately 8% to 53%. Recent studies (Van Bezooijen et al. 1984), also in female WAG/Rij rats, indicate that this major shift in ploidy has already taken place by 6 weeks of age. In addition, these authors found that the percentage of mononuclear tetraploid cells reached a peak (72%) at 12 months and declined thereafter to 37% at 36 months. The proportion of binuclear tetraploid cells increased gradually from 4% at 4

▶

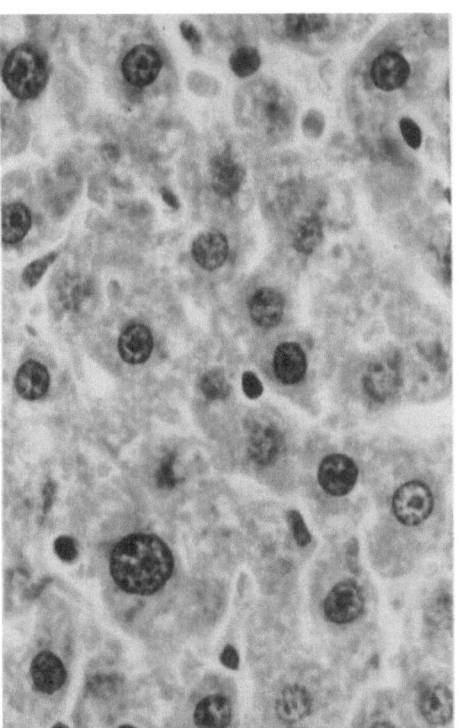

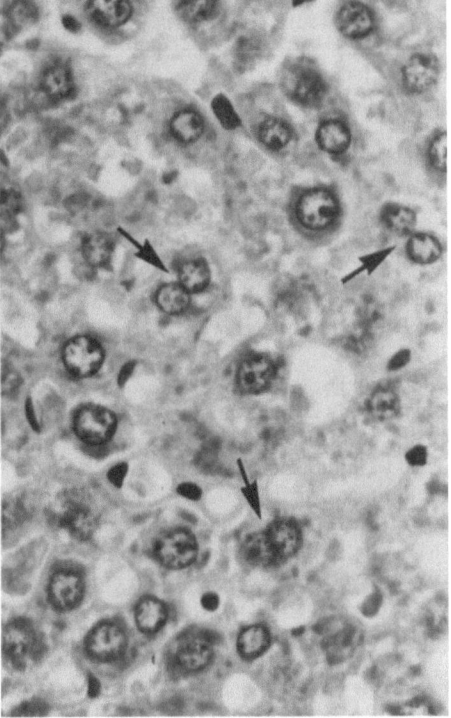

Fig. 103. (*above*) Polyploidy, liver, 8-month-old female BN/BiRij rat. Note the variation in nuclear size, with one large nucleus below center. Hematoxylin-phloxine-saffron, ×800

Fig. 104. (*below*) Binuclear hepatocytes (*arrows*), liver, same rat shown in Fig. 103. Note again the variation in nuclear size. Hematoxylin-phloxine-saffron, ×800

weeks of age to 30% at 36 months of age. The percentage of cells of the higher ploidy classes showed only small but gradual increases with age. Relative frequencies of the various ploidy classes and temporal relationships similar to those just described were found in female Long-Evans rats (Alfert and Geschwind 1958), in male Wistar rats (James et al. 1979), and in Wistar rats of unspecified sex (Nadal and Zajdela 1966). Female RU rats, on the other hand, showed the greatest shift from mononuclear diploid cells to mononuclear tetraploid cells between 3 and 12 months of age (De Leeuw-Israel et al. 1972; Van Bezooijen et al. 1973). Between 12 and 27 months, the relative frequencies of the mononuclear classes in this rat strain did not change appreciably.

The percentage of binuclear cells in isolated hepatocytes may also show strain-related differences. Several authors (reviewed by Meinders-Groeneveld 1969) observed a peak in the percentage of binuclear hepatocytes at 4–5 weeks of age, with a subsequent decline at older ages. Van Bezooijen et al. (1974, 1984) reported that, in female WAG/Rij rats, binuclear hepatocytes accounted for 2% of all hepatocytes at 2 weeks of age, 52% at 4 weeks, 32% at 3 months, 25% at 12 months, 32% at 24 months, 26% at 30 months, and 40% at 36 months. Engelmann et al. (1981) found that the percentage of binuclear hepatocytes in female F344 rats increased from 19% at 3–4 months to 28% at 29–30 months. It may be of interest here to point out that the number of binuclear cells found in histologic sections of liver is approximately one third of that found in suspensions of isolated hepatocytes (Harrison 1953).

Partial hepatectomy in young rats induces an accelerated shift to a higher ploidy state, whereas in older rats the ploidy state remains fairly constant following this procedure (Nadal and Zajdela 1966).

Interestingly, these authors found that partial hepatectomy in very young rats (less than 3 weeks of age) whose livers did not contain binuclear cells failed to induce polyploidy.

In addition to an increase in nuclear volume, there is evidence that polyploidy is associated with an increase in mean cell volume (Meinders-Groeneveld 1969; Van Bezooijen et al. 1974) and thus a reduction in the surface to volume ratio of hepatocytes (Epstein 1967). This may have important consequences for certain membrane-specific functions of the hepatocyte. Indeed, Van Bezooijen et al. (1973) concluded that an increase

in sulfobromophthalien (BSP) retention in RU rats correlated with the change to a higher ploidy level. Currently there is considerable interest in whether the rate of uptake of other drugs and compounds which are transported across the hepatocyte plasma membrane may be relatively decreased as a function of increasing polyploidy (Van Bezooijen 1984).

Comparison with Other Species

Polyploidy is a normal finding in many different species of animals and in several different organs, but it has been studied most extensively in the rodent liver. Despite scientific inquiry for many years, detailed information about the functional consequences of polyploidy is still lacking. It may be of interest to note that a great deal is known about polyploidization in plants, especially concerning its evolutionary role in allowing new species to emerge rapidly and to acquire a wide range of environmental tolerances. In the animal kingdom, hepatic polyploidization appears to be a typically mammalian phenomenon, since hepatocytes from birds, reptiles, and amphibians showed a lack of polyploidization (Gahan and Middleton 1982). These authors suggested that because polyploidization in the mammalian liver is accompanied by a reduced mitotic activity and an increase in cell volume, a physiologic advantage may be gained in terms of a higher efficiency in performing certain cellular functions and a less rapid accumulation of irreversible damage ultimately requiring cellular replacement.

Giant, hyperchromatic hepatic nuclei, often with multiple nucleoli, as well as binuclear hepatocytes, have been described in humans (Andrew et al. 1943; Carr et al. 1960). Nuclear volume and DNA content has been determined in human hepatocytes (Swartz 1956), and it was evident that the major shift to polyploid cells occurred at or around the time of puberty.

Giant nuclei have long been known to occur in mice (Andrew et al. 1943), and nuclear alterations associated with polyploidy tend to be much more prominent in mice than in rats. Indeed, nuclei with ploidy levels of 16n and 32n are not uncommon (Shima and Sugahara 1976). Although, as in rats, the major shift to higher ploidy levels occurs relatively early in life (within approximately 4 months), a prominent degree of polyploidization continues to an age of at least 28 months (Shima and Sugahara 1976). It has been shown in mice that partial hepatectomy induces an increase in the percentage of nuclei of higher ploidy level as compared with controls (Hollander and Thung 1966). Recently, it was reported that dietary protein restriction retarded the rate of polyploidization in mice (Enesco and Samborsky 1983). This finding may provide additional insights into the cellular mechanisms which may influence the expression of polyploidy.

References

Alfert M, Geschwind II (1958) The development of polysomaty in the rat liver. Exp Cell Res 15:230–232

Andrew W, Brown HM, Johnson JB (1943) Senile changes in the liver of mouse and man, with special reference to the similarity of the nuclear alterations. Am J Anat 72:199–221

Carr RD, Smith MJ, Keil PG (1960) The liver in the aging process. Histology. Arch Pathol 70:15–18

De Leeuw-Israel FR, van Bezooijen CFA, Hollander CF (1972) Ploidy as a possible explanation for the variation in liver function during the life span of the rat. Z Alternsforsch 26:29–33

Enesco HE, Samborsky J (1983) Liver polyploidy: influence of age and of dietary restriction. Exp Gerontol 18:79–87

Engelmann GL, Richardson A, Katz A, Fierer JA (1981) Age-related changes in isolated rat hepatocytes. Comparison of size, morphology, binucleation, and protein content. Mech Ageing Dev 16:385–395

Epstein CJ (1967) Cell size, nuclear content, and the development of polyploidy in the mammalian liver. Proc Natl Acad Sci USA 57:327–334

Gahan PB, Middleton J (1982) Hepatocyte euploidization is a typical mammalian physiological specialization. Comp Biochem Physiol 71A:345–348

Harrison MF (1953) Percentage of binucleate cells in the livers of adult rats. Nature 171:611

Hollander CF, Thung PJ (1966) Relations between regenerative growth and ageing in the mouse liver. In: Lindop PJ, Sacher GA (eds) Proceedings of the colloquium on radiation and ageing. Taylor and Francis, London, pp 3–14

James J, Tas J, Bosch KS, de Meere AJP, Schuyt HC (1979) Growth patterns of rat hepatocytes during postnatal development. Eur J Cell Biol 19:222–226

Meinders-Groeneveld J (1969) Enige Kwantitatieve Aspecten der Polyploidie. Een experimentel onderzoek bij ratten. Doctoral dissertation, University of Amsterdam

Nadal C, Zajdela F (1966) Polyploidie somatique dans le foie de rat. I. Le role des cellules binuclees dans la genese des cellules polyploides. Exp Cell Res 42:99–116

Shima A, Sugahara T (1976) Age-dependent ploidy class changes in mouse hepatocyte nuclei as revealed by Feulgen-DNA cytofluorometry. Exp Gerontol 11:193–203

Swartz FJ (1956) The development in the human liver of multiple desoxyribose nucleic acid (DNA) classes and their relationships to the age of the individual. Chromosoma (Berl) 8:53–72

Van Bezooijen CFA (1984) Influence of age-related changes in rodent liver morphology and physiology on drug metabolism. A review. Mech Ageing Dev 25:1–22

Van Bezooijen CFA, de Leeuw-Israel FR, Hollander CF (1973) On the role of hepatic cell ploidy in changes in liver function with age and following partial hepatectomy. Mech Ageing Dev 1:351–356

Van Bezooijen CFA, Van Noord MJ, Knook DL (1974) The viability of parenchymal liver cells isolated from young and old rats. Mech Ageing Dev 3:107–119

Van Bezooijen CFA, Bukvic SJ, Sleyster E Ch, Knook DL (1984) Bromsulfophthalein storage capacity of rat hepatocytes separated into ploidy classes by centrifugal elutriation. In: van Bezooijen CFA (ed) Proceedings of the Eurage workshop on pharmacological, morphological and physiological aspects of liver aging. EURAGE, Rijswijk, pp 115–120

Intranuclear and Intracytoplasmic Inclusions, Liver, Rat

Matthew J. van Zwieten and Carel F. Hollander

Synonyms. Inclusion bodies, intranuclear cytoplasmic invagination, cytoplasmic inclusions, acidophilic inclusions, hyaline droplets, hyaline bodies

Gross Appearance

Intranuclear and intracytoplasmic inclusions are not visible grossly.

Microscopic Features

Intranuclear inclusions are seen relatively infrequently in rat hepatocytes; however, they are distinctive and not readily overlooked. They are round, usually eccentrically located, and fill about one third to one half of the nucleus (Fig. 105). The inclusion body is sharply outlined and the exterior of its limiting membrane is sometimes studded with fine chromatin granules. The contents of the

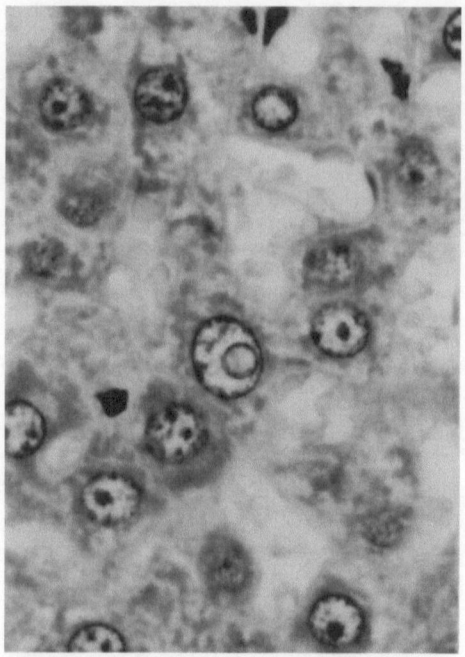

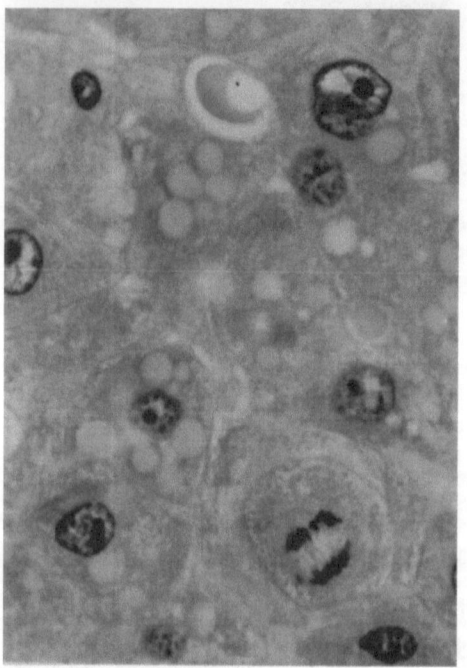

inclusion body are acidophilic and often somewhat granular or flocculent, with a morphological appearance closely resembling that of the cytoplasm in the same cell. Occasionally, an empty vacuole can be discerned within the inclusion body.

Intracytoplasmic inclusion bodies may occasionally be found in normal or neoplastic rat hepatocytes. They are usually round, homogeneous, nonrefractile, acidophilic structures, occurring singly or in clusters within the cytoplasm (Fig. 106). Such inclusions, which resemble hyaline droplets or globules, are often strongly periodic acid-Schiff (PAS) positive and diastase resistant (Popper et al. 1960; Ruebner et al. 1965). A narrow halo may sometimes be present. Their diameter may range from a few micrometers to that of the cell nucleus or larger.

Ultrastructure

The ultrastructural appearance of intranuclear inclusions in rat hepatocytes has been described by Kleinfeld et al. (1956) and summarized by Jones (1967). By electron microscopy, it is clear that the inclusions represent invaginations of cytoplasm into the nucleus (Figs. 107, 108). Cytoplasmic organelles, including mitochondria and endoplasmic reticulum, as well as lipid vacuoles and glycogen particles, can often be identified within the inclusion body. In some of the larger inclusion bodies, identification of cytoplasmic organelles may be difficult due to their apparent degeneration. The limiting membrane is a double membrane identical to that of the nuclear envelope and, in fortuitous sections, the latter can sometimes be seen to be continuous with the inclusion body membrane (Fig. 109).

Information on the electron microscopy features of intracytoplasmic inclusions in rat hepatocytes is scanty. It is not known, for example, whether the inclusions found in various experimental situations have a similar ultrastructural substrate. One type of hyaline inclusion found in rats exposed

◄

Fig. 105. (*above*) Intranuclear inclusion in a hepatocyte of an untreated 8-month-old female BN/BiRij rat. Hematoxylinphloxine-saffron, ×1000

Fig. 106. (*below*) Acidophilic hyaline intracytoplasmic inclusions in neoplastic hepatocytes of a 13-month-old male Sprague-Dawley rat. The animal was administered nitrosomorpholine (200 mg/l) in the drinking water for 3 weeks starting at 3 months of age. (Histologic section courtesy of Prof. P. Bannasch, University of Heidelberg, Germany) H&E,

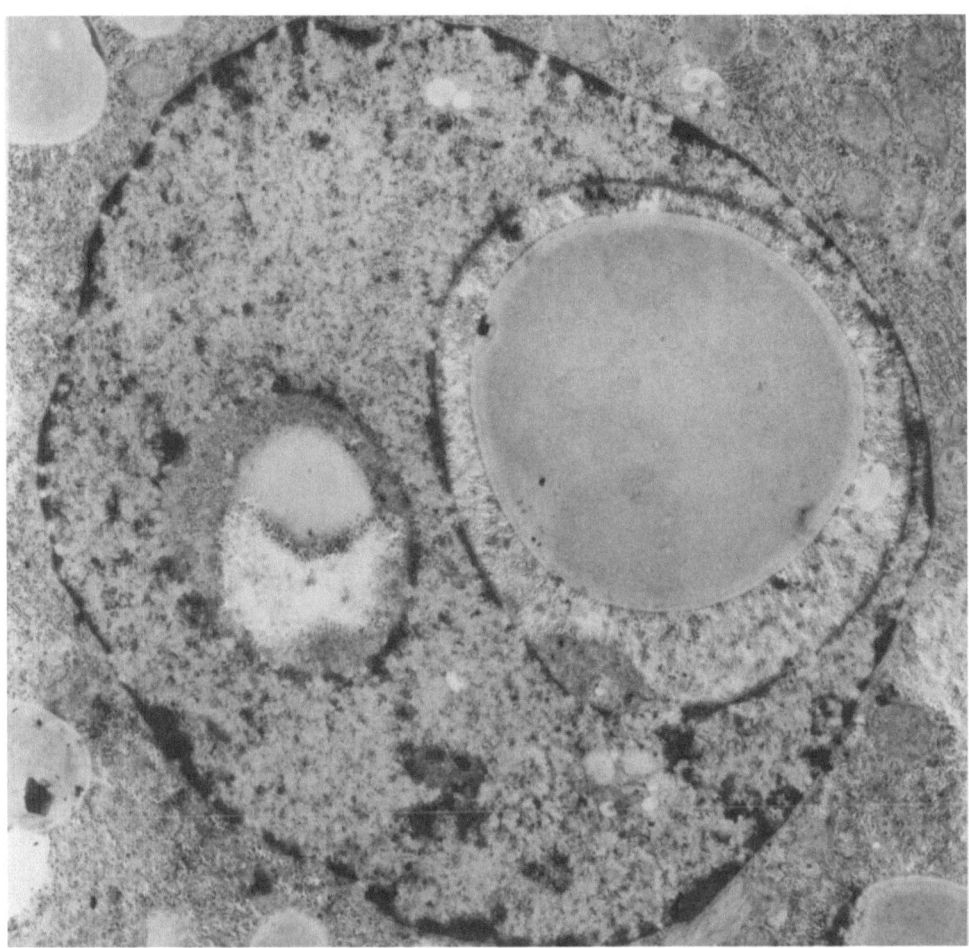

Fig. 107. Two intranuclear inclusions in a hepatocyte of a 9-month-old male WAG/Rij rat which had been irradiated with 10 Gy X-rays of the liver at the age of 3 months. Each inclusion is composed of a portion of cytoplasm containing glycogen particles and a lipid vacuole. (Specimen courtesy of Dr. A.M. de Leeuw, Institute for Experimental Gerontology, TNO, Rijswijk) TEM, ×12 720

to a chemical compound was composed of a network of vesicles and tubules and was regarded as arising from hypertrophied smooth endoplasmic reticulum (Schauer and Kunze 1976).

Differential Diagnosis

Both intranuclear and intracytoplasmic inclusions in rat hepatocytes are distinctive cytologic alterations. Their characteristic light microscopy features, aided if necessary, by appropriate histochemical or electron microscopy studies, should preclude misinterpretation.

A small intranuclear inclusion may at times be mistaken for a nucleolus; the intense basophilia of the latter and the lack of a well-defined limiting membrane are reliable distinguishing criteria. Similarly, intranuclear inclusions may sometimes be regarded as being viral in origin. Of the common rat viruses, only a parvovirus infection in suckling or weanling rats is likely to result in intranuclear inclusion bodies in hepatocytes (Jacoby et al. 1979). Such viral inclusion bodies are basophilic and large, often filling the nucleus, and are associated with hepatocellular necrosis.

Since none of the common rat viruses are associated with hepatocytic intracytoplasmic inclusion bodies (Jacoby et al. 1979), the most important consideration regarding these structures is to attempt to define the nature of the substance accumulated within the inclusion. Rarely, partial or

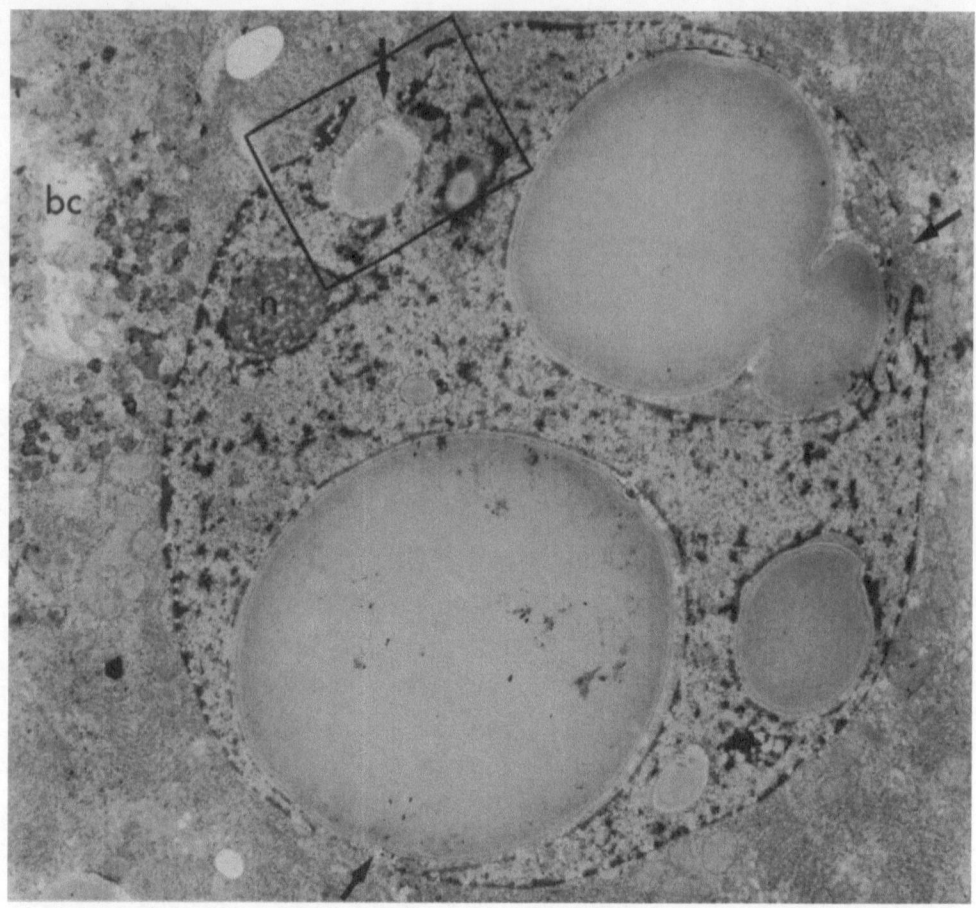

Fig. 108. Multiple lipid-containing cytoplasmic inclusions in another hepatocyte nucleus from the same rat in Fig. 107. At several locations (*arrows*), a continuity exists between the cell cytoplasm and the contents of the inclusion bodies, representing the points of invagination of the nuclear membrane. *bc*, Bile canalicus; *n*, nucleolus. (Specimen courtesy of Dr. A.M. de Leeuw, Institute for Experimental Gerontology TNO, Rijswijk.) TEM, ×5760

total coagulation necrosis of single hepatocytes, resulting in Councilman-like bodies, may be misinterpreted as intracytoplasmic inclusions, but, unlike the latter, these frequently replace the nucleus of the hepatocyte. In addition, remnants of cytoplasmic organelles can usually be demonstrated by electron microscopy.

Biologic Features

Beyond descriptive studies, details on the incidence and morphogenesis of intranuclear inclusions in rat hepatocytes are scarce. Considerably more work has been carried out on the intranuclear inclusions in mouse hepatocytes, which resemble those in the rat in most respects. Presumably, the biologic features of the "lesion" in both species may be considered comparable. Thus Herbst (1976) has described in detail the various stages seen in the formation of intranuclear inclusions in Chbi:NMRI (SPF) mice of different ages. The earliest stage consisted of a simple invagination of the nuclear membrane introducing a portion of cytoplasm into the nucleus. The invaginated portion subsequently became isolated within the nucleus as a membrane-bound structure, at which time the cytoplasmic organelles showed signs of degeneration. Ultimately, the inclusion body membrane became fragmented and disappeared altogether, often resulting in a free "glycogen body" within the nu-

cleus. This investigator concluded that the incidence of intranuclear inclusions in mice increased
with advancing age and that the presence of inclusions was not correlated with pathologic changes
in the cells harboring such inclusions or in the
surrounding parenchyma. The author speculated
that the intranuclear cytoplasmic invaginations
may occur as a result of reduced intranuclear pressure due not to cytoplasmic swelling, but to
nuclear atony, perhaps as a result of specific metabolic changes in the animal. However, the exact
etiology and significance remain unknown.

In rats, intranuclear inclusions are also found in
apparently normal hepatocytes, although no information on a possible age dependency is available.
Intranuclear inclusions have also been described
in rat hepatocellular carcinomas (Schauer and
Kunze 1976).

Intracytoplasmic inclusions in both normal and
abnormal rat hepatocytes were described in detail
by Popper et al. (1960). The PAS-positive inclusions were nonglycogenic in nature and were composed in part of mucopolysaccharides, lipids, and
other unidentified substances. The authors suggested that some of these structures, at least, appeared to be related to lysosomes and may have
resulted from alterations in the metabolic activities of the hepatocyte. Ruebner et al. (1965) described similar PAS-positive, diastase-resistant
hyaline globules in liver cells of a normal Sprague-
Dawley rat.

Other studies (reviewed by Jones 1967) have related the development of intracytoplasmic inclusion bodies in rat hepatocytes to experimental
procedures such as partial hepatectomy, median
lobe ligation, and parenteral administration of albumin, serum, whole blood, and other substances.
The resulting inclusion bodies were shown to contain various types of proteins, including serum
immunoglobulins, albumin, and others, depending
on the study. Some of these cytoplasmic inclusions
apparently resulted from a disturbed permeability
of the liver cell membrane.

Acidophilic cytoplasmic inclusion bodies have
also been found in hepatocellular neoplasms of
the rat (Edwards and White 1941; Schauer and
Kunze 1976), which develop following treatment
with a variety of chemical carcinogens. Based on
ultrastructural examination, some of these were
regarded as arising from hypertrophied smooth
endoplasmic reticulum.

It is apparent that cytoplasmic inclusions may be
rather heterogeneous in their origin, and conclu-

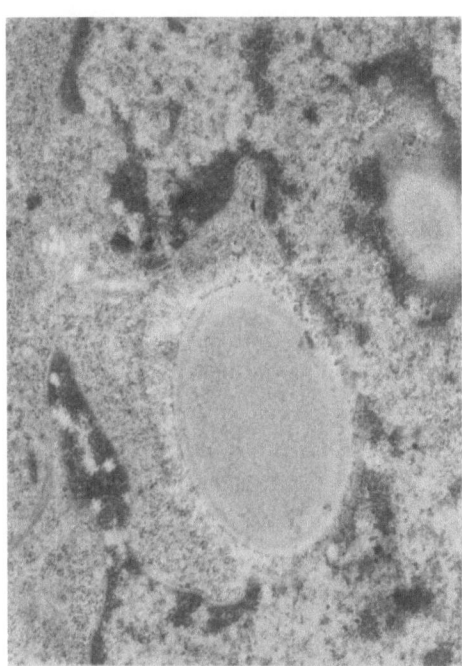

Fig. 109. Higher magnification of invaginated nuclear mem-
brane seen in the *rectangle* in Fig. 108. (Specimen courtesy of
Dr. A.M. de Leeuw, Institute for Experimental Gerontology
TNO, Rijswijk.) TEM, ×17 000

sions about their significance or pathogenesis must
be made with caution.

Comparison with Other Species

As mentioned previously, intranuclear inclusions
have been well documented in mice (Andrew
1962; Andrew et al. 1943; Frith and Ward 1980;
Herbst 1976; Ruebner et al. 1965). Several authors
have indicated that the occurrence of such inclusions may be age related (Andrew et al. 1943;
Andrew 1962; Hollander and Thung 1966). The
incidence of intranuclear inclusions in hepatocytes
of normal, 23.5-month-old male CBA mice was
reported to range from 0.25% to 0.72% (Hollander 1967). In mice of the same age and strain
partially hepatectomized 7–8 months earlier, the
incidence ranged from 1.31% to 2.93%. This was
regarded as indicating that the regenerative
growth following partial hepatectomy induced
the appearance of certain changes associated with
aging. Similar intranuclear inclusions were found
in 3-month-old male BCBA mice as early as 8h

after partial hepatectomy (Hollander and Thung 1966).

Intranuclear inclusions have also been found in neoplastic mouse hepatocytes (Frith and Ward 1980). In male CBA mice with liver tumors, the incidence of intranuclear inclusions in tumor cells was not different from that in normal liver cells from mice of the same age (Hollander and Thung 1966).

Intranuclear inclusions similar to those reported in rats and mice have been found in normal human hepatocytes (Andrew et al. 1943; Carr et al. 1960) as well as in neoplastic cells of a human hepatocellular carcinoma (Hollander 1967). Similar intranuclear inclusions have also been described in hamsters (Lussier and Pavilanis 1968). Of an apparently different nature are the acidophilic crystalline inclusions not uncommonly found in canine hepatocytic nuclei (Richter et al. 1965). These inclusions are known to be composed of protein, although their origin and significance remain unclear.

Intracytoplasmic inclusions have been described in normal and neoplastic mouse hepatocytes (Frith and Ward 1980; Wharton and Wright 1977). These inclusions, which occur infrequently in normal hepatocytes and much more commonly in benign hepatocellular tumors, have been well characterized morphologically. The ultrastructural features of intracytoplasmic inclusions in hepatocellular neoplasms of male CBA/H-T6T6 mice have been described (Helyer and Petrelli 1978), and three types of inclusion bodies were distinguished. The most common type was composed of a dense reticulated substance associated with a diffuse granulofibrillary material and arose within dilated cisternae of rough endoplasmic reticulum. The second type also occurred within rough endoplasmic reticulum, but was larger than the first and was composed mainly of a fine granular matrix. The third type was least common and was characterized as a non-membrane-bound accumulation of dense granulofibrillar material in the cytoplasm. The inclusions were shown to contain protein and phospholipids, although their exact composition was not defined. In another morphological study of similar intracytoplasmic inclusion bodies in diethylnitrosamine-induced hepatocellular tumors in C57BL × C3HF$_1$ mice, evidence was obtained by immunofluorescence methods that the inclusions contained a glycoprotein which reacted positively with antisera against human α_1-antitrypsin

(Rijhsinghani et al. 1980). To date, no evidence of viral material in such inclusions has been obtained, and their precise etiology and significance remain unknown.

Intracytoplasmic hyaline inclusions have also been reported in normal (Nakanuma et al. 1982) and neoplastic (Dekker and Krause 1973) human hepatocytes. Some of these were composed of a granular protein-containing substance and located within rough endoplasmic reticulum (Dedkker and Krause 1973). Based on their ultrastructural and histochemical features, they were felt to be distinct from the so-called Mallory's alcoholic hyline, and they closely resembled the inclusions found in mouse hepatocytes. Other inclusions in human hepatocytes not characterized electron microscopically were shown by immunohistochemistry to contain a variety of plasma proteins (Nakanuma et al. 1982). The authors postulated that these inclusions may have formed as a result of altered permeability of the hepatocytes, perhaps related to hypoxia or circulatory disturbances in the liver.

Cytoplasmic inclusions in hepatocytes of other species, including dogs, rabbits, and monkeys, have been described and their characteristics have been summarized by McClure et al. (1978).

References

Andrew W (1962) An electron microscope study of age changes in the liver of the mouse. Am J Anat 110:1–18

Andrew W, Brown HM, Johnson JB (1943) Senile changes in the liver of mouse and man, with special reference to the similarity of the nuclear alterations. Am J Anat 72:199–221

Carr RD, Smith MJ, Keil PG (1960) The liver in the aging process. Histology. Arch Pathol 70:15–18

Dekker A, Krause JR (1973) Hyaline globules in human neoplasms. A report of three autopsy cases. Arch Pathol 95:178–181

Edwards JE, White J (1941) Pathologic changes, with special reference to pigmentation and classification of hepatic tumors in rats fed p-dimethylaminoazobenzene (butter yellow). JNCI 2:157–183

Frith CH, Ward JM (1980) A morphologic classification of proliferative and neoplastic hepatic lesions in mice. J Environ Pathol Toxicol 3:329–351

Helyer BJ, Petrelli M (1978) Cytoplasmic inclusions in spontaneous hepatomas of CBA/H-T6T6 mice. Histochemistry and electron microscopy. JNCI 60:861–869

Herbst M (1976) Glycogenous hepatonuclear inclusions in the aged mouse – an electron microscopic study of the histogenesis of nuclear inclusions. Pathol Eur 11:69–79

Hollander CF (1967) Preliminary note on the relations between regenerative growth, ageing and tumor formation in the mouse liver. Epatologia 13:447–454

Hollander CF, Thung PJ (1966) Relations between regenerative growth and ageing in the mouse liver. In: Lindop PJ, Sacher GA (eds) Proc colloq on radiation and ageing. Taylor and Francis, London, pp 3–14

Jacoby RO, Bhatt PN, Jonas AM (1979) Viral diseases. In: Baker HJ, Lindsey JR, Weisbroth SH (eds) The laboratory rat, vol I. Biology and diseases. Academic, New York, chap 11, pp 271–306

Jones TC (1967) Pathology of the liver of rats and mice. In: Cotchin E, Roe FJC (eds) Pathology of laboratory rats and mice. Blackwell Scientific, Oxford, chap 1, pp 1–23

Kleinfeld RG, Greider MH, Frajola WJ (1956) Electron microscopy of intranuclear inclusions found in human and rat liver parenchymal cells. J Biophys Biochem Cytol 2 [Suppl]:435–438

Lussier G, Pavilanis V (1968) Nuclear inclusions in the liver cells of the golden hamster. Can J Comp Med 32:568–570

McClure HM, Chapman WL Jr, Hooper BE, Smith FG, Fletchert OJ (1978) The digestive system. In: Benirschke K, Garner FM, Jones TC (eds) Pathology of laboratory animals, vol 1. Springer, New York Heidelberg Berlin, chap 4, pp 175–317

Nakanuma Y, Ohta G, Matsubara F, Wantanabe K, Doishita K (1982) Cytoplasmic blood plasma inclusions in human hepatocytes. Liver 2:212–221

Popper H, Paronetto F, Barka T (1960) PAS-positive structures of nonglycogenic character in normal and abnormal liver. Arch Pathol 70:300–313

Richter WR, Stein RJ, Rdzok EJ, Moize SM, Bischoff MB (1965) Ultrastructural studies of intranuclear crystalline inclusions in the liver of the dog. Am J Pathol 47:587–599

Rijshsinghani K, Krakower C, Swerdlow M, Abrahams C, Ghose T (1980) Alpha-1-antitrypsin in intracellular inclusions of diethylnitrosamine induced hepatomas of C57BL × C3HF₁ mice. Carcinogenesis 1:473–479

Ruebner BH, Lindsey JR, Melby EC Jr (1965) Hepatitis and other spontaneous liver lesions of small experimental animals. In: Ribelin WE, McCoy JR (eds) The pathology of laboratory animals. Thomas, Springfield, chap 7, pp 160–182

Schauer A, Kunze E (1976) Tumours of the liver. In: Turusov VS (ed) Pathology of tumours in laboratory animals, vol I: tumours of the rat, part 2. IARC, Lyon, pp 41–72 (IARC scientific publications no 6)

Wharton FP, Wright DJM (1977) Observations on a new liver inclusion in the mouse. Lab Anim 11:109–111

Extramedullary Hematopoiesis, Liver, Rat

Matthew J. van Zwieten and Carel F. Hollander

Synonyms. Extramedullary hemopoiesis, hepatic hematopoiesis, ectopic hematopoiesis, myeloid metaplasia

Gross Appearance

Extramedullary hematopoiesis in the liver is usually not discernible macroscopically. In severe cases, the liver may be slightly enlarged.

Microscopic Features

In late prenatal and early postnatal life, rat liver sinusoids contain numerous hematopoietic cells (Fig. 110). Immature myeloid and erythroid cells, as well as megakaryocytes, are spread diffusely throughout the liver of fetal and newborn rats. Variable numbers of hematopoietic cells may also be found in adult rat livers under certain circum-

stances (Fig. 111). They appear as small foci or clusters of typical hyperchromatic erythroid and/ or myeloid cells with cytologic features identical to those in bone marrow and spleen. Each individual cluster tends to be composed of a relatively pure population of either erythroid or myeloid cells, but not a mixture of the two. When myeloid cells predominate in a particular case, the term granulopoiesis is applied, while erythropoiesis indicates a predominance of erythrocytic precursors. Megakaryocytes may also be found in association with extramedullary hematopoiesis in the adult rat liver.

Foci of hematopoietic cells are often randomly distributed throughout a section of liver, although at time they may be confined to a particular region, such as a nodule of proliferating hepatocytes or an area of altered vascular architecture, e.g., dilated sinusoids.

It is important to keep in mind that when extramedullary hematopoiesis is found in the

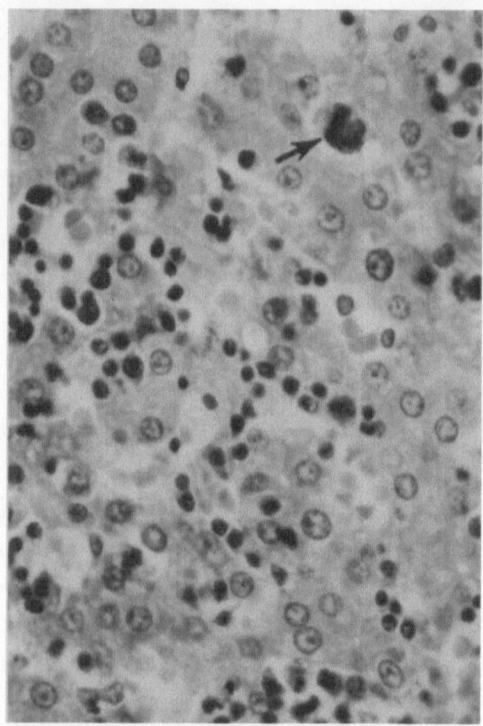

liver, splenic hematopoiesis is frequently extensive, and foci of hematopoietic cells can be found in other tissues including lymph nodes (medullary region), adrenal glands, and perirenal adipose tissue.

Ultrastructure

One report (Enomoto et al. 1978) describes the foci of extramedullary hematopoiesis which appear in the adult rat in the course of carcinogenesis caused by 3'-methyl-4-(dimethylamino) azobenzene (3'-Me-DAB). The hematopoietic cells, identified as erythroblasts, were located between the endothelial cells and hepatocytes and therefore in the spaces of Disse. The plasma membranes of hepatocytes were deeply indented on the surface adjacent to the erythroblasts, providing lacunae in which the erythropoietic cells were situated.

Certain ultrastructural features of hematopoietic cells in fetal rat liver, and especially their intimate contact with the developing sinusoidal and parentchymal liver cells, have been reported (Naito and Wisse 1977; Bankston and Pino 1980). Some of the more extensive data available on the ultrastructure of hematopoietic cells in fetal mouse livers (e.g., Medlock and Haar 1983a,b) may be applicable in part to the rat.

Differential Diagnosis

The major conditions which must be distinguished from extramedullary hematopoiesis in the liver include inflammation and involvement of the liver by hematologic neoplasms, such as myeloid leukemia. Inflammatory cell infiltrates generally are composed predominantly of mature granulocytes, often accompanied by other cell types, including lymphocytes and macrophages, as opposed to foci of granulopoiesis, which are composed of myeloid cells in various stages of maturation. In addition, ample evidence of hepatocellular

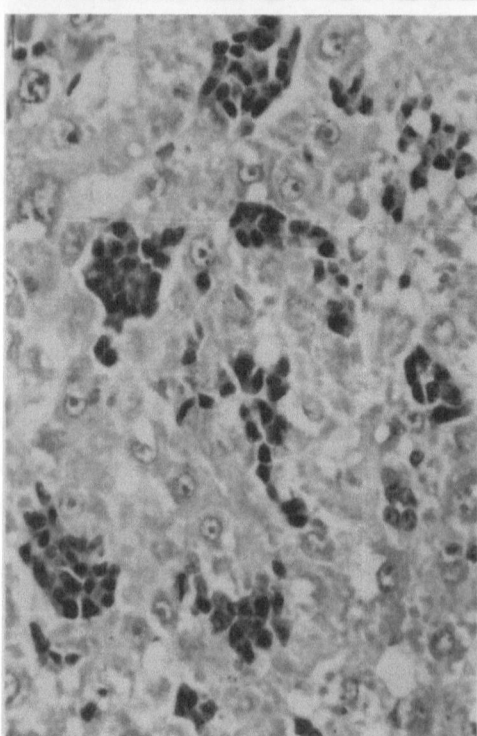

◄

Fig. 110. (*above*) Extramedullary hematopoiesis, liver of a 1-day-old BN/BiRij rat. Note the megakaryocyte (*arrow*). Hematoxylin-phloxine-saffron, ×600

Fig. 111. (*below*) Clusters of hematopoietic cells, liver of a 12-month-old female WAG/Rij rat. The erythropoietic cell foci in this case were confined to an area of hepatocellular proliferation. Hematoxylin-phloxine-saffron, ×500

degeneration and necrosis is usually present in inflammatory liver lesions. Leukemic infiltrates usually exhibit a maturation arrest, and such cells are frequently also present in numerous other tissues, including the peripheral blood. The presence of erythropoietic foci and megakaryocytes in the liver in addition to granulopoietic elements aids in the differential diagnosis. Nonetheless, at times it may be difficult, if not impossible, to distinguish with certainty between early leukemic infiltrates and foci of hematopoiesis, especially granulopoiesis.

Biologic Features

The embryonic yolk sac is the first hematopoietic organ to develop in mammals, but the liver is the main site of hematopoiesis during the latter half of fetal development. In the rat, hematopoiesis in the liver begins between the 12th and 13th day of gestation (Metcalf and Moore 1971) and by day 15 hematopoietic cells account for about 37% of the liver volume (Greengard et al. 1972). At birth, about 10% of the liver volume is occupied by hematopoietic cells and the bone marrow takes over as the primary hematopoietic site, although in rodents the spleen also remains a hematopoietic organ throughout much of life. During the first few weeks of postnatal life, the number of hematopoietic cells in the liver diminishes sharply, and by day 28 only about 1% of the liver volume is filled by hematopoietic cells (Greengard et al. 1972). Although not readily appreciable in standard light microscopy sections, electron microscopy studies have shown that hematopoiesis in the liver is predominantly an extravascular event, i.e., the immature blood-forming cells are located in the subendothelial space in close contact with the hepatocytes (Bankston and Pino 1980; Medlock and Haar 1983a). The endothelial cells are involved in the transmural migration of erythroid cells and megakaryocytic processes which pass through the sinusoidal lining by a process of diapedesis (Bankston and Pino 1980).

A variety of conditions can result in the reappearance of hematopoietic activity in the adult liver. The most common conditions are those which result in a particular demand for erythrocytes, leukocytes, or thrombocytes and in which the bone marrow is unable to meet this demand (Jones 1967). Anemia, inflammation, or extensive necrosis, such as may be associated with neo-

plasms, abscesses, or infarcts, are examples of conditions which may result in extramedullary hematopoiesis. An embryonic-type hemoglobin has been detected in hematopoietic cells in the liver of rats during hepatocarcinogenesis (Enomoto et al. 1980). This finding suggests that the appearance of such foci of hematopoiesis in the adult rat liver could be explained on the basis of a fetal environment in the liver during the carcinogenic process. Experimental infection of rats with a type C helper virus (Nooter et al. 1981) also resulted in marked hepatic erythropoiesis.

Comparison with Other Species

Extramedullary hematopoiesis involving the liver has been described in many animal species, including humans. The circumstances which cause a return of hematopoietic activity to the adult liver are generally quite similar in most species. In the human fetus, the period of hepatic hematopoiesis begins by about the third month of gestation, reaches a peak during the fifth and sixth month, and declines by the tenth month (Miale 1971). It has been reported that granulopoietic cells in the human fetal liver are located almost exclusively within the portal areas in close association with connective tissue fibers (Calvo and Carbonell 1980), a site also preferentially involved initially by leukemic cells of the myeloid series. Erythropoietic cells, on the other hand, are found beneath the sinusoidal lining in close contact with fetal hepatocytes.

In the fetal canine liver, hematopoiesis begins on about day 33 of gestation and consists primarily of erythropoiesis and thrombopoiesis until day 57, when granulopoiesis is also observed (W. Calvo, personal communication). Also in the dog, granulopoietic cells are seen predominantly in the portal areas, whereas erythropoietic cells are found extravascularly in close contact with hepatocytes (W. Calvo, personal communication). It is presumed that hepatocytes, or possibly Kupffer cells, provide an essential environment for the developing erythroid cells, whereas the mesenchymal tissues of the portal tract serve a similar function for the granulocytic precursors.

Extramedullary hematopoiesis has been described in the mouse liver by a number of authors (Dunn 1954; Frith and Ward 1980; Hollander 1975). Dunn (1954) provided detailed morphological criteria to aid in differentiating non-malignant extramedullary myelopoiesis from myeloid leukemia. In mice, the tissues adjacent to the portal vein were also reported to be the site of greatest granulopoietic activity (Dunn 1954). Cosgrove and Upton (1965) referred to extramedullary hematopoiesis in mice as either a leukemoid reaction or erythroblastosis, depending on the types of cells predominating. They also pointed out that features such as the degree of maturation of granulocytic elements and the presence of megakaryocytes, erythroid cells, and lymphoid cells are of importance in distinguishing a leukemoid reaction from leukemia.

Hepatic extramedullary hematopoiesis in adult mice occurs in response to repeated blood loss as well as spontaneous (Dunn 1954) or phenylhydrazine-induced (Ploemacher and van Soest 1977) hemolytic anemia. The latter investigators described the ultrastructural features of erythropoietic islands in the livers of phenylhydrazine-treated CBA/Rij mice, the possible origin of two types of centrally located macrophages consistently present in such islands, and some differences between erythropoiesis in adult liver and in fetal liver.

Active erythropoiesis is also commonly found in mice with histiocytic sarcoma (reticulum cell sarcoma, Dunn's type A; Dunn 1965), in which clumps of hematopoietic cells are scattered among the neoplastic histiocytes (Lemon 1967). Extensive extramedullary granulopoiesis may develop in mice with transplanted submaxillary gland myoepitheliomas or other neoplasms, including fibrosarcomas and squamous cell carcinomas (Dunn 1965).

References

Bankston PW, Pino RM (1980) The development of the sinusoids of fetal rat liver: morphology of endothelial cells, Kupffer cells, and the transmural migration of blood cells into the sinusoids. Am J Anat 159:1–15

Calvo W, Carbonell F (1980) The development of liver granulopoiesis in the human fetus. In: Lucarelli G, Fliedner TM, Gale RP (eds) Fetal liver transplantation. Excerpta Medica, Amsterdam, pp 14–18

Cosgrove GE, Upton AC (1965) Pathology of the reticuloendothelial system. In: Ribelin WE, McCoy JR (eds) The pathology of laboratory animals. Thomas, Springfield, chap 2, pp 21–28

Dunn TB (1954) Normal and pathologic anatomy of the reticular tissue in laboratory mice, with a classification and discussion of neoplasms. JNCI 14:1281–1433

Dunn TB (1965) Spontaneous lesions of mice. In: Ribelin WE, McCoy JR (eds) The pathology of laboratory animals. Thomas, Springfield, chap 11, pp 303–329

Enomoto K, Dempo K, Mori M, Onoe T (1978) Histopathological and ultrastructural study on extramedullary hematopoietic foci in early stage of 3'-methyl-4-dimethyl-aminoazobenzene hepatocarcinogenesis. Gann 69:249–254

Enomoto K, Dempo K, Mori M, Onoe T (1980) Demonstration of embryonic-type hemoglobin in extramedullary hematopoietic cells in the liver during experimental liver carcinogenesis by 3'-methyl-4-dimethylaminoazobenzene. Cancer Res 40:1769–1773

Frith CH, Ward JM (1980) A morphologic classification of proliferative and neoplastic hepatic lesions in mice. J Environ Pathol Toxicol 3:329–351

Greengard O, Federman M, Knox WE (1972) Cytomorphometry of developing rat liver and its application to enzymic differentiation. J Cell Biol 52:261–272

Hollander CF (1975) Embryology and ageing effects. In: Butler WH, Newberne PM (eds) Mouse hepatic neoplasia. Elsevier Scientific, Amsterdam, chap 2, pp 7–19

Jones TC (1967) Pathology of the liver of rats and mice. In: Cotchin E, Roe FJC (eds) Pathology of laboratory rats and mice. Blackwell Scientific, Oxford, chap 1, pp 1–23

Lemon PG (1967) Hepatic neoplasms of rats and mice. In: Cotchin E, Roe FJC (eds) Pathology of laboratory rats and mice. Blackwell Scientific, Oxford, chap 2, pp 25–56

Medlock ES, Haar JL (1983a) The liver hemopoietic environment: I. Developing hepatocytes and their role in fetal hemopoiesis. Anat Rec 207:31–41

Medlock ES, Haar JL (1983b) The liver hemopoietic environment: II. Peroxidase reactive mouse fetal liver hemopoiesis. Anat Rec 207:43–53

Metcalf D, Moore MAS (1971) Haemopoietic cells. Elsevier/North Holland, Amsterdam

Miale JB (1971) Hemopoietic system: reticuloendothelium, spleen, lymph nodes, blood, and bone marrow. In: Anderson WAD (ed) Pathology, 6th edn, vol II. Mosby, St Louis, chap 31, pp 1297–1386

Naito M, Wisse E (1977) Observations on the fine structure and cytochemistry of sinusoidal cells in fetal and neonatal rat liver. In: Wisse E, Knook DL (eds) Kupffer cells and other liver sinusoidal cells. Elsevier/North-Holland Biomedical, Amsterdam, pp 497–505

Nooter K, Dubbes R, Jore J, Zurcher C (1981) Induction of haemopoietic tumours in rats by the type-C helper virus of the woolly monkey sarcoma virus. Leuk Res 5:97–99

Ploemacher RE, van Soest PL (1977) Morphological investigation on phenylhydrazine-induced erythropoiesis in the adult mouse liver. Cell Tissue Res 178:435–461

Nutritional Fatty Liver, Cirrhosis, and Hepatocellular Carcinoma, Rat, Mouse

Paul M. Newberne and Adrianne E. Rogers

Synonyms. Alcoholic cirrhosis, nutritional cirrhosis, nutritional hepatocellular carcinoma

Gross Appearance

In methyl (lipotrope or choline) deficiency, as lipid accumulates the liver loses its mahogany-brown color and becomes beige or yellowish. The weight of the liver is reduced, and it is buoyant when immersed in aqueous solution. The cut surface is friable and oily in texture, and increased fibrous connective tissue gives the surface of the liver a roughened appearance, dividing the parenchyma into grossly visible lobules or sublobules of various size (Fig. 112).

Regeneration of liver cells eventually results in the presence of many distinct, individual nodules of varied size that present a botryoid or grape-cluster appearance (Figs. 113, 114). Maturation of fibrous tissue contributes to contraction and enhances the lobular appearance. The color varies among these nodules; many are light-brown to tan, while others are bright-yellow, depending on lipid content and degree of parenchymal regeneration. Large, circumscribed masses may be present in later stages as the result of neoplastic growth.

Microscopic Features

The development of cirrhosis proceeds through a series of histologically identifiable changes ranging from fat accumulation, first in the centrilobular zone and then extending throughout the lobules, leading to fibrosis. Extensive fibrosis and nodular regeneration of hepatic cells follow to become the essential hallmarks of cirrhosis. Some of the rats (Mikol et al. 1983) and mice (Newberne et al. 1982) later develop hepatocellular carcinoma that may metastasize to the lung. Fatty fibrosis and cirrhosis evolve in a well-defined sequence of histologic changes in the animal model, similar in many respects to changes observed in nutritional or alcoholic cirrhosis in humans. Figure 113 illustrates the gross and Fig. 116 the microscopic ap-

pearance of fatty liver, fibrosis, and early cirrhosis in a rat fed a methyl-deficient diet for 29 weeks. Figure 114 demonstrates the mature cirrhotic liver. Figure 116 shows the microscopic appearance of the liver of a rat fed the deficient diet for 3 months, demonstrating the changes typical of this intermediate stage. Here, islands of parenchymal cells are separated by connective tissue, proliferating bile duct cells, inflammation, and an accumulation of ceroid pigment in the bands of connective tissue coursing through the nodular hyperplastic liver. This is typical of early cirrhosis, but the lesions have not yet fully matured into the final, full-blown cirrhotic organ; however, the proliferating hepatocytes have largely excluded fat from the nodular areas. Figures 115 and 117 illustrate progressive stages in the deficient liver, and Fig. 118 illustrates the end stage of cirrhotic liver of a rat maintained on the deficient diet for more than 1 year. The organ is grossly nodular and much of the fat has disappeared, although some nodules still retain varying amounts of large-droplet fat. The fibrous tissue bands have matured and contracted, and the regenerative, nodular nature of the lesion is emphasized.

Biologic Features

History

The nutritional fatty liver and cirrhosis model had its origin in experiments with dogs and rats during the isolation, purification, and identification of insulin in 1921–1922 (Fisher 1923/1924). Impaired assimilation of fat and protein in the depancreatized dog was recognized during the various experiments leading to the discovery of insulin, and it was noted that the diabetes induced in the dog by pancreatectomy was reversed by feeding raw pancreas. Later, crude pancreatic extract or purified insulin served a similar purpose. However, despite this, the dogs did not survive more than a few months and died with an enlarged fatty liver. At first, early demise and fatty liver was attributed to a lack of the external secretion of the pancreas (Bliss 1922), but this hypothesis was not accepted by all investigators. Attention was then focused on the failure of liver function and disturbance of liver fat metabolism in the depancreatized, insulin-supplemented dog (Hershey and Soskin 1931). Hershey proposed that the deterioration in physical condition and depressed liver function of the dogs was related to phospholipids. Testing this hypothesis led to the

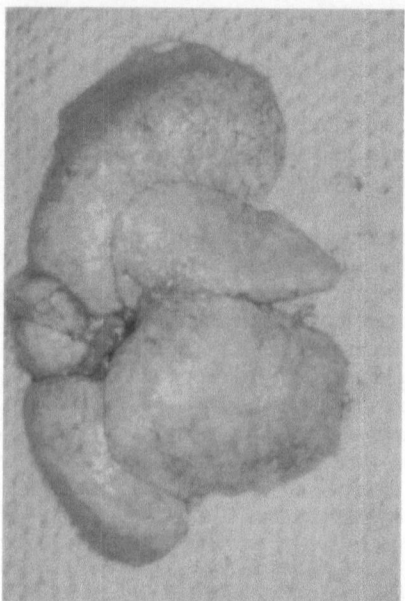

Fig. 112. Liver, rat. The pale, fatty organ of an animal fed a choline-deficient diet for 3 months. The surface is irregular with a fine nodularity resulting from fibrosis and parenchymal cell regeneration

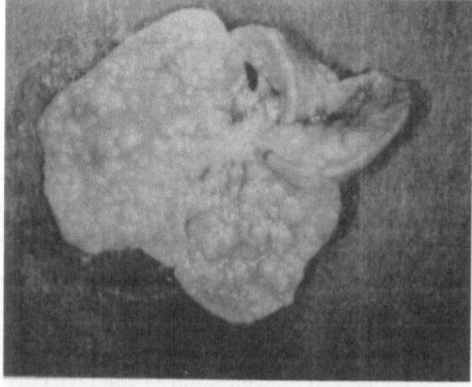

Fig. 113. Gross appearance of the liver of a rat fed a choline-deficient diet for 29 weeks. Note the nodular surface typical of a cirrhotic liver

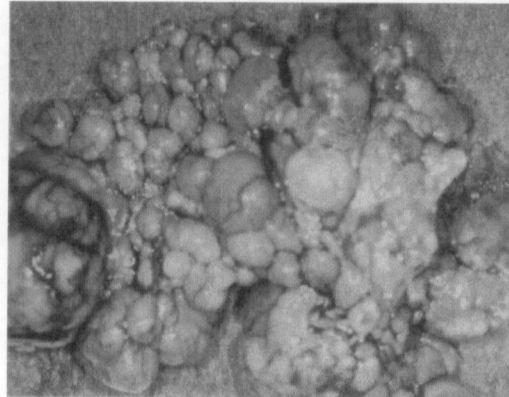

Fig. 114. Cirrhotic liver, fully mature rat. Most fat has disappeared; nodular islands of hepatocytes have been compressed by the contracted fibrous tissue to form the typical advanced, nodular lesions

discovery that lecithin, contained in phospholipids, prevented the condition and, further, that the only component of lecithin that prevented the fatty liver was choline (Best and Huntsman 1932).

Following the discoveries of Hershey and Best, a number of investigators joined the search for metabolic effects of choline and its relation to fatty liver. A large number of studies culminated in the concept of transmethylation, one-carbon metabolism, and elucidated the significance of choline, methionine, folic acid, and vitamin B_{12}, collectively designated as lipotropes (Best and Huntsman 1935; Channon et al. 1938; du Vigneaud et al. 1939; Bennett 1950). The relation of these nutrients to overall tissue metabolism, particularly the liver, is continuing to emerge with increasing use of more sophisticated methods and technology (Newberne 1993a).

Concomitant to and in parallel with some of the studies noted above, there was intense interest in the mechanism(s) of fatty liver associated with choline (methyl) deficiency. With respect to this model, however, the report of Copeland and Salmon (1946) was the seminal paper describing the effects of choline (methyl) deficit on the liver. In the studies conducted in Salmon's laboratory, it was shown that cirrhosis in the rat liver proceeded through a series of changes from fat infiltration through fibrosis and cirrhosis; a significant portion of these livers developed hepatocellular carcinoma. A most careful investigator, Stanley Hartroft, contributed significantly to the elucidation of fatty liver (Hartroft 1954). These interesting findings are detailed elsewhere and make for fascinating reading (Salmon and Copeland 1954; Salmon and Newberne 1963; Newberne 1986).

The methyl-deficient (CD) rat model was first described in detail more than 40 years ago in a series of reports from the laboratory of W.D. Salmon, Auburn University, where its early development took place (Engel and Salmon 1941). Rats maintained on choline-deficient diets for long periods reportedly developed hepatocellular carcinomas, an observation that attracted widespread attention in scientific circles because, in this case, omitting a substance (choline) from the diet, rather than adding one, resulted in neoplasia (Copeland and Salmon 1946; Salmon and Copeland 1954; Newberne et al. 1982). Hepatocellular carcinoma associated with choline deficiency was considered to be a result of a methyl group deficiency, which, over time, produced a series of alterations including fatty liver, rupture of fat cells with release of fat into extracellular space, parenchymal cell necrosis and hyperplasia, fibrosis, cirrhosis, and ultimately liver cancer.

While choline deficiency alone is unlikely to be encountered under usual conditions of veterinary or human clinical experience, protein deficiency and low dietary concentrations of methionine, folic acid, and vitamin B_{12} are not uncommon (Newberne and McConnell 1980a); such deficits result in methyl deficiency in humans and lower animals.

Etiology

All strains of rats and mice examined thus far are sensitive to the deficiency and its enhancing effects on chemically induced cancer. Generally, the animals have been fed a choline-free diet that is also marginal in methionine (Table 12) from weaning until cirrhosis develops, usually 6–8 months. Rats or mice must be placed on the diet from weaning and kept on it. It appears, however, that the initiation process for neoplasia is estab-

▶

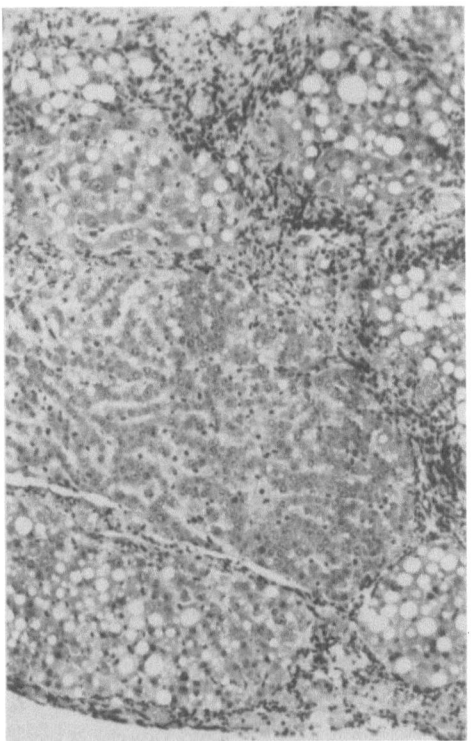

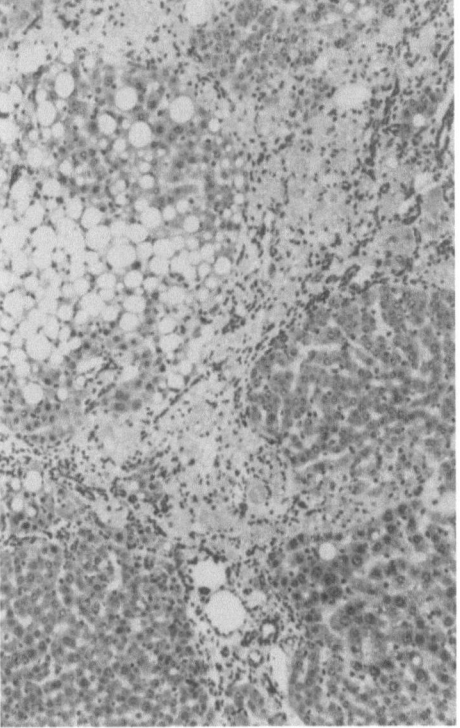

Fig. 115. (*above*) Liver, rat, after a choline-deficient diet for 29 weeks. Fibrous connective tissue bands separate islands of liver cells and contain bile duct cells, inflammatory infiltrate, and yellow-brown pigment (ceroid). The proliferated islands of hepatocytes have largely excluded fat. H&E, ×100

Fig. 116. (*below*) Liver, same rat as in Fig. 112. Fat, fibrosis, parenchymal cell, and bile duct hyperplasia are characteristic of this stage of development toward cirrhosis. H&E, ×100

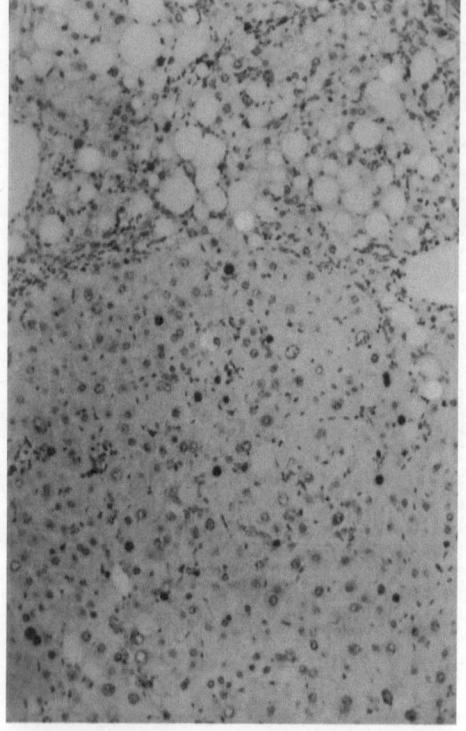

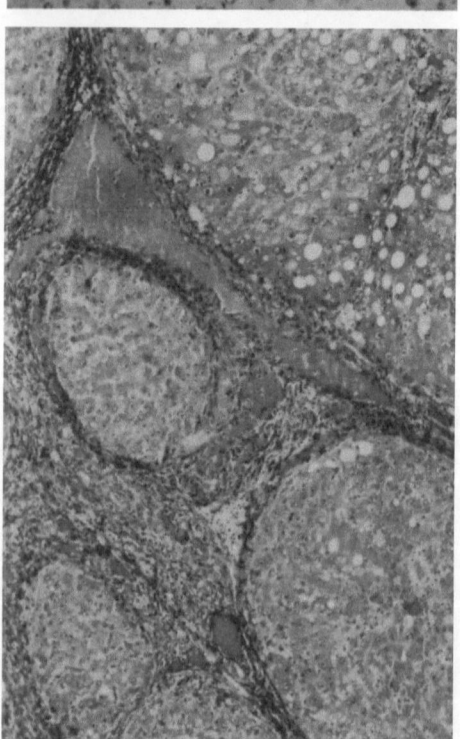

lished early in the deficiency, possible in the first 3 months of dietary exposure (Lombardi and Smith 1994). A few losses occur, usually between days 8 and 10, due to the hemorrhagic kidney syndrome and uremia associated with choline deficiency. The deprived rodent liver proceeds to cirrhosis with very little in the way of clinical signs aside from a modest inhibition in weight gain.

The severity of deficiency and the rapidity with which fatty liver and cirrhosis develop are governed by dietary levels of these nutrients and by the rats' age at the time of initiating the dietary treatment.

While most liver tumors in untreated mice and rats have been considered "spontaneous" in the past, it is now clear that some of these may be a result of dietary contamination (Newberne and McConnell 1980a, b) or of enhancement by dietary deficiencies or imbalances.

The development of the model, elucidation of the dietary factors involved, and description of the histopathology and its comparison to human cirrhosis have been reported by many investigators (Hartroft 1954; Rogers and MacDonald 1965;

Table 12. Composition of diets

Ingredient	Control (%)	Choline-free (%)
Casein, alcohol extracted	6.0	6.0
Peanut meal, alcohol extracted[a]	25.0	25.0
Sucrose	46.7	47.0
Salts mix, Rogers and Harper	5.0	5.0
Vitamin mix[b]	2.0	2.0
Fat (corn oil, Wesson oil, beef tallow)	15.0	15.0
Choline	0.3	0.0
Vitamin B_{12}	5.0 µg	0.0 µg
	100.0	99.0

[a] Assayed, free from detectable aflatoxins at a level of 0.2 ppb.
[b] Complete vitamin requirements for rats except for choline and vitamin B_{12}, which are added at the time of mixing the diet.

◀

Fig. 117. (*above*) Liver, choline-deficient rat injected with tritiated thymidine 2 h prior to death. All cell types label, an indiciation of DNA synthesis and incorporation of thymidine into the cell nucleus. H&E, ×100

Fig. 118. (*below*) Cirrhotic liver, same rat as in Fig. 114. Islands of hepatocytes are separated by bands of mature connective tissue, present as a bridging fibrosis with accompanying bile duct cells, inflammatory cells, and ceroid. Most of the lipid has disappeared and the parenchymal cells have lost their normal architecture. H&E, ×100

Newberne et al. 1966, 1969, 1982, 1983; Rogers and Newberne 1969, 1980; Newberne 1986). It should be pointed out that a severe deficiency that induces cirrhosis is not essential to predisposing the liver to carcinogenesis. Marginal methyl deficiency, which induces fatty liver but not cirrhosis, is also useful in long-term studies of chemical toxicity and carcinogenicity (Rogers and Newberne 1980), because it supports nearly normal growth and longevity while inducing many of the biochemical aberrations found in severe deficiency. It can be induced by deleting choline (0.2%) and vitamin B12 (5 µg%) from the diet shown in Table 12. Table 13 illustrates effects of marginal deficiency of lipotropes on induction of hepatocellular carcinoma by a variety of chemical carcinogens. With only one exception, dimethylnitrosamine (DMN), the marginal deficit enhanced cancer induction.

Table 13. Marginal deficiency of lipotropes and hepatocellular carcinoma induction by a variety of chemical car-cinogens (from Rogers and Newberne 1980)

Treatment	Tumor incidence (%)	
	Control	Low lipotrope
AFB$_1$ – 3 experiments	6–15	22–87
DEN, 18 weeks	70	80
12 weeks	24–43	60–86
DBN	24	64
DMN	28	27
AAF	36	72
DDCP	29	63

AFB$_1$, aflatoxin B$_1$; DEN, diethylnitrosamine; DBN, dibutylnitrosamine; DMN, dimethylnitrosamine; AAF, acetylnitrosamine; DDCP, 3,3-diphenyl-3-dimethylcarbamoyl-1-propyn.

Table 14. Effects of choline-deficiency cirrhosis on aflatox-in-induced hepatocellular carcinoma in rats (from Newberne et al. 1966)

Treatment (dietary)	Duration of study (days)	Liver lesions	
		Cirrhosis	Tumors
Control	422	0/20	0/20
Choline deficiency	425	18/18	0/18
Control + AFB$_1$	380	0/20	12/20
Choline deficiency + AFB$_1$	400	18/18	19/20
Choline deficiency to cirrhosis, then AFB$_1$	465	20/22	22/22

AFB$_1$, aflatoxin B$_1$.

Pathogenesis

Cirrhosis in the rat predisposes the liver to the effects of a number of hepatocarcinogens (Rogers and Newberne 1980). We have shown that choline deficiency, with or without cirrhosis, enhanced the carcinogenicity of aflatoxin for rat liver (Table 14). The presence of the cirrhotic liver appears to have a unique effect on the induction of liver carcinoma, similar in many ways to the observations in human populations, where cirrhosis and hepatocellular carcinoma often coexist (Hutt 1971; Gibson et al. 1980; MacSween and Scott 1973; Okuda and MacKay 1982).

During the induction of cirrhosis and of hepatocellular carcinoma, total lipids, cholesterol, and phospholipids significantly increase in the liver and decrease in the serum of the deficient rats. Supplementation with choline reverses the process with mobilization of lipids from the tissues and a return of serum lipid fractions to normal. Figure 119 illustrates clearly that in the choline-deficient rat, as liver fat increases, the number of cells that are labeled also increases. This is a reflection of the increased DNA synthesis and cell turnover which results, in part, from the large amount of cellular fat, parenchymal cell necrosis,

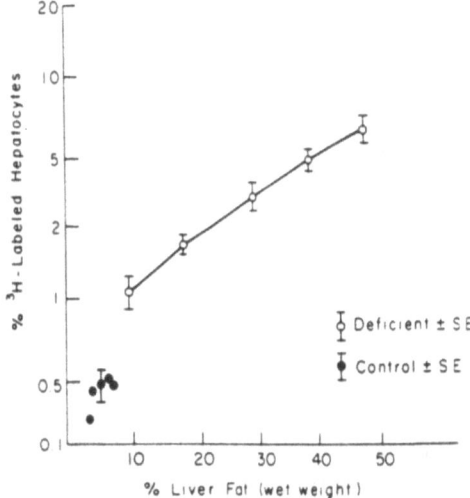

Fig. 119. Parallel increases in liver fat and in parenchymal cell DNA synthesis in choline-deficient rats. The increasing lipid concentration in hepatocytes, minimal necrosis of hepatocytes, and other undetermined factors "switch on" the process of cell proliferation which continues to nodular hyperplasia and, in some rats and mice, hepatocellular carcinoma

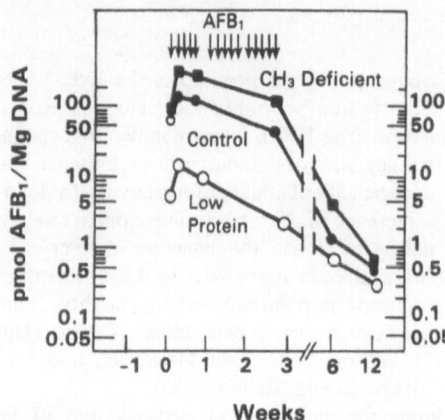

Fig. 120. Measurement of aflatoxin B_1 (AFB)$_1$–DNA adducts in the liver of rats 2 h following a dose of AFB$_1$, 3 weeks after initiating the methyl-deficient or protein-deficient diets. Methyl deficiency enhanced and protein deficiency inhibited adduct formation, parallel to observations in carcinogenesis studies

and regeneration. The liver attempts to accommodate to the large accumulation of fat in the parenchymal cells; some of them rupture and several will coalesce into a larger "cyst" (Hartroft 1954), which in the process allows escape of fat into the extracellular space; this is a likely source of significant irritation contributing to hepatitis. This phenomenon provides more target cells sensitive to attack by carcinogens; there is also the framework for methyl deficiency alone and its aberrant methylation, which also contributes to endogenous carcinogenesis (Rogers and MacDonald 1965; Rogers and Newberne 1969, 1971, 1975; Newberne et al. 1982). Reports of other investigators, which have been summarized (Newberne and Thurman 1981), have also described the importance of lipids and their metabolism to the regulation of cell proliferation.

The presence of fat in some way seems to enhance cell proliferation (Fig. 119), as determined by counts of labeled cells from animals treated with tritiated thymidine in vivo (Fig. 117).

Recently, we have shown that continuous necrogenic injury to the liver resulted in aberrant DNA methylation accompanied by increases in 7-methyl and O^6-methyl guanine which remained unrepaired (Bosan et al. 1987). The latter is associated with greater susceptibility to hepatocarcinogenesis. The aberrant methylation and lack of repair, in this case, was reflected in the induction of liver cancer in the hamster, a species resistant to carcinogenesis by hydrazine under

normal conditions in which DNA repair is achieved.

We have also demonstrated that the methyl-deficient liver, highly susceptible to a number of hepatocarcinogens, accumulates aflatoxin–DNA adducts (aflatoxin B_1–DNA) which are only slowly removed (repaired). Figure 120 illustrates this phenomenon and demonstrates also the greatly reduced aflatoxin B_1–DNA adduct formation in the liver of rats on a low-protein diet, the latter a protective regimen (Newberne 1974). Early exposure to aflatoxin B_1 and its metabolites, as shown in Fig. 120, differs depending to some extent on the availability of methyl groups. This interesting observation is supported further by the fact that increased dietary methyl groups significantly inhibited aflatoxin B_1-induced liver tumors in mice (Newberne et al. 1990). The nature of mechanisms for these manifestations is unclear but point to the value of this model in the investigations of basic biologic metabolism.

Comparisons with Other Species

Fatty liver and cirrhosis have been recognized for decades as sequelae of the excessive intake of alcohol and its accompanying nutritional problems (Davidson 1970). In humans, fatty liver can result from either nutritional deficiency or alcohol intake alone, but the development of cirrhosis apparently requires both alcohol and a nutritionally damaged liver (Porta et al. 1970; Lieber et al. 1971). Cirrhosis can be induced in rats and monkeys by nutritional deficiency alone or by certain hepatotoxins, but not by alcohol alone (Hartroft 1954; Hoffbauer 1959; Newberne and Conner 1980; Rogers et al. 1981).

In Western countries, particularly the United States, chronic alcoholism seems to be the most common agent associated with the genesis of both cirrhosis and liver cell cancer. The cirrhosis associated with alcohol is classically micronodular; it can progress to macronodular cirrhosis, followed by more coarse scarring and perhaps greater susceptibility to carcinogenic agents. In cases of human cancer occurring in cirrhotic livers, the cirrhosis is usually macronodular. The methyl-deficient liver of the rat follows a similar pattern. The relationship between cirrhosis and liver cell cancer in humans is unique and makes the animal model even more interesting for investigating this phenomenon. In no other tissue does there appear to be a counterpart to the nodular transformation produced by the regeneration of parenchymal cells with its high potential for malignant change. The malignant change occurring in the cirrhotic liver varies from one part of the world to another, but generally the coexistence is significant and the risk is high. Most estimates are in the range of 68%–98% malignant change occurring in cirrhotic livers. Liver cancer in patients from Japan and the Far East is usually solitary and often operable. The liver neoplasms in African patients are typically multicentric and inoperable (Okuda and MacKay 1982).

The fatty liver and cirrhosis induced in rats resemble closely the human disease in gross and microscopic lesions and in development of hepatic dysfunction (Hartroft 1954). The variation in biochemical function and pathology induced in rats by manipulation of the diet mimic the several patterns found in alcoholic patients.

These effects of methyl deficiency are equally applicable to rats and mice. However, most strains of mice are less sensitive than rats to the deficiency and its effects on tumor induction. It is of significance, however, that the B6C3F1 male mouse, widely used in safety evaluations of drugs and chemicals, develops liver nodules "spontaneously" with an incidence ranging between 7% and 58% (Tarone et al. 1981). It has now been documented that when this hybrid strain of mouse is fed a choline-deficient diet and/or is subjected to partial hepatectomy, the development of nodules is enhanced (Newberne et al. 1982) and some of the mice develop hepatocellular carcinomas (Newberne et al. 1982; L.A. Poirier, personal communication 1992). With respect to the B6C3F1 mouse, it is worth noting that the "spontaneous" liver nodules usually appear in animals older than

1 year (Ward et al. 1979) and that promoting conditions, such as the choline-deficient diet or phenobarbital, lowers the age to 6 months when the diet has been fed from weaning. These observations are even more impressive when one considers that the putative preneoplastic hyperplastic foci appear after only 3 months on a choline-deficient diet (Newberne et al. 1982). This suggests either that the liver of the B6C3F1 mouse is already "initiated" by the time the animals are put on the diet (at weaning) or that the experimental conditions in some way make the liver more susceptible to an "initiation" process at some point before the nodules or the preneoplastic foci appear. Data in Table 15 suggest that the liver of the B6C3F1 mouse is "initiated" by time of weaning and that nonspecific factors other than chemicals can enhance the formation of tumors in the liver that is already predisposed to neoplastic change. It will be useful to examine methyl deficiency in C3H and C57BL mouse strains, which appear to differ in carcinogenic susceptibility from the B6C3F1 hybrid.

An explanation of the phenomenon of liver cancer without a chemical carcinogen would be of value in understanding the initiation/promotion mechanisms that play a role in the choline-deficient mouse model. Such an understanding would also significantly improve our attempts at safety evaluation of chemicals considered risks for human populations and would help to further understand the enigmatic variation in incidence and types of human hepatocellular carcinoma around the world.

The methyl-deficient model has recently engendered a profound increase in interest and investigation throughout the oncologic community of scholars. Several lines of investigation are now

Table 15. Liver lesions in $B_6C_3F_1$ male mice after 13 months

Treatment	Liver lesions					
	Nodules		Adenoma		Hepatocarcinoma	
	(n)	(%)	(n)	(%)	(n)	(%)
Control	2/41	4.9	1/41	2.4	0/41	0.0
Control + partial hepatectomy[a]	2/41	4.9	6/41	14.6	1/41	2.4
Control + sham operation	1/23	4.3	2/23	8.7	0/18	0.0
Choline deficient	24/37	65.0	10/37	27.0	3/37	8.0
Choline deficient + partial hepatectomy[a]	20/34	58.8	9/34	26.5	5/34	14.7

[a] Partial hepatectomy performed after 3 weeks on diet.

ongoing in a number of laboratories, and promising results are beginning to emerge; a few comments regarding selected studies are included here. First, research in our laboratory (Newberne 1993b) continues to define some aspects of methyl deficiency and DNA adduct formation and removal, as it relates to the safety of the food supply. The greater sensitivity to toxic injury of the methyl-deficient liver of rodents permits detection of potential toxins with greater precision.

Lombardi and Smith (1994) failed to find *ras* gene mutations, but did find amplification of the c-*myc* protooncogene in all methyl-deficient tumors examined so far. Furthermore, these same investigators have found that mutations of the tumor suppressor gene *p53* are present with high frequency in the methyl-deficiency hepatocellular carcinoma (Smith et al. 1993).

Shinozuka and associates (Shinozuka et al. 1993) are pursuing the role of several growth factors which regulate liver cell growth; they have produced evidence that alterations in cell membrane receptors for some of these growth factors are present in methyl-deficient livers. These changes may relate to uncontrolled cell proliferation, which is extant in the methyl-deficiency hepatocellular carcinoma.

Christman and her group (Christman et al. 1993) have observed alterations in expression and methylation of specific genes (c-Ha-*ras*, c-*fos*, and c-*myc*) in the methyl-deficient liver which may participate in the cancer-causing properties of the methyl-deficient diet. Modifications in methylation of mRNA of genes involved in parenchymal cell growth regulation may lead to improved general understanding of basic metabolism in the liver and other tissues.

References

Bennett MA (1950) Utilization of homocystine for growth in presence of vitamin B_{12} and folic acid. J Biol Chem 87:751–756

Best CH, Huntsman ME (1932) The effects of components of lecithine upon deposition of fat in the liver. J Physiol (Lond) 75:405–412

Best CH, Huntsman ME (1935) The effects of choline on the liver fat of rats in various states of nutrition. J Physiol (Lond) 83:255–274

Bliss SW (1922) Effects of insulin on diabetic dogs. J Metab Res 2:385–401

Bosan W, Shank RC, MacEwen JD, Gaworski CL, Newberne PM (1987) Methylation of DNA guanine during the course

of induction of liver cancer in hamsters by hydrazine or dimethylnitrosamine. Carcinogenesis 8:439–444

Channon HJ, Manifold MC, Platt AP (1938) CXXXI. The action of cystine and methionine on liver fat deposition. Biochem J 32:969–975

Christman J, Chen ML, Sheikhnejad G, Dizik M, Abileah S, Wainfan E (1993) Methyl deficiency, DNA methylation, and cancer. J Nutr Biochem 4:672–680

Copeland DH, Salmon WD (1946) The occurrence of neoplasms in the liver, lungs and other tissues of rats as a result of prolonged choline deficiency. Am J Pathol 22:1059–1079

Davidson CS (1970) Nutrition, geography and liver diseases. Am J Clin Nutr 23:427–436

du Vigneaud V, Chandler JP, Meyer AW, Keppel DM (1939) The effect of choline on the ability of homocystine to replace methionine in the diet. J Biol Chem 131:57–76

Engel RW, Salmon WD (1941) Improved diets for nutritional and pathologic studies of choline deficiency in young rats. J Nutr 22:109–121 (+2 plates)

Fisher NF (1923/1924) Attempts to maintain the life of totally pancreatectomized dogs indefinitely by insulin. Am J Physiol 67:634–643

Gibson JB, Wu PC, Ho JC, Lauder IJ (1980) Hepatitis B surface antigen, hepatocellular carcinoma and cirrhosis in Hong Kong: a necropsy study: 1963–1976. Br J Cancer 42:370–377

Hartroft WS (1954) The sequence of pathological events in the development of experimental fatty liver and cirrhosis. Ann N Y Acad Sci 57:633–641

Hershey JM, Soskin S (1931) Substitution of lecithin for raw pancreas in the diet of the depancreatized dog. Am J Physiol 98:74–85

Hoffbauer FW (1959) Fatty cirrhosis in the rat. A method of grading specimens. Arch Pathol 68:160–170

Hutt MSR (1971) Epidemiology of human primary liver cancer. In: Liver cancer. IARC, Lyon, pp 21–29 (IARC scientific publications no 1)

Lieber CS, Rubin E, DeCarli LM (1971) Effects of ethanol on lipid, uric acid, intermediary and drug metabolism, including the pathogenesis of the alcoholic fatty liver. In: Kissin B, Begleiter H (eds) The biology of alcoholism, vol 1. Plenum, New York

Lombardi B, Smith ML (1994) Tumorigenesis, protooncogene activation and other gene abnormalities in methyl deficiency. J Nutr Biochem 5:2–9

MacSween RNM, Scott AR (1973) Hepatic cirrhosis: a clinicopathological review of 520 cases. J Clin Pathol 26:936–942

Mikol YB, Hoover KL, Creasia D, Poirier LA (1983) Hepatocarcinogenesis in rats fed methyl-deficient, amino acid-defined diets. Carcinogenesis 4:1619–1629

Newberne PM (1974) Mycotoxins: toxicity, carcinogenicity, and the influence of various nutritional conditions. Environ Health Perspect 9:1–32

Newberne PM (1986) Lipotropic factors and oncogenesis. In: Poirier L, Newberne P, Pariza M (eds) Essential nutrients in carcinogenesis. Plenum, New York, pp 223–251

Newberne PM (1993a) The methyl deficiency model: history characteristics and research directions. J Nutr Biochem 4:619–624

Newberne PM (1993b) Contribution of nutritional sciences to food safety: control of mycotoxins. J Nutr 123:289–293

Newberne PM, Connor M (1980) Effects of sequential exposure to aflatoxin B_1 and diethylnitrosamine on vascular and stomach tissue and additional target organs in rats. Cancer Res 40:4037–4042

Newberne PM, McConnell RG (1980a) Dietary nutrients and contaminants in laboratory animal experimentation. J Environ Pathol Toxicol 4:105–122

Newberne PM, McConnell RG (1980b) Toxicologic pathology: issues and uncertainties. In: Galli CL, Paoletti R (eds) The principles and methods in modern toxicology. Elsevier North Holland, Amsterdam, pp 223–258

Newberne PM, Thurman GB (1981) Working Group IV: lipids and the immune system. Report and recommendations. Cancer Res 41:3803–3804

Newberne PM, Harrington DH, Wogan GN (1966) Effects of cirrhosis and other liver insults on induction of liver tumors by aflatoxin in rats. Lab Invest 15:962–969

Newberne PM, Rogers AE, Bailey C, Young VR (1969) The induction of liver cirrhosis in rats by purified amino acid diets. Cancer Res 29:230–235

Newberne PM, deCamargo JLV, Clark AJ (1982) Choline deficiency, partial hepatectomy and liver tumors in rats and mice. Toxicol Pathol 10:95–109

Newberne PM, Rogers AE, Nauss KM (1983) Choline, methionine and related factors in oncogenesis. In: Butterworth CE Jr, Hutchinson ML (eds) Nutritional factors in the induction and maintenance of malignancy. Academic, New York, chap 15

Newberne PM, Suphiphat V, Locniskar M, deCamargo J (1990) Inhibition of hepatocarcinogenesis in mice by dietary methyl donors methionine and choline. Nutr Cancer 14:175–181

Okuda K, MacKay I (eds) (1982) Hepatocellular carcinoma. UICC, Geneva (UICC technical report ser, vol 74, rep no 17)

Porta EA, Koch OR, Hartroft WS (1970) Recent advances in molecular pathology: a review of the effects of alcohol on the liver. Exp Mol Pathol 12:104–132

Rogers AE, MacDonald RA (1965) Hepatic vasculature and cell proliferation in experimental cirrhosis. Lab Invest 14:1710–1726

Rogers AE, Newberne PM (1969) Aflatoxin B_1 carcinogenesis in lipotrope-deficient rats. Cancer Res 29:1965–1972

Rogers AE, Newberne PM (1971) Diet and aflatoxin B_1 toxicity in rats. Toxicol Appl Pharmacol 20:113–121

Rogers AE, Newberne PM (1975) Dietary effects on chemical carcinogenesis in animal models for colon and liver tumors. Cancer Res 35:3427–3431

Rogers AE, Newberne PM (1980) Lipotrope deficiency in experimental carcinogenesis. Nutr Cancer 2:104–112

Rogers AE, Fox JG, Murphy JC (1981) Ethanol and diet interactions in male rhesus monkeys. Drug Nutr Interact 1:3–14

Salmon WD, Copeland DH (1954) Liver carcinoma and related lesions in chronic choline deficiency. Ann N Y Acad Sci 57:664–667

Salmon WD, Newberne PM (1963) Occurrence of hepatomas in rats fed diets containing peanut meal as a major source of protein. Cancer Res 23:571–575

Shinozuka H, Masuhara M, Kubo Y, Katyal LL (1993) Growth factors and receptors in rat liver by choline-methionine deficiency. J Nutr Biochem 4:610–617

Smith ML, Yeleswarapu L, Scalamogna P, Locker J, Lombardi B (1993) p53 mutations in hepatocellular carcinomas induced by a choline-devoid diet in male Fischer-344 rats. Carcinogenesis 14:503–510

Tarone RE, Chu KC, Ward JM (1981) Variability in the rates of some common naturally occurring tumors in Fischer 344 rats and (C57BL/6N × C3H/HeN)F$_1$ (B6C3F$_1$) mice. JNCI 66:1175–1181

Ward JM, Goodman DG, Squire RA, Chu KC, Linhart MS (1979) Neoplastic and nonneoplastic lesions in aging (C57BL/6N × C3H/HeN)F$_1$ mice. JNCI 63:849–854

Cirrhosis, Mouse

Jerrold M. Ward

Synonyms. Hepatic fibrosis, nodular hyperplasia, chronic toxic hepatitis, postnecrotic, toxic, biliary, or pericellular cirrhosis

Gross Appearance

Diagnosis of cirrhosis in mice is relatively easy on gross examination. The liver is shrunken and nodular and has a scarred, uneven capsule (Fig. 121). Some lobes may be more severely affected than others, and large nodules (more than 1 cm in diameter) may be present.

Microscopic Features

Cirrhosis has not been well characterized morphologically and experimentally in the mouse (Edwards and Dalton 1942; Stowell et al. 1951; Ward et al. 1977, 1979). In chronic, persistent murine hepatitis virus infection in nude mice,

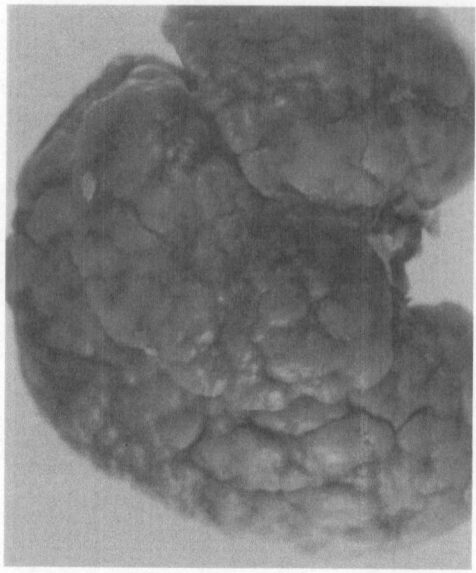

Fig. 121. Hepatic cirrhosis in liver of nude mouse infected with mouse hepatitis virus (MHV)

necrosis and loss of hepatocytes are replaced by a collapsed parenchyma, which condenses the reticulin network and appears to increase the amount of connective tissue (Figs. 122, 123). Nodularity is due to collapse more often than to hyperplasia.

In carbon tetrachloride-induced cirrhosis and cirrhosis induced by a nitrosamine, similar lesions are seen, but nodularity is more prominent (Figs. 124, 125). The nodularity in both cases is accompanied by some nodules containing hepatocytes with mitotic figures, degenerative changes, and megalocytosis. These may be true nodules of regenerative hyperplasia.

Biliary cirrhosis may occur from chronic lesions in the biliary tract, such as inflammatory bile duct and ductular hyperplasia (Fig. 126) and parasitic disease. All lesions are rare in mice. Pericellular fibrosis without nodularity (Fig. 123) appears more common in mice as a chronic toxic effect (Ward et al. 1979). Since hepatocytes can produce collagen, this lesion is not surprising.

Ultrastructure

The electron microscopy features of cirrhosis have not been described in mice.

Differential Diagnosis

The most difficult feature of hepatic cirrhosis in mice is distinguishing between neoplastic and hyperplastic nodular lesions. In general, the neoplastic lesions, termed adenomas or hyperplastic nodules, are composed of a uniform population of hepatocytes. The non-neoplastic nodular lesions in cirrhotic livers are usually composed of a pleomorphic population of hepatocytes with features of karyomegaly, cytomegaly, nuclear hyperchromatism, and mitotic figures. The nodule is usually lobular and delineated by a thin layer of fibrous tissue from other lobules (see "Hepatocellular Adenoma," p. 49).

Biologic Features

Cirrhosis in mice is usually the end stage of a continuous process of degeneration, necrosis, and an attempt at reparative hyperplasia. Thus, if severe enough, it is fatal; if not, it contributes to poor weight gain and decreased survival. It is not known whether chronic hepatic damage is reversible in mice or if it may, in itself, lead to neoplasia.

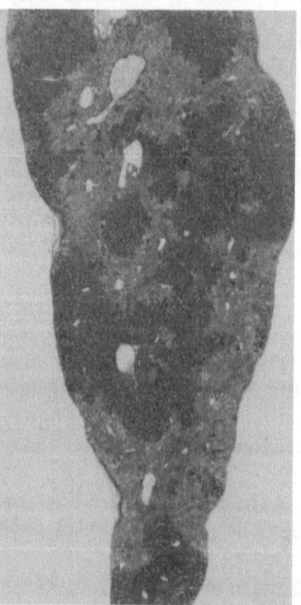

Fig. 122. Cirrhosis in mouse hepatitis virus (MHV) infection of nude mouse. Note lobular collapse, fibrosis, and focal nodules. H&E, ×16

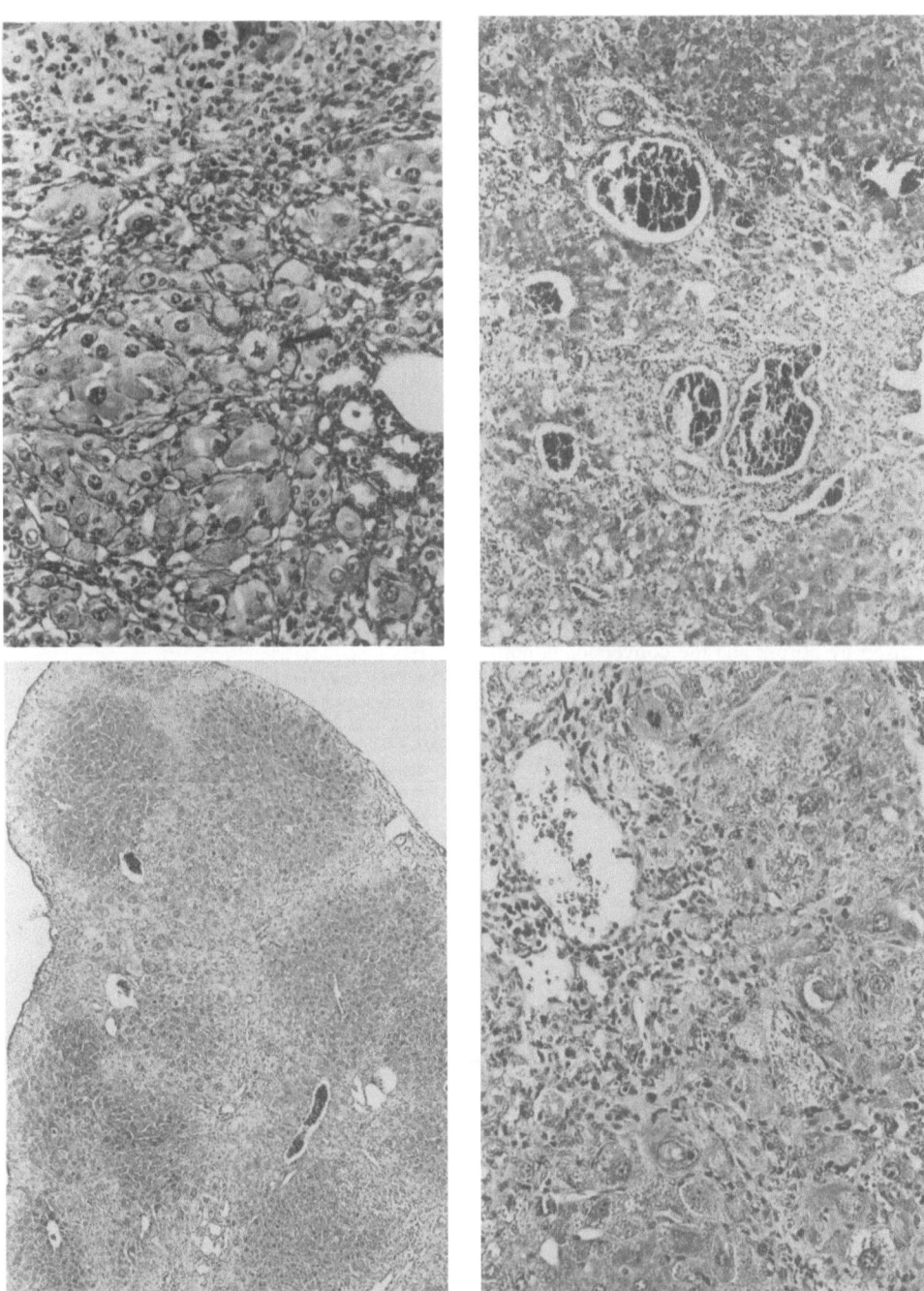

Fig. 123. (*upper left*) Mouse hepatitis virus (MHV) infection with pericellular reticulin fibers, degenerative hepatocytes, one mitotic figure (*arrow*). and chronic inflammation. Reticulin stain. H&E, ×220

Fig. 124. (*lower left*) Cirrhosis in mouse liver induced by diethanolnitrosamine. Note focal nodular lobules and irregular hepatic surface. H&E, ×54

Fig. 125. (*upper right*) Collapsed hepatic parenchyma and fibrosis in mouse exposed to carbon tetrachloride. H&E, ×80

Fig. 126. (*lower right*) Loss of hepatic parenchyma, fibrosis, cholangiolar hyperplasia, and mitotic figures (*arrow*) in liver of mouse given diethanol-nitrosamine. H&E, ×220

Cirrhosis is classified on the basis of etiology, pathogenesis, and morphology. It is a chronic response to chronic or severe liver damage. In mice, cirrhosis is poorly defined and rarely reported. Terms such as micronodular and macronodular cannot be used, as yet, in mice. At present, it seems more appropriate to use terms such as postnecrotic, toxic, pericellular, or biliary to describe murine cirrhosis.

Comparison with Other Species

Hepatic cirrhosis in mice is not well described. It is a rare lesion in control mice, but has been seen in nude mice infected with mouse hepatitis virus and in mice exposed to hepatotoxins. Cases in mice usually resemble cirrhosis grossly in other species, but generally lack the prominent nodularity and histologic features, especially the degree of fibrosis. In rats, cirrhosis has been induced by hepatotoxins as well (Tamayo 1983). In humans, cirrhosis is described as a diffuse process characterized by fibrosis and the conversion of normal liver architecture into structurally abnormal nodules, features representative of the sequelae of repeated episodes of hepatocellular necrosis, regeneration, collapse of the existing collagen network, and de novo collagen synthesis (Anthony et al. 1977; Tamayo 1983). The fibrosis usually results from necrosis and the nodules, surrounded by connective tissue, lack normal organization. These nodules have been termed "regenerative" or "hyperplastic."

Further research should be performed in mice to characterize fully the histogenesis and morphology of cirrhosis in mice and thus allow a more complete comparison with cirrhosis in other species.

References

Anthony PP, Ishak KG, Nayak NC, Poulsen HE, Scheuer PJ, Sobin LH (1977) The morphology of cirrhosis: definition, nomenclature, and classification. Bull WHO 55:521–540

Edwards JE, Dalton AJ (1942) Induction of cirrhosis of the liver and of hepatomas in mice with carbon tetrachloride. JNCI 3:19–41

Stowell RE, Lee CS, Tsuboi KK, Villasana A (1951) Histochemical and microchemical changes in experimental cirrhosis and hepatoma formation in mice by carbon tetrachloride. Cancer Res 11:345–354

Tamayo RP (1983) Is cirrhosis of the liver experimentally produced by CCl_4 an adequate model of human cirrhosis? Hepatology 3:112–120

Ward JM, Collins MJ Jr, Parker JC (1977) Naturally occurring mouse hepatitis virus infection in the nude mouse. Lab Anim Sci 27:372–376

Ward JM, Bernal E, Buratto B, Goodman DG, Strandberg JD, Schueler R (1979) Histopathology of neoplastic and nonneoplastic hepatic lesions in mice fed diets containing tetrachlorvinphos. JNCI 63:111–118

Peliosis Hepatis, Rodents

Peter Bannasch, Klaus Wayss, and Heide Zerban

Synonyms. Blood cyst, angiectasis, phlebectetic or sinusoidal peliosis hepatis, parenchymal peliosis hepatis

Gross Appearance

Macroscopically, peliosis hepatis is best visible at the surface of the liver. The lesion is seen as blood-filled, thin-walled cavities projecting above the liver surface (Fig. 127). The blood-filled spaces stand out clearly as dark areas against the otherwise brown liver tissue. Large peliotic lesions may also be identified at the cut surface. Frequently, however, it is extremely difficult or even impossible to detect peliosis with the naked eye after the animal has died spontaneously under conditions with a weak blood supply to the liver. When the liver is removed from anesthetized animals after laparotomy, peliosis is readily visible only as long as the liver is connected with the blood circulation, but usually becomes invisible when the blood is lost during removal of the organ.

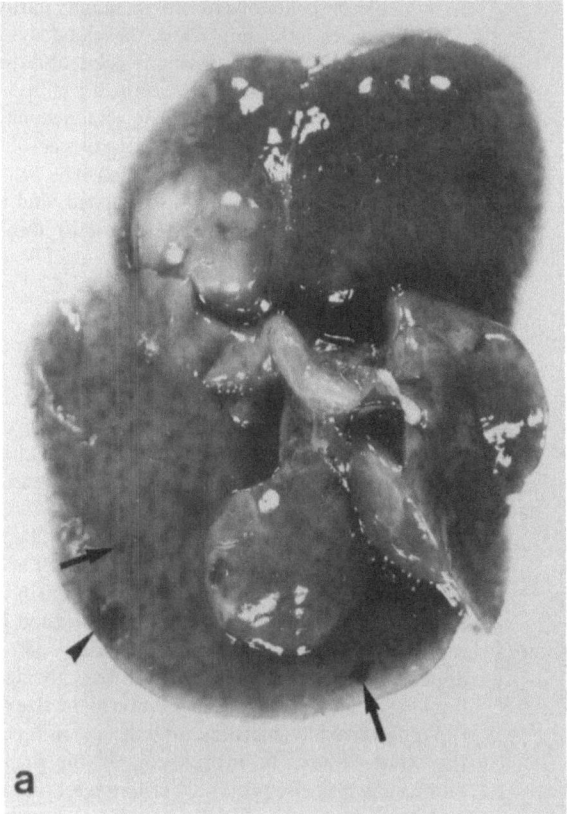
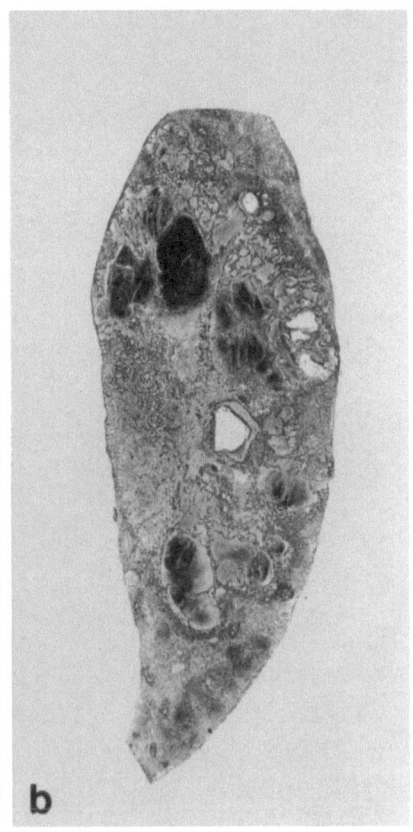

Fig. 127a,b. Vascular liver lesions induced in *Mastomys natalensis* by dimethylnitrosamine. **a** Macroscopic appearance of peliosis hepatis (*arrows*) and a benign hemangioendothelioma (*arrowhead*) at the liver surface. ×2. **b** Microscopic overview showing a large number of small peliotic lesions and large blood cysts (hemangioendotheliomatosis). ×7.5

Microscopic Features

Based on experience in human pathology (Yanoff and Rawson 1964), two types of peliosis hepatis are also distinguished in rodents, namely the "phlebectatic" and the "parenchymal" form (Ruebner et al. 1970). *Phlebectatic* or *sinusoidal peliosis hepatis* is characterized by irregular focal dilatations of the sinusoids (Trainin 1963; Popper et al. 1977; Wayss et al. 1979). It may occur as an isolated lesion, but frequently is multicentric. The cavernous lacunae of the peliotic lesions are densely packed with blood (Figs. 128, 129). They are separated from one another by cords of liver parenchyma and coated by a single layer of endothelium. As a rule, the cells of this endothelium appear to be unaltered morphologically except for an inconsistent enlargement and

sometimes also an increase in number of nuclei. Mitotic figures are very rare. The sinusoidal dilatations may be distributed quite unevenly in the liver, and a preferential location in particular regions is not usually seen. However, the occurrence of peliosis within hepatic adenomas (neoplastic nodules) is rather common (Popper et al. 1977; Ward 1981). As a rule, the tissue adjacent to the dilated sinusoids is well preserved and free from necrotic cells.

The irregular sinusoidal dilatations may also be located close to unequivocal benign hemangioendothelioma (Figs. 128b, 129b,c) and even to malignant angiosarcomas of the liver (Fig. 129d).

Parenchymal peliosis hepatis has also been described in rodent liver (Ruebner et al. 1970; Tuchweber et al. 1973; Schauer and Kunze 1976).

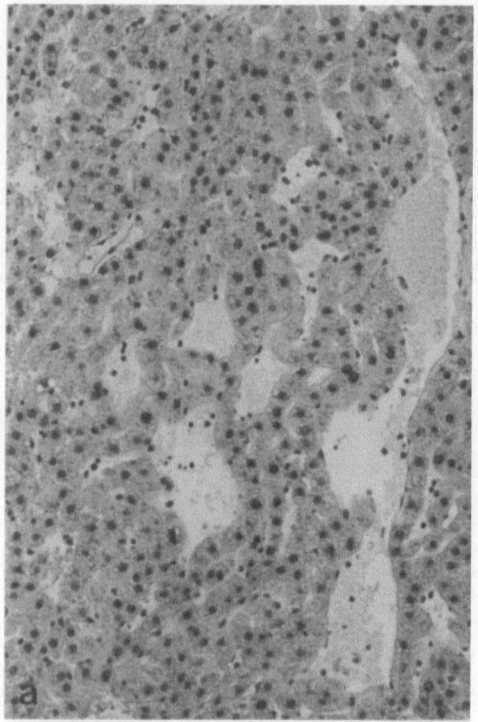

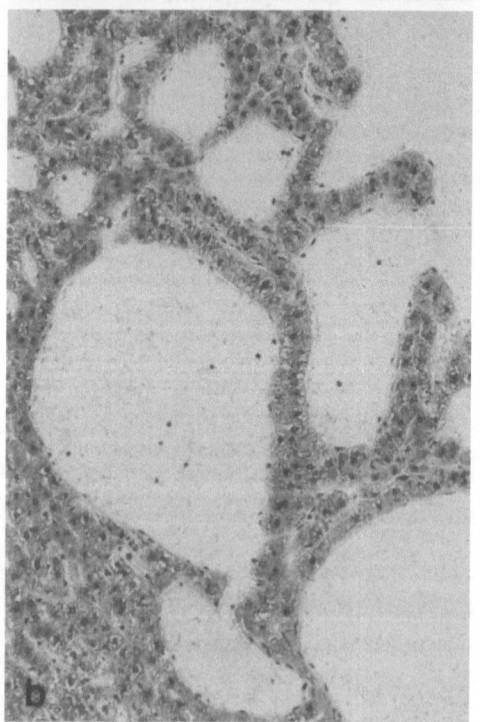

In this case, the peliotic cysts are not or only partly lined by an endothelium. Thus the blood cysts involve not only the sinusoidal lumen, but also the space of Disse, and the blood frequently is in direct contact with the neighboring parenchymal cells. The parenchyma undergoes focal necrotic changes without any zonal distribution. The cytoplasm of many hepatocytes is vacuolated, and it appears that blood may penetrate into these vacuoles (Ruebner et al. 1970).

Differential Diagnosis

The differential diagnosis of sinusoidal peliosis hepatis and benign hemangioendotheliomas of the liver is arbitrary; all possible transitions between these two lesions have been described in a number of animal models (Trainin 1963; Takayama 1968; Maltoni and Lefemine 1974; Popper et al. 1977; Wayss et al. 1979; Bannasch and Zerban 1990). However, as long as sinusoidal dilatation is not very pronounced and the lesion is relatively small, the diagnosis "peliosis hepatis" appears to be more appropriate than "benign hemangioendothelioma." In addition to the size of the individual blood lacunae and the extension of the whole lesion, the morphology of the lining endothelia may also provide a criterion for differential diagnosis, since their nuclei are usually more prominent in hemangioendotheliomas (Fig. 129c) than in the small peliotic cysts. The distinction of sinusoidal peliosis hepatis and liver angiosarcoma is less difficult, since cellular atypia and mitotic figures are frequent in malignant tumors and the cells are usually no longer confined to a single cell layer (Fig. 129d).

A hitherto largely neglected lesion which may be mistaken for peliosis hepatis (Lee 1983) has been described as *spongiosis* hepatis (Bannasch et al. 1981; see p. 104). In contrast to the lumina of peliosis, the cavities of spongiosis are not filled with blood but with a finely flocculent, acidophilic material rich in proteoglycans. Knowing this difference, the discrimination between peliosis and spongiosis hepatis is easy.

◄

Fig. 128a,b. Vascular liver lesions induced in rat liver by *N*-nitrosomorpholine. **a** Peliosis hepatis. H&E, ×130. **b** Benign hemangioendothelioma. H&E, ×150

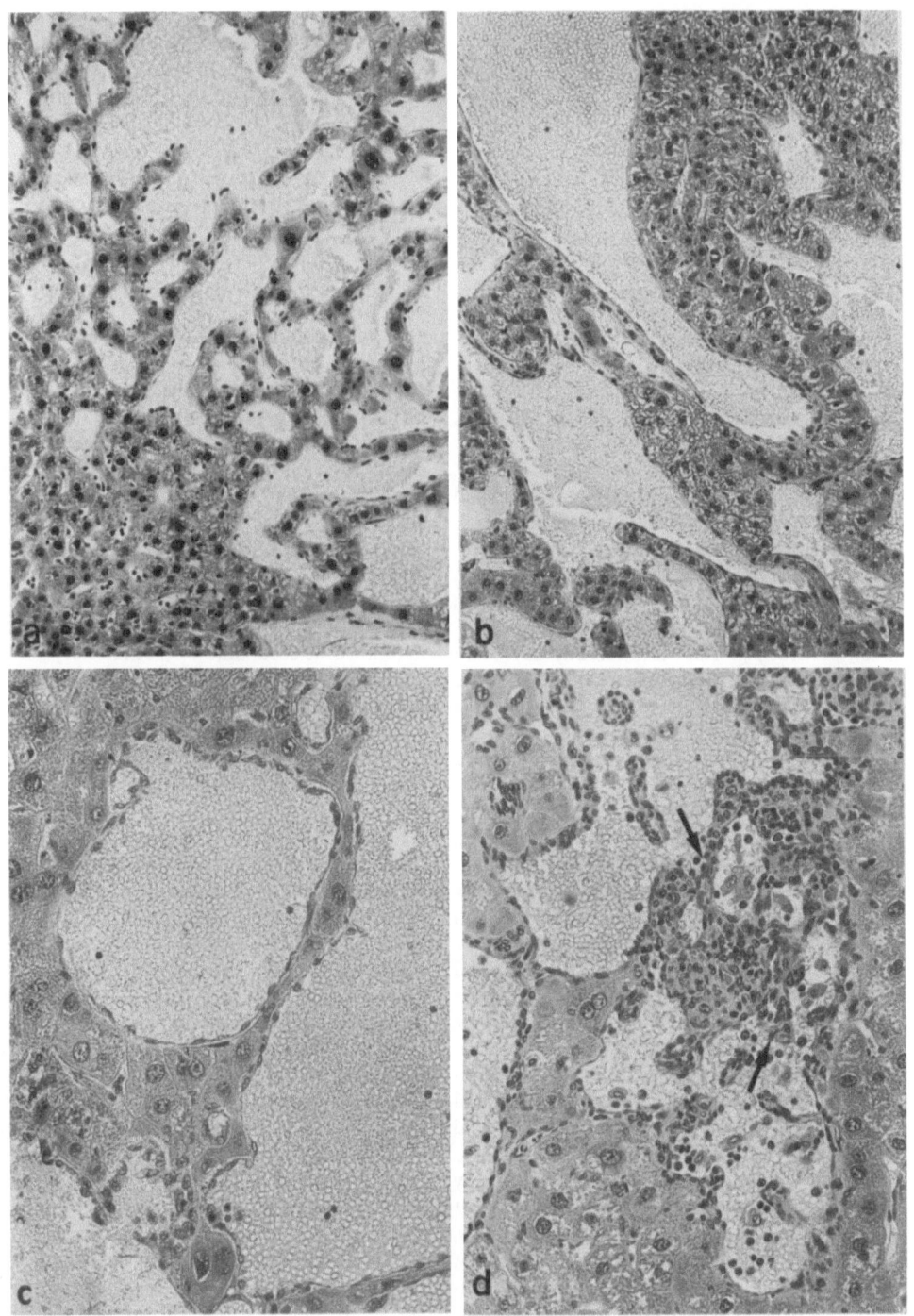

Fig. 129a–d. Vascular liver lesions induced in *Mastomys natalensis* by dimethylnitrosamine. **a** Peliosis hepatis. H&E, ×200. **b** Benign hemangioendothelioma. H&E, ×200. **c** Benign hemangioendothelioma with particularly prominent end-othelial lining cells. H&E, ×130. **d** Transitions from benign hemangioendothelioma into highly atypical cell populations (*arrows*) of an angiosarcoma. H&E, ×150

Pathogenesis

In rodent liver, the pathogenesis of both the phlebectatic and parenchymal form of peliosis hepatis has been followed in some detail. All authors agree that the sinusoidal form originates from sinusoidal lining cells, namely the endothelial cells (Popper et al. 1977; Wayss et al. 1979). It may occur "spontaneously" in rodents, but reliable data on the spontaneous incidence are not available. Focal sinusoidal dilatations corresponding to peliosis hepatis have been induced in rodents by vinyl chloride, vinyl carbamate, nitrosamines, nitrosamides, and nitrosomethylurea (mouse: Trainin 1963; Schmahl and Thomas 1965; Terracini et al. 1966; Takayama 1968; Maltoni and Lefemine 1974; Magee et al. 1976; Wright et al. 1991; rat: Druckrey et al. 1967; Hadjiolov and Markov 1973; Maltoni and Lefemine 1974; Magee et al. 1976; Popper et al. 1977; mastomy: Wayss et al. 1979; Chinese hamster: Reznik 1975; Reznik et al. 1976; European hamster: Mohr et al. 1974; Syrian golden hamster: Tomatis et al. 1964; Herrold 1967; Magee et al. 1976; guinea pig: Rao and Reddy 1977). In *Mastomys natalensis*, Wayss et al. (1979) have shown that sinusoidal peliosis hepatis as well as benign hemangioendotheliomas and angiosarcomas may be produced by administration of a relatively low single dose of dimethylnitrosamine (10 mg/kg body weight). Whereas sinusoidal peliosis hepatis appeared early in this experiment, benign hemangioendotheliomas developed later, and angiosarcomas were only seen after very long latent periods. These observations are in line with the suggestion that the carcinogen-induced peliotic lesions coated by a single-layered endothelium are the result of an autonomous growth process (Trainin 1963; Takayama 1968; Maltoni and Lefemine 1974; Popper et al. 1977; Wayss et al. 1979). The most important indication of the possible neoplastic nature of sinusoidal peliosis is its joint occurrence with unequivocal benign hemangioendotheliomas and even malignant angiosarcomas, into which they occasionally continue directly (Wayss et al. 1979). This might indicate that phlebectatic peliosis hepatis, benign hemangioendotheliomas, and malignant angiosarcomas are only different stages of a uniform pathogenetic series. Conceivably, however, the angiosarcoma may be produced independently of the benign lesions. Because of the occasional association of phlebectatic peliosis hepatis with hepatic adenomas (neoplastic nodules), some authors speculated that these two lesions might be causally related to each other (Popper et al. 1977). We do not share this view since peliosis and adenomas often appear independently of each other in the same liver. It is much more likely, therefore, that the occasional combination of the two lesions results from a parallel hit of both the endothelial cells and the hepatocytes by the same causative agent within a certain liver area.

The parenchymal form of peliosis hepatis has been produced in rodent liver by the pyrrolizidine alkaloid lasiocarpine (Ruebner et al. 1970) and by phalloidin (Tuchweber et al. 1973). The pathogenesis of this lesion induced in mouse liver by lasiocarpine has been investigated by light and electron microscopy (Ruebner et al. 1970). The authors suggested from their results that a lytic necrosis of vacuolated hepatocytes led to the development of the peliotic blood lakes. The pathogenesis of phalloidin-induced parenchymal peliosis in rats remained unclear (Tuchweber et al. 1973). From the observation of early hepatic changes resembling those seen after ischemia, the authors concluded that microcirculatory disturbances may play a role in the production of this liver injury.

Comparison with Other Species

The phlebectatic form of peliosis hepatis has attracted considerable interest in view of its increased incidence in humans in the past two decades (Bagheri and Boyer 1974; Kühböck et al. 1975; Nadell and Kosek 1977; Nissen et al. 1977; Popper et al. 1977; Balázs 1988; Soe et al. 1992). The lesion is frequently found in human livers bearing at the same time angiosarcomas induced by vinyl chloride (Thomas et al. 1975; Popper et al. 1977). Peliosis is also often associated with liver cell adenomas in patients who have taken oral contraceptives or anabolic androgenic steroids for long periods (Bagheri and Boyer 1974; Kühböck et al. 1975; Nadell and Kosek 1977; Nissen et al. 1977; Popper et al. 1977; Balázs 1988; Soe et al. 1992). In addition, peliosis hepatis has been observed in association with extrahepatic neoplastic diseases, such as Wilms' tumor (Björk et al. 1985) and hairy cell leukemia (Zafrani et al. 1987), in patients suffering from the acquired immunodeficiency syndrome (AIDS) (Czapar et al. 1986; Scoazec et al. 1988), after administration of immunosuppressant drugs such as azathioprine to renal transplant recipients (Takahara et al. 1987), and following diethylstilbestrol therapy for prostatic cancer, which likewise resulted in a multicentric hepatocellular carcinoma (Endo et al. 1987). Of particular interest are reports on the appearance of peliosis hepatis in two cases of the rare inborn hepatic glycogen storage disease (glycogenosis) type I (Spycher and Gitzelmann 1971; Schmidt et al. 1991), since peliosis hepatis in the rat is frequently accompanied by a focal hepatic glycogenosis, which has been shown to indicate an early event in hepatocarcinogenesis in various species (Bannasch et al. 1980; see "Foci of Altered Hepatocytes, Rat," p. 3 this volume). Views on the pathogenesis of peliosis hepatis in man are controversial (Taxy 1978; Spellberg et al. 1979; Zafrani et al. 1984). In cattle, peliosis hepatis has been described as a specific lesion in St. George's disease (Seawright and Francis 1971).

References

Bagheri SA, Boyer L (1974) Peliosis hepatis associated with androgenic-anabolic steroid therapy. A severe form of hepatic injury. Ann Int Med 81:610–618

Balázs M (1988) Sinusoidal dilatation of the liver in patients on oral contraceptives. Electron microscopical study of 14 cases. Exp Pathol 35:231–237

Bannasch P, Zerban H (1990) Tumours of the liver. In: Turusov VS, Mohr U (eds) Pathology of tumours in laboratory animals, vol I: tumours of the rat, 2nd edn. IARC, Lyon, pp 199–240 (IARC scientific publications no 99)

Bannasch P, Mayer D, Hacker HJ (1980) Hepatocellular glycogenosis and hepatocarcinogenesis. Biochim Biophys Acta 605:217–245

Bannasch P, Bloch M, Zerban H (1981) Spongiosis hepatis. Specific changes of perisinusoidal liver cells induced in rats by N-nitrosomorpholine. Lab Invest 44:252–264

Björk O, Eklöf O, Willi U, Åhström L (1985) Venoocclusive disease and peliosis of the liver complicating the course of Wilms' tumour. Acta Radiol Diagn 26:589–597

Czapar CA, Weldon-Linne CM, Moore DM, Rhone DP (1986) Peliosis hepatis in the acquired immunodeficiency syndrome. Arch Pathol Lab Med 110:611–613

Druckrey H, Preussmann R, Ivankovic S, Schmähl D (1967) Organotropic carcinogenic effects of 65 various N-nitrosocompounds on BD rats. Z Krebsforsch 69:103–201

Endo H, Murakami T, Nishimoto I, Sekine I, Yokoyama M (1987) Multicentric hepatocellular carcinoma following phosphate diethylstilbestrol therapy for prostatic cancer. Acta Pathol Jpn 37:795–806

Hadjiolov D, Markow D (1973) Fine structure of hemangioendothelial sarcomas in the rat liver induced with N-Nitrosodimethylamine. Arch Geschwulstforsch 42:120–126

Herrold KM (1967) Histogenesis of malignant liver tumors induced by dimethylnitrosamine. An experimental study in Syrian hamsters. J Natl Cancer Inst 39:1099–1111

Kühböck J, Radaszkiewicz T, Walek H (1975) Peliosis hepatis, eine Komplikation der Anabolikatherapie. Med Klin 70:1602–1607

Lee KP (1983) Peliosis hepatis-like lesions in aging rats. Vet Pathol 20:410–423

Magee PN, Montesano R, Preussmann R (1976) N-nitroso compounds and related carcinogens. In: Searle CE (ed) Chemical carcinogens. ACS Monogr Ser 173:491–625

Maltoni C, Lefemine G (1974) Carcinogenicity bioassays of vinylchloride 1. Research plan and early results. Environ Res 7:387–405

Mohr U, Haas H, Hilfrich J (1974) The carcinogenic effects of dimethylnitrosamine and nitrosomethylurea in European hamsters (Cricetus cricetus L.). Br J Cancer 29:359–364

Nadell J, Kosek J (1977) Peliosis hepatis. Twelve cases associated with oral androgen therapy. Arch Pathol Lab Med 101:405–410

Nissen ED, Kent DR, Nissen SE (1977) Etiologic factors in the pathogenesis of liver tumors associated with oral contraceptives. Am J Obstet Gynecol 127:61–66

Popper H, Selikoff U, Maltoni C, Squire RA, Thomas LB (1977) Comparison of neoplastic hepatic lesions in man and experimental animals. In: Hiatt HH, Watson JD, Winsten JA (eds) Human risk assessment. Cold Spring Harbor Conference on Cell Proliferation, vol 4: origins of human cancer. Cold Spring Harbor, Cold Spring Harbor Laboratory, pp 1359–1382

Rao MS, Reddy JK (1977) Induction of malignant vascular tumors of the liver in guinea pigs treated with 2,2'-dihydroxy-di-n- propylnitrosamine. J Natl Cancer Inst 58:387–392

Reznik G (1975) The carcinogenic effect of dimethylnitrosamine on the Chinese hamster (Cricetulus griseus). Cancer Lett 1:25–28

Reznik G, Mohr U, Kmoch N (1976) Carcinogenic effects of different nitroso-compounds in Chinese hamsters. 1. Dimethylnitrosamine and N-diethylnitrosamine. Br J Cancer 33:411–418

Ruebner BH, Watanabe K, Wand JS (1970) Lytic necrosis resembling peliosis hepatis produced by lasiocarpine in the mouse liver. A light and electron microscopic study. Am J Pathol 60:247–271

Schauer A, Kunze E (1976) Tumours of the liver. In: Turusov VS (ed) Pathology of tumours in laboratory animals, vol 1, Tumours of the rat, part 2. IARC, Lyon, pp 41–61 (IARC scientific publications no 6)

Schmähl D, Thomas C (1965) Dosis-Wirkungs-Beziehungen bei der Erzeugung von Hämangioendotheliomen der Leber bei Mäusen durch Diethylnitrosamin. Z Krebsforsch 66:533–535

Schmidt H, Ullrich K, von Lengerke HJ, Peters PE (1991) Peliosis hepatis with type 1 glycogen storage disease. J Inherited Metab Dis 14:831–832

Scoazec JY, Marche C, Girard PM, Houtmann J, Durand-Schneider AM, Saimot AG, Benhamou JP, Feldmann G (1988) Peliosis hepatis and sinusoidal dilatation during infection by the human immunodeficiency virus (HIV). An ultrastructural study. Am J Pathol 131:38–47

Seawright AA, Francis J (1971) Peliosis hepatis. A specific lesion in St. George's disease of cattle. Aust Vet J 47:91–99

Soe KL, Soe M, Gluud CN (1992) Liver pathology associated with the use of anabolic-androgenic steroids. Liver 12:73–79

Spellberg MA, Mirro J, Chowdhury L (1979) Hepatic sinusoidal dilatation related to oral contraceptives. A study of two patients showing ultrastructural changes. Am J Gastroenterol 72:248–252

Spycher MA, Gitzelmann R (1971) Glycogenesis type I (glucose-6-phosphatase deficiency): ultrastructural alterations of hepatocytes in a tumor bearing liver. Virchows Arch [B] Cell Pathol 8:133–142

Takahara S, Ihara H, Ichikawa Y, Nagano S, Fukunishi T, Sonoda T, Shinji Y (1987) Prospective study and long-term follow-up of liver damage in renal transplant recipients. Transplant Proc 19:2221–2224

Takayama S (1968) The histological and autoradiographical studies of mouse liver during the course of carcinogenesis by dimethylnitrosamine. Z Krebsforsch 71:246–254

Taxy JB (1978) Peliosis: a morphologic curiosity becomes an iatrogenic problem. Hum Pathol 9:331–340

Terracini B, Palestro G, Gigliardi MR, Montesano G (1966) Carcinogenicity of dimethylnitrosamine in Swiss mice. Br J Cancer 20:871–876

Thomas LB, Popper H, Berk PD, Selikoff I, Falk H (1975) Vinyl-chloride-induced liver disease. From idiopathic portal hypertension (Banti's syndrome) to angiosarcoma. N Engl J Med 292:17–22

Tomatis L, Magee PN, Shubik P (1964) Induction of liver tumors in the Syrian golden hamster by feeding dimethylnitrosamine. J Natl Cancer Inst 33:341–345

Trainin N (1963) Neoplastic nature of liver "blood cysts" induced by urethan in mice. J Natl Cancer Inst 31:1489–1499

Tuchweber B, Kovacs K, Khandekar JD, Garg BD (1973) Peliosis-like changes induced by phalloidin in the rat liver. A light and electron microscopic study. J Med 4:327–345

Ward JM (1981) Morphology of foci of altered hepatocytes and naturally-occurring hepatocellular tumors in F344 rats. Virchows Arch Pathol Anat 390:339–345

Wayss K, Bannasch P, Mattern J, Volm M (1979) Vascular liver tumors induced in Mastomys (Praomys) natalensis by single or twofold administration of dimethylnitrosamine. J Natl Cancer Inst 62:1199–1207

Wright JA, Marsden AM, Willets JM, Orton TC (1991) Hepatocarcinogenic effect of vinyl carbamate in the C57Bl/lOJ strain mouse. Toxicol Pathol 19:258–265

Yanoff M, Rawson AJ (1964) Peliosis hepatis. An anatomic study with demonstration of two varieties. Arch Pathol 77:159–165

Zafrani ES, Cazier A, Baudelot AM, Feldmann G (1984) Ultrastructural lesions of the liver in human peliosis. A report of 12 cases. Am J Pathol 114:349–359

Zafrani ES, Degos F, Guigui B, Durand-Schneider AM, Martin N, Flandrin G, Benhamou JP, Feldmann G (1987) The hepatic sinusoid in hairy cell leukemia: an ultrastructural study of 12 cases. Hum Pathol 18:801–807

Hyperplasia, Diffuse, Following Partial Hepatectomy, Mouse

Jerrold M. Ward

Synonyms. Mitotic figures, hyperplasia

Gross Appearance

About 40–60 h after surgical removal of the left and medial lobes of the liver, the remaining lobes are swollen and have rounded edges. The liver reaches its normal gross weight 3–5 days after two thirds of the liver has been removed. The shape of the liver, however, remains abnormal.

Microscopic Features

The peak numbers of hepatocytes in mitosis are seen at about 40–60 h after two-thirds partial hepatectomy. Nodularity of fibrosis does not occur during diffuse hyperplasia (Figs. 130, 131), but some fatty metamorphosis may be seen within hepatocytes (Ward 1984). Glycogen depletion is prominent, and nuclear chromatin may appear more dispersed than normal. As the liver reaches normal weight, the numbers of mitotic figures diminish to near normal levels.

Ultrastructure

The ultrastructure of this induced lesion appears not to have been described.

Differential Diagnosis

Many mitotic figures may be seen normally in young suckling mice; however, in adult control mice, few are observed. Increased numbers of mi-

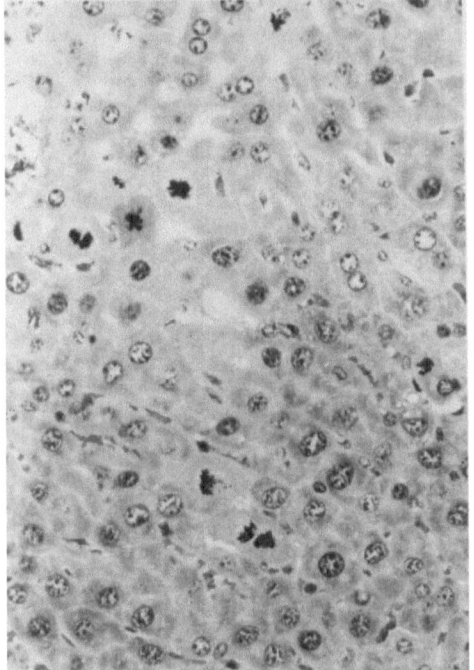

totic figures may be seen in livers of mice exposed to some liver toxins (Ward 1984). Similar numbers of mitoses may be seen in mouse livers during recovery from mouse hepatitis virus infection (p. 179). The absence of fibrosis and nodularity also serve to distinguish this lesion from cirrhosis (p. 151).

Biologic Features

Diffuse hyperplasia of hepatocytes after two-thirds partial hepatectomy (Higgins and Anderson 1931) is indicated by increased DNA synthesis in hepatocytes prior to the appearance of numerous mitotic figures (Chernozemski and War-wick 1970). α-fetoprotein synthesis may also be affected (Mohanty et al. 1978; Tuczek et al. 1981). Partial hepatectomy in mice usually results in a histologically normal liver by 3 days after surgery. In some mice, however, removal of the gallbladder with the median lobe may induce a chronic bile duct and ductular hyperplasia with hepatocyte megalocytosis and pleomorphism.

Comparison with Other Species

Partial hepatectomy in rats also results in a diffuse hyperplasia which peaks, however, approximately 20h prior to that in mice (Fabrikant 1968). The liver morphology returns to normal, but bile duct lesions are usually not seen, presumably because rats do not have a gallbladder.

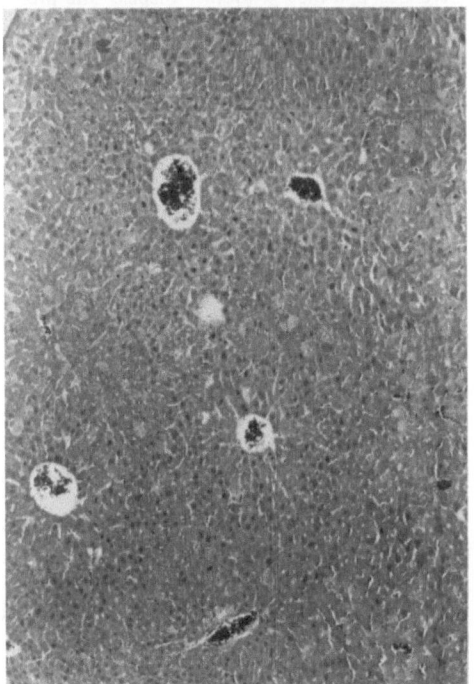

References

Chernozemski IN, Warwick GP (1970) Liver regeneration and induction of hepatomas in B6AF$_1$ mice by urethan. Cancer Res 30:2685–2690

Fabrikant JI (1968) The kinetics of cellular proliferation in regenerating liver. J Cell Biol 36:551–565

Higgins GM, Anderson RM (1931) Experimental pathology of the liver. I. Restoration of the liver of the white rat following partial surgical removal. Arch Pathol 12:186–202

Mohanty M, Das PK, Mittal A, Nayak NC (1978) Cellular basis of induced alpha-fetoprotein synthesis by hepatocytes of adult mouse after hepatotoxic injury and partial hepatectomy. Int J Cancer 22:181–188

Fig. 130. (*above*) Hepatocytes in mitoses in the liver of a 6-week-old mouse 46h after partial hepatectomy. H&E, ×330

Fig. 131. (*below*) Liver of mouse in Fig. 130. Note the lack fibrosis or nodule formation. Dark nuclei are mitotic figures. H&E, ×80

Tuczek HV, Fritz P, Wagner T, Braun U, Grau A, Wegner G (1981) Synthesis of alpha-fetoprotein (AFP) and cell proliferation in regenerating livers of NMRI mice after partial hepatectomy. An immunohistochemical and astoradiographic study with 3 H-thymidine. Virchows Arch (Cell Pathol) 38:229–237

Ward JM (1984) Pathology of toxic, preneoplastic and neo-plastic lesions. In: Douglas JF (ed) Carcinogenesis, mutagenesis. Humana, New York, pp 97–130

Oval Cells in Rodent Liver, Mouse, Rat

Natalia V. Engelhardt

Synonyms. Nonparenchymal epithelial cells, oval cell hyperplasia, ductular cell hyperplasia

Gross Appearance

Livers may exhibit a granular or nodular appearance due to the disruption in architecture accompanying oval cell proliferation. Alterations in the normal color of the liver may also occur.

Microscopic Features

A new population of small, oval to round epithelial cells with prominent nuclei and scanty basophilic cytoplasm is among the first cellular changes induced by most hepatocarcinogens and several toxic agents in the rat liver (Farber 1956; Inaoka 1967; Solt and Farber 1976; Sells et al. 1981; Lombardi 1982; Sell et al. 1987) and less frequently in the mouse (Hoover 1985; Uryvaeva and Faktor 1988; Radaeva and Factor 1990). On the basis of their morphology, these cells have been designated oval cells. They emerge in the portal regions and are usually arranged as small ducts and ductules located in close vicinity to the portal veins. Oval cells may proliferate actively and radiate further into the liver acinus in tortuous ducts (Fig. 132). At the time of their maximum expansion, oval cells can occupy more than half of the hepatic lobule. The connection of the newly formed ducts to the true bile ducts has been demonstrated in rats treated with different carcinogens (Schaffner and Popper 1961; Dunsford et al.

1985) and in Dipin-damaged liver of mice (Factor et al. 1994).

At the peak of oval cell proliferation, foci of small basophilic hepatocytes appear in the portal regions close to the oval cells. They proliferate actively, the size of the foci of small hepatocytes increases, and they gradually displace the preexisting damaged hepatocytes and oval cells.

There are three main questions related to oval cell proliferation in rodent liver: (1) What is the cell of their origin? (2) What is the fate of oval cells? (3) What is their relation to the emerging hepatocarcinomas?

Ultrastructure

Oval cells in the mouse (Factor et al. 1994) appear similar to those in the rat (Grisham and Hartroft 1961; Ogawa et al. 1974; Novikoff et al. 1991). When they first appear in Dipin-damaged mouse liver, they form small ducts consisting of two to five cells of primitive type with basal lamina and central lumen. Their nuclei are ovoid or irregular, contain densely scattered chromatin, and occupy approximately half of the cell volume. Free polysomes, individual cisternae of rough endoplasmic reticulum, a well-developed Golgi apparatus, and single mitochondria are detected in the cytoplasm. They have numerous microvilli, well-developed tight junctions, and desmosomes. In later stages, the mean cell diameter of the oval cells lining the ducts which have expanded into parenchyma increases, and the nucleus to cytoplasm ratio decreases, but their ultrastructure re-

Fig. 132a,b. Oval cell proliferation in the periportal regions of Dipin-damaged liver of F1 CBA/C57Bl6 mouse 4 months after treatment. **a** H&E, ×400. **b** Indirect immunoperoxidase staining with monoclonal antibody G7 reacting with common antigen of oval (*double-headed arrows*) and bile ductular cells (*arrows*). *PV*, portal vein. *Bar*, 70 µm. (From Engelhardt et al. 1990)

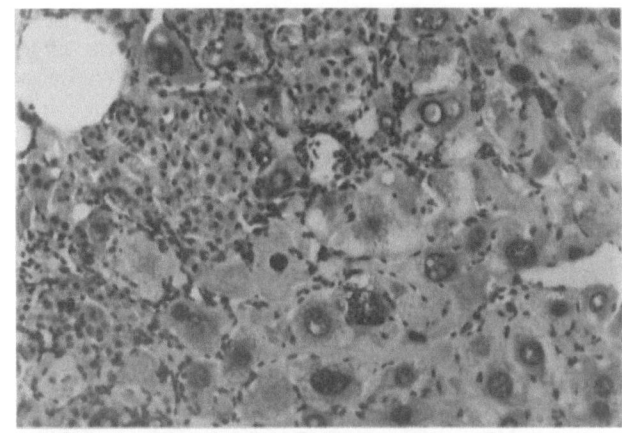

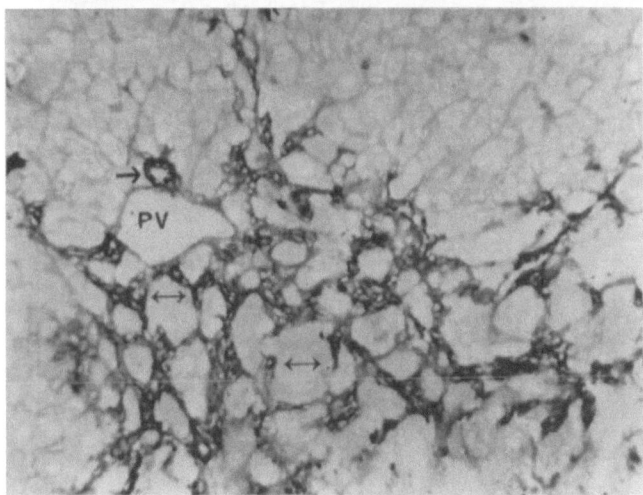

mains generally the same. The basal lamina surrounding the oval cell ducts is well defined. However, it is often lost along the surfaces attached to hepatocytes and desmosomes, and intermediate junctions develop between oval cells and hepatocytes.

Most interesting changes – gradual transition of oval cells to hepatocytes – take place in some oval cell ducts exclusively in the periportal areas of the liver lobules. The signs of hepatocytic differentiation include rounding of cell and nuclear shape, decreased nucleus to cytoplasm ratio, and increased number and size of mitochondria. Biliary epithelial cells proliferating in the liver after bile duct ligation are larger than oval cells and have more developed cytoplasm, and none of them show signs of hepatocytic differentiation (Factor et al. 1994).

Biologic Features

Interest in oval cells is provoked by their possible role as progenitors of hepatocarcinomas, as they regularly appear in various rat hepatocarcinogenic regimens (see above) and in mice transgenic for hepatitis B virus (Dunsford et al. 1990) or bearing simian virus 40 (SV40) large T antigen under the control of human antithrombin III regulatory sequence (Bennoun et al. 1993), which always develop hepatocarcinomas.

Massive proliferation of oval cells is triggered by severe hepatocellular injury in which hepatocytes cannot respond to a growth stimulus. Such a situation is created by galactosamine poisoning in rats (Lemire et al. 1991). Sudden and extensive induction of oval cells is found in the modification of the so-called resistant hepatocyte model of carcinogenesis in the rat (Solt and Farber 1976), in which the initial exposure to dimethylnitrosamine is omitted and partial hepatectomy is performed in the animals receiving 2-acetylaminoflurene, which prevents hepatocyte regeneration (Evarts et al. 1987). In mouse models, oval cell proliferation is less common. However, it is strongly induced after treatment with the genotoxic agent Dipin, followed by partial hepatectomy (Uryvaeva and Faktor 1988). Moreover, we found some small cells actively producing α-fetoprotein (AFP) in the regenerating liver of mice poisoned with carbon tetrachloride. They were ultrastructurally similar to oval cells and formed ductular structures with each other and hepatocytes (Baranov et al. 1982; Engelhardt et al. 1984; Fig. 133).

These data present the major support for the existence of hepatic stem cells in the adult liver (Marceau et al. 1989; Sell 1990; Fausto et al. 1992). The precise anatomic location of the hepatic stem cells is still unclear, but most authors believe that

they persist in the portal areas among the ductular cells of terminal bile ductules (Hering canals) (Wilson and Leduc 1958; Grisham 1980; Fausto et al. 1992) or as distinct populations of periductular cells (Sell and Salman 1984). The oval cells apparently correspond to the transient amplifying population.

The oval cell population is greatly heterogeneous. It contains cells with diverse levels of differentiation and developmental potential. Depending on the manner of induction and strain of the animals used, in addition to two hepatocytic lineages (hepatocytes and biliary epithelial cells) glandular intestinal-type epithelium (Tatematsu et al. 1985; Evarts et al. 1987) and pancreatic-type tissue (Rao et al. 1986) may be found.

Several attempts have been made to find specific markers of rat and mouse oval cells using the hybridoma technique. However, all the monoclonal antibodies proposed to distinguish them from hepatocytes revealed antigens in common with biliary epithelial cells (Hixson and Allison

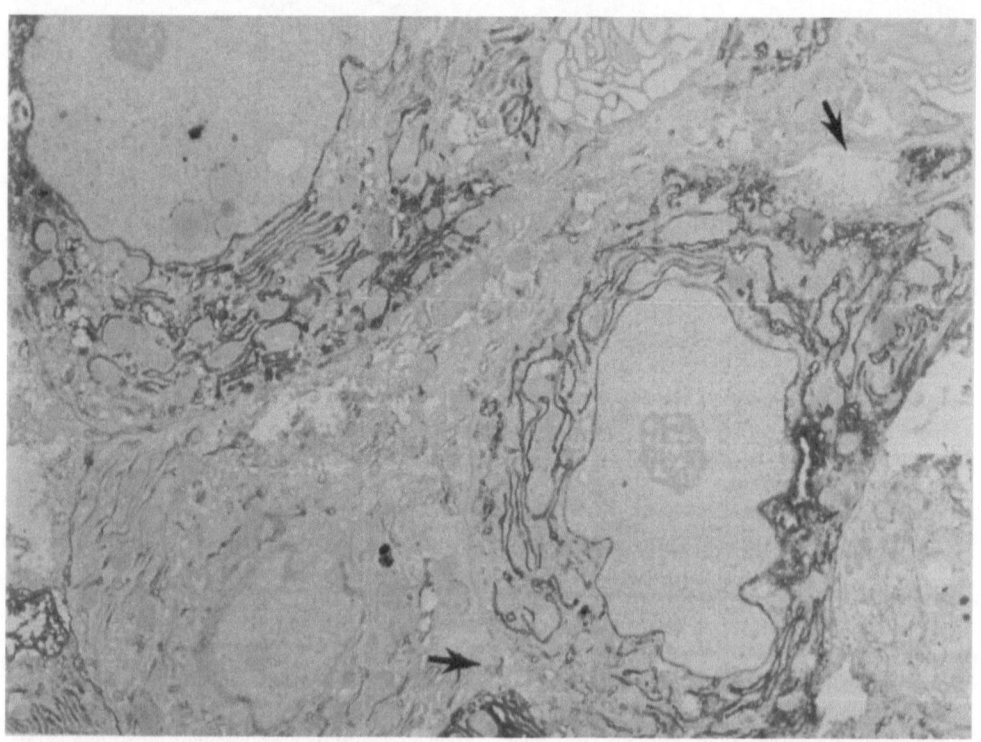

Fig. 133. Section of the regenerating liver of BALB/C/J mouse 72 h after intoxication with carbon tetrachloride. Indirect immunoperoxidase staining with monoclonal antibody to α-fetoprotein (AFP) revealed active AFP synthesis in a group of small, oval-shaped cells. *Arrows* indicate ductular structures formed by these cells. ×5400

1985; Marceau et al. 1986; Dunsford and Sell 1989; Engelhardt et al. 1990; Fig. 132). The protein of gap junctions, connexin 43 (Zhang and Thorgeirsson 1994), is also the same in oval and ductular cells. At the same time, rat oval cells are often capable of producing typical hepatocyte proteins (albumin and AFP), thus indicating a common progenitor for biliary epithelial and hepatocyte cell lineages. In the Dipin-damaged mouse liver, oval cells are AFP negative (V.S. Poltoranina and V.M. Factor, personal communication), but they contain AFP-RNA (V.M. Factor, personal communication). However, AFP-producing oval-like cells can be found in the regenerating mouse liver (see above). Similarities in the set of antigenic, functional, and structural markers permit the suggestion that oval cells are analogous to primitive intrahepatic bile ducts in the fetal rat liver (Shiojiri et al. 1991; Fausto et al. 1992). It is generally accepted that, in rats, oval cells directly develop to hepatocytes after azo-dye treatment. In other models, the interpretation of the results is contradictory, depending on the technical approach used. Some authors conclusively demonstrate that [^{3}H]thymidine is transferred from oval cells to immature hepatocytes and that oval cells give rise to hepatocytes (Evarts et al. 1987, 1989). Other authors suggest hepatocytes to be the cell of origin of newly formed hepatocytes and hepatocarcinomas in rats (Gerlyng et al. 1994; Farber 1992; Fausto et al. 1992). In most mouse hepatocarcinogen models, oval cells are not seen and foci seem to develop from hepatocytes. However, the sequence of morphological events, together with finding of intermediate cells and small hepatocytes within the oval cell ducts in Dipin-damaged mouse liver, suggests that there is transition of oval cells to hepatocytes. An important feature of this phenomenon is that it is not a massive process. Formation of the hepatocyte foci and their further growth seem to occur mostly as a result of active proliferation of a few newly formed hepatocytes. At the same time, most of the oval cells invading the parenchyma seem to constitute a temporary abortive population (Factor et al. 1994).

In various rat and mouse in vivo models, periportal zones are found to be the regions where the first foci of newly formed basophilic hepatocytes appear. Clearly, the direction of differentiation of the oval cells strongly depends on the microenvironment, which includes the intercellular contacts and extracellular matrix. For these reasons, in vitro studies of isolated oval cells seem to be inadequate.

The estimation of the exact mechanisms governing the differentiation of oval cells along different lineages offers an important and intriguing problem for future investigations.

Comparison with Other Species

Oval cell proliferation following liver damage seems to be a universal phenomenon. It has been found in various animals, including hamsters, rabbits, and dogs (Hoover 1985). In humans, small cells with ovoid nuclei resembling oval cells proliferate in the portal areas of submassive necrosis during acute liver failure (Koukoulis et al. 1992) and in hepatitis B virus-associated hepatocarcinogenesis (Hsia et al. 1992).

Acknowledgements. I would like to thank Ms. R.A. Essayan for her excellent assistance in preparing this paper. This work was supported by grant 93-04-21793 from the Russian Foundation for Fundamental Research and by W. Murray's "Russian Program."

References

Baranov VN, Engel'gardt NV, Lazareva MN, Gusev AI, Iasova AK, Shaklamov VA, Abelev GI (1982) Atypical alpha-fetoprotein-synthesizing cells in the regenerating mouse liver (in Russian). Biull Eksp Biol Med 94:82–84

Bennoun M, Rissel M, Engelhardt N, Guillouzo A, Briand P, Weber-Benarous A (1993) Oval cell proliferation in early stages of hepatocarcinogenesis in Simian virus 40 large T transgenic mice. Am J Pathol 143:1326–1336

Dunsford HA, Sell S (1989) Production of monoclonal antibodies to preneoplastic liver cell populations induced by chemical carcinogens in rat and to transplantable Morris hepatomas. Cancer Res 49:4887–4893

Dunsford HA, Maset R, Salman J, Sell S (1985) Connection of ductlike structures induced by a chemical hepatocarcinogen to portal bile ducts in the rat liver detected by injection of bile ducts with a pigmented barium gelatin medium. Am J Pathol 118:218–224

Dunsford HA, Sell S, Chisari FV (1990) Hepatocarcinogenesis due to chronic liver cell injury in Hepatitis B virus transgenic mice. Cancer Res 50:3400–3407

Engelhardt NV, Baranov VN, Lazareva MN, Goussev AI (1984) Ultrastructural localization of alpha-fetoprotein (AFP) in regenerating mouse liver poisoned with CCL4. 1. Reexpression of AFP in differentiated hepatocytes. Histochemistry 80:401–407

Engelhardt NV, Factor VM, Yazova AK, Poltoranina VS, Baranov VN, Lasareva MN (1990) Common antigens of mouse oval and biliary epithelial cells. Expression on newly formed hepatocytes. Differentiation 45:29–37

Evarts RP, Nagy P, Marsden E, Thorgeirsson S (1987) A precursor-product relationship exists between oval cells and hepatocytes in rat liver. Carcinogenesis 8:1737–1740

Evarts R, Nagy P, Nakatsukasa H, Marsden E, Thorgeirsson SS (1989) In vivo differentiation of rat liver oval cells into hepatocytes. Cancer Res 49:1541–1547

Factor VM, Radaeva SA, Thorgeirsson SS (1994) Origin and fate of oval cells in dipin-induced hepatocarcinogenesis in the mouse. Am J Pathol 145:409–422

Farber E (1956) Similarities in the sequence of early histologic changes induced in the liver of the rat by ethionine, 2-acetylaminofluorene, and 3-methyl-4-dimethylaminoazobenzene. Cancer Res 16:142–149

Farber E (1992) On cells of origin of liver cell cancer. In: Sirica AE (ed) The role of cell types in hepatocarcinogenesis. CRC, Boca Raton, pp 1–28

Fausto N, Lemire JM, Shiojiri N (1992) Oval cells in liver carcinogenesis: cell lineages in hepatic development and the identification of stem cells in normal liver. In: Sirica AE (ed) The role of cell types in hepatocarcinogenesis. CRC, Boca Raton, pp 89–108

Gerlyng P, Grotmol T, Stokke T, Erikstein B, Seglen PO (1994) Flow cytometric investigation of a possible precursor-product relationship between oval cells and parenchymal cells in the rat liver. Carcinogenesis 15:53–59

Grisham JW (1980) Cell types in long-term propagable cultures of rat liver. Ann N Y Acad Sci 349:128–137

Grisham JW, Hartroft WS (1961) Morphological identification by electron microscopy of oval cells in experimental hepatic degeneration. Lab Invest 10:317–332

Hixson DC, Alison JP (1985) Monoclonal antibodies recognizing oval cells induced in the liver of rats by N-2-fluorenylacetamide or ethionine in a choline-deficient diet. Cancer Res 45:3750–3760

Hoover KL (1985) Oval cell hyperplasia, liver, mouse, rat. In: Jones TC, Mohr U, Hunt RD (eds) Monographs on pathology of laboratory animals, digestive system. Springer, Berlin Heidelberg New York, pp 125–127

Hsia CC, Evarts RP, Nakatsukasa H, Marsden ER, Thorgeirsson SS (1992) Occurrence of oval-type cells in hepatitis B virus-associated human hepatocarcinogenesis. Hepatology 16:1327–1333

Inaoka Y (1967) Significance of the so-called oval cell proliferation during azo-dye hepatocarcinogenesis. Gann 58:355–366

Koukoulis G, Rayner A, Tan KC, Williams R, Portmann B (1992) Immunolocalization of regenerating cells after submassive liver necrosis using PCNA staining. J Pathol 166:359–368

Lemire J, Shiojiri N, Fausto N (1991) Oval cell proliferation and the origin of small hepatocytes in liver injury induced by D-galactosamine. Am J Pathol 139:535–552

Lombardi B (1982) On the nature, properties and significance of oval cells. In: Pani P, Feo F, Columbano A (eds) Recent trends in chemical carcinogenesis. ESA, Cagliari, Italy, pp 36–56

Marceau N, Germain L, Goyette R, Noel M, Gourdeau H (1986) Cell of origin of distinct cultured rat liver epithelial cells, as typed by cytokeratin and surface component selective expression. Biochem Cell Biol 64:788–802

Marceau N, Blouin MJ, Germain L, Noel M (1989) Role of different epithelial cell types in liver ontogenesis, regeneration and neoplasia. In Vitro Cell Dev Biol 25:336–341

Novikoff PM, Ikeda T, Hixon DC, Yam A (1991) Characterizations of and interactions between bile ductule cells and hepatocytes in early stages of rat hepatocarcinogenesis induced by ethionine. Am J Pathol 139:1351–1368

Ogawa K, Minase T, Onhoe T (1974) Demonstration of glucose-6-phosphatase activity in the oval cells of rat liver and the significance of the oval cells in azo dye carcinogenesis. Cancer Res 34:3379–3386

Radaeva SA, Factor VM (1990) Time course of development of oval cell population induced in the mouse liver by dipin and partial hepatectomy. Bull Exp Biol Med 109:514

Rao MS, Bendayan M, Kimbrough RD, Reddy JK (1986) Characterization of pancreatic-type tissue in the liver of rat induced by polychlorinated biphenyls. J Histochem Cytochem 34:197–201

Schaffner F, Popper H (1961) Electron microscopic studies of normal and proliferated bile ductules. Am J Pathol 38:393–410

Sell S (1990) Is there a liver stem cell? Cancer Res 50:3811–3815

Sell S, Salman J (1984) Light- and electron-microscopic autoradiographic analysis of proliferating cells during the early stages of chemical hepatocarcinogenesis in the rat induced by feeding N-2-fluorenylacetamide in a choline-deficient diet. Am J Pathol 114:287–300

Sell S, Hunt JM, Knoll BJ, Dunsford HA (1987) Cellular events during hepatocarcinogenesis in rats and the question of premalignancy. Adv Cancer Res 48:37–111

Sells MA, Katyal SL, Shinozuka H, Estes LW, Sell S, Lombardi B (1981) Isolation of oval cells and transitional cells from the livers of rats fed the carcinogen DL-ethionine. J Natl Cancer Inst 66:355–362

Shiojiri N, Lemire JM, Fausto N (1991) Cell lineages and oval cell progenitors in rat liver development. Cancer Res 51:2611–2620

Solt D, Farber E (1976) New principle for the analysis of chemical carcinogenesis. Nature 263:701–703

Tatematsu M, Kaku T, Medline A, Farber E (1985) Intestinal metaplasia as a common option of oval cells in relation to cholangiofibrosis in liver of rats exposed to 2-acetylaminofluorene. Lab Invest 52:354–362

Uryvaeva IV, Faktor VM (1988) Complete change of the cell composition of the liver parenchyma following exposure to dipin and partial resection (in Russian). Biull Eksp Biol Med 105:77–80

Wilson JW, Leduc EH (1958) Role of cholangioles in restoration of the liver of the mouse after dietary injury. J Pathol Bacteriol 76:441–449

Zhang M, Thorgeirsson SS (1994) Modulation of connexins during differentiation of oval cells into hepatocytes. Exp Cell Res 213:37–42

Herniation of Liver Through Esophageal Hiatus, Rat

Matthew J. van Zwieten and Carel F. Hollander

Synonym. Diaphragmatic hernia

Gross Appearance

Herniation of a portion of a liver lobe through the diaphragm appears grossly as a well-defined smooth or slightly dimpled nodule on the diaphragmatic surface of the liver (Figs. 134, 135). The nodule is frequently the same color as the liver or may be darker or lighter depending on the degree of vascular congestion or hepatocellular alteration and fatty change. The nodule can vary in size and shape depending on the diameter and nature of the diaphragmatic defect, i.e., whether a direct opening is present (complete hernia) or whether the defect results from a weakening of the diaphragmatic fascia with formation of a hernial sac (incomplete hernia). Nodules protruding through a diaphragmatic opening, especially if accompanied by some degree of constriction, such as the example illustrated in Fig. 134, can generally be identified properly at the time of necropsy. Smaller nodules (Fig. 135), especially those resulting from an incomplete hernia, may be mistaken for a hepatoproliferative lesion unless a careful gross inspection is carried out to identify a corresponding diaphragmatic defect (L. Emerson, personal communication).

Microscopic Features

The histologic appearance of the liver tissue within a nodule varies from case to case. The nodule may have a relatively normal hepatic architecture (Fig. 136) with, at most, signs of compression of tissue in the region of contact with the hernial ring. Foci of hepatocellular alteration (see also p. 3), resembling those found elsewhere in the liver, may occur in a nodule (Fig. 137).

Frequently, however, the histologic appearance of the nodule differs from that observed in the rest of the liver (Emerson, personal communication), being characterized predominantly by cellular changes suggestive of regeneration and hyperactivity (Ward 1981) as found in true hepatoproliferative lesions.

Ultrastructure

Electron microscopy studies of this lesion have not been done.

Differential Diagnosis

The primary liver lesions from which these nodules must be differentiated are the true hepatoproliferative lesions, including the so-called hyperplastic or neoplastic nodules, and discrete nodular benign or malignant hepatic neoplasms (see also p. 55). As noted previously, an adequate gross examination during necropsy will ensure the proper interpretation of a nodule resulting from a diaphragmatic hernia.

Biologic Features

The hepatic nodular lesion we have described results from herniation of a portion of a liver lobe through a cleft in the diaphragm. Diaphragmatic clefts are most often congenital and are formed by incomplete fusion of the pleuroperitoneal folds, or, more rarely, by the failure of one of these folds to fuse with the septum transversum. These diaphragmatic clefts are often present in the dorsal tendinous portions of the diaphragm, representing enlargement of the normal esophageal hiatus (Jubb and Kennedy 1970).

The diaphragmatic defect may be large enough to permit herniation of an entire liver lobe, which may become severely infarcted (Woodard and Montgomery 1978). Portions of the intestines and uterus may also herniate into the thoracic cavity in such cases (Andersen 1949). More frequently, the diaphragmatic cleft allows the herniation of only a part of a liver lobe, resulting in the formation of a small nubbin of liver tissue. When the defect is covered by a thin membrane derived from the diaphragmatic fascia, the herniated liver tissue is contained within a small sac, and the hernia is said to be incomplete (Andersen 1949).

Information on the frequency of liver herniation in rats is scarce. It is probably more common than is generally recognized, however, since it is likely

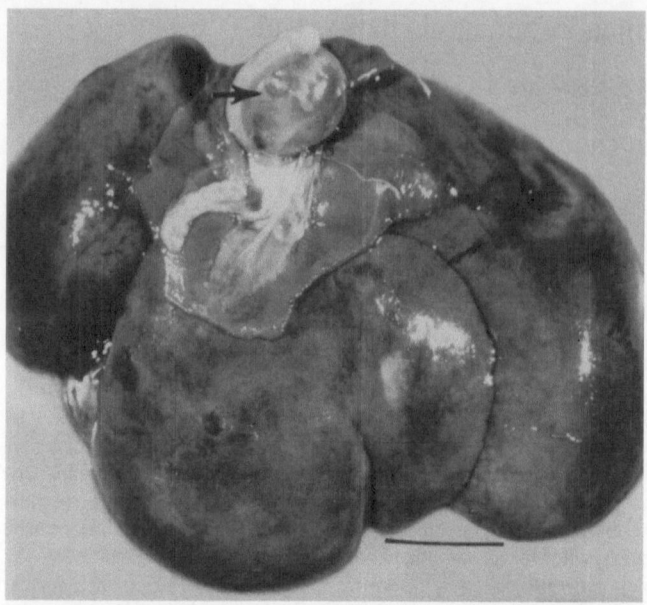

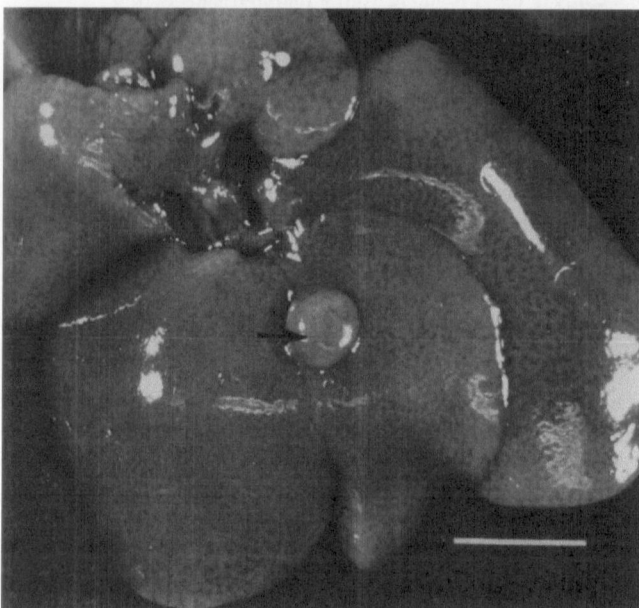

Fig. 134. (*above*) Incomplete diaphragmatic herniation of part of a liver lobe through the esophageal hiatus in an 18-month-old female WAG/Rij rat. The herniated part of the liver is present within a thin sac adjacent to the esophagus (*arrow*). *Bar*, 1 cm

Fig. 135. (*below*) Small hepatic nodule resulting from diaphragmatic hernia in a 20-month-old female F344 rat. *Bar*, 1 cm. (Courtesy of Dr. R.R. Maronpot, NTP, Research Triangle Park, NC)

often misdiagnosed as a nodular hepatopro-liferative lesion. Andersen (1949) observed a baseline incidence of 2.7% in an inbred albino strain of rats. The incidence was increased to 7.7% by selectively breeding females from litters with a high proportion of affected siblings. In 144 male F344 rats studied histopathologically, one case of diaphragmatic hernia (0.7%) was found (Coleman et al. 1977) and another investigator, who de-scribed liver lesions in 2700 male and female F344 rats, reports the occasional finding of this lesion without indicating its exact frequency (Ward 1981). In our own laboratory, liver herniation has been encountered infrequently in rats of two other inbred strains (WAG/Rij and BN/BiRij) used for aging research (M.J. van Zwieten, C. Zurcher, and C.F. Hollander, unpublished observations), al-though the incidence in these strains has not been tabulated. In the experience of some investigators, however, F344 rats appear to develop this lesion more frequently than do rats of other commonly used strains (Emerson, personal communication), although the paucity of data in the literature pre-cludes firm conclusions.

The cause of the diaphragmatic defect leading to liver herniation in rats is not known. Results of studies by Andersen (1949) indicate that genetic factors play an important role in the expression of this lesion. A genetic predisposition may also ex-plain the more common occurrence observed in F344 rats. Andersen (1949) has presented evi-dence, however, that nutritional factors during pregnancy, i.e., vitamin A deficiency, were in-volved in the expression of a genetic trait.

Comparison with Other Species

Diaphragmatic clefts, with or without herniation of abdominal viscera, occur uncommonly in a vari-ety of animals species, including humans. A he-reditary form has been described in rabbits (Fox and Crary 1973), and diaphragmatic defects were

▶

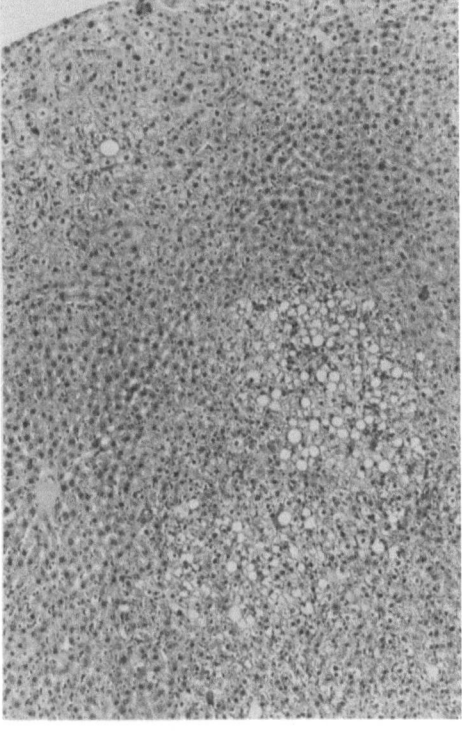

Fig. 136. (*above*) Portion of liver lobe which had herniated through the diaphragm of a F344 rat. The focus indicated by an *arrow* is shown in more detail in Fig. 137. (Section courtesy of Dr. G.A. Boorman, NTP, Research Triangle Park, NC) H&E, ×13

Fig. 137. (*below*) Focus of hepatocellular alteration and fatty change in a liver nodule resulting from diaphragmatic hernia (see arrow in Fig. 136). (Section courtesy of Dr. G.A. Boorman, NTP, Research Triangle Park, NC.) H&E, ×70

found in 1%–2% of rabbit fetuses showing a number of other anomalies (Wilson et al. 1978). In humans, the most commonly encountered type of diaphragmatic hernia, the so-called hiatal hernia, is characterized by protrusion of a portion of the stomach and abdominal esophagus into the thoracic cavity, with symptoms related to the reflux of gastric contents into the esophagus (Horn 1971).

References

Andersen DH (1949) Effect of diet during pregnancy upon the incidence of congenital hereditary diaphragmatic hernia in the rat. Am J Pathol 25:163–185

Coleman GL, Barthold SW, Osbaldiston GW, Foster SJ, Jonas AM (1977) Pathological changes during aging in barrier-reared Fischer 344 male rats. J Gerontol 32:258–278

Fox RR, Crary DD (1973) Hereditary diaphragmatic hernia in the rabbit. Genetics and pathology. J Hered 64:333–336

Horn RC Jr (1971) Alimentary tract. In: Anderson WAD (ed) Pathology, vol 2. Mosby, St Louis, chap 27, p 1124

Jubb KVF, Kennedy PC (eds) (1970) Pathology of domestic animals, 2nd edn, vol 2. Academic, New York, pp 455–456

Ward JM (1981) Morphology of foci of altered hepatocytes and naturally-occurring hepatocellular tumors in F344 rats. Virchows Arch (Pathol Anat) 390:339–345

Wilson JG, Kalter H, Palmer AK, Hoar RM, Shenefelt RE, Beck F, Earl FL, Nelson NS, Berman E, Stara JF, Selby LA, Scott WJ Jr (1978) Developmental abnormalities. In: Benirschke K, Garner FM, Jones TC (eds) Pathology of laboratory animals, vol II. Springer, New York Heidelberg Berlin, chap 20, pp 1851–1856

Woodard JC, Montgomery CA (1978) Musculoskeletal system. In: Benirschke K, Garner FM, Jones TC (eds) Pathology of laboratory animals, vol I. Springer, New York Heidelberg Berlin, chap 10, p 832

K Virus Infection, Mouse

John E. Greenlee

Synonyms. None. However, K virus is also referred to as K papovavirus

Gross Appearance

Gross pathologic changes associated with K virus infection are confined to the lungs and are most prominent in mice less than 8 days of age, in whom K virus produces a fatal interstitial pneumonia (Fisher and Kilham 1953; Holt 1959; Greenlee 1979, 1981); lungs of these fatally infected animals are congested and edematous on cut section. In older suckling animals, similar changes of lesser severity are present during the acute stages of infection (Fisher and Kilham 1953; Greenlee 1981). Gross changes are not evident in lungs of immunologically normal weanling or adult mice during acute K virus infection and do not occur during K virus infection in nude (*nu/nu*) mice or in reactivated K virus infection under conditions of immunosuppression (Fisher and Kilham 1953; Greenlee 1981, 1986; Greenlee and Dodd 1984). Acute K virus infection in older animals immunosuppressed with cyclophosphamide, however, resembles the infection seen in newborn mice and may be accompanied by pulmonary congestion (Mokhtarian and Shah 1980). K virus does not produce gross pathologic changes in other organs of newborn or older mice, and gross changes in other organs are not seen in immunosuppressed animals.

Microscopic Features

Acute Infection

Histopathologic changes due to K virus are most prominent in suckling mice with fatal K virus pneumonia (Fisher and Kilham 1953; Holt 1959; Greenlee 1979, 1981). In older animals, K virus produces a clinically inapparent infection in which the severity of pulmonary and systemic involvement depends upon the age of the animal at the time of inoculation (Fisher and Kilham 1953; Greenlee 1981). Even in newborn animals, however, identification of histologic findings depends heavily on methods of tissue processing, and many features of K virus infection are obscured in conventional formalin-fixed, paraffin-embedded sections (Holt 1959; Greenlee 1979; Greenlee and Keeney 1982). Identification of intranuclear inclusions within K virus-infected cells is greatly enhanced by fixation with Bouin's solution, 96% ethanol/1% glacial acetic acid, or paratoluenesulfonic acid (Greenlee 1981; Greenlee and Keeney 1982; Margolis et al. 1976). Preservation of antigens for immunohistochemistry may be achieved using 96% ethanol/1% glacial acetic acid or, less optimally, Bouin's fixative (Greenlee and Keeney 1982). Paraformaldehyde (2% or 4%) or periodate-lysine-paraformaldehyde-glutaraldehyde (PLPG) may also be used for immunohistologic studies, but these fixatives do not allow identification of intranuclear inclusions (Greenlee et al. 1991). Diagnostic accuracy is also improved by use of paraffin or glycol methacrylate-embedded sections prepared at 2-μm thickness. Optimal detection of infected cells requires immunohistologic staining of K virus early (T) or capsid (V) antigen, using either immunoperoxidase techniques or study of adjacent sections by histologic and fluorescent antibody methods (Greenlee and Keeney 1982). In situ nucleic acid hybridization methods can also be employed to identify infected cells (Greenlee et al. 1991, 1994). The extent of K virus infection in vascular endothelia and other cell types is always much more extensive in sections studied using immunohistologic methods or in situ nucleic acid hybridization than can be demonstrated by routine histologic methods.

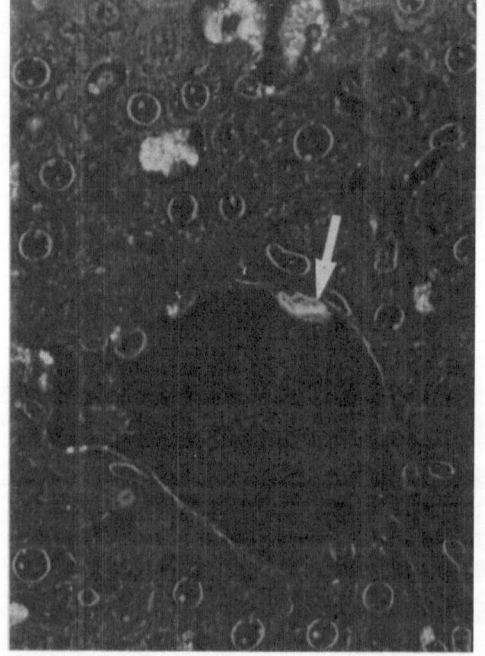

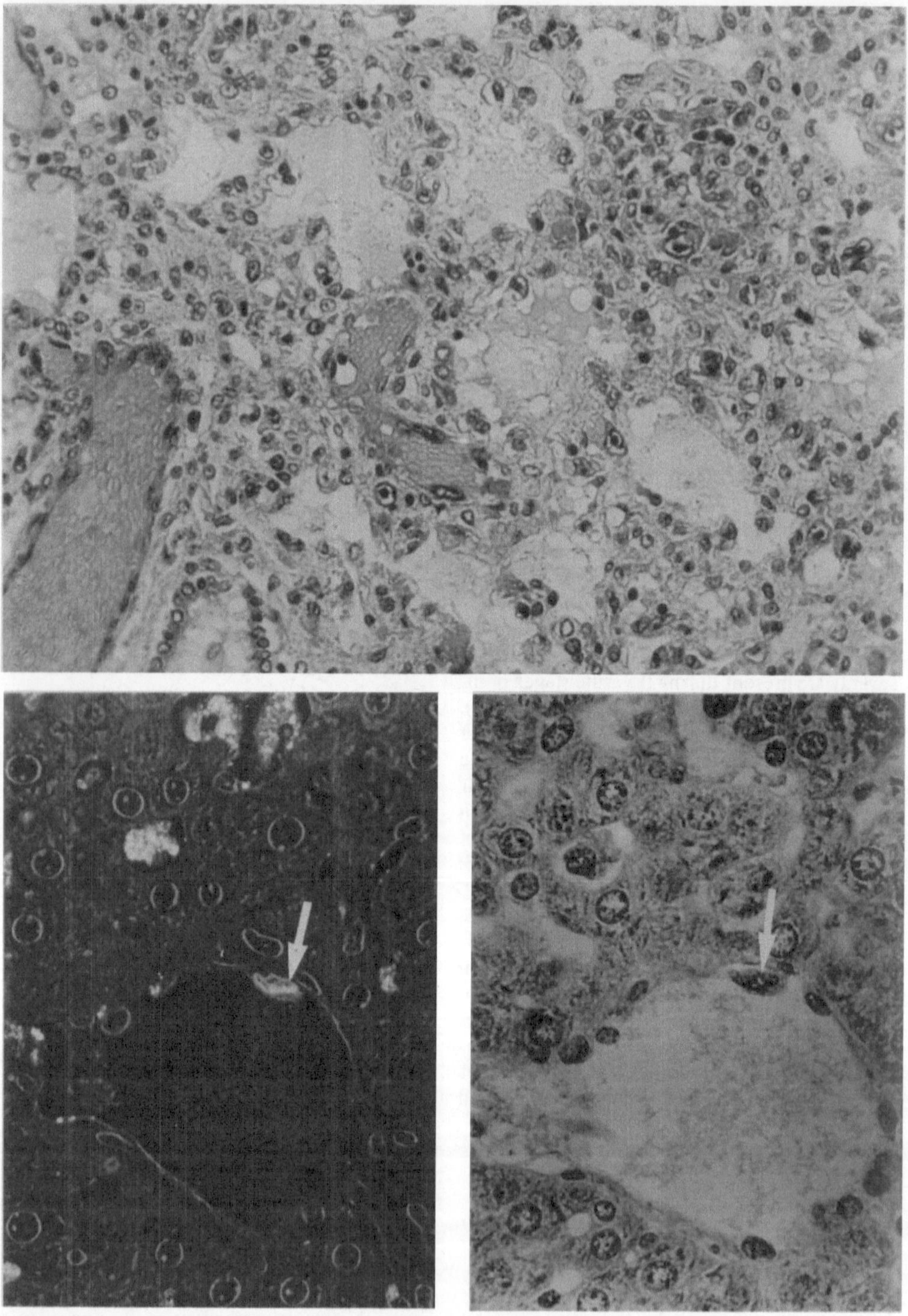

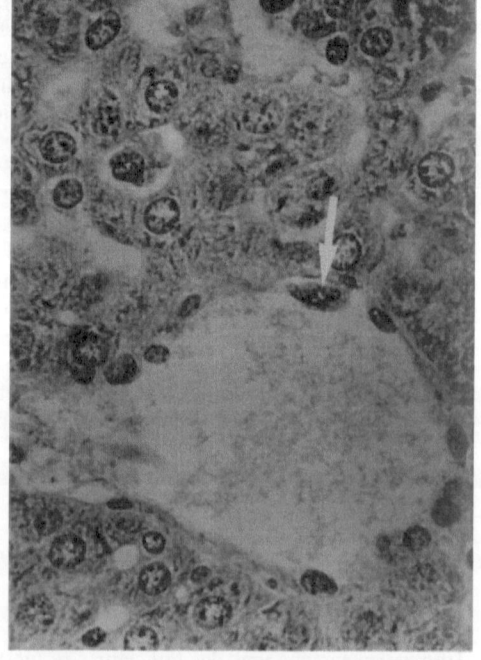

K virus pneumonia in newborn mice is characterized by thickening and lymphocytic infiltration of alveolar walls, transudation of fluid into alveolar spaces, and nuclear enlargement or development of Cowdry A-type intranuclear inclusions within endothelial cells of alveolar capillaries and larger pulmonary vessels (Fig. 138). Bronchial and bronchiolar endothelial cells are not involved, although involvement of rare pulmonary alveolar epithelial cells may be detected by immunofluorescence or immunoperoxidase methods. Histologic examination of livers from mice dying from K virus pneumonia reveals nuclear swelling or intranuclear inclusions within endothelial cells lining hepatic sinusoids. Examination of livers by immunohistologic or in situ hybridization methods reveals K virus antigens and nucleic acids in endothelial and Kupffer's cells, but not in hepatic parenchymal cells (Figs. 139, 140). Nuclei of endothelial cells are intensely stained. Kupffer's cells contain cytoplasmic but not nuclear viral antigens and nucleic acids, indicating phagocytosis of circulating virions without viral replication. Livers from many – but not all – fatally infected animals contain large numbers of membrane-bound, empty spaces. These can be shown by in situ nucleic acid hybridization methods to be surrounded by K virus nucleic acids and

by immunofluorescence or immunoperoxidase methods to be rimmed with viral antigen (Figs. 141, 142; Greenlee et al. 1994). The spaces are believed to represent clusters of adjacent infected cells which have undergone lytic infection within a given sinusoid. Involvement of scattered vascular endothelia can be observed in virtually all other organs. Inclusions are perhaps most easily detected in capillaries of intestinal villi (Figs. 143, 144), but are also found in spleens, kidneys, and adrenals (Fig. 145; Greenlee 1979, 1981, 1983). In situ nucleic acid hybridization methods reveal that acute K virus infection also results in extensive involvement of lymphocytes within Peyer's patches (Fig. 146) and spleens and in scattered renal tubular epithelial cells (Greenlee et al. 1991, 1994). Within brains, inclusions can be found in vascular endothelia and in occasional small cells which appear to be glia (Greenlee 1983; Greenlee et al. 1994). Microscopic findings in suckling animals between 7 and 14 days of age are similar to, but less severe than those observed in newborn mice. Inclusions are rarely identified by histologic methods in weanling or adult animals, although involvement of scattered pulmonary and extrapulmonary vascular endothelial cells can be identified by immunohistological or in situ hybridization methods (Holt 1959; Fisher and Kilham 1953; Greenlee 1981).

Persistent Infection

During acute infection, K virus infection involves vascular endothelial cells almost exclusively. Beginning approximately 2 months after initial infection (at death in suckling animals with overwhelming infection), however, viral nucleic acids and proteins can be detected in renal tubular epithelial cells (Greenlee et al. 1994). By 6 months after infection, renal tubular epithelia represent the major cell type involved (Greenlee et al. 1991). For the most part, infected tubular epithelia appear histologically normal, but occasional cells have enlarged nuclei or contain intranuclear inclusions. Infected tubules may be surrounded by mononuclear inflammatory cells (Greenlee et al. 1991). Cells found to be positive by nucleic acid hybridization are often present in clusters. These data suggest that the major site of K virus persistence is the renal tubular epithelial cell and that the virus may persist by cell-to-cell spread

◄

Fig. 138. (above) Lung from mouse dying of K virus pneumonia. Note prominent vascular congestion and transudation of fluid into alveoli. Alveolar walls are thickened and contain an infiltrate consisting predominantly of lymphocytes. Numerous Cowdry type A inclusions bodies are present in nuclei of alveolar capillaries and other pulmonary vessels. Fixed using Bouin's solution and sectioned at 3 μm. H&E, ×180

Fig. 139. (lower left) Liver from mouse with lethal K virus infection. Nuclear fluorescence indicative of viral replication is present in the nucleus of an endothelial cell (arrow). Numerous other cells contain cytoplasmic antigen without nuclear involvement, consistent with phagocytosis of circulating viral particles. The section was fixed in 96% ethanol plus 1% glacial acetic acid, embedded in glycol methacrylate, sectioned at 2-μm thickness, stained for K virus capsid (V) antigen using indirect immunofluorescence methods, and examined using phase fluorescence microscopy. ×300

Fig. 140. (lower right) Section adjacent to Fig. 139. The involved endothelial cell is indicated by the arrow. H&E, ×300

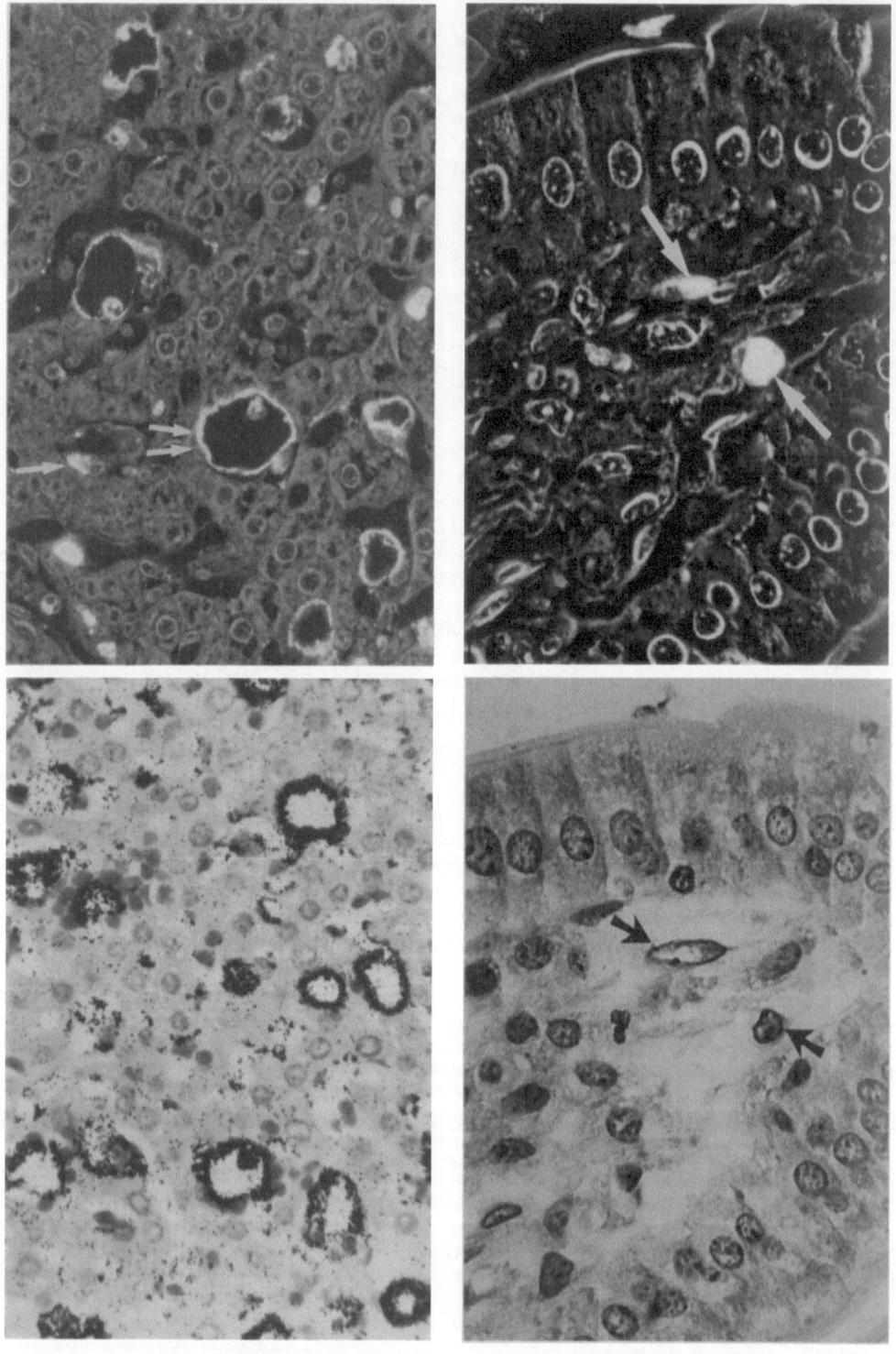

◀

Fig. 141. (*upper left*) Liver from mouse with lethal K virus infection. Multiple microcystic changes are present, rimmed with viral antigen (*double arrows*). Nuclear fluorescence is present in many endothelial cells, and scattered Kupffer's cells (*single arrow*) exhibit cytoplasmic fluorescence with nuclear sparing, indicating phagocytosis of circulating virus. The section was fixed in 96% ethanol plus 1% glacial acetic acid, embedded in glycol methacrylate, sectioned at 2-μm thickness, stained for K virus capsid (V) antigen using indirect immunofluorescence methods, and examined using phase fluorescence microscopy. ×280

Fig. 142. (*lower left*) Liver from mouse with lethal K virus infection. Numerous microcystic spaces are surrounded by dense collections of exposed emulsion grains, indicating the presence of K virus DNA and RNA. Cells lining hepatic sinusoids also exhibit specific hybridization. Probed for K virus nucleic acids using in situ hybridization methods. ×210

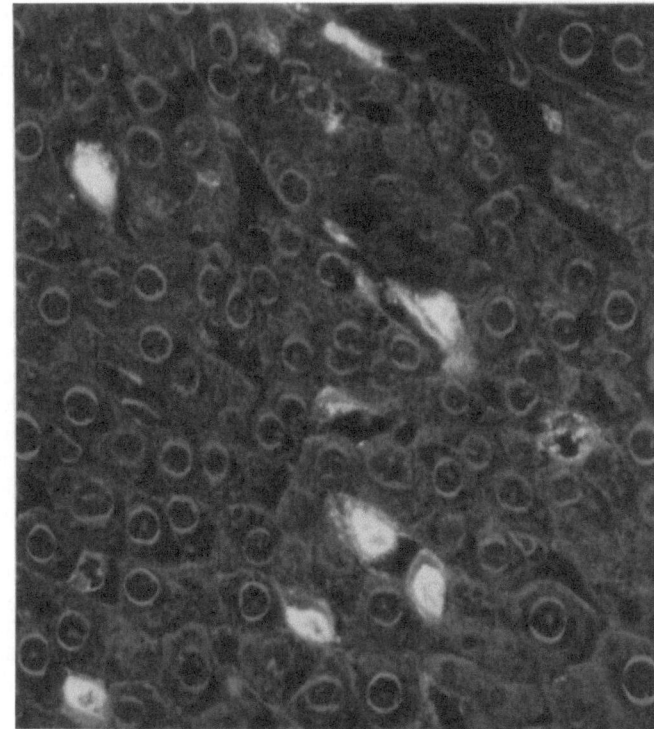

◀

Fig. 143. (*upper right*) Intestine, mouse, with lethal K virus infection. Nuclear fluorescence indicative of viral replication is present in the nuclei of two endothelial cells (*arrows*). The section was fixed in 96% ethanol plus 1% glacial acetic acid, embedded in glycol methacrylate, sectioned at 2-μm thickness, stained for K virus capsid (V) antigen using indirect immunofluorescence methods, and examined using phase fluorescence microscopy. ×300

Fig. 144. (*lower right*) Section adjacent to Fig. 143. The involved endothelial cells are indicated by *arrows*. H&E, ×300

▶

Fig. 145. (*above*) Adrenal from mouse with lethal K virus infection. Nuclei of numerous capillary endothelial cells are positively stained. The section was fixed in 96% ethanol plus 1% glacial acetic acid, embedded in glycol methacrylate, sectioned at 2-μm thickness, stained for K virus capsid (V) antigen using indirect immunofluorescence methods, and examined using phase fluorescence microscopy. ×525

Fig. 146. (*below*) Peyer's patch from mouse with lethal K virus infection probed for K virus DNA and RNA. Increased numbers of exposed emulsion grains overly a lymphocyte, indicating the presence of K virus DNA and RNA. ×650

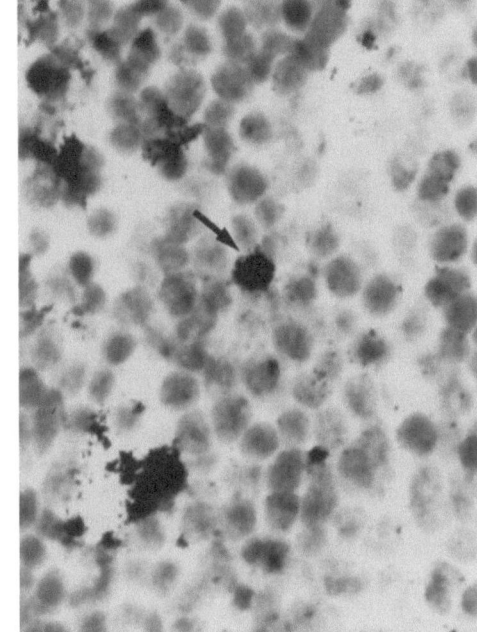

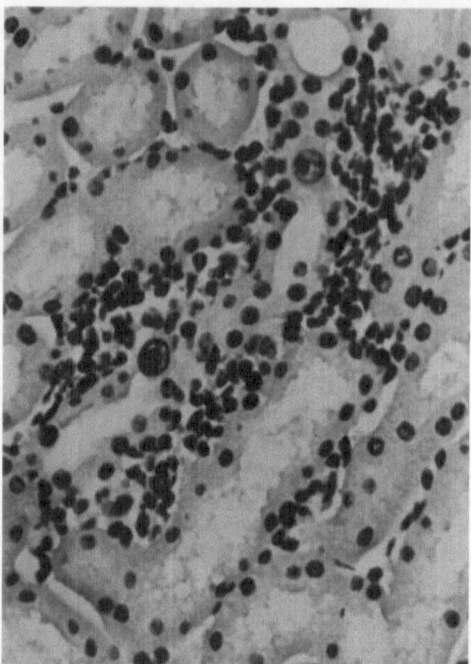

Fig. 147. Kidney from mouse with reactivated K virus infection during immunosuppression with cyclophosphamide, 6 months after initial infection. The nucleus of a renal tubular epithelial cell contains a Cowdry type A inclusion which contains viral antigen. The tubule is surrounded by an infiltrate of mononuclear cells. Stained for K virus V antigen using peroxidase–antiperoxidase methods. ×425

of lytic infection with wider dissemination of infection being prevented by the host's immune response.

Reactivated Infection Following
Immunosuppression

Immunosuppression of mice surviving acute K virus infection following inoculation at ages 8–12 days of age results in the reappearance of viral antigen and detectable titers of virus in multiple organs, including kidneys, intestines, lungs, livers, and brains (Greenlee and Dodd 1984). Viral titers and numbers of infected cells are highest in kidneys (Greenlee and Dodd 1984; Greenlee et al. 1991). Reactivated infection is not accompanied by detectable histopathologic changes except in

kidneys, where occasional renal tubular epithelial cells may contain intranuclear inclusions (Greenlee et al. 1991). Infected tubules are often surrounded by infiltrates of mononuclear inflammatory cells (Fig. 147).

K Virus Infection in Nude (*nu/nu*) Mice

K virus infection of nude mice results in a clinically silent, progressive infection which is partially contained by low levels of circulating antibody (Greenlee 1986; Mokhtarian and Shah 1983). Although occasional infected cells have histologically detectable nuclear enlargement, intranuclear inclusions are usually not found, and the infection is best studied using either immunohistologic or in situ hybridization methods. These techniques will indicate infection of vascular endothelial cells in multiple organs, including lungs, but with most prominent involvement of capillary endothelia in intestinal villi and spleens (Greenlee 1986). Extensive infection of renal tubular epithelial cells can also be found in animals with advanced infection (Greenlee 1986). Inflammatory infiltrates are not seen.

Ultrastructure

Limited ultrastructural studies of K virus infection have identified cytoplasmic virions within pulmonary alveolar capillaries and hepatic Kupffer's cells (Dalton et al. 1959, 1963; Gleiser and Heck 1972; Jordan and Doughly 1969). Extensive ultrastructural examination of other K virus-infected organs has not been reported.

Differential Diagnosis

Several other murine agents, including polyoma virus and murine cytomegalovirus, may produce systemic infection and result in intranuclear inclusions. K virus is unique, however, in its ability to produce a fatal interstitial pneumonia with predominant involvement of pulmonary and extrapulmonary vascular endothelial cells and with sparing of bronchial and alveolar epithelial cells. The microcystic lesions seen in livers of newborn mice with fatal infection are also unique to K virus infection. Specific histologic diagnosis of the renal tubular infection caused by K virus, as distinguished from polyoma virus, may require immunohistologic or in situ methods, particularly in nude mice, in whom pulmonary and hepatic changes are often subtle or absent.

Biologic Features

K virus is distinct from the only other known mouse papovavirus, polyoma virus, in its nucleic acid sequence and antigenic properties (Law et al. 1979; Bond et al. 1978; Mayer and Dorries 1991). K virus was initially identified in both wild and laboratory mice, but is now rarely found in laboratory mouse colonies. The virus is believed to be transmitted in nature by the oral route, with initial replication of virus in capillaries of intestinal villi and probably also Peyer's patches (Greenlee 1979; Greenlee et al. 1994). Epidemiologic studies indicate that infection with the virus occurs between 4 and 7 months of age. Two epidemiological patterns of infection have been observed: antiviral antibody has been detected in the majority of animals within some colonies, suggesting widespread infection; in other colonies, however, only a few animals have been found to have antibody (Parker et al. 1966; Tennant et al. 1966; Rowe et al. 1963). In newborn mice, K virus causes an overwhelming respiratory infection in which viral nucleic acids and intranuclear inclusions containing viral antigens can be identified in endothelial cells of alveolar capillaries and larger pulmonary vessels (Fisher and Kilham 1953; Holt 1959; Greenlee 1979; Greenlee et al. 1994). Smaller numbers of infected cells are present in capillary endothelia within virtually all other organs and tissues. Lungs, livers, spleens, and adrenals are most heavily involved. Rare alveolar epithelial cells may be involved in fatally infected animals. Immunohistologic studies of brains demonstrate infection of both vascular endothelial cells and scattered extravascular cells within brain parenchyma (Greenlee 1983; Ikeda et al. 1988). Studies employing immunohistochemical staining for glial fibrillary acidic protein and myelin basic protein as markers of astrocytes and oligodendrocytes, respectively, indicate that both of these cell types may be infected (J.E. Greenlee et al., unpublished data). Studies employing nucleic acid hybridization methods also indicate the presence of K virus DNA and RNA in Peyer's patches and other lymphoid tissues and in rare renal tubular epithelial cells (Greenlee et al. 1994). Recovery from acute K virus infection is heavily dependent on development of antiviral antibody, and newborn mice can be protected from lethal infection by passive transfer of anti-K virus antibody or by adoptive transfer of K virus-sensitized B cells (Greenlee 1981; Mokhtarian and Shah 1980, 1983). Containment of the virus during persistent infection, however, appears to require intact T lymphocyte function, in addition to circulating antiviral antibody (Greenlee and Dodd 1984; Greenlee et al. 1991).

Inoculation of older, suckling, weanling, and adult animals results in a protracted, clinically silent infection similar to that in newborn mice, but whose extent and distribution is determined by the age and immunologic status of the animal at the time of inoculation (Greenlee 1981). In older animals, vascular endothelia remain the major cell type supporting K virus replication during acute infection. By 2 months after inoculation, however, viral nucleic acids can be detected in scattered renal tubular epithelial cells, and by 6 months after acute infection, renal tubular cells rather than vascular endothelia are the major site of viral persistence (Greenlee et al. 1991). Renal tubular epithelial cells are also the major site of viral replication during reactivated infection under conditions of immunosuppression (Greenlee et al. 1991). In nude (nu/nu) mice, K virus causes a nonfatal infection with predominant involvement of intestinal capillaries and renal tubules (Greenlee 1986).

Comparison with Other Species

K virus differs from all other known polyomaviruses in its strong tropism for vascular endothelial cells during acute infection and in its ability to produce a lethal infection in its natural host. A single case has been reported in which SV40 virus produced a fatal pneumonia in a laboratory primate, but the pulmonary infection in this animal was characterized by involvement of epithelial cells rather than endothelial cells (Sheffield et al. 1980). Murine polyoma virus has been shown to involve occasional vascular endothelial cells in brains of persistently infected nude mice; however, vascular involvement is not normally a feature of polyoma virus infection (McCance et al. 1983). In situ nucleic acid hybridization studies have demonstrated that K virus, like virtually all other known members of the polyoma virus group, persists in renal tubular epithelial cells and causes productive infection of these cells during reactivated infection in immunosuppressed animals. K virus, like human JC virus, appears able to replicate in lymphoid tissue and, within brains, to infect both astrocytes and oligodendrocytes (Houff et al. 1988; Greenlee et al. 1991, 1994; J.E. Greenlee et al., unpublished observations). Unlike JC virus, however, K virus does not appear to cause destruction of oligodendrocytes, and K virus infection of brain does not result in demyelination or in the appearance of atypical, abortively infected astrocytes as is the case in human progressive multifocal leukoencephalopathy (PML).

Acknowledgments. The above description of the pathology of K virus infection is based in part upon work supported by the United States Department of Veterans Affairs and by Public Health Grant 7 R01 NS24925 from the National Institute of Neurological and Communicative Disorders and Stroke. Cloned K virus DNA was provided by Dr. Peter Howley and Dr. Ming Fan Law. The technical assistance of Mrs. Paula Keeney, Mr. Richard Phelps, and Mrs. Susan Clawson is gratefully acknowledged.

References

Bond SB, Howley PM, Takemoto KK (1978) Characterization of K virus and its comparison with polyoma virus. J Virol 28:337–343

Dalton AJ, Moore AE, Mottrain FC (1959) Electron microscopic observations on the "K" virus of mice. J Appl Phys 30:20–25

Dalton AJ, Kilham L, Zeigel RF (1963) A comparison of polyoma, "K", and Kilham rat viruses with the electron microscope. Virology 20:391–398

Fisher ER, Kilham L (1953) Pathology of a pneumotropic virus recovered from C3H mice carrying the Bittner milk agent. Arch Pathol 55:14–19

Gleiser CA, Heck FC (1972) The pathology of experimental K-virus infection in suckling mice. Lab Anim Sci 22:865–869

Greenlee JE (1979) Pathogenesis of K virus infection in newborn mice. Infect Immun 26:705–713

Greenlee JE (1981) Effect of host age on experimental K virus infection in mice. Infect Immun 33:297–303

Greenlee JE (1983) Pathogenesis of K virus infection in mice. In: Sever JL, Madden DL (eds) Polyomaviruses and human neurological disease. Liss, New York, pp 325–342

Greenlee JE (1986) Chronic infection of nude mice by murine K papovavirus. J Gen Virol 67:1109–1114

Greenlee JE, Dodd WK (1984) Reactivation of persistent papovavirus K infection in immunosuppressed mice. J Virol 51:425–429

Greenlee JE, Keeney PM (1982) Immunofluorescent labelling of K-papovavirus antigens in glycol methacrylate embedded material: a method for studying infected cell populations by fluorescence microscopy and histological staining of adjacent sections. Stain Technol 57:197–205

Greenlee JE, Phelps RC, Stroop WG (1991) The major site of murine K-papovavirus persistence and reactivation is the renal tubular epithelium. Microbial Pathol 11:237–247

Greenlee JE, Clawson SH, Phelps RC, Stroop WG (1994) Distribution of K-papovavirus in infected newborn mice. J Comp Pathol 111:259–268

Holt D (1959) Presence of K-virus in wild mice in Australia. Aust J Exp Biol 37:183–192

Houff SA, Major EO, Katz DA, Kufta CV, Sever JL, Pittaluga S, Roberts JR, Gitt J, Saini N, Lux W (1988) Involvement of JC virus-infected mononuclear cells from the bone marrow and spleen in the pathogenesis of progressive multifocal leukoencephalopathy. N Engl J Med 318:301–305

Ikeda K, Dorries K, ter Meulen V (1988) Morphological and immunohistochemical studies of the central nervous system involvement in papovavirus K infection in mice. Acta Neuropathol (Berl) 77:175–181

Jordan SW, Doughty WE (1969) Ultrastructural pathology of murine pneumonitis caused by K-papovavirus. Exp Mol Pathol 11:1–7

Law MF, Takemoto KK, Howley PM (1979) Characterization of the genome of the murine papovavirus K. J Virol 30:90–97

Margolis G, Jacobs LR, Kilham L (1976) Oxygen tension and the selective tropism of K-virus for mouse pulmonary endothelium. Am Rev Respir Dis 114:45–51

Mayer M, Dorries K (1991) Nucleotide sequence and genome organization of the murine polyomavirus, Kilham strain. Virology 181:469–480

McCance DJ, Sebesteny A, Griffin BE, Balkwill F, Tilly R, Gregson NA (1983) A paralytic disease in nude mice associated with polyoma virus infection. J Gen Virol 64:57–67

Mokhtarian F, Shah KV (1980) Role of antibody response in recovery from K-papovavirus infection in mice. Infect Immun 29:1169–1179

Mokhtarian F, Shah KV (1983) Pathogenesis of K-papovavirus infection in athymic nude mice. Infect Immun 41:434–436

Parker JC, Tennant RW, Ward TG (1966) Prevalence of K virus in mouse colonies. In: Holdenried R (ed) Viruses of laboratory rodents. Natl Cancer Inst Monogr 20:33

Rowe WP, Hartley JW, Huebner RJ (1963) Polyoma and other indigenous mouse viruses. Lab Anim Care 13:166–175

Sheffield WD, Strandberg JD, Braun L, Shah K, Kalter SS (1980) Simian virus 40-associated fatal interstitial pneumonia and renal tubular necrosis in a rhesus monkey. J Infect Dis 142:618–622

Tennant RW, Parker JC, Ward TG (1966) Respiratory virus infections of mice: K virus. In: Holdenried R (ed) Viruses of laboratory rodents. Natl Cancer Inst Monogr 20:100–104

Mouse Hepatitis Virus Infection, Liver, Mouse

Stephen W. Barthold

Synonyms. Hepatoencephalitis virus, murine hepatitis virus infection, mouse coronavirus infection

Gross Appearance

Gross lesions can occur in liver, intestine, and lymphoreticular organs. Intestinal lesions are described in detail elsewhere. Affected livers have random, small, pale or hemorrhagic foci to multiple confluent foci with depression of the capsular surface. The liver may be diffusely pale and covered with fibrinous peritoneal exudate. Infant mice can be runted, jaundiced, or may manifest neurologic signs, including tremor, incoordination, or convulsions (Piazza 1969). During the acute phase of infection, involution of lymph nodes, spleen, and thymus can occur. Recovered mice develop mild splenomegaly or lymphadenomegaly, particularly in cervical nodes. Athymic nude mice can become progressively cachectic (wasting disease). Their livers are contracted with rough, nodular surfaces (Ward et al. 1977), and splenomegaly can be pronounced (Ishida et al. 1978).

Microscopic Features

Depending on virus and host factors, foci of necrosis, leukocytic infiltration, and syncytium formation may be encountered in several organs. Acute focal hepatocellular necrosis is accompanied by hemorrhage and mild mixed leukocyte infiltration. Lesions in susceptible mice are more severe and often coalesce, with parenchymal collapse. Nuclei of degenerating cells often have characteristic dense, marginated chromatin, or chromatin is condensed in multiple dense bodies (Fig. 148). Syncytia arising from hepatocytes or other cells can be present (Fig. 149; Barthold 1985, 1986a; Barthold and Smith 1984; Jones and Cohen 1962; Piazza 1969). In athymic nude mice, parenchymal collapse, fibrosis, and syncytium formation are pronounced (Fujiwara et al. 1977; Ishida et al. 1978; Tamura et al. 1977; Ward et al. 1977). They often have marked myelopoiesis in

portal regions (Fig. 150) and spleen (Ishida et al. 1978). Hepatocellular mitotic activity is elevated in infected nude mice or immunocompetent mice recovering from mouse hepatitis virus infection (Fig. 151; Barthold 1985, 1986a; Carthew 1981; Jones and Cohen 1962). Discrete nodular foci of macrophage or leukocyte accumulations (microgranulomas) are often present in the liver of recovering mice (Fig. 152).

Two patterns of mouse hepatitis virus infection are seen: *respiratory* and *enteric*. The enteric pattern is described elsewhere. The respiratory pattern is most often associated with hepatitis. Mild necrosis of nasal epithelium, perivascular lymphocytic infiltration in the lung, and focal necrosis with syncytia in lymph nodes, spleen, brain, bone marrow, mesothelium, and other organs can also be present. In resistant adult hosts, lesions are restricted to upper respiratory mucosa, with minimal dissemination (Barthold 1985, 1986a; Barthold and Smith 1984, 1987). Athymic mice have endothelial syncytia in blood vessels of lung, heart base, brain, and other organs. Syncytia, necrosis, and marked myelopoiesis are frequent in bone marrow and spleen. Epithelial syncytia in the intestine are occasionally present (Barthold 1985, 1986a; Fujiwara et al. 1977; Ishida et al. 1978; Tamura et al. 1977; Ward et al. 1977), but most virus activity is in gut-associated lymphoid tissue and lamina propria (Barthold and Smith 1987).

Ultrastructure

Kupffer's cells and hepatocytes develop a number of nonspecific degenerative changes. Specific changes include dissociation of ribosomes from endoplasmic reticulum, aggregation of ribosomes, formation of discrete, compact arrays of electron-dense reticular structures (reticular inclusions), and virion formation. Virions bud into cytoplasmic cisternae and are usually dispersed in small numbers, but occasional compact aggregates can be found. Virions are pleomorphic with a corona of surface spikes and an average diameter of about 90nm. These changes, including reticular inclusions, have been observed in a variety of infected cells in vitro (NCTC 1469 cells) and in vivo

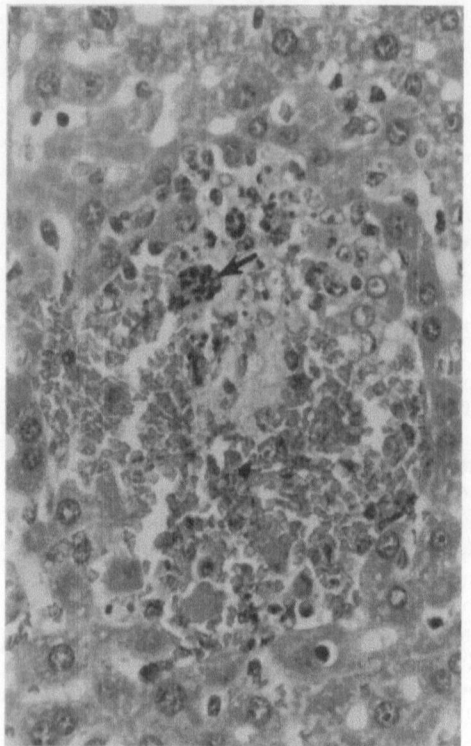

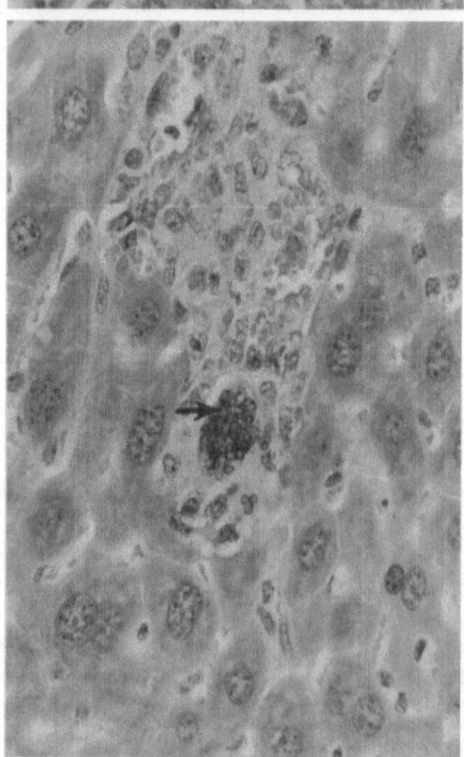

(oligodendroglia, astrocytes, Kupffer's cells, hepatocytes, and enterocytes) (Barthold et al. 1982; David-Ferreira and Manaker 1965; Lampert et al. 1973; Ruebner et al. 1967; Svoboda et al. 1962).

Differential Diagnosis

Necrotizing hepatitis and focal hepatitis mimic lesions induced by a number of other pathogens. Idiopathic focal hepatic necrosis can be encountered in normal mice in the absence of pathogens. Careful examination of liver lesions for multinucleate syncytia and examination of other organs, including nose, lung, and bowel helps to provide a definitive diagnosis. Antigen of the virus can be demonstrated in active lesions (Barthold and Smith 1984, 1987; Barthold et al. 1990; Brownstein and Barthold 1982), and seroconversion of recovered mice is confirmatory. Virus isolation is a difficult and insensitive means of diagnosis. Wasting disease in nude mice must be differentiated from other chronic infectious diseases, including those caused by polyoma virus, Sendai virus, pneumonia virus of mice, mouse adenovirus, and *Pneunocystis carinii*. Histologic findings are confirmatory.

Biologic Features

Natural History

Mouse hepatitis virus is represented by innumerable different strains that are generally highly contagious and spread by respiratory or orofecal routes. In utero transmission of the virus can be shown experimentally, but only under the somewhat artificial combination of highly virulent strain, susceptible host, and stage of pregnancy. These events are not likely to occur under natural conditions (Barthold et al. 1988). Respiratory strains vary widely in their virulence. Most of these virus strains are only mildly pathogenic,

◀
Fig. 148. (*above*) Mouse hepatitis virus infection, liver, mouse. Focal necrotizing hepatitis with hemorrhage. Note the large, degenerating cell with densely clumped chromatin (*arrow*), a frequent finding in these lesions. H&E, ×410

Fig. 149. (*below*) Liver, mouse. Focal hepatitis with a characteristic syncytium (*arrow*), probably of histiocytic origin. H&E, ×660

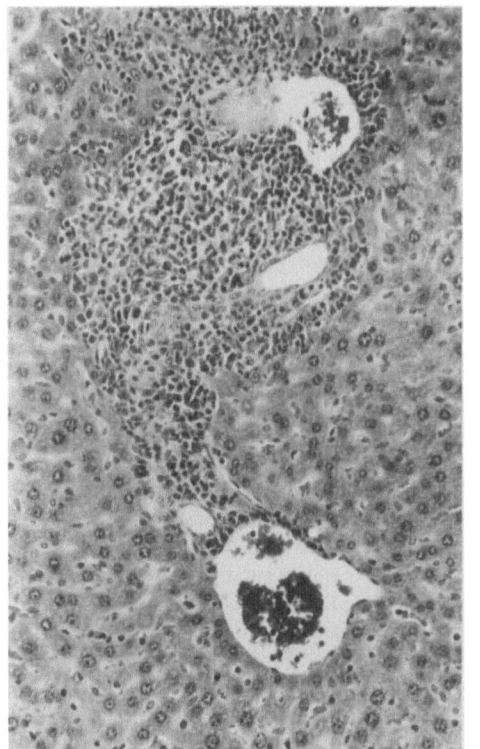

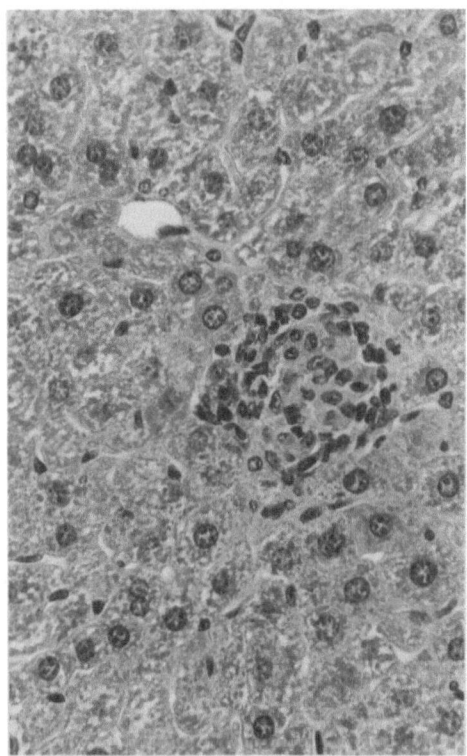

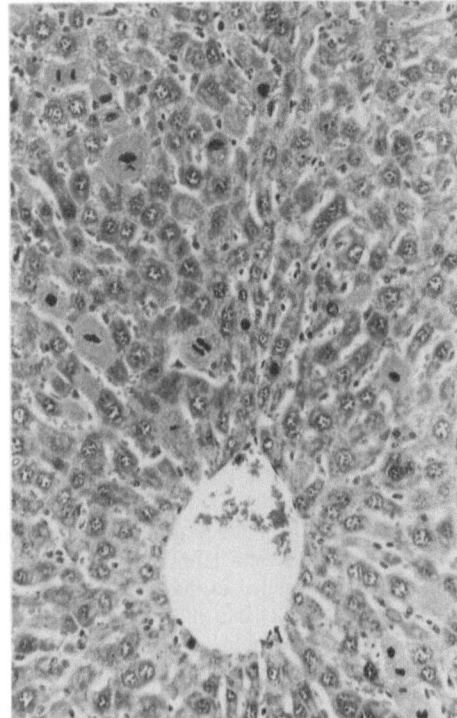

▶

Fig. 150. (*upper left*) Extramedullary myelopoiesis, liver, athymic nude mouse infected with mouse hepatitis virus. H&E, ×165

Fig. 151. (*upper right*) Liver, mouse recovering from mouse hepatitis virus infection. Increased hepatocellular mitotic activity. H&E, ×165

Fig. 152. (*below*) Microgranuloma, liver, mouse recovering from mouse hepatitis virus infection. H&E, ×410

even in nude mice (Hirano et al. 1975), while some can be highly virulent in adult mice (Le Prevost et al. 1975). In naturally infected, uncompromised adult mice, the infection is usually subclinical. Clinical disease can be precipitated during active infection by a variety of stressful situations, particularly immunosuppression, alteration of macrophage function, and tumor transplantation (Barthold 1985, 1986a). The respiratory pattern is usually subclinical, unless in infant or immuno-compromised mice, which manifest signs of encephalitis and hepatitis (Barthold 1985, 1986a; Barthold and Smith 1987). Athymic nude mice infected with this virus develop chronic wasting disease lasting up to several weeks (Hirano et al. 1975). The virus can be introduced in a population of laboratory mice by feral mice, subclinically infected mice, or infected biologic products. Transplantable tumors, particularly lymphoreticular and ascites tumors, commonly carry the virus, which can cause oncolysis or other abnormal host–tumor kinetics (Barthold 1985, 1986a).

Pathogenesis

The course of infection with mouse hepatitis virus is dependent on the strain of virus and host factors, which consist predominantly of genotype and lymphoreticular function (Barthold 1985, 1986a; Hirano et al. 1975, 1981; Le Prevost et al. 1975; Piazza 1969). Apparently, many viral strains exist, as mouse hepatitis virus is subject to a high rate of mutation and recombination, events which no doubt contribute to the survival of the virus in mouse populations. Hepatotropism and neurotropism are characteristic features in most strains when virus is inoculated intraperitoneally or intracerebrally (Piazza 1969). The role of these organs in natural disease is secondary and not essential for successful infection and transmission. Hepatitis, however, is a very common lesion in naturally infected mice. Strains of virus that cause respiratory infections seem to replicate in nasal or olfactory mucosa as a primary target and then disseminate to other internal organs if the host is susceptible (Barthold and Smith 1992). Strains of low virulence in resistant hosts cause asymptomatic infections limited to the nasal mucosa, but some extend directly into the brain through the olfactory tracts (Barthold et al. 1988). In susceptible hosts, vascular endothelial cells become infected, with hematogenous dissemination to

multiple organs (Barthold and Smith 1992). Intestine can be infected, but involvement is mild with minimal lesions compared to enteric strains (Barthold 1985, 1986a; Barthold and Smith 1983, 1984, 1987). Fecal transmission seems to play only a minor role with respiratory strains of virus. Virus is recoverable and viral antigen is demonstrable in most organs for only about 1–3 weeks after inoculation, after which time infection is cleared (Barthold 1985, 1986a; Barthold and Smith 1983, 1987; Fujiwara et al. 1977; Tamura et al. 1977). Infection in immunocompetent mice is short term with no carrier state (Barthold and Smith 1990), but selected low-virulence, neurotropic strains can persist in the brain, causing chronic demyelination when inoculated intracerebrally (Stohlman and Weiner 1981). This does not seem to play a crucial role in the natural history of the infection.

Much of what is known about host resistance to mouse hepatitis virus has been derived from studies on strains of respiratory type. Host genotype is a significant factor in resistance and seems to be mediated through a number of mechanisms (Bang 1978; Hirano et al. 1975; Taguchi et al. 1976, 1979; Piazza 1969). Lymphoreticular function is also an important factor (Bang 1978; Dupuy et al. 1975; Levy-Leblond and Dupuy 1977; Taguchi et al. 1979; Tardieu et al. 1980). Neonatal mice have immature lymphoreticular function and are thus highly susceptible to infection, usually dying of hepatitis or encephalitis. Natural resistance develops at about 2 weeks of age, but significant differences in susceptibility continue to be present in adults, depending on genotype (Bang 1978; Barthold and Smith 1987; Hirano et al. 1975; Taguchi et al. 1979). Maternal antibody seems to play a protective role through this period of age-related susceptibility. Antibody to respiratory strains of the virus appears to be transferred primarily through colostrum, and not in utero, and continues to be absorbed through the bowel of infant mice until around 2 weeks of age. Maternally derived antibody seems to protect pups against different mouse hepatitis virus strains (Barthold et al. 1988). Mice that have recovered from active infection with one strain of mouse hepatitis virus resist reinfection with the homologous strain, but are fully susceptible to infection with another strain of the virus. Furthermore, recovered mice can be actively reinfected with the same strain of virus after several months. Thus immunity to mouse hepatitis virus is highly strain specific and not long-lasting, analogous to common respiratory coronavirus infections (colds) in humans (Barthold and Smith 1989a,b).

Infection of mice with other agents, such as K virus and retroviruses, can increase susceptibility to mouse hepatitis virus, presumably through effects on macrophages (Barthold 1985, 1986b). Recovery from the infection appears to be a function of cell-mediated immunity. Passive transfer of sensitized lymphocytes and macrophages, but not antibody, is protective (Levy-Leblond and Dupuy 1977; Tardieu et al. 1980). Athymic, T cell-deficient nude mice are unable to recover, succumbing to chronic progressive infection of multiple organs, particularly liver and brain (Fujiwara et al. 1977; Tamura et al. 1977). Other strains of mouse hepatitis virus cause enteric infections. They are more easily recognized because of their more severe clinical effects and more obvious intestinal lesions. Enterotropic strains infect primarily the intestinal mucosa and are less likely to disseminate widely to other organs (see p. 379 for further discussion of enterotropic infections).

Etiology

Mouse hepatitis virus is a coronavirus with many strains that possess complex antigenic interrelationships (Barthold 1985, 1986a; Piazza 1969), virulence, and organotropism (Barthold and Smith 1984, 1987; Hirano et al. 1981; Piazza 1969). The antigenic composition of a strain does not predict its virulence or organotropism (Barthold 1985, 1986a; Barthold and Smith 1984). Coronaviruses can be divided into four distinct antigenic groups, with antigenic homology within each group, but no common antigen among groups. Mouse hepatitis virus belongs to a group containing rat coronaviruses, human coronaviruses (OC43 group), bovine coronaviruses, and porcine hemagglutinating encephalomyelitis virus. Neonatal mice can be experimentally infected with all of these viruses by oronasal inoculation, but adult mice can only be infected with rat and mouse coronaviruses. Despite susceptibility to infection, mice transmit rat coronavirues inefficiently (Barthold et al. 1990).

Frequency

Mouse hepatitis virus is an extremely prevalent virus among laboratory mice throughout the world. Confusion exists over the true frequency of this infection for a number of reasons. Most infected mice do not manifest clinical disease and have limited lesions that are difficult to discern. In enzootically infected populations, disease is often transient, since mice recover rapidly or signs may be obscured by partial protection with maternal antibody. Evidence of past infection is usually confirmed by indirect immunofluorescence and enzyme-lifted immunosorbent sero-assays, which are most sensitive (Smith 1983; Smith and Winograd 1986).

Comparison with Other Species

There was considerable interest in mouse hepatitis virus when it was first discovered, since it provided a potential model of viral hepatitis in humans (Piazza 1969). However, close similarities turn out to be minimal. Focal necrotizing hepatitis is a nonspecific lesion and is seen as a feature of many infectious diseases in the mouse and other species. Mouse hepatitis virus, like coronaviruses of other species, has either respiratory or enteric primary tropism, depending on virus strain.

References

Bang FB (1978) Genetics of resistance of animals to viruses. I. Introduction and studies in mice. Adv Virus Res 23:269–348

Barthold S (1985) Research complications and state of knowledge of rodent coronaviruses. In: Hamm TE Jr (ed) Complications of viral and mycoplasmal infections in rodents to toxicology research and testing. Hemisphere, Washington DC, pp 53–89

Barthold SW (1986a) Mouse hepatitis virus biology and epizootiology. In: Bhatt PN, Jacoby RO, Morse HC II, New AE (eds) Viral and mycoplasmal infections of laboratory rodents: effects on biomedical research. Academic, Orlando, pp 571–601

Barthold SW (1986b) Olfactory neural pathway in mouse hepatitis virus nasoencephalitis. Acta Neuropathol (Berl) 76:502–506

Barthold SW, Smith AL (1983) Mouse hepatitis virus S in weanling Swiss mice following intranasal inoculation. Lab Anim Sci 33:355–360

Barthold SW, Smith AL (1984) Mouse hepatitis virus strain-related pattern of tissue tropism in suckling mice. Arch Virol 81:103–112

Barthold SW, Smith AL (1987) Response of genetically susceptible and resistant mice to mice to intranasal inoculation with mouse hepatitis virus JHM. Virus Res 7:225–239

Barthold SW, Smith AL (1989a) Duration of challenge immunity to coronavirus JHM in mice. Arch Virol 107:171–177

Bartold SW, Smith AL (1989b) Virus strain specificity of challenge immunity to coronavirus. Arch Virol 104:187–196

Barthold SW, Smith AL (1990) Duration of mouse hepatitis virus infection: studies in immunocompetent and chemically immunosuppressed mice. Lab Anim Sci 40:133–137

Barthold SW, Smith AL (1992) Viremic dissemination of mouse hepatitis virus-JHM following intranasal inoculation of mice. Arch Virol 122:35–44

Barthold SW, Smith AL, Lord PF, Bhatt PN, Jacoby RO, Main AJ (1982) Epizootic coronaviral typhlocolitis in suckling mice. Lab Anim Sci 32:376–383

Barthold SW, Beck DS, Smith AL (1988) Mouse hepatitis virus and host determinants of vertical transmission and maternally-derived passive immunity in mice. Arch Virol 100:171–183

Barthold SW, de Souza MS, Smith AL (1990) Susceptibility of laboratory mice to intranasal and contact infection with coronaviruses of other species. Lab Anim Sci 40:481–485

Brownstein DG, Barthold SW (1982) Mouse hepatitis virus immunofluorescence in formalin- or Bouin's-fixed tissues using trypsin digestion. Lab Anim Sci 32:37–39

Carthew P (1981) Inhibition of the mitotic response in regenerating mouse liver during viral hepatitis. Infect Immun 33:641–642

David-Ferreira JF, Manaker RA (1965) An electron microscope study of the development of a mouse hepatitis virus in tissue culture cells. J Cell Biol 24:57–78

Dupuy J, Levy-Leblond E, Le Prevost C (1975) Immunopathology of mouse hepatitis virus type 3 infection. II. Effect of immunosuppression in resistant mice. J Immunol 114:226–230

Fujiwara K, Tamura T, Taguchi F, Hirano N, Ueda K (1977) Wasting disease in nude mice infected with facultatively virulent mouse hepatitis virus. Proceedings of the 2nd international workshop on nude mice, pp 53–60

Hirano N, Takenaka S, Fujiwara K (1975) Pathogenicity of mouse hepatitis virus for mice depending upon host age and route of infection. Jpn J Exp Med 45:285–292

Hirano N, Murakami T, Taguchi F, Fujiwara K, Matumoto M (1981) Comparison of mouse hepatitis virus strains for pathogenicity in weanling mice infected by various routes. Arch Virol 70:69–73

Ishida T, Tamura T, Ueda K, Fujiwara K (1978) Hepatosplenic myelosis in naturally occurring mouse hepatitis virus infection in the nude mouse. Nippon Juigaku Zasshi 40:739–743

Jones WA, Cohen RB (1962) The effect of murine hepatitis virus on the liver. An anatomic and histochemical study. Am J Pathol 41:329–347

Lampert PW, Sims JK, Kniazeff AJ (1973) Mechanism of demyelination in JHM virus encephalomyelitis. Acta Neuropathol (Berl) 24:76–85

Le Prevost C, Levy-Leblond E, Virelizier JI, Dupuy JM (1975) Immunopathology of mouse hepatitis virus type 3 infection. Role of humoral and cell-mediated immunity in resistance mechanisms. J Immunol 114:221–225

Levy-Leblond E, Dupuy JM (1977) Neonatal susceptibility to MHV 3 infection in mice. I. Transfer of resistance. J Immunol 118:1219–1222

Piazza M (1969) Experimental viral hepatitis. Thomas, Springfield

Ruebner BH, Hirano T, Slusser RJ (1967) Electron microscopy of the hepatocellular and Kupffer-cell lesions of mouse hepatitis, with particular reference to the effect of cortisone. Am J Pathol 51:163–189

Smith AL (1983) An immunofluorescence test for detection of serum antibody to rodent coronaviruses. Lab Anim Sci 33:157–160

Smith AL, Winograd DF (1986) Two enzyme immunoassays for the detection of antibody to rodent coronaviruses. J Virol Methods 14:335–343

Stohlman SA, Weiner LP (1981) Chronic central nervous system demyelination in mice after JHM virus infection. Neurology 31:38–44

Svoboda D, Nielson A, Werder A, Higginson J (1962) An electron microscopic study of viral hepatitis in mice. Am J Pathol 41:205–224

Taguchi F, Hirano N, Kiuchi Y, Fujiwara K (1976) Difference in response to mouse hepatitis virus among susceptible mouse strains. Jpn J Microbial 20:293–302

Taguchi F, Yamada A, Fujiwara K (1979) Factors involved in the age-dependent resistance of mice infected with low-virulence mouse hepatitis virus. Arch Virol 62:333–340

Tamura T, Taguchi F, Ueda K, Fujiwara K (1977) Persistent infection with mouse hepatitis virus of low virulence in nude mice. Microbial Immunol 21:683–691

Tardieu M, Hery C, Dupuy JM (1980) Neonatal susceptibility to MHV3 infection in mice. II. Role of natural effecter marrow cells in transfer of resistance. J Immunol 124:418–423

Ward JM, Collins MJ Jr, Parker JC (1977) Naturally occurring mouse hepatitis virus infection in the nude mouse. Lab Anim Sci 27:372–376

Rat Parvovirus Infection, Liver

Robert O. Jacoby

Synonyms. Rat virus infection, Kilham rat virus infection, H-1 virus infection

Gross Appearance

Gross lesions normally occur only in rats infected as infants or fetuses. Mechanical or toxic injury may, however, facilitate virus-induced necrosis in adults (Margolis et al. 1968; Ruffolo et al. 1966). During acute infection, the liver may be soft and pale brown with rounded edges and contain gray–white foci (necrosis) or red foci (laked blood or hemorrhage). These lesions can be accompanied by ascites and icterus (Ruffolo et al. 1966; Coleman et al. 1983; Jacoby et al. 1987). Mild lesions resolve uneventfully, but if necrosis is severe the liver may become firm or nodular due to stromal collapse, fibrosis, and compensatory hepatocytic and biliary hyperplasia. Small, red capsular cysts or elevations resembling those of peliosis hepatis may also develop (Bergs and Scotti 1967).

Microscopic Features

Viral-induced hepatocytic necrosis is the central lesion. Basophilic type A parvoviral inclusions may develop in hepatocytic nuclei as early as 24 h after infection and can persist for up to 3 weeks (Fig. 153; Margolis et al. 1968). Inclusions also may be found in vascular endothelium, Kupffer cells, bile duct epithelium, and connective tissue fibroblasts. They vary in size and may fill the nucleus or can be separated from the nuclear membrane by a halo. Immunohistochemical staining and in situ molecular hybridization have revealed viral antigen and DNA, respectively, in these cell types (Jacoby et al. 1987; Gaertner et al. 1993). Nuclear chromatin in infected cells is often concentrated at the nuclear envelope. The cytoplasm of infected hepatocytes often becomes increasingly dense and eosinophilic or may undergo ballooning degeneration. Cell nuclei become pyknotic and karyorrhectic.

Necrosis can occur among individual cells or groups of cells, but is in random distribution with respect to lobular zones. During severe necrosis, large segments of one or more lobules may be destroyed. Inflammation develops during convalescence (see below), especially in portal triads, and consists primarily of mononuclear cells and some polymorphonuclear leukocytes (Ruffolo et al. 1966). Bile stasis can be observed in icteric livers and may be accompanied by bile thrombi.

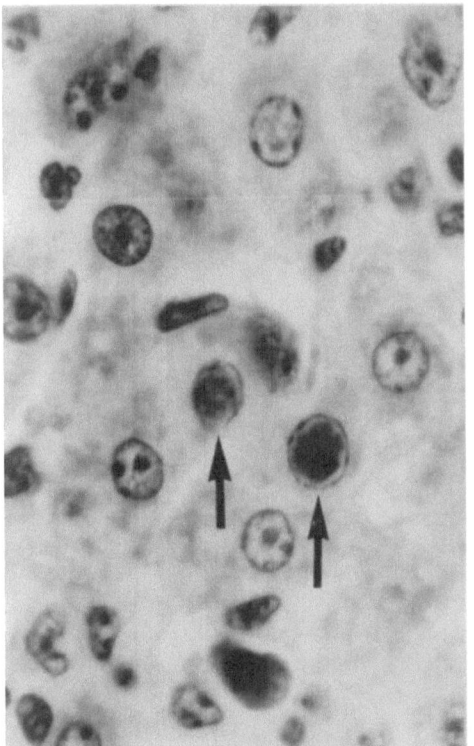

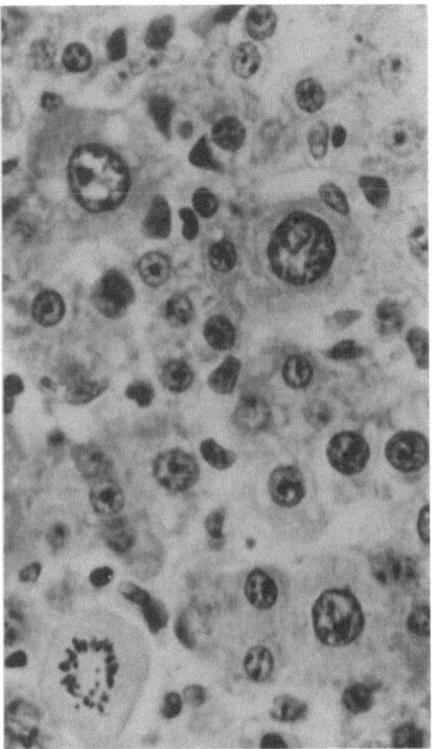

Fig. 153. Liver of a 4-day-old rat with natural RV infection. Several phases of inclusion body formation in hepatocytes are demonstrated (*arrows*). A necrotic hepatocyte is located at the *center bottom* of the field. (Courtesy of Dr. G. Margolis). H&E, ×900

Fig. 154. Giant hepatocytes in the liver of a rat with rat virus infection. (From Margolis et al. 1968, with permission of Dr. G. Margolis and *Experimental and Molecular Pathology*) H&E, ×910

Liver lesions in convalescent rats occur in four patterns, either individually or in varying combinations (Margolis et al. 1968). These include the following: (1) giant cell transformation, (2) nonsuppurative portal hepatitis and biliary hyperplasia, (3) sinusoidal dilatation, and (4) postnecrotic stromal collapse, fibrosis, and nodular hyperplasia. These changes usually develop over several months. Giant cell transformation is characterized by cytomegaly, nuclear enlargement, and polyploidy (Fig. 154). Enlarged cells may contain multiple nuclei. Lesions of the type pictured in Fig. 154 have been detected at 16–43 days in rats inoculated experimentally as neonates or sucklings. Nonsuppurative portal hepatitis and biliary hyperplasia can begin as early as 8 days after infection and affect primarily small ducts (Fig. 155). Sinusoidal dilatation peliosis hepatis (Yanoff and Rawson 1964) is characterized by irregular blood-filled spaces enclosed by distorted plates of hepatocytes that may be one cell thick (Fig. 156). Postnecrotic stromal collapse and fibrosis, together with nodular hepatocytic hyperplasia, produce irreversible distortion of hepatic architecture (Fig. 157).

Ultrastructure

The ultrastructure of naturally occurring rat parvovirus infection has not been thoroughly described. Ruffolo and coworkers (Ruffolo et al. 1966) have, however, studied responses of partially hepatectomized rats to H-1 virus (Fig. 158). Viral particles were found primarily in nuclei of hepatocytes and Kupffer cells. Nuclear degeneration was marked by increased density and confluence of chromatin, especially adjacent to the

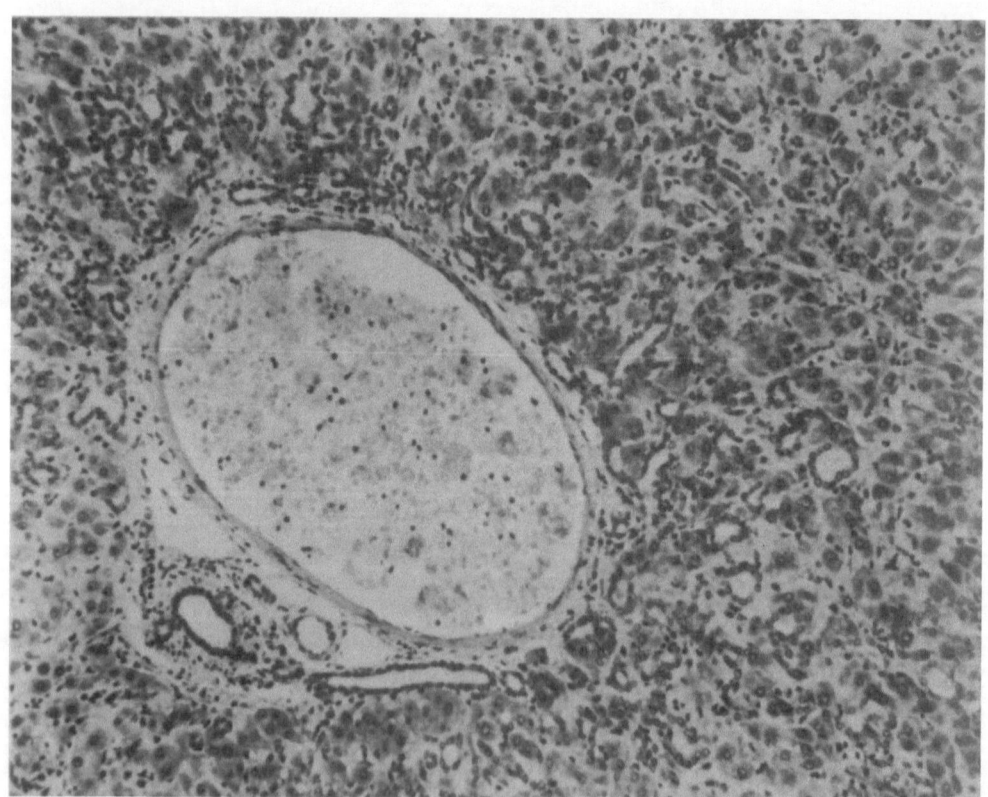

Fig. 155. Liver of a 19-day-old rat inoculated 8 days postnatally with H-1 virus. Numerous small, dilated, proliferated biliary ducts extend from the portal areas deep into the hepatic lobules. The hypocellular areas are remnants of necrotic foci. Note the chronic inflammatory responses. (Courtesy of Dr. G. Margolis.) H&E, ×80

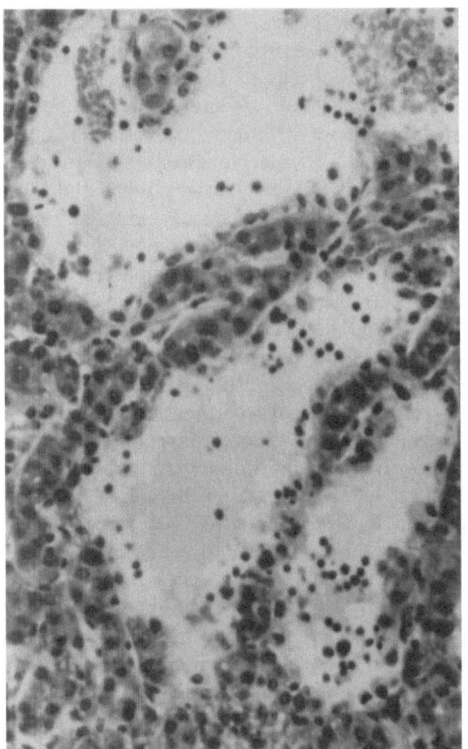

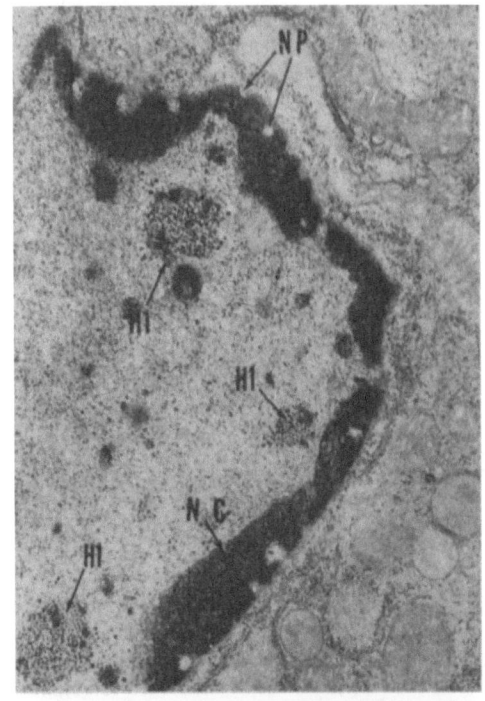

▶

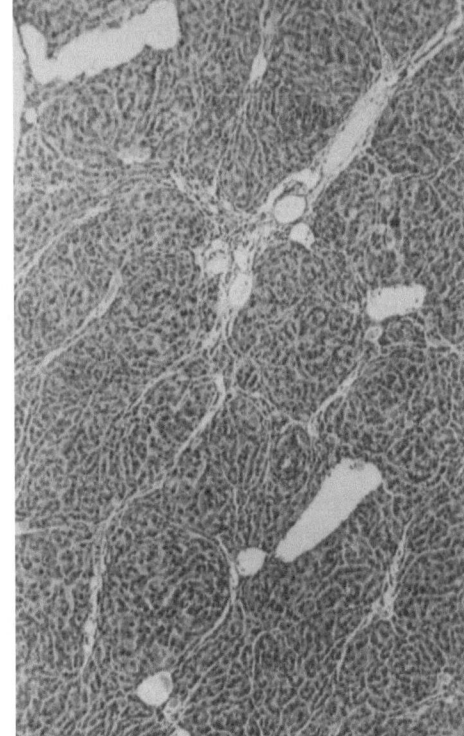

Fig. 156. (*upper left*) Peliosis hepatis in a rat inoculated with rat parvovirus. (From Margolis et al. 1968, with permission of Dr. G. Margolis and *Experimental and Molecular Pathology*) H&E, ×250

Fig. 157. (*below*) Nodular hyperplasia and fibrosis in the liver of a rat infected with rat virus about 7 weeks previously. (From Margolis et al. 1968, with permission of Dr. G. Margolis and *Experimental and Molecular Pathology*) H&E, ×46

Fig. 158. (*upper right*) Early changes in a hepatocyte of a partially hepatectomized rat infected with H-1 virus. Clusters of H-1 virus (*H1*) and nuclear pores (*NP*). The nuclear chromatin (*NC*) is denser than normal and condensed at the nuclear envelope. The cytoplasm is devoid of glycogen. (From Ruffolo et al. 1966, with permission of Dr. G. Margolis and the *American Journal of Pathology*.) TEM, ×20300

nuclear membrane, and by clear spaces and membrane-bound vesicles containing nuclear debris. Cytoplasmic changes occurred when nuclear degeneration was advanced. There was a relative increase in smooth endoplasmic reticulum. Cells were shrunken and the number of phagolysosomes was increased. Mitochondrial cristae were lost and intracristal spaces widened. Amorphous or granular material accumulated in cell matrices, and the cellular limiting membrane underwent lysis.

Differential Diagnosis

The hepatic lesions of parvovirus infection are not found in other viral infections of rats. Liver necrosis can occur in Tyzzer's disease (Jonas et al. 1970), but the causative organism (Clostridium piliforme) can usually be demonstrated with silver impregnation stains at the margin of necrotic foci. Liver necrosis and its sequelae can potentially follow inadvertent exposure to hepatotoxins that may contaminate the environment (e.g., food, bedding), but such lesions are rare among laboratory rats. Biliary hyperplasia, portal fibrosis, and peliosis hepatis occur in aged rats, but the etiology and pathogenesis of these lesions are not clear.

Physical or chemical hepatic injury can render even adult rats susceptible to parvoviral hepatitis, as noted above. Conversely, infection could potentially exacerbate effects of hepatotoxins and complicate interpretation of drug-induced lesions. In addition, the predilection of parvoviruses for mitotically active cells (Margolis and Kilham 1965) has the potential for altering the kinetics of hepatic neoplasia and immunologic responses (Campbell et al. 1977; McKisic et al. 1995).

Biologic Features

Prenatal infection can cause fetal deaths and resorption, but the propensity for in utero infection appears to depend on virus strain as well as dose and route of inoculation (Kilham and Margolis 1969; Margolis and Kilham 1972; Jacoby et al. 1988). During natural infection, virus appears to enter through the respiratory tract and is widely disseminated by viremia (Gaertner et al. 1993). Clinical disease is rare, but can be severe or lethal. It occurs most commonly among sucklings

as the result of hepatic necrosis, granuloprival cerebellar hypoplasia from cytolytic infection of external germinal cells, and hemorrhagic infarction from infection of vascular endothelium and megakaryocytes (Kilham and Margolis 1966; Jacoby et al. 1987; Gaertner et al. 1993). Infection of adult rats is usually asymptomatic, although hemorrhagic infarcts can occur, especially in the central nervous system (Coleman et al. 1983). Chemical immunosuppression appears to be at least one factor that predisposes adults to hemorrhagic disease (Eldadah et al. 1967).

Infection of the liver results from viremia. The highly fenestrated sinusoidal endothelium of rat liver may facilitate direct infection of hepatocytes (Burkel and Low 1966; Margolis and Kilham 1965). Additionally, mitotic activity among rat hepatocytes remains high for up to 6 weeks postpartum (Steiner et al. 1966). Because rat parvovirus has a predilection for dividing cells, susceptibility to liver damage is predictably greater in young rats (Margolis et al. 1968). A predilection for mitotically active cells has also been used to explain prolonged hepatic infection, i.e., mitotic activity in response to virus-induced liver injury could initiate repeated cycles of hepatocytic infection (Margolis et al. 1968).

Rat parvoviruses are small (18–30 nm), single-stranded, negative-sense DNA viruses that hemagglutinate guinea pig erythrocytes and, to some extent, erythrocytes of other species (Siegl 1976). Three serogroups are distinguishable by hemagglutination inhibition or neutralization serology: the RV type, which incorporates a number of strains, including the prototype strain described by Kilham and Olivier (1959); the H-1 type discovered by Toolan (1968); and a recently discovered serotype named rat parvovirus-1 (RPV-1; Ball-Goodrich et al., unpublished data). The open reading frame encoding nonstructural proteins, which is located on the left side of the rat parvovirus genome, is highly conserved among rat parvoviruses. Thus generic assays such as enzyme-linked immunosorbent assay (ELISA), which detect both nonstructural and structural proteins do not distinguish among the viruses serologically.

Both RV and H-1 can be cultivated in primary monolayer cultures of rat embryo cell or in continuous cell lines from other species. RV, for example, replicates in 324K cells, a line of SV40-transformed human embryonic kidney cells (Smith 1983; Jacoby et al. 1987). Productively infected cells express viral hemagglutinin, develop intranuclear inclusions, and undergo lysis (Kilham and Oliver 1959).

RV is excreted in urine, milk, feces, and possibly in expired air (Jacoby et al. 1988; Kilham 1966; Lipton et al. 1972). Infection is therefore communicable by contact with infected rats or by airborne transmission. Prevalence rates of 100%, based on seroconversion, are not unusual in enzootically infected colonies (Robinson et al. 1971). The varied expression of infection is influenced by host age and immunological status and by virus strain. RV can persist in euthymic rats exposed as infants (Jacoby et al. 1991) or in athymic rats exposed as infants or adults (Gaertner et al. 1989, 1993). Anatomic sites that harbor persistent virus appear to include endothelium, lymphoid cells, and smooth muscle fibers. Preexisting immunity, such as maternal immunity, can protect naive rats from acute and persistent infection, but immunity is not protective once infection has been established (Jacoby et al. 1988; Gaertner et al. 1991).

Comparison with Other Species

Rat parvovirus infection among rat colonies is widespread in many areas of the world, but there is no firm evidence that species other than the rat (laboratory or wild) are naturally infected. Experimental infections have, however, been demonstrated in other species (Siegl 1976; Jacoby et al. 1979). The most notable historical model is the hamster, in which RV and H-1 produce osteolytic lesions that result in dental and skeletal deformities, giving affected animals a mongoloid appearance (Toolan 1960).

Necrotizing viral hepatitis occurs in young or fetal animals of several species. These conditions include poxviral, coronaviral, and reoviral hepatitides in the mouse, infectious canine hepatitis, equine viral rhinopneumonitis, and exotic diseases such as Rift valley fever and Wesselbron disease in sheep (Jubb and Kennedy 1970).

A number of viruses can cause necrotizing hepatitis in humans (Edington 1979; Ishak et al. 1982), including the classic syndromes of human viral hepatitis (Ishak 1976; Koff and Galambos 1982; MacSween 1980; Poulsen 1976). Because these conditions are associated with necrosis and inflammation and are variably associated with viral inclusions or chronic degenerative sequelae, they have some morphological similarities to rat parvoviral hepatitis. However, they do not appear to be sufficiently similar, etiologically or pathogenetically, to warrant using parvoviral hepatitis as a model. The rat disease has also been suggested as a model for hepatitis of intrauterine or neonatal onset (Margolis et al. 1968) and has been compared with neonatal jaundice of humans accompanied by giant cell transformation (Margolis et al. 1968; Smetana et al. 1965).

References

Bergs VV, Scotti TM (1967) Virus-induced peliosis hepatis in rats. Science 158:377–378

Burkel WE, Low FN (1966) The fine structure of rat liver sinusoids, space of Disse and associated tissue space. Am J Anat 118:769–783

Campbell DA Jr, Staal SP, Menders EK, Bonnard GD, Oldham RK, Salzman LA, Herberman RB (1977) Inhibition of in vitro lymphoproliferative responses by in vivo passaged rat 13762 mammary adenocarcinoma cells. II. Evidence that Kilham rat virus is responsible for the inhibitory effect. Cell Immunol 33:378–391

Coleman GL, Jacoby RO, Bhatt PN, Smith AL, Jonas AM (1983) Naturally occurring lethal parvovirus infection of juvenile and young-adult rats. Vet Pathol 20:49–56

Edington GM (1979) Other viral and infectious diseases. In: MacSween RNM, Anthony PP, Scheuer PJ (eds) Pathology of the liver. Churchill Livingstone, New York

Eldadah AH, Nathanson N, Smith KO, Squire RA, Santos GW, Melby EC (1967) Viral hemorrhagic encephalopathy of rats. Science 156:392–394

Gaertner DJ, Jacoby RO, Smith AL, Ardito RB, Paturzo FX (1989) Persistence of rat parvovirus in athymic rats. Arch Virol 105:259–268

Gaertner DJ, Jacoby RO, Paturzo FX, Johnson EA, Brandsma JL, Smith AL (1991) Modulation of lethal and persistent rat parvovirus infection by antibody. Arch Virol 118:1–9

Gaertner DJ, Jacoby RO, Johnson EA, Paturzo FX, Smith AL, Brandsma JL (1993) Characterization of acute rat parvovirus infection by in situ hybridization. Virus Res 28:1–18

Ishak KG (1976) Light microscopic morphology of viral hepatitis. Am J Clin Pathol 65:787–827

Ishak KG, Walker DH, Coetzer JA, Gardner JJ, Gorelkin L (1982) Viral hemorrhagic fevers with hepatic involvement: pathologic aspects with clinical correlations. In: Popper H, Schaffner F (eds) Progress in liver diseases, vol 7. Grune and Stratton, New York, chap 29

Jacoby RO, Bhatt PN, Jonas AM (1979) Viral diseases. In: Baker HJ, Lindsey JR, Weisbroth SH (eds) The laboratory rat, vol 1. Academic, New York, chap 11

Jacoby RO, Bhatt PN, Gaertner DJ, Smith AL, Johnson EA (1987) The pathogenesis of rat virus infection in infant and juvenile rats after oronasal inoculation. Arch Virol 95:251–270

Jacoby RO, Gaertner DJ, Bhatt PN, Paturzo FX, Smith AL (1988) Transmission of experimentally induced rat virus infection. Lab Anim Sci 38:11–14

Jacoby RO, Johanson EA, Paturzo FX, Gaertner DJ, Brandsma JL, Smith AL (1991) Persistent rat parvovirus infection in individually housed rats. Arch Virol 117:193–205

Jonas AM, Percy DH, Craft J (1970) Tyzzer's disease in the rat. Its possible relationship with megaloileitis. Arch Pathol 90:516–521

Jubb KVF, Kennedy PC (1970) The liver and biliary system. The nervous system. In: Pathology of domestic animals, 2nd edn, vol 2. Academic, New York, chaps 3, 7

Kilham L (1966) Viruses of laboratory and wild rats. Natl Cancer Inst Monogr 20:117–146

Kilham L, Margolis G (1966) Spontaneous hepatitis and cerebellar hypoplasia in suckling rats due to congenital infections with rat virus. Am J Pathol 49:457–475

Kilham L, Margolis G (1969) Transplacental infection of rats and hamsters induced by oral and parenteral inoculations of H-1 and rat viruses (RV). Teratology 2:111–123

Kilham L, Olivier LJ (1959) A latent virus of rats isolated in tissue culture. Virology 7:428–437

Koff RS, Galambos J (1982) Viral hepatitis. In: Schiff L, Schiff ER (eds) Diseases of the liver, 5th edn, Lippincott, Philadelphia, chap 15

Lipton H, Nathanson N, Hodous J (1972) Enteric transmission of parvoviruses: pathogenesis of rat virus infection in adult rats. Am J Epidemiol 96:443–446

MacSween RNM (1980) Pathology of viral hepatitis and its sequelae. Clin Gastroenterol 9:23–45

Margolis G, Kilham L (1965) Rat virus, an agent with an affinity for the dividing cell. In: Gadjusek DC, Gibbs CJ, Alpers M (eds) Slow, latent and temperate virus infections. US Dept Health Education and Welfare, pp 361–367 (NINDB monographs 2)

Margolis G, Kilham L (1972) Rat virus infection of megakaryocytes: a factor in hemorrhagic encephalopathy? Exp Mol Pathol 16:326–340

Margolis G, Kilham L, Ruffolo PR (1968) Rat virus disease, as an experimental model of neonatal hepatitis. Exp Mol Pathol 8:1–20

McKisic MD, Paturzo FX, Gaertner DJ, Jacoby RO, Smith AL (1995) Nonlethal rat parvovirus infection suppresses rat T-lymphocyte effector functions. J Immunol 155(8):3979–3986

Poulsen H (1976) Histological features of acute viral hepatitis. Ann Clin Res 8:139–150

Robinson GW, Nathanson N, Hodous J (1971) Seroepidemiological study of rat virus infection in a closed laboratory colony. Am J Epidemiol 4:91–100

Ruffolo PR, Margolis G, Kilham L (1966) The induction of hepatitis by prior partial hepatectomy in resistant adult rats injected with H-1 virus. Light and electron microscopy and virologic studies. Am J Pathol 49:795–824

Siegl G (1976) The parvoviruses. Springer, Vienna New York (Virology monograph, vol 15)

Smetana HF, Edlow JB, Glunz PR (1965) Neonatal jaundice. A critical review of persistent obstructive jaundice in infancy. Arch Pathol 80:553–574

Smith AL (1983) Response of weanling random-bred mice to inoculation with minute virus of mice. Lab Anim Sci 33:37–39

Steiner JW, Perz ZM, Taichman LB (1966) Cell population dynamics in the liver. A review of quantitative morphological techniques applied to the study of physiological and pathological growth. Exp Mol Pathol 5:146–181

Toolan HW (1960) Experimental production of mongoloid hamsters. Science 131:1446–1448

Toolan HW (1968) The picodna viruses, H, RV, and AAV. Int Rev Exp Pathol 6:135–180

Yanoff M, Rawson AJ (1964) Peliosis hepatis. An anatomical study with demonstration of two varieties. Arch Pathol 77:159–165

Mousepox, Liver, Mouse

Robert O. Jacoby

Synonyms. Infectious ectromelia

Gross Appearance

The liver is a major site of viral replication in mousepox, but gross lesions, even during acute disease, are not readily apparent until shortly before death. Severely affected livers are usually swollen and friable and may occupy up to half the volume of the peritoneal cavity, whereas mildly affected livers may remain grossly normal or have sparse focal necrosis. Necrotic areas appear first as pinpoint yellow–white foci, but increase rapidly in size and number. Confluent areas of necrosis can produce a reticulated pattern of yellow–brown to pink discoloration on the surface and throughout the parenchyma. Areas of hemorrhage also may develop. The pale hue of severely affected livers is in part due to fatty change, and the fat content of such livers can be as much as four times normal. Livers from mice that survive acute infection usually have a normal gross appearance. A few small scars may be present, however, especially at the margins (Fenner 1948d, 1949b: Allen et al. 1981).

Microscopic Features

The major liver lesion is coagulation necrosis, which begins in a random fashion among individual or small groups of hepatocytes approximately 5 days after infection (Fig. 159). In highly susceptible mice, necrotic areas enlarge rapidly and in 2–4 days lead to extensive necrosis and variable degrees of hemorrhage.

Hepatocytes in early lesions or at the margins of advanced lesions can undergo ballooning degeneration or can shrink and develop intensely eosinophilic cytoplasm. Hepatocytes with clear cytoplasm can occur at some distance from necrotic foci and have reportedly undergone glycogen depletion (De Burgh 1950). Nuclei of infected cells may be transiently enlarged, but quickly become pyknotic or karyorrhectic.

Two types of intracytoplasmic inclusions occur in mousepox: the early or type B inclusion and the late or type A inclusion. Type B inclusions commonly develop in the liver, whereas type A inclusions are rare. Type B inclusions are basophilic to amphophilic and can occur singly or in groups (Fig. 160). They may be surrounded by a thin "halo", but this is often hard to see in formalin-fixed sections. Type B inclusions also are difficult to detect unless the hematoxylin staining time is at least doubled (Allen et al. 1981). Alternatively, because they are sites of viral replication (Cairns 1960; kameyama et al. 1959), they can be detected in formalin-fixed liver by immunohistochemical methods. Immunoperoxidase staining of infected hepatocytes will reveal intracytoplasmic antigen in particulate and diffuse distribution (Fig. 161). Viral antigen also can be found in cells lining vascular channels, especially during early phases of infection. Type A inclusions are large and intensely eosinophilic. They can more readily be detected in other tissues, particularly skin (Fig. 162), mucous membranes, and intestinal epithelium, to help confirm the etiology of hepatic lesions.

The inflammatory response during fulminating hepatic necrosis is minimal. Some mononuclear cells and polymorphonuclear leukocytes may infiltrate portal triads or the margin of necrotic lesions. In milder or more prolonged hepatic infection, such as occurs in genetically resistant mice, inflammatory infiltrates are more prominent. The mononuclear cell response is compatible with host defenses mediated by cellular immunity (Blanden 1974).

The histologic sequelae of mousepox in the liver have not been described in detail, but appear to be unremarkable. This is probably due to the fact that extensive hepatic involvement commonly leads to rapid death, whereas mild hepatic lesions are quickly repaired. Large or multinucleated hepatocytes can accumulate at the margin of necrotic lesions during repair. Chronic hepatitis has not been reported. "Hyalinized" areas have been found and may represent local postnecrotic scarring, but widespread fibrosis or cirrhosis does not occur.

Ultrastructure

The most complete observations have been made by Leduc and Bernhard (1962) on livers of naturally infected mice. The earliest signs of infection are seen among periportal hepatocytes. Cytoplasmic changes are characterized by loss of glycogen, vesicular swelling of the endoplasmic reticulum, mitochondria, and Golgi apparatus and sparse lipid droplets (Fig. 163). Irregular dense bodies develop and are found occasionally in nucleoplasm. Nucleolar fragmentation is also observed. In moribund or necrotic cells, the cytoplasm is dense and, in addition to viral particles, contains membranous whorls, vesicles of various sizes, vacuoles with cell debris, and collapsed, distorted mitochondria. Nuclear chromatin is dense and marginated, and blisters form in the nuclear envelope. Although cells lining hepatic sinusoids support viral replication, they do not undergo necrosis.

Virus formation begins in granular to reticular cytoplasmic matrices of low electron density (Fig. 163). Matrix zones displace the cytoplasm and lack sharp boundaries. They are thought to be the ultrastructural correlates of type B inclusions. Three early forms of virus emerge from the matrix: (1) 220 mu, membrane-bound, oval particles with uniform viroplasm similar in appearance to matrix (Fig. 164 arrow), (2) membrane-bound oval particles with a dense nucleoid surrounded by a halo (Fig. 164), and (3) lobulated masses of matrix-like viroplasm partially enclosed by C-shaped membranes which probably represent incompletely developed viral particles (Fig. 164, arrow).

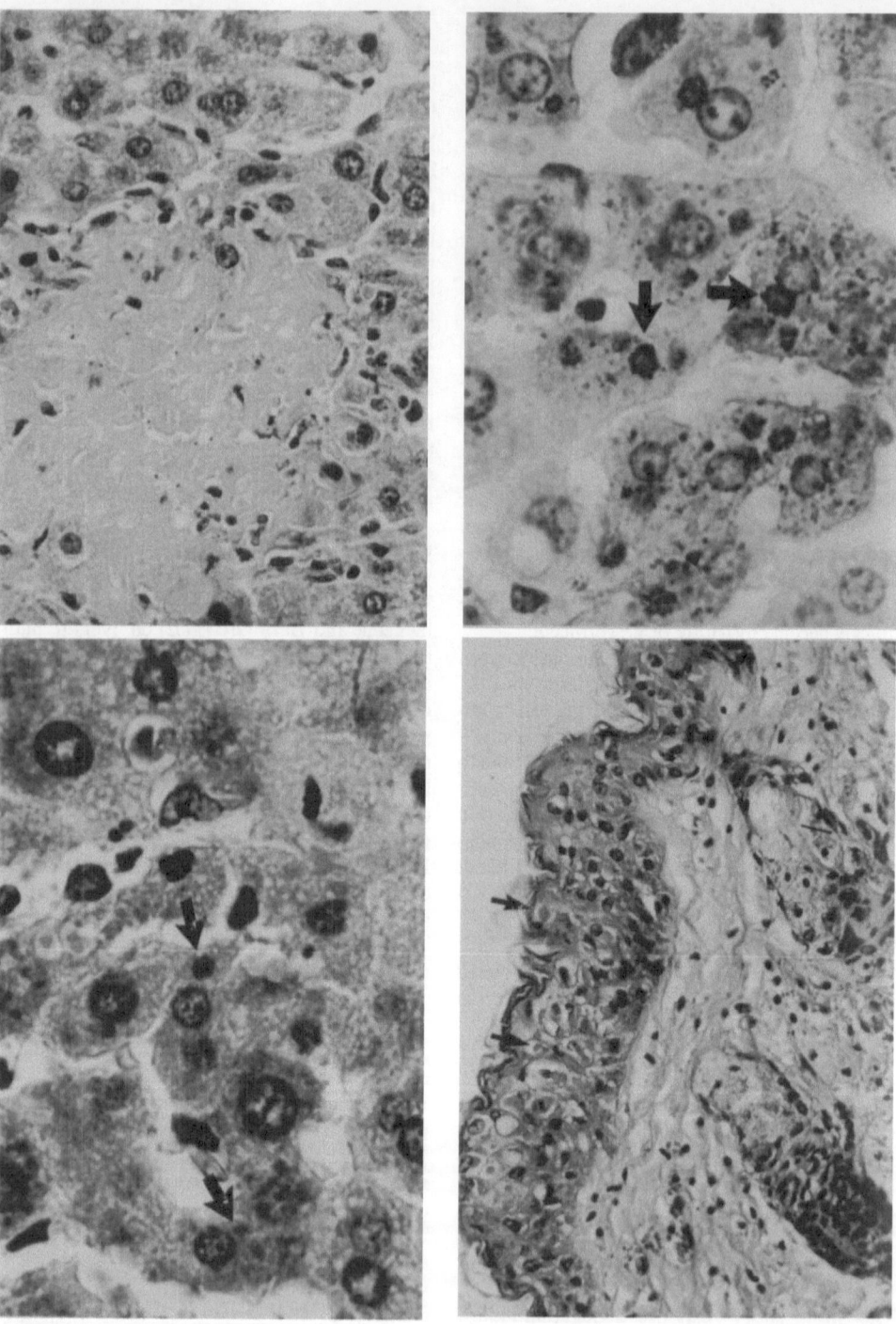

Fig. 159. (*upper left*) Focal coagulative necrosis in the liver of a mouse during the early stages of acute mousepox. H&E ×720

Fig. 160. (*lower left*) Type B inclusions (*arrows*) in hepatocytes of a mouse infected with ectromelia virus. Double stained with Harris' hematoxylin and counterstained with eosin, ×1800

Fig. 161. (*upper right*) Ectromelia viral antigen in hepatocytes. Intracytoplasmic bodies of various size correspond to type B inclusions (*arrows*). Avidin-biotin-conjugate, immunoperoxidase method, ×1800

Fig. 162. (*lower right*) Type A inclusions of ectromelia virus (*arrows*) in the epidermis of a mouse. H&E, ×450

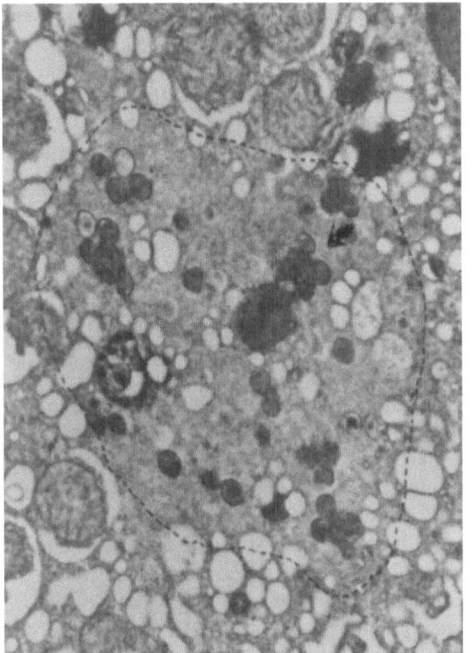

Fig. 163. Hepatocyte infected with mousepox. Matrix zone (encircled by *dotted line*) contains developing virus particles. Incomplete C shaped particles presumably represent immature virions (*arrow*). (From Leduc and Bernhard 1962, with permission of *Journal of Ultrastructural Research.*) TEM, ×17 000

mousepox, whereas viral inclusions are not found in coronaviral hepatitis. Extrahepatic manifestations, especially necrosis of spleen and other tissues, are also more severe in mousepox. Several strains of mouse coronavirus are neurotropic, whereas mousepox does not cause significant lesions of the central nervous system unless virus is inoculated intracerebrally.

Reovirus-3 can cause necrotizin hepatitis in infant mice, but not in adults (see p. 196) Necrosis tends to begin in centrilobular zones and to spread peripherally (Walters et al. 1963). Mice that recover may develop chronic active hepatitis that has immunopathologic overtones expressed by a significant, persistent inflammatory response (Stanley 1974). The disease in infant mice is frequently accompanied by necrosis in other organs; most frequently among these is brain, a finding that is not compatible with mousepox. Mouse cytomegalovirus and lymphocytic choriomeningitis virus can cause hepatic necrosis after experimental inoculation, but infection with these agents is not prevalent among well-managed mouse colonies.

Differentiation of the various forms of viral hepatitis in mice can be difficult. Therefore, in the absence of specific morphological changes, ancillary data should be collected to confirm the diagnosis. These include one or more of the following:

Mature virus particles are seen only in necrotic hepatocytes, but they also occur in viable Kupffer's cells and sinusoidal endothelium.

Differential Diagnosis

Several naturally occurring murine viruses cause necrotizing hepatitis in mice. Hepatotropic strains of mouse coronaviruses (mouse hepatitis viruses) may induce acute multifocal hepatic necrosis, but lesions, especially in adult mice, are seldom as severe as those of lethal mousepox (see p. 194). Because the course of coronaviral hepatitis is often longer than that of mousepox, signs of repair (mitotic activity, large or binucleated hepatocytes) can be found more commonly at the margins of necrotic lesions. Inflammation also may be more prominent. Syncytia, which are common in mouse coronavirus infection, are not typical of

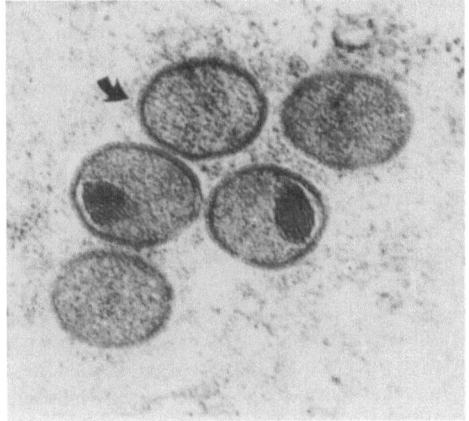

Fig. 164. Two types of mousepox viral particles. Oval particles with uniform viroplasm (*arrow*) and particles with nucleoids of precisely aligned filaments. (From Leduc and Bernhard 1962, with permission of *Journal of Ultrastructural Research.*) TEM, ×53 000

viral serology, immunohistochemical demonstration of viral antigen in tissue, electron microscopy for viral particles, and, when feasible, viral isolation.

Tyzzer's disease should be considered whenever hepatic necrosis is found in mice. The causative organism, *Bacillus piliformis*, is difficult to find in hematoxylin and eosin (H&E)-stained sections, but can be demonstrated with special stains (Giemsa, periodic acid-Schiff, Warthin-Starry; see p. 201). Organisms are generally found in the cytoplasm of hepatocytes bordering necrotic areas. They may be present in small numbers, making it necessary to examine several sections to locate them. Although the acute form of other bacterial infections such as salmonellosis may cause necrotizing hepatitis, these organisms are rarely found in modern vivariums. Nevertheless, it is advisable to obtain bacterial cultures from mice with hepatic necrosis.

Biologic Features

The clinical expression of mousepox can vary and is influenced strongly by genotype (Briody et al. 1956; Briody 1966; O'Neill and Blanden 1983; Bhatt and Jacoby 1985). Broadly speaking, three clinical courses are recognized. *Highly susceptible mice* develop acute, fatal disease wherein apparently healthy mice may die within several hours from the onset of illness. This form of mousepox is closely associated with severe hepatic necrosis. Clinical signs are relatively nonspecific and include hunched posture, rough haircoat, and diarrhea. If infection has occurred through skin abrasion, careful examination of the skin may reveal a "primary lesion," which indicates the initial site of viral replication (Fenner 1947). *Moderately susceptible mice* often develop a chronic form of mousepox with variable mortality. It is characterized by erosive or ulcerative dermatitis which is most easily observed on the ears, tail, and feet, but which also may occur as a general exanthema. These mice can also have edema of the face and extremities and conjunctivitis. Amputation of part or all of one or more limbs or the tail can occur, a lesion that gave rise to the term "infectious ectromelia." Skin lesions among survivors of chronic mousepox heal as hairless scars. *Resistant mice* commonly have asymptomatic infection (Fenner 1982).

From an epizootiologic aspect, acute, lethal mousepox is a major hazard to a mouse colony, but is self-limiting provided that additional susceptible animals are not introduced during an outbreak. Chronic or asymptomatically infected mice are a relatively greater hazard in the long run, because they can perpetuate enzootic infection (Fenner 1984a,b). Introduction of highly susceptible mice to an enzootically infected colony may initiate a clinically explosive and devastating outbreak.

The prevalence and biologic significance of latent, persistent mousepox infection (carrier state) is unsettled. Evidence for a carrier state is sparse (Gledhill 1962; Horzinek and Höpken 1965), but has not been thoroughly or systematically discounted. If a carrier state occurs, its expression is likely to be influenced by mouse genotype. Briody and coworkers (Briody et al. 1956) and Briody (1959), for example, studied responses of mice to natural epizootics of mousepox and found that some inbred strains such as DBA/1, A and C3H were highly susceptible to lethal infection, whereas other strains such as C57BL/6, AKR, and BALB/c were highly resistant. Although these findings underscore the significance of genotype for the outcome of infection, they should be extrapolated cautiously, because recent evidence suggests that susceptibility to lethal mousepox can vary even among sublines of a given inbred strain (Bhatt and Jacoby 1985).

Fenner (1947, 1948a–d, 1949a,b) is largely responsible for deciphering the pathogenesis of mousepox. Although ectromelia virus can enter the body by several routes, including the respiratory tract, the most common mode of entry is thought to be through abraded skin. Virus replicates at the site of entry and infects the draining lymph node. It reenters the lymphatics and finally reaches the blood to produce a primary viremia. During primary viremia, virus invades parenchymal organs, including the liver. Immunofluorescent studies by Mims (1964) disclosed that blood-borne virus first infects littoral cells lining vascular channels and that infected cells seed virus to hepatocytes. Mice that survive initial infection of parenchymatous tissues develop a secondary viremia during which virus is widely distributed to skin and causes a typical pox rash. Recovery from mousepox and resistance to lethal infection depends on intact cellular immunity (Blanden 1974). Evidence that humoral responses are critical to host survival are less compelling (Schell 1960a,b).

Ectromelia virus is an orthopoxvirus that is morphologically identical and antigenically similar to vaccinia virus, a relationship which enables vaccinia virus to elicit protective immunity to mousepox (Briody 1959). The virus grows in a number of continuous cell lines, including L cells and Vero cells, and classic pox lesions are produced if virus is grown on the chorioallantoic membrane of embryonated hen eggs (Fenner 1982). A line of kidney cells (BS-C-1), derived from African green monkeys, is highly susceptible to ectromelia virus (Bhatt and Jacoby 1985).

Several strains of ectromelia virus have been studied intensively. The Moscow strain is highly virulent, whereas the Hampstead strain is less so (Fenner 1949b). Avirulent strains of Hampstead virus replicate poorly in hepatic Kupffer's cells, but readily invade hepatocytes. It has been proposed from this that virulence may depend on the ability of virus to infect Kupffer's cells and that survival of the host may depend on the ability of Kupffer's cells to protect hepatocytes from viral invasion (Roberts 1964). A strain of ectromelia virus (NIH-79) was isolated from a recent mousepox epizootic in the United States (Allen et al. 1981). Its behavior in mice is under scrutiny, but is seems to be moderately to highly virulent (Bhatt and Jacoby 1985).

All strains of ectromelia virus studied thus far express an immunogenic hemagglutinin. Infected animals develop hemagglutinin-inhibiting antibody, which can be detected by routine serologic methods (Briody 1966). The sensitivity and specificity of hemagglutination inhibition has been the center of some controversy, and current evidence suggests that the hemagglutination inhibition (HAI) test, although still widely used, may elicit occasional false-positive results (Collins et al. 1981; Wallace et al. 1981). Other serologic tests used to detect infection include complement fixation (relatively insensitive), immunofluorescence (Christensen et al. 1966) virus neutralization. and enzyme-linked immunoadsorbent assay (ELISA). ELISA is highly sensitive, but it must be refined before it can be used to discriminate ectromelia-infected from vaccinated mice (Buller et al. 1983; Collins et al. 1981). HAI is still the test of choice for this purpose. The strain of vaccinia virus used to immunize mice (IHD-T) is hemagglutinin deficient (Briody 1959). Thus vaccinia-immune mice should remain free of HAI antibody to either vaccinia virus or ectromelia virus.

Naturally occurring mousepox is limited to mice. Outbreaks in laboratory mice have occurred in Europe, Asia, Australia, and the United States. The risk of introducing infection in a susceptible population is increased by extensive exchange of mice and mouse tissues among laboratories. The spread of mousepox in a newly infected colony depends on a combination of factors such as mouse genotype, husbandry, viral virulence, and experimental manipulation. Although mousepox is an infectious disease, it can spread slowly even among mice housed in the same room (Wallace et al. 1981). *In utero* infection has been produced experimentally (Schwanzer et al. 1975; Wylekshanin 1935), but its prevalence and significance in naturally occurring disease have not been established.

Comparison with Other Species

The skin form of mousepox has been studied extensively as a model of a viral exanthcm because it resembles human smallpox (Fenner 1948c). Hepatic lesions, however, are not characteristic of smallpox or of poxvirus diseases of domestic mammals. A notable exception is rabbit pox, which tends to be severe and generalized and is frequently accompanied by hepatic necrosis (Greene 1934).

References

Allen AM, Clarke GL, Ganaway JR, Lock A, Werner RM (1981) Pathology and diagnosis of mousepox. Lab Anim Sci 31:599–608

Bhatt PN, Jacoby RO (1985) The pathogenesis of mousepox in genetically resistant and genetically susceptible inbred mice (in preparation)

Blanden RV (1974) T cell response to viral and bacterial infection. Transplant Rev 19:56–88

Briody BA (1959) Response of mice to ectromelia and vaccinia viruses. Bacteriol Rev 23:61–95

Briody BA (1966) The natural history of mousepox. Natl Cancer Inst Monogr 20:105–115

Briody BA, Hauschka TS, Mirand EA (1956) The role of genotype in resistance to an epizootic of mouse pox (ectromelia). Am J Hyg 63:59–68

Buller RM, Bhatt PN, Wallace GD (1983) Evaluation of an enzyme-linked immunosorbent assay for the detection of

196   S.W. Barthold

ectromelia (mousepox) antibody. J Clin Microbiol 18:1220–1225

Cairns J (1960) The initiation of vaccinia infection. Virology 11:603–623

Christensen LR, Weisbroth S, Matanic B (1966) Detection of ectromelia virus and ectromelia antibodies by immunofluorescence. Lab Anim Care 16:129–141

Collins MJ Jr, Peters RL, Parker JC (1981) Serological detection of ectromelia virus antibody. Lab Anim Sci 31:595–598

De Burgh PM (1950) Cytochemical changes in early ectromelia infection of mice. Aust J Exp Biol 28:214–218

Fenner F (1947) Studies in infectious ectromelia of mice. II. Natural transmission: the portral of entry of the virus. Aust J Exp Biol 25:275–282

Fenner F (1948a) The epizootic behavior of mouse-pox (infectious ectromelia). Br J Exp Pathol 29:69–91

Fenner F (1948b) The epizootic behavior of mousepox (infectious ectromelia of mice). II. The course of events in long-continued epidemics. J Hyg 46:383–393

Fenner F (1948c) The pathogenesis of the acute exanthems. An interpretation based upon experimental investigations with mousepox (infectious ectromelia of mice). Lancet 2:915–920

Fenner F (1948d) The clinical features and pathogenesis of mousepox (infectious ectromelia of mice). J Pathol Bacteriol 60:429–552

Fenner F (1949a) Studies in mousepox (infectious ectromelia of mice). VII. The effect of the age of the host upon the response to infection. Aust J Exp Biol 27:45–53

Fenner F (1949b) Mouse-pox (infectious ectromelia of mice): a review. J Immunol 63:341–373

Fenner F (1982) Mousepox. In: Foster HL, Small JD, Fox JG (eds) The mouse in biomedical research, vol II, diseases. Academic, New York, chap 11

Gledhill AW (1962) Latent ectromelia. Nature 196:298

Greene HSN (1934) Rabbitpox. I. Pathology of the epidemic disease. J Exp Med 60:441–457

Horzinek M, Höken W (1965) Untersuchungen Uber die inapparente Infektion mit dem Ektromelievirus. Arch Ges Virusforsch 17:125–138

Kameyama S, Takahashi M, Toyoshima K, Kato S, Kamahora J (1959) Studies on the inclusion bodies of ectromelia virus using the fluorescent antibody technique. Biken J 2:341–344

Leduc EH, Bernhard W (1962) Electron microscope study of mouse liver infected by ectromelia virus. J Ultrastruct Res 6:466–488

Mims CA (1964) Aspects of the pathogenesis of virus diseases. Bact Rev 28:30–71

O'Neill HC, Blanden RV (1983) Mechanisms determining innate resistance to ectromelia virus infection in C57BL mice. Infect Immun 41:1391–1394

Roberts JA (1964) Growth of ectromelia virus in the liver parenchymal cells of different strains of mouse. Nature 202:1140–1141

Schell K (1960a) Studies on the innate resistance of mice to infection with mousepox. I. Resistance and antibody production. Aust J Exp Biol Med Sci 38:271–288

Schell K (1960b) Studies on the innate resistance of mice to infection with mousepox. II. Route of inoculation and resistance; and some observations on the inheritance of resistance. Aust J Exp Biol Med Sci 38:289–300

Schwanzer V, Deerberg F, Frost J, Liess B, Schwanzerova I, Pitterman W (1975) Zur intrauterinen Infektion der Maus mit Ektromelie-virus. Z Versuchstierkd 17:110–120

Stanley NF (1974) The reovirus murine models. Prog Med Virol 18:257–272

Wallace GD, Werner RM, Golway PL, Hernandez DM, Alling DW, George DA (1981) Epizootiology of an outbreak of mousepox at the National Institutes of Health. Lab Anim Sci 31:609–615

Walters MN, Joske RA, Leak PJ, Stanley NF (1963) Murine infection with reovirus. I. Pathology of the acute phase. Br J Exp Pathol 44:427–436

Wylekshanin AJ (1935) Transplancental transmission of the filterable virus of infectious ectromelia. Z Ihikrobiol (Moscow) 15:433

Reovirus Type 3 Infection, Liver, Mouse

Stephen W. Barthold

Synonyms. Reo-3 virus infection, hepato-encephalomyelitis virus (HEV) infection, ECHO 10 virus infection

Gross Appearance

Natural infection of adult mice with reovirus type 3 is asymptomatic. Infection of neonatal mice can result in runting, emaciation, jaundice, bilirubinuria, conjunctivitis, incoordination, tremor, paralysis, and oily hair effect. Survivors are runted with transient alopecia, which is most marked along the dorsum of the head, neck, and rear legs. Yellow or blood-tinged peritoneal exudate can be present. Livers are enlarged and dark, with multiple sharply demarcated yellow foci up to 3 mm in diameter on all surfaces. Gallbladders

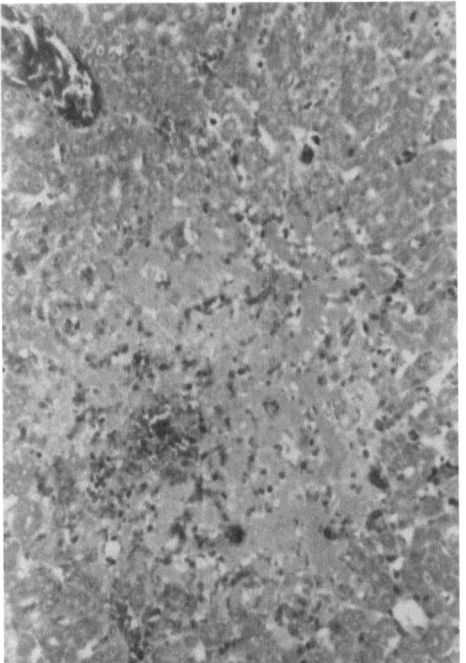

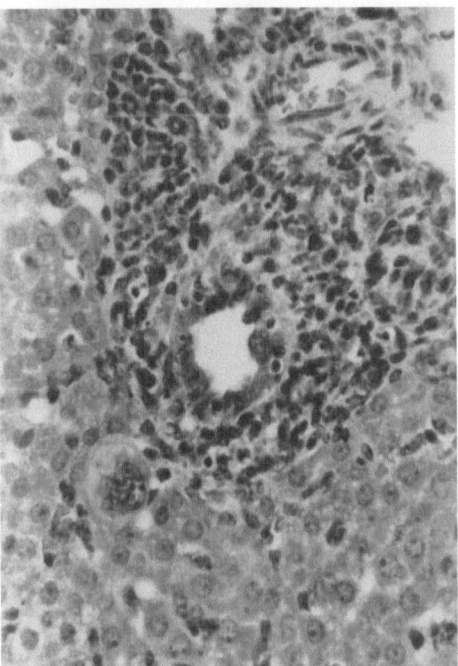

are distended with dark bile. The intestines are reddened and distended with yellow digesta. Fibrinous exudate and small gray foci have been observed on the epicardium. Other lesions include general pallor, focal pulmonary hemorrhages, pale kidneys, and atrophy of thymus, lymph nodes, and spleen. Chronically infected mice can have recrudescence of jaundice and liver lesions (Bennette et al. 1967a; Cook 1963; Stanley et al. 1953, 1954; Walters et al. 1963).

Microscopic Features

Lesions are generalized in mice exposed as neonates. Sharply demarcated foci of acute hepatocellular coagulation necrosis surrounded by variable leukocytic infiltrates are evident, particularly in the subcapsular parenchyma and near centrilobular veins (Fig. 165). This is followed by proliferation of Kupffer's cells and sinusoidal accumulations of mixed leukocytes. Initially, bile ducts are dilated and contain eosinophilic and necrotic cellular debris, followed by infiltration of portal areas with leukocytes (Fig. 166). Parenchymal mitotic activity is increased. Vascular congestion, hemorrhage, foci of neuronal degeneration and necrosis, perivascular leukocytic infiltrates, and nonsuppurative meningitis are present in the brain and spinal cord. Foci of necrosis and inflammation are also present in lymphoreticular organs, pancreas, salivary glands, kidney, brown fat, skin, myocardium, and skeletal muscle. Lungs have patchy areas of alveolar hemorrhage and edema (Barthold et al. 1993; Bennette et al. 1967a; Cook 1963; Papadimitriou 1968; Stanley et al. 1953, 1954; van Tongeren 1957; Walters et al. 1963). Intestinal changes are subtle or absent. Small intestinal villi can be blunted, with dilatation of lymphatics and mild mononuclear leukocyte infiltration (Branski et al. 1980b).

The chronic phase of infection is characterized by the presence of nonsuppurative inflammatory and degenerative changes in many organs. Reduction

◄

Fig. 165. (*above*) Focal necrotizing hepatitis in an infant mouse experimentally infected with reovirus type 3. H&E, ×100

Fig. 166. (*below*) Portal hepatitis and cholangitis in an infant mouse experimentally infected with reovirus type 3. Individual bile ductal epithelial cells are undergoing pyknosis. H&E, ×375

in hair follicles, edema, and subepidermal leukocytic infiltrates are found in alopecic skin. Depletion of lymphocytes occurs in thymus, spleen, and lymph nodes. Foci of hepatic necrosis, mixed leukocytic infiltrates, nonsuppurative portal inflammation, and bile ductular proliferation can be present. There may also be pancreatic acinar atrophy with adipose replacement, focal necrosis, and inflammation (Bennette et al. 1967a,b; Joske et al. 1966; Phillips et al. 1969; Stanley et al. 1964; Walters et al. 1963).

Ultrastructure

Within 4–5 days of intraperitoneal inoculation of neonatal mice, monocytes and neutrophils infiltrate hepatic sinusoids and the space of Disse. Monocytic phagocytosomes contain collections of virus particles, which are also scattered throughout the cytoplasm of degenerating monocytes. By 5–6 days, sinusoids contain cellular debris intermixed with viral particles. These particles are within phagocytic and pinocytotic vacuoles of Kupffer's cells and are free within the space of Disse. Virus entry by pinocytosis and virus replication is also evident in hepatocytes, with viral crystalloids in the paranuclear and sometimes the pericanalicular cytoplasm. Viral particles measure 650–700 Å and consist of a 300-Å dense, inner nucleoid and a less osmiophilic capsule. Some particles do not contain nucleoids. Virus crystalloids sometimes form in areas containing a loose mesh of fine tubules measuring 200 Å and lined by finely granular membranous walls approximately 50–60 Å thick. The nuclei of infected hepatocytes contain one or more enlarged peripheral nucleoli and occasional clear vacuoles. Virus release occurs by cytolysis. Necrosis of hepatocytes and Kupffer's cells in the absence of discernible viral replication also occurs (Papadimitriou 1965). Following hepatocellular virus replication, biliary epithelial cells undergo lytic, replicative infection or display increased osmiophilia without virus, as seen in hepatocytes (Papadimitriou 1968). In pancreas, monocytes containing membrane-bound and free virus particles in their cytoplasm are evident in small venules by day 4. By days 5–8, acinar epithelial cells possess autophagosomes containing cellular debris, dense bodies, and structures 250–300 Å in diameter, suggesting viral nucleoids. Crystalloids of virus are evident only in terminal pancreatic duct epithelium, with exfoliation of necrotic cells into the lumen (Papadimitriou and

Walters 1967). In brain, vessels contain virus-bearing mononuclear leukocytes, followed sequentially by virus replication in vascular endothelium, perivascular astrocytes, and then neurons (Papadimitriou 1967). In the chronic hepatitis of mice that survive the acute phase, there is increased osmiophilia and irregularity of mitochondria, vesiculation and degranulation of endoplasmic reticulum, an increase in lysosomes, and blunting of microvilli within affected hepatocytes. Sinusoids contain thrombocytes, many of which are adherent to sinusoidal lining cells, and focal areas of multilayered Kupffer's cells, some with deposition of distinct basement membrane material. In a few hepatocytes, stacks of viral filaments or microtubules occur, but there is a remarkable absence of replicating virus (Papadimitriou 1966).

Differential Diagnosis

The diverse lesions of reovirus type 3 infection are similar to those caused by mouse hepatitis virus and a number of other less common infectious agents. Reoviral hepatic lesions do not have pathognomonic features at the light microscopy level. Oily hair effect should not be considered specific for reovirus type 3 infection, since it can also occur in other enteritides of the young mouse, including murine rotavirus and enteric mouse hepatitis virus infection.

Biologic Features

Natural History

Reovirus type 3 infection in laboratory mice is usually not clinically manifested under natural conditions. Mice are susceptible to infection at all ages, but exposure or inoculation in the early neonatal period (3 or less days of age) is necessary for induction of lesions and disease (Barthold et al. 1993; Stanley et al. 1954). Furthermore, infant mice may be protected from experimental inoculation during this susceptible period by maternal antibody (Bennette et al. 1967a,b). However, all of the signs and lesions of experimental reovirus type 3 infection can be present in susceptible, naturally infected neonates (Cook 1963; Stanley et al. 1954). The most likely means of reovirus type 3 excretion and transmission is the orofecal route (Branski et al. 1980b; Stanley et al. 1954; Walters et al. 1963). The duration of virus excretion is not known. Virus replication and excretion are diminished or absent in mice that survive the acute phase, but can be reactivated under certain circumstances (Stanley 1974).

Pathogenesis

Reovirus type 3 can be transmitted via a number of routes, including intraperitoneal, intrahepatic, intracerebral, subcutaneous, intradermal, oronasal, and contact. The outcome of infection is dependent on age, virus strain, mouse genotype, and presence of maternal antibody (Bennette et al. 1967a,b; Stanley 1974; Stanley et al. 1953, 1954; van Tongeren 1957; Walters et al. 1963). If mice are infected at 3 or less days of age, they develop a generalized, multisystemic infection. Older mice are apparently not susceptible to generalized infection and develop no discernible lesions (Barthold et al. 1993; Bennette et al. 1967a; Stanley et al. 1953, 1954; van Tongeren 1957). Within a few days after inoculation, neonatal mice have a plasma- and leukocyte-associated viremia (Papadimitriou 1965, 1967, 1968; Papadimitriou and Walters 1967; Stanley et al. 1953, 1954). Viral antigen is readily discernible, and virus is recoverable in high titer from multiple organs for up to 1 month, but lesions are absent in many virus-positive organs (Barthold et al. 1993; Stanley 1974; Stanley et al. 1953, 1954). Tropism of virus for neurons and lymphocytes is mediated through the viral hemagglutinin (Weiner et al. 1977, 1980). Mice can develop steatorrhea, with a many fold increase in fecal volume and fecal fat (Stanley et al. 1954), due to pancreatic insufficiency or intestinal malabsorption, since function of both organs is altered (Branski et al. 1980a,b). Biliary tract occlusion can also contribute (Papadimitriou 1968; Phillips et al. 1969). Clinical signs during this acute phase of infection thus include signs referable to central nervous system infection (tremor, ataxia), hepatitis (jaundice), and steatorrhea (oily hair effect). Hepatic, pancreatic, and intestinal lesions are apparently variable, either because of differences in virus strain, route and dose of inoculum, mouse genotype, or coinfection with other pathogens.

Mice that survive have prolonged retardation of growth. They have atrophy of lymphoreticular organs, normocytic normochromic anemia, leukopenia, and monocytosis (Bennette et al. 1967a; Joske et al. 1966). There are chronic inflammatory, degenerative, and necrotizing lesions in multiple organs. As duration of infection increases, virus recovery from various organs is increasingly difficult (Joske et al. 1966; Walters et al. 1973). The necrotizing hepatitis in this phase of infection, unlike the acute phase, is not associated with viral replication (Papadimitriou 1966; Walters et al. 1973). It has been suggested that the chronic disease and runting syndrome mimics a graft-versus-host reaction (Bennette et al. 1967b; Joske et al. 1966; Papadimitriou 1966; Stanley 1974; Stanley et al. 1964, 1966) and can be reproduced by transfer of homologous spleen cells from mice with chronic disease into neonates (Joske et al. 1966; Stanley 1974; Stanley et al. 1966). The late chronic disease also appears to predispose to the development of lymphoma (Joske et al. 1966; Stanley et al. 1966).

Etiology

Mammalian reoviruses share common antigens, but three distinct antigenic types of reovirus can be differentiated serologically (reovirus types 1, 2, and 3). They are ubiquitous, infecting at least 60 different species of mammals (including humans), birds, and reptiles (Hartley et al. 1961; Stanley 1974). Although all three types produce infection in experimentally inoculated mice, reovirus type 3 is associated with natural laboratory and wild mouse infections and clinical disease (Bennette et al. 1967a; Cook 1963; Hartley et al. 1961). The pathology in mice induced by human isolates of reovirus type 3 is indistinguishable from that caused by murine isolates (Bennette et al. 1967a; Cook 1963; Stanley et al. 1953, 1954, 1964; Walters et al. 1963). Neonatal mice inoculated oronasally with reovirus type 1 develop lesions similar to reovirus type 3, whereas infection with reovirus type 2 results in intestinal lesions reminiscent of rotaviral disease (Barthold et al. 1993). Natural seroconversion to reovirus type 3 is observed in laboratory rats, hamsters, guinea pigs, and rabbits, but is not associated with clinical disease (Barthold et al. 1993).

Frequency

Because of the usual absence of clinical signs, the prevalence of reovirus type 3 infection can only be assessed serologically. Hemagglutination inhibition assays of mouse colonies have suggested a high prevalence of infection in the past (Parker et al. 1966). Seroconversion to reovirus type 3 is common among contemporary mouse, rat, and hamster colonies, and the virus is a frequent contaminant of transplantable tumors (S.W. Barthold, unpublished observations).

Comparison with Other Species

The host range is broad, but the nature of infection in species other than the mouse has not been well studied. Rats and hamsters experimentally inoculated with virus became clinically ill with hepatitis (Bennette et al. 1967a), but three species of nonhuman primates *(Macaca nemestrina, M. iris,* and *M. mulatta)* did not. Fecal virus excretion and liver lesions were demonstrated in those primates (Stanley et al. 1954). In humans, reoviruses (including reovirus type 3) have been isolated from feces of normal children as well as from patients with respiratory and intestinal disease, steatorrhea, hepatitis, conjunctivitis, alopecia, and neurologic disease (Papadimitriou 1967; Stanley et al. 1954; van Tongeren 1957; Walters et al. 1963).

References

Barthold SW, Smith AL, Bhatt PN (1993) Infectivity, disease patterns, and serologic profiles of reovirus serotypes 1, 2, and 3 in infant and weanling mice. Lab Anim Sci 43:425–430

Bennette JG, Bush PV, Steele RD (1967a) Characteristics of a newborn runt disease induced by neonatal infection with an oncolytic strain of reovirus type 3 (RE03MH). 1. Pathological investigations in rats and mice. Br J Exp Pathol 48:251–266

Bennette JG, Bush PV, Steele RD (1967b) Characteristics of a newborn runt disease induced by neonatal infection with an oncolytic strain of reovirus type 3 (RE03MH). II. Immunological aspects of the disease in mice. Br J Exp Pathol 48:267–284

Branski D, Lebenthal E, Faden HS, Hatch TF, Krasner J (1980a) Reovirus type 3 infection in a suckling mouse: the effects on pancreatic structure and enzyme content. Pediatr Res 14:8–11

Branski D, Lebenthal E, Faden H, Hatch TF, Krasner J (1980b) Small intestinal epithelial brush border enzymatic changes in suckling mice infected with reovirus type 3. Pediatr Res 14:803–805

Cook I (1963) Reovirus type 3 infection in laboratory mice. Aust J Exp Biol 41:651–659

Hartley JW, Rowe WP, Huebner RJ (1961) Recovery of reoviruses from wild and laboratory mice. Proc Soc Exp Biol Med 108:390–395

Joske RA, Leak PJ, Papadimitriou JM, Stanley NF, Walters MN (1966) Murine infection with reovirus. IV. Late chronic disease and the induction of lymphoma after reovirus type 3 infection. Br J Exp Pathol 47:337–346

Papadimitriou JM (1965) Electron micrographic features of acute murine reovirus hepatitis. Am J Pathol 47:565–585

Papadimitriou JM (1966) Ultrastructural features of chronic murine hepatitis after reovirus type 3 infection. Br J Exp Pathol 47:624–631

Papadimitriou JM (1967) An electron microscopic study of murine reovirus 3 encephalitis. Am J Pathol 50:59–75

Papadimitriou JM (1968) The biliary tract in acute murine reovirus 3 infection. Light and electron microscopic study. Am J Pathol 52:595–611

Papadimitriou JM, Walters MN (1967) Studies on the exocrine pancreas. II. Ultrastructural investigation of reovirus pancreatitis. Am J Pathol 51:387–403

Parker JC, Tennant RW, Ward TG (1966) Prevalence of viruses in mouse colonies. Natl Cancer Inst Monogr 20:25–45

Phillips PA, Keast D, Papadimitriou JM, Walters MN, Stanley NF (1969) Chronic obstructive jaundice induced by Reovirus type 3 in weanling mice. Pathology 1:193–203

Stanley NF (1974) The reovirus murine models. Prog Med Virol 18:257–272

Stanley NF, Dorman DC, Ponsford J (1953) Studies on the pathogenesis of a hitherto undescribed virus (hepatoencephalomyelitis) producing unusual symptoms in suckling mice. Aust J Exp Biol 31:147–159

Stanley NF, Dorman DC, Ponsford J (1954) Studies on the hepatoencephalomyelitis virus (HEV). Aust J Exp Biol 32:543–562

Stanley NF, Leak PJ, Walters MN, Joske RA (1964) Murine infections with reovirus: II. The chronic disease following reovirus type 3 infection. Br J Exp Pathol 45:142–149

Stanley NF, Walters MN, Leak PJ, Keast D (1966) The association of murine lymphoma with reovirus type 3 infection. Proc Soc Exp Biol Med 121:90–93

van Tongeren HAE (1957) A familial infection with hepatoencephalomyelitis virus in the Netherlands. Study on some properties of the infective agent. Arch Virusforsch 7:429–448

Walters ML, Stanley NF, Dawkins RL, Alpers MP (1973) Immunological assessment of mice with chronic jaundice and runting induced by reovirus 3. Br J Exp Pathol 54:329–345

Walters MN, Joske RA, Left PJ, Stanley NF (1963) Murine infection with reovirus. I. Pathology of the acute phase. Br J Exp Pathol 44:427–436

Weiner HL, Drayna D, Averill DR Jr, Fields BN (1977) Molecular basis of reovirus virulence: role of the S1 gene. Proc Natl Acad Sci USA 74:5744–5748

Weiner HL, Ault KA, Fields BN (1980) Interaction of reovirus with cell surface receptors. I. Murine and human lymphocytes have a receptor for hemagglutinin of reovirus type 3. J Immunol 124:2143–2148

BACTERIAL INFECTIONS

Tyzzer's Disease, Rat, Mouse, and Hamster

Lela K. Riley and Craig L. Franklin

Gross Appearance

Gross lesions may vary from none to severe and are characteristically found in the liver, lower intestine (ileum, cecum, and proximal colon), and less frequently in the heart. The most consistent finding is hepatomegaly with multiple gray to white foci scattered throughout all lobes of the liver (Fig. 167). While hepatic lesions are a hallmark of Tyzzer's disease, the liver is not always affected. Intestinal lesions are usually evident in acutely infected animals and consist of varying degrees of serosal edema with or without obvious hemorrhage. Cardiac lesions consisting of white streaks within the myocardium have been reported in hamsters, rats, and mice.

Microscopic Features

Lesions of Tyzzer's disease are characterized by necrosis with varying degrees of inflammation in response to the necrosis. Acute hepatic lesions consist of necrotic foci surrounded by minimal, primarily neutrophilic inflammation (Fig. 168). As the disease progresses, the inflammatory response may increase, but rarely becomes a predominant feature. Chronic foci may become mineralized or fibrotic. Because the causative agent of Tyzzer's disease, *Clostridium piliforme* (previously known as *Bacillus piliformis*), stains faintly with hematoxylin and eosin (H&E), silver stains such as Steiner or Warthin-Starry are used to enhance detection of bacteria. In acute to subacute lesions, intracellular bacteria are found in viable cells adjacent to necrotic foci. Within the cell, *C. piliforme* are found lying next to each other in a unique, almost pathognomonic arrangement which has been described as a "bundle of sticks" or "pick up sticks" arrangement (Fig. 169; Ganaway et al. 1971).

Acute lesions of the intestine consist of necrosis of primarily the luminal enterocytes. The lamina propria and submucosa may be edematous and contain a mild neutrophilic infiltrate. As the lesion progresses, necrosis becomes more extensive and the inflammatory response shifts to lymphocytic or granulomatous. Occasionally, necrotic foci are evident in the tunica muscularis of the ileum or cecum. These foci may be accompanied by either typical necrotizing epithelial lesions or minimal mucosal pathology characterized by a mild lymphocytic infiltrate (Fig. 170). With severe lesions, complete loss of mucosal architecture may be observed (Fig. 171). Healing and repair are evidenced by hyperplasia of the crypt or gland epithelium. Tyzzer's bacilli are usually present in viable enterocytes or myocytes of the muscularis (Fig. 172). Myocardial lesions consist of mild inflammation associated with small necrotic foci (Fig. 173).

Ultrastructure

Tyzzer's bacillus is a filamentous bacterium with peritrichous flagella (Fig. 174). The vegetative form is large, ranging from 8 to 40 μm in length. Spores are 0.5 μm wide and 3 μm long and may be found either as terminal endospores within the bacteria or free within host cells.

The intracellular events in *C. piliforme* infections have been examined with an in vitro model which utilizes a Caco-2 colonic carcinoma cell line that possesses a polarized monolayer with a functional brush border similar to intestinal epithelium. This model closely resembles in vivo infection, since the initial site of bacterial invasion in animals is the intestinal epithelium. Ultrastructural examination of infected Caco-2 monolayers reveals a putative sequence of events that occur in *C. piliforme* infections (Fig. 175). *C. piliforme* apparently enter

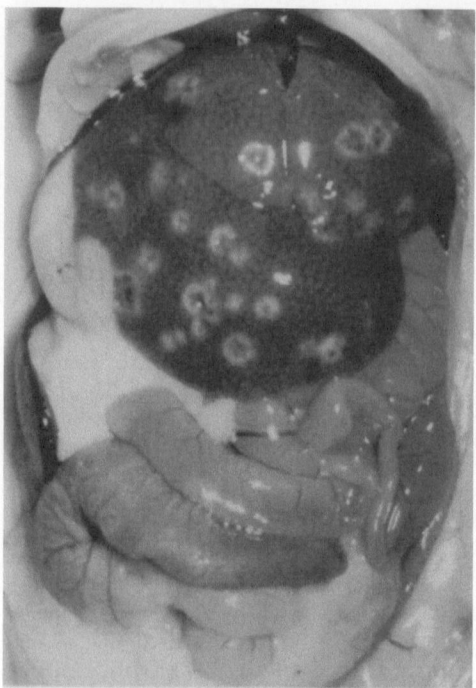

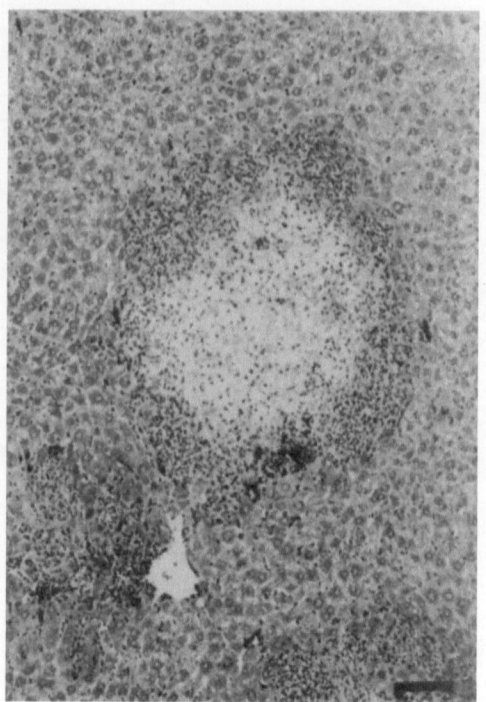

Fig. 167. Multiple foci of hepatic necrosis in a mouse with Tyzzer's disease

the host epithelial cell through the apical surface by a phagocytic-like mechanism (Fig. 175a). On rare occasions, *C. piliforme* can be seen in membrane-bound vacuoles, usually subjacent to the microvillar surface (Fig. 175b). However, the majority of intracellular *C. piliforme* are located within the cytoplasm, with bacterial numbers ranging from one to 50 per cell (Fig. 175c). Both vegetative and spore forms can be found within host cells. Because nearly all of the bacteria are found free within the cytoplasm, *C. piliforme* are believed to rapidly escape from the phagocytic vacuole. The ability to exit the phagosome is

►

Fig. 168. (*above*) Liver from a mouse infected with *Clostridium piliforme*. A focal area of necrosis is surrounded by a thin band of infiltrating neutrophils. Also note the sharp demarcation between necrosis and viable hepatocytes. No bacteria are evident. *Bar*, 100 μm. H&E

Fig. 169. (*below*) Liver from the same animal as in Fig. 168. Hepatocytes at the periphery of the necrotic foci contain intracellular bundles of long filamentous argyrophilic bacteria arranged in a "pick up stick" pattern. Individual bacteria are also evident in hepatocytes and necrotic debris. *Bar*, 10 μm. Steiner silver stain

thought to be an important virulence determinant, allowing the organism to survive intracellularly and evade host-mediated killing mechanisms. Once inside the host cell, the organism replicates (Fig. 175d). In addition to cytoplasmic localization of the organism, the bacteria may also be found in the nucleus or bridging the nuclear membrane (Fig. 175e). The morphology of lightly infected cells is comparable to that of uninfected cells, whereas heavily infected cells, i.e., cells with more than 20 bacteria per cell, exhibit marked vacuolation (Fig. 175c,f). Eventual cell lysis results in release of the bacteria for dissemination to new host cells (Fig. 175f).

Differential Diagnosis

A presumptive diagnosis of Tyzzer's disease can be made upon gross examination of the liver. Differential diagnoses for hepatic lesions include infections with *Corynebacterium kutscheri* (rats and mice), mouse hepatitis virus (mice), ectromelia virus (mice), or *Salmonella* (all species) or hepatic dissemination of enteric bacteria or bacterial toxins (all species). A tentative histologic diagnosis of Tyzzer's disease can be made based on the necrotizing nature of the lesions. Definitive diagnosis is made by demonstrating typical intracellular bacteria in hepatocytes adjacent to necrotic foci.

A diagnosis of Tyzzer's disease was once thought to require hepatic involvement. However, experimental studies have shown that infection of hepatocytes is a transient phenomenon, lasting only 3–7 days (unpublished data). In contrast, intestinal involvement appears to be maintained for a longer period of time, up to 13 days postinfection. Thus histologic examination of intestinal tissues may be critical in establishing a definitive diagnosis.

Serologic assays have been used diagnostically to detect *C. piliforme* infections. Experimental stud-

▶

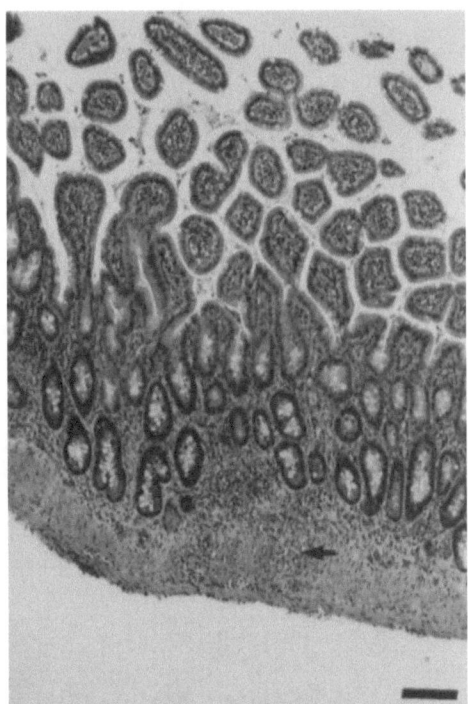

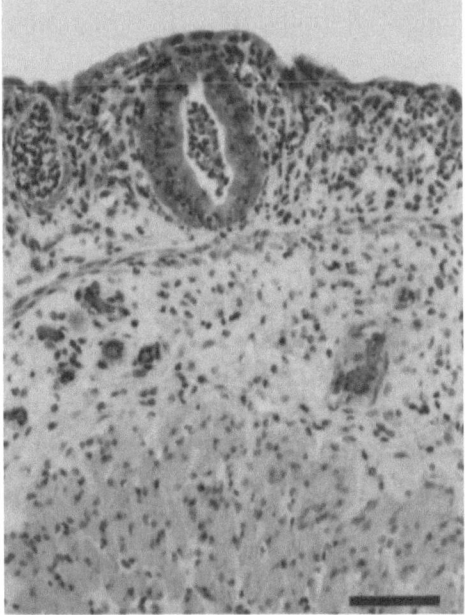

Fig. 170. (*above*) Terminal ileum from a hamster infected with *Clostridium piliforme*. The *arrow* indicates focal necrosis in the tunica muscularis. Note the minimal inflammation associated with the necrotic foci and lack of lesions in enterocytes. *Bar*, 100 μm. H&E

Fig. 171. (*below*) Cecum from a hamster infected with *Clostridium piliforme*. Note the loss of mucosal architecture, moderate pyogranulomatous infiltrate, mucosal and submucosal edema and focal hyperplasia of remaining cecal gland epithelial cells. *Bar*, 60 μm. H&E

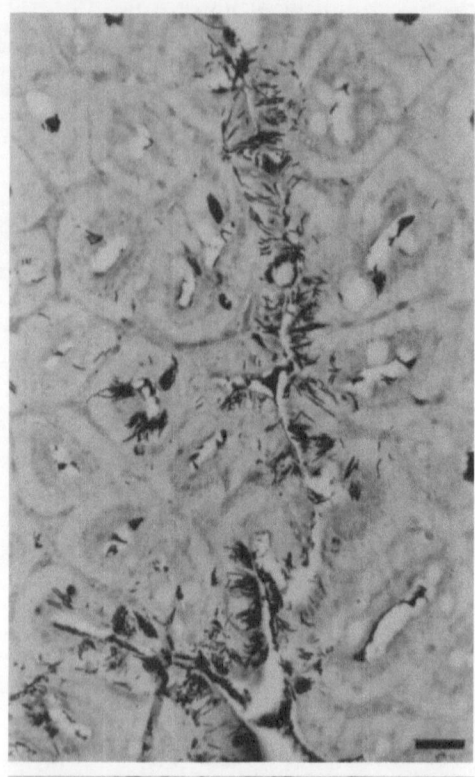

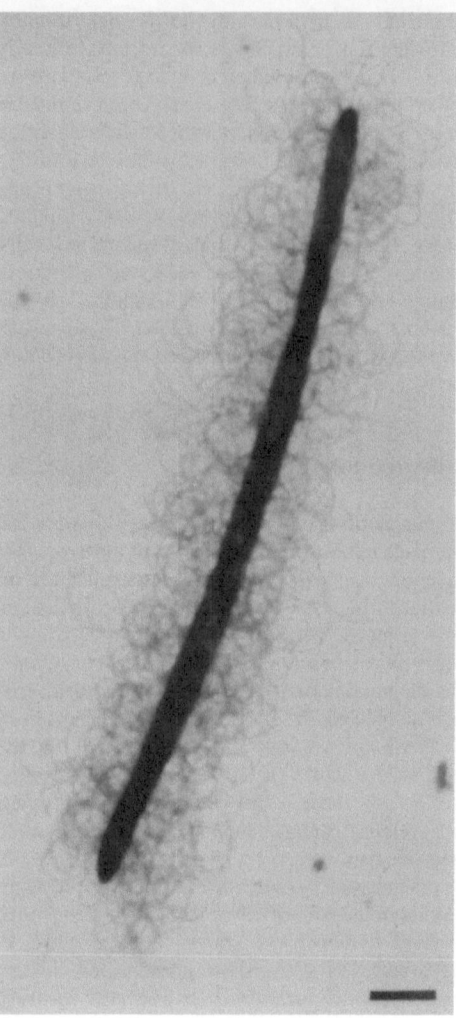

Fig. 174. *Clostridium piliforme* bacterium. The bacillus is surrounded by large numbers of peritrichous flagella. TEM. *Bar,* 2 μm

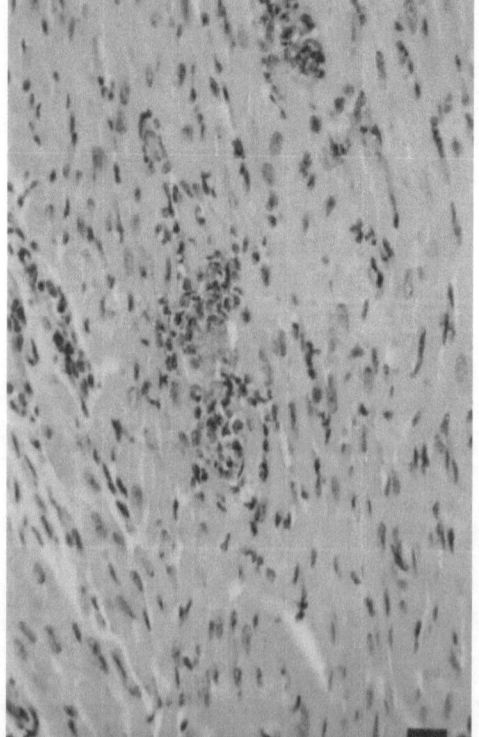

◄

Fig. 172. (*above*) Cecal epithelium from a mouse infected with *Clostridium piliforme*. Packets of argyrophilic intracellular bacteria are present within viable enterocytes. Individual bacteria are also occasionally evident within enterocytes. *Bar,* 20 μm. Steiner silver stain

Fig. 173. (*below*) Heart from a hamster infected with *Clostridium piliforme*. Note focal myofiber necrosis with associated inflammation. *Bar,* 20 μm. H&E

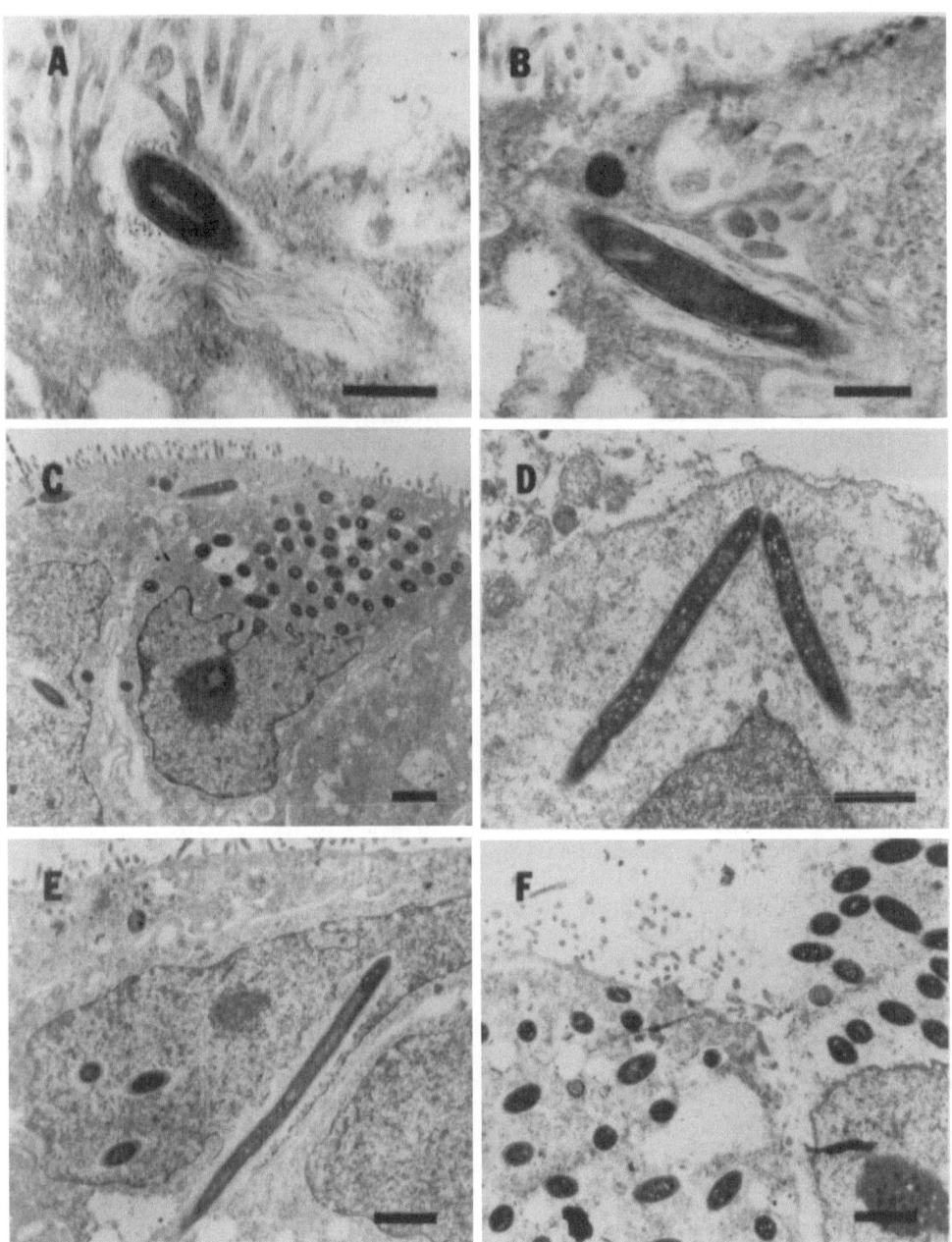

Fig. 175a–f. Caco-2 cell monolayers infected with *Clostridium piliforme*. **a** Bacterium within an apical surface invagination of a Caco-2 cell. **b** Bacterium within a membrane-bound vacuole subjacent to the microvillar surface of a Caco-2 cell. **c** Two infected Caco-2 cells. The cell on the *left* contains a few intracytoplasmic bacteria and a single intranuclear bacterium, whereas the cell on the *right* contains abundant intracytoplasmic bacteria and is mildly vacuolated, suggesting dete- rioration. **d** Bacterium undergoing replication within the cyto- plasm of an infected cell. **e** Infected Caco-2 cell containing four intranuclear bacteria, with one bacterium bridging the nuclear membrane. **f** Two heavily infected Caco-2 cells. The cell on the *left* is vacuolated and contains abundant bacteria, whereas the cell on the *right* appears to have lysed and is releasing bacteria. TEM. *Bars,* 0.5 μm (**a,b**), 1.0 μm (**c–f**)

ies have shown that induction of serum antibody to *C. piliforme* requires viable organisms; simple exposure to killed bacteria is insufficient to generate an immune response (Motzel and Riley 1992). These findings suggest that the presence of specific serum antibodies to *C. piliforme* indicates that the animal has been infected. In early studies, complement fixation tests were used, but these lacked sensitivity (Fujiwara et al. 1971, 1973, 1974). Subsequently, indirect immunofluorescence assays (IFA) and enzyme-linked immunosorbent assays (ELISA) were developed using whole organisms or extracts from infected tissues or embryonated eggs (Savage and Lewis 1972; Waggie et al. 1987). However, nonspecific cross-reactivity of antibodies to other organisms often yielded false-positive results in these assays. Recent advancement in techniques for propagation of *C. piliforme* in mammalian cell cultures has provided more defined antigen preparations for ELISA testing. These refinements have lessened, but not eliminated nonspecific cross-reactivity and false-positive testing results. More purified and specific antigens will be required to enhance the specificity of ELISA assays and resolve these problems. Until such reagents are developed, definitive serologic diagnosis of *C. piliforme* infections will remain difficult.

Recently, a specific monoclonal-based competitive inhibition ELISA has been developed which allows identification of the specific isolate involved in *C. piliforme* infections of laboratory animals (Boivin et al. 1994). The ability to delineate the specific isolate may provide a tremendous epidemiologic tool for the control and prevention of subsequent disease outbreaks in laboratory animals.

Biologic Features

Natural History

The natural history of Tyzzer's disease is not thoroughly understood. The disease was named after Ernest Tyzzer, who initially described the syndrome in 1917. Tyzzer reported a fatal epizootic of necrotizing hepatitis and hemorrhagic enteritis that resulted in the death of an entire colony of Japanese waltzing mice (Tyzzer 1917). He observed gram-negative, pleomorphic, spore-forming, intracellular bacteria adjacent to necrotic hepatic and intestinal lesions and postu-

lated that the bacterium was the causative agent. For many years, Tyzzer's disease was believed to be restricted to mice, until Allen et al. (1965) reported the disease in laboratory rabbits in 1965. Since then, the disease has been reported in a wide range of mammalian animal species, including domestic, wild, and laboratory animals. In laboratory animals, the disease has been found in rats (Jonas et al. 1979), mice (Tyzzer 1917; Gibson et al. 1987), gerbils (Carter et al. 1969; White and Waldron 1969), hamsters (Zook et al. 1977), rabbits (Allen et al. 1965), guinea pigs (McLeod et al. 1977), and rhesus monkeys (Niven 1968). Outbreaks of the disease have occurred worldwide.

The causative agent was initially named *B. piliformis* by Tyzzer, based strictly on its morphology. However, recent studies by Duncan and colleagues (1993), in which they sequenced the 16S rRNA gene of an isolate obtained from a rabbit with Tyzzer's disease, indicated that the organism is phylogenetically more closely related to the *Clostridium* genus than to the *Bacillus* genus. Based on these findings, the Tyzzer's bacillus has been reclassified as *C. piliforme*.

Despite numerous attempts by many investigators, the organism has not been cultured on cell-free medium. In one study alone, over 190 different microbiologic media were tested for their ability to support growth of the bacillus (Thunert 1984). Based on the universal lack of success in culture of the bacterium on cell-free media and the apparent requirement of viable host cells for propagation of the organism (Ganaway et al. 1971), *C. piliforme* is believed to be an obligate intracellular bacterium. Early studies of Tyzzer's disease required cultivation of the bacteria in susceptible animals. However, in 1971, Ganaway and colleagues reported successful culture of the bacteria in yolk sacs of embryonated eggs (Ganaway et al. 1971). The organism was subsequently grown on primary cultures of mouse hepatocytes (Kawamura et al. 1983). In recent years, the organism has been cultivated on a number of immortal mammalian cell lines (Spencer et al. 1990; Riley et al. 1990; Franklin et al. 1993).

A number of factors may predispose animals to Tyzzer's disease. Stresses such as overcrowding, change in environmental conditions, and immunosuppression may contribute to disease outbreaks in laboratory animals. Weanling animals are particularly susceptible to acute disease.

A high-protein diet has also been shown to predispose animals to Tyzzer's disease (Maejima et al. 1965).

Pathogenesis

Transmission of the organism is believed to occur primarily via ingestion of spores from contaminated feces. Because the vegetative form is particularly labile (Craigie 1966), it is likely that it does not play a major role in natural infections. In contrast, spores are stable and can remain infectious for 1–2 years in contaminated bedding. Thus it is thought that spores are the primary means of transmission among animals. Experimental studies have documented that spores are shed into feces for 1–2 weeks after infection and that susceptible animals exposed to contaminated fecal material can develop Tyzzer's disease (Itoh et al. 1988; Motzel and Riley 1992; Waggie et al. 1984). Ingested spores are believed to germinate in the intestine and colonize the intestinal epithelium. Dissemination of bacteria to the liver and heart occurs via the blood during later stages of the disease (Takagaki and Fujiwara 1968). Transplacental transfer has been reported in experimentally infected mice and immunosuppressed rats, although the relative importance of this mode of transmission in natural infections is not known (Fries 1978, 1979).

Virulence determinants of *C. piliforme* are not well defined. The bacterium's ability to adhere to and invade epithelial cells provides it with a mechanism to evade host immunity. In addition, recent studies have shown that certain isolates produce cytotoxins that may function as virulence determinants (Riley et al. 1992). Ten of 13 isolates examined in our laboratories to date produce a high molecular weight, nonprotease, temperature-sensitive protein molecule that is cytotoxic for mammalian cells. While the role of toxins in pathogenesis of disease has not been determined, preliminary studies indicate that animals infected with toxigenic isolates exhibit strikingly more severe clinical signs and histopathologic lesions and have markedly higher mortality rates than do animals infected with nontoxigenic isolates. Based on these observations, it is tempting to speculate that outbreaks with severe clinical disease and high mortality rates may be due to toxigenic isolates, whereas asymptomatically infected animals may be infected with nontoxigenic isolates.

Etiology

Organisms designated as *C. piliforme* represent a heterogeneous collection of bacterial isolates that have indistinguishable morphology, but are antigenically distinct. All isolates are weakly gram negative, pleomorphic, spore forming, rod shaped, and motile by means of peritrichous flagella. Spores are relatively heat labile in that they survive 60°C, but not 70°C for 30 min (Ganaway 1980). Evidence of diversity among isolates came initially from vaccination studies in mice (Fujiwara et al. 1971, 1973, 1974). Vaccination was most effective when the challenge isolate of *C. piliforme* was the same as the isolate used for vaccination. Further evidence of differences among isolates was demonstrated by agar gel diffusion, complement fixation tests, and western blots (Riley et al. 1990; Fujiwara et al. 1971, 1973, 1974). These tests revealed that both conserved and isolate-specific antigens were present in the various *C. piliforme* isolates. Subsequent studies revealed antigenic and size variation among immunodominant flagellin proteins from *C. piliforme* isolates (Motzel and Riley 1991).

Host specificity of *C. piliforme* was initially suggested by Fujiwara and colleagues (1971, 1973, 1974). Their findings have been further supported by studies in which well-characterized, antigenically diverse isolates were used to infect rats, mice, and hamsters (Franklin et al. 1994). Results of these investigations indicate that isolates are predominately host specific, but some cross-infectivity may occur among animal species, particularly between rats and mice. The molecular mechanism or mechanisms responsible for this host specificity are not known, but it is likely that specific bacterial adhesions or host receptors may control initial events in infection that result in host specificity of isolates.

Frequency

Mortality rates in excess of 50% have been reported during epizootics in mice, hamsters, gerbils, and rabbits. However, outbreaks of clinical disease are relatively rare. Subclinical infections with *C. piliforme* are more widespread (Fries 1980), and animals latently infected may develop overt disease when subjected to stress. Subclinically infected animals are capable of transmitting *C. piliforme* to naive animals (Motzel and Riley 1992), allowing insidious spread of the infection throughout an animal facility without evidence of clinical signs of disease.

Comparison with Other Species

In general, hamsters, gerbils, and guinea pigs experience more severe clinical disease and a higher mortality rate than do rats and mice. Although the mechanisms governing disease severity are not known, both bacterial factors, such as cytotoxin production, and host factors, such as immune status, are thought to play a role. Despite the difference observed in clinical signs among various host species, the tissues involved and lesions associated with the disease are similar among affected host species.

References

Allen AM, Ganaway JR, Moore TD, Kinard RF (1965) Tyzzer's disease syndrome in laboratory rabbits. Am J Pathol 46:859–882

Boivin GP, Hook RR Jr, Riley LK (1994) Development of a monoclonal antibody-based competitive inhibition enzyme-linked immunosorbent assay for detection of Bacillus piliformis isolate-specific antibodies in laboratory animals. Lab Anim Sci 44:153–158

Carter GR, Whitenack DL, Julius LA (1969) Natural Tyzzer's disease in Mongolian gerbils (Meriones unguiculatus). Lab Anim Care 19:648–651

Craigie J (1966) Bacillus piliformis (Tyzzer) and Tyzzer's disease of the laboratory mouse. I. Propagation of the organism in embryonated eggs. Proc R Soc Lond [B] Biol Sci 165:35–60

Duncan AJ, Carmen RJ, Olsen GJ, Wilson KH (1993) Assignment of the agent of Tyzzer's disease to Clostridium piliforme comb. nov. on the basis of 16S rRNA sequence analysis. Int J Syst Bacteriol 43:314–318

Franklin CL, Kinden DA, Stogsdill PA, Riley LK (1993) In vitro model of adhesion and invasion by Bacillus piliformis. Infect Immun 61:876–883

Franklin CL, Motzel SL, Besch-Williford CL, Hook RR Jr, Riley LK (1994) Tyzzer's infection: host specificity of Clostridium piliforme isolates. Lab Anim Sci 44:568–572

Fries AS (1978) Demonstration of antibodies to Bacillus piliformis in SPF colonies and experimental transplacental infection by Bacillus piliformis in mice. Lab Anim 12:23–26

Fries AS (1979) Studies on Tyzzer's disease: transplacental transmission of Bacillus piliformis in rats. Lab Anim 13:43–46

Fries AS (1980) Antibodies to Bacillus piliformis (Tyzzer's disease) in sera from man and other species. In: Spiegel A, Erichsen S, Solleveld HA (eds) Animal quality and models in biomedical research. 7th symposium of the ICLAS, 21–23 Aug 1979. Fischer, Stuttgart, pp 249–252

Fujiwara K, Yamada A, Ogawa H, Oshima Y (1971) Comparative studies on the Tyzzer's organisms from rats and mice. Jpn J Exp Med 41:125–133

Fujiwara K, Kurashina H, Magaribuchi T, Takenaka S, Yokoiyama S (1973) Further observation on the difference between Tyzzer's organisms from mice and those from rats. Jpn J Exp Med 43:307–315

Fujiwara K, Takasaki Y, Kubokawa K, Takenaka S, Kubo M, Sato K (1974) Pathogenic and antigenic properties of the Tyzzer's organisms from feline and hamster cases. Jpn J Exp Med 44:365–372

Ganaway JR (1980) Effect of heat and selected chemical disinfectants upon infectivity of spores of Bacillus piliformis (Tyzzer's disease). Lab Anim Sci 30:192–196

Ganaway JR, Allen AM, Moore TD (1971) Tyzzer's disease of rabbits: isolation and propagation of Bacillus piliformis (Tyzzer) in embryonated eggs. Infect Immun 3:429–437

Gibson SV, Waggie KS, Wagner JE, Ganaway JR (1987) Diagnosis of subclinical Bacillus piliformis infection in a barrier-maintained mouse production colony. Lab Anim Sci 37:786–788

Jonas AM, Percy DH, Craft J (1979) Tyzzer's disease in the rat. Arch Pathol 90:516–528

Itoh T, Kagiyama N, Fujiwara K (1988) Production of Tyzzer's disease in rats by ingestion of bacterial spores. Jpn J Exp Med 59:9–15

Kawamura S, Taguchi F, Ishida T, Nakayama M, Fujiwara K (1983) Growth of Tyzzer's organism in primary monolayer cultures of adult mouse hepatocytes. J Gen Microbiol 129:277–283

Maejima K, Fujiwara K, Takagaki Y, Naiki M, Kurashina H, Tajima Y (1965) Dietetic effects on experimental Tyzzer's disease. Jpn J Exp Med 35:1–10

McLeod CG, Stookey JL, Harrington DG, White JD (1977) Intestinal Tyzzer's disease and spirochetosis in a guinea pig. Vet Pathol 14:229–235

Motzel SL, Riley LK (1991) Bacillus piliformis flagellar antigens for serodiagnosis of Tyzzer's disease. J Clin Microbiol 29:2566–2570

Motzel SL, Riley LK (1992) Subclinical infection and transmission of Tyzzer's disease in rats. Lab Anim Sci 42:439–443

Niven JS (1968) Tyzzer's disease in laboratory animals. Z Versuchstierkd 10:168–174

Riley LK, Besch-Williford C, Waggie KS (1990) Protein and antigenic heterogeneity among isolates of Bacillus piliformis. Infect Immun 58:1010–1016

Riley LK, Caffrey CJ, Musille VS, Meyer JK (1992) Cytotoxicity of Bacillus piliformis. J Med Microbiol 37:77–80

Savage NL, Lewis DH (1972) Application of immunofluorescence to detection of Tyzzer's disease agent (Bacillus piliformis) in experimentally infected mice. Am J Vet Res 33:1007–1011

Spencer TH, Ganaway JR, Waggie KS (1990) Cultivation of Bacillus piliformis (Tyzzer) in mouse fibroblasts (3T3 cells). Vet Microbiol 22:291–297

Takagaki Y, Fujiwara K (1968) Bacteremia in experimental Tyzzer's disease of mice. Jpn J Microbiol 12:129–143

Thunert A (1984) Is it possible to cultivate the agent of Tyzzer's disease (Bacillus piliformis) in cellfree media? Z Versuchstierkd 26:145–150

Tyzzer EE (1917) A fatal disease of the Japanese waltzing mouse caused by a spore-bearing bacillus (Bacillus piliformis NSP). J Med Res 37:307–338

Waggie KS, Ganaway JR, Wagner JE, Spencer TH (1984) Experimentally induced Tyzzer's disease in Mongolian gerbils (Meriones unguiculatus). Lab Anim Sci 34:53–57

Waggie KS, Spencer TH, Ganaway JR (1987) An enzyme-linked anti-Bacillus piliformis serum antibody in rabbits.

Lab Anim Sci 37:176–179

White DJ, Waldron MM (1969) Naturally-occurring Tyzzer's disease in the gerbil. Vet Rec 85:111–114

Zook BC, Huang K, Rhorer RG (1977) Tyzzer's disease in Syrian hamsters. J Am Vet Med Assoc 171:833–836

Corynebacterium kutscheri Infection, Liver, Mouse and Rat

Stephen W. Barthold

Synonyms. Pseudotuberculosis, corynebacteriosis, *Corynebacterium murium* infection

Gross Appearance

Infected animals often have no gross lesions, but clinically affected animals may be emaciated with starry coats, naso-ocular discharge, and dyspnea. Submandibular lymph nodes are enlarged. Lungs contain randomly distributed, single large to numerous miliary cream-colored foci surrounded by red zones. Single to several variably sized caseopurulent foci protrude from the hepatic capsule or lie deep within the parenchyma. Adhesions to pleura, pericardium, and peritoneum are frequent. Less frequently, similar foci can occur in kidney, heart, spleen, joints, skin, lymph nodes, prepucial glands, and middle ears. Lesions are often restricted to lung and liver in rats, but liver and kidney lesions predominate in mice (Brownstein et al. 1985; Fauve et al. 1964; Ford and Joiner 1968; Giddens et al. 1968; Kutscher 1894; Matheson et al. 1955; Nelson 1973; Tadokoro et al. 1961; Weisbroth 1979; Weisbroth and Scher 1968a).

Microscopic Features

Regardless of tissue, lesions consist of microscopically small to grossly large foci of inflammation. The centers of each focus are necrotic, surrounded by a zone of leukocytic infiltration consisting initially of neutrophils and later of macrophages, lymphocytes, plasma cells, and fibroplasia (Fig. 176). Colonies of numerous small, gram-positive rods, arranged in irregular palisades, are present in the junctional zones between the necrotic centers and the reactive periphery (Figs. 177, 178; Weisbroth 1979; Giddens et al. 1968).

Ultrastructure

The ultrastructural features of this infection have not been reported.

Differential Diagnosis

Differential diagnosis of pulmonary abscesses must distinguish primarily those caused by *Mycoplasma, Streptococcus*, and *Mycobacterium*. The bacterial etiology of disseminated abscesses in other organs, particularly liver, include *Pseudomonas aeruginosa, Salmonella*, and *Streptococcus*. Gram stains and microbiologic examinations of abscesses provide definite diagnosis of *C. kuscheri* infection. Subclinical infections can be detected serologically or by culture of submandibular lymph nodes (Brownstein et al. 1985; Fox et al. 1987).

Biologic Features

Natural History

C. kutscheri infection was one of the earliest recognized rodent infectious diseases (Kutscher 1894) and still occurs today. The term "pseudotuberculosis" is an accepted synonym, al-

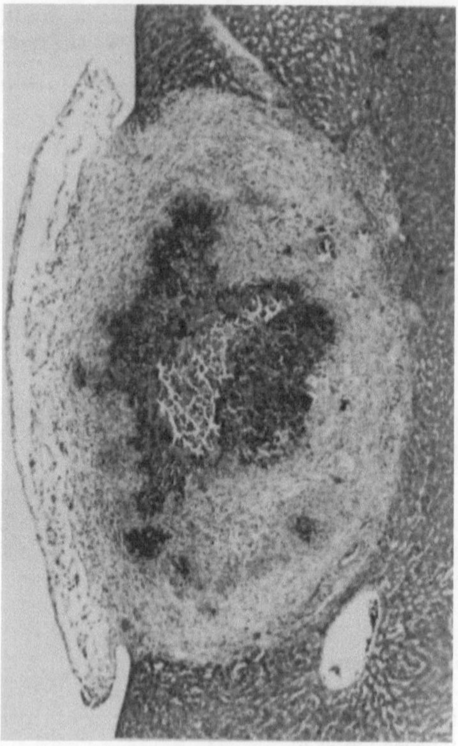

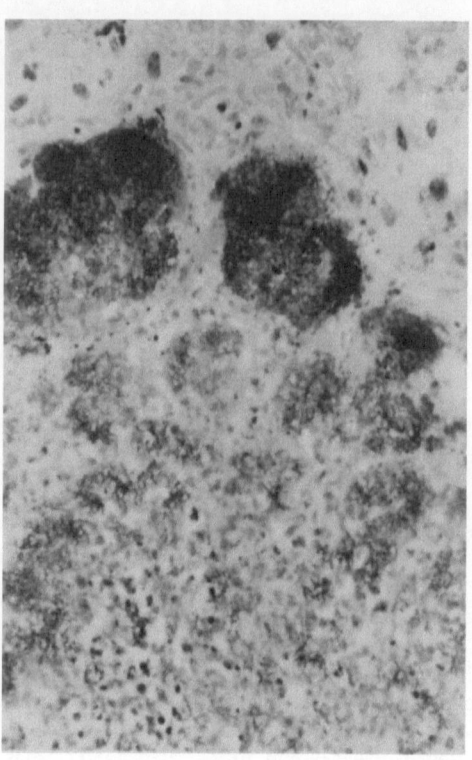

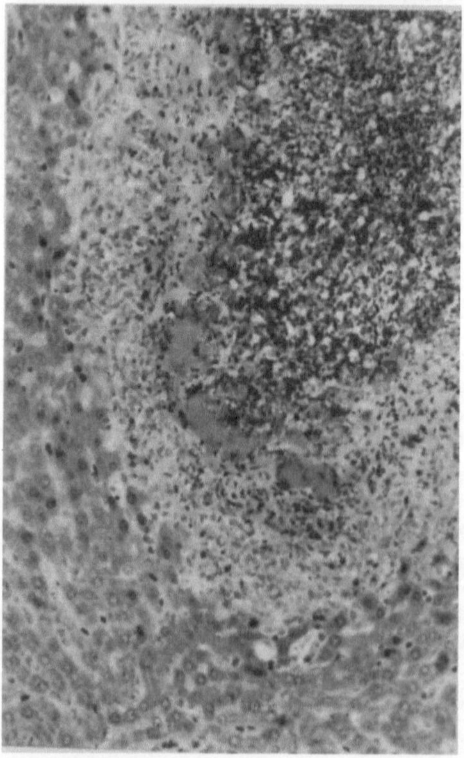

◄

Fig. 176. (*upperleft*) Liver, rat naturally infected with *Corynebacterium kutscheri*. A granulomatous lesion elevates the hepatic capsule. H&E, ×56

Fig. 177. (*below*) *Corynebacterium kutscheri* granuloma, liver, rat. The center of the lesion (*upper right*) contains necrotic debris, surrounded by a zone of mixed leukocyte infiltration. Colonies of *C. kutscheri* are visible at the junction between the necrotic center and outer reactive zone. H&E, ×175

Fig. 178. (*upper right*) Gran-positive *Corynebacterium kutscheri* organisms in the center and in dense colonies (*near top*) of a hepatic lesion in a rat. Brown and Brenn, ×440

though it may cause confusion with other similarly named bacterial infections, such as *C. pseudotuberculosis* and *Yersinia pseudotuberculosis*, which seem not to be natural laboratory rodent pathogens (Weisbroth 1979). *C. kutscheri* has been associated with spontaneous epizootics of disease without apparent provocation (Ford and Joiner 1968; Giddens et al. 1968; Kutscher 1894; Weisbroth 1979; Weisbroth and Scher 1968a; Wolff 1950), but it is more commonly associated with subclinical infections that are precipitated into clinical disease by some other predisposing factor, such as irradiation, corticosteroid treatment, nutritional deficiency, neoplasia, and experimental stress of coinfection with other pathogens (Antopol et al. 1953; Barthold and Brownstein 1988; Fauve and Pierce-Chase 1967; Fauve et al. 1964; Ford and Joiner 1968; Lawrence 1957; LeMaistre and Tompsett 1952; Pierce-Chase et al. 1964; Robinson et al. 1968; Seronde et al. 1956; Shechmeister and Adler 1953; Sokoloff 1965; Tadokoro et al. 1961; Wolff 1950). The means by which the agent is transmitted within infected rodent colonies has not been elucidated, but it does not appear to be highly contagious.

Pathogenesis

Pseudotuberculosis has been reproduced in rodents following subcutaneous, intraperitoneal, oral, intranasal, and intravenous inoculation of *C. kutscheri* (Giddens et al. 1968; Pierce-Chase et al. 1964). Until recently, the primary site of *C. kutscheri* colonization and the site of entry have been unknown. A carrier state exists in the absence of lesions or culturable *C. kutscheri* in internal organs, but pseudotuberculosis can be precipitated following immunosuppression. Once precipitated, *C. kuscheri* is readily isolated from caseopurulent lesions (Fauve and Pierce-Chase 1967; Fauve et al. 1964; Giddens et al. 1968; LeMaistre and Tompsett 1952; Pierce-Chase et al. 1964; Weisbroth and Scher 1968a,b). It was hypothesized that *C. kutscheri* existed in an altered, avirulent form during latent infection and reverted to virulence as a result of host modification (Fauve et al. 1964, 1966; Fauve and Pierce-Chase 1967; Pierce-Chase et al. 1964). This hypothesis has not been supported by subsequent investigations (Bruce et al. 1969; Hirst and Campbell 1977; Hirst and Olds 1978a,b; Hirst and Wallace 1976). Following oronasal inoculation, *C. kutscheri* colo-

nized the oral cavity, from which it was readily recovered for several weeks. The bacterium was also often cultured from submandibular lymph nodes in these carrier rats, but was infrequently isolated from nasal cavity, lung, liver, spleen, mesenteric lymph node, cecum, and blood. Corticosteroid treatment resulted in conversion from localized, oral colonization to disseminated disease (Brownstein et al. 1985). Evaluation of naturally infected rats has confirmed that the submandibular lymph nodes are preferential tissues for recovery of *C. kutscheri* in rats without clinical signs or gross lesions (Fox et al. 1987). The vascular distribution of lesions, particularly in the lung, in both natural and experimental infections suggests hematogenous dissemination of the bacterium (Giddens et al. 1968). It remains to be determined whether the oral cavity is the site of localized colonization in the mouse.

Etiology

Pseudotuberculosis in rodents is usually caused by *C. kutscheri*, a gram-positive rod. Older synonyms for this agent are *Bacillus pseudotuberculosis murium*, *Corynethrix pseudotuberculosis murium*, *Corynebacterium pseudotuberculosis murium*, *Bacillus muris*, and *Corynebacterium murium* (Weisbroth 1979). *C. pseudodiphtheriticum* and possibly other *Corynebacterium* species can also produce disease in rodents indistinguishable from that caused by *C. kutscheri* (Lee and Lang 1979).

Frequency

Liver lesions are common in rats and mice with pseudotuberculosis, especially mice. Historically, *C. kutscheri* has been very common in laboratory rodent stocks (Hirst and Campbell 1977; Hirst and Olds 1978a,b; Hirst and Wallace 1976; Pierce-Chase et al. 1964). Changes in husbandry practices and cesarean rederivation of most commercial stock has reduced the prevalence of infection substantially in the United States, but it still occurs sporadically, particularly in rats. Manifestation of infection within a population varies from subclinical infections to unprovoked epizootics of pseudotuberculosis. Usually, but not always, a predisposing factor is involved. An accurate assessment of the true prevalence of infection is difficult to obtain because of the frequency of "carrier" infections without lesions and the lack of application of available serologic tests. Agglutination, precipitin, immunofluorescence, and enzyme-linked immunosorbent assays (ELISA) have all been used for detection of *C. kutscheri* antibody (Ackerman et al. 1984; Brownstein et al. 1985; Weisbroth and Scher 1968b). Surveillance for infection by treatment with cortisone is possible, but is not a practical approach under most circumstances (Weisbroth 1979). Culture of submandibular lymph nodes is the most sensitive means of detecting active subclinical infections (Brownstein et al. 1985; Fox et al. 1987).

Comparison with Other Species

C. kutscheri usually produces subclinical carrier infections in mice and rats that may disseminate under appropriate conditions, but the agent has also been isolated from pulmonary lesions of a guinea pig (Vallee et al. 1969). Oral localization and dissemination of *C. kutscheri* resemble nosocomial *Corynebacterium* infections in humans, especially in immunocompromized or cancer patients. In these individuals, corynebacteria have been isolated from the nose, throat, skin, and rectum, as well as from lesions of internal organs (Young et al. 1981).

References

Ackerman JI, Fox JG, Murphy JC (1984) An enzyme linked immunosorbent assay for detection of antibodies to Corynebacterium kutscheri in experimentally infected rats. Lab Anim Sci 34:38–43

Antopol W, Quittner H, Saphra I (1953) "Spontaneous" infections after the administration of cortisone and ACTH. Am J Pathol 29:599–600

Barthold SW, Brownstein DG (1988) The effect of selected viruses on Corynebacterium kutscheri infection in rats. Lab Anim Sci 38:580–583

Brownstein DG, Barthold SW, Adams RL, Terwilliger GA, Aftosmis JG (1985) Experimental Corynebacterium kutscheri infection in rats: bacteriology and serology. Lab Anim Sci 35:135–138

Bruce DL, Bismanis JE, Vickerstaff JM (1969) Comparative examinations of virulent Corynebacterium kutscheri and its presumed avirulent variant. Can J Microbial 15:817–818

Fauve RM, Pierce-Chase CH (1967) Comparative effects of corticosteroids on host resistance to infection in relation to chemical structure. J Exp Med 125:807–821

Fauve RM, Pierce-Chase CH, Dubos R (1964) Corynebacterial pseudotuberculosis in mice. II. Activation of natural and experimental latent infections. J Exp Med 120:283–304

Fauve RM, Bouanchaud D, Delaunay A (1966) Resistance cellulaire à l'infection bacterienne. Ann Inst Pasteur (Paris) 110 [Suppl]:106–117

Ford TM, Joiner GN (1968) Pneumonia in a rat associated with Corynebacterium pseudotuberculosis. A case report and literature survey. Lab Anim Care 18:220–223

Fox JG, Niemi SM, Ackerman J, Murphy JC (1987) Comparison of methods to diagnose an epizootic of Corynebacterium kutscheri pneumonia in rats. Lab Anim Sci 37:72–75

Giddens WE Jr, Keahey KK, Carter GR, Whitehair CK (1968) Pneumonia in rats due to infection with Corynebacterium kutscheri. Pathol Vet 5:227–237

Hirst RG, Campbell R (1977) Mechanisms of resistance to Corynebacterium kutscheri in mice. Infect Immun 17:319–324

Hirst RG, Olds RJ (1978a) Corynebacterium kutscheri and its alleged avirulent variant in mice. J Hyg (Lond) 80:349–356

Hirst RG, Olds RJ (1978b) Serological and biochemical relationships between the alleged avirulent variant of Corynebacterium kutscheri and streptococci of group N. J Hyg (Lond) 80:356–363

Hirst RG, Wallace ME (1976) Inherited resistance to Corynebacterium kutscheri in mice. Infect Immun 14:475–482

Kutscher D (1894) Ein Beitrag zur Kenntnis der bacillaren Pseudotuberculose der Nagetiere. Z Hyg Infektkr 18:327–342

Lawrence JJ (1957) Infection of laboratory mice with Corynebacterium murium. Aust J Sci 20:147

Lee KJ, Lang CM (1979) The pathogenicity of Corynebacterium pseudodiptheriticum in rats. Am Assoc Lab Anim Sci Abstract 105

Le Maistre C, Tompsett R (1952) The emergence of pseudotuberculosis in rats given cortisone. J Exp Med 95:393–408

Matheson BH, Grice HC, Connell MRE (1955) Studies of middle ear disease in rats. I. Age of infection and infecting organisms. Can J Comp Med Vet Sci 19:91–97

Nelson JB (1973) Response of mice to Corynebacterium kutscheri on footpad injection. Lab Anim Sci 23:370–372

Pierce-Chase CH, Fauve RM, Dubos R (1964) Corynebacterial pseudotuberculosis in mice. I. Comparative susceptibility of mouse strains to experimental infection with Corynebacterium kutscheri. J Exp Med 120:267–281

Robinson HJ, Phares HF, Graessle OE (1968) Effects of indomethacin on acute, subacute, and latent infections in mice and rats. J Bacterial 96:6–13

Seronde J Jr, Zucker TF, Zucker LM (1956) Thaimine, pyridoxine and pantothenic acid in the natural resistance of the rat to a Corynebacterium infection. J Nutr 59:287–298

Shechmeister IL, Adler FL (1953) Activation of pseudotuberculosis in mice exposed to sublethal total body irradiation. J Infect Dis 92:228–239

Sokoloff L (1965) Musculoskeletal lesions in experimental animals. In: Ribelin WF, McCoy JR (eds) Pathology of laboratory animals. Thomas, Springfield, chap 1

Tadokoro S, Kurihara Y, Kurihara N, Ogawa H, Shibata K (1961) Emergence of infection in rats after administration of corticosteroids. 1. Symptoms, autopsy findings and bacteriological observations. Gunma J Med Sci 10:245–269

Vallee A, Guillon JC, Cayeux P (1969) Isolement d'une souche de Corynebacterium kutscheri chez un cobaye. Bull Acad Vet France 42:797–800

Weisbroth SH (1979) Bacterial and mycotic diseases. In: Baker HJ, Lindsey JR, Weisbroth SH (eds) The laboratory rat, vol 1, biology and diseases. Academic, New York, chap 9

Weisbroth SH, Scher S (1968a) Corynebacterium kutscheri infection in the mouse. 1. Report of an outbreak, bacteriology, and pathology of spontaneous infections. Lab Anim Care 18:451–458

Weisbroth SH, Scher S (1968b) Corynebacterium kutscheri infection in the mouse. II. Diagnostic serology. Lab Anim Care 18:459–468

Wolff HL (1950) On some spontaneous infections observed in mice; Corynebacterium kutscheri and Corynebacterium pseudotuberculosis. Antonie van Leeuwenhoek 16:105–110

Young VM, Meyers WF, Moody MR, Schimpff SC (1981) The emergence of coryneform bacteria as a cause of nosocomial infections in compromised hosts. Am J Med 70: 646–650

Idiopathic Focal Hepatic Necrosis in Inbred Mice

John P. Sundberg, James G. Fox, Jerrold M. Ward, and Hendrick G. Bedigian

Synonyms. Noninfectious focal hepatic necrosis, acute hepatic coagulative necrosis

Gross Appearance

Focal hepatic necrosis is identified as an incidental finding at the time of necropsy. Livers have solitary or multiple small, irregular, light-brown foci that may coalesce (Fig. 179). These foci may be areas in which only the color has changed or may form small depressions in the surface of the liver, depending upon the age of the lesion.

Microscopic Features

With the light microscope, the liver lesions are seen as single or multiple foci (often only one to three per liver) consisting of clusters of hepatocytes undergoing coagulative necrosis (Fig. 180). The foci are sharply demarcated. Scattered individual necrotic hepatic cells may be seen in livers that also contain prominent foci of necrosis. Affected hepatocytes are usually swollen, have rounded edges, bright eosinophilic cytoplasm, and pyknotic or karyorrhectic nuclei or no nuclear remnants (Fig. 181). Cytoplasmic borders are usually evident. Based upon evaluation of multiple cases or lesions with various degrees of inflammation within an individual (Fig. 181), there appears to be a progression of changes that lead to healing and resolution. Inflammation is absent in peracute lesions (Fig. 179). Early migration of neutrophils occurs into the necrotic focus (Fig. 180). Subacute lesions have mild infiltrations of neutrophils and some lymphocytes that circumscribe the necrotic focus (Fig. 181). In chronic lesions, various degrees of peripheral fibrosis may be seen (Fig. 181). Thrombi can occasionally be identified in adjacent vessels, but extensive serial sectioning may be necessary to identify them (Fig. 179). No pathogenic organisms have been identified in tissue sections in any of the cases diagnosed at the Jackson Laboratory using a variety of special stains, including Warthin-Starry, Steiner, Brown and Brenn, Gomori methenamine silver, and Giemsa stains (Fig. 179).

Ultrastructure

An abrupt change from viable hepatocytes to cells undergoing coagulative necrosis is evident in both transmission electron and light microscopy studies. No recognizable organism has been identified in the junctional regions.

Differential Diagnosis

Focal hepatic necrosis in the mouse is a nonspecific lesion associated with mouse adenovirus (Percy and Barthold 1993), mouse hepatitis virus (Frith and Ward 1988; Percy and Barthold 1993), *Clostridium* (*Bacillus*) *piliforme* (Tyzzer's disease;

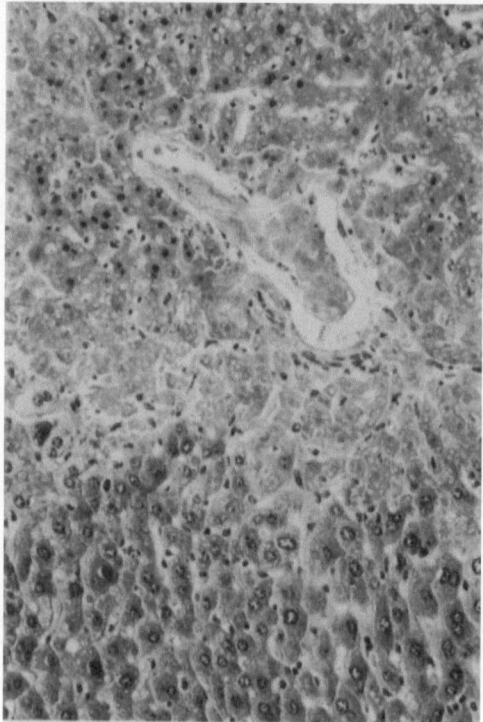

Fig. 179. Acute, multifocal hepatic coagulative necrosis in a 10-month-old female C57BL/6J mouse consists of multiple foci of pale eosinophilic necrotic "ghost" cells around a thrombosed central vein. H&E, ×100

Frith and Ward 1988; Percy and Barthold 1993; Franklin et al. 1994), *Proteus mirabilis* infection in severe combined immunodeficiency (*scid*) mice (Percy and Barthold 1993), *Streptobacillus moniliformis* (Percy and Barthold 1993), two new *Helicobacter* species (Fox et al. 1994, 1995; Ward et al. 1994a,b), chemical agents (Clapp 1973; Ward 1985), and ischemia (Frith and Ward 1988). Hepatic necrosis may be associated with extensive metastasis or multicentric involvement of a variety of spontaneous neoplastic diseases that spread to the liver. *Salmonella* species may also cause thrombosis and focal hepatic necrosis, although the hepatic lesions are often granulomatous in nature (Percy and Barthold 1993).

▶

Fig. 180a,b. Two foci from a single 7-month-old female BALB/cJ mouse with various degrees of hepatocyte necrosis and inflammation. Necrotic foci may be at various stages of resolution within an individual. H&E, ×200

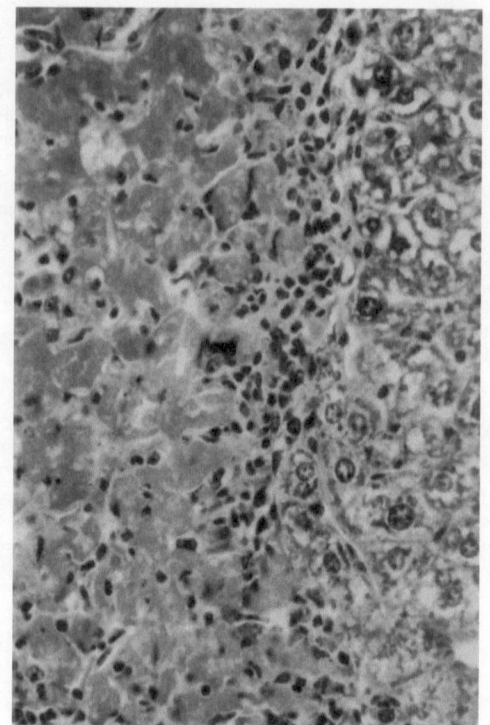

a

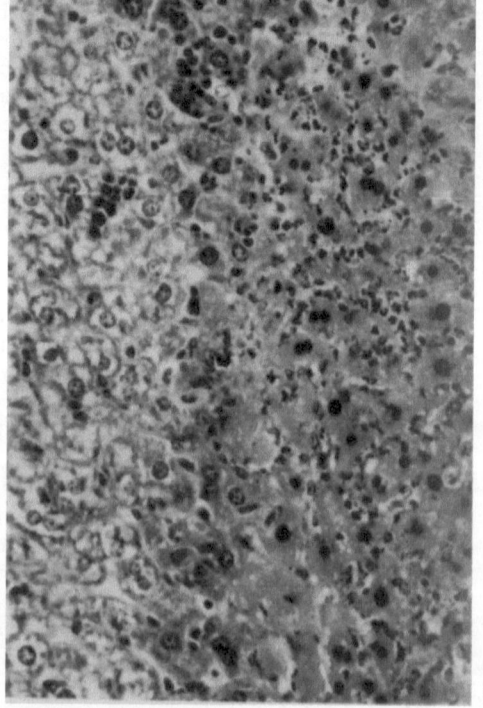

b

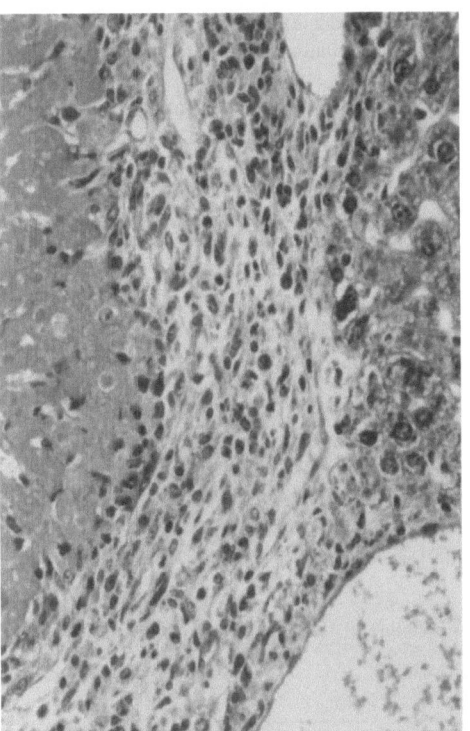

Fig. 181. As lesions heal, the inflammatory cells surrounding the necrotic foci become primarily lymphocytes, plasma cells, and some macrophages within the fibrotic area. This results in a depression on the surface of the liver. This lesion occurred in an 8-month-old male DW.C3H dw+/J mouse. H&E, ×200

Hepatitis caused by *Helicobacter hepaticus* consists initially of small foci of necrosis with nonsuppurative inflammation (Fig. 182). Lesions may be single or multifocal throughout the liver. Most are small, but as the disease progresses, large areas of necrosis are occasionally found. These areas are visible at the gross level as white spots or streaks. The lesions progress to involve the entire liver, and chronic lesions include oval (ductular) hyperplasia (Fig. 182), hepatocytomegaly, cholangitis, and occasional focal necrosis. Bacteria are seen between hepatocytes with silver stains. On ultrastructural examination, these organisms

▶

Fig. 182. *Helicobacter hepaticus* infection in an A/JCr mouse. **a** Acute, focal, necrotizing hepatitis. H&E, ×160. **b** Chronic hepatitis with oval cell hyperplasia. H&E, ×250. **c** Note pleomorphic hepatocyte nuclei, liver demonstrating *H. hepaticus* (thin dark bacteria) between hepatocytes. Steiner's stain, ×360. Bacteria are often found away from areas of inflammation or in the parenchyma adjacent to necrotic foci

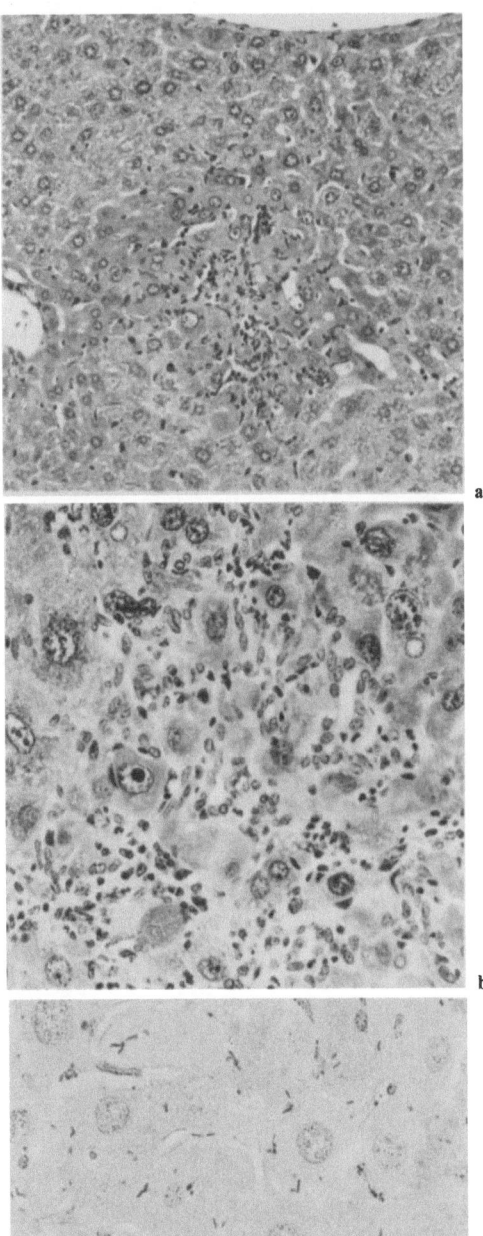

are found within bile canaliculi (Ward et al. 1994a,b). The disease is transmissible and in some strains, such as A/JCr, liver tumors are associated with the disease process (Ward et al. 1994b).

Biologic Features

The pathogenesis of focal hepatic necrosis in inbred laboratory mice appears to involve ischemia secondary to vascular occlusion, usually a thrombus in a vessel supplying blood to the affected area (Fig. 179). Several strains of mice have a propensity for developing auricular mural thrombi within their hearts (Clapp 1973; Frith and Ward 1988; Meier and Hoag 1966), which may be the source of emboli that ultimately lodge in the liver.

No pathogenic aerobic bacteria have been isolated, nor have antibodies to mouse pathogens been detected in routine serologic surveys of colonies or evaluation of individuals with hepatic lesions.

Focal hepatic necrosis is a rarely described incidental lesion observed in a variety of inbred strains of mice not exposed to conventional pathogens or chemical irritants (Table 16; Sundberg and Bedigian 1994). Since the mice do not exhibit clinical signs of disease, it is difficult to establish frequency rates in large colonies without systematic strain characterization studies.

Comparison with Other Species

Many of the infectious agents listed above that cause focal hepatic necrosis in mice also infect other mammalian species. Those pathogenic organisms that are species specific have other members of the genus that induce similar lesions in the

Table 16. Summary of cases of focal hepatic necrosis identified as an incidental finding at the time of necropsy during an 18-month period (1993–1994) in colonies at the Jackson Laboratory

Strain	Females				Males			
	Mice with liver necrosis (n)	Mice examined (n)[a]	Age range (days)	Mean age (days)	Mice with liver necrosis (n)	Mice examined (n)[a]	Age range (days)	Mean age (days)
A/J	1	9	90	90	1	9	601	601
AKR/J	1	32	88	88	0	7	–	–
B10.AKM/SnJ	1	1	222	222	0	0	–	–
B10.R111(71NS)/SnJ	0	1	–	–	1	1	124	124
BALB/cByJ	1	53	127	127	0	8	–	–
BALB/cByJ nu/nu	0	2	–	–	2	11	270–329	300
BXA-7	1	1	86	86	0	0	–	–
BXSB/MpJ	0	2	–	–	3	54	123–218	179
C3H/HeJ	3	50	161–205	179	1	31	399	399
C57BL/10SnJ	1	2	101	101	0	0	–	–
C57BL/6J	2	107	196–322	259	0	69	–	–
C58/J	0	7	–	–	1	8	54	54
CD-1	1	4	196	196	0	0	–	–
CX5129	1	1	120	120	0	0	–	–
CXJ-9/SalkMob	1	1	332	332	0	0	–	–
DBA/1J	2	8	261–271	266	0	0	–	–
FVB/NJ	2	5	203–232	218	0	1	–	–
FVB/nMob	1	4	258	258	0	2	–	–
LP/J	1	1	293	293	0	0	–	–
MRL/MpJ	0	1	–	–	1	1	146	146
NOD/LtJ	1	11	172	172	0	0	–	–
NU/J nu/nu	6	27	161–312	251	2	13	300–338	319
NU/J nu/+	45	47	29–345	193	2	2	313	313
NZB/B1NJ	2	3	179–193	186	0	0	–	–
SM/J	0	2	–	–	1	1	208	208
Total	75	382	29–345	198	17	218	54–601	250

[a] Number of mice of that strain examined for unrelated reasons.

appropriate host. Infarction due to thrombosis with fibrin emboli is well known in all species.

Acknowledgement. This work was supported by grants from the National Cancer Institute (CA 34196, JPS, HGB; RFP 594–69 and RO1 CA67529, JGF), National Institutes of Health (RR01046 and RR07036, JGF), and funds from The Jackson Laboratory (JPS, HGB). The authors thank J. Miller for technical assistance with necropsies and gross photography, M.E. Hogan for electron microscopy, and P. Jewett for the histologic work.

References

Clapp NK (1973) An atlas of RF mouse pathology: disease descriptions and incidences. United States Atomic Energy Commission, Springfield, Virginia, p 9

Fox JG, Dewhirst FE, Tully JG, Paster BJ, Yan L, Taylor NS, Collins MJ Jr, Gorelick PL, Ward JM (1994) Helicobacter hepaticus sp. nov., a microaerophilic bacterium isolated from livers and intestinal mucosal scrapings from mice. J Clin Microbiol 32:1238–1245

Fox JG, Yan LL, Dewhirst FE, Paster BJ, Shames B, Murphy JC, Hayward A, Belcher JC, Mendes EN (1995) Helicobacter bilis sp. nov., a novel Helicobacter species isolated from bile, livers, and intestines of aged, inbred mice. J Clin Microbiol 33:445–454

Franklin CL, Motzel SL, Besch-Williford CL, Hook RR Jr, Riley LK (1994) Tyzzer's infection: host specificity of Clostridium piliforme isolates. Lab Anim Sci 44:568–572

Frith CH, Ward JM (1988) Color atlas of neoplastic and non-neoplastic lesions in aging mice. Elsevier, Amsterdam, pp 3–11

Meier H, Hoag WG (1966) Blood coagulation. In: Green EL (ed) Biology of the laboratory mouse, 2nd edn. McGraw-Hill, New York, pp 373–376

Percy DH, Barthold SW (eds) (1993) Pathology of laboratory rodents and rabbits. Iowa State University Press, Ames, pp 3–69

Sundberg JP, Bedigian H (1994) Focal hepatic necrosis in inbred laboratory mice. Jax Notes 456:2–4

Ward JM (1985) Cirrhosis, mouse. In: Jones TC, Mohr U, Hunt RD (eds) Monographs on pathology of laboratory animals, digestive system. Springer, Berlin Heidelberg New York, pp 107–110

Ward JM, Anver MR, Haines DC, Benveniste RE (1994a) Chronic active hepatitis in mice caused by Helicobacter hepaticus. Am J Pathol 145:959–968

Ward JM, Fox JM, Anver MR, Haines DC, George CV, Collins MJ Jr, Gorelick PL, Nagashima K, Gonda MA, Gilden RV, Tully JG, Russell RJ, Benveniste RE, Paster BJ, Dewhirst FE, Donovan JC, Anderson LM, Rice JM (1994b) Chronic active hepatitis and associated liver tumors in mice caused by a persistent bacterial infection with a novel Helicobacter species. J Natl Cancer Inst 86:1222–1227

Multifocal Inflammation, Liver, Rat

A.J. Spencer, Raymond Everett, and James A. Popp

Synonyms. Chronic inflammation with or without associated single cell hepatocellular necrosis, lymphohistiocytic cell infiltration, hepatic microgranuloma

Gross Appearance

Multifocal inflammation in the liver is not discernable macroscopically.

Microscopic Features

Multiple foci of inflammation may be found in the liver of control laboratory rats, in the absence of demonstrable infection with viruses or bacteria (Greaves 1990; Greaves and Faccini 1992). In these lesions, hepatocytes are typically replaced by an infiltration of macrophages and, possibly, a few lymphocytes, but with no abnormalities in surrounding hepatocytes (Fig. 183). The predominance of macrophages in the infiltrate can be confirmed by using immunohistochemical stains for antigens such as ED1 (A.J. Spencer, unpublished observation). Compression of surrounding hepatocytes is not a feature, indicating previous loss of hepatocytes. Less commonly, lymphocytes may predominate in the cell infiltrate or there is a more acute lesion where necrotic hepatocytes and occasional polymorphonuclear leukocytes are present (Fig. 184). Giant cells, mineralization, and

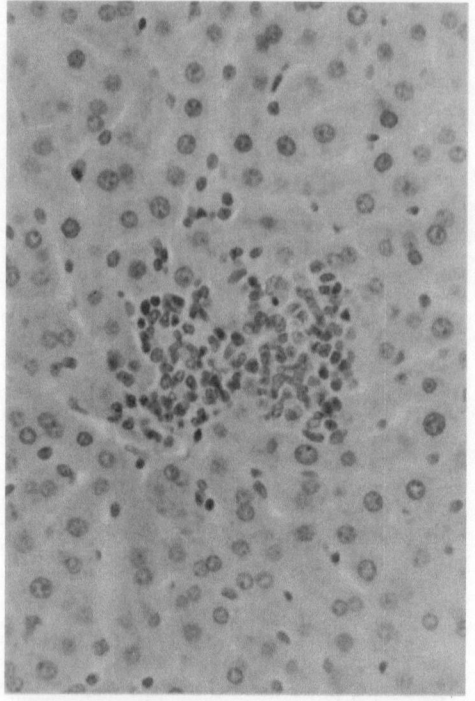

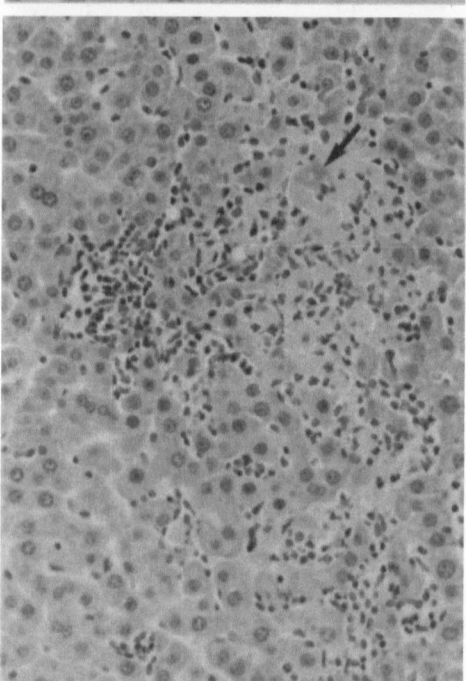

fibrosis are not features. These foci may occur singly or in small clusters. They can be present in any liver lobe and in any part of the lobule. However, no studies have been reported describing their relative frequency in each lobe or the different zones of the lobule. Most lesions are smaller than 100 μm in diameter, but lesions up to 300 μm in diameter may be present in our experience.

Ultrastructure

Electron microscopy examination of foci confirms the histologic appearance (Fig. 185). The majority of inflammatory cells present are macrophages; some may contain cell debris in lysosomes, while others have the features of monocytes. Most surrounding hepatocytes are normal, but occasional degenerate or necrotic hepatocytes may be seen. We have not identified microorganisms in these lesions.

Differential Diagnosis

Multifocal inflammation must be distinguished from extramedullary hematopoiesis and invasion of the liver by neoplasms (myeloid leukemia, lymphoma, histiocytic sarcoma; Van Zwieten and Hollander 1985; Gal et al. 1990; Squire 1985). Foci of extramedullary hematopoiesis are distinguished by the lack of hepatocellular necrosis and the presence of megakaryocytes and immature hematopoietic cells, both erythroid and myeloid, in varying stages of development. Neoplasms occur predominantly in older rats, have atypical cytologic features, usually are associated with a much greater extent of infiltration and destruction of hepatocytes, and may be present in other tissues.

Inflammatory foci in the liver are also seen as part of certain gastrointestinal infections, e.g., Tyzzer's disease (*Clostridium piliforme*), salmonellosis, and systemic infections, e.g., *Corynebacterium kutcheri* (Ganaway 1985a,b; Barthold 1985; see p. 209, this volume). In immunosuppressed non-

◄

Fig. 183. (*above*) Liver, rat, multifocal inflammation. Focus of mononuclear cells surrounded by normal hepatocytes. H&E, ×430

Fig. 184. (*below*) Liver, rat, multifocal inflammation. Necrotic hepatocytes (*arrow*) and associated inflammatory cells. H&E, ×215

barrier-reared animals, the presence of rat parvovirus can also cause necrosis and inflammation (Jacoby 1985).

Inflammatory foci may follow hepatocellular necrosis induced by xenobiotics. Typically, such lesions have a zonal pattern and can thus be distinguished from preexisting lesions. However, some xenobiotics such as galactosamine and β-napthylisocyanate induce multifocal necrosis and inflammation in all zones of the lobule (Kuhlmann and Wurster 1980; Leonard et al. 1981). Differential diagnosis in such cases is difficult and rests on identifying additional effects, e.g., giant cells in lesions induced by β-napthylisocyanate, proliferation of bile ducts with galactosamine.

Biologic Features

The cause of multifocal hepatic inflammation in barrier-maintained rats, where known pathogenic microorganisms are absent, is unknown. Lesions are thus often referred to as "spontaneous," "idiopathic," or "background" lesions. Suggestions have been made that endotoxins or bacteria from the intestine are the cause, but no evidence has been published to support this. No studies have been published detailing any supplier, strain, age, or sex predilection in the incidence and severity of multifocal inflammation. However, the authors have seen these lesions in Sprague-Dawley, Fischer 344, and Long-Evans rats. Foci have been seen in rats as young as 6 weeks old and appear to increase in number and size with age and be more prevalent in females than in males. In toxicity studies, administration of some (unspecified) compounds has been reported to exacerbate these lesions in the absence of obvious systemic toxicity (Greaves and Faccini 1992). As described above, some xenobiotics produce similar necrosis and inflammation per se (Kuhlmann and Wurster 1980; Leonard et al. 1981). On the other hand, administration of immunosuppressive antineoplastic agents such as doxorubicin, 5-fluorouracil, and etoposide have been found to reduce the incidence of lesions (Spencer, unpublished observation).

Comparison with Other Species

Scattered foci of inflammation can been seen in the liver of other rodents and nonrodents kept

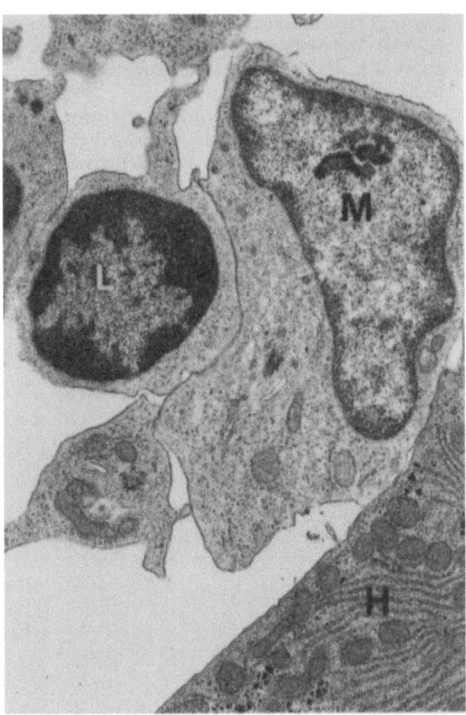

Fig. 185. Liver, rat, multifocal inflammation. Part of an inflammatory focus with a lymphocyte (*L*), macrophage (*M*), and cytoplasm of a neighboring hepatocyte (*H*). TEM, ×8850

under barrier conditions. As in the rat, the initiating cause is usually unknown (Greaves and Faccini 1992). In the mouse, an inflammatory disease called chronic/active hepatitis has recently been described from one laboratory associated with the presence of a newly described organism *Helicobacter hepaticus* (Ward et al. 1994). This organism has not, at the time of this writing, been described in rats.

References

Barthold SW (1985) Corynebacterium kutcheri infection, liver, mouse and rat. In: Jones TC, Mohr U, Hunt RD (eds) Monographs on pathology of laboratory animals, digestive system. Springer, Berlin Heidelberg New York

Gal F, Sugar J, Csuka O (1990) Granulocytic leukemia, rat. In: Jones TC, Ward JM, Mohr U, Hunt RD (eds) Monographs on pathology of laboratory animals, hematopoietic system. Springer, Berlin Heidelberg New York

Ganaway JR (1985a) Tyzzer's disease, liver, mouse, rat, hamster. In: Jones TC, Mohr U, Hunt RD (eds) Monographs on pathology of laboratory animals, digestive system. Springer, Berlin Heidelberg New York

Ganaway JR (1985b) Salmonellosis, liver, mouse, rat, hamster. In: Jones TC, Mohr U, Hunt RD (eds) Monographs on pathology of laboratory animals, digestive system. Springer, Berlin Heidelberg New York

Greaves P (1990) Digestive system. In: Greaves P (ed) Histopathology of preclinical toxicity studies: interpretation and relevance in drug safety evaluation. Elsevier, Amsterdam, chap 8

Greaves P, Faccini (1992) Digestive system. In: Greaves P, Faccini J (eds) Rat histopathology: a glossary for use in toxicity and carcinogenicity studies. Elsevier, Amsterdam, chap 6

Jacoby RO (1985) Rat parvovirus infection, liver. In: Jones TC, Mohr U, Hunt RD (eds) Monographs on pathology of laboratory animals, digestive system. Springer, Berlin Heidelberg New York

Kuhlmann WD, Wurster K (1980) Correlation of histology and alpha1-fetoprotein resurgence in rat liver regeneration after experimental injury by galactosamine. Virchows Arch (Pathol Anat) 387:47–57

Leonard TB, Popp JA, Graichen ME, Dent JG (1981) Beta-Naphthylisothiocyanate-induced alterations in hepatic drug metabolism and liver morphology. Toxicol Appl Pharmacol 60:527–534

Squire RA (1985) Histiocytic sarcoma, rat. In: Jones TC, Ward JM, Mohr U, Hunt RD (eds) Monographs on pathology of laboratory animals, hematopoietic system. Springer, Berlin Heidelberg New York

Van Zwieten MJ, Hollander CF (1985) Extramedullary hematopoiesis, liver, rat. In: Jones TC, Mohr U, Hunt RD (eds) Monographs on pathology of laboratory animals, digestive system. Springer, Berlin Heidelberg New York

Ward JM, Anver MR, Haines DC, Benveniste RE (1994) Chronic active hepatitis in mice caused by Helicobacter hepaticus. Am J Pathol 145:959–968

The Salivary Glands

Histology and Ultrastructure, Salivary Glands, Mouse

Charles H. Frith and James W. Townsend

Synonyms. Parotid, sublingual, submaxillary (submandibular) salivary glands

Gross Appearance

The mouse, as well as many other species, has three major paired salivary glands – submaxillary (submandibular), parotid, and sublingual – and minor salivary glands (Smith 1966). The minor salivary glands are located in the oral submucosa and tongue, but are not visible grossly. All three major glands are closely associated and located in the subcutaneous tissue of the ventral neck area (Fig. 186). The submaxillary (submandibular) are the largest and easiest of the salivary glands to identify grossly. They are lobulated and extend posteriorly to the sternum and clavicle, anteriorly to the hyoid bone, and medially to overlap slightly on the median line. Submaxillary salivary glands of male mice are larger and more opaque than those of females (Fig. 186). A single excretory duct from the anterior dorsal surface of each gland opens on the floor of the oral cavity posterior to the incisor teeth (Hummel et al. 1966).

The much smaller sublingual glands are closely associated with the anterolateral surface of the submaxillary salivary glands. Each is a single lobe, and each gland has a single excretory duct that opens in the oral cavity close to the opening of the submandibular duct (Hummel et al. 1966).

The parotid salivary glands are lobulated and are the most diffuse of the three salivary glands. They extend from the base of the ears, where they overlie the morphologically similar exorbital lacrimal gland, posteriorly to the clavicle. A single duct from each gland opens in the oral cavity opposite the molar teeth of the lower jaw (Jacoby and Leeson 1959; Hummel et al. 1966).

Microscopic Features

The minor salivary glands, located in the oral submucosa or the tongue, may be either mucous or serous (Fig. 187). All three pairs of major salivary glands are classified as compound tubuloalveolar glands. They are separated into lobules by connective tissue septa, and the alveoli are lined by secretory pyramidal cells. The alveoli are continuous with terminal intercalated ducts, which form larger intralobular ducts. Interlobular and main excretory ducts are lined with columnar or stratified columnar epithelium except at the orifices, where they are lined by stratified squamous epithelium.

The parotid salivary glands of the mouse are classified as serous glands (Gude et al. 1982), and the alveoli are small and composed of three or four secretory pyramidal cells. Nuclei are basally located and the cytoplsm is basophilic with hematoxylin and eosin (H&E) staining (Fig. 188). The parotid salivary gland is similar morphologically to and must be distinguished from the exorbital lacrimal gland (Fig. 189). Pyramidal secretory cells of this lacrimal gland are also basophilic, but larger than those of the parotid salivary gland. The exorbital lacrimal gland secretes a serous fluid, which moistens the cornea of the eye. The small sublingual glands are divided into lobules by connective tissue septa projecting from the capsule. These lobules are composed of acini containing tall pyramidal cells with basally located nuclei and pale- to blue-staining cytoplasm with H&E (Fig. 190). The sublingual salivary glands are mucous glands (Gude et al. 1982) and stain intensely with periodic acid-Schiff (PAS) stain (Fig. 191).

Submaxillary (submandibular) salivary glands are classified as mixed glands (Gude et al. 1982), and the tall pyramidal cells produce both mucous and serous fluid. These glands are sexually dimorphic in adult male and female mice. The pyramidal cells of male adult mice are larger than in females and stain intensely eosinophilic with H&E, a result of the large number of cytoplasmic zymogen granules (Fig. 192). In adult male mice, the alveoli and the intercalated ducts have tall columnar cells with basally located nuclei. In the female, the cells are much shorter and the nuclei are centrally located. Pyramidal cells of the male mouse contain fewer granules than in the female and consequently are much less intensely eosinophilic with H&E stain (Fig. 193).

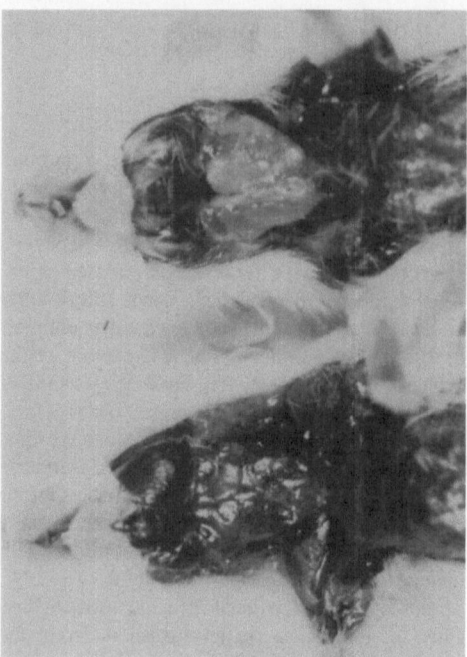

Fig. 186. Salivary glands of a male (*top*) and female mouse (*bottom*). The submaxillary salivary gland of the male mouse is larger and paler than that of the female

Ultrastructure

The acini of the parotid gland consist of pyramidally shaped cells with nuclei located toward the basilar portion of each cell (Fig. 194).
These acini may reveal a remarkable variation in appearance from one cell or acinus to the next. The nucleus of the acinar cell has no distinguishing characteristics: chromatin is more or less evenly distributed, and a reticulated nucleolus is occasionally encountered. Acinar cell endoplasmic reticulum is well developed, and the Golgi network, usually situated at the apices of the acinar cells, is prominent. The complex consists of a series of flattened and dilated lamellae whose interiors are clear. Associated with them are small arrays of poorly defined vesicles. The secretory granules have a consistent appearance. The membrane-bound droplets have a dense periphery and a small dense central zone. Mitochondria are randomly distributed and are not particularly numerous in parotid acinar cells. Myoepithelial cells have been described in the mouse and rat parotid

glands (Leeson and Jacoby 1959; Scott and Pease 1959; Leeson 1967).
The acini of the sublingual gland are composed primarily of pyramidally shaped cells whose major component consists of lobules of a slightly electron-dense substance with a delicate open network of fine dense strands (Fig. 195). These presumably represent formed mucin. The nucleus, mitochondria, endoplasmic reticulum, and Golgi membranes are crowded into the base of the cell. The density of this portion is often so extreme that the formed elements are difficult to visualize. The cell resembles closely the mature goblet cells of the intestinal mucosa, except for the pyramidal shape and the extreme cytoplasmic denstiy. The lobules of mucin are delineated by compressed stacks of membranes, cytoplasmic granules, and occasional mitochondria. Figure 196 is an electron micrograph of a myoepithelial cell in the sublingual salivary gland.
Acinar cells of the submaxillary gland have several features that differ from those of parotid cells. The nuclei of the serous portion of the gland of male mice have a shrunken appearance with a loose, wrinkled membrane and lie toward the basal aspect of the cells. They are surrounded by numerous dense secretory granules (Fig. 197).
Endoplasmic reticulum and mitochondria are not numerous. Scott and Pease (1959) stated that membrane-bound secretory granules are first seen in the basilar portions of the cells and that, as they move toward the apex, the membranes are lost. In the mucous portion of the cell, endoplasmic reticulum and mitochondria are much more prominent and the mucous granules of secretion are interspersed with groups of Golgi membranes and vesicles (Fig. 198). Scott and Pease (1959) also reported that the Golgi complex is intimately associated with secretion, but did not exclude the possibility that the endoplasmic reticulum is also involved. In the female mouse, secretory granules are much smaller and less numerous (Fig. 199) and the mucous portion more prominent.

Biologic Features

The salivary glands secrete saliva, which has both a chemical and mechanical function. The mechanical function is provided primarily by the mucous secretion, which aids in swallowing and keeps the mouth moist. The chemical function, provided by the serous secretion, adjusts pH, di-

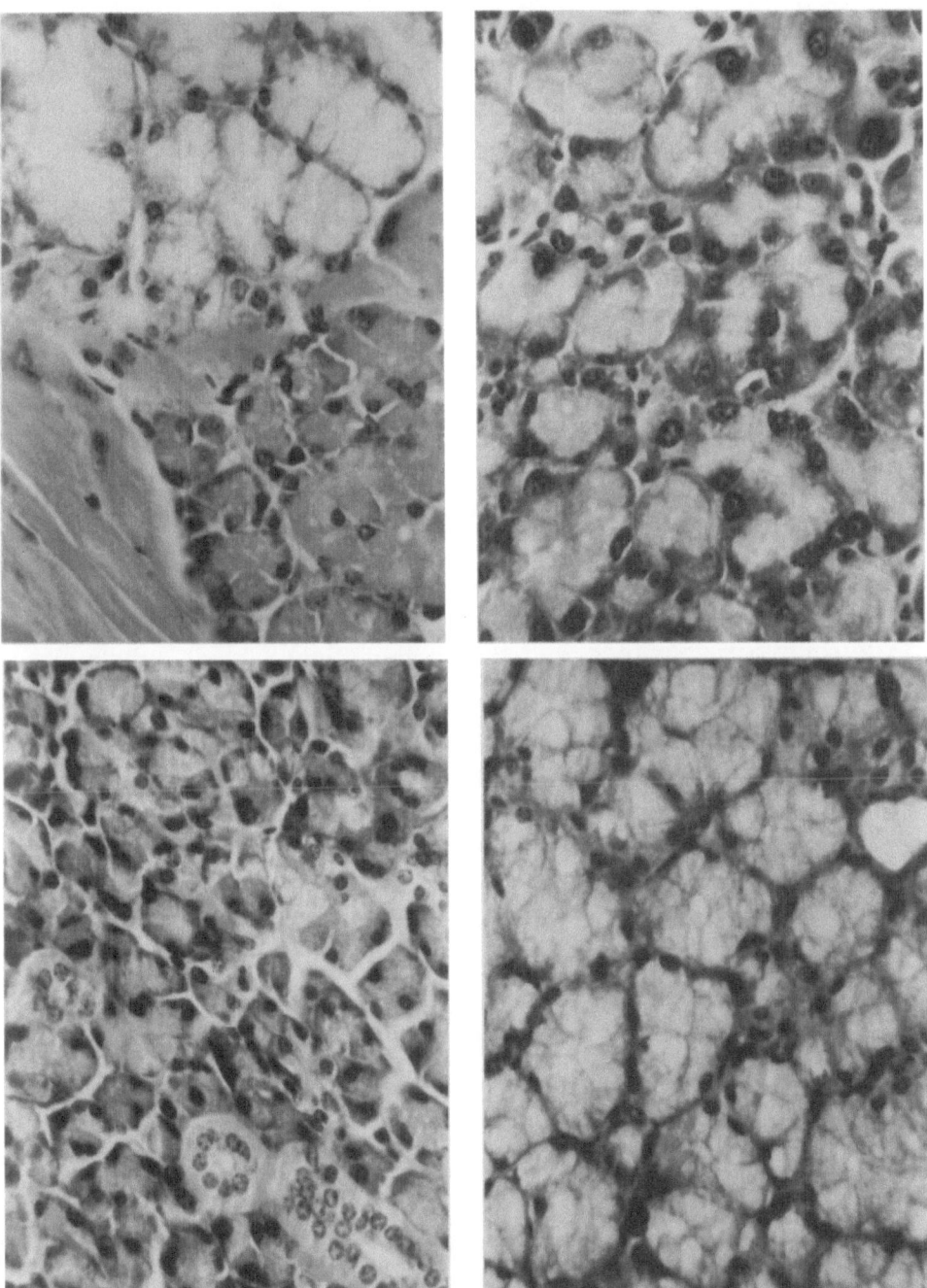

Fig. 187. (*upper left*) Minor salivary glands in the tongue mouse. Both mucous (*top*) and serous glands (*bottom*) are present. H&E, ×300

Fig. 188. (*lower left*) Parotid salivary gland. H&E, ×300

Fig. 189. (*upper right*) Exorbital lacrimal gland. The acini of the lacrimal gland are larger than those of the parotid (Fig. 188). H&E, ×300

Fig. 190. (*lower right*) Sublingual salivary gland. H&E, ×300

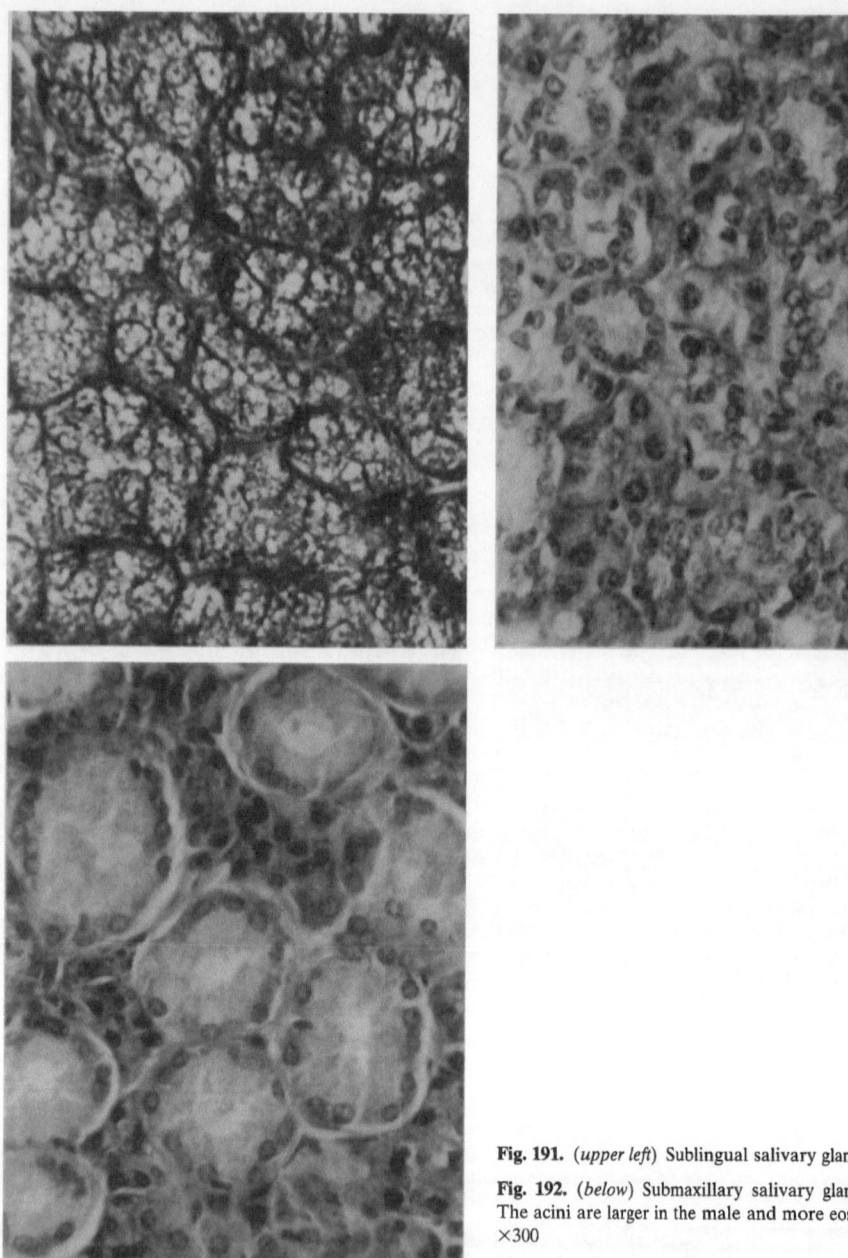

Fig. 191. (*upper left*) Sublingual salivary gland. PAS, ×300

Fig. 192. (*below*) Submaxillary salivary gland, male mouse. The acini are larger in the male and more eosinophilic. H&E, ×300

Fig. 193. (*upper right*) Submaxillary salivary gland, female mouse. The acini are smaller and less eosinophilic in the female. H&E, ×00

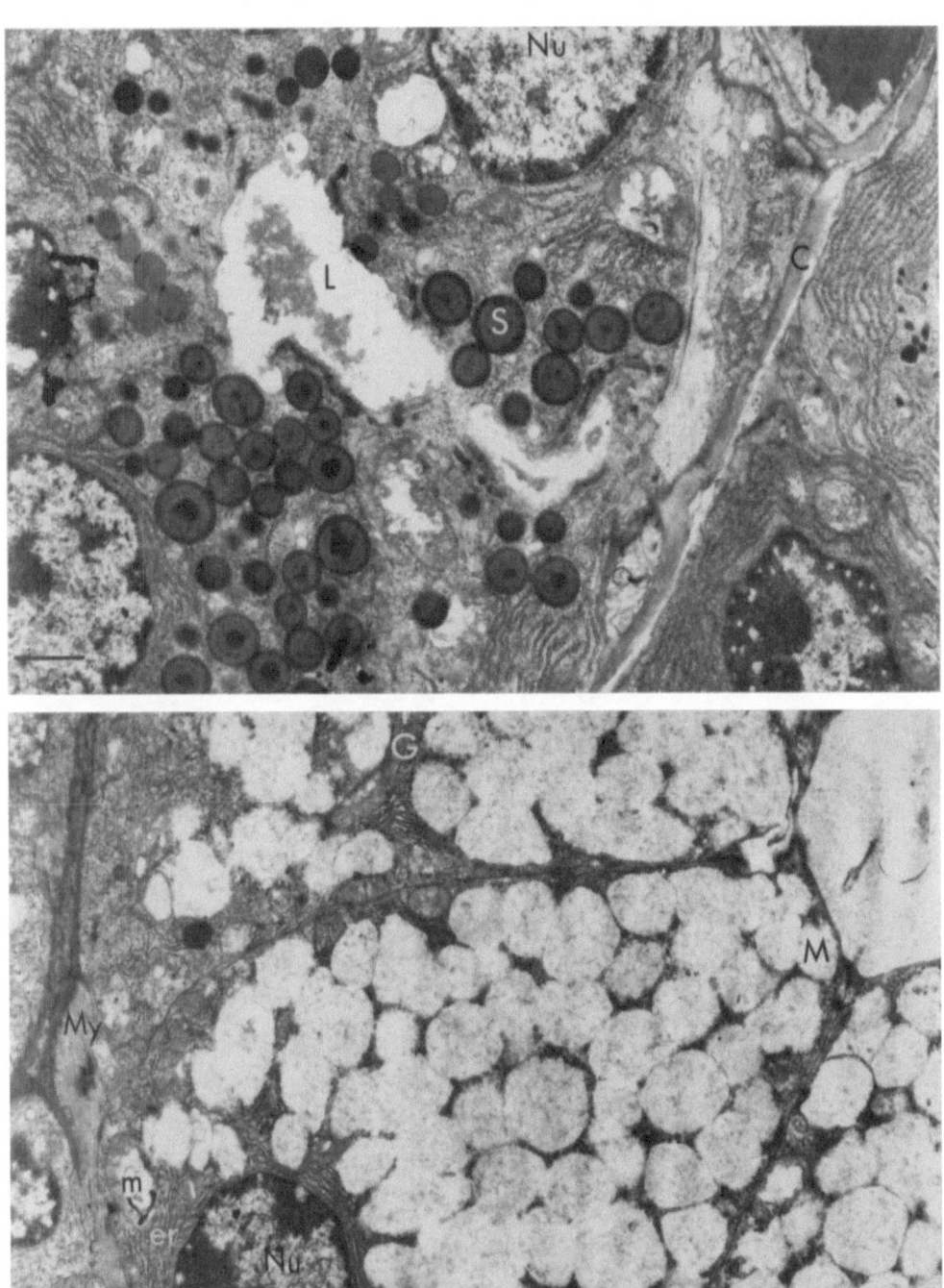

Fig. 194. (*above*) Parotid salivary gland. *S*, serous secretory droplets with dense periphery and core; *L*, acinar lumen; *C*, collagen fibrils of the septum: *Nu*, nucleus. TEM. *Bar*, 1.0 μm

Fig. 195. (*below*) Sublingual salivary gland. Note the acinar lumen (*L*) with the adjacent apical mucin granules (*M*). Mitochondria (*m*), endoplasmic reticulum (*er*), Golgi (*G*), and the nucleus (*Nu*) are located in the basal portion of the cell. A cell process of a myoepithelial cell (*My*) is also seen in this view. TEM. *Bar*, 1.0 μm

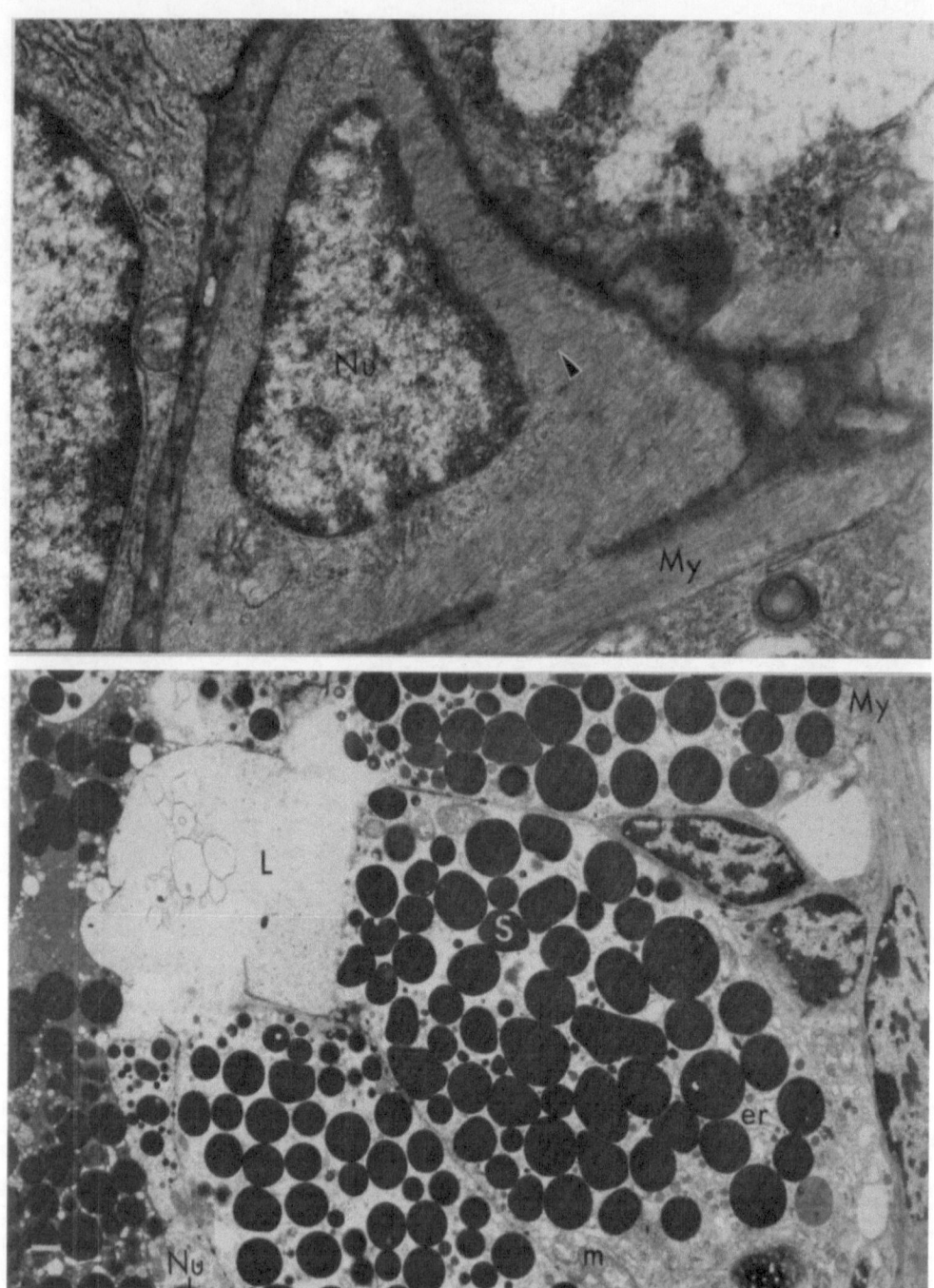

Fig. 196. (*above*) Myoepithelial cell in sublingual salivary gland. The centrally located nucleus (*Nu*) is surrounded by highly oriented myofilaments (*arrowhead*) which exhibit characteristic periodic densities. A myoepithelial cell process (*My*) is also seen cut in nearlongitudinal section. TEM. *Bar*, 1.0 μm

Fig. 197. (*below*) Submaxillary salivary gland, male mouse. The apical serous granules (*S*) are very prominent. Note the distinct difference in cytoplasmic density among the cells. *m*, Mitochondrion; *er*, endoplasmic reticulum; *Nu*, nucleus; *L*, acinar lumen; *My*, myoepithelial cell. TEM. *Bar*, 1.0 μm

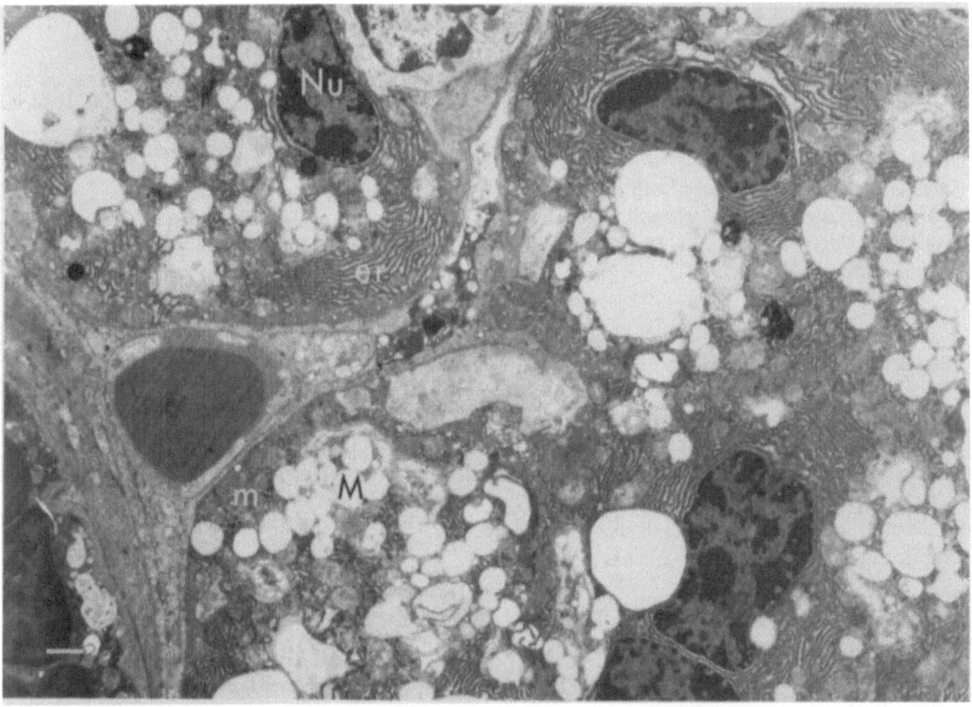

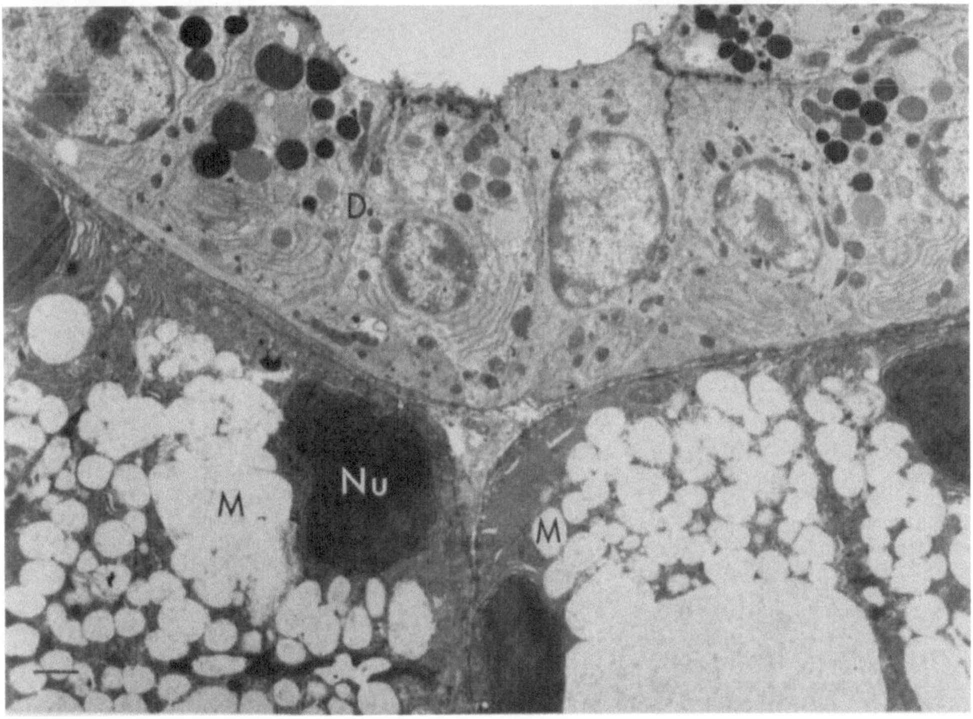

Fig. 198. (*above*) Submaxillary salivary gland, male mouse. These mucous cells have more conspicuous endoplasmic reticulum (*er*) and mitochondria (*m*) than seen in serous cells. Interspersed among these components are the mucous secretory granules (*M*). *Nu*, Nucleus. TEM. *Bar*, 1.0 μm

Fig. 199. (*below*) Submaxillary salivary gland, female mouse. Mucous granules (*M*) are more numerous than in the male submaxillary gland. The nucleus (*Nu*) is crowded into the base of the cell, along with most of the cytoplasmic-formed elements. Secretory duct (*D*) in cross-section. TEM. *Bar*, 1.0 μm

lutes food, makes it possible to taste food, and hydrolyzes carbohydrates (Young and van Lennep 1978).

Epidermal growth factor (EGF), renin, and protease A have been demonstrated immunocytochemically in the sublingual gland of the mouse (Gresik and Barka 1983), and both nerve growth factor (NGF) and epidermal growth factor (EGF) have been demonstrated in the mouse submaxillary gland (Walker 1982).

Comparison with Other Species

Salivary glands of the mouse are similar to those of other species; however, the classification as to a mucous/serous type gland may vary from species to species (Trautmann and Fiebiger 1957). Pecularities of the mouse include the sexual dimorphism of the submaxillary salivary gland.

Acknowledgements. We wish to acknowledge the excellent contributions of Ralph Nichols for specimen preparation and of Grace Miekina for photography.

References

Gresik EW, Barka T (1983) Epidermal growth factor, renin and protease in hormonally responsive duct cells of the mouse sublingual gland. Anat Rec 205:169–175

Gude WD, Cosgrove GE, Hirsch GP (1982) Histological atlas of the laboratory mouse. Plenum, New York, pp 17–18

Hummel KP, Richardson FL, Fekete E (1966) Anatomy. In: Green EL (ed) Biology of the laboratory mouse, 2nd edn. McGraw-Hill, New York, chap 13

Jacoby F, Leeson CR (1959) The post-natal development of the rat submaxillary gland. J Anat 93:201–216

Leeson CR (1967) Structure of salivary glands. In: Code CF (ed) Handbook of physiology, section 6, alimentary canal, vol 2. Secretion. American Physiological Society, Washington DC, chap 32, pp 463–495

Leeson CR, Jacoby F (1959) An electron microscopic study of the rat submaxillary gland during its post-natal development and in the adult. J Anat 93:287–295

Scott BL, Pease DC (1959) Electron microscopy of the salivary and lacrimal glands of the rat. Am J Anat 104:115–161

Smith JF (1966) Histopathology of salivary gland lesions. Lippincott, Philadelphia

Trautmann A, Fiebiger J (1957) Fundamentals of the histology of domestic animals. Comstock, New York, chap 11

Young JA, van Lennep EW (1978) The morphology of salivary glands. Academic, New York

Walker P (1982) The mouse submaxillary gland: a model for the study of hormonally dependent growth factors. J Endocrinol Invest 5:183–196

NEOPLASMS

Myoepithelioma, Salivary Glands, Mouse

Gary T. Burger, Charles H. Frith, and James W. Townsend

Synonyms. None

Gross Appearance

Myoepitheliomas usually occur in the major salivary glands and appear in the subcutis of the ventral neck. They are seen as reddish-yellow masses, which may be solid or a mixture of solid and cystic. The cystic structures are filled with mucoid material and necrotic debris. The parotid and submaxillary are most commonly involved, whereas sublingual involvement is rare. We are not aware of any reports of myoepitheliomas in the minor salivary glands (Dawe 1979).

Microscopic Features

Microscopically, the tumors are composed of large cells which vary in size and shape; some have elongated nuclei suggestive of a sarcoma (Fig. 200). Some cells form compact bundles or are arranged in a swirl around larger, more rounded cells with large vesicular nuclei. The cytoplasm of the neoplastic cells is darkly eosinophilic to amphophilic. These compact areas may become necrotic and form pseudocysts, which contain a mucoid material suggestive of a secretion (Fig. 201). True glands do not occur in the tumor, as an epithelial lining for the cysts is lacking. Intracytoplasmic fibers are visible with high dry or oil immersion using the phosphotungstic acid hematoxylin (PTAH) stain. The tumors occasionally metastasize to the lungs (Fig. 202).

Ultrastructure

The tumor cells vary in morphology; some are electron dense and others are electronlucent (Fig. 203) with irregular, centrally located nuclei.

Neoplastic cells are joined by prominent desmosomes (Fig. 204). The tumor cells are seen to contain filaments and fibrils within the cytoplasm (Fig. 205). These filaments are usually arranged in dense parallel bundles around the nucleus and exhibit the periodic densities characteristic of smooth muscle fibrils. Collagen fibers are present in the intercellular space (Fig. 206).

Differential Diagnosis

Pleomorphic tumors induced by the polyoma virus may be differentiated from myoepithelioma by (a) the tendency of the polyoma tumors to be multicentric and (b) the fact that their multiple morphological variants include mixed mesenchymal/epithelial components and pure mesenchymal or solely epithelial elements (see p. 239, this volume; Dawe 1979).

Adenomas and adenocarcinomas (p. 236) are readily distinguished by their tendency to form acini in their more differentiated parts.

Evidence that myoepitheliomas are derived from myoepithelial cells is somewhat tenuous. The presence of fibrils and filaments with periodic densities in the cytoplasm of the neoplastic cells and the presence of desmosomes is suggestive of myoepithelial origin. Another possible diagnosis is a poorly differentiated carcinoma, based upon interpretation of the fibrils as tonofilaments or keratin. However, tonofilaments do not have periodic density. The filaments appear to resemble myofilaments and not keratin. Immunocytochemistry may help clarify the true identity of the cell of origin.

Biologic Features

The authors have reviewed 37 myoepitheliomas from 14726 control mice of several strains ob-

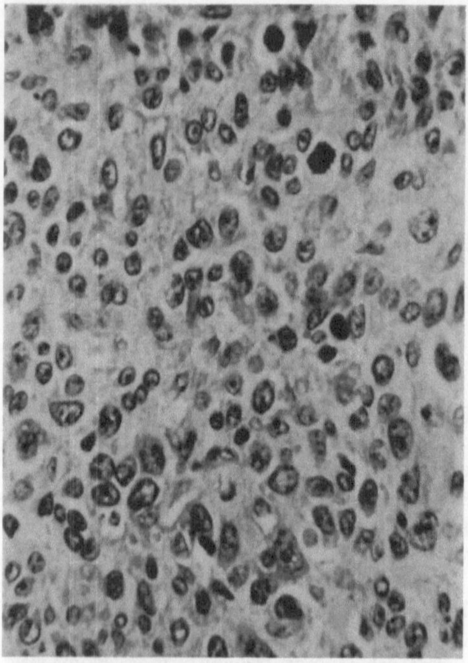

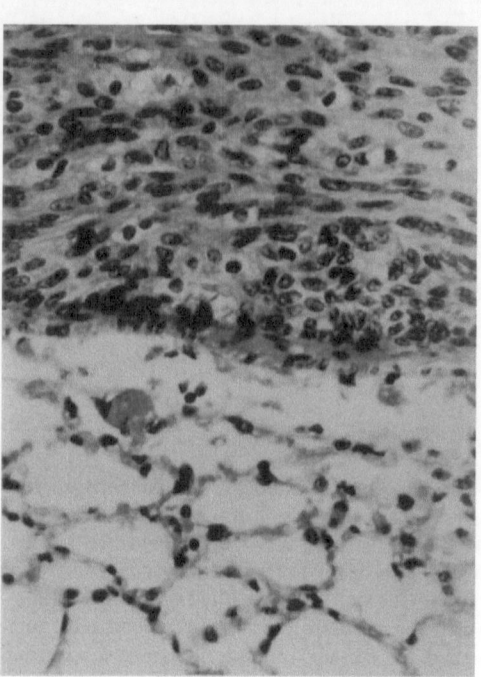

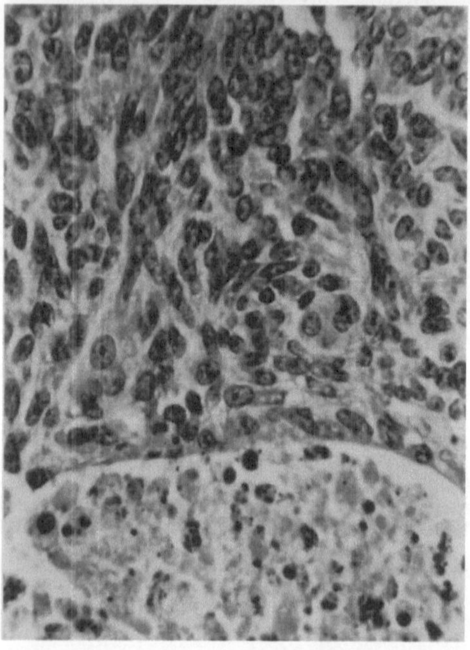

Fig. 200. (*upper left*) Myoepithelioma, mouse. Pleomorphic cells. H&E, ×300

Fig. 201. (*below*) Myoepithelioma, mouse. Note the pseudocyst filled with mucoid material and necrotic debris. H&E, ×300

Fig. 202. (*upper right*) Pulmonary metastasis of a myoepithelioma, mouse. H&E, ×300

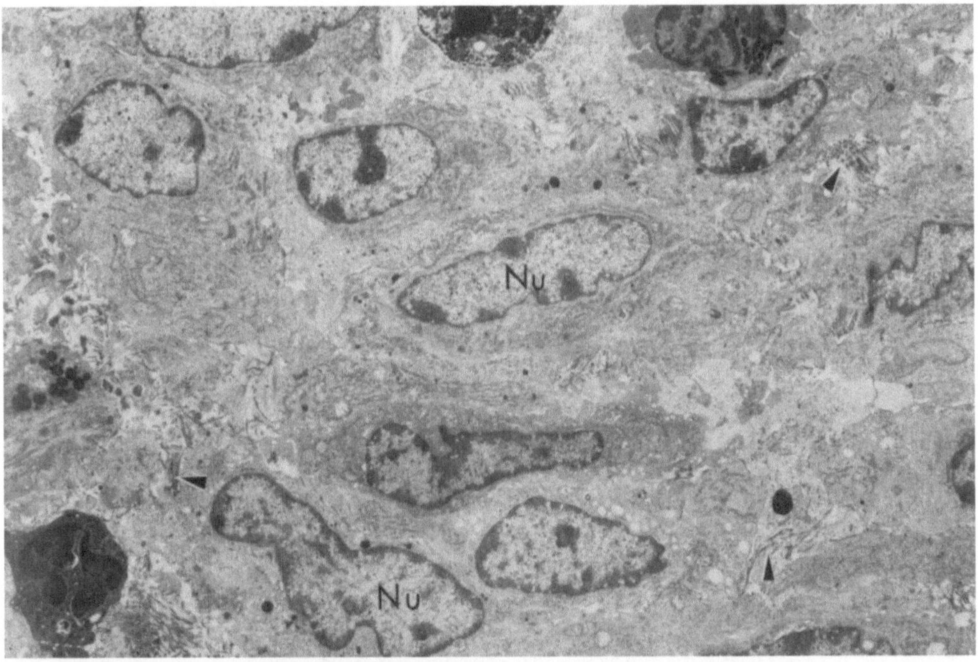

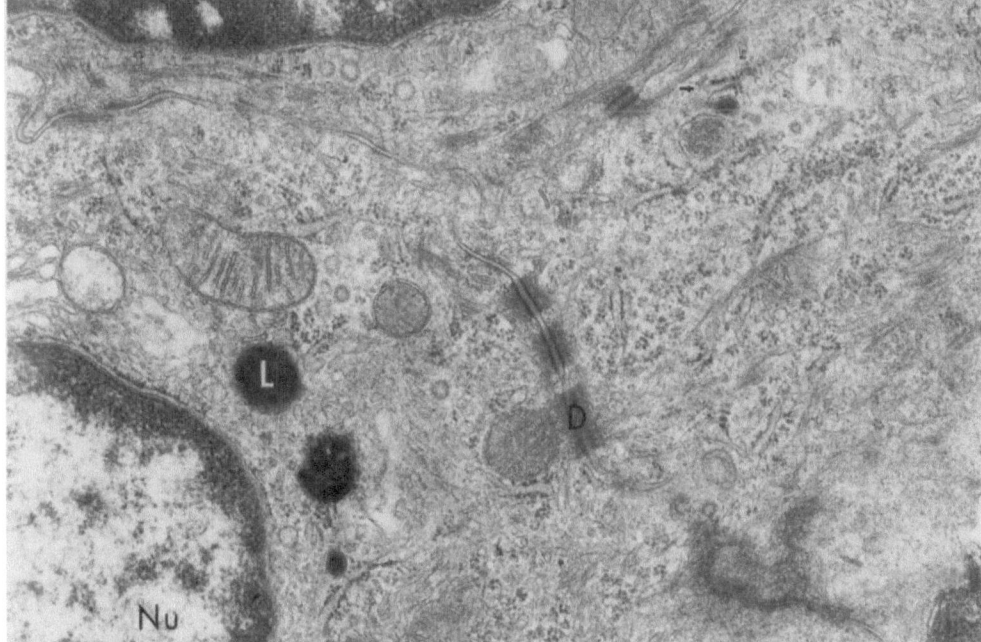

Fig. 203. (*above*) A typical low-magnification electron micrography view of a myoepithelioma. The centrally located, irregular nuclei (*Nu*) are characteristic. Collagen (*arrowheads*) is interspersed among the myoepithelial cells. Uranyl acetate-lead citrate, TEM, ×1500

Fig. 204. (*below*) Myoepithelioma, mouse. Tumor cells are joined by prominent desmosomes (*D*) with fine tonofilaments (*t*). The tonofilaments are not as highly ordered and do not exhibit the periodic densities seen in myofilaments of the myoepithelial cells. *Nu*, nucleus; *L*, lysosome. Uranyl acetate-lead citrate, TEM, ×1200

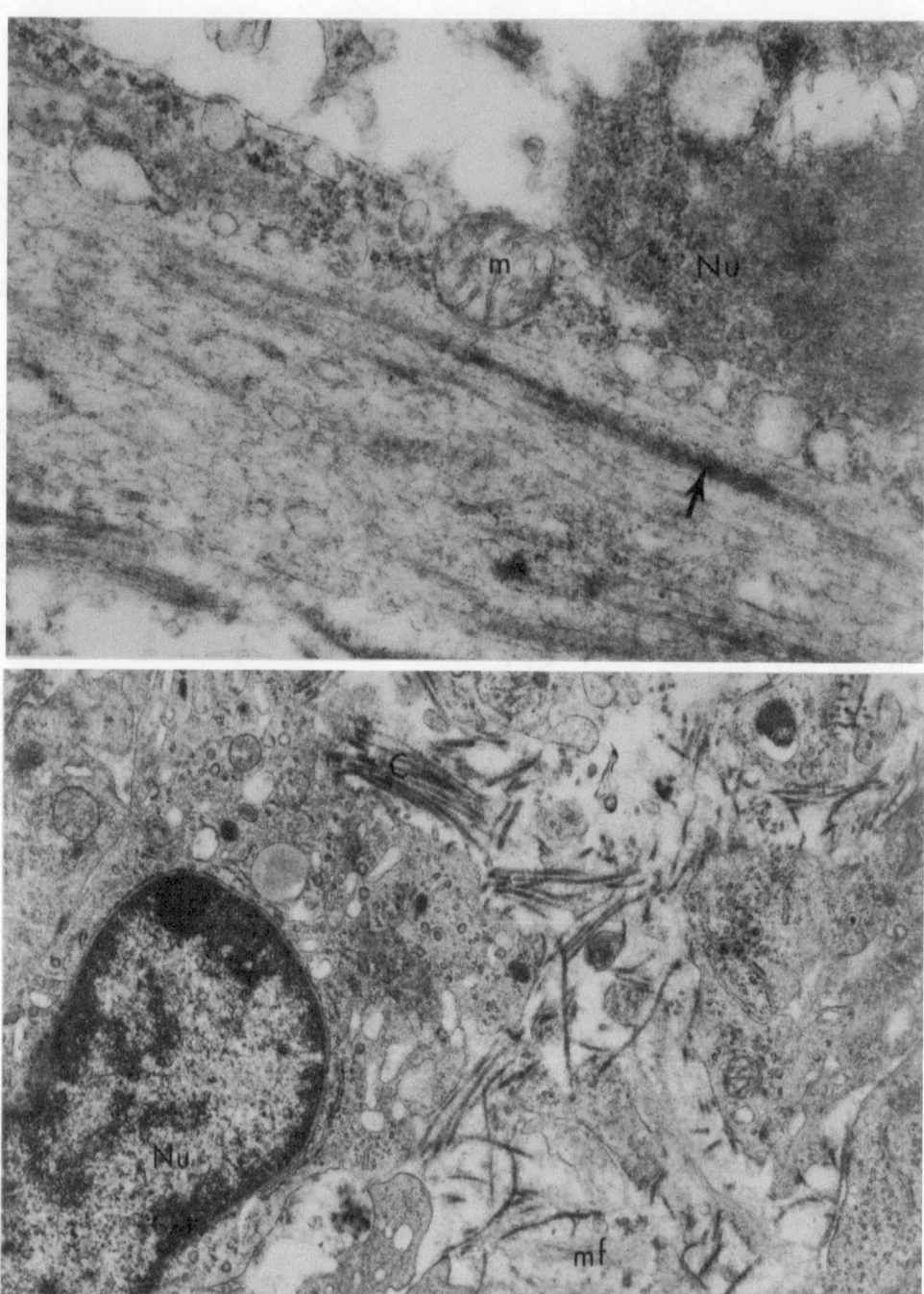

Fig. 205. (*above*) Myoepithelioma, mouse. The myofila-
mentous component of the tumor cells has a highly ordered
arrangement and the characteristic periodic density (*arrow*)
seen in smooth muscle fibrils. *Nu*, nucleus; *m*, mitochondria.
Uranyl acetate-lead citrate, TEM, ×15 000

Fig. 206. (*below*) Myoepithelioma, mouse. Note the intercel-
lular matrix composed of randomly arranged collagen fibers
(*C*). *Nu*, nucleus; *mf*, myofibrils. Uranyl acetate-lead citrate,
TEM, ×7000

tained from the control data base of the National Center for Toxicological Research (NCTR). Myoepitheliomas were only seen in the BALB/c strain and were predominantly in females. Only one of the tumors occurred in a male BALB/c mouse. Of the 14726 control mice, 6543 were BALB/c mice, of which 5090 were females and 1543 were males. The incidence was 0.7% in the BALB/c females and 0.07% in the BALB/c males, occurring in ages ranging from 100 to 776 days, with the largest percentage of tumors in mice 12–18 months old. Myoepitheliomas have also been reported in the (CBA/x BALB/c) F-1 hybrid, (BALB/c x C3H) F-1 hybrid, and the A strain. Of the 37 myoepitheliomas observed in the NCTR control BALB/c mice, 32 occurred in the subcutis of the ventral neck, presumably arising from one of the major salivary glands. Of those tumors in which a specific site could be identified, 13 involved the submaxillary salivary gland, three involved the parotid salivary gland, and seven involved the mammary glands.

The remaining 14 tumors involved more than one of these sites and a single specific site could not be determined. Three of the 37 myoepitheliomas in these animals metastasized to the lungs. The tumors are locally invasive and destructive, and a 10% incidence of pulmonary metastases in BALB/c female mice has been reported (Frith et al. 1981) They are also reported to be readily transplantable (Delaney 1977).

Cosgrove et al. (1978) reported five myoepitheliomas in 2928 (0.17%) control BALB/cCR mice necropsied before 18 months of age; Madison et al. (1968) reported eight myoepitheliomas from 2088 (0.39%) mice killed at 18 months of age; and Peters et al. (1972) reported 30 myoepitheliomas (0.52%) in 5800 BALB/cCR mice ranging in age from 10 to 28 months of age.

Comparison with Other Species

Myoepitheliomas have been reported in the parotid gland of humans (Leifer et al. 1974).

Ultrastructurally, they were composed of a single cell population characterized by numerous cytoplasmic filaments and the cytoplasmic appearance of myoepithelial cells. One of the most common salivary gland tumors in humans, the pleomorphic adenoma, has been reported to be ultrastructurally composed of a modified myoepithelial cell (Dardick et al. 1983).

Acknowledgements. We wish to acknowledge the excellent contributions of Annette Andrews for specimen preparation, Alan Warbritton for photography, and Ruby Franklin for her technical assistance.

References

Cosgrove GE, Satterfield LC, Bowles ND, Klima WC (1978) Diseases of aging untreated virgin female RFM and BALB/c mice. J Gerontol 33:178–183
Dardick I, Van Nostrand AWP, Jeans MT, Rippstein P, Edwards V (1983) Pleomorphic adenoma. II. Ultrastructural organization of "stromal" regions. Hum Pathol 14:398–809
Dawe CJ (1979) Tumours of the salivary and lachrymal glands, nasal fossa and maxillary sinuses. In: Turusov VS (ed) Pathology of tumors in laboratory animals, vol II. Tumors of the mouse. IARC, Lyon, pp 91–133 (IARC scientific publications no 23)
Delaney WE (1977) Transplantable murine salivary gland carcinoma (myoepithelioma). I. Biologic behavior and ultrastructural features. JNCI 58:(1)61–65
Frith CH, Littlefield NA, Ulmholtz R (1981) Incidence of pulmonary metastases for various neoplasms in BALB/cStCrlfC3H/Nctr female mice fed N-2-fluorenylacetamide. JNCI 66:703–712
Leifer C, Miller AS, Putong PB, Harwick RD (1974) Myoepithelioma of the parotid gland. Arch Pathol 98: 312–319
Madison RM, Rabstein LS, Bryan WR (1968) Mortality rate and spontaneous lesions found in 2928 untreated BALB/cCR mice. JNCI 40:683–685
Peters RL, Rabstein LS, Spahn GJ, Madison RM, Huebner RJ (1972) Incidence of spontaneous neoplasms in breeding and retired breeder BALB/cCr mice throughout the natural life span. Int J Cancer 10:273–282

Adenoma, Adenocarcinoma, Salivary Gland, Mouse

James E. Heath

Synonyms. None

Gross Appearance

Neoplasms of the parotid, submaxillary (submandibular), and sublingual salivary glands of the mouse are rare events, but when encountered appear as reddish-yellow masses in the subcutis of the ventral neck. Identification of the specific salivary gland involved is not always possible grossly and sometimes not even microscopically. Consistency of the mass may vary from firm to soft. Soft tumors may result in the formation of cysts, undergo necrosis, and rupture.

Microscopic Appearance

Adenoma. Although rarely encountered, a few tumors of major salivary glands identified as adenomas have the following microscopic features: small circumscribed lesions in the submaxillary gland partially or completely surrounded by a thin capsule. The cells in the mass have lost their zymogen granules, have amphophilic cytoplasm, and mitotic figures may be present (Fig. 207). The neoplastic cells may be large, well differentiated, and arranged in an orderly papillary pattern. The cytoplasm may have a mucoid appearance (Fig. 208), and nuclei are usually hyperchromic and basally oriented.

Adenocarcinoma. These lesions have been described as being composed of large pleomorphic polyhedral cells with amphophilic cytoplasm, and large vesicular nuclei with multiple nucleoli (Fig. 209). Necrosis and mitotic figures may be prominent (Fig. 210). Solid acinar and papillary areas may be seen. Local invasion of adjacent tissues is encountered in some cases, as is metastases to the lungs (Fig. 211).

Ultrastructure

Ultrastructural studies of mouse salivary gland adenomas and adenocarcinomas have been reported infrequently. Spontaneous carcinomas arising from the parotid salivary gland of transgenic mice were reported by Dardick et al. (1992) to contain secretory granules within the apical cytoplasm of the tumor cells that resembled the zymogen granules of the normal parotid acinar cell. Some of the tumor cells also had a prominent complement of rough endoplasmic reticulum.

Differential Diagnosis

the most frequent diagnostic problem is to differentiate adenomas or adenocarcinomas of salivary glands (paratid or submaxillary) from tumors arising in the anterior mammary glands, which in the mouse are located in the subcutis of the neck adjacent to the major salivary glands. Careful gross dissection followed by embedding these glands together in their normal relationship is helpful. Histologic identification of the tumor within the salivary or mammary gland is often definitive. The origin of undifferentiated carcinomas which involve both types of glands may be difficult or impossible to determine.

Biologic Features

Spontaneous salivary gland neoplasms of all types are rare in the mouse. Among these, pleomorphic polyoma virus tumors (see p. 239) and myoepitheliomas are most frequent (see p. 231). Adenomas and adenocarcinomas are least frequent. A spontaneous pleomorphic adenoma of the submandibular salivary gland in an aged male B6C3F1 mouse was described by Hagiwara et al. (1993), and a highly anaplastic scirrhous salivary gland carcinoma in a female B6C3F1 mouse from a National Toxicology Program bioassay was documented by Frith et al. (1994). Several reports in the literature indicate that, in addition to the above, squamous cell carcinoma, adenoacanthoma, giant cell carcinoma, undifferentiated carcinoma, and sarcoma can be induced in the major salivary glands with carcinogens.

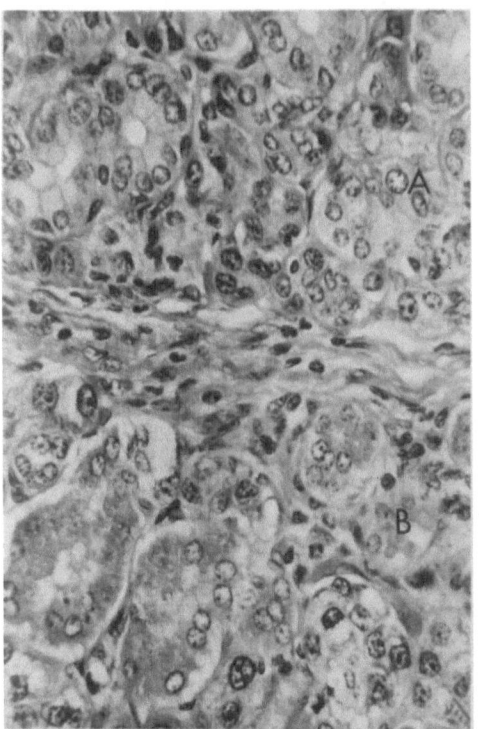

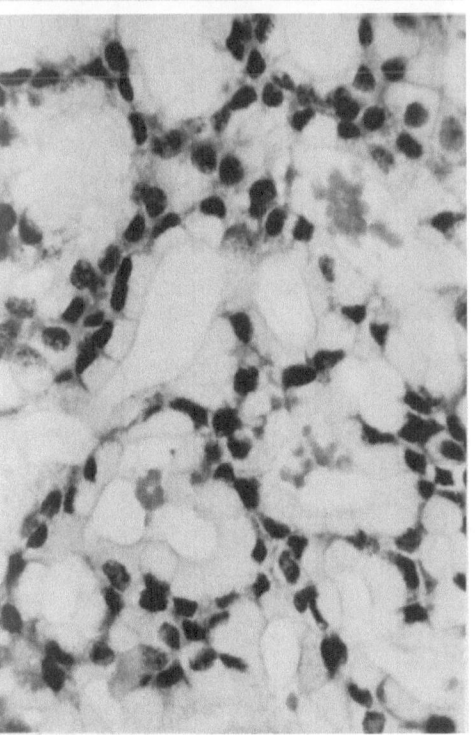

Rusch et al. (1940) induced squamous cell carcinomas of the salivary glands of mice by injecting benzpyrene directly into the salivary glands. The tumors were invasive and their development was preceded with metaplasia of the glandular tissue into a stratified squamous epithelium followed by the development of squamous cell carcinoma. The illustrations in the report of Rusch et al. (1940) were of a poorly differentiated squamous cell carcinoma with little evidence of keratin.

Franseen et al. (1941) introduced methylcholanthrene in the form of a pellet into the submaxillary gland of strain A mice. Seventeen tumors occurred in 30 mice and 12 were verified histologically as squamous cell carcinomas. The earliest tumor developed in 64 days and the latest in 215 days. Bauer and Byrne (1950) induced 58 salivary gland tumors with 7,12-dimethylbenz-(a)anthracene (DMBA), most of which were adenoacanthomas.

Sarcomas of the salivary glands of mice have also been induced with chemical carcinogens (dibenz-(a,K)anthracene, 3-methylcholanthrene, and benzo[a]pyrene; Benecke and Schroder 1939; Rusch et al. 1940; Franseen et al. 1941).

Dawe (1979) described pleomorphic tumors induced with polyoma virus, indicating a pure mesenchymal tumor, a pure epithelial tumors, and a mixed mesenchyma–epithelial tumor (see Dawe 1979; p. 239, this volume).

Comparison with Other Species

Spontaneous salivary gland neoplasms appear to be relatively rare in other species as well as mice. They are equally as rare in the rat (Glucksmann and Cherry 1973). The pleomorphic adenoma, which is the most common salivary gland neoplasm in humans (Shklar 1979), bears some resemblance to the pleomorphic polyoma tumor in the mouse.

◄

Fig. 207. (*above*) Adenoma, submaxillary salivary gland, BALB/c StCrlfC3H/NCTR, male mouse. Normal submaxillary salivary gland (*A*) and adenoma (*B*). H&E, ×300

Fig. 208. (*below*) Adenoma, submaxillary salivary gland of a 686-day-old female BALB/c StCrlfC3H/NCTR mouse. Note the mucoid appearance of the cells. H&E, ×300

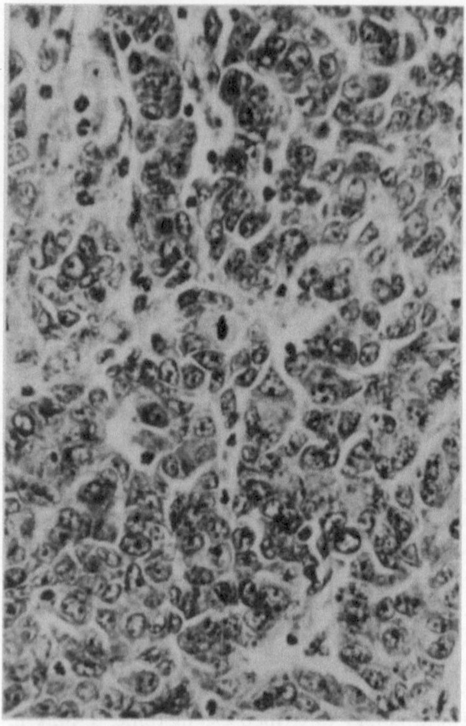

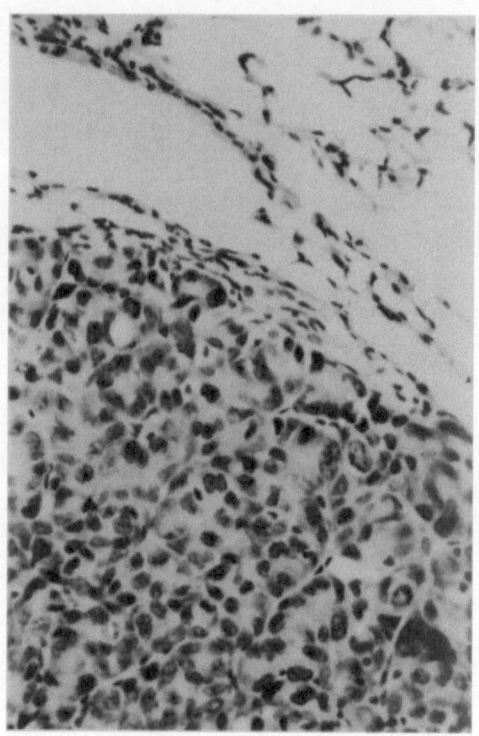

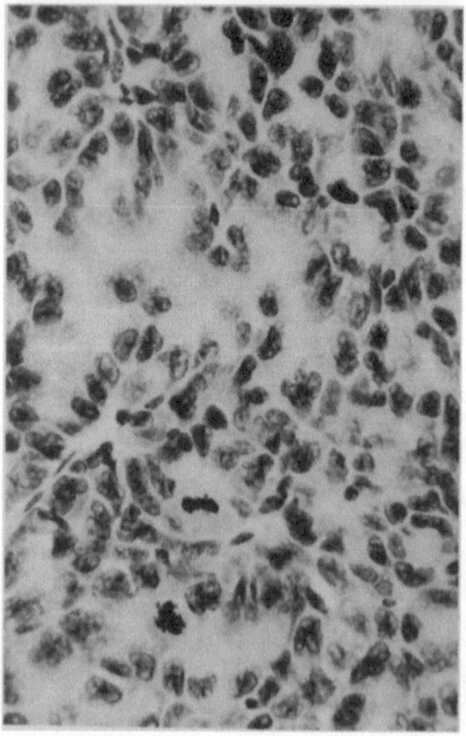

◀

Fig. 209. (*upper left*) Adenocarcinoma, parotid salivary gland of a 550-day-old C37BL/6JfC3H/NCTR × BALB/cStCrlfC3H/NCTRF₁ hybrid female mouse. Note the numerous mitotic figures. H&E, ×300

Fig. 210. (*below*) Adenocarcinoma, parotid salivary gland of a 685-day-old female B6C3Fl hybrid female mouse. Note the plemorphism of the neoplastic cells and mitotic figures. (Courtesy of Dr. Daniel R. Farnell, Southern Research Institute, Birmingham, Alabama.) H&E, ×300

Fig. 211. (*upper right*) Pulmonary metastasis of parotid adenocarcinoma shown in Fig. 210. H&E, ×300

References

Bauer WH, Byrne JJ (1950) Induced tumors of the parotid gland. Cancer Res 10:755–761

Benecke E, Schroder J (1939) Über experimentelle Geschwulsterzeugung in Speicheldrüsen. Z Krebsforsch 49:505–514

Dardick I, Burford-Mason AP, Garlick DS, Carney WP (1992) The pathobiology of salivary gland. II. Morphological evaluation of acinic cell carcinomas in the parotid gland of male transgenic (MMTV/v-Ha-ras) mice as a model for human tumors. Virchows Arch [A] Pathol Anat Histopathol 421(2):105–113

Dawe CJ (1979) Tumours of the salivary and lachrymal glands, nasal fossa and maxillary sinuses. In: Turusov VS (ed) Pathology of tumours in laboratory animals, vol II. Tumours of the mouse. IARC, Lyon, pp 91–134 (IARC scientific publications no 23)

Franseen CC, Aub JC, Simpson CL (1941) Experimental tumors in lymph nodes and in endocrine and salivary glands. Cancer Res 1:489–493

Frith CH, Heath JE (1994) Tumours of the salivary gland. In: V.S. Turosov (ed) Pathology of tumors in laboratory animals. II. Tumours of the mouse. IARC, Lyon, pp 115–139 (IARC scientific publications no 111)

Glucksmann A, Cherry CP (1973) Tumours of the salivary gland. In: Turusov VS (ed) Pathology of tumours in laboratory animals, vol I. Tumours of the rat, part 1. IARC, Lyon, pp 75–86 (IARC scientific publications no 5)

Hagiwara A, Ogiso T, Shibata, M-A, Shirai T (1993) Spontaneous pleomorphic adenoma in the submandibular salivary gland of an aged male (C57BL/6N × C3H/HeN)F$_1$ (B6C3F$_1$) mouse. Vet Pathol 30:394–396

Rusch HP, Baumann CA, Maison GL (1940) Production of internal tumors with chemical carcinogens. Arch Pathol 29:8–19

Shklar G (1979). The oral cavity, jaws and salivary glands. In: Robbins SL, Cotran RS (eds) Pathologic basis of disease, 2nd edn. Saunders, Philadelphia, pp 886–917

Polyoma Virus Infection, Salivary Glands, Mouse

Stephen W. Barthold

Synonyms. Stewart-Eddy (SE) polyoma virus, parotid tumor virus infection

Gross Appearance

Naturally infected mice have no gross lesions. Tumors arise in many organs, particularly parotid salivary glands, following experimental inoculation of neonatal mice. Submaxillary and sublingual salivary glands, lacrimal glands, and accessory mucous and serous glands of the head and neck are also affected. Salivary tumors are bilateral or unilateral, multilobular, usually well circumscribed, and up to 5 cm in diameter. Small tumors are soft, tan to gray, and bulge on cut surface. Larger tumors can have pseudocystic, mucoid centers and hemorrhage, and they occasionally ulcerate the overlying skin. Other common tumor sites are renal cortex, thymus, mammary gland, skin, subcutis, bone, mesothelium, adrenal gland, and, less commonly, else-

where. Liver, lung, and pancreas are frequent sites of metastases (Dawe 1979; Stewart 1960). Prior to tumor development, pups are runted with thymic atrophy. Small nodules on the costochondral and costovertebral junctions are often seen (Buffet and Levinthal 1962). Athymic nude mice develop a wasting disease with paralysis and multiple tumors, particularly of the uterus and bone (McCance et al. 1983; Sebesteny et al. 1980; Vandeputte et al. 1974).

Microscopic Features

Within 2–4 weeks after inoculation of neonates, epithelial cells in many organs enlarge, and nuclei swell with inclusions. Adjacent cells are mitotically active, and lymphocytic infiltrates are often present. Several weeks later, multicentric microtumors arise. Initially, microtumors contain enlarged, inclusion-bearing cells, as in the acute infection, but cytopathic changes become less conspicuous by 90 days. Rib nodules consist of periosteal fibroblastic proliferation, with invasion of adjacent bone, cartilage, muscle, and marrow suggesting fibrosarcomas, but many of these foci ossify and heal at later intervals (Buffet and Levinthal 1962; Stanton et al. 1959).

In order of decreasing frequency, mixed mesenchymoid–epithelial, pure mesenchymoid, and pure epithelial tumor types occur in salivary glands. Foci of several variants are often seen in the same mouse. Initially, there is loss of secretory activity of acinar cells within a lobule, with a change from large pyramidal cells with basal nuclei to low cuboidal cells with central nuclei. Clusters of mitotically active round or fusiform cells then appear between basement membrane and the altered, cuboidal epithelium. These clusters progressively enlarge, filling the glandular lumen and breaking out of the confining basement membrane. Many salivary tumor emboli are found in pulmonary arteries, but few develop as metastatic tumors.

Tumors of accessory glossal salivary glands, pharyngeal glands, lacrimal glands, and nasal glands are identical to those induced in the major salivary glands. An additional type of lesion, termed sclerosing nodule, also occurs in which hypocellular collagenous tissue and nonneoplastic intralobular ducts are present with only a few neoplastic cells present in the matrix (Dawe 1979; Stanton et al. 1959).

Other organs can develop benign and malignant tumors of mesenchyme (bone, subcutis, renal medulla, mesothelium, and uterus), epithelium (submucosal glands of nose, pharynx, trachea, lacrimal glands, thymic medulla, thyroid, mammary gland, dental organ, stomach, buccal mucosa, hair follicles, and cutaneous sweat and sebaceous glands), and neuroepithelium (olfactory mucosa, adrenal medulla). Nonsalivary digestive system tumors are rare, but include ameloblastomas and squamous carcinomas of buccal and gastric mucosa (for details, please see Dawe 1972, 1979; Stanton et al. 1959; Stewart 1960; Vandeputte et al. 1974).

Athymic nude mice have conspicuous cytolytic lesions and a variety of tumors, as in neonatally inoculated mice. However, osseous and uterine tumors predominate (Vandeputte et al. 1974). Nude mice also develop infection of oligodendroglia, demyelination, and paralysis. Cytopathic and proliferative changes are present in the bronchial mucosa and in epithelium of the renal pelvis and ureter (McCance et al. 1983; Sebesteny et al. 1980).

Ultrastructure

Cells in replicative infection develop crystalline arrays of closely packed, 40- to 45-nm viral particles in their nuclei. Phagocytes may contain phagosomes with aggregates of virus derived from lysed cells. Early tumors may contain virus, but replicative infection diminishes with time. In salivary tumors, both type B and C retrovirus particles can also be encountered. Salivary tumor cells, including mesenchymoid types, are often joined by desmosomes and lie on a basement membrane; some contain intracytoplasmic fibrils, suggesting epithelial and myoepithelial origin (Imamura 1968; Dawe 1979; Howatson et al. 1960; Stewart 1960).

Differential Diagnosis

Natural infections in immunologically competent mice are diagnosed by seroconversion. Early polyoma virus infections in neonatal mice must be differentiated from other disseminated virus infections that cause intranuclear inclusions (mouse adenovirus, cytomegalovirus), but the likelihood of encountering such infections is remote. Polyoma virus-induced salivary tumors do not resemble spontaneous or chemically induced salivary neoplasms. Mammary tumors can arise in the cervical region and must be differentiated from salivary tumors (Dawe 1979). Polyoma virus infection in athymic nude mice has been shown to cause posterior paralysis, but other, more likely etiologic agents include trauma, mouse hepatitis virus infection, and mouse encephalomyelitis virus infection.

Biologic Features

Natural History

Under natural conditions, polyoma virus infections are subclinical in laboratory and wild mice. This is because mice must be infected under specific conditions for neoplasia, the most obvious sign of infection, to occur. Chances of tumor development are greatest if a mouse is infected within the first 24 h of life, does not receive maternal antibody, is exposed to a relatively high dose of virus, and is inoculated parenterally with a virus strain selected for its oncogenicity. Nevertheless, polyoma viral tumors can occur rarely under natural conditions. Polyoma virus is highly resistant and is transmitted by contamination with infected urine. Wild mouse infection is maintained in contaminated nesting sites, which are used repeatedly for multiple generations. The opportunity for virus contamination and transmission to take place is limited in laboratory mouse colonies because of husbandry practices. Neonates are unlikely to develop infections because the statistical chance of acquiring early infection is small and they can be protected by maternal antibody. Infection is best maintained in laboratory mice when there is direct contact among weanlings. Although weanlings excrete virus inefficiently, protection by maternal antibody has waned by that age. Laboratory colonies can be infected by accidental exposure to experimental virus or to contaminated transplantable tumor lines (McCance et al. 1983; Rowe 1961; Sebesteny et al. 1980).

Pathogenesis

Upon infection of susceptible cells in vitro, polyoma virus causes: (a) lytic infection in which a cell is destroyed by virus replication, (b) transformation, in which the viral genome is permanently incorporated into the host cell without virus replication and cytolysis, or (c) abortive infection, in which transformed cells lose virus after several cell divisions. These are not exclusive, since virus replication has been seen in mitotic tumor cells in vivo (Allison 1980; Imamura 1968). Polyoma virus induces several virus-directed cellular antigens, including nuclear T antigens and cell membrane tumor-specific transplantation antigens (TSTA). Virus structural antigens are also produced in replicative infections. The outcome of infection in vivo depends largely on host immune response to virus or virus-induced tumor neoantigens. Fully immunocompetent adult mice mount an efficient and rapid immune response to both virus and tumor, preventing both generalized infection and tumor development. Neonatal mice have an ineffective, delayed immune response, allowing generalization of infection, virus excretion, and growth of transformed cells to the point that neoplastic growth can no longer be controlled

(Allison 1980). Susceptibility to the oncogenic effects of polyoma virus is most pronounced during the first 24 h of life, after which time it rapidly declines. This short period of susceptibility suggests a nonimmune, age-related factor in tumorigenesis, but susceptibility can persist into adulthood in athymic nude mice or if mice are neonatally thymectomized or irradiated (Dawe 1972; McCance et al. 1983; Sebesteny et al. 1980). Naturally infected athymic nude mice develop persistent infections with neoplasia. These mice can develop posterior paralysis due to vertebral bone tumors, as well as demyelination and focal leukoencephalopathy due to infection of oligodendroglia (McCance et al. 1983; Sebesteny et al. 1980).

Within 5 h of intranasal inoculation of a neonate, primary replication occurs in the nasal cavity, submaxillary salivary gland, and lungs, followed by viremia with dissemination to multiple organs. Viremia ceases with the appearance of serum-neutralizing antibody between days 3 and 6. Secondary virus replication occurs predominantly in liver, spleen, kidney, and colon. Neonatal pups experience high mortality during this phase. By day 12, polyoma virus DNA is partially cleared from most organs, coinciding temporally with the appearance of lymphocytic infiltrates and a cell-mediated immune response. By day 22, there is persistence of viral DNA only in lungs and kidneys, with a gradual decline over the next 3 months. At about 2 months, microtumors evolve in multiple organs and are grossly visible in up to 100% of mice surviving between 2 and 4 months. Urinary excretion is the major means of elimination of the virus, and the respiratory tract appears to be the primary portal of entry. Virus persists in mice for 2–5 months in trace amounts after infection of young adults, but is found in higher titers and for longer periods in mice infected as neonates. After 4 months, virus is localized to the kidneys (Allison 1980; Buffet and Levinthal 1962; Dubensky and Villarreal 1984; Dubensky et al. 1984; Rowe 1961).

Transplacental transmission can be effected experimentally in which fetal and newborn mice have renal infection, but this does not seem to occur naturally (McCance and Mims 1977). In young adult mice infected as neonates, there is reactivation of virus in the kidneys, but not other organs, late in gestation. Prior to pregnancy, little or no virus can be detected (McCance and Mims 1979).

Etiology

Polyoma virus is the type species of the polyoma virus subgroup (B) of the DNA papovavirus group. Polyoma virus has been used extensively as a model of viral oncogenesis. All strains of mice are susceptible to tumor induction, but genotypic differences in susceptibility occur (Stewart 1960). An uncharacterized subgroup B papovavirus has been reported to cause infection but not tumors of parotid, lacrimal, and laryngeal glands and respiratory epithelium in athymic nude rats (Ward et al. 1984).

Frequency

Polyoma virus was once common among both laboratory and wild mice (Rowe 1961), but has been largely eliminated from laboratory mouse stocks. Transplantable tumor lines can be contaminated with this agent. Tumor lines and experimental virus stocks are the most frequent sources of laboratory mouse contamination. Tumor development under natural conditions is rare, due to complex epizootiologic factors. Salivary gland tumors are the most common experimentally induced tumors and are the most likely to occur, albeit rarely, in natural infections (Rowe 1961; Stewart 1960).

Comparison with Other Species

Polyoma virus of mice resembles several viruses of the polyoma virus subgroup in that it generally produces subclinical infections with chronic urinary excretion, including rabbit kidney vacuolating virus, hamster papovavirus, SV40, STMV, SA12, and human viruses BK and JC. Several viruses of the polyoma virus subgroup are oncogenic under experimental conditions, but are generally not associated with naturally occurring neoplasia (Howley 1980). Experimental infection of the central nervous system of athymic nude mice with polyoma virus produces lesions resembling those of progressive multifocal leukoencephalopathy in immunocompromised nonhuman primates and humans caused by subgroup B papovaviruses (Sebesteny et al. 1980). Polyoma virus-induced pleomorphic salivary gland tumors have been compared with salivary mixed tumors or pleomorphic adenomas in hu-

mans, but the resemblance is only superficial (Dawe 1979).
Polyoma virus also induces mesenchymal tumors in experimentally infected hamsters, rats, ferrets, guinea pigs, rabbits, and Mastomys. In contrast to mice, hamsters are highly susceptible to tumor induction after intranasal inoculation of polyoma virus of mice, and there is no sharp development of resistance with age (Rowe 1961). Likewise, hamsters of all ages can be naturally infected with hamster papovavirus (a subgroup B papovavirus) and develop persistent infections with tumor formation, particularly lymphosarcomas. Lymphosarcomas can occur in epizootic form among young hamsters infected with this agent (Barthold et al. 1987).

References

Allison AC (1980) Immune responses to polyoma virus and polyoma virus-induced tumors. In: Klein G (ed) Viral oncology. Raven, New York, pp 481–487

Barthold SW, Bhatt PN, Johnson EA (1987) Further evidence for papovavirus as the probable etiology of transmissible lymphoma of Syrian hamsters. Lab Anim Sci 37:283–288

Buffet RF, Levinthal JD (1962) Polyoma virus infection in mice. Arch Pathol 74:513–526

Dawe CJ (1972) Epithelial-mesenchymal interactions in relation to the genesis of polyoma virus-induced tumours of mouse salivary gland. In: Tarin D (ed) Tissue interactions in carcinogenesis. Academic, New York, chap 10

Dawe CJ (1979) Tumours of the salivary and lachrymal glands, nasal fossa and maxillary sinuses. In: Turusov VS (ed) Pathology of tumours in laboratory animals, vol II. Tumours of the mouse. IARC, Lyon, pp 91–133 (IARC scientific publication, no 23)

Dubensky TW, Villarreal LP (1984) The primary site of replication alters the eventual site of persistent infection by polyomavirus in mice. J Virol 50:541–546

Dubensky TW, Murphy FA, Villarreal LP (1984) Detection of DNA and RNA virus genomes in organ systems of whole mice: patterns of mouse organ infection by polyomavirus. J Virol 50:779–783

Howatson AF, McCulloch EA, Almeida JD, Siminovich L, Axelrad AA, Ham AW (1960) Studies in vitro, in vivo, and by electron microscope of a virus recovered from a C3H mouse mammary tumor: relationship to polyoma virus. J Natl Cancer Inst 24:1131–1151

Howley PM (1980) Molecular biology of SV40 and the human polyomaviruses BK and JC. In: Klein G (ed) Viral oncology. Raven, New York, pp 489–550

Imamura M (1968) Electron microscopic study of polyoma-induced salivary gland tumors, with special reference to cell-virus interactions. J Natl Cancer Inst 41:1265–1283

McCance DJ, Mims CA (1977) Transplacental transmission of polyoma virus in mice. Infect Immun 18:196–202

McCance DJ, Mims CA (1979) Reactivation of polyoma virus in kidneys of persistently infected mice during pregnancy. Infect Immun 25:998–1002

McCance DJ, Sebesteny A, Griffin BE, Balkwill F, Tilly R, Gregson NA (1983) A paralytic disease in nude mice associated with polyoma virus infection. J Gen Virol 64:57–67

Rowe WP (1961) The epidemiology of mouse polyoma virus infection. Bacterial Rev 25:18–31

Sebesteny A, Tilly R, Balkwill F, Trevan D (1980) Demyelination and wasting associated with polyomavirus infection in nude (nu/nu) mice. Lab Anim 14:337–345

Stanton MF, Stewart SE, Eddy BE, Blackwell RH (1959) Oncogenic effect of tissue-culture preparations of polyomavirus on fetal mice. J Natl Cancer Inst 23:1441–1475

Stewart SE (1960) The polyoma virus. Adv Virus Res 7:61–90

Vandeputte M, Eyssen H, Sobis H, De Somer P (1974) Induction of polyoma tumors in athymic nude mice. Int J Cancer 14:445–450

Ward JM, Lock A, Collins MJ Jr, Gonda MA, Reynolds CW (1984) Papovaviral sialoadenitis in athymic nude rats. Lab Anim 18:84–89

NON-NEOPLASTIC LESIONS

Cytomegalovirus Infection, Salivary Glands, Mouse, Rat, and Hamster

Stephen W. Barthold

Synonyms. Salivary gland virus infection, cytomegalic inclusion disease

Gross Appearance

Under most circumstances there are no grossly visible lesions. During acute, generalized infections in mice, the liver may be enlarged and pale with hemorrhagic foci. Intestinal serosa can be irregularly reddened and mucosa congested (McCordock and Smith 1936). Mineralization of skeletal muscle and brown fat (Lussier 1975) and blood-tinged ascitic fluid (Olding et al. 1976) have also been observed. Mice infected as neonates are runted with thymic and splenic involution (Schwartz et al. 1975).

Microscopic Features

Acute generalized infections in susceptible (especially if immunosuppressed) mice are characterized by focal necrosis, cytomegaly, intracytoplasmic and intranuclear (type A) inclusion bodies, and nonsuppurative inflammation of multiple organs. Lesions can be found in salivary and lacrimal glands, brain, liver, spleen, thymus, lymph nodes, paritoneal connective and adipose tissue, lungs, skin, renal glomeruli, bowel, pancreas, adrenals, skeletal and cardiac muscle, cartilage, and brown fat (Brody and Craighead 1974; Gardner et al. 1974; Jordan 1978; Lussier 1975; McCordock and Smith 1936; Mims and Gould 1979; Olding et al. 1976; Schwartz et al. 1975). Lesions may be restricted to salivary glands in natural infections of mice, rats, and hamsters (Gardner et al. 1974; Lussier 1975; Lyon et al. 1959; Priscott and Tyrrell 1982). Submaxillary salivary glands are preferentially infected, the sublingual glands less so, and the parotids least of all (Mims and Gould 1979; Ruebner et al. 1966).

In mice, inclusions are most apt to be found in acinar epithelium (Fig. 212), whereas they occur in ductal epithelium in rats, guinea pigs, humans, and most other species (Bruggeman et al. 1985). Eosinophilic intracytoplasmic and intranuclear inclusions are present in acinar and ductal epithelial cells. Intranuclear inclusions are Feulgen positive, and intracytoplasmic inclusions are periodic acid-Schiff (PAS) and Feulgen positive. Infected cells can become atypically large (cytomegaly; Fig. 212). This is accompanied by infiltration of the surrounding interstitium with lymphocytes and plasma cells. In the acute phase of generalized infections, other organs, particularly liver (Fig. 213), are more apt to possess lesions than salivary glands, which develop lesions later in the course of infection (Brodsky and Rowe 1958; Gardner et al. 1974; Henson and Strano 1972; Ruebner et al. 1966).

Ultrastructure

Initially, nuclei of acinar cells of salivary glands become enlarged with chromatin uniformly dispersed. Nucleoli enlarge and are partially surrounded by fibrillar structures identical to those of the virus membrane. Pleomorphic dense cores become surrounded by these fibrils, followed by the appearance of cores, fibrils, and virions throughout the nucleoplasm. Virus particles acquire a second membrane by passing through the nuclear membrane into the cytoplasm. Particles make contact with membranes of dilated Golgi vesicles and, to a lesser extent, endoplasmic reticulum, acquiring a third membrane or envelope by the process of invagination. Vesicles enlarge and are filled with virus particles as they approach the cell apex and are extruded into the lumen. Intranuclear inclusions observed by light microscopy correspond to aggregates of granular and fibrillar material intermixed with virus parti-

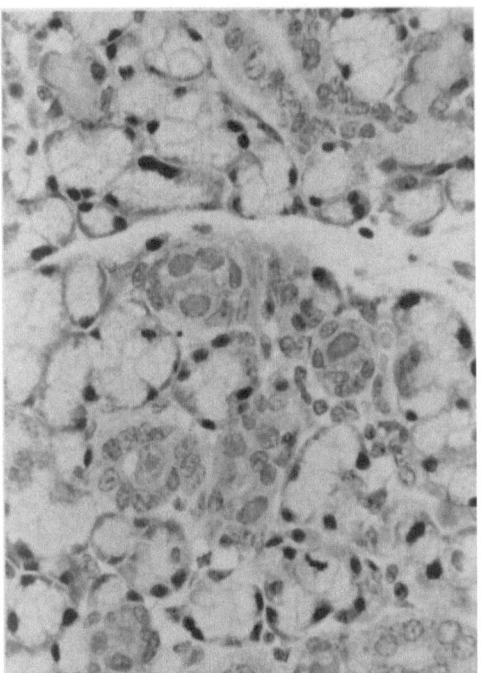

Fig. 212. Intranuclear inclusions in submaxillary salivary gland of a mouse experimentally infected with mouse cytomegalovirus. H&E, ×485

cles. Intracytoplasmic inclusions correspond to virus-filled vesicles (Fig. 214; Henson and Strano 1972; Ruebner et al. 1966). Similar viral replicative changes have been described in pulmonary macrophages (Brody and Craighead 1974), splenic reticulum cells, and hepatocytes (Ruebner et al. 1964, 1966).

Differential Diagnosis

In mice, sialoadenitis is also caused by polyoma virus and reovirus type 3. Polyoma virus induces intranuclear inclusions, but not cytomegaly or intracytoplasmic inclusions, and preferentially infects the parotid salivary gland. Reovirus type 3 can cause necrotizing sialoadenitis without inclusion bodies. Furthermore, mouse salivary gland is a primary target organ for another herpesvirus, mouse thymic virus, but salivary gland lesions have not been described (Cross et al. 1979). Murine mammary tumor virus also replicates in

and is shed from salivary glands, but is not associated with light microscopy changes (Bentvelzen and Hilgers 1980). Sialoadenitis can be caused by coronavirus and polyoma virus in rats. Coronavirus produces necrotizing lesions without inclusions in submandibular and parotid salivary glands, as well as lacrimal glands. Polyoma virus induces inclusions in parotid salivary glands. Papovavirus also induces very similar inclusions and sialoadenitis in parotid salivary glands of athymic nude rats (Ward et al. 1984).

Biologic Features

Natural History

Infections with cytomegalovirus in mice under natural conditions are almost invariably subclinical. Virus is transmitted by direct contact through inhalation or ingestion and is excreted in saliva, tears, and urine (Brodsky and Rowe 1958; Lussier 1975). In utero transmission does not play

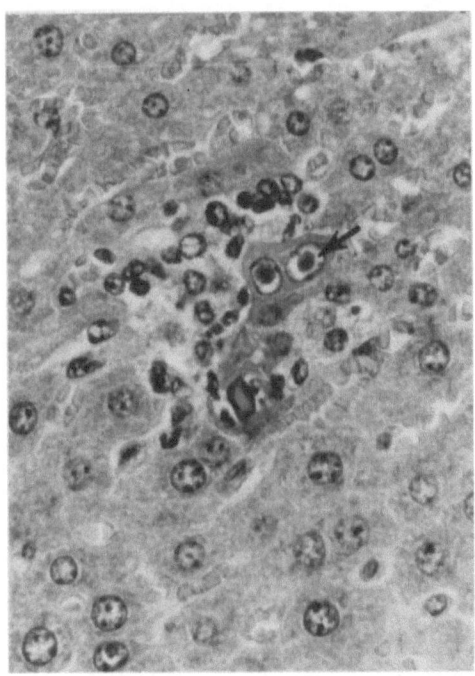

Fig. 213. Focal hepatitis, liver, mouse. Intranuclear inclusions (*arrow*); experimental mouse cytomegalovirus infection in a neonatal mouse. H&E, ×440

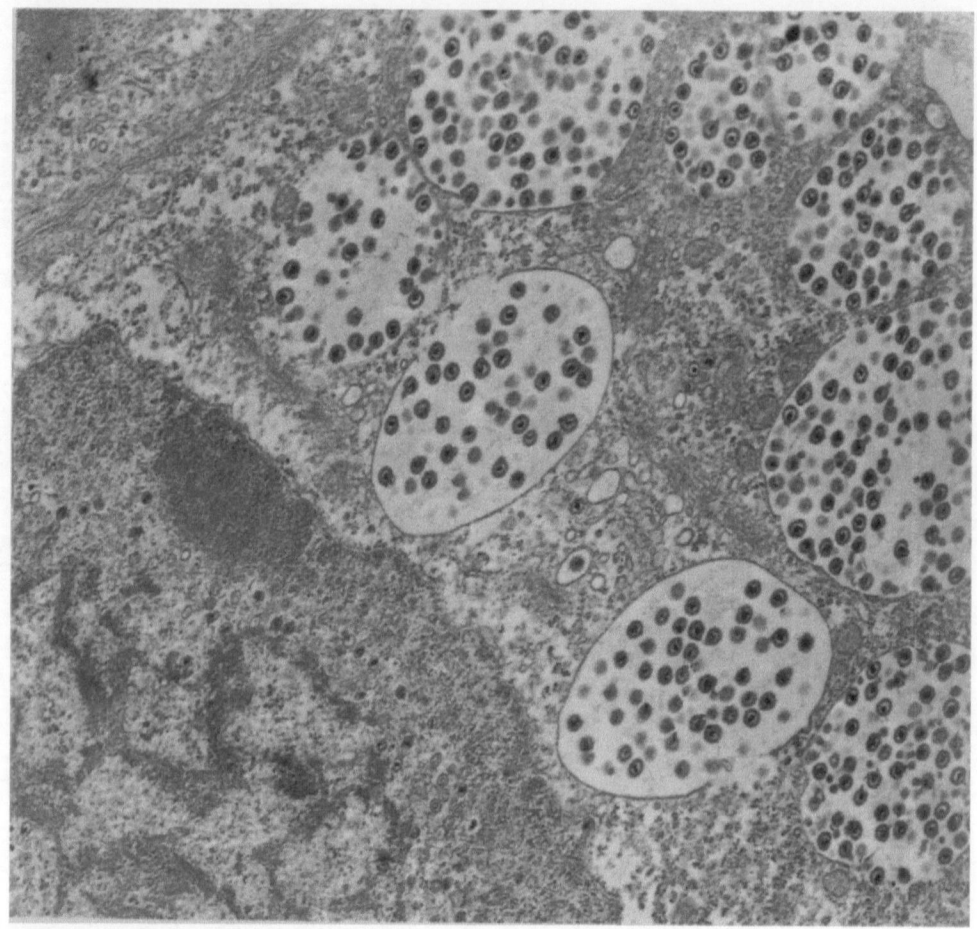

Fig. 214. Salivary gland, mouse, 16 days after infection with mouse cytomegalovirus. Viral particles are in the nucleus and in cytoplasmic vesicles. (Courtesy of Ruebner et al. 1964 and *American Journal of Pathology*) Uranylacetate, TEM, ×16300

an important role in natural murine infections, but transmission from dam to fetus can occur in latent form under experimental conditions (Chantler et al. 1979). Suckling pups are protected from infection by maternally derived antibody (Mannini and Medearis 1961; Medearis 1964). In susceptible mice, infection is followed by one of four outcomes: death, recovery, chronic localized replication in salivary glands, or latency, in which virus is harbored in tissues in a nonreplicative state (Olding et al. 1976). Considerably less is known about cytomegaloviruses of hamster and rat. In these species, infection appears to be largely restricted to salivary glands (Lussier 1975; Lyon et al. 1959; Priscott and Tyrrell 1982).

Pathogenesis

Most studies on pathogenesis of cytomegaloviruses have been done with the mouse virus. The course of infection is dependent on route of inoculation, host age and genotype, virulence of the agent, and immune status of the host (Hudson 1979; Lussier 1975; Mannini and Medearis 1961). Within a week of inoculation, young mice develop leukocytosis and viremia with dissemination to multiple organs. Hematocrit, leukocyte, and platelet counts decrease, but return to normal by 15 days (Lussier 1975; McCordock and Smith 1936; Osborn and Shahidi 1973). Intranasal or oral inoculation of low doses into young adult mice results in subclinical pulmonary infection, followed by viremia and dissemination. Virus replicates in monocytes and alveolar macrophages, with septal thickening and edema (Brody and Craighead 1974; Jordan 1978). This probably represents the natural course of infection. Salivary gland infection occurs regularly, but virus replication does not take place until later than other organs and infection persists for many months in this location after virus is cleared from other sites. Replication and excretion of salivary virus can occur in the presence or absence of discernible lesions. The affinity of the virus for salivary glands is underscored by the fact that this site is preferentially infected regardless of host age, and natural infections are localized to this site alone (Lussier 1975; Brodsky and Rowe 1958; Mims and Gould 1979; Olding et al. 1976).

Macrophages are important in early virus clearance and restriction (or replication) (Brautigam et al. 1979). Neutralizing antibody appears early after infection and provides a protective effect against challenge, but it does not seem to be involved in recovery (Hudson 1979). Cell-mediated immunity is important in host recovery (Ho 1980; Hudson 1979; Lussier 1975), and lymphocytic infiltration heralds termination of infection (Henson and Strano 1972). Immunity can result in recovery, but more often results in two types of persistent infections. In chronic infections, the virus is localized to salivary glands, where it replicates and is orally excreted for up to 1 year after infection (Brodsky and Rowe 1958). More often, latent infections develop, in which replicating virus cannot be detected (Olding et al. 1976). Salivary glands, macrophages, bone marrow-derived (B) lymphocytes, and reproductive tissues of latently infected mice harbor cytomegalovirus but not brain, thymus, liver, or kidney (Brautigam et al. 1979; Cheung et al. 1980; Olding et al. 1976). Latent infection can be reactivated in vivo and in vitro by immunosuppression and allogeneic reactions (Gardner et al. 1974; Hudson 1979; Lussier 1975; Montplaisir 1979).

Cytomegalovirus has been shown to have a number of immunosuppressive effects on infected mice due to alterations in T cell, B cell, and macrophage functions and interferon response (Hudson 1979). A synergistic effect on mouse mortality occurs in combined cytomegalovirus and *Pseudomonas aeruginosa* infections (Hamilton and Overall 1978). Chronically and latently infected mice can develop immune com-

plex renal glomerular lesions and antinuclear antibodies (Olding et al. 1976).

Experimental inoculation of neonatal rats with rat cytomegalovirus results in disseminated infection, but without the mortality and liver disease observed in mice infected with mouse cytomegalovirus. Infection in rats is also persistent, with the salivary gland being the preferential target for virus replication, shedding, and persistence (Bruggeman et al. 1985).

Etiology

Cytomegaloviruses are herpesviruses that are species specific in vivo. Several strains have been isolated from laboratory and wild mice and vary in virulence. Different cytomegaloviruses have been described in mice, rats, hamsters, guinea pigs, nonhuman primates, moles, pigs, dogs, feral rodents, sheep, horses, and humans (Lussier 1975). The mouse and rat cytomegaloviruses are antigenically and genetically distinct viruses (Bruggeman et al. 1985; Priscott and Tyrrell 1982).

Frequency

Based on observation of salivary gland inclusions or isolation of virus, infection is not common among laboratory mice and is rare among hamsters and rats. With appropriate techniques, infection is found to be very common among wild mice and rats (Bruggeman et al. 1985; Lussier 1975; Mannini and Medearis 1961). Molecular hybridization techniques have revealed latent cytomegaloviral DNA in specific pathogen-free mice, in which virus was undetectable by other means. It is likely many laboratory mice may be latently infected, but presence of viral DNA does not necessarily mean that the virus can be reactivated or will interfere with experimental procedures (Cheung et al. 1980).

Comparison with Other Species

Cytomegaloviruses frequently infect salivary glands in many species of animals and lesions are often present as incidental findings. Human infection mimics the murine disease in many ways, including its propensity for generalized infection following immunosuppression and allogeneic reactions. As in mice, fatal cytomegalovirus infections in humans are characterized by pneumonia and disseminated cytomegalic inclusion disease (Hudson 1979; Lussier 1975; Mims and Gould 1979; Montplaisir 1979).

References

Bentvelzen P, Hilgers J (1980) Murine mammary tumor virus. In: Klein G (ed) Viral oncology. Raven, New York, pp 311–355

Brautigam AR, Dutko EJ, Olding LB, Oldstone MB (1979) Pathogenesis of murine cytomegalovirus infection: the macrophage as a permissive cell for cytomegalovirus infection, replication and latency. J Gen Virol 44:349–359

Brodsky I, Rowe WP (1958) Chronic subclinical infection with mouse salivary gland virus. Proc Soc Exp Biol Med 99:654–655

Brody AR, Craighead JE (1974) Pathogenesis of pulmonary cytomegalovirus infection in immunosuppressed mice. J Infect Dis 129:677–689

Bruggeman CA, Meijer H, Bosman F, van Boven CPA (1985) Biology of rat cytomegalovirus infection. Intervirology 24:1–9

Chantler JK, Misra V, Hudson JB (1979) Vertical transmission of murine cytomegalovirus. J Gen Virol 42:621–625

Cheung KS, Huang ES, Lang DJ (1980) Murine cytomegalovirus: detection of latent infection by nucleic acid hybridization technique. Infect Immun 27:851–854

Cross SS, Parker JC, Rowe WP, Robbins ML (1979) Biology of mouse thymic virus, a herpesvirus of mice, and the antigenic relationship to mouse cytomegalovirus. Infect Immun 26:1186–1195

Gardner MB, Officer JE, Parker J, Estes JD, Rongey RW (1974) Induction of disseminated virulent cytomegalovirus infection by immunosuppression of naturally chronically infected wild mice. Infect Immun 10:966–969

Hamilton JR, Overall JC Jr (1978) Synergistic infection with murine cytomegalovirus and Pseudomonas aeruginosa in mice. J Infect Dis 137:775–782

Henson D, Strano AJ (1972) Mouse cytomegalovirus. Necrosis of infected and morphologically normal submaxillary gland acinar cells during termination of chronic infection. Am J Pathol 68:183–202

Ho M (1980) Role of specific cytotoxic lymphocytes in cellular immunity against murine cytomegalovirus. Infect Immun 27:767–776

Hudson JB (1979) The murine cytomegalovirus as a model for the study of viral pathogenesis and persistent infections. Arch Virol 62:1–29

Jordan MC (1978) Interstitial pneumonia and subclinical infection after intranasal inoculation of murine cytomegalovirus. Infect Immun 21:275–280

Lussier G (1975) Murine cytomegalovirus (MCMV). Adv Vet Sci Comp Med 9:223–247

Lyon HW, Christian JJ, Miller CW (1959) Cytomegalic inclusion disease of lacrimal glands in male laboratory rats. Proc Soc Exp Biol Med 10:164–166

Mannini A, Medearis DN Jr (1961) Mouse salivary gland virus infections. Am J Hyg 73:329–343

McCordock HA, Smith MG (1936) The visceral lesions produced in mice by the salivary gland virus of mice. J Exp Med 63:303–310

Medearis DN Jr (1964) Mouse cytomegalovirus infection. III. Attempts to produce intrauterine infections. Am J Hyg 80:113–120

Mims CA, Gould J (1979) Infection of salivary glands, kidneys, adrenals, ovaries and epithelia by murine cytomegalovirus. J Med Microbial 12:113–122

Montplaisir S (1979) Latency and activation of cytomegalovirus in man and in mice. Can J Microbiol 25:261–266

Olding LB, Kingsbury DT, Oldstone MB (1976) Pathogenesis of cytomegalovirus infection. Distribution of viral products, immune complexes and autoimmunity during latent murine infection. J Gen Virol 33:267–280

Osborn JE, Shahidi NT (1973) Thrombocytopenia in murine cytomegalovirus infection. J Lab Clin Med 81:53–63

Priscott PK, Tyrrell DA (1982) The isolation and partial characterization of a cytomegalovirus from the brown rat Rattus norvegicus. Arch Virol 73:145–160

Ruebner BH, Miyai K, Slusser RJ, Wedemeyer P, Medearis DN Jr (1964) Mouse cytomegalovirus infection. An electron microscopic study of hepatic parenchymal cells. Am J Pathol 44:799–821

Ruebner BH, Hirano T, Slusser R, Osborn J, Medearis DN Jr (1966) Cytomegalovirus infection. Viral ultrastructure with particular reference to the relationship of lysosomes to cytoplasmic inclusions. Am J Pathol 48:971–989

Schwartz JN, Daniels CA, Klintworth GK (1975) Lymphoid cell necrosis, thymic atrophy, and growth retardation in newborn mice inoculated with murine cytomegalovirus. Am J Pathol 79:509–522

Ward JM, Lock A, Collins MJ, Gonda MA, Reynolds CW (1984) Papovaviral sialoadenitis in athymic nude rats. Lab Anim 18:84–89

Sialodacryoadenitis Virus Infection, Rat

Robert O. Jacoby

Synonym. Rat coronavirus infection

Gross Appearance

Lesions develop during the first week after infection and occur primarily in the submandibular and parotid salivary glands, which are located on the anteroventral and anterolateral aspects of the neck, respectively (Fig. 215). Affected glands are unilaterally or bilaterally enlarged and pale yellow to white in contrast to their normal tan color (Fig. 216). Periglandular connective tissue is often edematous and, together with glandular enlargement, may cause clinically detectable cervical swelling. The cervical lymph nodes may also be enlarged and edematous. They are frequently congested, especially early in infection, and occasionally flecked with red spots.

Microscopic Features

The glandular lesions of sialodacryoadenitis (SDA) consist of necrosis and inflammation during acute disease and squamous metaplasia as tissues are repaired (Jonas et al. 1969; Jacoby et al. 1975). Serous and mixed salivary glands in the oropharynx and neck are affected, but mucous salivary glands are resistant. Infection begins as rhinotracheitis. Earliest glandular lesions are detected in the submandibular salivary glands, where necrosis of ductular epithelium can begin by 4 days postinfection (Fig. 217) and is associated with intracytoplasmic viral antigen. Infection and necrosis of adjacent acinar parenchyma begins shortly thereafter and quickly involves large areas of the parenchyma (Fig. 218). Necrosis is accompanied by inflammatory edema, which also engulfs periglandular connective tissues. The inflammatory cells consist of small and large mononuclear cells and polymorphonuclear leukocytes. Necrosis and inflammation in the parotid and oropharyngeal salivary glands develop in a similar pattern. The sublingual salivary glands, which are mucin-secreting glands at the anterior pole of the submandibular glands

(Greene 1959), are not affected (Fig. 219). Cervical lymph nodes are congested during the first 4–5 days of infection, but remain enlarged from ensuing hyperplasia. By the end of the first week, active germinal centers appear in the cortex and medullary cords begin to fill with plasma cells. Mild focal necrosis and focal hemorrhage may occur during this period.

Repair begins during the second week of infection and is normally complete by the end of the third week. It proceeds efficiently because acinar and ductular basement membrane remain intact through acute infection. The characteristic sign of repair is squamous metaplasia of ductular epithelium (Figs. 219, 220), a transient lesion that is replaced by normal columnar epithelium. Acinar regeneration entails the maturation of clusters or rosettes of small hyperchromatic cells into normal acini. Inflammation and edema subside coincidently with reconstitution of glandular parenchyma, but interstitial foci of nonsuppurative inflammation may linger after parenchymal repair is complete. Fibrosis is not a significant sequel of SDA, but small interstitial scars may develop.

Ultrastructure

The ultrastructure of SDA infection has not been thoroughly described. Jonas and coworkers (1969) found two types of viral particles in ductal epithelium of submaxillary glands of experimentally infected rats. One type was 60–70 nm in diameter and had an electron-dense core. The second type was about 10–20 nm in diameter and occurred as cytoplasmic aggregates. Two types of cytoplasmic vacuoles also were observed. One had an electron-lucent matrix, was membrane bound, and contained cytoplasmic debris, membrane fragments, and cell organelles. The other also was membrane bound and contained the electron-dense viral particles. Parker and coworkers (1970) described rat coronavirus as a pleomorphic virus 60–200 nm in diameter. The variation in reported measurements appears to be due to differences in the preparations used.

Differential Diagnosis

Sialoadenitis in the rat is characteristic of rat coronavirus infection. Two isolates were described in the early 1970s; SDA virus (Bhatt et al. 1972), rat coronavirus (Parker et al. 1970), and others have since been added (Jacoby 1986). The various isolates have comparable pathogenicity for rats and thus far cannot be distinguished serologically. Cytomegalovirus infection in the salivary glands of rats is associated with enlarged ductal epithelial cells which may contain intranuclear inclusions (Lyon et al. 1959). An unclassified cytopathic agent has been isolated from the submaxillary salivary glands of rats, but it is apparently nonpathogenic (Ashe 1969). Salivary gland tumors are rare in rats, and no other classes of microbial agents have been consistently associated with sialoadenitis. It is worth noting that rat

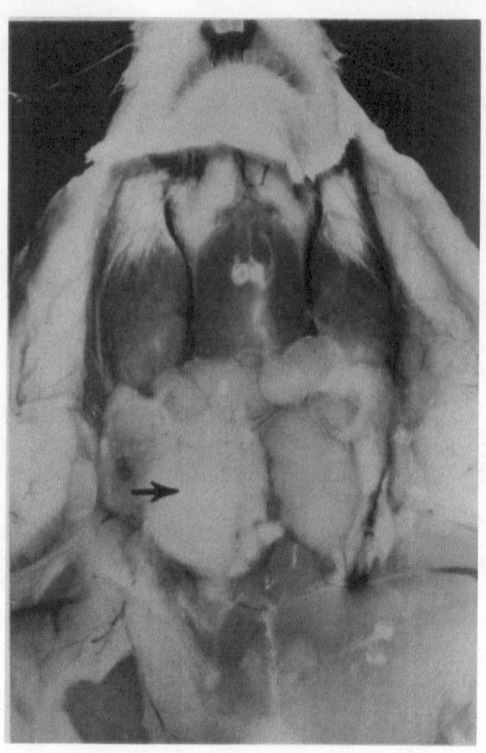

Fig. 216. Submaxillary gland (*arrow*), rat inoculated intranasally with sialodacryoadenitis virus 5 days previously. The gland is swollen and pale. Cervical lymph nodes are also enlarged. (From Jacoby et al. 1979, with permission of Academic Press)

▶

Fig. 217. (*upper left*) Submaxillary salivary gland, rat, 4 days after intranasal inoculation of sialodacryoadenitis virus. Note necrosis of a large salivary duct and mild interstitial inflammation. Early signs of degeneration in adjacent acini are indicated by intracytoplasmic vacuoles. (From Jacoby et al. 1975, with permission of *Veterinary Pathology*.) H&E, ×400

Fig. 218. (*lower left*) Submaxillary gland, rat infected intranasally with sialodacryoadenitis virus 5 days previously. Note widespread necrosis and inflammation. (From Jacoby et al. 1975, with permission of *Veterinary Pathology*.) H&E, ×160

Fig. 219. (*upper right*) Submaxillary gland, rat, effaced by severe sialoadenitis 6 days after infection with sialodacryoadenitis virus. Squamous metaplasia is evident in a salivary duct. The adjacent sublingual gland is normal. (From Jacoby et al. 1975, with permission of *Veterinary Pathology*.) H&E, ×160

Fig. 220. (*lower right*) Submaxillary gland, rat inoculated with sialodacryoadenitis virus 9 days previously. Note the prominent squamous metaplasia of ductal epithelium and clusters of regenerating acini, especially in the lower right quadrant. (From Jacoby et al. 1975, with permission of *Veterinary Pathology*.) H&E, ×160

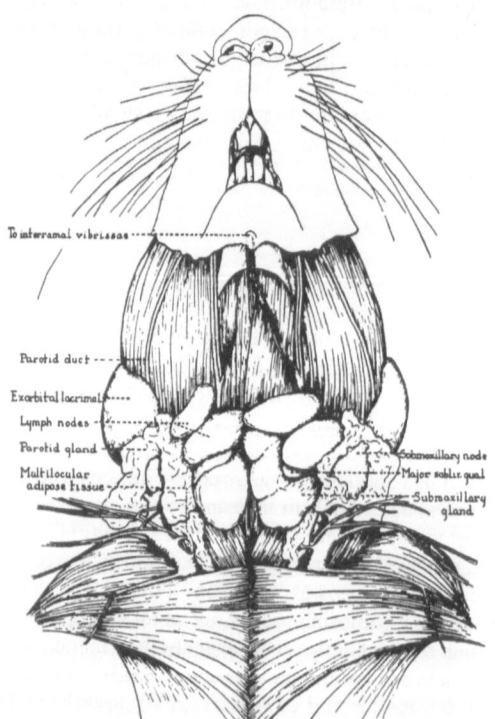

Fig. 215. Gross anatomy, ventral aspect of the neck, rat. (Adapted from Greene 1959, with permission of Hafner Publ. Co.)

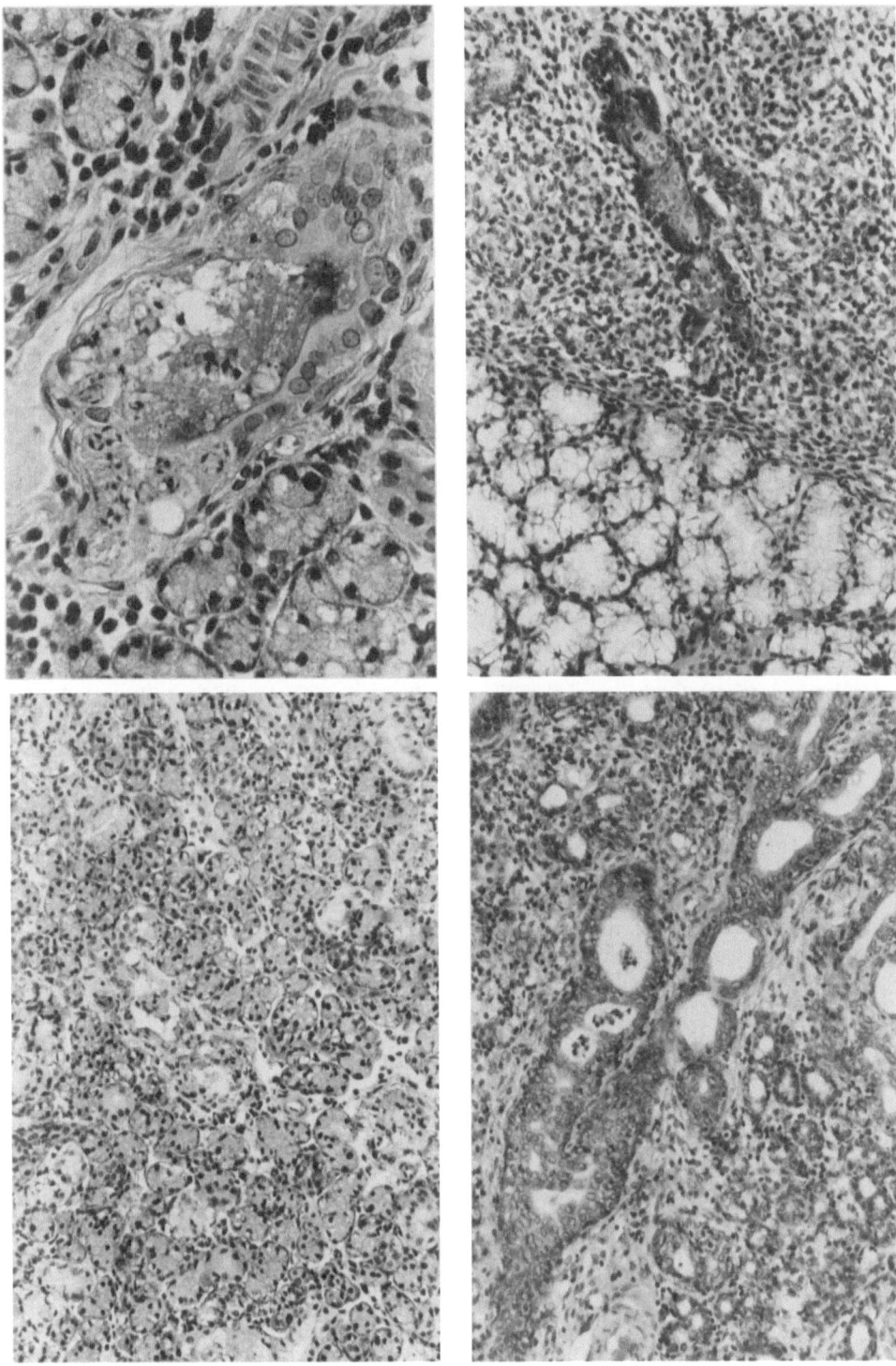

coronaviruses are antigenically related to mouse coronaviruses commonly grouped under the name mouse hepatitis virus (MHV) (Bhatt et al. 1972; Parker et al. 1970). Because under experimental conditions SDA virus can infect mice (Bhatt et al. 1977) and MHV can infect rats (Taguchi et al. 1979), some caution in serologic techniques and interpretation is advised when investigating coronavirus infections in these species, especially when they are housed near one another (Jacoby et al. 1979).

Biologic Features

SDA is a common infectious disease of rats. It is acute and self-limiting and can affect rats at any age. It usually occurs, however, in weanlings that have recently lost maternal immunity or in sucklings of nonimmune dams. Morbidity is high, but mortality is low. Clinical disease develops by 1 week after exposure to the virus and is characterized by one or more of the following signs: sneezing; photophobia; lacrimation, including excessive secretion of porphyrin pigment (so-called chromodacryorrhea), which imparts a red stain to periocular and perinasal skin; cervical swelling due to palpably enlarged submaxillary salivary glands and edema; transient anorexia and weight loss; and reduced breeding performance. The severity of clinical disease can vary among rat strains, and mild disease may be asymptomatic (Bhatt and Jacoby 1985). Seroconversion, which typically occurs about 1 week after exposure to virus, marks the cessation of infection and transmission and the initiation of repair. Although immunity to systemic infection is long-lived, seropositive rats can be transiently reinfected and transmit virus to naive rats for several days (Percy et al. 1990; Weir et al. 1990b).

Convalescence is associated with humoral immunity, which can be detected serologically by neutralization (Bhatt et al. 1972), immunofluorescence (Smith 1983), or enzyme-linked immunoabsorption (Peters and Collins 1981). Infection of athymic rats, by contrast, is persistent and results in chronic active inflammation and prominent scarring of the salivary glands. Affected rats also transmit infection for prolonged periods and may develop a wasted appearance (Weir et al. 1990a). SDA has been detected in laboratory rats in North America, Europe, and

Asia and is probably one of the most common infections of laboratory rats worldwide. The infection is difficult to eliminate until susceptible animals have become immune or until rooms with infected rats have been depopulated. Virus is spread rapidly by contact or aerosol and infects the respiratory tract before reaching the salivary glands and lacrimal glands. The route of spread from the nasopharynx has not been confirmed, but might involve retrograde infection of salivary and lacrimal excretory ducts. Viremia, if it occurs, is likely to be transient (Jacoby et al. 1975). Virus spread in immunocompetent rats is limited to the respiratory tract, salivary glands, lacrimal glands, and regional lymph nodes. Infection of the urinary tract has been demonstrated, however, in athymic rats (Weir et al. 1990a). Keratoconjunctivitis associated with infection has been attributed to corneal dehydration from interrupted tear production rather than to viral invasion of the cornea (Lai et al. 1976).

As previously stated, SDA virus is closely related to a coronavirus isolated from rats by Parker and coworkers (Parker et al. 1970; Bhatt and Jacoby 1977). SDA virus is relatively large (60–200nm) and pleomorphic and can be grown in monolayer cultures of primary rat kidney (Bhatt et al. 1972) or in mouse L2 fibroblasts (Percy et al. 1989; Gaertner et al. 1991). The cytopathic effect of rat coronaviruses is syncytia formation. Infection is typically detected serologically by enzyme-linked immunosorbent assay (ELISA) or immunofluorescence assay.

SDA may affect the results in research using rats (Barthold 1985). The most significant interference stems from chronic ophthalmologic lesions secondary to dacryoadenitis, which are described elsewhere in these monographs. During acute disease, rats may consume less food and have transient weight loss and reduced breeding performance (Utsumi et al. 1980). Infected rats may also be at greater risk during general anesthesia, especially if they are restrained with head and neck extended. Nasal exudates from acute rhinitis (see p. 249), and swelling of cervical tissues may combine to narrow or obstruct the upper airways. Biochemical or morphological studies on affected glands would not be suitable during acute infection or convalescence (Percy et al. 1988). Whether squamous metaplasia of healing glands can affect the course of carcinogenicity studies has not been clarified. There is no published evidence, however, that such lesions increase the prevalence of salivary tumors in rats treated with potential or known carcinogens.

Comparison with Other Species

The lesions of SDA in rats are not duplicated in natural disease of other laboratory or domestic animal species. Sialoadenitis in humans may be infectious (e.g., mumps) or noninfectious (Vickers and Gorling 1977), but the pathogenesis, distribution of lesions, and other manifestations of human sialoadenitis do not closely resemble SDA.

References

Ashe WK (1969) Properties of the rat submaxillary gland virus hemagglutinin and antihemagglutinin and their incidence in apparently healthy gnotobiotic and conventional rats. J Gen Virol 4:1–7

Barthold SW (1985) Research complications and state of knowledge of rodent coronaviruses. In: Hamm TA (ed) Complications of viral and mycoplasmal infections in rodents to toxicology research and testing. Hemisphere, Washington DC

Bhatt PN, Jacoby RO (1977) Experimental infection of adult axenic rats with Parker's rat coronavirus. Arch Virol 54:345–352

Bhatt PN, Jacoby RO (1985) Epizootiological observations of natural and experimental infection with sialodacryoadenitis virus in rats. Lab Anim Sci 35:129–134

Bhatt PN, Percy DH, Jonas AM (1972) Characterization of the virus of sialodacryoadenitis of rats: a member of the coronavirus group. J Infect Dis 126:123–130

Bhatt PN, Jacoby RO, Jonas AM (1977) Respiratory infection in mice with sialodacryoadenitis virus, a coronavirus of rats. Infect Immun 18:823–827

Gaertner DJ, Smith AL, Paturzo FX, Jacoby RO (1991) Susceptibility of rodent cell lines to rat coronaviruses and differential enhancement by trypsin or DEAE-dextran. Arch Virol 118:57–66

Greene EC (1959) Anatomy of the rat. Hafner, New York, p 93 (Transactions of American Philosophical Society, vol 27)

Jacoby RO (1986) Rat coronavirus. In: Bhatt PN, Jacoby RO, Morse HC, New AE (eds) Viral and mycoplasmal infections of laboratory rodents: effects on biomedical research. Academic, Orlando, pp 625–638

Jacoby RO, Bhatt PN, Jonas AM (1975) Pathogenesis of sialodacryoadenitis in gnotobiotic rats. Vet Pathol 12:196–209

Jacoby RO, Bhatt PN, Jonas AM (1979) Viral disease. In: Baker H, Lindsey JR, Weisbroth SH (eds) The laboratory rat, vol 1: biology and diseases. Academic, New York, chap 11

Jonas AM, Craft J, Black L, Bhatt PN, Hilding D (1969) Sialodacryoadenitis in the rat. A light and electron microscopic study. Arch Pathol 88:613–622

Lai YL, Jacoby RO, Bhatt PN, Jonas AM (1976) Keratoconjunctivitis associated with sialodacryoadenitis in rats. Invest Ophthalmol 15:538–541

Lyon HW, Christian JJ, Miller CW (1959) Cytomegalic inclusion disease of lacrimal glands in male laboratory rats. Proc Soc Exp Biol Med 101:164–166

Parker JC, Cross SS, Rowe WP (1970) Rat coronavirus (RCV): a prevalent, naturally occurring pneumotropic virus of rats. Arch Gesamte Virusforsch 31:293–302

Percy DH, Hayes MA, Kocal TE, Wojcinski ZW (1988) Depletion of salivary gland epidermal growth factor by sialodacryoadenitis virus infection in the Wistar rat. Vet Pathol 25:183–192

Percy DH, Bond S, MacInnes J (1989) Replication of sialodacryoadenitis virus in mouse L-2 cells. Arch Virol 104:323–333

Percy DH, Bond SJ, Paturzo FX, Bhatt PN (1990) Duration of protection from reinfection following exposure to sialodacryoadenitis virus in Wistar rats. Lab Anim Sci 40:144–149

Peters RL, Collins MJ Jr (1981) Use of mouse hepatitis virus antigen in an enzyme-linked immunosorbent assay for rat coronaviruses. Lab Anim Sci 31:472–475

Smith AL (1983) An immunofluorescence test for detection of serum antibody to rodent coronaviruses. Lab Anim Sci 33:157–160

Taguchi F, Yamada A, Fujiwara K (1979) Asymptomatic infection of mouse hepatitis virus in the rat. Arch Virol 59:275–279

Utsumi K, Ishikawa T, Maeda T, Shimizu S, Tatsumi H, Fujiwara K (1980) Infectious sialodacryoadenitis and rat breeding. Lab Anim 14:303–307

Vickers RA, Gorlin RJ (1977) Face, lips, teeth, mouth, jaws, salivary glands and neck. In: Anderson WAD, Kissane JM (eds) Pathology, 7th edn, vol 2. Mosby, St Louis, chap 29

Weir EC, Jacoby RO, Paturzo FX, Johnson EA, Ardito RB (1990a) Persistence of sialodacryoadenitis virus in athymic rats. Lab Anim Sci 40:138–143

Weir EC, Jacoby RO, Paturzo, FX, Johnson EA (1990b) Infection of SDAV-immune rats with SDAV and rat coronavirus. Lab Anim Sci 40:363–366

The Exocrine Pancreas

Embryology, Histology, and Ultrastructure of the Exocrine Pancreas

Scot L. Eustis and Gary A. Boorman

Introduction

The primary function of the exocrine pancreas is the synthesis and secretion of the enzymes necessary to digest and assimilate ingested foods. Digestive enzymes are synthesized in acinar cells and stored within the cells as zymogen granules. A second function of the exocrine pancreas is the maintenance of homeostasis within the intestinal lumen by secretion of fluid containing high levels of bicarbonate. The epithelium comprising the duct system of the exocrine pancreas appears to be the major contributor to electrolyte secretion. In the following paragraphs, the embryologic development, histology, ultrastructure, and structural–functional relationships of the exocrine pancreas are reviewed.

Overview of Embryonic Development

The mammalian pancreas develops by fusion of dorsal and ventral pancreatic rudiments derived from the gut endoderm. The dorsal pancreatic rudiment emerges from the gut at the 26-somite stage (approximately day 10.5 of gestation in the rat and day 9.5 in the mouse), and the ventral rudiment follows about 12 h later (28–30 somites; Pictet et al. 1972). From day 11 to 14 of gestation in the rat, there is continued cell proliferation with cell-to-cell interaction, which produces the characteristic organ shape and histologic architecture (Rutter et al. 1968a–c). Organization of the pancreatic epithelium into acini and ducts is well developed by days 16–17 in the rat, when the two rudimentary glands merge.

Cellular differentiation resulting in the typical cytologic features of the adult exocrine cell begins about day 15 in the rat. It is during the process of differentiation that acinar cells acquire the cellular organelles and enzyme systems necessary for the synthesis, intracellular transport, and secretion of digestive enzymes. Corresponding to these changes, acinar cell volume doubles between days 15 and 19 of gestation (Uchiyama and Watanabe 1984). Mesenchymal cells, which comprise as much as 55% of the embryonic pancreas on day 18, are essential for the growth and differentiation of the exocrine component. A substance (mesodermal factor) produced by mesenchymal cells apparently binds to epithelial membrane receptors to stimulate endogenous production of cyclic adenosine monophosphate (cAMP) and, ultimately, acinar cell differentiation and proliferation (Rutter and Pictet 1976).

This pattern of development is correlated with the transcription and translation of pancreas-specific genes and the accumulation of exocrine proteins (e.g., carboxypeptidase, amylase, chymotrypsin) within the embryonic pancreas (Rutter et al. 1968c; Doyle and Jamieson 1978; Rutter et al. 1964; Sanders and Rutter 1974). Very low levels of specific activity for carboxypeptidase, amylase, and chymotrypsin are detected in the pancreatic rudiment between gestation days 12–14, at which stage no zymogen granules are seen. With the rapid development of rough endoplasmic reticulum (RER) beginning on day 15 and the accumulation of zymogen granules in acinar cells, there is an increase in transcription of exocrine-specific genes, an increase in enzyme-specific mRNA, and a 10^3- to 10^4-fold increase in the specific activity of these enzymes.

In the rat, acinar cells begin to respond to cholinergic stimuli (secretagogues such as carbamylcholine) on gestation day 21, shortly before birth, although full responsiveness does not occur until the first postnatal day or later (Doyle and Jamieson 1978; Werlin and Grand 1979; Werlin and Stefaniak 1982). Muscarinic (cholinergic) receptors can be detected on acinar cells as early as day 15, but the number of receptors per cell increases until postnatal day 60 (Dumont et al. 1981).

Developmental Histology and Ultrastructure

Prior to formation of the pancreatic rudiment, the endodermal tube comprising the gut is flattened in the dorsal–ventral direction. The dorsal rudiment is formed by an evagination of the endodermal epithelium at the 26-somite stage (Pictet et al. 1972). As cells proliferate, there is progressive thickening of the rudiment and progressive constriction at the junction of the diverticulum with the gut endoderm. A cap of mesodermal cells surrounds the epithelial component of the rudimentary pancreas and is essential to further morphogenesis of the gland and development of acini (Spooner et al. 1977). Under the influence of a mesodermal factor or factors, the epithelial cells of the pancreatic rudiment undergo mitosis and assume a pseudostratified appearance. Continued cell proliferation apparently causes an increase in lateral pressure until a slight deformation occurs with a change in orientation of the cells with respect to the others. Continued mitoses at these sites result in branching of the pancreatic bud and eventual lobulation of the primitive gland. Acinar structures with ducts are present by days 16–17 in the rat (Fig. 221) and day 15 in the mouse (Pictet et al. 1972).

The epithelium of the developing pancreatic rudiment is arranged in branching tubules in which the lumen exists as a narrow or potential space delineated by the cell apices (Fig. 222). Throughout

morphogenesis, the epithelial cells comprising the pancreatic rudiment maintain the degree of structural organization and polarity shown by the gut epithelium from which the rudiment arises. The cells are joined by junctional complexes and are disposed in a single layer. A continuous basal lamina is present at the cell–mesenchyme interface. The polarity and specific relationship of the cells are perpetuated by the pattern of cell division, where the axis of the mitotic spindle is always parallel to the lumen. New junctional complexes are formed prior to completion of cytokinesis. Thus each daughter cell receives about half the apical surface of the mother cell and retains the same junctional relationship with adjacent cells (Pictet et al. 1972).

From gestation days 11–14 in the rat, all exocrine pancreatic cells of the rudiment are identical to the endodermal cells of the gut. The luminal surface of each cell is covered by microvilli, and the cytoplasm contains abundant free ribosomes. The

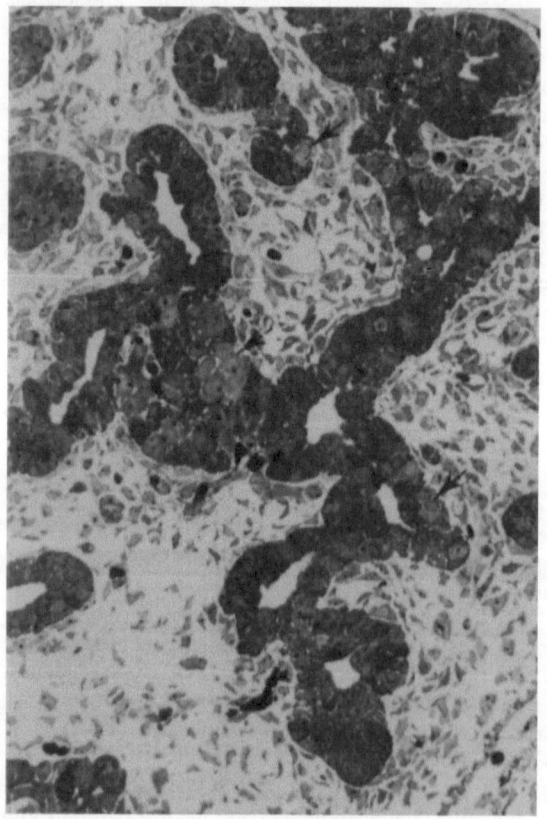

Fig. 221. Fetal rat pancreas, gestation day 16. The exocrine pancreas consists of branching tubules in a loose mesenchymal stroma. A lumen is visible in some parts of the tubules. Cells with lightly staining cytoplasm (*arrows*) are endocrine cells. Toluidine blue, ×960

Fig. 222. Fetal rat pancreas, gestation day 16. Columnar epithelial cells are arranged in a compact tubule with a narrow lumen (*L*) defined by the cell apices and the junctional complexes (*arrows*) that attach adjacent cells. Short microvilli project into the lumen. Epithelial cell nuclei are located in the basal region of the cells. Note the interstitial (mesenchymal) cell (*IC*). TEM, ×2700

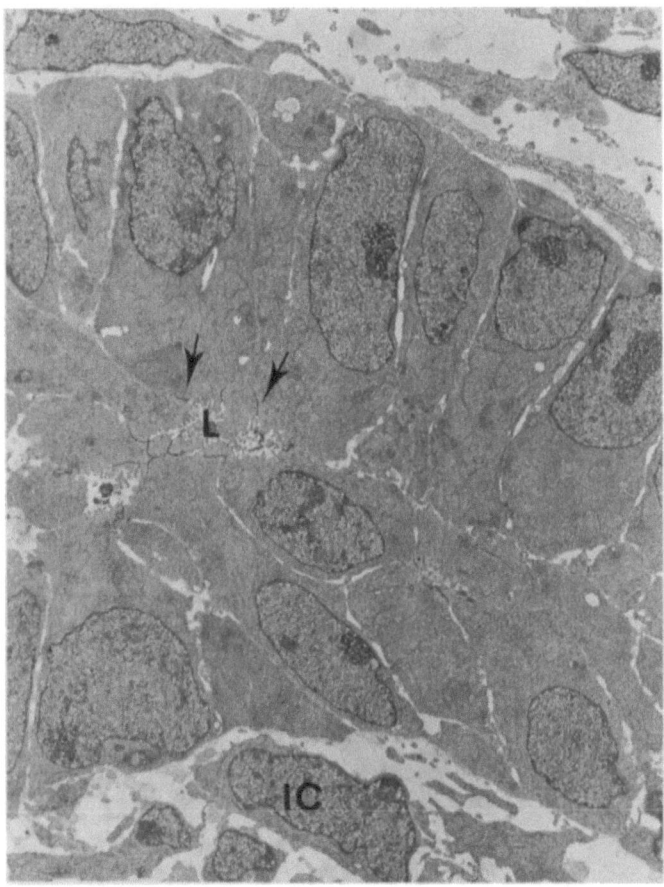

RER is sparse relative to that of differentiated acinar cells and consists of flattened cisternae that are not oriented with respect to the nuclear and plasma membranes. Zymogen granules are not present (Fig. 223). Cellular differentiation begins about day 15 (rat) as the acinar cells develop the ultrastructural features of adult secretory cells. Differentiation is not synchronous and it varies in time of onset among acini (Pictet et al. 1972). There is a rapid increase in the amount of RER as the cisternae lengthen and become oriented parallel to the circumference of the nucleus. The appearance of dense material in dilated saccules of the Golgi complex is associated with the formation of zymogen granules. Small numbers of granules located adjacent to the apical membrane are apparent in a substantial number of acinar cells by day 17 (Figs. 224, 225). Between days 17 and 21, zymogen granules accumulate and may be seen throughout the cytoplasm (Fig. 226). Zymogen granules become progressively larger during prenatal development. In contrast to acinar cells, duct cells show little ultrastructural change during this period to correlate with the development of specific functional capabilities.

Postnatal Development

The weight of the rat pancreas doubles about every 5 days for the first 3–4 weeks after birth (Sesso et al. 1973), and the pancreas reaches its adult size at about 2 months of age. This period of rapid growth is correlated with the proliferative rate of acinar cells, as demonstrated by mitotic and tritiated thymidine labeling indices, which are high during the first 2 weeks and decline rapidly thereafter (Sesso et al. 1973; Sidorova and

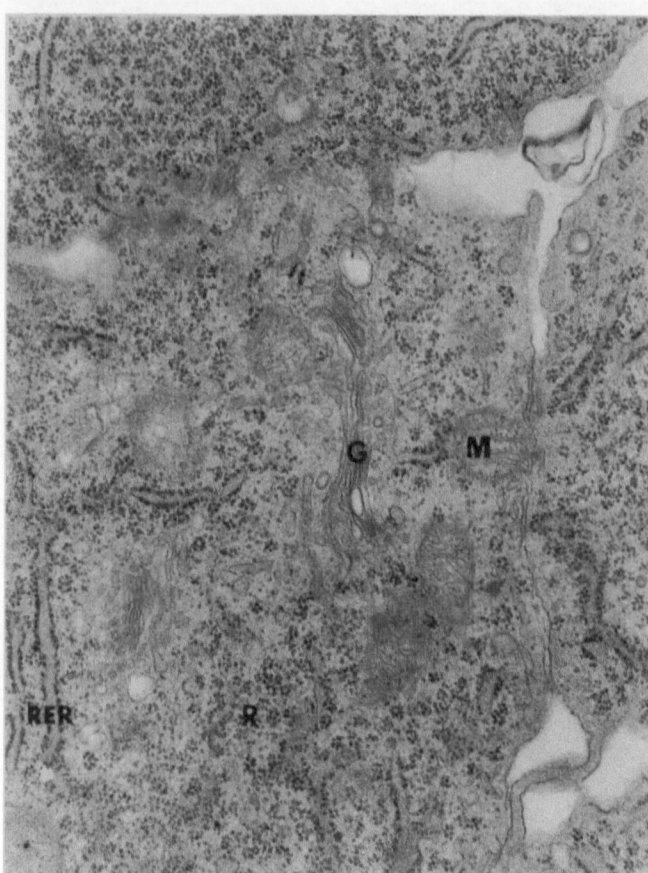

Fig. 223. Fetal rat pancreatic epithelial cell, gestation day 16. The cell has scant rough endoplasmic reticulum (*RER*), free ribosomes (*R*), Golgi apparatus (*G*), and mitochondria (*M*). TEM, ×24 000

Babaeva 1968). The labeling index of acinar cells in the rat drops from approximately 13%–14% on the last day of gestation to 4%–8% on the second day after birth (Wenzel et al. 1972). This level is maintained during the first 2–3 postnatal weeks, but the labeling index decreases to below 1% between the second and third week, when weaning occurs in suckling rats. The labeling index of acinar cells declines further to about 0.1%–0.2% between the first and second month of age and remains relatively unchanged thereafter.

The mitotic indices of centroacinar and duct cells are similar to that of acinar cells during the immediate postnatal period. However, since acinar cells comprise the vast majority of cells in the pancreas, the development of new acini seems to result from the proliferation of preexisting acinar cells (Sesso et al. 1973). There is little evidence to suggest that centroacinar cells differentiate into acinar cells

during normal postnatal growth. Subsequent growth of the pancreas is associated with enlargement of individual acini as well as continued cell proliferation. In 40-day-old rats, the mean area of one acinus is about twice that of rats aged 22 days (Sidorova and Babaeva 1968).

▶

Fig. 224. (*above*) Fetal rat pancreas, gestation day 18. The acinus has a prominent lumen (*L*), and electron-dense zymogen granules have accumulated in the apical cytoplasm of acinar cells. Note the acinar cell in mitosis (*M*). A cluster of five cells in the *lower left corner* of the micrograph comprise a developing acinus. Notice the cytoplasmic process of the interstitial cell (*IC*) that extends between the acinar bud and the developed acinus. TEM, ×2700

Fig. 225. (*below*) Fetal rat pancreatic epithelial cell, gestation day 18. Cell has zymogen granules, mitochondria, Golgi apparatus, microvilli, and junctional complexes (*arrows*). TEM, ×9600

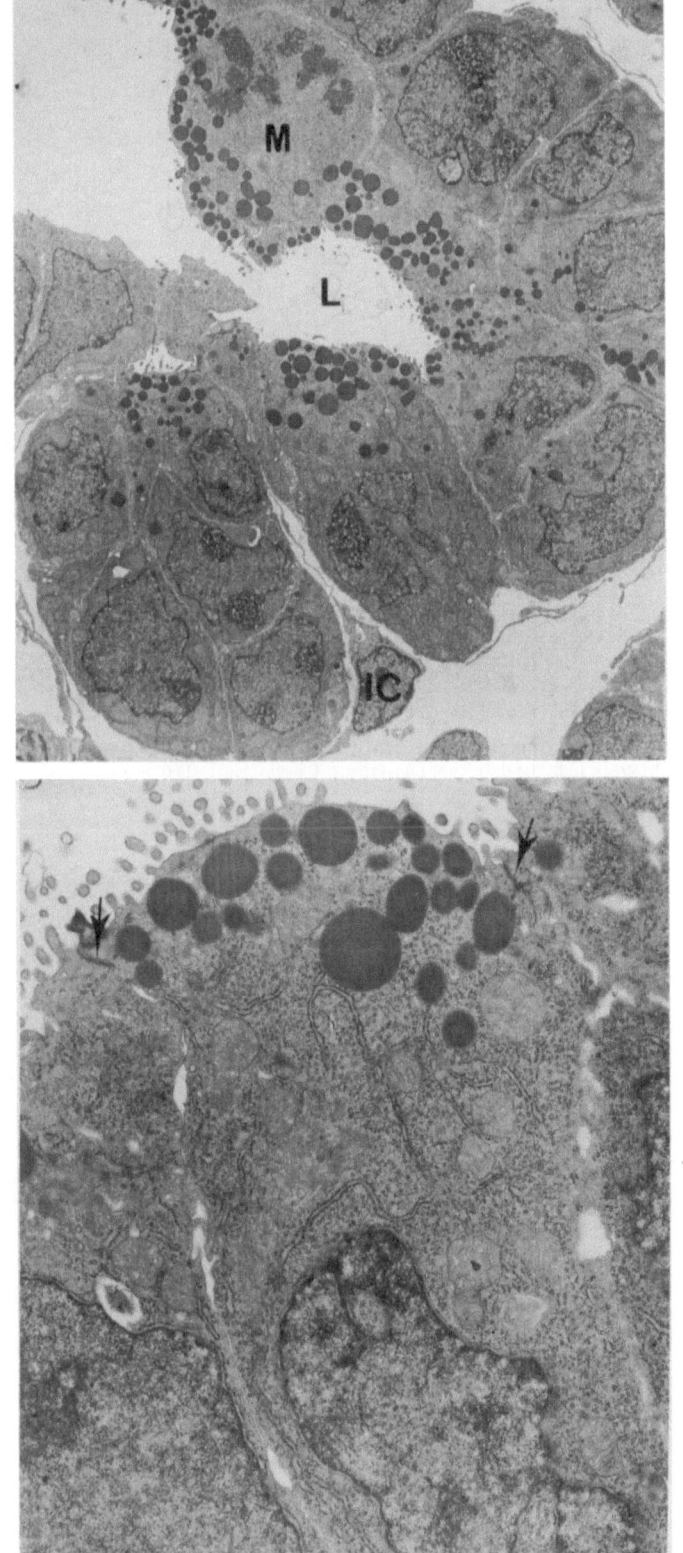

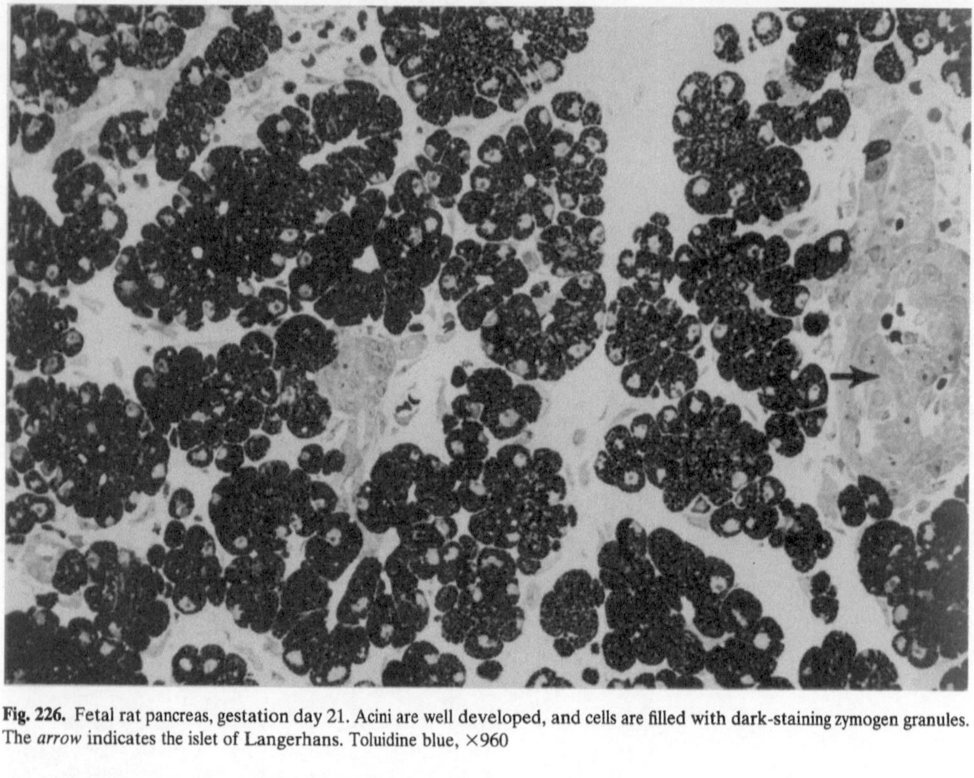

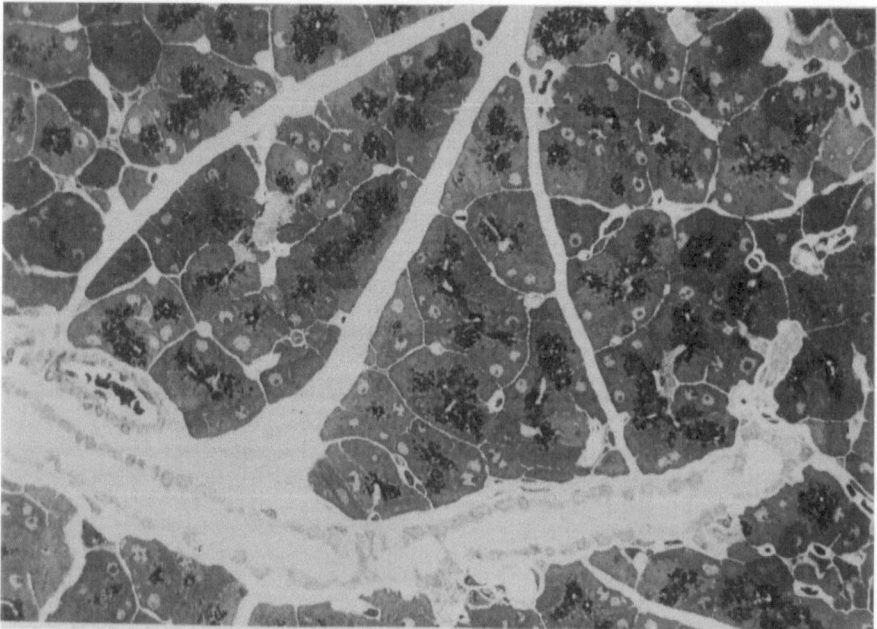

Fig. 226. Fetal rat pancreas, gestation day 21. Acini are well developed, and cells are filled with dark-staining zymogen granules. The *arrow* indicates the islet of Langerhans. Toluidine blue, ×960

Fig. 227. Adult rat pancreas. Acinar lumens are narrow, and dark-staining zymogen granules fill the apical cytoplasm of acinar cells. Note the interlobular duct. Toluidine blue, ×960

Fig. 228. Adult rat pancreas. Note the pyramidal-shaped acinar cells (*AC*), centroacinar cell (*CC*), and interstitium (*arrows*). TEM, ×2100

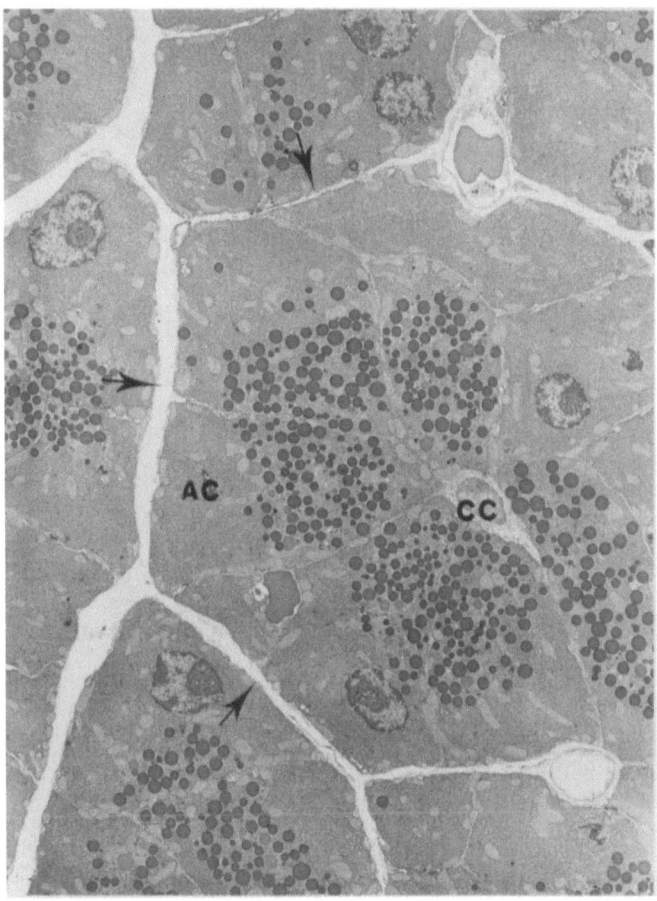

Postnatal growth of the exocrine pancreas is associated with an increase in diploid binucleate acinar cells, and in the adult pancreas a small proportion of binucleate acinar cells become tetraploid (Oates and Morgan 1986). These authors reported that, from 5 to 17 days after birth, 95% of the cells are mononucleate diploid and 5% are binucleate diploid. By 56 days after birth and thereafter, approximately 64% are binucleate diploid. In adults, 4%–6% of mononucleate cells and approximately 3% of binucleate cells are tetraploid.

Adult Pancreas

The exocrine pancreas is a lobular, compound tubuloalveolar gland (Dixon 1979). The lobules are separated by thin connective tissue septa that are continuous with the capsule. The connective tissue is more abundant around the main and interlobular ducts.

Acini are irregular, oval or elongated glands comprising a single layer of pyramid-shaped cells. The narrow apical ends of the cells define the acinar lumina, and the broad, basal portions of the cells rest on a basal lamina and scant reticular stroma (Figs. 227, 228). The nucleus of each acinar cell is spherical and located in the basal region of the cytoplasm. One or several nucleoli are present within the nucleus. The cytoplasm surrounding the nucleus is basophilic in sections of pancreas stained with hematoxylin and eosin (H&E), and the apical cytoplasm contains numerous brightly acidophilic, zymogen granules. Flattened epithelial cells with scant pale-staining cytoplasm, the centroacinar cells, protrude into the acinar lumens to a variable extent and overlie the apical surface of some acinar cells. These cells represent

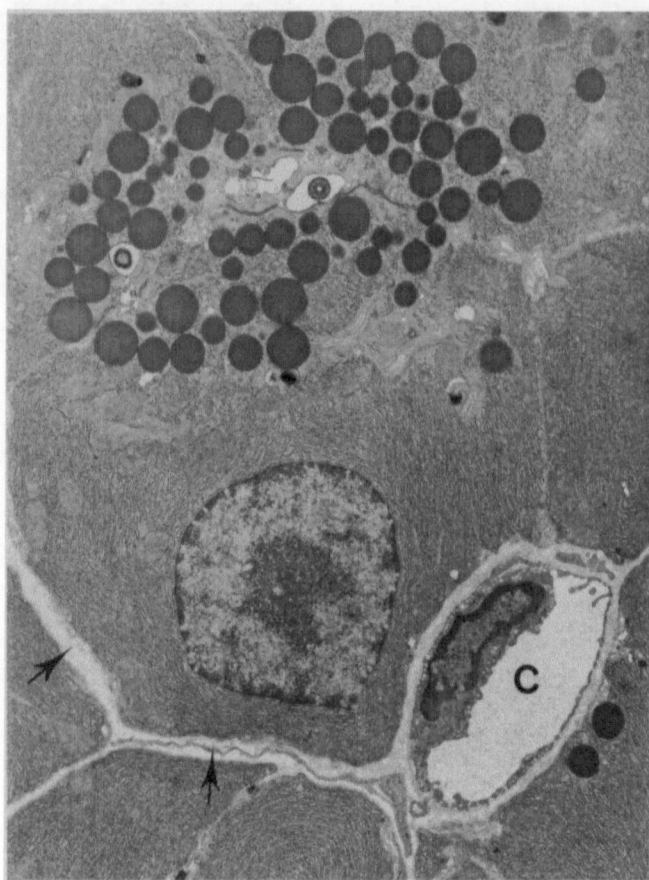

Fig. 229. Adult rat pancreas. Acinar cell contains abundant rough endoplasmic reticulum arranged in concentric lamellar arrays parallel to the plasma and nuclear membranes. The nucleus has a basal location, and zymogen granules are accumulated in the apical cytoplasm. A basal lamina is present at the cell–mesenchyme interface (*arrows*), and reticular fibers are present in the interstitial space. Note the capillary with fenestrated endothelial cell (*C*). TEM, ×6000

the beginning of the duct system and are continuous with the small intercalated ducts that are interposed between the acini and the intralobular ducts. Intercalated and intralobular ducts are comprised of a low-cuboidal epithelium. Interlobular ducts located in the connective tissue septa consist of simple cuboidal or columnar epithelium. Generally, ducts that are greater in diameter have taller epithelium.

The characteristic ultrastructural features of acinar cells are the abundant lamellar arrays of RER, which are responsible for the cytoplasmic basophilia, and the spherical membrane-bound zymogen granules containing homogenous electron-dense secretory protein (Fig. 229). Zymogen granules normally have an average diameter of 0.8–1.0 μm and occupy about 10%–20% of the cell volume (Nadelhaft 1973; Liebow and Rothman

1973). In fasted rats (deprived of food for 16 h), mean acinar cell volume has been estimated as 1670 μm³ with an average of 450 granules per cell, while after feeding the mean cell volume was 1300 μm³ with 190 granules per cell (Aughsteen and Cope 1987). Acinar cells surrounding islets of Langerhans tend to contain more and larger zymogen granules.

The Golgi apparatus is located between the nucleus and cell apex (Fig. 230). The Golgi apparatus or stack is a continuous ribbon-like structure consisting of a highly fenestrated, osmiophilic, tubular membranous network known as the *cis* element, several layers of fenestrated, flattened saccules, and saccules containing spheroidal dilatations or prosecretory granules (Rambourg et al. 1988). Detached prosecretory granules (condensing vacuoles) are seen along the *trans* aspect of the

Fig. 230. Adult rat pancreatic acinar cell. Note the Golgi apparatus (*G*), condensing vacuoles (*CV*), zymogen granules (*Z*), mitochondria (*M*), and rough endoplasmic reticulum (*RER*). TEM, ×18 900

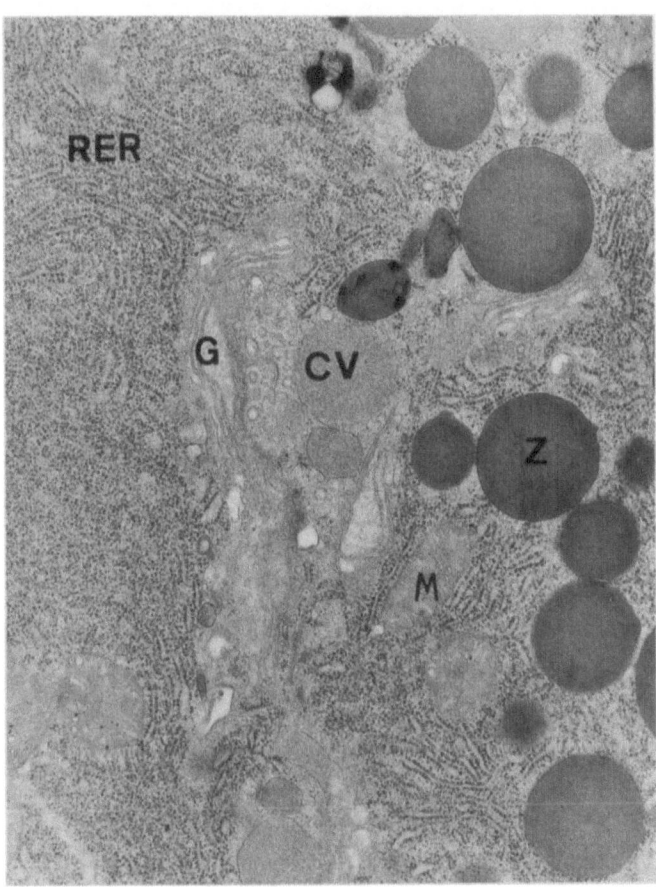

Golgi stack. Accumulation of moderately electron-dense material can occasionally be seen in dilated portions of the Golgi saccules and prosecretory granules on the trans-face. Morphological and autoradiographic studies after administration of tritium-labeled amino acids indicate that the secretory proteins are synthesized by ribosomes of the RER, segregated within the lumen of the RER, and transferred to the Golgi apparatus, where they are "packaged" into prosecretory and zymogen granules (Scheele 1979). Mitochondria are oval or elongated and generally occur more frequently in the basal two thirds of the acinar cell. A few microvilli project from the apical surface into the acinus lumen. Adjacent acinar cells are attached near their luminal borders by junctional complexes consisting of zonula accludens (tight junctions), zonula adherens, and macula adherens (desmosomes). Zonulae occludentes and adherentes completely encircle the acinar cells, whereas the desmosomes form distinct plaques on opposing membranes. The zonulae occludentes may function to seal off the luminal surface and prevent macromolecules (secretory proteins) from escaping, while the primary function of desmosomes seems to be cell adhesion. The ultrastructural appearance of centroacinar cells and cells of the intercalated and larger ducts are similar. The luminal surfaces of these cells have a few microvilli, and their cytoplasm has relatively few organelles in comparison to acinar cells (Figs. 231, 232). There are a few scattered mitochondria, a small Golgi complex, a scant amount of endoplasmic reticulum, and occasional free ribosomes. Occasional cilia are seen. Scattered mucin-producing goblet cells are present in the columnar epithelium of the main ducts.

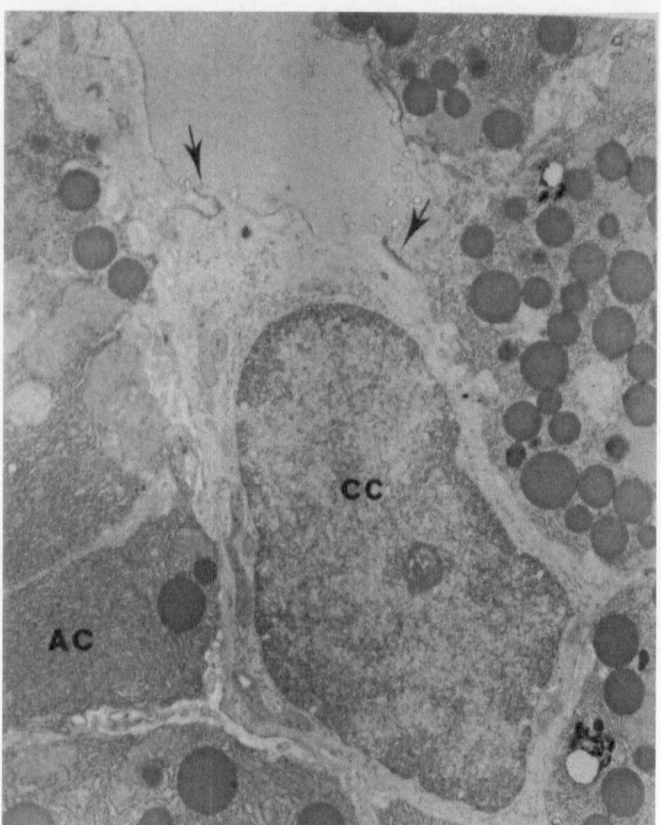

Fig. 231. (*above*) Adult rat pancreas. A centroacinar cell (*CC*) protrudes into the acinar lumen and overlays the apical surface of several acinar cells (*AC*). Adjacent centroacinar cells are attached by junctional complexes (*arrows*). Centroacinar cells have scant cytoplasm with few mitochondria, occasional profiles of rough endoplasmic reticulum, and occasional free ribosomes. TEM, ×7500

Fig. 232. (*below*) Intercalated duct, adult rat pancreas. Duct lumen (*L*) contains electron-dense secretory material. Junctional complexes attach adjacent epithelial cells, and a basal lamina is present at the cell–mesenchyme interface. The *arrow* indicates a cilium. TEM, ×7500

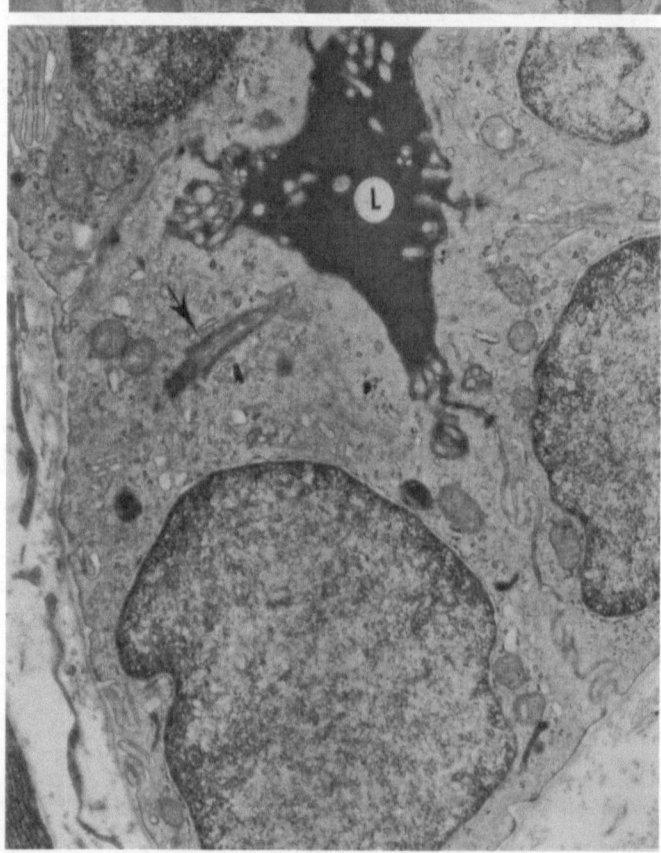

Acknowledgments. The authors gratefully acknowledge the expert assistance of Fred Talley and John Horton for electron microscopy and Sharon Walker for manuscript preparation.

References

Aughsteen AA, Cope GH (1987) Changes in the size and number of secretion granules in the rat exocrine pancreas induced by feeding or stimulation in vitro. Cell Tissue Res 249:427–436

Dixon JS (1979) Histology: ultrastructure. In: Howat HT, Sarles H (eds) The exocrine pancreas. Saunders, Philadelphia, pp 30–47

Doyle CM, Jamieson JD (1978) Development of secretagogue response in rat pancreatic acinar cells. Dev Biol 65:11–27

Dumont Y, Larose L, Morisset J, Poirier GG (1981) Parallel maturation of the pancreatic secretory response to cholinergic stimulation and the muscarinic receptor population. Br J Pharmacol 73:347–354

Liebow C, Rothman SS (1973) Distribution of zymogen granule size. Am J Physiol 225:258–262

Nadelhaft I (1973) Measurement of the size distribution of zymogen granules from rat pancreas. Biophys J 13:1014–1029

Oates PS, Morgan RG (1986) Changes in pancreatic acinar cell nuclear number and DNA content during aging in the rat. Am J Anat 177:547–554

Pictet RL, Clark WR, Williams RH, Rutter WJ (1972) An ultrastructural analysis of the developing embryonic pancreas. Dev Biol 29:436–467

Rambourg A, Clermont Y, Hermo L (1988) Formation of secretion granules in the Golgi apparatus of pancreatic acinar cells of the rat. Am J Anat 183:187–199

Rutter WJ, Pictet RL (1976) Hormone-like factor(s) in mesenchymal epithelial interactions during embryonic development. Ciba Found Symp 40:259–272

Rutter WJ, Wessells NK, Grobstein C (1964) Control of specific synthesis in the developing pancreas. Natl Cancer Inst Monogr 13:51–65

Rutter WJ, Ball WD, Bradshaw WS, Clark WR, Sanders TG (1968a) Levels of regulation in cytodifferentiation. In: Hagan E, Wechsler W, Zilliken F (eds) Experimental biology and medicine, vol 1. Karger, Basel, pp 110–124

Rutter WJ, Clark WR, Kemp JD, Bradshaw WS, Sanders TG, Ball WD (1968b) Multiphasic regulation in cytodifferentiation. In: Billingham RE, Fleischmajer R (eds) Epithelial-mesenchymal interactions. Williams and Wilkins, Baltimore, pp 114–131

Rutter WJ, Kemp JD, Bradshaw WS, Clark WR, Ronzio RA, Sanders TG (1968c) Regulation of specific protein synthesis in cytodifferentiation. J Cell Physiol 72 [Suppl 1]:1–18

Sanders TG, Rutter WJ (1974) The developmental regulation of amylolytic and proteolytic enzymes in the embryonic rat pancreas. J Biol Chem 249:3500–3509

Scheele GA (1979) The secretory process in the pancreatic exocrine cell. Mayo Clin Proc 54:420–427

Sesso A, Abrahamsohn PA, Tsanaclis A (1973) Acinar cell proliferation in the rat pancreas during early postnatal growth. Acta Physiol Lat Am 23:37–50

Sidorova VF, Babaeva AG (1968) Postnatal development of the pancreas in albino rats. Biull Eksp Biol Med 65:566–569

Spooner BS, Cohen HI, Faubion J (1977) Development of the embryonic mammalian pancreas: the relationship between morphogenesis and cytodifferentiation. Dev Biol 61:119–130

Uchiyama Y, Watanabe M (1984) A morphometric study of developing pancreatic acinar cells of rats during prenatal life. Cell Tissue Res 237:117–122

Wenzel G, Stocker E, Heine WD (1972) Zur Zellproliferation im exokrinen Pankreas der Ratte. Autoradiographische Untersuchungen mit H-Thymidin. Virchows Arch B Zellpathol 10:118–126

Werlin SL, Grand RJ (1979) Development of secretory mechanisms in rat pancreas. Am J Physiol 236:E446–E450

Werlin SL, Stefaniak J (1982) Maturation of secretory function in rat pancreas. Pediatr Res 16:123–125

Acinar Cell Carcinoma, Pancreas, Rat

Gary A. Boorman, Robert C. Sills, and Scot L. Eustis

Synonyms. Adenocarcinoma, carcinoma

Gross Appearance

Spontaneous acinar cell carcinoma of the rat pancreas has rarely been described grossly. In one case, confluent yellow to white nodules formed a flattened mass 10 × 5 × 7 mm in the pancreas adjacent to the stomach (Benitz and Roth 1980), while in another case the tumor consisted of firm, glistening, grape-like nodules in the mesentery (Naumann and Kunstyr 1982).

Microscopic Features

The sampling of the pancreas plays an important role in the number of proliferative lesions recognized (Boorman et al. 1987), and in most toxicology and carcinogenicity studies only a single sample of pancreas is examined. At low magnification, acinar cell carcinomas have a lobular appearance, frequently with varying morphological patterns within the tumor. These tumors may be partially encapsulated with lobules of neoplastic cells extending through the fibrous capsule. Vascular areas with dilated blood-filled spaces separated by poorly formed lobules of tumor cells are a common feature (Fig. 233). Some tumors may elicit a scirrhous response with clusters of acinar cells separated by a dense fibrous stroma (Fig. 234). Neoplastic acinar cells arranged in acini, tubular structures, or solid sheets are the varying patterns found in acinar cell carcinomas (Figs. 234, 235). In the rat, well-differentiated acinar cells containing abundant zymogen granules often give the tumor cell cytoplasm a red, granular appearance. The tumor cells have large irregular vesicular nuclei and abundant basophilic to eosinophilic cytoplasm and in some areas will assume a typical acinar pattern with red, granular cytoplasm near

the apex of the cell and cytoplasmic basophilia near the base of the cell (Fig. 235).

Ultrastructure

Spontaneous acinar cell carcinomas of the rat pancreas are characterized by cells with abundant rough endoplasmic reticulum, prominent Golgi complex, and numerous zymogen granules (Fig. 236). The tumor cell nuclei are often irregular, with indentations of the nuclear membrane and margination of the chromatin. A transplantable pancreatic acinar cell carcinoma from a rat, induced by a carcinogen, had similar features after 12–18 transplantations, except that there was more morphological variation in the secretory granules (Iwanij et al. 1982). Secretory granule protein content and membrane phospholipid composition also differed in the transplantable rat acinar cell carcinoma when compared to normal acinar cells (Hansen et al. 1983).

Differential Diagnosis

A cell carcinoma must be differentiated from carcinomas of other origins and from acinar cell adenoma. In the rat, acinar cell carcinoma usually has areas with formation of a typical acinar pattern and prominent zymogen granules, making it easy to distinguish from pancreatic islet cell tumors and metastatic carcinomas. The features distinguishing acinar cell adenoma from acinar cell carcinoma are less clear. Features that suggest carcinoma are a variable growth pattern (nodule within nodules), highly vascular areas, areas of scirrhous response, and glandular patterns (Boorman and Eustis 1984; Eustis et al. 1990). Loss of acinar structure and a heterogeneous growth pattern are useful criteria for distinguishing carcinomas from adenomas. Local invasion

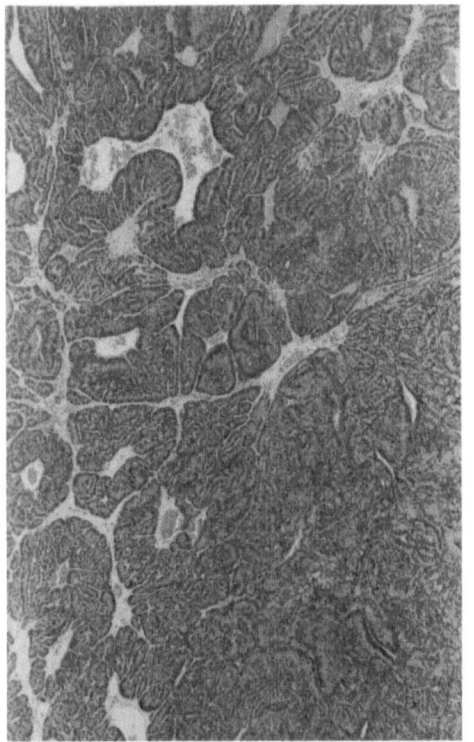

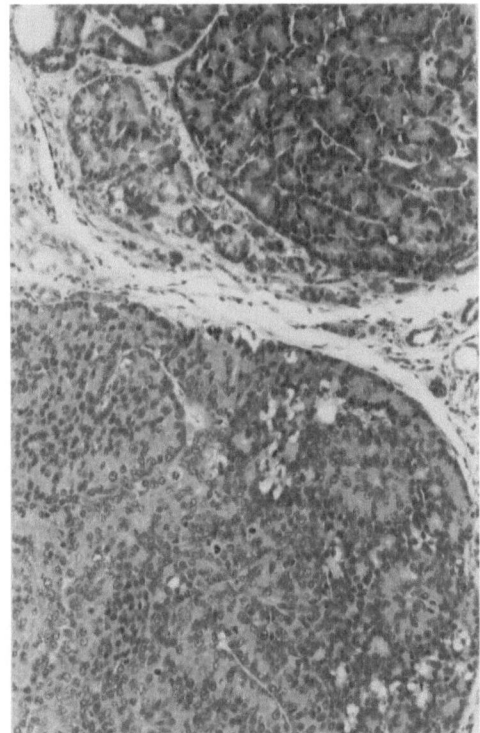

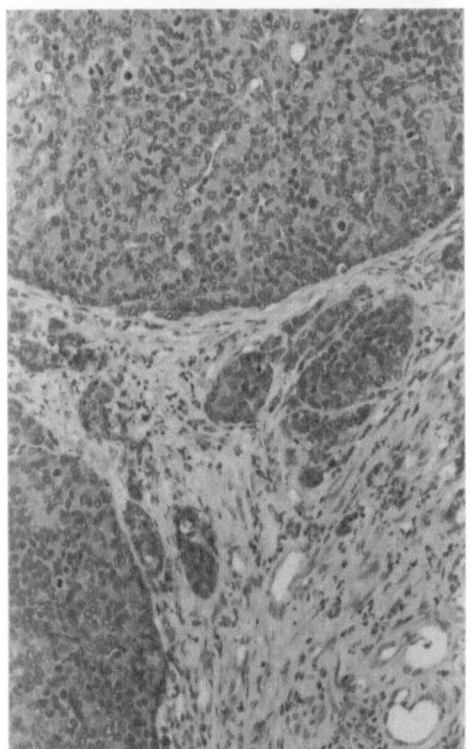

◄

Fig. 233. *(upper left)* Pancreatic acinar-cell carcinoma (spontaneous) in a 2-year-old male F344 rat. The neoplastic cells are well differentiated but are arranged in tubular or trabecular structures separated by scant stroma with dilated vascular channels. H&E, ×100

Fig. 234. *(below)* Pancreatic acinar cell carcinoma (spontaneous) in a male F344 rat. Note the solid growth pattern and small clusters of neoplastic cells separated by an immature fibrous connective tissue stroma that contains inflammatory cells. H&E, ×260

Fig. 235. *(upper right)* Acinar cell carcinoma, pancreas, (spontaneous) in a male F344 rat. Note the solid and glandular growth patterns. Acinar cells arranged in glandular structures are well differentiated and the apical cytoplasm contains abundant zymogen granules. H&E, ×260

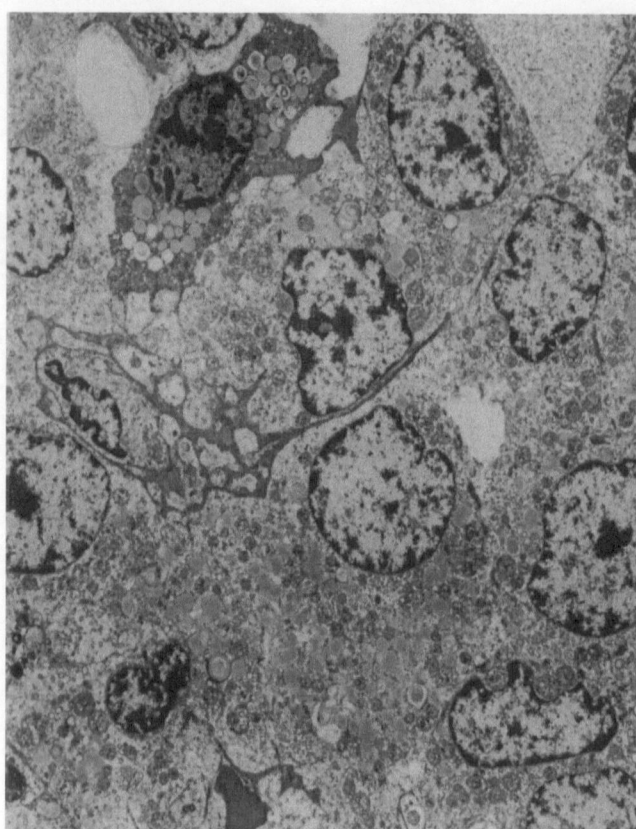

Fig. 236. Acinar cell carcinoma, in a male F344 rat pancreas, illustrating the ultrastructural characteristics of the neoplastic cells. Note the membrane-bound zymogen granules. TEM, ×3299

provides confirmation of the malignant nature of the lesions, since distant metastases are rarely found (Eustis et al. 1990). Acinar cell carcinoma should not be confused with hepatocyte metaplasia, which may occasionally occur around pancreatic islets (McDonald and Boorman 1989); these normal-appearing hepatocytes are an occasional naturally occurring lesion or may occur after chemical exposure.

Biologic Features

There are sex and strain differences in sensitivity to spontaneous and chemically induced neoplasms of the exocrine pancreas. Spontaneous acinar cell carcinomas are rare in rats. In a review of pancreata from 1041 untreated male F344 rats from 2-year toxicity and carcinogenicity studies, no carcinomas were found in untreated controls,

while two out of 992 vehicle (corn oil) controls had pancreatic carcinomas (Boorman and Eustis 1984). A similar pattern was found in a more recent review, with none out of 1868 acinar cell carcinomas in untreated male rats and five out of 1865 (0.3%) acinar cell carcinomas in male rats receiving corn oil vehicle (Haseman et al. 1990). There were no acinar cell carcinomas in over 3800 female F344/N rats, approximately half of which received corn oil as a vehicle control (Haseman et al. 1990). Generally, female rats appear to be less sensitive to the development of neoplasms of the exocrine pancreas. Testosterone appears to promote the experimental induction of pancreatic tumors in the rat (Longnecker 1991; Longnecker and Sumi 1990).

There is some evidence for strain variability in susceptibility to pancreatic cancer in rats (Table 17). Sprague-Dawley rats have the highest spontaneous incidence for carcinomas of the exocrine pancreas. Nitrofen caused carcinomas of the exocrine pancreas in female Sprague-Dawley rats (National Cancer Institute 1978), but did not induce pancreatic tumors in Fischer 344 rats, even though tested by the same route and at a higher dose (National Cancer Institute 1979). Furthermore, Longnecker (1981) reported that Lewis rats are more susceptible to azaserine-induced pancreatic cancer than are random-bred Wistar rats. In 100 female and 98 male Wistar rats examined, no acinar cell tumors were reported (Maekewa et al. 1983).

A number of chemicals have been found to produce acinar cell carcinomas, including azaserine (Bax et al. 1990; Longnecker et al. 1979b, 1984; Roebuck et al. 1987), 7,12-dimethylbenzene (Bockman et al. 1976), and 4-hydroxyaminoquinoline (Konishi et al. 1990). In the NCI/NTP Toxicity and Carcinogenicity Testing Program, in which over 450 chemicals were tested, only nitrofen produced tumors of the exocrine pancreas in F344/N rats (National Cancer Institute 1978). Long-term feeding of raw soya flour to rats is also reported to be associated with acinar cell carcinoma (McGuinness et al. 1980, 1981). Dietary factors may also influence the development of pancreatic neoplasms in rats. In azaserine-induced pancreatic cancer in rats, high dietary levels

of unsaturated fat increased the incidence, while lowered food intake, retinoids, and phenolic antioxidants all seem to have a protective effect on the pancreas (Longnecker 1990; Longnecker et al. 1984; Roebuck et al. 1985). The mammalian gastrin-releasing-peptide (GRP) also has been shown to be a trophic factor for the transplanted pancreatic carcinoma in the rat (Hajri et al. 1992).

Comparison with Other Species

In humans, the vast majority of exocrine pancreatic tumors have a glandular pattern with morphological features that have been assumed to indicate a ductal cell origin (Longnecker et al. 1984; Morohoshi et al. 1983; Pour et al. 1982). Others have challenged this viewpoint and suggested that some may be of acinar cell origin (Bockman 1981; Flaks et al. 1982a,b). Oncocytic and giant-cell variants of pancreatic carcinoma have been described in humans (Hunkatroon 1983). Several cases have been described in which tumor cells were demonstrated to contain both polypeptide (endocrine) and zymogen granules, suggesting that the tumors may have been of an intermediate cell origin (Morrison et al. 1984;

Table 17. Carcinoma of the exocrine pancreas, rats

Strain	Sex	Study type	Exocrine carcinomas (n) (%)	No. of rats examined	Reference
F344	M	2-year	1 (<0.1%)	2320	Solleveld et al. (1984)
F344	M	Life span	3 (0.6%)	526	Solleveld et al. (1984)
F344	F	2-year	1 (<0.1%)	2370	Solleveld et al. (1984)
F344	F	Life span	1 (0.2%)	529	Solleveld et al. (1984)
F344	M	2-year	0	1868[a]	Haseman et al. (1990)
F344	M	2-year	5	1865[b]	Haseman et al. (1990)
F344	F	2-year	0	3809[c]	Haseman et al. (1990)
O-M[d]	M	2-year	0	975	Goodman et al. (1980)
O-M	F	2-year	0	970	Goodman et al. (1980)
S-D[e]	M/F[d]	2-year	0	223	MacKenzie and Garner (1973)
S-D, Holtzman	M/F	2-year	0	200	MacKenzie and Garner (1973)
S-D, Charles River	M/F	2-year	0	446	MacKenzie and Garner (1973)
S-D, (Diablo)	M/F	2-year	1 (0.5%)	196	MacKenzie and Garner (1973)
S-D, (Spartan)	M/F	Life span	5 (1.1%)[e]	447	Leong et al. (1981)

O-M, Osborne-Mendel strain; S-D, Sprague-Dawley strain.
[a] Untreated controls.
[b] Corn oil vehicle controls.
[c] Untreated and vehicle controls combined.
[d] Males and females combined in approximately equal numbers.
[e] Treated and controls combined (two of 112 controls and three of 335 treated had carcinomas).

Ulich et al. 1982). Some of these cancers have arisen in children and been called pancreatoblastoma (Buchino et al. 1984). In the rat, the nearly identical phenotype patterns of the hepatic and pancreatic cell lines suggest that they may be derived from a common primitive epithelial cell type present in both rat liver and pancreas (Bisgaard and Thorgeirsson 1991).

In primary adenocarcinomas of the pancreas in humans, more than 80% had point mutations in codon 12 of the K-*ras* oncogene (Hruben et al. 1993; Wakita et al. 1992). This change is found in small carcinomas, suggesting that c-K-*ras* mutation occurs early, perhaps playing a role in initiation of human pancreatic adenocarcinoma (Schaeffer et al. 1994). In pancreatic neoplasms examined for K-*ras* mutations, genetic alterations were detected frequently in pancreatic carcinomas in Syrian hamsters, but were absent in pancreatic neoplasms from rats (van Kranen et al. 1991).

In humans, overexpression of the p53 protein has been reported to occur in 40% of primary pancreatic tumors and 35% of hyperplastic pancreatic lesions, which suggests that p53 protein overexpression is a common genetic alteration that occurs early in the development of these tumors (Boschman et al. 1994; Lee et al. 1993). The c-*erbB2* oncogene was also found to be amplified and overexpressed in 15 cases (17%) of human pancreatic adenocarcinomas (Hall et al. 1990).

Similar to the rat, exocrine pancreatic tumors are uncommon in mice (Cavaliere et al. 1981; Corbett et al. 1984; Prejean et al. 1973), but occur spontaneously or can be induced in a variety of species, including guinea pigs (Reddy et al. 1974), rabbits (Roudebush 1977), domestic animals (Banner et al. 1978, 1979; Popp 1990), coyotes (Sundberg et al. 1982), and donkeys (Kerr et al. 1982). Both the hamster and rat have been used extensively as models for experimental induction of acinar cell cancer (Birt et al. 1983, 1988; Longnecker 1990; Longnecker et al. 1979a; Scarpelli et al. 1992), and experimentally induced exocrine pancreatic tumors are discussed further in this volume (see pp. 274–288).

References

Banner BF, Alroy J, Pauli BU, Carpenter JL (1978) An ultrastructural study of acinic cell carcinomas of the canine pancreas. Am J Pathol 93:165–182

Banner BF, Alroy J, Kipnis RM (1979) Acinar cell carcinoma of the pancreas in a cat. Vet Pathol 16:543–547

Bax J, Schippers-Gillissen C, Woutersen RA, Scherer E (1990) Kinetics of induction and growth of putative precancerous acinar cell foci in azaserine-induced rat pancreas carcinogenesis. Carcinogenesis 11:245–250

Benitz KF, Roth RN (1980) A spontaneous metastasizing exocrine adenocarcinoma of the pancreas in the rat. Lab Anim Sci 30:64–66

Birt DF, Stepan KR, Pour PM (1983) Interaction of dietary fat and protein on pancreatic carcinogenesis in Syrian golden hamsters. J Natl Cancer Inst 71:355–360

Birt DF, Julius AD, Runice CE, White LT, Lawson T, Pour PM (1988) Enhancement of BOP-induced pancreatic carcinogenesis in selenium-fed Syrian golden hamsters under specific dietary conditions. Nutr Cancer 11:21–33

Bisgaard HC, Thorgeirsson SS (1991) Evidence for a common cell of origin for primitive epithelial cells isolated from rat liver and pancreas. J Cell Physiol 147:333–343

Bockman DE (1981) Cells of origin of pancreatic cancer: experimental animal tumors related to human pancreas. Cancer 47:1528–1534

Bockman DE, Black O Jr, Mills LR, Mainz DL, Webster PD III (1976) Fine structure of pancreatic adenocarcinoma induced in rats by 7,12-dimethylbenz(a)anthracene. J Natl Cancer Inst 57:931–936

Boorman GA, Eustis SL (1984) Proliferative lesions of the exocrine pancreas in male F344/N rats. Environ Health Perspect 56:213–217

Boorman GA, Banas DA, Eustis SL, Haseman JK (1987) Proliferative exocrine pancreatic lesions in rats. The effect of sample size on the incidence of lesions. Toxicol Pathol 15:451–456

Boschman CR, Stryker S, Reddy JK, Rao MS (1994) Expression of p53 protein in precursor lesions and adenocarcinoma of human pancreas. Am J Pathol 145:1291–1295

Buchino JJ, Castello FM, Nagaraj HS (1984) Pancreatoblastoma. A histochemical and ultrastructural analysis. Cancer 53:963–969

Cavaliere A, Bacci M, Fratini D (1981) Spontaneous pancreatic adenocarcinoma in a mouse (Mus musculus) Lab Anim Sci 31:502–503

Corbett TH, Roberts BJ, Leopold WR, Peckham JC, Wilkoff LJ, Griswold DJ, Schabel FJ (1984) Induction and chemotherapeutic response of two transplantable ductal adenocarcinomas of the pancreas in C57BL/6 mice. Cancer Res 44:717–726

Eustis SL, Boorman GA, Hayashi Y (1990) Exocrine pancreas. In: Boorman GA, Eustis SL, Elwell MR, Montgomery CAJ, MacKenzie WF (eds) Pathology of the Fischer rat: reference and atlas, chap 8. Academic, San Diego, pp 95–108

Flaks B, Moore MA, Flaks A (1982a) Ultrastructural analysis of pancreatic carcinogenesis. V. Changes in differentiation of acinar cells during chronic treatment with N-nitrosobis(2hydroxypropyl)amine. Carcinogenesis 3:485–498

Flaks B, Moore MA, Flaks A (1982b) Ultrastructural analysis of pancreatic carcinogenesis. VI. Early changes in hamster acinar cells induced by N-nitroso-bis(2-hydroxypropyl)amine. Carcinogenesis 3:1063–1070

Goodman DG, Ward JM, Squire RA, Parton MB, Reichardt WD, Chu KC, Linhart MS (1980) Neoplastic and nonneoplastic lesions in aging Osborne-Mendel rats. Toxicol Appl Pharmacol 55:433–447

Hajri A, Balboni G, Koenig M, Garaud JC, Damge C (1992) Gastrin-releasing peptide: in vivo and in vitro growth effects on an acinar pancreatic carcinoma. Cancer Res 52:3726–3732

Hall PA, Hughes CM, Staddon SL, Richman PI, Gullick WJ, Lemoine NR (1990) The c-erb B-2 proto-oncogene in human pancreatic cancer. J Pathol 161:195–200

Hansen LJ, Reddy MK, Reddy JK (1983) Comparison of secretory protein and membrane composition of secretory granules isolated from normal and neoplastic pancreatic acinar cells of rats. Proc Natl Acad Sci USA 80:4379–4383

Haseman JK, Eustis SL, Arnold J (1990) Tumor incidences in Fischer 344 rats: NTP historical data. In: Boorman GA, Eustis SL, Elwell MR, Montgomery CAJ, MacKenzie WF (eds) Pathology of the Fischer rat, chap 35. Academic, San Diego, pp 555–564

Hruban RH, van Mansfeld AD, Offerhaus GJ, van Weering DH, Allison DC, Goodman SN, Kensler TW, Bose KK, Cameron JL, Bos JL (1993) K-ras oncogene activation in adenocarcinoma of the human pancreas. A study of 82 carcinomas using a combination of mutant-enriched polymerase chain reaction analysis and allele-specific oligonucleotide hybridization. Am J Pathol 143:545–554

Huntrakoon M (1983) Oncocytic carcinoma of the pancreas. Cancer 51:332–336

Iwanij V, Hull BE, Jamieson JD (1982) Structural characterization of a rat acinar cell tumor. J Cell Biol 95:727–733

Kerr OM, Pearson GR, Rice DA (1982) Pancreatic adenocarcinoma in a donkey. Equine Vet J 14:338–339

Konishi N, Ward JM, Waalkes MP (1990) Pancreatic hepatocytes in Fischer and Wistar rats induced by repeated injections of cadmium chloride. Toxicol Appl Pharmacol 104:149–156

Lee CS, Rush M, Charalambous D, Rode J (1993) Immunohistochemical demonstration of the p53 tumor suppressor gene product in cancer of the pancreas and chronic pancreatitis. J Gastroenterol Hepatol 8:465–469

Leong BK, Kociba RJ, Jersey GC (1981) A lifetime study of rats and mice exposed to vapors of bis(chloromethyl)ether. Toxicol Appl Pharmacol 58:269–281

Longnecker DS (1981) Carcinoma of the pancreas in azaserinetreated rats. Am J Pathol 105:94–96

Longnecker D (1990) Experimental pancreatic cancer: role of species, sex and diet. Bull Cancer (Paris) 77:27–37

Longnecker DS (1991) Hormones and pancreatic cancer. Int J Pancreatol 9:81–86

Longnecker DS, Sumi C (1990) Effects of sex steroid hormones on pancreatic cancer in the rat. Int J Pancreatol 7:159–165

Longnecker DS, Curphey TJ, French JI, Lilja HS (1979a) Response of the Syrian golden hamster to a nitrosourea amino acid carcinogen. Cancer Lett 8:163–168

Longnecker DS, Lilja HS, French J, Kuhlmann E, Noll W (1979b) Transplantation of azaserine-induced carcinomas of pancreas in rats. Cancer Lett 7:197–202

Longnecker DS, Wiebkin P, Schaeffer BK, Roebuck BD (1984) Experimental carcinogenesis in the pancreas. Int Rev Exp Pathol 26:177–229

MacKenzie WF, Garner FM (1973) Comparison of neoplasms in six sources of rats. J Natl Cancer Inst 50:1243–1257

Maekawa A, Onodera H, Tanigawa H, Furuta K, Kodama Y, Horiuchi S, Hayashi Y (1983) Neoplastic and non-neoplastic lesions in aging Slc:Wistar rats. J Toxicol Sci 8:279–290

McDonald MM, Boorman GA (1989) Pancreatic hepatocytes associated with chronic 2,3-dichloro-p-phenylene diamine administration in Fischer 344 rats. Toxicol Pathol 17:1–6

McGuinness EE, Morgan RG, Levison DA, Frape DL, Hopwood D, Wormsley KG (1980) The effects of long-term feeding of soya flour on the rat pancreas. Scand J Gastroenterol 15:497–502

McGuinness EE, Morgan RG, Levison DA, Hopwood D, Wormsley KG (1981) Interaction of azaserine and raw soya flour on the rat pancreas. Scand J Gastroenterol 16:49–56

Morohoshi T, Held G, Kloppel G (1983) Exocrine pancreatic tumours and their histological classification. A study based on 167 autopsy and 97 surgical cases. Histopathology 7:645–661

Morrison DM, Jewell LD, McCaughey WT, Danyluk J, Shnitka TK, Manickavel V (1984) Papillary cystic tumor of the pancreas. Arch Pathol Lab Med 108:723–727

National Cancer Institute (NCI) (1978) Bioassay of nitrofen for possible carcinogenicity, NCI-CG-TR-26. Department of Health, Education, and Welfare publ no (NIH) 78:826, Washington

National Cancer Institute (NCI) (1979) Bioassay of nitrofen for possible carcinogenicity, NCI-CG-TR-184. Department of Health, Education, and Welfare publ no (NIH) 79:1740, Washington

Naumann S, Kunstyr I (1982) Adenokarzinom des exokrinen Pankreas bei Ratte. Z Versuchstierkd 24:181–183

Popp JA (1990) Tumors of the liver, gall bladder, and pancreas. In: Moulton JE (ed) Tumors in domestic animals. University of California Press, Berkeley, pp 436–457

Pour PM, Sayed S, Sayed G (1982) Hyperplastic, preneoplastic and neoplastic lesions found in 83 human pancreases. Am J Clin Pathol 77:137–152

Prejean JD, Peckham JC, Casey AE, Griswold DP, Weisburger EK, Weisburger JH (1973) Spontaneous tumors in Sprague-Dawley rats and Swiss mice. Cancer Res 33:2768–2773

Reddy JK, Svoboda DJ, Rao MS (1974) Susceptibility of an inbred strain of guinea pigs to the induction of pancreatic adenocarcinoma by N-methyl-N-nitrosourea. J Natl Cancer Inst 52:991–993

Roebuck BD, Longnecker DS, Baumgartner KJ, Thron CD (1985) Carcinogen-induced lesions in the rat pancreas: effects of varying levels of essential fatty acid. Cancer Res 45:5252–5256

Roebuck BD, Baumgartner KJ, Longnecker DS (1987) Growth of pancreatic foci and development of pancreatic cancer with a single dose of azaserine in the rat. Carcinogenesis 8:1831–1835

Roudebush P (1977) Pancreatic adenocarcinoma in a domestic rabbit. Vet Med Small Anim Clin 72:467–468

Scarpelli DG, Cerny WL, Mangold KA (1992) Gene alterations in rodent and human tumors of the exocrine and endocrine pancreas. Prog Clin Biol Res 376:223–243

Schaeffer BK, Glasner S, Kuhlmann E, Myles JL, Longnecker DS (1994) Mutated c-K-ras in small pancreatic adenocarcinomas. Pancreas 9:161–165

Solleveld HA, Haseman JK, McConnell EE (1984) Natural history of body weight gain, survival, and neoplasia in the F344 rat. J Natl Cancer Inst 72:929–940

Sundberg JP, Reilly MJ, Wyand DS, Fichandler PD (1982) Pancreatic adenocarcinoma in a coyote-dog cross. J Wildl Dis 18:513–515

Ulich T, Cheng L, Lewin KJ (1982) Acinar-endocrine cell tumor of the pancreas. Report of a pancreatic tumor containing both zymogen and neuroendocrine granules. Cancer 50:2099–2105

van Kranen HJ, Vermeulen E, Schoren L, Bax J, Woutersen RA, van Iersel P, van Kreijl C, Scherer E (1991) Activation

of c-K-ras is frequent in pancreatic carcinomas of Syrian hamsters, but is absent in pancreatic tumors of rats. Carcinogenesis 12:1477–1482

Wakita K, Ohyanagi H, Yamamoto K, Tokuhisa T, Saitoh Y (1992) Overexpression of c-Ki-ras and c-fos in human pancreatic carcinomas. Int J Pancreatol 11:43–47

Experimental Carcinogenesis, Exocrine Pancreas, Hamster and Rat

Dante G. Scarpelli

Introduction

Ten years have passed since rodent models of experimental pancreatic carcinogenesis were last reviewed in the first edition of *Digestive System* in this monograph series. At the time of that review (Scarpelli et al. 1985), it was clear that the exocrine pancreas of rodent species was susceptible to a variety of chemical carcinogens and that these might be models of utility for pathogenetic and other analyses of the carcinogenic process. A remarkable feature of these models is that in various rodent species different types of exocrine cells appear to be susceptible to the neoplastic transformation. For example, in the rat and guinea pig, the majority of tumors induced are acinar cell adenomas and carcinomas, as contrasted to the hamster, in which almost exclusively ductal adenocarcinomas develop. In mice, on the other hand, both acinar cell carcinomas and duct-like carcinomas have been obtained.

By way of introduction in the next few pages, the major thrust of the previous review, namely the pathogenesis of the experimentally induced neoplasms of exocrine pancreas, will be summarized. For a more detailed discussion of acinar and duct cell carcinogenesis and numerous photographs of specific lesions, the reader is referred to that review. The rat and hamster models of chemically induced pancreatic carcinogenesis have been the most studied and best characterized. In both models, pathogenesis consists of a spectrum of preneoplastic epithelial proliferative lesions which progress to noninvasive, invasive, and eventually

metastatic carcinomas. In the rat, the genesis of acinar cell carcinomas induced by the diazoketone azaserine and other carcinogens can be readily followed from focal nests of proliferating acinar cells to larger nodules, some of which have limited growth potential, as contrasted to those that are more aggressive and eventually evolve into carcinomas. The appearance to varying degrees of a duct-like component in some rat (Fig. 237) and guinea pig acinar cell carcinomas is a finding that was both confounding and little understood. The pathogenesis of ductal adenocarcinomas by the carcinogen N-nitrosobis(2-oxopropyl)amine (BOP) in the hamster is also not entirely clear. The development of ductal carcinomas by BOP consists of proliferative epithelial lesions that arise in pancreatic ducts of all sizes from the main duct and its various-sized branches to the smallest ductules. In larger ducts, the lesions begin as focal epithelial proliferations (Fig. 238) which become progressively larger, papillary, and cytologically atypical. Such lesions, with continued growth, evolve to in situ (Fig. 239), invasive, and eventually metastatic carcinomas, most of which are well differentiated (Fig. 240). The epithelium of preexisting ductules undergoes the same proliferative, preneoplastic, and neoplastic alterations described above, except that papillary lesions develop rarely or not at all in ductules due to their small size. Most of the tumors induced produce mucin to varying degrees, and a few secrete large amounts, which accumulate in large pools that also contain exfoliated tumor cells and are lined by neo-

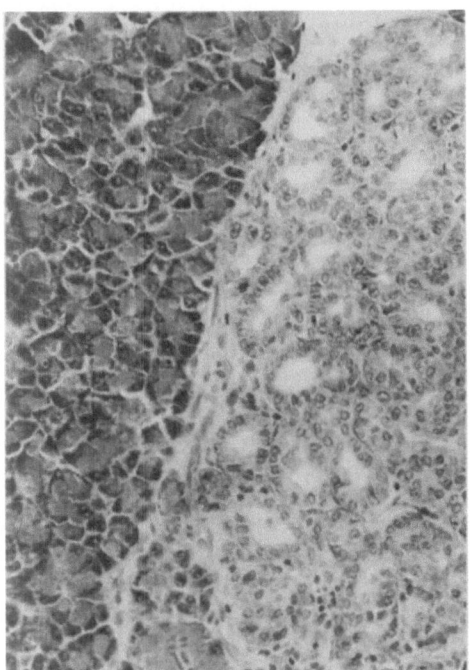

Fig. 237. A tumor nodule containing duct-like profiles is shown in the right half of the photomicrograph of pancreas from a Lewis rat treated with azaserine. H&E, ×200. Courtesy of D.S. Longnecker

plastic epithelium characteristic of mucinous adenocarcinoma.

The apparent relatively straightforward sequences of lesions encountered during pancreatic carcinogenesis in all rodent species are complicated by the capacity of their adult acinar cells to readily undergo metaplastic transformation to duct-like cells in response to pancreatic carcinogens. In the hamster, sustained exposure to BOP causes the alteration of acinar cells to duct-like cells. The fact that this change, aptly named neoduct or pseudoduct formation, can also be experimentally induced by focal blunt physical trauma to the hamster pancreas suggests that the transformation may be a nonspecific response to sublethal injury. The newly formed ductular cells (Fig. 241) are stable, especially with continued BOP treatment, show an augmented rate of proliferation, and persist for many months after induction. Several weeks after their development, neoducts are indistinguishable from normal preexisting ductules. The appearance, with continued BOP treatment, of preneoplastic and in situ carcinomas in some neoductules suggests that they,

too, may serve as a substrate for the induction of ductal adenocarcinoma.

In the hamster, the high sensitivity of preexisting ductules to BOP stimulates their proliferation throughout the pancreas, including the usually obscure and largely imperceptible ductular elements present in and around the islets of Langerhans. These small peri- and intrainsular ductular profiles proliferate under the influence of metabolites of BOP to the point at which they largely replace the entire islet (Pour 1978). The epithelium lining these ductules becomes atypical and may progress to in situ carcinomas. Thus it was not entirely clear whether the BOP-induced epithelial atypias and in situ carcinomas documented in intrainsular ducts evolved to invasive carcinomas, and to what extent they contributed overall to the pathogenesis of pancreatic ductal adenocarcinomas in the hamster.

In both the rat and hamster pancreatic cancer models, manipulation of the composition of diets significantly altered the development of chemically induced pancreatic carcinomas. For example, diets high in unsaturated fats such as corn oil or safflower oil substantially increased the incidence of azaserine-induced pancreatic acinar cell neoplasms in rats and of BOP-induced ductal carcinomas in hamsters, respectively (Birt et al. 1991). The enhancement of acinar cell carcinogenesis in rats fed a diet containing raw soy flour initiated studies which focused on the role of trypsin inhibitors and the gastrointestinal hormone cholecystokinin (CCK) in pancreatic carcinogenesis.

The current review deals with progress made from 1985 to the present in our understanding of the pathogenesis of acinar cell carcinomas in the rat and ductular/ductal carcinomas in the hamster. Special attention is given to the possible role of duct formation from acinar cells and within pancreatic islets in the genesis of pancreatic duct carcinoma, the presence of pancreatic stem cells, the modulatory effects of diet and hormones on pancreatic carcinogenesis, and the molecular analysis of experimental rat and hamster pancreatic carcinomas.

Genesis of Neoductules from Pancreatic Acinar and Islet Cells and Their Possible Role in Pancreatic Duct Carcinogenesis

Studies on the formation of neoducts in hamsters (Pour 1988; Meijer et al. 1989) have identified and

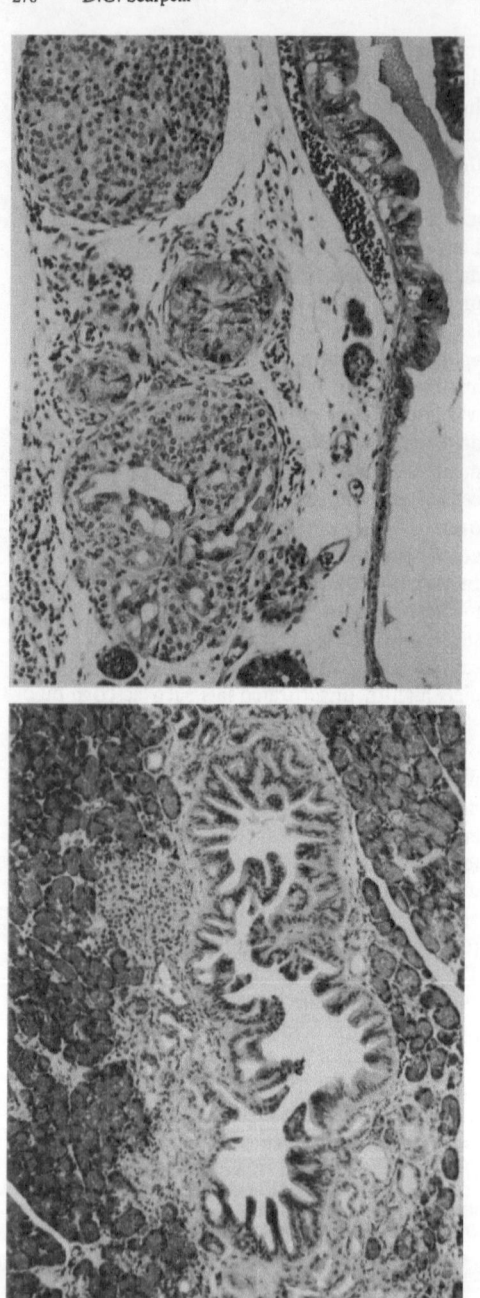

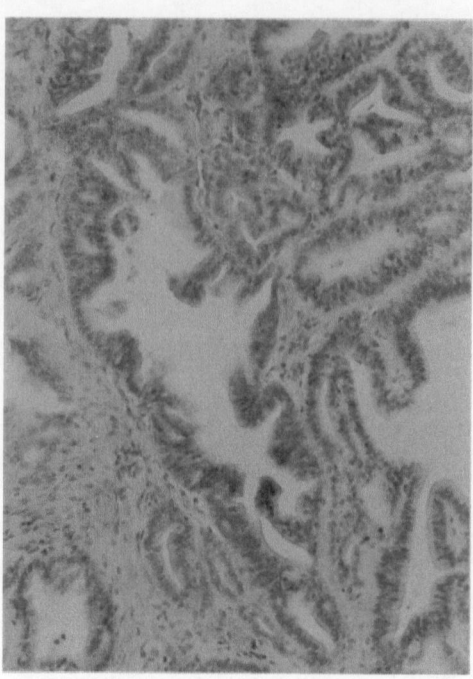

Fig. 238. (*upper left*) Along the right border, part of a large pancreatic duct lined by a focus of hyperplastic epithelium is present (*upper half*), as contrasted with normal-appearing epithelium (*lower half*). *Bottom left,* an islet with numerous duct-like structures induced by *N*-nitrosobis(2-oxopropyl)amine (BOP). H&E, ×250

Fig. 239. (*below*) A medium-sized pancreatic duct lined by severely dysplastic epithelium displaying a markedly papillary pattern; some papillae are so extended that they almost bridge the lumen. This is an early carcinoma in situ. H&E, ×150

Fig. 240. (*upper right*) A well-differentiated duct adenocarcinoma, pancreas, hamster. H&E, ×190

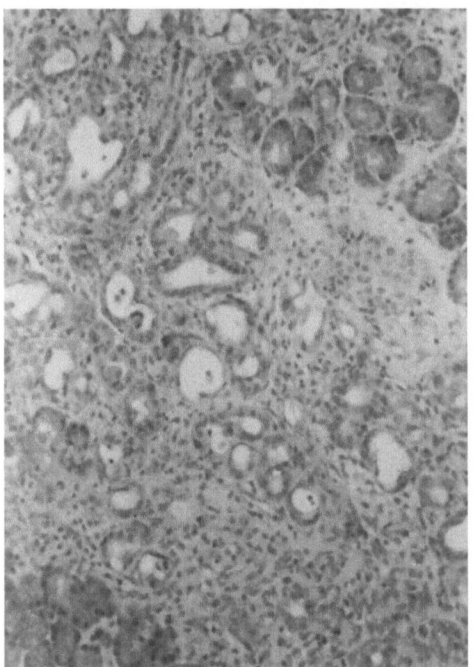

Fig. 241. A focus of neoduct formation from acinar cells that have undergone metaplasia from treatment with *N*-nitrosobis(2-oxopropyl)amine (BOP). H&E, ×190

documented an interesting mechanism involving centroacinar cells that represents a hitherto unrecognized pathway. Hamsters exposed to a single dose of BOP to obviate the possible confounding effects of repeated toxic insults showed cytologic alterations of centroacinar and subsequently acinar cells. These consisted of the elongation of centroacinar cell cytoplasmic processes, which eventually come to overlie and surround adjacent acinar cells. Following BOP treatment, centroacinar cells express type-A blood group and cytokeratin antigen, respectively, which the authors suggest serve as ductal/ductular cell markers that identify the phenomenon in acini by light microscopy. Acinar cells enveloped by centroacinar cells undergo degenerative changes and subsequently die. The authors interpret this to mean that the resulting ductule-like structures are derived from hyperplasia of centroacinar cells that transform into cells indistinguishable from preexisting ductules (Scarpelli et al. 1991).

Although the foregoing observations are convincing, other equally compelling studies indicate that pancreatic acinar cells in vitro can readily undergo direct transformation to duct-like cells. This was first demonstrated when isolated malignant acinar cells derived from a transplantable tumor in the rat were cultured on a substrate of basement membrane derived from seminiferous tubules. The neoplastic acinar cells aggregated into duct-like profiles and lost their complement of zymogen granules to varying degrees (Reddy et al. 1986). More recently, acinar cell to duct cell transformation has also been documented with normal pancreatic acinar cells isolated from the mouse (DeLisle and Logsdon 1990), human (Hall and Lemoine 1992), rat (Arias and Bendayan 1993), and more recently the Syrian golden hamster (SGH) (S. Hubchak and D.G. Scarpelli, unpublished observations; Hubchak et al. 1990).

Branched interconnecting tubules develop in acinar cell monolayers in vitro that are covered by basement membrane matrix. In each instance, the duct-like profiles consist of cells containing residual zymogen granules linking them to precursor acinar cells. Of even greater significance are the findings that the alteration to duct-like cells is accompanied by a progressive loss of the enzyme amylase- and *Helix pomatia* lectin-binding affinity, both of which are phenotypic characteristics of acinar cells. These phenotypic changes indicate that the transformation of acinar cells to duct cells does indeed occur. One could argue that the transformation is limited to cultured acinar cells; however, the relative ease with which it occurs in vivo in diverse species makes this explanation quite unlikely. Despite the strong evidence that at least two pathways exist for neoductule formation, it is not yet clear whether either or both pathways contribute significantly to the pathogenesis of pancreatic ductal adenocarcinoma. Pour (1988) has continued to study the possible significance of intrainsular pancreatic duct formation in BOP-induced duct carcinogenesis in hamster pancreas. The experimental system consists of transplantation of isolated homologous pancreatic islets to extrapancreatic sites such as the submandibular gland (SMG) and beneath the renal capsule in the hamster. In a recent paper, Ishikawa et al. (1995) report on preliminary results which, though promising, do not unequivocally establish that duct lesions induced by BOP in pancreatic islets transplanted to the SMG progress to invasive duct carcinomas. They found that BOP treatment induced duct epithelial hyperplasia, atypia, and lesions consistent with in situ carcinomas which expressed blood group A antigen, as determined by immunochemical staining with specific antibody. The latter finding, which has been previously demonstrated in BOP-induced duct carcinomas including invasive ones (Pour et al. 1986), suggests that these in situ lesions in islets may indeed contribute to the pathogenesis of duct carcinogenesis in hamster pancreas. However, in view of the well-established frequency of cancer development in preexisting pancreatic ducts and the infrequency of severe atypia and in situ carcinoma in neoducts (Fig. 242), the contribution of ducts arising from metaplasia of acinar cells and from islets in the hamster in the usual carcinogenesis protocol must at present be considered as being relatively minor and of secondary significance until proven otherwise.

The possibility that metaplastic acinar cells may also be involved in the pathogenesis of ductal adenocarcinoma in human pancreas is suggested by the coexpression of acinar and duct cell antigens in duct-like profiles in the non-neoplastic portions of the pancreas in patients with pancreatic ductal adenocarcinoma (Parsa et al. 1985). The phenotypic plasticity (metaplasia) of gut endoderm-derived epithelia of stomach, liver, and pancreas in adulthood is well known (Scarpelli 1985) and has been documented to accompany the development of human gastric carcinoma (Lauren 1965; Matsukara et al. 1980) and experimental hepatic cancer (Yoshida et al. 1978). Since the phenomenon of neoduct formation in the human pancreas has not been well documented, more studies are needed before its role in pancreatic carcinogenesis can be ascertained.

About 5 years ago, an interesting transgenic mouse model of pancreatic acinar cell carcinogenesis and a model of in vitro rat acinar cell cancer were developed which may contribute to a clearer understanding of the acinar to duct cell conversion. The first model bears a transgene consisting of the elastase promoter (Ela-1), which targets expression to acinar cells, and the *Myc* oncogene, which stimulates cell growth and induces carcinogenesis. The transgene is designated Tg(Ela-1, Myc)Bril 59 (Sandgren et al. 1991). These mice, as expected, develop diffuse hyperplasia of acinar cells and eventually acinar cell carcinomas. In about half of the acinar cell carcinomas, areas of ductal differentiation appear that are associated with stromal hyperplasia. Less frequently, some of the ducts undergo squamous

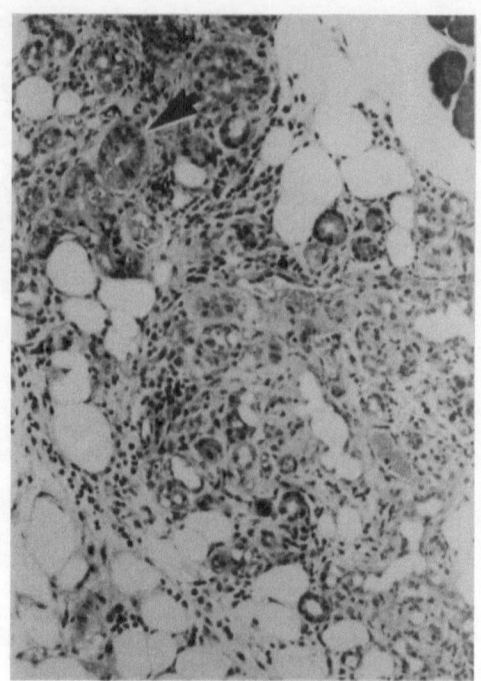

Fig. 242. Note at the *upper left* a focus of neoduct formation from acinar cells during *N*-nitrosobis(2-oxopropyl)amine (BOP)-induced carcinogenesis. Within the focus (*arrow*), there is a ductule with a carcinoma in situ, hamster. H&E, ×155

metaplasia. The ductal components grow more slowly than acinar cell tumors and do not evolve into pure pancreatic carcinomas. The second model consists of a rat pancreatic acinar carcinoma cell line derived from an in vivo azaserine-induced acinar cell carcinoma. This cell line underwent metaplastic conversion to ducts which retained the malignant phenotype during alternating serial in vitro culture followed by subcutaneous transplantation. During conversion to duct cells, acinar cells gradually lost their capacity to secrete amylase as they assumed the ductal phenotype. The tumor regularly grows as a ductal adenocarcinoma with foci of adenosquamous differentiation when transplanted in syngeneic Lewis rats (Longnecker et al. 1991). These models of acinar to duct cell conversion may lead to the identification of the genetic and other factors that are involved in the metaplastic phenomenon which appears to be prevalent in both the rodent and human pancreas.

Presence of Putative Stem Cells
in the Rodent Pancreas

An understanding of the cellular events in pancreatic growth and carcinogenesis would be incomplete without a consideration of whether pancreatic stem cells exist and may be involved in the development of cancer of the exocrine pancreas. In the immediate postnatal period, epithelial cells of the pancreas, like those of liver and kidney, are capable of proliferation, a property that decreases with increasing age to levels of about 0.1% or less in the adult animal, but they are stimulated to increased growth when organ mass is significantly decreased by cell injury and death. These are referred to as expanding cell populations and lie intermediate between static cells such as neurons or cardiocytes, which lose their capacity to divide shortly after birth and normally persist through the life of the animal, and renewing cell populations, in which there is a continuous replacement of cells, the prime example being the epithelia of the digestive tract (Leblond 1964).

Studies of massive acinar cell loss by sustained dietary copper deficiency in the adult rat pancreas (Rao et al. 1986, 1988) have identified populations of undifferentiated single cells during copper restitution that resemble stem cells in that they aggregate to form ductules (Ide et al. 1993) and differentiate into hepatocytes (Rao et al. 1990; Reddy et al. 1991). These cells resembling stem cells appear to originate from pancreatic ductal and periductal cells in the rat and are comparable to similar, though less numerous cells present in the ductules of regenerating adult hamster pancreas, which also differentiate into functional hepatocytes (Scarpelli and Rao 1981; Makino et al. 1990). It is well known that the three epithelial cell types of the pancreas are derived from ductules arising early in embryonic pancreatic development and persisting into adult life; these cell types appear to have the properties characteristic of stem cells. The widespread loss of acinar cells in the copper-depleted/replenished rat model appears to unmask a second stem cell that is one step removed from the ductular cell in the intact pancreas and may represent a backup pathway for pancreatic restitution activated in the event of massive pancreatic injury. The pathway suggested for cellular restitution of the severely atrophic pancreas resembles that seen in fulminant hepatitis, in which there is a massive loss of hepatocytes, and so-called oval cells, putative stem cells, proliferate and differentiate into liver cells in an attempt to reconstitute the organ. The capacity of stem cells for sustained proliferation and their importance as possible target cells in carcinogenesis makes their continued study in this regard warranted. Further elucidation of the possible role of stem-like cells in cell replacement of the injured pancreas and in pancreatic carcinogenesis is awaited with great interest.

Modulation of Experimental Pancreatic
Acinar and Ductal Carcinogenesis
by Dietary Manipulation

Although epidemiologic studies have implicated diet and nutrition in the etiology of pancreatic cancer in humans (Rowlatt 1967; Wynder et al. 1973; Howe et al. 1990), they have not unambiguously identified specific high-risk factors. This is necessarily, so because epidemiologists obtain data regarding etiologic factors for groups rather than individuals and at best can only suggest etiologic relationships rather than recognize particular factors. The epidemiologic suggestion that diets high in fat may be involved in the pathogenesis of pancreatic carcinoma in humans continues to stimulate the use of rat and hamster models to experimentally test this hypothesis.

Earlier studies clearly showed that diets high in saturated fats enhanced carcinogenesis of the exocrine pancreas in both rats and hamsters. A more recent study in which the caloric intake of experimental and control animals was carefully controlled also showed a significant promotional effect of a high-fat diet on carcinogenesis, establishing that the effect is not simply due to the increased calorie content of high-fat diets (Birt et al. 1989). In the absence of mechanistic studies on how unsaturated fats exert their promotional effect on pancreatic acinar and duct cell carcinogenesis, it is noteworthy that in chemical carcinogenesis of skin (Verma et al. 1980) and colon (Reddy and Sugie 1988) the effect has been linked to prostaglandins, cyclic metabolites of unsaturated fatty acids that enhance tumor promotion. Elegant studies (Verma et al. 1980) have established that prostaglandins potentiate the induction of ornithine decarboxylase, an enzyme essential for tumor progression. Experiments in hamsters fed diets containing high levels of beef tallow following BOP treatment suggested for the first time that saturated animal fat also promoted pancreatic duct carcinogenesis (Woutersen et al. 1987). A recent and more definitive study with beef tallow (Birt et al. 1990), in which experimental animals consumed the same number of calories as control animals fed the high-corn oil diet, has yielded interesting results. The incidence and the number of ductal adenomas per animal (multiplicity) were increased by the beef tallow diet as compared to controls; carcinoma in situ multiplicity was also increased in hamsters fed a high-fat diet regardless of the nature of the fat, and the multiplicity of pancreatic ductal adenocarcinoma was elevated in hamsters fed either the low- or high-tallow diet compared to those that ingested the low- or high-corn oil diets. Feeding a diet containing a mixture of unsaturated and saturated fats yielded an intermediate number of tumors. The latter finding suggests that the mechanism by which saturated fat mediates its promotional effect probably differs from that of unsaturated fat. The enhancing effect of high-fat diets, especially those consisting of saturated fat, on experimental pancreatic duct carcinogenesis, coupled with studies documenting similar results with high-protein diets (Pour and Birt 1983; Birt and Pour 1987), is consonant with numerous epidemiologic investigations that suggest increased rates of pancreatic cancer in populations that consume diets high in fat and/or protein (Rowlett 1967; Hirayama 1972; Wynder et al. 1973; Wynder 1975, 1976; Falk et al. 1988; Mills et al. 1988; Farrow and Davis 1990).

Influence of Hormones on the Growth of Normal Pancreas and Pancreatic Carcinoma

It is well established that growth of the normal pancreas is controlled by peptide hormones such as bombesin, CCK, epidermal growth factor (EGF), and secretin, which all stimulate growth, and somatostatin, which inhibits growth. Growth-modulatory hormones and factors mediate their effects by binding to specific receptors on the surface of cells and sending transmembrane signals that trigger cytoplasmic and nuclear events that are vital for cell proliferation (Townsend et al. 1986; Longnecker 1991). During the process of carcinogenesis, cell surface receptors of initiated target cell populations are frequently qualitatively and quantitatively altered and new growth factors

expressed, affording the neoplastic cells a selective growth advantage over their noninitiated counterparts (Smith et al. 1987; Jhappan et al. 1990; Okamura et al. 1990; Barton et al. 1991; Shepherd et al. 1993). Such changes are considered important components of carcinogenesis and a major focus of study, since they quite likely contribute significantly to tumor promotion and progression. The observation that dietary fat and protein are potent secretagogues for the release of CCK from the duodenum, together with the experimental demonstration that the hormone stimulates both hypertrophy and hyperplasia of the exocrine pancreas in the rat, led to the suggestion that it might be a factor in the development of pancreatic cancer. Stimulation of normal pancreatic growth (Oates and Morgan 1982), the development of spontaneous acinar cell adenomas and carcinomas (McGuiness et al. 1980), and the promotion of azaserine-induced acinar cell carcinogenesis (McGuiness et al. 1981) in rats by feeding a diet containing raw soy flour rich in trypsin inhibitors tended to support the thesis that CCK might be involved in pancreatic tumorigenesis. It was also well established that trypsin inhibitors cause increased levels of CCK in plasma (Fölsch et al. 1984), because they prevent the usual tryptic inactivation of duodenal monitor peptide, which stimulates secretion of CCK by the duodenum (Lu et al. 1989). Feeding a diet containing camostate (Foy-305), a synthetic trypsin inhibitor, to azaserine-treated rats gave results similar to those seen in rats treated with azaserine and fed a raw soy flour diet (Lhoste et al. 1988). Direct studies with CCK administered weekly by subcutaneous injections to rats after a single injection of azaserine showed that CCK increased the yield of atypical acinar cell neoplastic (AACN) foci eightfold and that treatment with the CCK receptor antagonist CR-1409 nullified the increase completely (Douglas et al. 1989); a second study showed that CCK treatment increased AACN to even greater levels of eight- to 20-fold (Bell et al. 1992). Further, the latter study established that the stimulatory effect of CCK on the growth and development of azaserine-induced AACN in rats was closely linked to an overexpression of CCK receptors on the surface of the cells comprising AACN. It is of interest to note that, once bound to its receptor, CCK increases the level of ornithine decarboxylase activity (Scemama et al. 1989), an important event obligatory for cell proliferation. Experiments with selective CCK agonists of the

CCK-A and CCK-B receptors have shown that the growth of azaserine-induced pancreatic acinar cell lesions is mediated specifically by CCK-A receptors. The preceding studies, together with others involving bombesin and caerulein (Lhoste and Longnecker 1987) and bombesin alone (Meijers et al. 1992a), showed that pancreaticotrophic hormones that control nominal pancreatic growth serve a similar role in acinar cell carcinogenesis. In the hamster model of BOP-induced pancreatic duct carcinogenesis, on the other hand, CCK was not effective in promoting either the development or growth of pancreatic duct tumors (Pour et al. 1988; Meijers et al. 1990). Epidemiologic studies have documented that the age-adjusted incidence of pancreatic cancer in humans is higher in males than in females. A similar sex difference exists in the response of rats to azaserine with higher numbers of AACN and acinar cell neoplasms (Longnecker et al. 1981). Orchiectomy alone or in combination with the administration of 17β-estradiol inhibited growth of a azaserine-induced transplantable acinar cell carcinoma DSL-2 in inbred male Lewis rats. Castration together with estradiol had an additive inhibitory effect (Sumi et al. 1989b). Further studies (Sumi et al. 1989a) showed that estradiol and castration were both highly effective in inhibiting the development and growth of azaserine-induced AACN, indicating that estrogen inhibits and testosterone promotes acinar cell carcinogenesis at an early stage of the process. In addition to confirming the studies just described, Meijers et al. (1991) extended their experiments to include treatment of azaserine-treated rats and BOP-treated hamsters by castration, alone and in combination with the aromatase inhibitor aminoglutethimide. None of these treatments inhibited either the development of AACN in rats or early preneoplastic pancreatic ductular lesions in hamsters, despite the lowering of plasma testosterone levels in both species. Subsequent studies employing orchiectomy, alone or in combination with testosterone and the antiandrogen cyproterone acetate, were done to assess their effect on carcinogenesis of the exocrine pancreas in rats and hamsters (Meijer et al. 1992). The results showed that testosterone enhanced acinar cell neoplasia in rats to a slight degree and was without effect on duct cell carcinogenesis in hamsters. These findings are in support of earlier clinical trials which demonstrated that antiestrogen therapy of pancreatic ductal adenocarcinomas were not effective (Andrén-Sandberg 1989; Bakkevold et al. 1990).

Another approach to hormonal modulation of pancreatic carcinogenesis introduced by Schally and his associates in 1989 (Zalatinai and Shelly 1989a,b) involved the use of hormones such as somatostatin and luteinizing hormone release hormone (LHRH) that inhibit hormone secretion from a variety of endocrine cells. They demonstrated that both experimental rodent and spontaneous human malignant pancreatic ductal tumor cells possess high-affinity receptors for both peptides (Fekete et al. 1989). The growth-modulatory efficacy of such receptors was exploited in subsequent experiments that showed treatment with the somatostatin analogue RC-160 and/or the LHRH superagonist Dtrp-6-LHRH (LHRH-A) administered 18 weeks after BOP treatment decreased both the incidence and weight of pancreatic duct carcinomas (Szende et al. 1990a,b). It is noteworthy that malignant pancreatic ductal tumors in hamsters were susceptible to the inhibitory effects of LHRH-A, as contrasted to preneoplastic ductal lesions, in which only minor effects were present (Zalatnai and Schally 1989a,b). This was further supported by the subsequent finding that LHRH-A receptors are present in duct adenocarcinomas and absent in normal pancreas of hamsters, suggesting that the receptors appear as part of the malignant phenotypes (Zalatnai and Schally 1989a). In other studies with the more potent LHRH agonist SB-75, treatment with the agonist caused regression of established BOP-induced ductal carcinomas (Szende et al. 1990a). It has been found that human pancreatic ductal carcinomas do not express somatostatin receptors (Reubi et al. 1988), which may explain the lack of a therapeutic response to the somatostatin analogue sandostatin in patients with metastatic pancreatic carcinomas (Klijn et al. 1990). A recent development which may eventually lead to effective therapy of pancreatic cancer with somatostatin and possibly other growth-inhibiting peptides is the identification of subtypes of somatostatin receptors with different affinities for somatostatin analogues (Buscail et al. 1994).

Molecular Biology of Pancreatic Carcinomas

The multistep pathogenesis of cancer postulated and established by early experimental studies (Berenblum and Shubik 1947; Farber 1984) has been corroborated and extended by recent molecular analyses of human malignant neoplasms. The exponential relationship between the incidence of human cancers and age of onset suggests that the carcinogenic process requires between three and seven genetic mutations to yield a malignant neoplasm (Miller 1980). The foregoing and the close resemblance of pancreatic duct carcinomas induced in SGH by BOP stimulated us to undertake the genetic analysis of the model.

Our initial studies began with an attempt to determine whether SGH pancreatic duct carcinomas contained transforming DNA and, if so, to identify the gene responsible and to determine the molecular basis for its activation. High molecular weight DNA was isolated from two transplantable SGH pancreatic duct carcinomas, one induced by BOP, the other occurring spontaneously. The DNA was transfected into NIH/3T3 cells and resulted in cells that developed into fibrosarcomas when injected subcutaneously into nude mice. Analysis of the DNA isolated from the fibrosarcomas by Southern blotting revealed a band specific for hamster K-ras. Polymerase chain reaction (PCR) amplification of the codon 12–13 region of K-ras in genomic DNA from the transplantable tumors produced a 117-bp product that was analyzed by allele-specific oligonucelotide hybridization and DNA sequencing. Both methods identified a G to A transversion in the second position of codon 12 in DNA from the BOP-induced duct carcinoma, and an identical mutation in the second position of codon 13 of the spontaneous carcinoma (Cerny et al. 1990). These mutations paralleled those found in a high proportion of human pancreatic duct carcinomas (Bos 1989). An almost simultaneous report by Fuji et al. (1990) of analysis of the status of K-ras in pancreatic duct carcinomas induced by BOP and a later study by van Kranen et al. (1991) corroborated our findings. We next attempted to determine the time point during the process of SGH pancreatic duct carcinogenesis at which the mutational activation of the K-ras oncogene occurs. In these experiments, animals treated weekly with BOP were serially killed after 8, 12, 14, 16, and 24 weeks of treatment. Hyperplasias, carcinoma in situ, and invasive and metastatic ductal carcinomas were identified, marked, and scraped from 10-μm-thick H&E sections. The DNA was isolated and the first exon of K-ras amplified by PCR and probed by oligonucleotide-specific hybridizations

for mutations at either codon 12 or 13. Analysis of 186 samples detected mutations in codon 12 of K-ras in 26% of hyperplasias, 46% of papillary hyperplasias, 76% of carcinoma in situ, 80% of invasive adenocarcinomas, and 43% of duct carcinomas metastatic to lymph nodes. Codon 13 mutations were much rarer, with only nine detected in 142 samples assayed (Cerny et al. 1992). These results established that the mutational activation of the K-ras oncogene is an early event in SGH BOP-induced carcinogenesis and may in fact be at least one of the genetic alterations that occur during the initiation phase of carcinogenesis. Using fine-needle aspirates of pancreatic masses, Shibata et al. (1990) found activated K-ras in 25% of patients with atypical cytology as compared to 72% in patients with malignant cytology, lending support to our findings.

More recently, we extended the genetic analysis of SGH pancreatic duct carcinogenesis to include the tumor suppressor genes p53, deleted in colon cancer (DCC), and RB-1 and the mouse double minute (mdm-2) oncogene (Chang et al. 1995). Twenty-one invasive pancreatic ductal adenocarcinomas greater than 0.4 cm in diameter derived from 17 different BOP-treated SGH were analyzed. Single-strand conformation polymorphism (SSCP) analysis of hamster p53 exons 5–8 showed no evidence of p53 mutations in 21 tumors as did immunohistochemistry of (mAb) D07. Although these results were consonant with the low frequency of p53 mutations in rodent tumors (Greenblatt et al. 1994), we studied the status of the mdm-2 oncogene to determine whether it might be involved when the p53 gene is not mutationally deregulated. The protein of amplified or overexpressed mdm-2 binds p53 protein, inactivates it (Momand et al. 1992), and thereby removes the tumor-suppressive function of the gene (Finlay 1993). RNAse protection assays (RPA) showed overexpression of mdm-2 in five of 19 (26%) tumors (Fig. 243). Thus p53, though not mutated, is involved indirectly as a tumor suppressor in SGH duct carcinogenesis.

Semiquantitative reverse transcriptase PCR analysis of the DCC tumor suppressor gene showed either a complete or partial loss of expression in ten of 19 (53%) of duct neoplasms (Fig. 244). The DCC gene codes for a cell adhesion molecule (CAM) which, in contrast to other CAM's, is normally expressed at relatively low levels. In human colon carcinomas, high levels of DCC expression have been found in tumor vari-

Fig. 243. RNase protection assay (RPA) analysis of mdm-2 expression in Syrian golden hamster (SGH) *N*-nitrosobis(2-oxopropyl)amine (BOP)-induced pancreatic duct carcinomas. RPA using antisense RNA probes against SGH *mdm*-2 and G3PDH. *N*, normal pancreatic control; *T*, tumor. The *numbers* indicate tumor samples. From Chang et al. (1995)

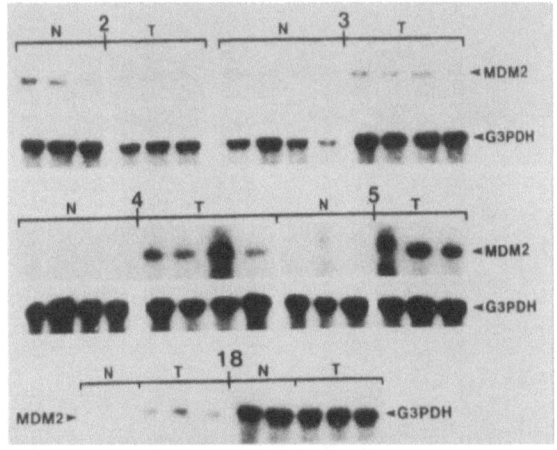

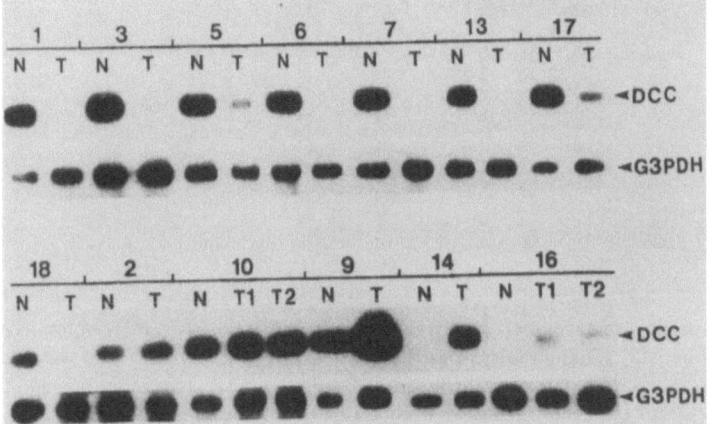

Fig. 244. Reverse transcriptase–polymerase chain reaction (RT-PCR) semiquantitation of *deleted in colon cancer* (*DCC*) gene expression. Autoradiograms of 13 representative sample pairs obtained in one experiment. *N*, normal pancreatic control; *T*, tumors; *T1* and *T2*, two separate tumors induced in the same animal, nos. 10 and 16. There is a decreased expression of DCC in tumors 1, 3, 5–7, 13, 17 and 18. From Chang et al. (1995)

ants called mucinous adenocarcinomas (Hedrick et al. 1994). These tumors produce abundant mucin, a marker of differentiation. This has led to the suggestion that it may play a role in cell differentiation rather than adhesion (Hedrick et al. 1994). In our series of duct tumors, DCC expression was low in tumors 16-2 and 17, which were mucinous ductal adenocarcinomas; conversely, in tumors 9 and 10, DCC expression was quite high (Fig. 244). While sample size in our study is too small to draw a firm conclusion, our findings suggest that, in SGH pancreatic ducts, DCC protein may serve a function other than controlling cell differentiation. Loss of expression

and/or loss of heterozygosity of DCC also has been found in a variety of human epithelial neoplasms (Huang et al. 1992; Uchino et al. 1992; Scheck and Coons 1993; Gao et al. 1993), including pancreatic duct carcinoma (Hohne et al. 1992; Hedrick 1994; Seymour et al. 1994; Simon et al. 1994).

Semiquantitative reverse transcriptase (RT)-PCR analysis showed a significant decrease or loss of RB-1 expression in tumors 1, 4, 5, and 17 when compared to matched normal pancreas controls (Fig. 245b). Although RB-1 was expressed in controls 6, 11-1, 11-2, and 18, there was a complete absence of expression in the corresponding

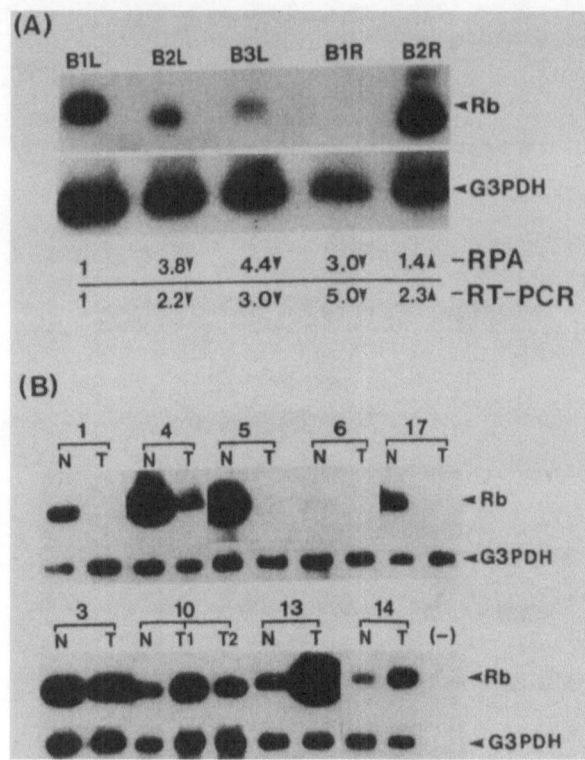

Fig. 245a,b. Semiquantitation of RB-1 expression. **a** Comparison of reverse transcriptase–polymerase chain reaction (*RT-PCR*) to RNase protection assay (RPA) analysis in five in vitro *N* nitrosobis(2-oxopropyl)amine (BOP)-induced tumor cell cultures designated *B1L, B2L, B3L, B1R,* and *B2R. Top two rows,* Southern hybridization of PCR products. The RT-PCR values obtained by image analysis and RPA values generated previously are compared and confirm the accuracy of RT-PCR semiquantitation. The *arrowheads* indicate an increase or decrease of values in other samples relative to B1L. **b** Rearranged autoradiogram of representative sample pairs obtained from a single experiment. *N,* matched normal pancreatic control; *T,* tumors; *T1* and *T2,* tumors 10-1 and 10-2; *line,* negative control. There is decreased expression of RB-1 in tumors 1, 4–6, and 17. From Chang et al. (1995)

tumors (see representative pair 6 as an example in Fig. 245). There was increased level of expression over control pancreas in tumors 2, 9, 10, 10-1, 13, and 14 (see representative pairs 13 and 14). The remaining six tumors showed no differences in RB-1 expression between matched controls and tumors (see representative pairs 3 and 10-2). The levels of RB-1 mRNA were less in tumors than in matched controls in eight of 19 sample pairs (42%). The RB-1 gene encodes a nuclear protein which, when unphosphorylated, regulates cell growth by preventing the passage of cells from G to the S phase of the cell cycle (Goodrich et al. 1991). Phosphorylation of the protein interferes with this control and permits such passage. Mutations of RB-1 that inactivate the protein are detected in virtually all retinoblastomas and most small lung carcinomas (Lee et al. 1987; Levine 1993). Lower frequencies of RB-1 mutation have been identified in urinary bladder and breast carcinomas (Levine 1993; Varley et al. 1989). However, the role of RB-1 in human pancreatic carcinomas is not firmly established, since only

two immunohistochemical reports have been published with evidence of RB-1 underexpression in two of ten and none of seven primary human pancreatic carcinomas, respectively (Seymour et al. 1994).

Incidences of 90% for K-*ras* mutations, of 53% for complete or partial loss of expression of DCC, of 42% for complete or partial loss of expression of RB-1, and the absence of p53 mutations except for 26% of cases in which the p53 gene is indirectly involved due to overexpression of the *mdm-2* oncogene are characteristic genetic features of SGH pancreatic adenocarcinomas. It is noteworthy that, in this relatively small series of 21 different pancreatic duct cancers, six showed either a solitary K-*ras* mutation or the absence of any alterations for the five genes that were analyzed, suggesting that other as yet unidentified genes are also involved in the pathogenesis of SGH pancreatic duct carcinoma. With the exception of the K-*ras* mutation, which we believe to be an early event in carcinogenesis, we have no similar information on the timing or sequence of any of the genetic alterations presented above. In situ studies for the detection and localization of gene products or transcripts will be needed to time and sequence the appearance of the gene alterations.

In rat pancreas acinar cell carcinomas induced by azaserine, numerous attempts to identify the relevant molecular lesions have not been fruitful. Early studies (Schaeffer et al. 1990) reported the absence of K-*ras* mutations in rat acinar cell adenomas and carcinomas. A more recent study confirmed their findings and extended the genetic analysis to include the status of the H-*ras* gene, which was also found to be unaltered (Van Kranen et al. 1991). These results indicate that azaserine acinar cell carcinogenesis is mediated by a pathway involving not *ras* genes, but rather genes that remain to be identified.

References

Andrén-Sandberg A (1989) Androgen influence on exocrine pancreatic cancer. Int J Pancreatol 4:363–369

Arias AE, Bendayan M (1993) Differentiation of pancreatic acinar cells into duct-like cells in vitro. Lab Invest 69:518–530

Bakkevold KE, Pettersen A, Amesjo B, Espehaug B (1990) Tamoxifen therapy in unresectable adenocarcinoma of the pancreas and the papilla of vater. Br J Surg 77:725–730

Barton C, Hall PA, Hughes CM, Gullick WJ, Lemoine NR (1991) Transforming growth factor α and epidermal growth factor in human pancreatic cancer. J Pathol 163:111–116

Bell RH Jr, Kuhlmann ET, Jensen RT, Longnecker DS (1992) Overexpression of cholecystokinin receptors in azaserine-induced neoplasms of the rat pancreas. Cancer Res 52:3295–3299

Berenblum I, Shubik P (1947) A new quantitative approach to the study of the stages of chemical carcinogenesis in the mouse's skin. Br J Cancer 1:383–391

Birt DF, Pour PM (1987) Pancreatic cancer enhancement in the hamster model by diets high in fat and/or protein. In: Scarpelli DG, Reddy JK, Longnecker DS (eds) Experimental pancreatic carcinogenesis. CRC Press, Boca Raton, FL, pp 175–186

Birt DF, Stepan KR, Pour PM (1983) Interaction of dietary fat and protein on pancreatic carcinogenesis in Syrian golden hamsters. J Natl Cancer Inst 71:355–360

Birt DF, Julius AD, White LT, Pour PM (1989) Enhancement of pancreatic carcinogenesis in hamsters fed a high-fat diet ad libitum and at a controlled calorie intake. Cancer Res 49:5848–5851

Birt DF, Julius AD, Dwork E, Hanna T, White LT, Pour PM (1990) Comparison of the effects of dietary beef tallow and corn oil on pancreatic carcinogenesis in the hamster model. Carcinogenesis 11:745–748

Bos JL (1989) ras oncogene in human cancer: a review. Cancer Res 49:4682–4689

Buscail L, Delesque N, Estève JP, Saint-Laurent N, Prats H, Clerc P, Rofferecht P, Bell GI, Liebow C, Schally AV, Voysse N, Susini C (1994) Stimulation of tyrosine phosphatase and inhibition of cell proliferation by somatostatin analogues: mediation by human somatostatin receptor subtypes SSTR1 and SSTR2. Proc Natl Acad Sci USA 91:2315–2319

Cerny WL, Marigold KA, Scarpelli DG (1990) Activation of K-ras in transplantable pancreatic ductal adenocarcinomas of Syrian golden hamsters. Carcinogenesis 11:2075–2079

Cerny WL, Marigold KA, Scarpelli DG (1992) K-ras mutation is an early event in pancreatic duct carcinogenesis in the Syrian golden hamster. Cancer Res 52:4507–4513

Chang K-W, Laconi S, Mangold KA, Hubchak S, Scarpelli DG (1995) Multiple genetic alterations in hamster pancreatic ductal adenocarcinomas. Cancer Res 55:2560–2568

DeLisle RC, Logsdon CD (1990) Pancreatic acinar cells in culture: expression of acinar and ductal antigens in a growth-related manner. Eur J Cell Biol 51:64–75

Douglas BR, Woutersen RA, Jansen JB, de Jong AJL, Rovati LC, Lamers CB (1989) Modulation by CR-1409 (Lorglumide), a cholecystokinin receptor antagonist, of trypsin inhibitor-enhanced growth of azaserine-induced putative preneoplastic lesions in rat pancreas. Cancer Res 49:2438–2441

Falk RT, Pickle LW, Fontham ET, Correa P, Fraumeni JF Jr (1988) Life-style risk factors for pancreatic cancer in Louisiana: a case control study. Am J Epidemiol 128:324–336

Farber E (1984) The multistep nature of cancer development. Cancer Res 44:4217–4223

Farrow DC, Davis S (1990) Diet and the risk of pancreatic cancer in men. Am J Epidemiol 132:423–431

Fekete M, Zalatnai A, Comaru-Schally AM, Schally AV (1989) Membrane receptors for peptides in experimental and human pancreatic cancers. Pancreas 4:521–528

Finlay CA (1993) The mdm-2 oncogene can overcome wild type p53 suppression of transformed cell growth. Mol Cell Biol 13:301–306

Fölsch UR, Mustroph D, Schafmayer A, Becker HD, Creutzfeld W (1984) Elevated CCK plasma concentrations during acute and chronic feeding of soybean flour. Digestion 30:88

Fuji H, Egami H, Chaney W, Pour P, Pelling J (1990) Pancreatic ductal adenocarcinomas induced in Syrian hamsters by N-nitrosobis(2-oxopropyl)amine contain a c-K-ras oncogene with a point-mutated codon 12. Mol Carcinog 3:296–301

Gao X, Honn KV, Grignon D, Sake W, Chen YQ (1993) Frequent loss of expression and loss of heterozygosity of the putative tumor suppressor gene DCC in prostatic carcinomas. Cancer Res 53:2723–2727

Goodrich DW, Wang NP, Qian YW, Lee EY, Lee WH (1991) The retinoblastoma gene product regulates progression through the G1 phase of the cell cycle. Cell 67:293–302

Greenblatt MS, Bennett WP, Hollsteins M, Harris CC (1994) Mutations in the p53 tumor suppressor gene: clues to cancer etiology and molecular pathogenesis. Cancer Res 54:4855–4878

Hall PA, Lemoine NR (1992) Rapid acinar to ductal transdifferentiation in cultured human exocrine pancreas. J Pathol 166:97–103

Hedrick L, Cho KR, Fearon ER, Wu TC, Kinzler KW, Voglstein B (1994) The DCC gene product in cellular differentiation and colorectal tumorigenesis. Genes Dev 8:1174–1183

Hirayama T (1972) Smoking in relation to the death rates of 265,118 men and women in Japan. A report on five years of follow-up. Presented at the American Cancer Society's 14th Science Writers Seminar. Clearwater Beach FL

Hohne MW, Halatch M-E, Kahl GF, Weinel RJ (1992) Frequent loss of expression of the potential tumor suppressor gene DCC in ductal pancreatic adenocarcinoma. Cancer Res 52:2616–2619

Howe GR, Jain M, Miller AB (1990) Dietary factors and risk of pancreatic cancer: results of a Canadian population-based case-control study. Int J Cancer 45:604–608

Huang Y, Boynton RF, Blount PL, Silverstein RJ, Yin J, Tong Y, McDaniel TK, Newkirk C, Resau JH, Sridhara R, Reid BJ, Meltzer SJ (1992) Loss of heterozygosity involves multiple tumor suppressor genes in human esophageal cancers. Cancer Res 52:6525–6530

Hubchak S, Mangino MM, Reddy MK, Scarpelli DG (1990) Characterization of differentiated Syrian golden hamster pancreatic duct cells maintained in extended monolayer culture. In Vitro Cell Dev Biol 26:889–897

Ide H, Subbarao V, Reddy JK, Rao MS (1993) Formation of ductular structures in vitro by pancreatic epithelial oval cells. Exp Cell Res 209:38–44

Ishikawa O, Ohigashi H, Imaoka S, Nakai I, Mitsuo M, Weide L, Pour P (1995) The role of pancreatic islers in experimental pancreatic carcinogenicity. Am J Pathol 147:1458–1464

Jhappan C, Staahle C, Harkins RN, Fausto N, Smith GH, Merlino GT (1990) TGFalpha overexpression in transgenic mice induces liver neoplasia and abnormal development of the mammary gland and pancreas. Cell 61:1137–1146

Klijn JG, Hoff AM, Planting AS, Verweij J, Kok T, Lamberts SWJ, Portengen H, Foekens JA (1990) Treatment of patients with metastatic pancreatic and gastrointestinal tumours with the somatostatin analogue Sandostatin: a phase II study including endocrine effects. Br J Cancer 62:627–630

Lauren P (1965) The two histological main types of gastric carcinoma: diffuse and so-called intestinal-type carcinoma. An attempt at a histo-clinical classification. Acta Pathol Microbiol Scand 64:31–49

Leblond CP (1964) Classification of cell populations on the basis of their proliferative behavior. Natl Cancer Inst Monogr 14:119–150

Lee WH, Bookstein R, Hong F, Young LJ, Shew JY, Lee EY (1987) Human retinoblastoma susceptibility gene; cloning, identification and sequence. Science 235:1394–1399

Levine AJ (1993) The tumor suppressor genes. Annu Rev Biochem 62:623–651

Lhoste EF, Longnecker DS (1987) Effect of bombesin and caerulein on early stages of carcinogenesis induced by azaserine in the rat pancreas. Cancer Res 47:3273–3277

Lhoste EF, Roebuck BD, Longnecker DS (1988) Stimulation of the growth of azaserine-induced nodules in the rat pancreas by dietary camostate (FOY-305). Carcinogenesis 9:901–906

Longnecker DS (1991) Hormones and pancreatic cancer. Int J Pancreatol 9:81–86

Longnecker DS, Roebuck BD, Yager JD, Lilja HS, Siegmund B (1981) Pancreatic carcinoma in azaserine-treated rats, induction, classification and dietary modulation of incidence. Cancer 47:1562–1572

Longnecker DS, Faris RA, Bell RH Jr, Kuhlmann ET, Pettengill OS (1991) Ductal metaplasia in cell lines derived from an acinar cell carcinoma of the rat pancreas. Pancreas 6:710

Lu L, Louie D, Owyang CA (1989) A cholecystokinin releasing peptide mediates feedback regulation of pancreatic secretion. Am J Physiol 256:G430–G435

Makino T, Usuda N, Rao S, Reddy JK, Scarpelli DG (1990) Transdifferentiation of ductular cells into hepatocytes in regenerating hamster pancreas. Lab Invest 62:522–561

Matsukara N, Suzuki K, Kawochi T, Aoyai M, Sugimura T, Kitaoka H, Numajiri H, Shirota A, Itaboshi M, Hirota T (1980) Distribution of marker enzymes and mucin in intestinal metaplasia in human stomach and relation of complete and incomplete types of metaplasia to minute gastric carcinomas. J Natl Cancer Inst 65:231–240

McGuiness EE, Morgan RGH, Levison DA, Frape DL, Hopewood D, Wormsley KG (1980) The effects of longterm feeding of soya flour on the rat pancreas. Scand J Gastroenterol 15:497–502

McGuiness EE, Morgan RGH, Levison DA, Hopewood D, Wormsley KG (1981) Interaction of azaserine and raw soya flour on the rat pancreas. Scan J Gastroenterol 16:49–56

Meijers M, Bruijntjes JP, Hendriksen EG, Woutersen RA (1989) Histogenesis of early preneoplastic lesions induced by N-nitrosobis-(2-oxopropyl)amine in exocrine pancreas of hamsters. Int J Pancreatol 4:127–137

Meijers M, van Garderen-Hoetmer A, Lamers CB, Rovati LC, Jansen JBMJ, Woutersen RA (1990) Role of cholecystokinin in the development of BOP-induced pancreatic lesions in hamsters. Carcinogenesis 11:2223–2226

Meijers M, Woutersen RA, van Garderen-Hoetmer A, Bakker GH, de Jong FH, Foekens JA, Klijn JG (1991) Effects of castration, alone and in combination with amino glutethimide, on growth of preneoplastic lesions in exocrine pancreas of rats and hamsters. Carcinogenesis 12:1707–1713

Meijers M, Appel MJ, van Garderen-Hoetmer A, Lamers CB, Rovati LC, Jansen JB, Woutersen RA (1992a) Effects of cholecystokinin and bombesin on development of azaserine-induced pancreatic tumours in rats: modulation by the cholecystokin receptor antagonist lorgiumide. Carcinogenesis 13:1525–1528

Meijers M, Visser CJ, Klijn JG, Lamberts SWJ, van Garderen-Hoetmer A, de Jong FH, Foekens JA, Woutersen RA (1992b) Effects of orchiectomy, alone or in combination with testosterone, and cyproterone acetate on exocrine pancreatic carcinogenesis in rats and hamsters. Int J Pancreatol 11:137–146

Miller DG (1980) On the nature of susceptibility to cancer. The presidential address. Cancer 46:1307–1318

Mills PK, Beeson WL, Abbey DE, Fraser GE, Phillips RL (1988) Dietary habits and past medical history as related to fatal pancreas cancer risk among Adventists. Cancer 61:2578–2585

Momand J, Zambetti GP, Olson DC, George D, Levine AJ (1992) The mdm-2 oncogene product forms a complex with the p53 protein and inhibits p53-mediated transactivation. Cell 69:1237–1245

Oates PS, Morgan RG (1982) Pancreatic growth and cell turnover in the rat fed raw soya flour. Am J Pathol 108:217–224

Okamura E, Okoda M, Onoda N, Kamiya Y, Murakami H, Tsuhima T, Shizume K (1990) Insulin-like growth factor I and transforming growth factor α as autocrine growth factors in human pancreatic cancer cell growth. Cancer Res 50:103–107

Parsa I, Longnecker DS, Scarpelli DG, Pour P, Reddy JK, Lefkowitz M (1985) Ductal metaplasia of human exocrine pancreas and its association with carcinoma. Cancer Res 45:1285–1290

Pour PM (1978) Islet cells as a component of pancreatic ductal neoplasms. 1. Experimental study: ductular cells including islet cell precursors and primary progenitor cells of tumors. Am J Pathol 90:295–316

Pour PM (1988) Mechanism of pseudoductular (tubular) formation during pancreatic carcinogenesis in the hamster model. An electron-microscopic and immune-histochemical study. Am J Pathol 130:335–344

Pour PM, Birt DF (1983) Modifying factors in pancreatic carcinogenesis in the hamster model. IV. Effects of dietary protein. J Natl Cancer Inst 71:347–353

Pour PM, Uchida E, Burnett DA, Steplewski Z (1986) Blood-group antigen expression during pancreatic cancer induction in hamsters. Int J Pancreatol 1:327–340

Pour PM, Lawson T, Hegelson S, Donnerly T, Stepan K (1988) Effect of cholecystokinin on pancreatic carcinogenesis in the hamster model. Carcinogenesis 9:597–601

Rao MS, Subbarao V, Reddy JK (1986) Induction of hepatocytes in the pancreas of copper-depleted rats following copper repletion. Cell Differ 18:109–117

Rao MS, Dwivedi RS, Subbarao V, Usman MI, Scarpelli DG, Nemali MR, Yeldandi A, Thangada S, Kumar S, Reddy JK (1988) Almost total conversion of pancreas to liver in the adult rat: a reliable model to study transdifferentiation. Biochem Biophys Res Commun 156:131–136

Rao MS, Yeldandi AV, Reddy JK (1990) Stem cell potential of ductular and periductular cells. Cell Differ Dev 29:155–163

Reddy BS, Sugie S (1988) Effect of different levels of omega-3 and omega-6 fatty acids on azoxymethane-induced colon carcinogenesis in F344 rats. Cancer Res 48:6642–6647

Reddy JK, Kanwar YS, Rao MS, Watanabe TK, Reddy MK, Parsa I, Longnecker DS, Tafuri S (1986) Duct-like morphogenesis of Longnecker pancreatic acinar carcinoma cells maintained in vitro on seminiferous tubular basement membrane. Cancer Res 46:347–354

Reddy JK, Rao MS, Yeldandi AV, Tan X, Dwivedi RS (1991) Pancreatic hepatocytes. An in vivo model for cell lineage in pancreas of adult rat. Dig Dis Sci 36:502–509

Reubi JC, Horisberger U, Essed CE, Jeekel J, Klijn JG, Lamberts SWJ (1988) Absence of somatostatin receptors in human exocrine pancreatic adenocarcinomas. Gastroenterology 95:760–763

Rowlatt U (1967) Spontaneous epithelial tumours of the pancreas of mammals. Br J Cancer 21:82–107

Sandgren EP, Quaife CJ, Paulovich AG, Palmiter RD, Brimster RL (1991) Pancreatic tumor pathogenesis reflects the causative genetic lesion. Proc Natl Acad Sci USA 88:93–97

Scarpelli DG (1985) Editorial. Multipotent developmental capacity of cells in the adult animal. Lab Invest 52:331–333

Scarpelli DG, Rao MS (1981) Differentiation of regenerating pancreatic cells into hepatocyte-like cells. Proc Natl Acad Sci USA 78:2577–2581

Scarpelli DG, Rao MS, Reddy JK (1985) Experimental carcinogenesis of exocrine pancreas: animal models, neoplasms, and current understanding of pathogenesis. In: Jones

TC, Mohr U, Hunt RD (eds) Digestive system. Springer, Berlin Heidelberg New York, pp 224–238

Scarpelli DG, Rao MS, Reddy JK (1991) Are acinar cells involved in the pathogenesis of ductal adenocarcinoma of the pancreas? Cancer Cells 3:275–277

Scemama JL, DeVries L, Pradayrol L, Seva C, Tronchere H, Vaysse N (1989) Cholecystokinin and gastrin peptides stimulate ODC activity in a rat pancreatic cell line. Am J Physiol 256:G846–G850

Schaeffer BK, Zurlo J, Longnecker DS (1990) Activation of c-K-ras not detectable in adenomas or adenocarcinomas arising in rat pancreas. Mol Carcinog 3:165–170

Scheck AC, Coons SW (1993) Expression of the tumor suppressor gene DCC in human gliomas. Cancer Res 53:5605–5609

Seymour AB, Hruban RH, Redston M, Caldas C, Powell SM, Kinzler KW, Yeo CJ, Kern SE (1994) Allelotype of pancreatic adenocarcinoma. Cancer Res 54:2761–2764

Shepherd JG, Chen JR, Tsao M-S, Duguid WP (1993) Neoplastic transformation of cultured rat pancreatic duct epithelial cells by azaserine and streptozotocin. Carcinogenesis 14:1027–1033

Shibata D, Almoguera C, Forrester K, Dunitz J, Martin SE, Cosgrove MM, Perucho M, Amheim N (1990) Detection of c-K-ras mutations in fine needle aspirates from human pancreatic adenocarcinomas. Cancer Res 50:1279–1283

Simon B, Weinel R, Hohne M, Watz U, Schmidt J, Kortner G, Arnold R (1994) Frequent alterations of the tumor suppressor genes p53 and DCC in human pancreatic carcinoma. Gastroenterology 106:1645–1651

Smith JJ, Derynck R, Korc M (1987) Production of transforming growth factor α in human pancreatic cancer cells: evidence for a superagonist autocrine cycle. Proc Natl Acad Sci USA 84:7567–7570

Sumi C, Longnecker DS, Roebuck BD, Brinck-Johnsen T (1989a) Inhibitory effects of estrogen and castration on the early stage of pancreatic carcinogenesis in Fischer rats treated with azaserine. Cancer Res 49:2332–2336

Sumi C, Brinck-Johnsen T, Longnecker DS (1989b) Inhibition of a transplantable pancreatic carcinoma by castration and estradial administration in rats. Cancer Res 49:6687–6692

Szende B, Srkalovic G, Schally AV, Lapis K, Groot K (1990a) Inhibitory effects of analogs of leutinizing hormone-releasing hormone and somatostatin on pancreatic cancers in hamsters. Events which accompany tumor regression. Cancer 65:2279–2290

Szende B, Srkalovic G, Groot K, Lapis K, Schally AV (1990b) Regression of nitrosamine-induced pancreatic cancers in hamsters treated with luteinizing hormone-releasing hormone antagonists or agonists. Cancer Res 50:3716–3721

Townsend CM, Singh P, Thompson SC (1986) Gastrointestinal hormones and gastrointestinal and pancreatic carcinomas. Gastroenterology 91:1002–1006

Uchino S, Tsuda H, Noguchi M, Yokota J, Terada M, Saito T, Kobayashi M, Sugimura T, Hirohashi S (1992) Frequent loss of heterozygosity at the DCC locus in gastric cancer. Cancer Res 52:3099–3102

van Kranen HJ, Vermeuien E, Schoren L, Bax J, Woutersen RA, van Iersel P, van Kreije CF, Scherer E (1991) Activation of c-K-ras is frequent in pancreatic carcinomas of Syrian hamsters, but is absent in pancreatic tumors of rats. Carcinogenesis 12:1477–1482

Varley JM, Armour J, Swallow JE, Jeffrey AJ, Ponder BA, Walker RA (1989) Retinoblastoma gene is frequently altered leading to loss of expression in primary breast tumors. Oncogene 4:725–729

Verma AK, Ashendel CL, Boutwell RK (1980) Inhibition by prostaglandin synthesis inhibitors of the induction of epidermal orothine decarboxylase activity, the accumulation of prostaglandins, and tumor promotion caused by 12-O-tetradecanoylphorbol-13-acetate. Cancer Res 40:308–315

Woutersen RA, van Garderen-Hoetmer A, Longnecker DS (1987) Characterization of a 4-month protocol for the quantitation of BOP-induced lesions in hamster pancreas and its application in studying the effect of dietary fat. Carcinogenesis 8:833–837

Wynder EL (1975) An epidemiological evaluation of the causes of cancer of the pancreas. Cancer Res 35:2228–2233

Wynder EL (1976) Nutrition and cancer. Fed Proc 35:1309–1315

Wynder EL, Mabuchi K, Maruchi N, Former JG (1973) Epidemiology of cancer of the pancreas. J Natl Cancer Inst 50:645–667

Yoshida Y, Kaneko A, Chisaka N, Onoé T (1978) Appearance of intestinal type of tumor cells in hepatoma tissue induced by 3'-methyl-4-dimethylaminoazobenzene. Cancer Res 38:2753–2758

Zalatnai A, Schally AV (1989a) Treatment of N-nitrosobis(2-oxopropl)amine-induced pancreatic cancer in Syrian hamsters with D-Trp-6-LH-RH and somatostatin analog RC-160 microcapsules. Cancer Res 49:1810–1815

Zalatnai A, Schally A (1989b) Responsiveness of the hamster pancreatic cancer to treatment with microcapsules of D-Trp-6-LH-RH and somatostatin analog RC-160. Histological evidence of improvement. Int J Pancreatol 4:149–160

Atrophy, Exocrine Pancreas, Rat

Gary A. Boorman and Scot L. Eustis

Synonyms. Pancreatic microcystic transformation, pancreatic degeneration, pancreatic involution

Gross Appearance

Naturally occurring atrophy of the exocrine pancreas in the rat was described by Kendrey and Roe (1969) as having a macroscopic granular appearance. In experimentally induced progressive atrophy, the pancreas appeared fatty in rats after 6 weeks on a copper-deficient diet (Smith et al. 1982).

Microscopic Features

Atrophy of the exocrine pancreas varies from small discrete lesions to diffuse changes involving almost the entire pancreas. The atrophy often has a lobular pattern with affected lobules sharply demarcated from the more normal appearing adjacent pancreas (Fig. 246). The atrophic changes are frequently accompanied by an infiltrate of inflammatory cells consisting predominantly of lymphocytes and macrophages (Fig. 247). Within the area of atrophy, the pancreatic ducts are often dilated (Fig. 248) and the acini consist of small glands lined by cuboidal epithelium (Fig. 249). Other changes occasionally seen are dilated glands lined by flattened epithelium, interstitial fibrosis (Fig. 250) and stromal infiltration by fat cells (Fig. 251). In spite of severe atrophic changes, there is preservation of the islet tissue (Fig. 251).

Ultrastructure

Ultrastructural examination of pancreatic tissue from 30-month-old male F344 rats revealed changes in the interstitial stroma and in exocrine cells. There was a relative increase in collagen fibers and focal accumulations of finely granular or flocculent material resembling the basal lamina. This material was often confluent with the basal lamina of acini. Focally, the basal lamina was thrown into complex folds extending away from the acinar or duct cells. Affected acinar cells were reduced in size and contained fewer zymogen granules and less abundant rough endoplasmic reticulum than normal cells. In addition, the atrophic cells sometimes contained cytosegrasomes or autophagic vacuoles and lipid droplets. Occasional mitochondria were swollen and contained disrupted and small lamellar arrays of membranous material. Some ducts had dilated lumina with extruded whorls of membranous material, and epithelial cells contained autophagic vacuoles (Figs. 252, 253).

Ultrastructural examination of the pancreas from zinc-deficient rats revealed a decrease and rupture of zymogen granules, increased lipid droplets, nuclear pyknosis, and prominent lysosome-like bodies (Koo and Turk 1977). Pancreatic atrophy in rats induced by copper deficiency was characterized by prominent hypertrophy of mitochondria in the acinar cells (Fell et al. 1982). There was extensive degeneration of the rough endoplasmic reticulum and failure of zymogen granule synthesis. In contrast, ultrastructural examination of pancreatic tissue from magnesium-deficient rats revealed increased zymogen granules. However, that study failed to produce clear lesions of the exocrine pancreas, as has been reported for humans (The et al. 1980).

Differential Diagnosis

The lesions of pancreatic atrophy are common and easily recognized. Acute pancreatitis, which is rare in the rat, can be recognized by the acute

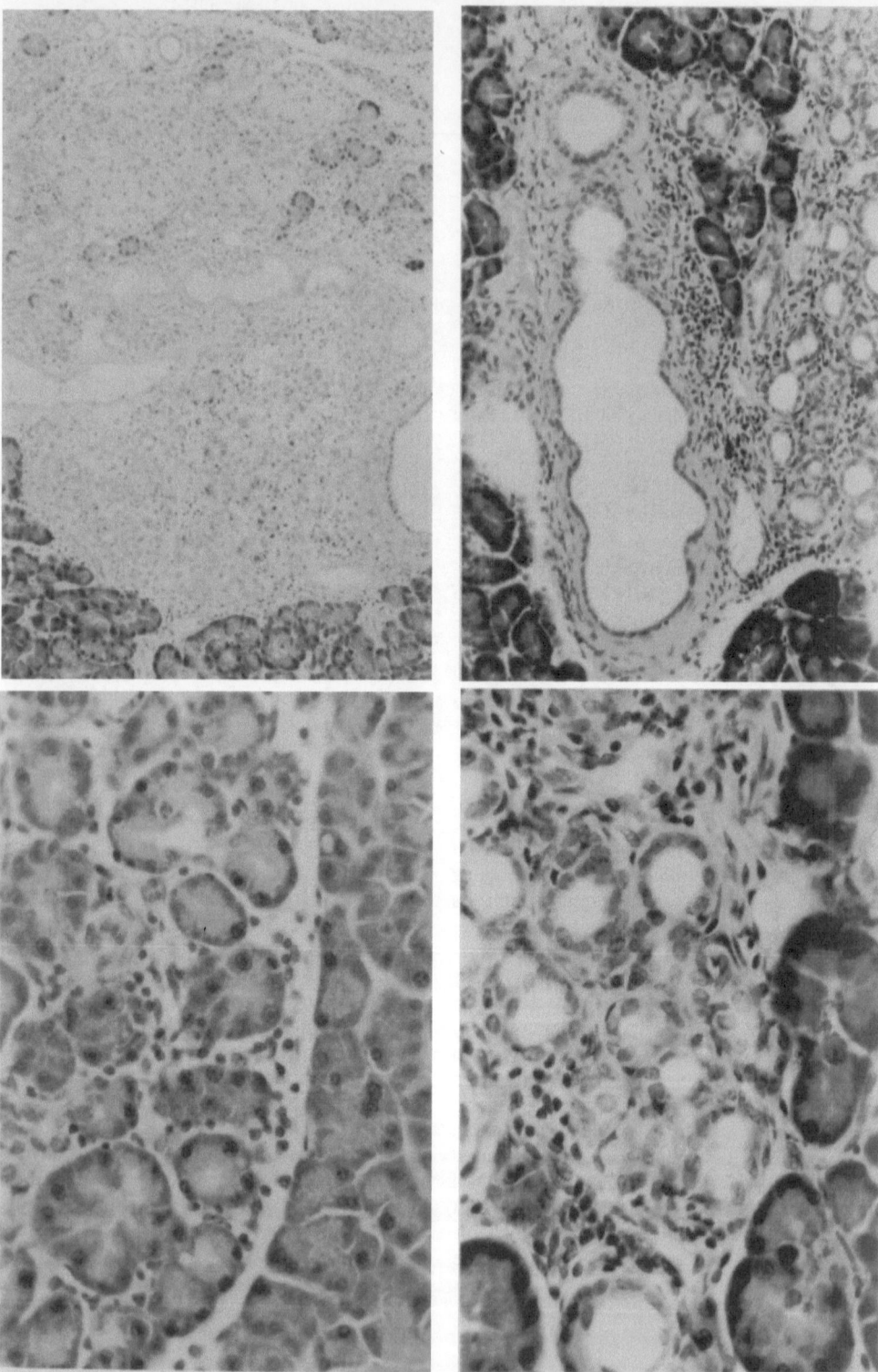

Fig. 246. (*upper left*) Exocrine pancreas, rat. Focal area of atrophy sharply demarcated from more normal appearing adjacent pancreas. H&E, ×100

Fig. 247. (*lower left*) Exocrine pancreas, rat. Margin of normal (*right*) and atrophic pancreas (*left*), with influx of mononuclear inflammatory cells. H&E, ×250

Fig. 248. (*upper right*) Exocrine pancreas, rat. Dilated pancreatic ducts in an atrophic area. H&E, ×200

Fig. 249. (*lower right*) Exocrine pancreas, rat. Ducts in the atrophic area lined by cuboidal epithelial cells. H&E, ×300

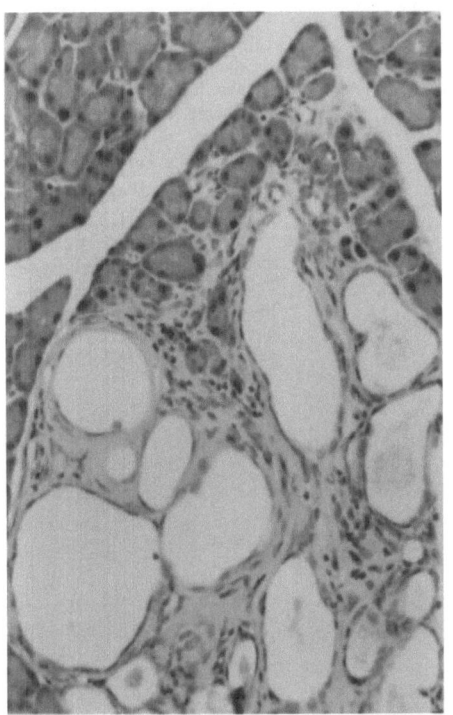

inflammatory changes characterized by hemorrhage, edema, necrosis, and influx of polymorphonuclear granulocytes.

Biologic Features

Natural History

Atrophy of the exocrine pancreas is uncommon in animals less than 3 months of age and, when found, is not extensive or severe (Andrew 1944). As animals age beyond 1 year, the lesion becomes more common and extensive (Berg 1967; Boorman and Hollander 1973). In one study, nearly all rats older than 2 years had some degree of exocrine pancreatic atrophy (Kendrey and Roe 1969). Both the frequency and severity of the lesion appear age related (Anver and Cohen 1979). Rats with pancreatic atrophy have normal islet tissue (Burek 1978), and clinical disease in the rat resulting from naturally occurring pancreatic atrophy has not been reported (Kendrey and Roe 1969).

Pathogenesis

The pathogenesis of naturally occurring pancreatic atrophy of aging rats is unknown (Andrew 1944; Kendrey and Roe 1969). In copper-deficient rats, there is marked loss and atrophy of acinar cells (Fell et al. 1982; Muller 1970; Smith et al. 1982). However, these changes are much more diffuse and affect the entire gland, in contrast to the spontaneous lesion, which may be focal and quite circumscribed. Pancreatic changes in magnesium-deficient rats bear little resemblance to the naturally occurring lesion (The et al. 1980). Manganese exposure in rats by intraperitoneal injection produces a selective toxicity on pancreatic tissue characterized by destruction of acinar cells, influx of neutrophils, lymphocytes, and macrophages, and increased fibrous tissue (Scheuhammer 1983; Scheuhammer and Cherian 1983). Ductal ligation in the rat produced rapid

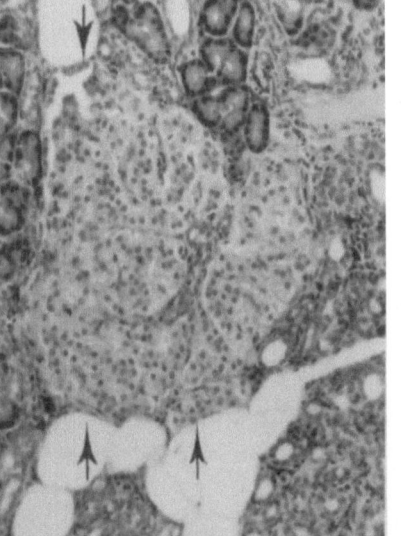

◄

Fig. 250. (*above*) Exocrine pancreas, rat. Focal area of interstitial fibrosis and cystically dilated ducts lined by flattened epithelial cells. H&E, ×150

Fig. 251. (*below*) Pancreas, rat. Normal appearing islet (*arrows*) within an area of severe atrophy and fat infiltration. H&E, ×145

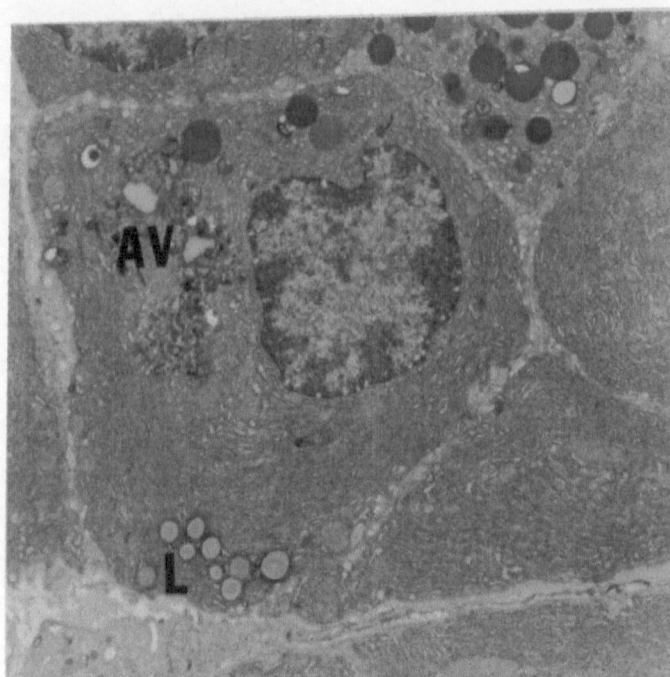

Fig. 252. (*above*) Pancreatic acinar cell, 30-month-old male F344 rat. Lipid droplets are present in the basal region of the cell (*L*) and a large cyto-segrasome (autophagocytic vacuole) containing membranous material is present (*AV*). TEM, ×5850

Fig. 253. (*below*) Pancreatic duct, 30-month-old male F344 rat. The lumen of the duct contains whorled arrays of membranous material and a cytoplasmic bleb (*CB*) with diminished electron density. Duct cells contain membrane-bound vacuoles filled with debris (*arrows*). TEM, ×5850

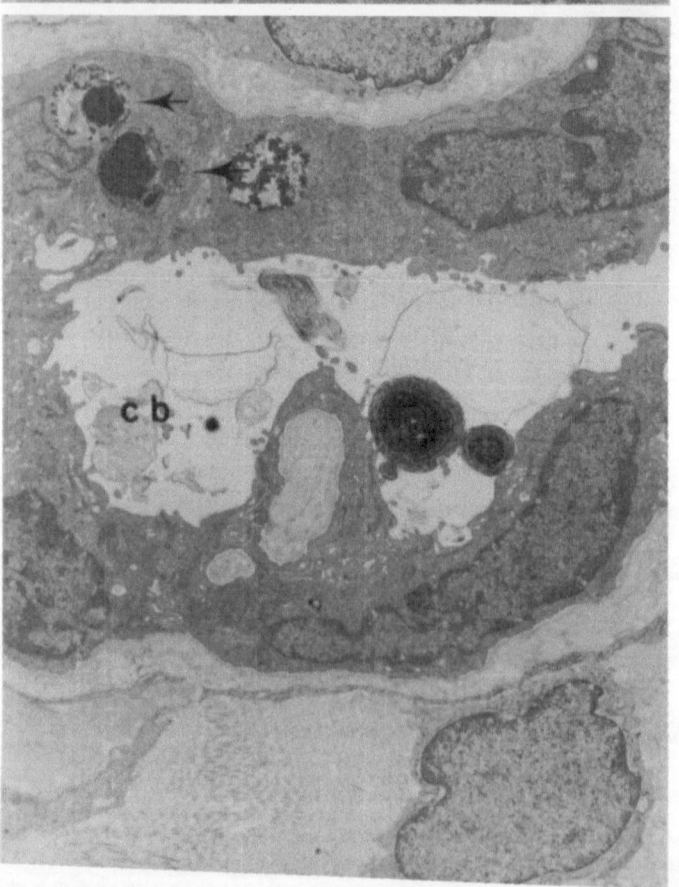

loss of acinar cells with the acini forming duct-like structures, influx of mononuclear cells, and fibrosis (Hultquist and Jonsson 1965; Mann et al. 1979; Pound and Walker 1981). Autoradiographic studies show that DNA synthesis in acinar cells ceases and there is death of acinar cells. There is proliferation of small duct cells in the area distal to the ligation (Walker and Pound 1983).

The morphological changes found in rats 3 months after duct ligation, i.e., small, dilated, duct-like structures, interstitial fibrosis, and influx of mononuclear cells, are somewhat similar to the naturally occurring pancreatic atrophy. However, evidence of duct obstruction was not reported in naturally occurring pancreatic atrophy (Kendrey and Roe 1969) and, similarly, we have not found evidence of ductal obstruction in F344 rats. In some cases, eosinophilic material is present in the duct lumen in affected areas, but this was reported to be an artifact of fixation of high-protein pancreatic secretion and not a physiologic "plug" (Papp et al. 1982). Vascular lesions, such as polyarteritis nodosa, which is common in rats, also merit consideration, but an association between vascular disease and pancreatic atrophy could not be shown for the rat (Kendrey and Roe 1969). In summary, the pathogenesis of naturally occurring pancreatic atrophy in the rat is still unknown.

Incidence

In eight chronic studies reported by the National Toxicology Program (Research Triangle Park, NC, United States) in 1983, exocrine pancreatic atrophy was diagnosed in 49 out of 393 (12%) male F344 rats and in 36 out of 387 (9%) female

F344 rats. These studies were terminated at 2 years. The incidence appears higher in rats allowed to complete their life span than in 2-year-old F344 rats, but there may be strain differences in the incidence of pancreatic atrophy as well (Table 18).

Comparison with Other Species

Necrosis and atrophy of the exocrine pancreas in mice can be induced by feeding DL-ethionine in protein-deficient diets (Herman and Fitzgerald 1962), choline-deficient diets (Lombardi et al. 1975), and prolonged fasting (Nevalainen and Janigan 1974). Pancreatic exocrine cell necrosis and atrophy can also be induced in mice by a variety of viruses, including strain E encephalomyocarditis virus (EMC) (Craighead 1966), group B coxsackievirus (Pappenheimer et al. 1951), and others (Angiolelli and Rio 1972; Gorelkin and Jahrling 1974). Coxsackievirus-induced pancreatic insufficiency in mice has many similarities to a genetic mutation found in CBA/J mice (Lansdown 1974; Pivetta and Green 1973). However, the disease cannot be propagated as an autosomal recessive mutation, and the genetic factors predisposing to the disease are unknown (Eppig and Leiter 1977; Leiter and Cunliffe-Beamer 1977). In mice, this disease is characterized by the breakdown of zymogen granules, release of zymogen contents into the cell, and destruction of acinar cells (Leiter and Cunliffe-Beamer 1977). Pancreatic degenerative atrophy has been described in German shepherd dogs, and it appears to be inherited as an autosomal recessive trait (Westermarck 1980; Rimaila-Parnanen

Table 18. Incidence of acinar cell pancreatic atrophy

Strain	Age	Males			Females			Reference
		Affected (n)	Total (n)	(%)	Affected (n)	Total (n)	(%)	
F344	2 years	49	393	12	36	387	9	NTP (1983)
BN/Bi	Life span	30	74	40	81	236	34	Burek (1978)
WAG/Rij	Life span	14	124	11	16	101	16	Burek (1978)
BN/Bi/WAG/Rij(Fl)	Life span	31	67	46	4	68	6	Burek (1978)
Sprague-Dawley (HAP)	6–29 months	12	71	17	–			Anver et al. (1982)
Sprague-Dawley/ Crl/COBS/CD	12–38 months	43	108	40	–			Anver et al. (1982)

NTP, National Toxicology Program, National Institute of Environmental Health Services, Research Triangle Park, NC, United States.

and Westermarck 1982; Hashimoto et al. 1979; Szabo et al. 1978). The disease has also been reported in young beagle dogs (Prentice et al. 1980). Atrophy of the exocrine pancreas in dogs can be induced by complete or incomplete duct obstruction (Churg and Richter 1971; Leong et al. 1982) and by duct blockage with a silicone adhesive (White et al. 1981). The morphological lesions of exocrine pancreatic atrophy in the dog are similar to those reported for the rat (Anderson and Low 1965; Hill et al. 1971). As in the rat, the pathogenesis of the disease is unknown (Pfister et al. 1980). Exocrine pancreatic lesions similar to those described in rats and dogs have also been reported for dasyurid marsupials (Attwood and Woolley 1980).

Pancreatic atrophy occurs quite frequently in older people and bears some resemblance to the disease reported in rats (Andrew 1944).

References

Anderson N, Low D (1965) Juvenile atrophy of the canine pancreas. Anim Hosp 1:101–109

Andrew W (1944) Senile changes in the pancreas of Wistar Institute rats and of man with special regard to the similarity of locale and cavity formation. Am J Anat 74:97–126

Angiolelli RF, Rio GJ (1972) Infectious pancreatic necrosis virus-induced pancreatic lesions in Swiss albino mice: elecctron microscopy. Am J Vet Res 33:1513–1520

Anver MR, Cohen BJ (1979) Lesions associated with aging. In: Baker HJ, Lindsey JR, Weisbroth SH (eds) The laboratory rat, vol I. Biology and diseases. Academic, New York, pp 377–399

Anver MR, Cohen BJ, Lattuada CP, Foster SJ (1982) Ageassociated lesions in barrier-reared male Sprague-Dawley rats: a comparison between Hap: (SD) and Crl:COBS[R]CD[R](SD) stocks. Exp Aging Res 8:3–24

Attwood HD, Woolley PA (1980) Pancreatic pathology in dasyurid marsupials. J Wildl Dis 16:245–249

Berg BN (1967) Longevity studies in rats. II. Pathology of ageing rats. In: Cotchin E, Roe FJC (eds) Pathology of laboratory rats and mice. Blackwell Scientific, Oxford, chap 23

Boorman GA, Hollander CF (1973) Spontaneous lesions in the female WAG/Rij (Wistar) rat. J Gerontol 28:152–159

Burek JD (1978) Pathology of aging rats. CRC, Boca Raton FL

Churg A, Richter WR (1971) Early changes in the exocrine pancreas of the dog and rat after ligation of the pancreatic duct. Am J Pathol 63:521–546

Craighead JE (1966) Pathogenicity of the M and E variants of the encephalomyocarditis (EMC) virus. II. Lesions of the pancreas, parotid and lacrimal glands. Am J Pathol 48:375–386

Eppig JJ, Leiter EH (1977) Exocrine pancreatic insufficiency syndrome in CBA/J mice. I. Ultrastructural study. Am J Pathol 86:17–30

Fell BF, King TP, Davies NT (1982) Pancreatic atrophy in copper-deficient rats: histochemical and ultrastructural evi-

dence of a selective effect on acinar cells. Histochem J 14:665–680

Gorelkin L, Jahrling PB (1974) Pancreatic involvement by Venezuelan equine encephalomyelitis virus in the hamster. Am J Pathol 75:349–362

Hashimoto A, Kita I, Okada K, Fujimoto Y (1979) Juvenile acinar atrophy of the pancreas of a dog. Vet Pathol 16:74–80

Herman L, Fitzgerald PJ (1962) The degenerative changes in pancreatic acinar cells caused by DL-ethionine. J Cell Biol 12:277–296

Hill FWG, Osborne AD, Kidder DE (1971) Pancreatic degenerative atrophy in dogs. J Comp Pathol 81:321–330

Hultquist GT, Jonsson LE (1965) Ligation of the pancreatic duct in rats. Acta Soc Med Uppsala 70:82–88

Kendrey G, Roe FJC (1969) Histopathological changes in the pancreas of laboratory rats. Lab Anim 3:207–220

Koo SI, Turk DE (1977) Effect of zinc deficiency on the ultrastructure of the pancreatic acinar cell and intestinal epithelium in the rat. J Nutr 107:896–908

Lansdown ABG (1974) Exocrine pancreatic insufficiency: a comparison of the clinical findings with the epi/epi mutation and Coxsackie virus B infection in mice. J Hered 65:378

Leiter EH, Cunliffe-Beamer T (1977) Exocrine pancreatic insufficiency syndrome in CBA/J mice. Pathological and genetic analysis. Gastroenterology 73:260–266

Leong ASY, Slavotinek AH, Deakin EJ, Nance SH, Elmslie RG (1982) The pathology of experimental chronic fibrosing pancreatitis – light microscopic and ultrastructural observations. Pathology 14:363–368

Lombardi B, Estes LW, Longnecker DS (1975) Acute hemorrhagic pancreatitis (massive necrosis) with fat necrosis induced in mice by DL-ethionine fed with a choline-deficient diet. Am J Pathol 79:465–480

Mann JR, Slater DN, Boyle P, Managnall Y, Fox M (1979) The effects of steroids and glucagon on the morphological changes in the duct-ligated pancreas of the rat. Br J Exp Pathol 60:423–433

Muller HB (1970) Der Einfluss kupferarmer Kost auf das Pankreas. Virchows Arch (Pathol Anat) 350:353–367

Nevalainen TJ, Janigan DT (1974) Degeneration of mouse pancreatic acinar cells during fasting. Virchows Arch (Cell Pathol) 15:107–118

Papp M, Fodor I, Varga G, Folly G (1982) Pancreatic edema: its effect on the function and morphology of the pancreas in dogs and rats. Mt Sinai J Med (NY) 49:456–464

Pappenheimer AM, Kunz LJ, Richardson S (1951) Passage of Coxsackie virus (Connecticut-5 strain) in adult mice with production of pancreatic disease. J Exp Med 94:45–64

Pfister K, Rossi GL, Freudiger U, Bigler B (1980) Morphological studies in dogs with chronic pancreatic insufficiency. Virchows Arch (Pathol Anat) 386:91–105

Pivetta OH, Green EL (1973) Exocrine pancreatic insufficiency: a new recessive mutation in mice. J Hered 64:301–302

Pound AW, Walker NI (1981) Involution of the pancreas after ligation of the pancreatic ducts. I. A histological study. Br J Exp Pathol 62:547–558

Prentice DE, James RW, Wadsworth PF (1980) Pancreatic atrophy in young Beagle dogs. Vet Pathol 17:575–580

Rimaila-Parnanen E, Westermarck E (1982) Pancreatic degenerative atrophy and chronic pancreatitis in dogs. A comparative study of 60 cases. Acta Vet Scand 23:400–406

Scheuhammer AM (1983) Chronic manganese exposure in rats: histological changes in the pancreas. J Toxicol Environ Health 12:353–360

Scheuhammer AM, Cherian MG (1983) The influence of manganese on the distribution of essential trace elements. II. The tissue distribution of manganese, magnesium, zinc, iron and copper in rats after chronic manganese exposure. J Toxicol Environ Health 12:361–370

Smith PA, Sunter JP, Case RM (1982) Progressive atrophy of pancreatic acinar tissue in rats fed a copper-deficient diet supplemented with D-penicillamine or triethylene tetramine: morphological and physiological studies. Digestion 23:16–30

Szabo T, Greenstein AJ, Geller SA, Dreiling DA (1978) Pancreatic atrophy in the canine: an entity of exocrine-endocrine dissociation. Mt Sinai J Med (NY) 45:503–508

The TL, Maxwell WL, Thirumalai C (1980) Light microscopic and ultrastructural changes of the exocrine pancreas in the magnesium deficient rats. Exp Pathol (Jena) 18:245–253

Walker NI, Pound AW (1983) An autoradiographic study of the cell proliferation during involution of the rat pancreas. J Pathol 139:407–418

Westermarck E (1980) The hereditary nature of canine pancreatic degenerative atrophy in the German shepherd dog. Acta Vet Scand 21:389–394

White DC, Sutherland DE, Najarian JS (1981) Endocrine function and histology of the canine pancreas after exocrine ablation by ductal injection of silicone rubber adhesive. J Surg Res 31:371–374

Exocrine Pancreas of Hypophysectomized Rats

Yoichi Konishi

Gross Appearance

Transauricular hypophysectomy was performed on male Wistar rats weighing approximately 100 g. The average body and pancreas weights 52 weeks after hypophysectomy were 133 ± 25 g and 0.25 ± 0.04 g, respectively. The pancreatic weight was 0.19% of the body weight, which was less than that of nonhypophysectomized rats (0.25%). The pancreas appeared slightly thin when fresh and in situ, but normal in color.

Microscopic Features

Histologic changes of the nonendocrine pancreas are clearly seen by staining with toluidine blue. Compared with acini of the normal pancreas (Fig. 254), most acini of hypophysectomized rats are partially involuted with loss of cellular size. The nuclei of acinar cells are irregular or invaginated in shape and contain increased chromatin. In most of the pancreas, the zymogen granules appear larger in diameter and fewer in number. Empty vacuoles are often seen in the basal side of the cytoplasm (Fig. 255). The centriacinar and ductal cells retain their normal appearance.

Ultrastructure

This is probably the first demonstration of the ultrastructure of pancreatic acinar cells in hypophysectomized rats after 1 year. Acinar cell nuclei are invaginated or lobulated and contain abundant perinuclear heterochromatin. The size of zymogen granules varies, and large zymogen granules can be seen. A well-developed Golgi complex and augmented condensing vacuoles are present (Fig. 256). Some acinar cells contain huge mitochondria with long tubular cristae (Fig. 257). The nuclei in some parts of the pancreas are compressed by zymogen granules, which diffusely fill they cytoplasm (Fig. 258).

Biologic Features

It has been reported that hypophysectomy results in a loss of pancreatic weight (Baker 1958, Mayston and Barrowman 1973), depletion of zymogen granules (Baker 1958), and a fall in the levels of both protease (Sesso and Valeri 1962) and lipase (Bicknell and Baker 1962). These depressions of acinar cell functions are overcome by the combined administration of somatotropin,

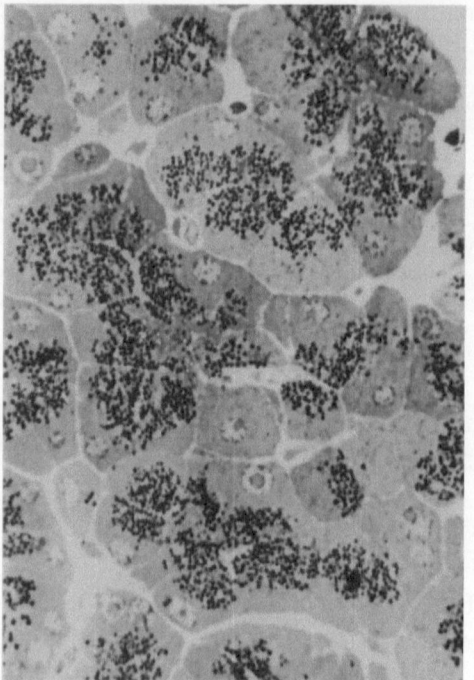

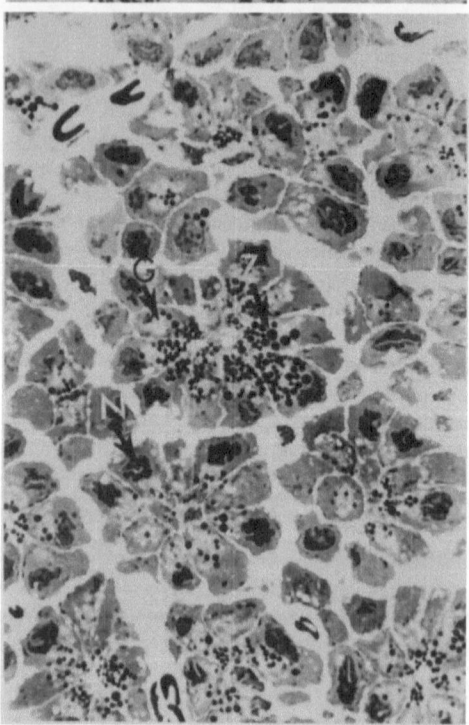

corticosterone, and thyroxin (Baker 1958; Baker et al. 1961) or administration of pentagastrin (Mayston and Barrowman 1973). Pituitary hormones also influence acinar cell function through the adrenal, thyroid, and endocrine cells of the gastrointestinal tract. The ultrastructural features we describe suggest that synthesis is decreased, as fewer large zymogen granules are present in hypophysectomized rats.

Ultrastructural studies have been carried out on various tissues of rats, mice, and humans. Hypophysectomy leads to disorganization of rough endoplasmic reticulum of rat hepatocytes (Lando et al. 1980). The differing effects of hypophysectomy on the ultrastructure of proximal tubules in the rat kidney are sex dependent. Reduction of the tubule circumference and ribosomes and decrease of Golgi area are the predominant changes (Daigeler 1981). Number and size of thyroid follicular cells are decreased, and their microvilli become flattened after hypophysectomy in rats (Gaal et al. 1976). Long, thin mitochondria appear in zona fasciculata cells of rat adrenals (Sharawy and Penney 1973). The larger granulated vesicles in the nerve terminalis of the vascular organ of the lamina terminalis almost disappear 2 months after hypophysectomy (Wenger 1976). Myocardial cell atrophy after hypophysectomy has also been reported (Smith and Page 1976).

Comparison with Other Species

In the mouse, an accumulation of dense secretory vesicles and conspicuous development of rough endoplasmic reticulum are observed in the B cells of the synovial membrane after hypophysectomy. This has been interpreted as being the result of secretory stimulation of the B cells, which are intimal polypeptide-producing secretory elements specific to the synovial membrane (Linck and Porte 1981). In humans with acromegaly, ultrastructural examination of the skeletal muscles reveals altered mitochondria (pleomorphism,

◄

Fig. 254. (*above*) Pancreatic acini of an unoperated rat. Toluidine blue, ×800

Fig. 255. (*below*) Pancreatic acini, hypophysectomized rat. Note the invaginated nuclei (*N*), reduced number and variably sized zymogen granules (*Z*), and prominent Golgi area (*G*). Toluidine blue, ×800

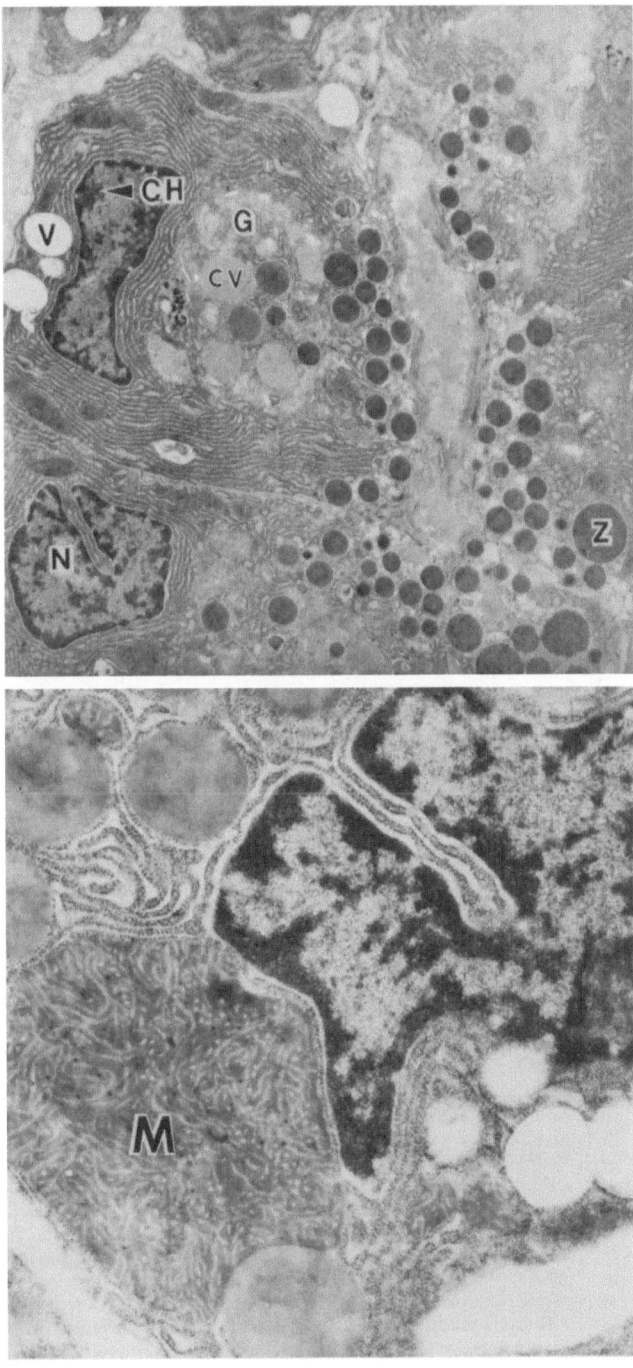

Fig. 256. (*above*) Pancreatic acinar cell, hypophysectomized rat. Invaginated nucleus (*N*), abundant heterochromatin (*CH*), large zymogen granules (*Z*), Golgi complex (*G*), augmented condensing vacuoles (*CV*), and vacant vacuoles (*V*). TEM, ×7000

Fig. 257. (*below*) Pancreatic acinar cell, hypophysectomized rat, containing huge mitochondria with long tubular cristae (*M*). TEM, ×20 000

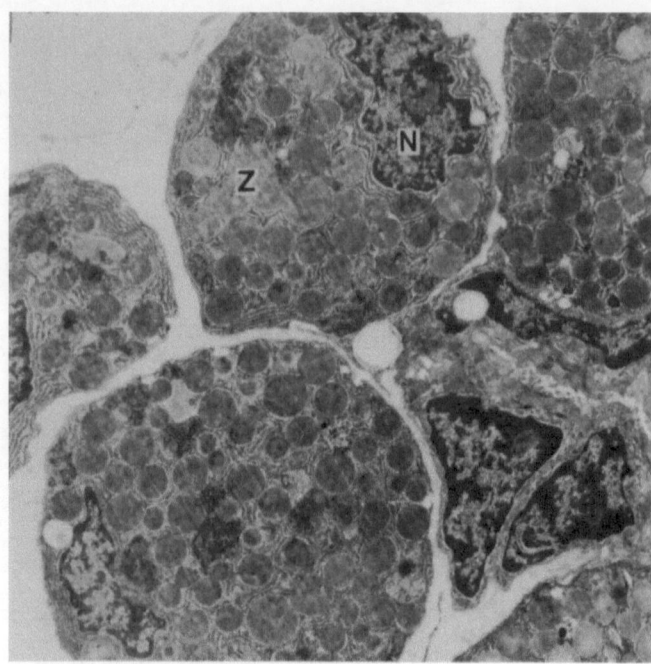

Fig. 258. Pancreatic acinar cells, hypophysectomized rat. Zymogen granules (*Z*) and compressed nucleus (*N*). TEM, ×7000

elongation, matrical pallor, and cristae abnormalities); these changes are diminished after hypophysectomy (Cheah et al. 1975). The effects of hypophysectomy on various tissue are varied, and further detailed morphological studies related to the function of organs are needed.

References

Baker BL (1958) Restoration of involuted zymogenic cells in hypophysectomized rats by replacement therapy. Anat Rec 131:389–403

Baker BL, Clapp HW, Annable CR, Dewey MM (1961) Elevation of proteolytic activity in the pancreas of hypophysectomized rats by hormonal therapy. Proc Soc Exp Biol Med 108:238–242

Bicknell JM, Baker BL (1962) Influence of hypophysectomy on lipolytic activity of the rat pancreas. Endocrinology 71:853–856

Cheah JS, Chua SP, Ho CL (1975) Ultrastructure of the skeletal muscles in acromegaly – before and after hypophysectomy. Am J Med Sci 269:183–187

Daigeler R (1981) Sex-dependent changes in the rat kidney after hypophysectomy. Cell Tissue Res 216:423–443

Gaal JM, Kovacs K, Sellers EA (1976) Effect of hypophysectomy and short- and long-term propylthiouracil treatment on the rat thyroid. Transmission and electronmicroscopy. Acta Anat 96:356–365

Lando D, Secchi J, Roche J, Raynaud JP (1980) Adrenocorticotropin analogs and glucocorticoids in the hypophysectomized rat. I. Effects on liver polyribosomes and rough endoplasmic reticulum. Endocrinology 107:2055–2062

Linck G, Porte A (1981) B-cells of the synovial membrane. IV. Ultrastructural evidence of secretory variations in hypophysectomized or propylthiouracyl-treated mice. Cell Tissue Res 218:123–128

Mayston PD, Barrowman JA (1973) Influence of chronic administration of pentagastrin on the pancreas in hypophysectomized rats. Gastroenterology 64:391–399

Sesso A, Valeri V (1962) Effect of hypophysectomy on the amylase and trypsin. Activatable protease activities in the pancreatic juice and pancreatic acinar cell of the rat. Pflugers Arch Physiol 274:345–355

Sharawy M, Penney DP (1973) Unusual mitochondrial morphology in the rat adrenal cortex following hypophysectomy. Am J Anat 136:395–401

Smith HE, Page E (1976) Morphometry of rat heart mitochondrial subcompartments and membranes: application to myocardial cell atrophy after hypophysectomy. J Ultrastruct Res 55:31–41

Wenger T (1976) Ultrastructural changes in the vascular organ of the lamina terminalis following ovariectomy and hypophysectomy in the rat. Brain Res 101:95–102

Necrotizing Pancreatitis Induced by 4-Hydroxyaminoquinoline, Rat

Yoichi Konishi

Synonyms. Acute pancreatitis, pancreatic necrosis

Gross Appearance

Twenty-four and 48 h after intraperitoneal injection of the rat with 20 mg 4-hydroxyaminoquinoline (4-HAQO)/kg body weight, the abdominal cavity contained a small amount of serosanguinous fluid. The pancreas was whitish and edematous and contained areas of fat necrosis and small hemorrhagic foci. Gross evidence of edema was apparent by 24 h after intravenous injection. Treatment with lower doses of 4-HAQO (7 mg/kg intraperitoneally or 10 mg/kg intravenously) did not change the normal appearance of the pancreas.

Microscopic Features

Different patterns of pancreatic damage, depending upon the route and dose of 4-HAQO administered, were observed microscopically. Extensive interstitial edema and fat and acinar cell necrosis with severe acute inflammatory cell infiltration were seen at 24 and 48 h after intraperitoneal injection of 20 mg/kg. Necrosis of acinar cells frequently extended deep into the center of the acinus, considerably disrupting the lobular architecture. At a dosage of 7 mg/kg, these effects were confined to the subcapsular regions and periphery of the lobules. Islets and pancreatic ducts appeared normal.

Twenty-four hours after the intravenous injection of 20 mg/kg, necrosis of multiple single acinar cells was seen (Fig. 259).

At 72 h, extensive acinar cell necrosis and interstitial edema with slight inflammatory cell infiltration was present (Fig. 260). Hemorrhage was not seen, nor were fat necrosis and capsular edema prominent. At this dose (20 mg/kg), although some islets were necrotic, the ductal epithelial cells were well preserved. At lower dosages of 7 or 13 mg/kg, injury was confined to acinar cells. A single intraperitoneal or intravenous injection of 7, 13, or 20 mg 4-HAQO/kg induced dose-dependent pancreatic acinar cell necrosis by 24 and 48 h. Eleven days after the intravenous injection of 14 mg 4-HAQO/kg, regenerating cells, which have already formed part of the acinar architecture, appear; they contain abundant cytoplasm with perinuclear basophilia and zymogen granules. At 14 days progressive restoration of the acinar cells, increased numbers of regenerating acini, and numerous mitotic figures are observed. The acinar and lobular architecture is almost restored by 4 weeks after the injection, and there is a slight adipose tissue infiltration. The acinar cells are eosinophilic; however, occasional mitotic figures can still be seen.

Differential Diagnosis

Acute pancreatic necrosis may be histologically diagnosed as interstitial (edematous) pancreatitis or necrotizing (hemorrhagic) pancreatitis (Baggenstoss 1973). However, an acute diffuse interstitial inflammatory process characterized by edema and a polymorphonuclear leukocytic reaction dominates the histologic picture of interstitial pancreatitis. Necrotizing pancreatitis is characterized by acinar cell necrosis, hemorrhage, foci of fat necrosis, and a mild inflammatory reaction to the fat necrosis. Pancreatic necrosis induced in rats by intraperitoneal or intravenous injections of 4-HAQO histologically resembles interstitial pancreatitis seen in humans. However, reasons for the lack of hemorrhage and only the slight degree of fat necrosis in rats remain obscure. The differential diagnosis of chronic pancreatitis can be made on the basis of the lack of inter- or intralobular fibrosis and ductal lesions.

Biologic Features

Ethionine (Herman and Fitzgerald 1962), azaserine (Hruban et al., 1965a), triparanol (Hruban et al. 1965b), 3-furylalanine (Hruban et al. 1965c), and puromycin (Longnecker and Farber 1967) induce pancreatic acinar cell necrosis accompanied by considerable liver cell damage. However, 4-HAQO induces pancreatic acinar cell

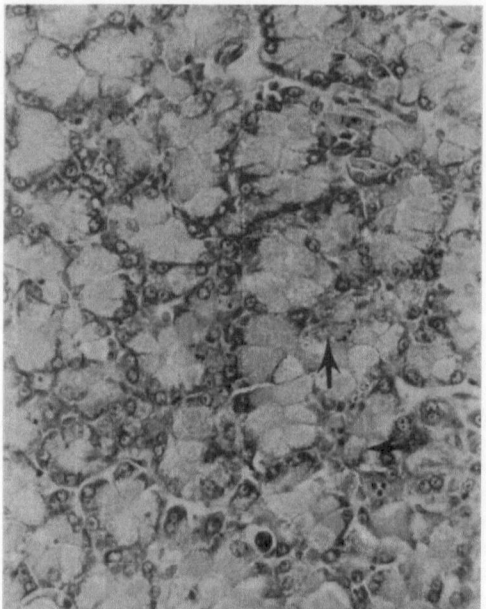

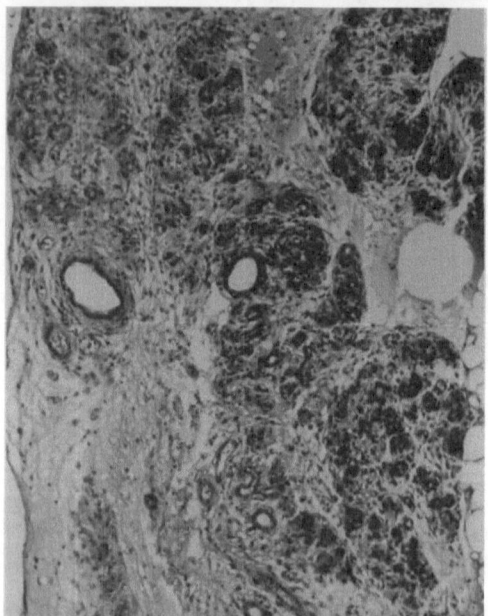

necrosis without morphological alterations in the liver (Hayashi et al. 1972; Konishi et al. 1974). The mechanism by which 4-HAQO causes a selective necrosis of the exocrine pancreas is not known. 4-HAQO is known to form a complex with DNA (Kawazoe et al. 1972), which might interfere with DNA-directed RNA synthesis. However, this DNA-binding mechanism cannot adequately explain the selective pancreaticotoxic action in the absence of a hepatotoxic effect.

After subcutaneous injection of 4-HAQO into mice, the liver metabolizes 4-HAQO rapidly to 4-AQO and 4-QO (Kawazoe et al. 1970). It is possible, therefore, that the selective pancreatic necrosis produced by 4-HAQO may be attributed to increased uptake and concentration of this compound by the pancreas. The liver would rapidly metabolize 4-HAQO to nontoxic metabolites, whereas the pancreas, with its slow metabolic processes, would accumulate a sufficient amount of toxic metabolites to produce necrosis. Animal death caused by necrotizing pancreatitis depends upon the severity of tissue damage; if the damage is not sufficient to upset the hemodynamic balance and the pancreas per se possesses the ability of restoration, the damage is not lethal.

Serum amylase levels after intraperitoneal and intravenous injection of 4-HAQO are shown in Fig. 261. The main amylase component in the serum of nontreated (control) rats is salivary-type amylase, indicated by its isozyme. The serum amylase level is statistically elevated 16 h after intravenous injection of 20 mg/kg, but not after intraperitoneal injection of this amount, suggesting that this elevation may be the reflection of increased pancreatic-type amylase. In fact, isozyme studies indicated that pancreatic-type amylase increased in the serum of rats after intravenous treatment with 4-HAQO. Pancreatic necrosis induced by the intravenous route of injections can be diagnosed by the elevation in serum amylase levels. Isozme studies also show that, after either intravenous or intraperitoneal injection, the ratio of salivary- and pancreatic-type amylase is changed in the serum of rats.

Fig. 259. (*above*) Necrotizing pancreatitis, rat, 24 h after intravenous injection of 20 mg 4-HAQO/kg body weight. Multiple single acinar cell necrosis (*arrows*). H&E, ×200

Fig. 260. (*below*) Necrotizing pancreatitis, rat, 72 h after intravenous injection of 20 mg 4-HAQO/kg. Diffuse acinar cell necrosis and interstitial edema with slight inflammatory cell infiltration. H&E, ×40

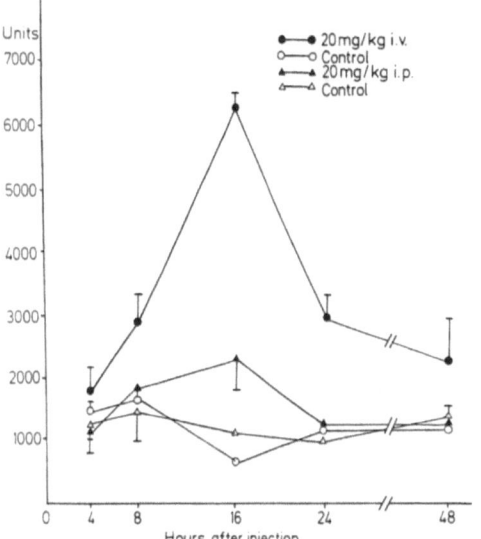

Fig. 261. Serum amylase in rats injected intravenously and intraperitoneally with 4-HAQO

Comparison with Other Species

Experimental pancreatic necrosis has been induced in several species of laboratory animals by the systemic administration of chemicals, by the intraductal administration of toxic substances, and by surgical procedures. Necrotizing pancreatitis was induced in guinea pigs by the intravenous injection of 4-HAQO (Reddy et al. 1975). It has been induced in rats by intraperitoneal injections of ethionine (Fitzgerald and Alvizouri 1952) and in mice by the intraperitoneal injection of human serum (Janigan et al. 1975), as well as by feeding choline-deficient diets containing 0.5% ethionine (Lombardi et al. 1975; Rao et al. 1982). Pancreatic damage caused by 4-HAQO in guinea pigs, ethionine in rats, and human serum in mice is histologically comparable to that induced by the intravenous injection of 4-HAQO in rats (Hayashi et al. 1972; Konishi et al. 1974). The damage induced in mice by choline-deficient diets containing 0.5% ethionine is accompanied by hemorrhage and can be diagnosed as acute hemorrhagic pancreatitis. Injections of bacterial toxin (Thal and Brackney 1954; Thal and Molestina 1955), bile salt (Waterman et al. 1969; Korbova et al.

1977), and bile (Herva 1970) into the pancreatic duct or directly into the parenchyma of dogs, rabbits, goats, guinea pigs, and rats cause pancreatic necrosis. Pancreatic necrosis can also be induced in cats and dogs by the surgical procedure of duct ligation (Menguy et al. 1957).

References

Baggenstoss AH (1973) Pathology of pancreatitis. In: Gambill EE (ed) Pancreatitis. Mosby, St Louis, pp 179–212

Fitzgerald PJ, Alvizouri M (1952) Rapid restitution of the rat pancreas following acinar cell necrosis subsequent to ethionine. Nature 170:929–930

Hayashi Y, Furrkawa H, Hasegawa T (1972) Pancreatic tumors in rats induced by 4-nitroquinoline 1-oxide derivatives. Proc 2nd Int Symp Princess Takamatsu Cancer Res Fund, pp 53–72

Herman L, Fitzgerald PJ (1962) The degenerative changes in pancreatic acinar cells caused by DL-ethionine. J Cell Biol 12:277–296

Herva P (1970) Experimental biliary pancreatitis in dogs. I. Agar electrophoretic study of the degradation of pancreatic tissue proteins. Scand J Gastroenterol 5 [Suppl 8]:14–22

Hruban Z, Swift H, Slesers A (1965a) Effect of azaserine on the fine structure of the liver and pancreatic acinar cells. Cancer Res 25:708–723

Hruban Z, Swift H, Slesers A (1965b) Effect of triparanol and diethanolamine on the fine structure of hepatocytes and pancreatic acinar cells. Lab Invest 14:1652–1672

Hruban Z, Swift H, Dunn FW, Lewis, Lewis DE (1965c) Effect of beta-2-furylalanine on the ultrastructure of the hepatocytes and the pancreatic acinar cells. Lab Invest 14:70–80

Janigan DT, Nevalainen TJ, MacAulary MA, Vethamany VG (1975) Foreign serum-induced pancreatitis in mice. I. A new model of acute pancreatitis. Lab Invest 33:591–607

Kawazoe Y, Tomura M, Araki M (1970) Metabolism of carcinogenic 4-hydroxyaminoquinoline 1-oxide in mice. Gann 61:593–596

Kawazoe Y, Araki M, Huang GF (1972) Chemical aspects of carcinogenesis by 4-nitroquinoline 1-oxide. In: Nakahara W, Takayama S, Sugimura T, Odashima S (eds) Topics in chemical carcinogenesis. University Park, Baltimore, pp 1–13

Konishi Y, Popp JA, Shinozuka H (1974) Pancreatic acinar cell damage induced by 4-nitroquinoline-1-oxide and 4-hydroxyaminoquinoline-1-oxide. JNCI 52:917–920

Korbova L, Kohout J, Malis F, Balas V, Cizkova J, Marek J, Cihak A (1977) Inhibitory effect of various cytostatics and cycloheximide on acute experimental pancreatitis in rats. Gut 18:913–918

Lombardi B, Estes LW, Longnecker DS (1975) Acute hemorrhagic pancreatitis (massive necrosis) with fat necrosis induced in mice by DL-ethionine fed with a choline-deficient diet. Am J Pathol 79:465–480

Longnecker DS, Farber E (1967) Acute pancreatic necrosis induced by puromycin. Lab Invest 16:321–329

Menguy RB, Hallenbeck GA, Bollman JL, Grindlay JH (1957) Ductal and vascular factors in etiology of experimentally induced acute pancreatitis. Arch Surg 74:881–889

Reddy JK, Rao MS, Svoboda DJ, Prasad JD (1975) Pancreatic necrosis and regeneration induced by 4-hydroxyaminoquinoline-1-oxide in the guinea pig. Lab Invest 32:98–104

Rao KN, Eagon PK, Okamura K, Van Thiel DH, Gavaler JS, Kelly RH, Lombardi B (1982) Acute hemorrhagic pancreatic necrosis in mice. Induction in male mice treated with estradiol. Am J Pathol 109:8–14

Thal A, Brackney E (1954) Acute hemorrhagic pancreatic necrosis produced by local Shwartzman reaction: experimental study on pancreatitis. JAMA 155:569–574

Thal A, Molestina JE (1955) Studies on pancreatitis. III. Fulminating hemorrhagic pancreatic necrosis produced by means of staphylococal toxin. Arch Pathol 60:212–220

Waterman NG, Walsky R, Moore R, Howell RS (1969) Acute pancreatitis: an experimental model Surgery 55:746–750

The Oral Cavity

Squamous Cell Carcinoma, Tongue, Rat

Richard J. Kociba

Gross Appearance

The gross appearance of a squamous cell carcinoma in the tongue of a rat is seen in Fig. 262 and may be compared with a normal rat tongue (Fig. 263). These tumors typically occur on the dorsal posterior aspect of the base of the tongue where the semilunar groove surrounds the vallate papilla. They appear as firm, usually white, papillary or fungiform masses. These masses can partially or toally occlude the lumen of the oral cavity and oropharynx and prevent ingestion of food. Smaller lesions may be present as papillary projections from the mucosal surface.

Microscopic Features

The semilunar groove, as noted in Fig. 264 surrounds the vallate papilla and is normally lined by invaginations of the squamous epithelium of the tongue. Nerve endings (taste buds) extend into the epithelium surrounding the semilunar groove. Normally a small complement of mononuclear cells are present in the adjacent submucosal interstitium.

Hair shafts and debris penetrate into the depths of the semilunar groove and adjacent submucosa (Fig. 265), eliciting a foreign body inflammatory reaction. A proliferative epithelial reaction and chronic inflammation are evoked by the embedded hair shafts (Figs. 266, 267).

Squamous cell carcinoma may arise from the epithelium of the semilunar groove (Fig. 268) in the presence of the proliferative and inflammatory lesions described previously. Squamous cell carcinoma at this site is recognized by the downgrowth of the squamous epithelium in a disorganized manner below the level of the basement membrane. The epithelium forms irregular masses often containing a central core of keratin (epithelial pearl). Individual cells often lose their polarity and fail to differentiate in an ordered fashion. Although invasion of adjacent tissues is apparent, these tumors do not appear to spread by metastasis.

Ultrastructure

At this time we are not aware of any studies on the ultrastructure of this series of lesions.

Differential Diagnosis

True neoplasms of the rat tongue, such as squamous cell carinoma, must be differentiated from the hyperplasia secondary to inflammation of the foreign body reaction. Squamous cell carcinomas at this site have the features of those that arise from squamous epithelium elsewhere.

Biologic Features

The first clinical signs noted are usually consistent with loss of body weight and loss of body condition secondary to the increasing inability to swallow food normally. As the lesion progresses in size, increased salivation is frequently noted. Most of the tongue neoplasms usually occur during the second year of life in the rat.

The pathogenesis of the series of lesions encountered in the region of the vallate papilla suggests a connection between foreign matter such as hair shafts and food particles, which is frequently found embedded in the semilunar groove or deeper into the excretory ducts and submucosal glandular structures, and the development of chronic inflammatory or foreign body reactions. The possibility exists that long-term chronic inflammation induces neoplasia through epigenetic mechanisms. In our experience, these neoplasms have been more common in dietary studies where the rat chow is pulverized. Tongue neoplasms have been found less frequently in inhalation studies where the rat chow is supplied in pelleted form.

In earlier toxicity or geriatric studies, tissues of the oral cavity of rodents were not always examined histologically on a routine basis unless a grossly recognized mass was noted at necropsy. During the past decade, most laboratories have been routinely examining the tongue both grossly and by light microscopy. Research in our laboratory has

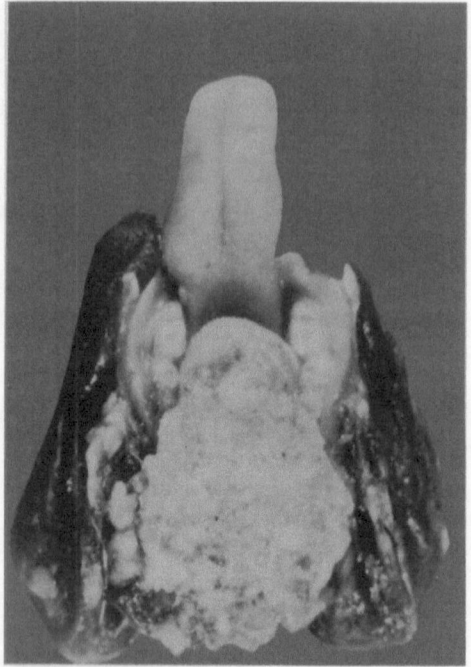

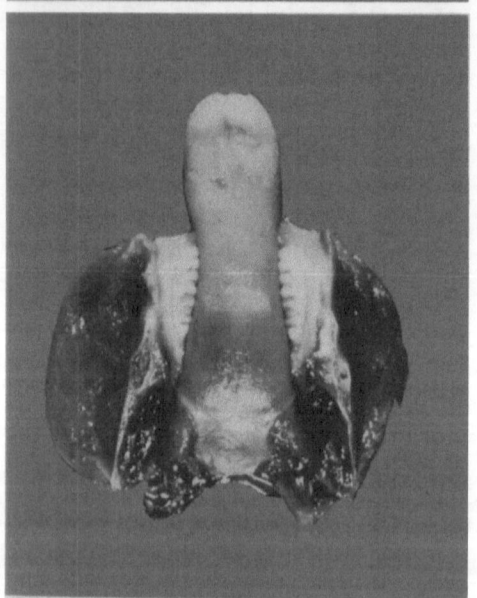

Fig. 262. (*above*) Mass originating at the base of the tongue, rat, in the region of the vallate papilla

Fig. 263. (*below*) Normal rat tongue

revealed that the single vallate papilla, located near the base of the tongue of the rat, frequently undergoes various age-associated inflammatory and sometimes proliferative lesions at this site (Keyes et al. 1980; Kociba et al. 1975, 1977, 1978, 1979).

The tongue of the rat contains various mechanical and gustatory papillae located on the dorsal surface (Hebel and Stromberg 1976; Weichert 1958). The rostral portion of the tongue is covered by well-developed filiform papillae, while fungiform papillae are located between the tip and dorsal prominence of the tongue. Paired foliate papillae are located on the lateral surfaces of the tongue at the level of the third molar. A single vallate papilla is located approximately 4 mm rostral to the epiglottis adjacent to a semilunar groove containing numerous taste buds (Greene 1963). Numerous ducts are located at the bottom of this groove. Examination of this groove frequently reveals submucosal inflammatory infiltrate and often the presence of hair shafts or other debris which have become embedded within this groove or the excretory ducts at the base of the tongue.

Table 19 lists the chronology within our laboratory regarding the diagnosis of spontaneous tongue neoplasms in control groups of animals. It is evident from the table that neoplasms of the tongue of rats were not recognized until the tongues were examined histologically on a routine

Table 19. Spontaneous tongue neoplasms occurring in control rats and mice used in 2-year diet studies

Strain of rat (mouse)	Study	Tongue examined routinely	Incidence of epithelial tongue neoplasms (%)	
			Male	Female
S-D rat	A	No	0	0
S-D rat	B	No	0	0
S-D rat	C	No	0	0
S-D rat	D	Yes	3.2	2.8
S-D rat	E	Yes	1.3	0
S-D rat	F	Yes	2	2.1
S-D rat	G	Yes	0	1.3
S-D rat	H	Yes	1.2	1.2
F344 rat	I	Yes	0	2
F344 rat	J	Yes	0.8	0.8
B6C3F1 mouse	K	Yes	0	1.6
B6C3F1 mouse	L	Yes	0	0
B6C3F1 mouse	M	Yes	0	0

S-D, Sprague-Dawley strain.

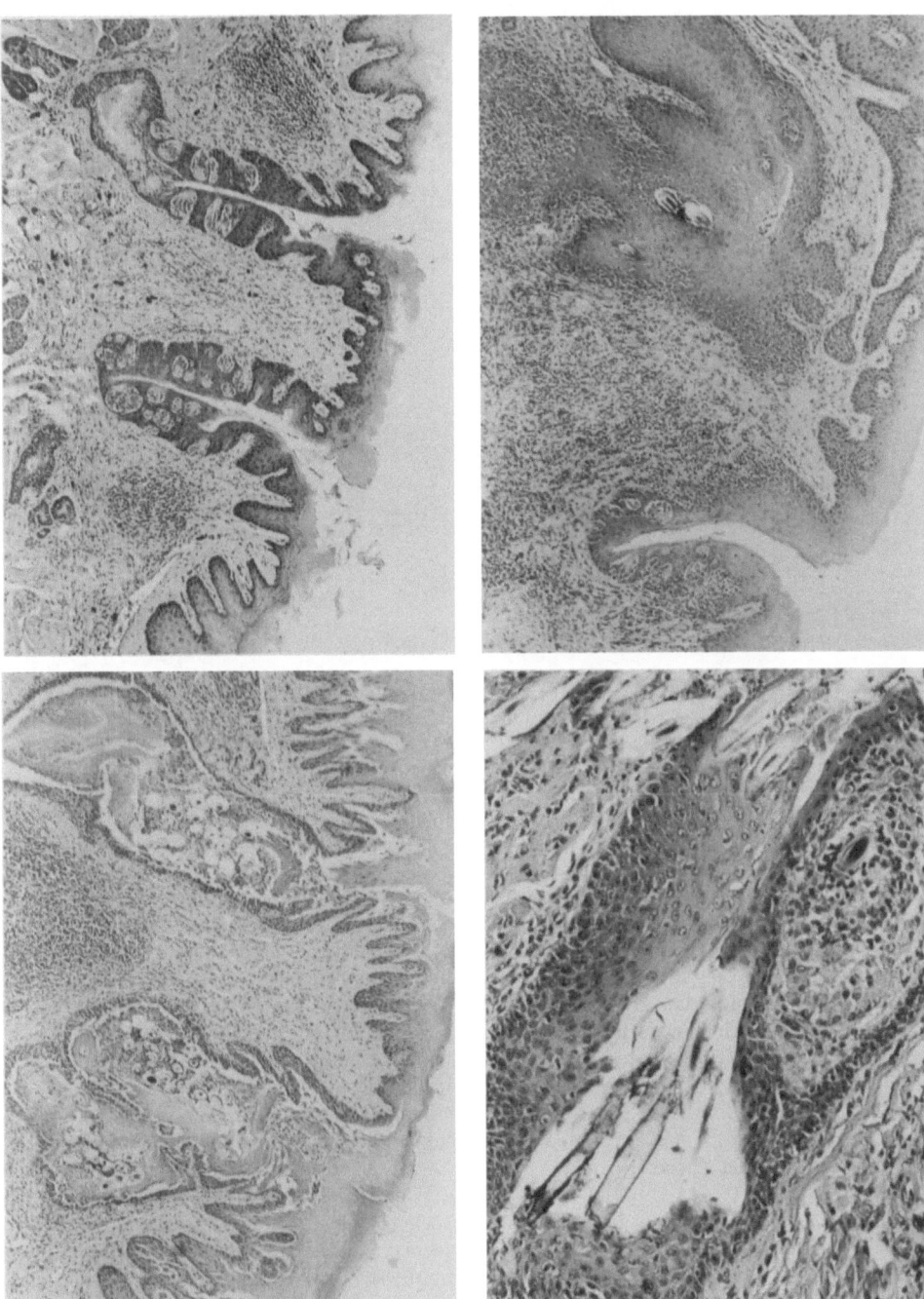

Fig. 264. (*upper left*) Cross-sectional histologic appearance of the vallate papilla and surrounding semilunar groove. H&E, ×40

Fig. 265. (*lower left*) Vallate papilla, tongue, rat, with penetrating hair shafts and debris in the semilunar groove eliciting a foreign body reaction. The mucosal surface has undergone acanthosis and hyperkeratosis in response to this reaction. H&E, ×40

Fig. 266. (*upper right*) Tongue, rat. Proliferative epithelial reaction to hair shafts embedded in submucosal tissue adjacent to semilunar groove. H&E, ×40

Fig. 267. (*lower right*) Tongue, rat. Higher magnification depicting inflammatory and epithelial reaction to foreign bodies (hair). H&E, ×100

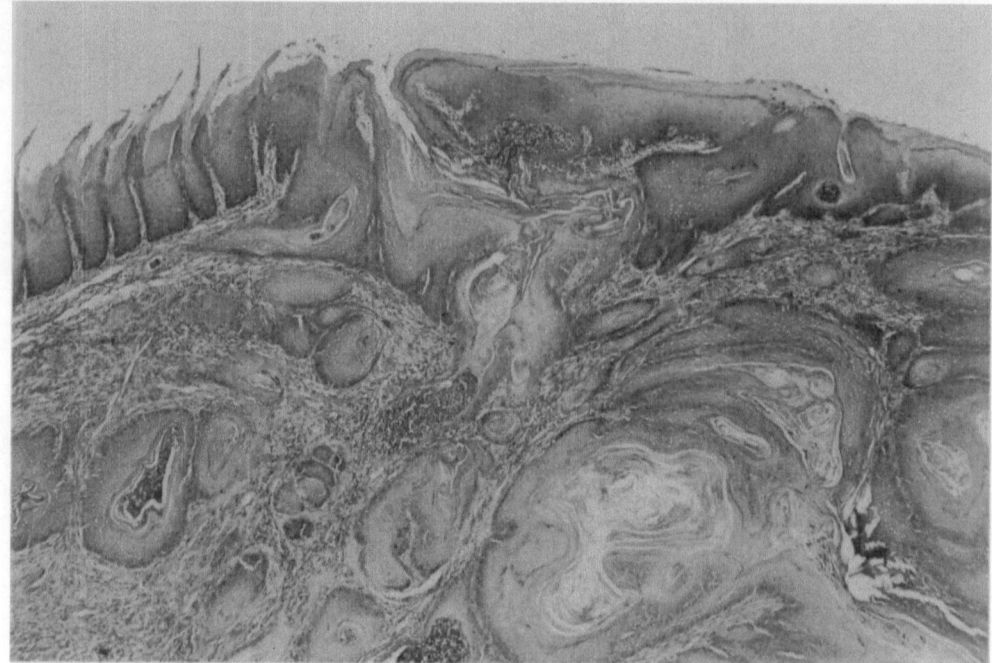

Fig. 268. Squamous cell carcinoma, tongue, rat, originating from the region of the vallate papilla and semilunar groove. H&E, ×40

basis. Beginning with study D, tongues were routinely subjected to histologic examination, and neoplasms of the tongue were diagnosed at the incidence rates noted in the table. In studies A–H, the Sprague-Dawley strain of rat was used routinely. During subsequent years (studies I, J), the Fisher 344 rat has been used. Neoplasms of the tongue have also been diagnosed in control groups of this rat strain. Limited data from three studies with the B6C3F1 hybrid mouse indicate a lower incidence of spontaneous neoplasms of the tongue in this species.

Comparison with Other Species

The inflammatory and proliferative reactions of the rat tongue discussed herein are similar in both pathogenesis and outcome to intramandibular cysts and carcinomas described for mice by van Rijssel and Muhlbock (1955). In their study, most old mice were found to have hair shafts embedded within their mandibular incisor sockets. The incidence rate varied among the different strains of mice, but was usually higher in males than females. These penetrating hair shafts initially elicited an epithelial cyst reaction. This was followed by epithelial growth that progressed to carcinoma. Some of these carcinomas were of the epidermoid type, while others contained either areas of squamous keratinization or anaplastic areas of a more undifferentiated morphology. Embedded hair shafts were found in the alveoli of all 59 mice in which carcinoma was found. Hollander and van Rijssel (1963) subsequently described the experimental production of intramandibular carcinoma in mice by foreign body mechanical damage. Chemical composition of the foreign body did not influence the development of the carcinomas. Nylon thread, steel wire, or the animal's own hair all elicited a foreign body reaction within the alveolar socket. This reaction included an initial cyst formation followed by a chronic metaplastic epithelial reaction that progressed to carcinoma. These authors concluded that chronic irritation of the epithelium provided the conditions necessary for the development of carcinoma.

It appears the same conditions can be created by foreign bodies (e.g. hair shafts) penetrating the epithelial structures within the semilunar groove surrounding the vallate papilla of the rat tongue.

References

Greene EC (1963) Anatomy of the rat. Hafner, New York (Transactions of the American Philosophical Society, vol 27)

Hebel R, Stromberg MW (1976) Anatomy of the laboratory rat. Williams and Wilkins, Baltimore

Hollander CF, Van Rijssel TG (1963) Experimental production of intramandibular carcinoma in mice by mechanical damage. JNCI 30:337–359

Keyes DG, Kociba RJ, Schwetz RW, Wade CE, Dittenber DA, Quinn T, Gorzinski SJ, Hermann EA, Momany JJ, Schwetz BA (1980) Results of a two-year toxicity and oncogenci study of rats ingesting diets containing dibromoneopentyl glycol (FR-1138). J Combustion Toxicol 7:76–98

Kociba RJ, Frauson LO, Humiston CG, Norris JM, Wade CE, Lisowe RW, Quast JF, Jersey GC, Jewett GL (1975) Results of a two year dietary feeding study with decabromodiphenyl oxide (DBDPO) in rats. J Combustion Toxicol 12:267–285

Kociba RJ, Keyes DG, Jersey GC, Ballard JJ, Ditenber DA, Quast JF, Wade CE, Humiston CG, Schwetz BA (1977) Results of a two-year chronic toxicity study with hexachloro-butadiene in rats. Am Ind Hyg Assoc J 38:589–602

Kociba RJ, Keyes DG, Beyer JE, Carreon RM, Wade CE, Dittenber DA, Kalnins RP, Frauson LE, Park CN, Barnard SD, Hummel RA, Humiston CG (1978) Results of a two-year chronic toxicity and oncogencity study of 2,3,7,8-tetrachlorodibenzo-p-dioxin (TCDD) in rats. Toxicol Appl Pharmacol 46:279–303

Kociba RJ, Keyes DG, Lisowe RW, Kalnins RP, Dittenber DA, Wade CE, Gorzinski SJ, Mahle NH, Schwetz BA (1979) Results of a two-year chronic toxicity and oncogenic study of rats ingesting diets containing 2,4,5,-trichlorophenoxyacetic acid (2,4,5-T). Food Cosmet Toxicol 17:205–221

van Rijssel TG, Mühlbock O (1955) Intramandibular tumors in mice, JNCI 16:659–689

Weichert CK (ed) (1958) Organ systems in vertebrates. In: Anatomy of chordates. McGraw Hill, New York

The Esophagus

Squamous Cell Papilloma, Esophagus, Rat

Maria Yolanda Ovelar and Antonio Cardesa

Synonyms. Keratotic papilloma, hyperkeratotic papilloma, nonkeratizing papilloma, papillomatosis

Gross Appearance

Papillomas appear as wart-like lesions, varying in size from 0.1 to 0.5 cm in diameter. They may be either peduncular with a long, thin stalk or sessile with a short, broad stalk. They usually appear as multiple lesions disseminated throughout the entire mucosal surface of the esophagus, although they have a tendency to be located more commonly at the areas of anatomic narrowing of the esophagus: the pharyngeo-esophageal junction, the anulus at the bifurcation of the trachea, and the cardial region (Fig. 269).

Microscopic Features

Papillomas are characterized by branching filiform, well-vascularized supporting cores of connective tissue that are covered by acanthotic, usually hyperkeratotic stratified squamous epithelium, imparting an overall tree-like appearance (Fig. 270). Keratinization is present in variable amounts. The thickening of the stratified squamous epithelium (acanthosis) is usually prominent. In addition, the basal cell layer of the cells is well demarcated. Atypical cells are not seen (Fig. 271).

Ultrastructure

Although electron microscopy studies have been performed in experiments with rats in which papillomas of the esophagus were induced, the ultrastructural investigations were primarily concerned with the simultaneously coexisting carcinomas (Ito et al. 1971; Levison et al. 1979).

Differential Diagnosis

Papillary lesions of the esophagus in the rat may often be difficult to classify properly. The differen-

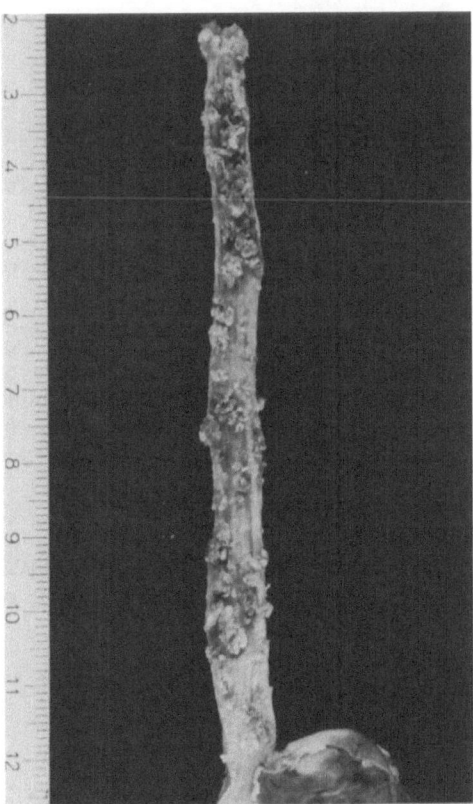

Fig. 269. Papillomas in a Sprague-Dawley rat, widely distributed throughout the upper, middle, and lower third of the esophagus

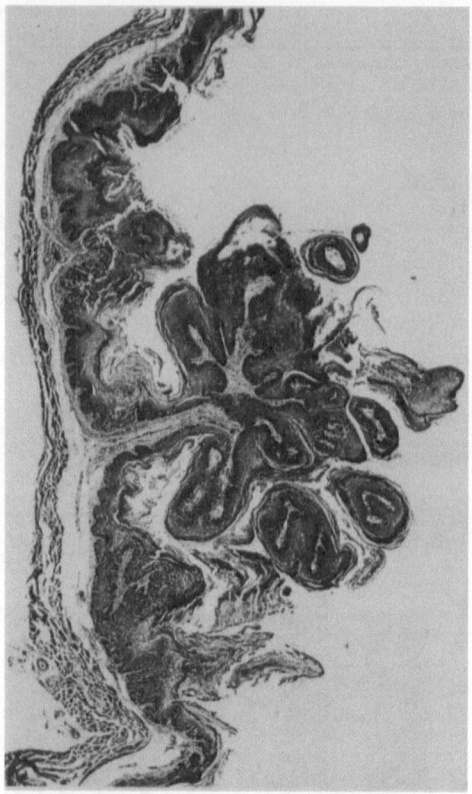

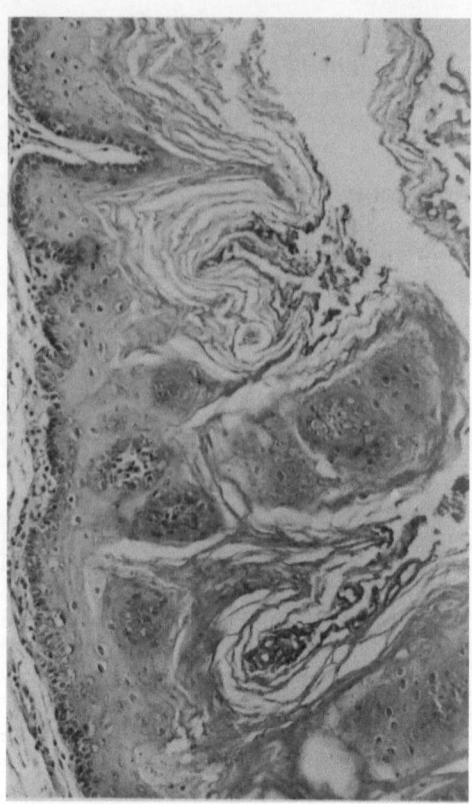

Fig. 270. Papilloma with a well-developed, peripherally branched stalk covered by acanthotic squamous epithelium. (From Cardesa et al. 1982, with permission of the publisher, CRC Press Inc., Boca Raton.) H&E, ×40

Fig. 271. Early papillary formations on the mucosal surface of the esophagus of a Sprague-Dawley rat. Notice the acanthosis, papillary growth of the epithelium, hyperkeratosis, and the well-demarcated basal cell layer. H&E, ×100

tial diagnosis has to be established between benign papillomas and others with varying grades of dysplasia. On occasions, several or even multiple sections of lesions measuring about 0.5 cm in diameter have to be performed in order to detect small squamous cell carcinomas, particularly in cases of sessile papillomas. Frank squamous cell carcinoma is not difficult to distinguish when its cells have such features as loss of polarity, atypical mitotic figures, epithelial pearls, and invasion of esophageal muscle.

Biologic Features

Papillomas are benign tumors. In our study (Cardesa et al. 1982), they were first detected after 18 weeks of treatment, appearing as small papillary projections on the mucosal surface (Fig. 271). In the majority of such experiments, papillomas

are only a part of the spectrum of tumor development and may progress to atypical papillomas and to invasive carcinomas. Ito et al. (1971) reported the regression of papillomas. Spontaneous papillomas of the esophagus are a rarity.

Comparison with Other Species

The experimental induction of papillomas of the esophagus has been reported in mice (Horie et al. 1965) and hamsters (Herrold 1966). In humans, papilloma of the esophagus is observed only rarely (Ming 1973).

References

Cardesa A, Ovelar MY, Mohr U (1982) Experimental esophageal neoplasms in the rat: histogenetic pathways and comparison with the developmental stages of cancer of the esophagus in man. In: Pfeiffer CJ (ed) Cancer of the esophagus, vol II. CRC, Boca Raton, Fl, chap 13, pp 199–213

Herrold KM (1966) Epidermoid carcinomas of esophagus and forestomach induced in Syrian hamsters by N-nitroso-N-methylurethan. JNCI 37:389–394

Horie A, Hohchi S, Kuratsune M (1965) Carcinogenesis in the esophagus. II. Experimental production of esophageal cancer by administration of ethanolic solution of carcinogens. Gann 56:429–441

Ito N, Kamamoto Y, Hiasa Y, Makiura S, Marugami M, Yokota Y, Sugihara S, Hirao K (1971) Histopathological and ultrastructural studies on esophageal tumors in rats treated with N-nitrosopiperidine. Gann 62:445–451

Levison DA, Hopwood D, Morgan RGH, Coghill G, Milne GA, Wormsley KG (1979) Oesophageal neoplasia in male Wistar rats due to parenteral di(2-hydroxylpropyl)-nitrosamine (DHPN): a combined histopathological, histochemical and electron microscopic study. J Pathol 129:31–361

Ming S-C (1973) Tumors of the esophagus and stomach. In: Atlas of tumor pathology, 2nd ser, fasc 7. Armed Forces Institute of Pathology, Washington DC, p 11

Carcinoma In Situ, Esophagus, Rat

Antonio Cardesa and Maria Yolanda Ovelar

Synonyms. Noninvasive carcinoma, keratinizing squamous cell carcinoma in situ, nonkeratinizing squamous cell carcinoma in situ

Gross Appearance

There are generally no gross findings that are characteristic of carcinoma in situ. These lesions in the esophagus of the rat may be undetectable to the naked eye or may be part of a noninvasive exophytic or of a leukoplakic lesion. Papillomas bearing carcinoma in situ have an average diameter of 0.5–1 cm and may cause stenosis of the esophageal lumen (Fig. 272). In endophytic lesions, carcinoma in situ may be present as slightly elevated, small white patches measuring a few millimeters in diameter.

Microscopic Features

Carcinoma in situ of the esophagus is characterized by the presence of abundant atypical neoplastic cells within the full thickness of its covering squamous epithelium (Fig. 273). The nuclei of these cells are abnormal in shape and enlarged with altered polarity and irregular distribution of the chromatin. Mitotic figures are frequently located several layers above the basal cell row (Fig. 274). Depending upon the production of keratin,

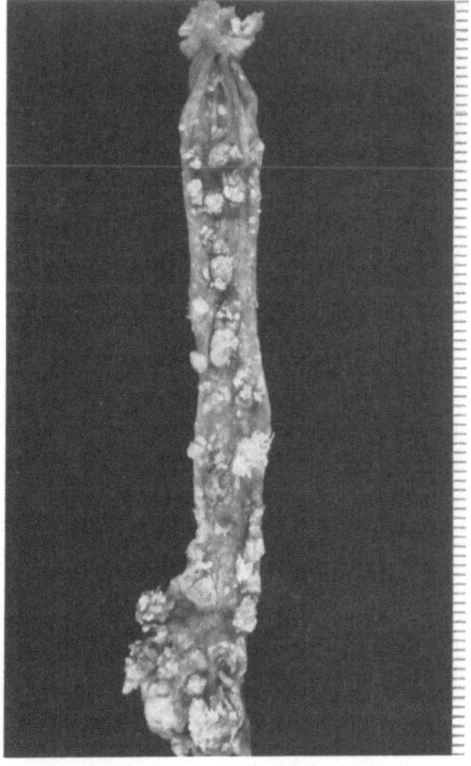

Fig. 272. Multiple papillomas in the esophagus of a Sprague-Dawley rat. The larger lesions at the *bottom* correspond to exophytic carcinoma in situ

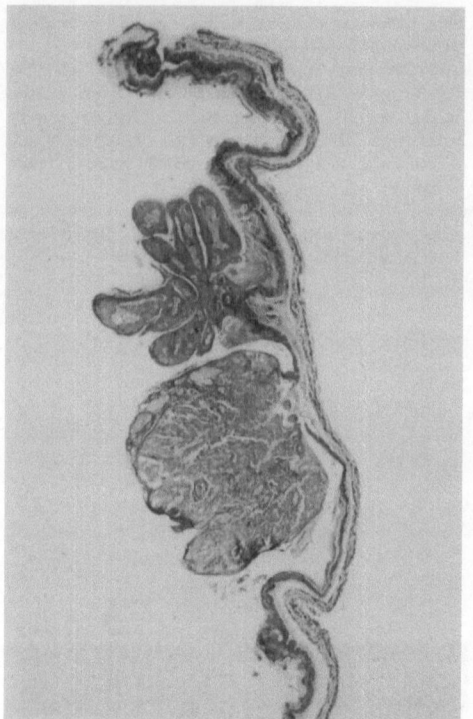

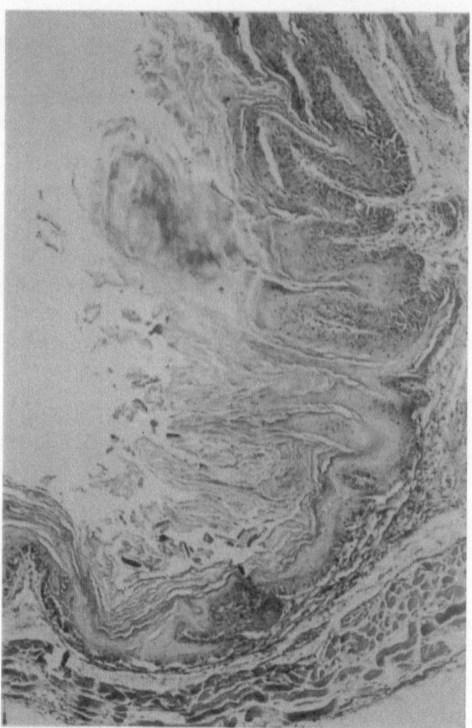

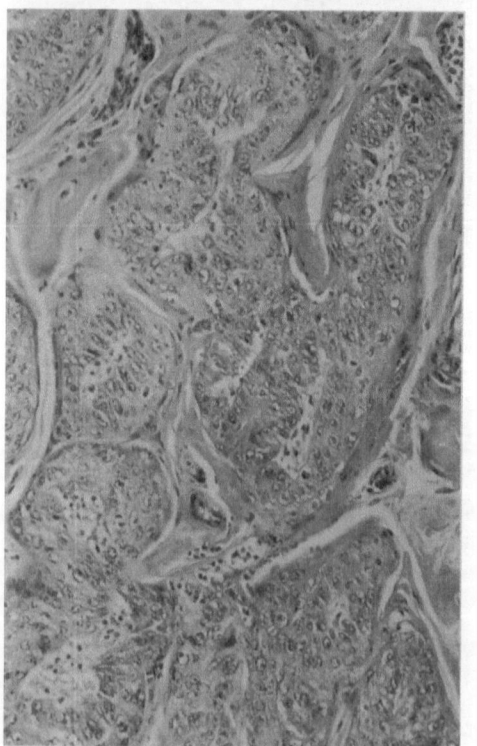

◀

Fig. 273. (*upper left*) Benign papilloma (*top*) coexisting with markedly atypical papilloma (*bottom*), with the features of carcinoma in situ. (From Cardesa et al. 1982, with permission of the CRC Press Inc.) H&E, ×10

Fig. 274. (*below*) Exophytic papilloma with atypical epithelial features from the basal layer up to the level of keratinization. These changes in the different foci range from moderate dysplasia to carcinoma in situ. H&E, ×200

Fig. 275. (*upper right*) Leukoplakia of the esophagus with marked superficial keratinization at the *bottom* and *center*. Progressive transition to carcinoma in situ with intact basal lamina at the top. H&E, ×40

these tumors may be divided into keratinized and nonkeratinized varieties. The cytoplasm of nonkeratinized cells is basophilic, whereas in the keratinized cells it is eosinophilic. The cell membrane usually is nondistinct. Both exophytic papillomatous lesions and endophytic hyperplastic growths may be the site of carcinoma in situ. In all instances of carcinoma in situ, the basal lamina separating the epithelium from the underlying supporting connective tissue appears intact (Fig. 275).

Ultrastructure

We are not aware of electron microscopy studies of carcinoma in situ of the esophagus in the rat. Nevertheless, the finding of a well-formed and intact basal lamina must be a prerequisite for establishing the diagnosis of carcinoma in situ.

Differential Diagnosis

The diagnosis of carcinoma in situ relies on the observation of full-thickness atypical changes within the covering epithelium of the esophagus plus the presence of an intact basal lamina. Keratinizing carcinoma in situ of the esophagus must be distinguished from other leukoplastic lesions in which the atypical cells do not represent full-thickness changes of the epithelium (Fig. 276). These are examples of dysplasia, but not of carcinoma in situ. Nonkeratinizing carcinoma in situ of the esophagus should not be confused with basal cell carcinoma in situ. In the later, a palisading arrangement of the basal cell row is characteristic, whereas in the former no palisading is observed at the basal cell layer.

Biologic Features

In Sprague-Dawley rats, exophytic squamous cell carcinoma in situ has been found to originate within atypical papillomas of the esophagus after 21 weeks of treatment with 2–6 dimethylnitrosamine (Cardesa et al. 1982). The dysplastic changes start at the basal epithelial layer, sometimes causing broadening of the stalk. Atypia extends upward, eventually involving the entire epithelial thickness of the papilloma. A similar sequence of events has been described by Pozharisski (1990).

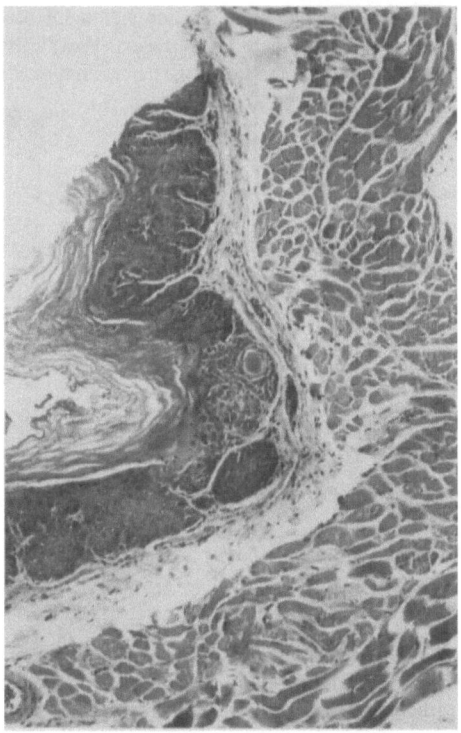

Fig. 276. Dysplastic squamous epithelium of the esophagus with prominent keratinization of the surface and increased basal cell layer thickness at the bottom, where keratin pearl formation is seen. H&E, ×60

Endophytic squamous cell carcinoma in situ originates from benign hyperplastic endophytic lesions. These lesions evolve from nondysplastic to dysplastic and then to carcinoma in situ, giving rise later on to invasive squamous cell carcinoma. In our study, the endophytic carcinomas in situ were first observed after 20 weeks of treatment with 2–6 dimethylnitrosamine.

Comparison with Other Species

Endophytic carcinoma in situ of the rat esophagus does not differ morphologically in its keratinizing and nonkeratinizing varieties from similar lesions in humans (Cardesa et al. 1982). The exophytic carcinoma in situ is a rare entity in humans. In studies of the hamster, Herrold (1966) reported the existence of atypical foci of exophytic growths in the vicinity of epidermoid carcinomas of the esophagus. With regard to mice, we are unaware of any description of these lesions, although Horie et al. (1965) reported, in the same experiment, the induction of papillomas and two carcinomas of the esophagus.

References

Cardesa A, Ovelar MY, Mohr U (1982) Experimental esophageal neoplasms in the rat: histogenetic pathways and comparison with the developmental stages of cancer of the esophagus in man. In: Pfeiffer CJ (ed) Cancer of the esophagus, vol II. CRC, Boca Raton, pp 199–213

Herrold KM (1966) Epidermoid carcinomas of esophagus and forestomach induced in Syrian hamsters by N-nitroso-N-methylurethan. JNCI 37:389–394

Horie A, Hohchi S, Kuratsune M (1965) Carcinogenesis in the esophagus. II. Experimental production of esophageal cancer by administration of ethanolic solution of carcinogens. Gann 56:429–441

Pozharisski KM (1990) Tumours of the oesophagus. In: Turusov VS (ed) Pathology of tumours in laboratory animals, vol I. Tumours of the rat, 2nd edn. IARC Sci Publ no 99, Lyon, pp 109–128

Squamous Cell Carcinoma, Esophagus, Rat

Antonio Cardesa, Maria Yolanda Ovelar, and Manuel Pera

Synonyms. Keratinizing squamous cell carcinoma, epidermoid carcinoma, squamous cell epithelioma, spinocellular carcinoma, nonkeratinizing squamous cell carcinoma

endophytic growth is observed. Both of the patterns may be so intermingled that an objective distinction of the type of growth is no longer possible.

Gross Appearance

According to their pattern of growth, squamous cell carcinomas of the esophagus are classified as exophytic and endophytic varieties (Cardesa et al. 1982). Exophytically growing carcinomas have a cauliflower-like shape, with markedly irregular contours, friable consistency, and whitish color. Occasionally, foci of hemorrhage and necrosis are seen on the surface. They give rise to protruding, broad-based, confluent, nodular formations measuring an average of 1 cm in diameter (Fig. 277). Due to their large size, they usually cause obstruction of the esophagus with dilatation of its proximal part. In endophytic carcinomas, there is a limited intraluminal growth. The tumor surface is irregular, necrotic, and hemorrhagic, showing erosions and defects that, in advanced carcinomas, give rise to wide annular ulcerations, reaching up to 1.5 cm in length (Fig. 278). The cut surface is whitish to whitish-red, showing conspicuous thickening of the esophageal wall due to the inward growing of the tumor. In some advanced carcinomas, however, a mixed pattern of exophytic and

Microscopic Features

Microscopically, exophytic carcinomas have a papillomatous or verrucous appearance (Fig. 279). They are formed by the proliferation of outward-growing malignant squamous epithelial cells, showing varying grades of differentiation and supported by thin, papillary, fibrous connective tissue stalks. The most external squamous epithelial cell layers may be markedly keratinizing. Deeply invasive exophytic carcinomas show nests and cords of squamous epithelial cells that destroy the basal lamina of the epithelium. They invade the different layers of the esophagus, compressing and progressively destroying its walls. In general, the depth of tumor invasion is less extensive in peduncular than in sessile forms of exophytic carcinoma.

Endophytically growing squamous cell carcinomas are formed by cords and sheets of inward-growing squamous epithelium with varying grades of dedifferentiation. The superficial part of these tumors shows a minimal or absent intraluminal growth, which is covered by a thick keratin layer.

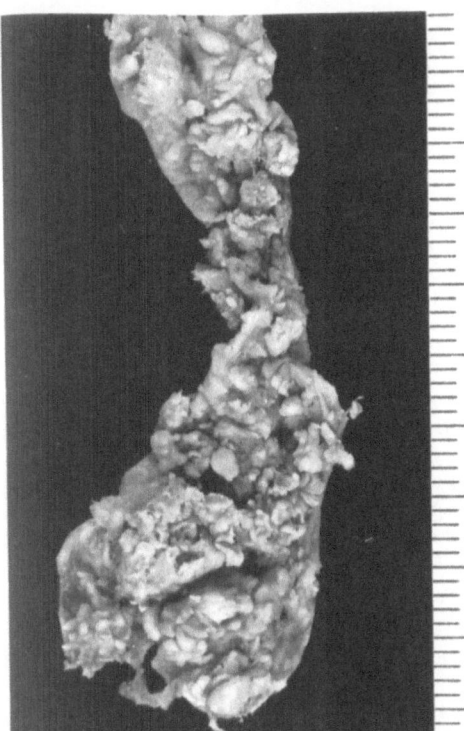

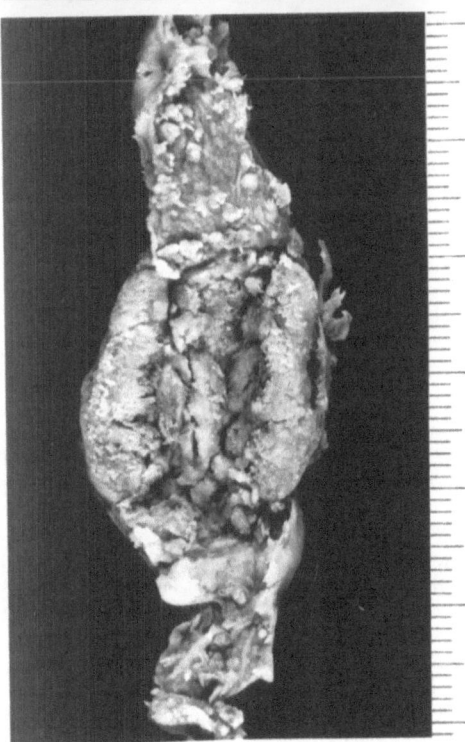

Endophytic squamous cell carcinomas are markedly invasive tumors infiltrating all or part of the thickness of the esophageal wall (Figs. 280, 281). A lateral growth takes place by invasion of the subepithelial connective tissue; therefore, the surface of this region is covered by normal squamous epithelium. Endophytic squamous cell carcinomas extend into the neighboring structures; however, in spite of evident marked cellular anaplasia, we have never seen distant tumor metastases from this carcinoma in the rat (Cardesa et al. 1982; Pera et al. 1989). In between the areas of growth of exophytic and endophytic squamous cell carcinomas, which usually coexist in the same animal, there are occasionally other fields which present nests and plaques of rather superficially invading squamous cell carcinomas. They are characterized by a basal cell layer of malignant cells, which is followed upward by two or three layers of nonkeratinizing squamous cells that abruptly give rise to marked keratinization. In the center of these fields, there is even pearl formation. The cells of these squamous carcinomas are generally of medium size, measuring about 30 μm. They appear slightly pleomorphic, with predominance of polygonal forms. The cytoplasm is homogeneous, abundant, and eosinophilic in the superficial and intermediate cell layers, whereas it is basophilic and scanty in the deep cell layers. The cell membrane appears, in general, poorly defined. The nuclei are picnotic or absent in the superficial cell layers. In the deep and intermediate cell layers, the nuclei are oval to round, measuring 15–20 μm. They have a smooth and reinforced nuclear membrane with occasional discrete ondulations. The internal structure of these nuclei is clear and vesicular, with a prominent eosinophilic nucleolus, which may be double. Exceptionally gigantic nuclei measuring more than 40 μm are observed. Also multinucleated tumor cells may be encountered. The number of mitotic figures varies from zero to five per high-power field (40 × 10), depending upon the tumor growth tendency in the different areas.

◄

Fig. 277. (*above*) Exophytic squamous cell carcinoma of the esophagus in a Sprague-Dawley rat. Multiple confluent verrucous lesions on the entire mucosal surface

Fig. 278. (*below*) Endophytic squamous cell carcinoma of the esophagus in a Sprague-Dawley rat. Involvement of the middle esophageal third by an inward-growing neoplasm

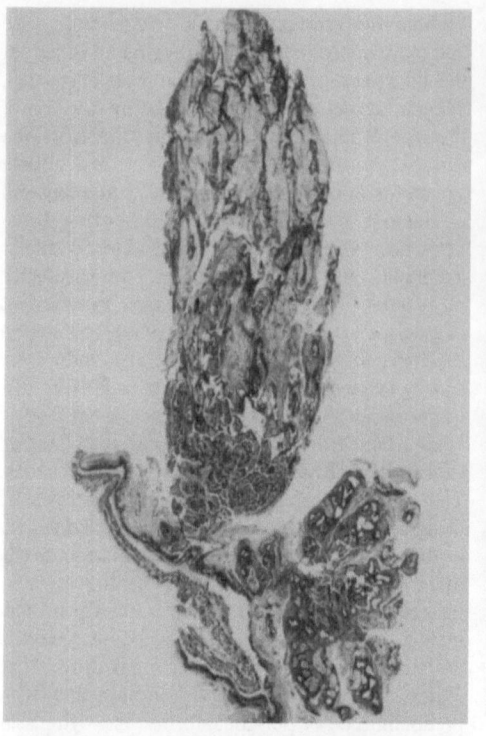

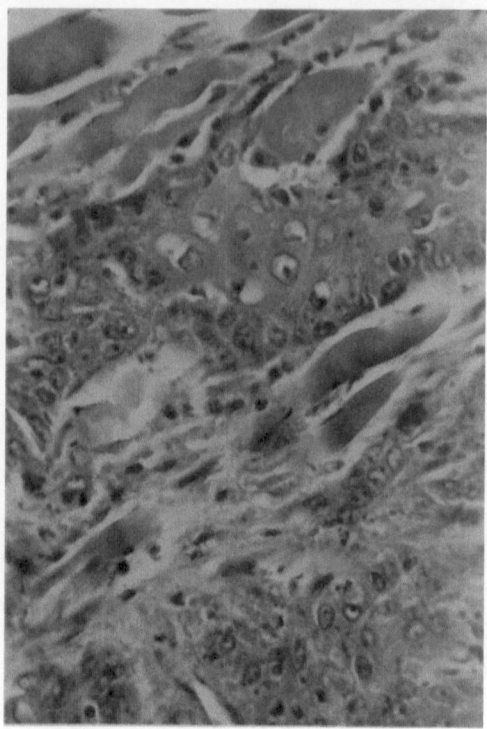

◀

Fig. 279. (*upper left*) Exophytic squamous cell carcinoma of the esophagus in a Sprague-Dawley rat. Heavily keratinized verrucous formation at the tip with malignant changes near the thin-stalked base. The smaller lesion at the *lower right* is a papilloma. H&E, ×60

Fig. 280. (*below*) Endophytic squamous cell carcinoma of the esophagus in a Sprague-Dawley rat. Invasion of the muscular layer of the wall by cords and nest of keratinizing squamous epithelial cells. H&E, ×100

Fig. 281. (*upper right*) Squamous cell carcinoma of the esophagus in a Sprague-Dawley rat. Higher magnification of Fig. 280. H&E, ×400

Ultrastructure

The electron microscopy features of squamous cell carcinoma of the esophagus have been described in the rat by Ito et al. (1971) and by Levison et al. (1979). The most characteristic features found were the presence of intercellular desmosomes, as well as tonofilaments and keratin granules in the cytoplasm. The nuclei had an oval or irregularly shaped nuclear membrane, and aggregated chromatin was seen in the nucleoplasm. At the ultrastructural level, these tumors are not different from those observed in the esophagus of other species of laboratory animals and humans.

Differential Diagnosis

The differential diagnosis of exophytically growing squamous cell carcinomas has to be mainly made with the benign squamous papillomas and the atypical papillomas of the esophagus. The endophytically growing invasive squamous cell carcinomas must be distinguished from the in situ squamous cell carcinomas and from squamous cell carcinoma with focal glandular differentiation (Cardesa et al. 1994).

Benign papillomas of the esophagus, due to their orderly pattern, bland nuclei, and preservation of the basal membrane, do not offer special differential diagnostic problems. The distinction may be more subtle between atypical squamous cell papillomas and exophytic squamous cell carcinomas, since atypical papillomas have dysplastic cells with atypical nuclei, mainly at the base of the epithelial growths. However, in these instances, the integrity of the basal lamina and the lack of invasion at the base of the lesion precludes the diagnosis of carcinoma. This diagnostic distinction becomes easier with pedunculated atypical papillomas than in cases of sessile atypical papillomas. In the presence of broad-based papillomas with atypical features at the base, the search for either rupture of the basal lamina or invasion is necessary in order to establish or rule out the diagnosis of invasive carcinoma. This may require cutting the tumor in graded or even serial sections. The distinction of endophytically growing squamous cell carcinoma from "in situ" carcinoma is also based on the integrity of the basal membrane of the epithelium. Nonkeratinizing squamous cell carcinoma with areas of undifferentiation at the base should not be con-

fused with basal cell carcinoma. Stains for periodic acid-Schiff (PAS), alcian blue, and mucicarmine will detect mucin production in cases of squamous cell carcinoma with focal glandular differentiation (Cardesa et al. 1994).

Finally, the distinction between squamous cell carcinoma originating in the esophagus and a metastatic squamous cell carcinoma has also to be kept in mind. In this regard, a careful dissection of the organs at autopsy and a detailed description of the gross findings may be of great help. In case of doubt, repeated tissue sections demonstrating the origin of the neoplasm from the covering squamous epithelium of the esophagus are mandatory. In the absence of convincing evidence of the origin of the tumor from the esophageal squamous epithelium, the diagnosis of a primary esophageal carcinoma is difficult to establish.

Biologic Features

In the rat, squamous cell carcinomas of the esophagus are usually multifocal in origin. At early stages of tumor development, they are seen preferentially to arise at the points of natural anatomic narrowing of the esophagus: the pharyngeo-esophageal junction, the anulus at the bifurcation of the trachea, and the cardial region. Among these three foci of narrowing, the pharyngeo-esophageal junction appears to be the most frequent site of tumor origin. At advanced stages of tumor development, esophageal carcinomas may arise from almost any point on the entire esophageal mocosal surface. They invade the wall of the esophagus, cause shortening of its length, and at the same time widening of its transverse diameter. On average, a normal adult esophagus of rat measures about 7–9 cm in length and 0.7 cm in width. In contrast, the carcinomatous esophagus measures approximately 5–6 cm in length and 1.2 cm in width.

With regard to the pathogenesis of squamous cell carcinoma of the esophagus in rats, emphasis has been placed on the description of the developmental stages of exophytic lesions leading to papillary carcinomas (Napalkov and Pazhärisski 1969; Ito et al. 1971; Stinson et al. 1978). On the other hand, Cardesa et al. (1982) have investigated the sequential steps of progressive preneoplastic and neoplastic changes leading not only to exophytic, but also to endophytic squamous cell carcinoma. The earliest lesions observed were hyperplastic epithelial proliferations, which were followed by the formation of prominent fibrotic submucosal esophageal ridges. From this stage on, the development of the esophageal lesions followed two different pathways. The exophytic growth path resulted in papillomas, atypical papillomas, and papillary carcinomas. The endophytic growth path gave rise to benign endophytic growths, atypical endophytic growths, and endophytic carcinomas.

No suitable models of esophageal carcinogenesis were available until the discovery by Druckrey et al. (1961) that N-methyl-N-nitrosoaniline given orally to rats resulted in a high incidence of squamous cell carcinoma of the esophagus. Since then, most of the experimentally induced squamous cell carcinomas of the esophagus have been due to the exposure of rats to several nitrosamines by various routes of administration (Druckrey et al. 1967). A rate of up to 100% of tumors of the esophagus was detected in Sprague-Dawley rats which received s.c. 2,6-DMNM at one tenth and $^1/_{20}$ LD_{50} (Ovelar et al. 1981). Approximately one third of these tumors were squamous cell carcinomas, mainly exophytically growing (Cardesa et al. 1982). If a reflux esophagitis was produced, by means of an esophagojejunostomy, prior to the administration of 2,6-DMNM, a predominance of endophytic squamous cell carcinomas was observed (Pera et al. 1989).

Spontaneous squamous cell carcinoma of the esophagus are rare in rats. In an extensive literature review, only two reported cases were found (Pozharisski 1990).

Comparison with Other Species

The induction of squamous cell carcinomas of the esophagus has been mainly reported in the rat, but also with less frequency in hamsters, mice, and other species (Iizuka et al. 1982). Squamous cell carcinoma is the most frequent malignant neoplasm of the esophagus in humans. Although they have been described, verrucous or true papillary carcinomas of the human esophagus are rare. Nonpapillary squamous carcinomas of the fungating, ulcerative, and infiltrating type, on the other hand, represent the overwhelming majority found in humans (Ming 1973). The endophytic pathway of carcinoma development in experimental esophageal carcinogenesis is thus significant, since it shows stages of progression that are histologically comparable to those of the human counterpart.

References

Cardesa A, Ovelar MY, Mohr U (1982) Experimental esophageal neoplasms in the rat: histogenetic pathways and comparison with the developmental states of cancer of the esophagus in man. In: Pfeiffer CJ (ed) Cancer of the esophagus, vol II. CRC, Boca Raton, chap 13, pp. 199–213

Cardesa A, Bombí JA, Fernandez PL, Campo E, Pera C, Mohr U (1994) Spectrum of glandular differentiation in experimental carcinoma of the esophagus induced by 2,6-dimethyl nitrosomorpholine under the influence of esophagojejunostomy. Exp Toxicol Pathol 46:41–49

Druckrey H, Preussmann R, Schmahl D, Blum G (1961) Carcinogene Wirkung von N-Methyl-N-nitroso-Anilin. Naturwissenschaften 48:722–723

Druckrey H, Preussmann R, Ivankovic S, Schmahl D (1967) Organotrope carcinogene Wirkung bei 65 verschiedenen N-Nitroso-Verbindungen an BD-Ratten. Z Krebsforsch 69:103–201

Iizuka T, Kato H, Ichimura S, Kawachi T (1982) Experimental esophageal carcinoma in rats, rabbits, dogs and other species. In: Pfeiffer CJ (ed) Cancer of the esophagus, vol II. CRC, Boca Raton, chap 12

Ito N, Kamamoto Y, Hiasa Y, Makiura S, Marugami M, Yokota Y, Sugihara S, Hirao K (1971) Histopathological and ultrastructural studies on esophageal tumors in rats treated with N-nitrosopiperidine. Gann 62:445

Levison DA, Hopwood D, Morgan RGH, Coghill G, Milne GA, Wormsley KG (1979) Oesophageal neoplasia in male Wistar rats due to parenteral di(2-hydroxylpropyl)-nitrosamine (DHPN): a combined histopathological, histochemical and electron microscopic study. J Pathol 129:31

Ming S-C (1973) Tumors of the esophagus and stomach. In: Atlas of tumor pathology, 2nd ser, fasc 7. Armed Forces Institute of Pathology, Washington, DC, p 11

Napalkov N, Pozharisski KM (1969) Morphogenesis of experimental tumors of the esophagus. JNCI 42:922

Ovelar MY, Cardesa A, Mohr U (1981) Carcinogenic effect of chronic subcutaneous injections of 2-6-di-methylnitrosomorpholine in Sprague-Dawley rats. Cancer Lett 13:159–163

Pera M, Cardesa A, Bombí JA, Erust H, Pera C, Mohr U (1989) Influence of esophagojejunostomy on the induction of adenocarcinoma of the distal esophagus in Sprague-Dawley rats by subcutaneous injection of 2,6-dimethylnitrosomorpholinee Cancer Res 49:6803–6808

Pozharisski KM (1990) Tumors of the esophagus. In: Turusov VS (ed) Tumours of the rat, vol I, 2nd edn. IARC Sci Publ no 99, Lyon, pp 109–128

Stinson SF, Squire RA, Sporn MB (1978) Pathology of esophageal neoplasms and associated proliferative lesions induced in rats by n-methyl-n-benzylnitrosamine. JNCI 61:1471

Papillary and Nonpapillary Squamous Cell Carcinoma, Esophagus, Rat (Zinc Deficiency, Alcohol, and Methylbenzylnitrosamine)

Paul M. Newberne

Synonyms. Keratinizing squamous papilloma, keratinizing squamous cell carcinoma in situ, epidermoid carcinoma of the esophagus

Gross Appearance

Initial gross changes are foci of leukoplakia followed shortly by irregular thickening of the epithelium, particularly the basal layer, along with a mixture of hyperkeratosis and parakeratosis. As tumors first begin to appear within 6–10 weeks after initiation of carcinogen exposure, their shape follows two patterns: papillary and nonpapillary. The appearance, both grossly and microscopically, depends on a number of factors, including the schedule of dosing (concentration, frequency, and total dose) and dietary factors such as zinc deficiency, vitamin A, selenium, and alcohol (ethanol, fusel alcohols) (Nauss et al. 1987). The predominant type of tumor in rats treated with methylbenzlnitrosamine (MBN) may be a single papilloma, but more often they are multiple, verrucous, and hyperkeratinized; some are sessile and endophytic with little tendency for papillary growth. While the papillary exophytic type of neoplasm predominates in rats given a standard complete diet and exposed to MBN only, the sessile endophytic form tends to predominate when the regimen is modified by a zinc-deficient diet or when some forms of alcohol are included as a part of the protocol. Descriptions and identifications of esophageal lesions in rats presented in this section emphasize the effects on tumor induction by the superimposition of zinc deficiency, ethanol, and/or 3-methylbutanol (3-MB), in concert with exposure to the environmental carcinogen MBN (Nauss et al. 1986; Newberne et al. 1994). Papillomas usually appear as single peduncular masses attached to the mucosa by a narrow stalk and extend into the esophageal lumen; the other form is a sessile (flat) mass. Zinc-deficient rats treated with MBN develop numerous, often confluent tumors which may obstruct the esophagus, leading to the rapid physical deterioration of the animal. Figure 282 illustrates the gross appearance of a papillary exophytic form of esophageal neoplasm in a control (Fig. 282a) and a zinc-deficient rat (Fig. 282b). Each rat was treated identically with MBN and killed at the same time following MBN exposure. There was a marked increase in the number of tumors and in their confluence in the deficient animal. However, confluence of the multiple tumors obscures many of the nonpapillary, endophytic types which are numerous in zinc-deficient rats.

Microscopic Features

The initial feature of the rat esophagus exposed to MBN is the leukoplakia noted grossly (Fig. 283). There is a thickening of the epithelium, a result of an increase in the number and size of cells, acanthosis, hyperkeratosis, and parakeratosis, the latter particularly marked in rats fed zinc-deficient diets. At about this time, dysplasia and a significant increase in mitotic figures usually develop in the basal cell layer of the epithelium with a variable inflammatory infiltrate into the subepithelial connective tissue. Subsequent lesions include papillary projections of the epithelium (Fig. 284), which later develop into true papillomas (Fig. 285). Some of the lesions are malignant squamous cell carcinomas, which may remain in situ or, more often, invade the esophageal wall (Fig. 286). Metastasis to regional lymph nodes and elsewhere is rare, but the primary tumors are locally quite destructive. Endophytic growth which develops along with papillary types can be observed as early as 3 weeks after commencing exposure to the carcinogen (Fig. 287) and progresses rapidly to malignancy (Fig. 288). The invasive carcinoma, consisting of ingrowths of atypical epithelium, extends through the lamina propria of the mucosa into the submucosa. This appears to be a de novo malignancy developing without proceeding through papillary growth or papilloma. The overlying epithelium remains essentially intact in the early stages, but within a short time is ulcerated. The tumors which develop via endophytic growth are more common with zinc deficiency or combined zinc deficiency plus alcohols, ethanol, and a fusel alcohol (3-MB). Typically, they de-

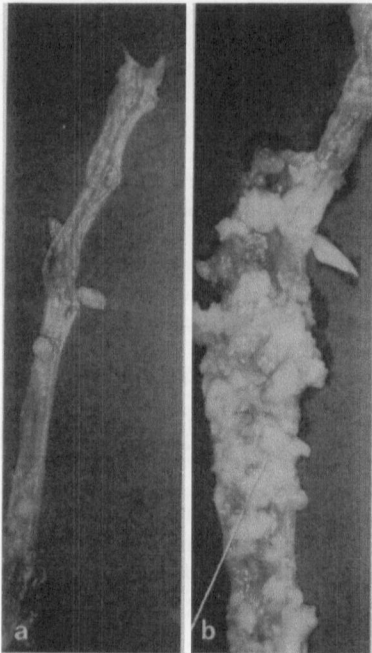

Fig. 282a,b. Esophagus **a** of rat treated with methylbenzlnitrosamine (MBN) and **b** of rat depleted of zinc and treated with MBN. The deficiency markedly enhances the induction, increases the number, and decreases the lag time for the appearance of tumors

velop rapidly and ulcerate early in the invasive process.

Biologic Features

Natural History

Observations by Lin et al. (1977) in human subjects (see Table 20) led us to develop the MBN rat model and to test the hypothesis that zinc deficiency, which alone damages the esophagus, enhances chemically induced esophageal cancer. In the Western world, esophageal cancer is strongly associated with alcohol consumption (Chilvers et al. 1979; Pottern et al. 1981; Blot et al. 1991). Epidemiologic evidence suggests that the incidence and severity of this type of neoplasm may be increasing worldwide in parallel to increased consumption of alcohol and tobacco products (Blot et al. 1991; Field et al. 1994). Clinical evidence in human patients of an enhanced excretion

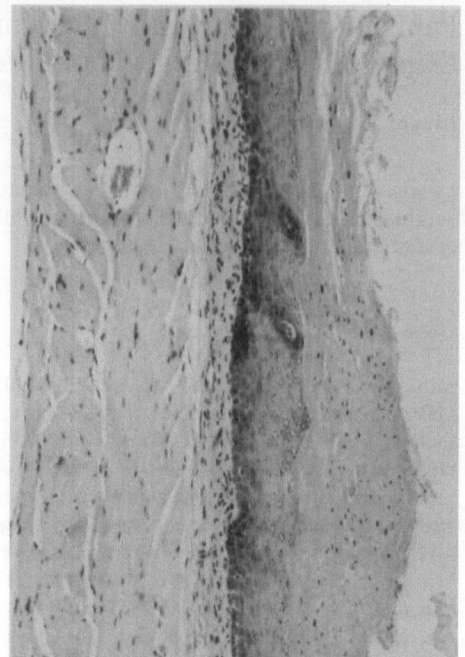

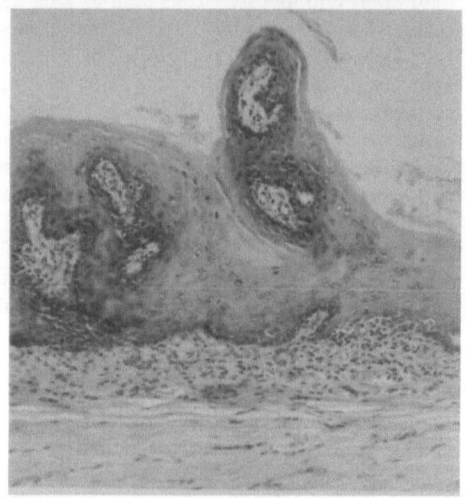

Fig. 283. (*above*) Esophagus, rat. Leukoplakia, acanthosis, hyperkeratosis, and parakeratosis typical of early lesions following exposure to methylbenzlnitrosamine (MBN). H&E, ×110

Fig. 284. (*below*) Papillary proliferation and hyperkeratosis of epithelium of rat esophagus. The subepithelial proliferation of connective tissue and variable reactive cell infiltrate is typical 4–6 weeks after dosing. H&E, ×250

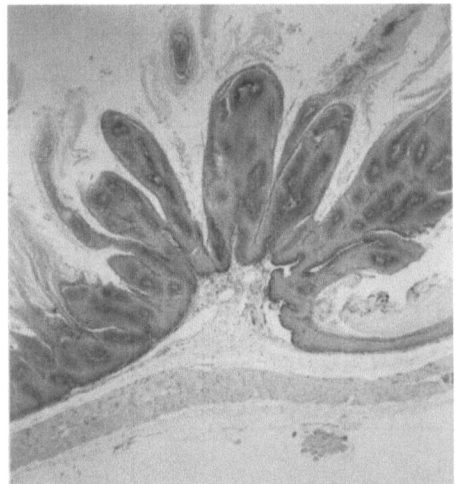

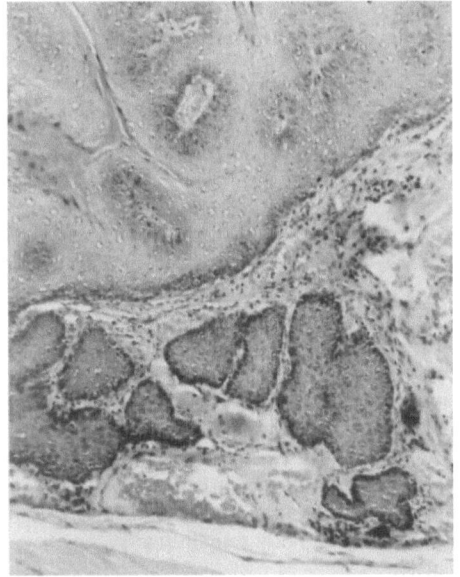

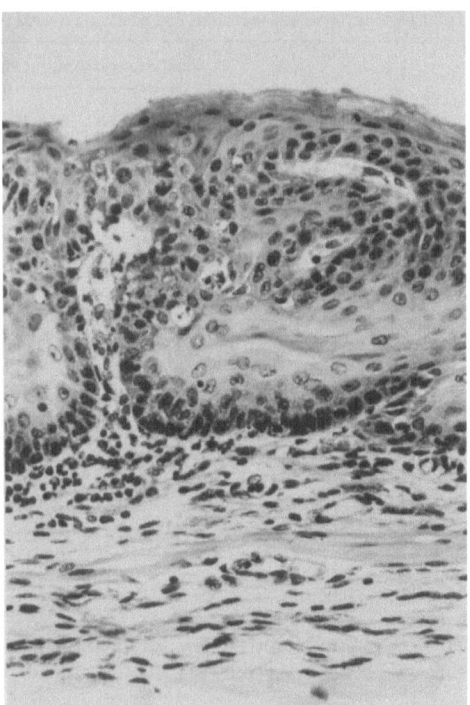

Fig. 287. Esophagus, rat. Early development of endophytic growth. The pattern of cellular growth and epithelium is distorted. H&E, ×400

Fig. 285. (*above*) Papilloma, esophagus of methylbenzylnitrosamine (MBN)-treated rat. Invasion of the stalk may occur, but at a later time. H&E, ×80

Fig. 286. (*below*) Squamous cell carcinoma of the esophagus of a rat treated with methylbenzylnitrosamine (MBN). The tumor has infiltrated the underlying connective tissue and muscle. H&E, ×250

of zinc following consumption of alcohol was reported by Russell (1980) and subsequently supported by studies in rats (Ahmed and Russell 1982). In a series of studies in rats using MBN as the carcinogen, we have clearly shown that dietary zinc deficiency in rats increases the con-

centration of zinc in serum, but decreases it in hair and esophageal tissues. These events markedly enhance the induction of esophageal cancer (Fong et al. 1978; Gabrial et al. 1982; Newberne and Schrager 1983). In addition, MBN alone also modifies serum and tissue zinc concentrations in a manner similar to the dietary deficiency (Table 21).

Etiology

Esophageal tumors induced by MBN are histologically similar to, but more predictable than, those induced by a number of other nitrosamines, since MBN is highly specific for the esophagus. Other chemicals also produce a variety of tumor types at different anatomic sites (Magee and Schoental 1964; Napalkov and Pozharisski 1969; Pozharisski 1990).

The methodology for developing the MBN rat model is as follows: Weanling rats (Sprague-Dawley, Fischer 344, Wistar, male or female) are

Table 20. Zinc levels in serum, hair, and esophageal tissues from esophageal cancer patients (from Lin et al. 1977)

	Zinc concentrations (μg/100 ml or g)			
	Serum	Hair	Esophageal	
			Tumor	Esophagus
Normal subjects	102.7 ± 18.5	195.0 ± 29.0	110.0 ± 22.4	160.0 ± 28.7
Patients with:				
Esophageal cancer	78.0 ± 14.9	162.0 ± 33.0	–	–
Other cancers	114.4 ± 31.8	169.0 ± 37.0	–	–
Other disorders[a]	96.2 ± 15.0	212.0 ± 48.0	149.0 ± 18	248.0 ± 17.0

[a] Taken from data on accidental deaths unrelated to the esophageal cancer study. Data expressed as mean ± SD.

Table 21. Zinc levels in rats fed a zinc-deficient diet and treated with methylbenzylnitrosamine (MBN)

Diet	Number of animals	Serum (μg/100 ml)	Hair (ppm)	Esophagus (ppm)
Control	8	101.9 ± 12.7	223.7 ± 25.5	211.3 ± 38.8
Control + MBN	8	124.4 ± 12.9	202.8 ± 11.4	156.2 ± 12.3
Zinc deficient	11	38.3 ± 11.6	165.2 ± 15.2	136.6 ± 16.8
Zinc deficient + MBN	10	46.8 ± 15.5	160.8 ± 15.6	126.0 ± 37.2

Differences in zinc concentrations of serum and hair of control and zinc-deficient groups significant at $p < 0.001$; esophageal levels significant at $p < 0.01$. Data expressed as mean ± SD.

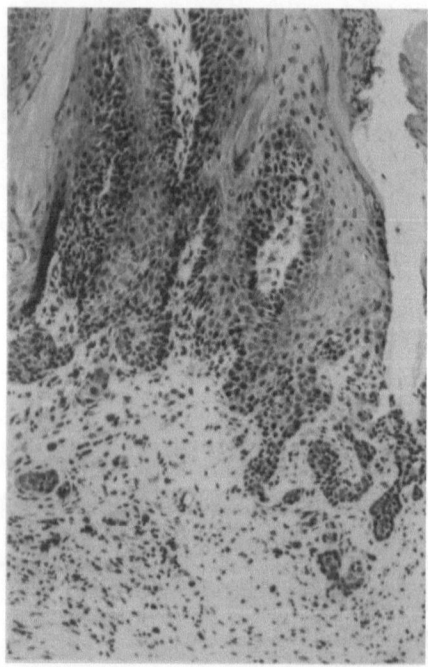

Fig. 288. Esophagus, rat. Infiltrating carcinoma which developed in a rat exposed to methylbenzylnitrosamine (MBN). This is the endophytic form of nonpapillary lesion. H&E, ×300

placed on a standard diet for a period of time, usually 3–4 weeks, for adjustment prior to dosing with the carcinogen. When the animals are ready for dosing, MBN (Illinois Institute of Technology, Chicago) is administered by gastric intubation at a dose of 2.5 mg/kg body weight, twice weekly for 5 weeks (total dose, 25 mg/kg). The incidence and the lag time for tumor development can be modified by reducing or increasing the number of doses and the total amount of MBN administered. It is important to keep the individual dose at no more than 2–2.5 mg/kg body weight to avoid significant toxicity.

Following the above regimen, papillomas and carcinomas are observed within a few weeks after the last dose of MBN. If zinc deficiency is to be a concomitant factor, it must be induced prior to exposure, as noted above. Other factors can also be introduced during and after exposure to MBN.

Pathogenesis

Studies (Gabrial et al. 1982) with zinc deficiency and alcohol confirmed earlier observations and revealed additional significant findings. Table 22

Table 22. Esophageal tumor incidence in control and zinc-deficient rats exposed to methylbenzylnitrosamine (MBN)

Diet		mg MBN/kg	First dose of MBN to death (days)	Carcinoma affected		Papilloma affected		Total (n)
				(n)	(%)	(n)	(%)	
Group 1	Zinc deficient ad libitum	34	58	5	83	6	100	6
	Control, pair-fed	34	58	0	0	2	33	6
Group 2	Zinc deficient ad libitum	34	93	4	33	12	100	12
	Control, pair-fed	34	93	0	0	8	66	12

Table 23. Tumor incidence induced by methylbenzylnitrosamine (MBN) in rats deficient in zinc and given ethyl alcohol and 13-*cis* retinoic acid (from Gabrial et al. 1982)

Group	Treatment, zinc content	MBN	4% alcohol in drinking water	13-*cis* retinoic acid	Rats with tumors		Total (n)
					(n)	(%)	
1	Control, 60 ppm	−	−	−	0	0	12
2	Control, 60 ppm	+	−	−	14	40.0	35
3	Deficient, 7 ppm	+	−	−	25	75.7	33
4	60 ppm control, deficient 7 ppm for postdosing	+	−	−	18	51.4	35
5	Deficient	+	+	−	29	85.3	34
6	Deficient	+	+	+	33	94.3	35

lists the major findings of this study. Zinc deficiency significantly enhanced carcinogenesis. Alcohol tended to further enhance tumor incidence, but 13-*cis* retinoic acid offered no protection. Switching the zinc-deficient group from deficient (7 ppm Zn) to the control diet (60 ppm) after exposure to six doses of 2.5 mg each of the carcinogen (21 days) afforded significant protection, indicating protective effects of zinc during both the initiation and the promotion stages of carcinogenesis.

A partial explanation for the effects of alcohol on zinc status may reside in the following observations: (a) alcohol increases zinc excretion (Russell 1980) and (b) increased zinc excretion precipitates a crisis, resulting in a conditioned deficiency in the esophagus and potentially a greater susceptibility to carcinogens. Comparisons of the tissue zinc concentration in rats fed either a marginally zinc-deficient diet or zinc-deficient diet plus alcohol (Mobarhan et al. 1984) indicate that alcohol precipitates an acute deficiency with accompanying esophageal lesions (Table 23). A marginal deficiency of zinc, when interacting with alcohol consumption, quickly progresses to an acute deficiency with the appearance of the typical histologic lesion of zinc deficiency, esophageal parakeratosis. This is not present with the marginal deficiency alone (see Table 24).

In a number of additional studies in our laboratory, the effects of ethanol have been variable, leading us to hypothesize that additional factors or conditions have been involved, such as the presence of high molecular weight alcohols which are present in many alcoholic beverages. In attempts to further define the nature of the effects of alcohol on esophageal cancer, we have modified the MBN rat model to include both ethanol and a fusel alcohol, 3-MB. Alcoholic beverages which contain significant amounts of 3-MB are consumed by subsets of the population at high risk for head and neck cancer and are considered by some to increase the risk of esophageal cancer. When ethanol is used as a variable in this model, a liquid diet facilitates control of alcohol intake, and up to 50% of calories in the diet can be substituted with ethanol or other forms of alcohol. The liquid diet, described earlier (Barch et al. 1987; Rogers et al. 1981), has been used to test effects of ethanol and the fusel oil 3-MB together on basal cell proliferation and, further, on the induction of esophageal tumors by MBN. Table 25 lists results of some of our experiments. These results illustrate the effects of a complex exposure (zinc deficiency,

Table 24. Relation of zinc concentration in tissues, zinc and alcohol in the diet, and changes in the esophageal epithelium, Rats. From Mobarhan et al. 1984

Diet	Zinc concentration (μg/g wet weight)		Histologic findings esophagus (Parakeratosis)
	Liver	Hair	
Control	25.5 ± 2.9[a]	212.4 ± 23.8	
Zinc deficient	11.6 ± 0.5	146.4 ± 9.0	1/9 (11%)
Zinc deficient + alcohol	9.0 ± 0.5	110.0 ± 18.3	7/9 (78%)

[a] Mean ± SD.

Table 25. Effects of zinc deficiency and alcohol on DNA synthesis, cell proliferation, and esophageal tumorigenesis in rats fed liquid diets

Diet	Treatment			Basal cell hyperplasia[a] (% of control)	Tumor type and incidence[b]			
	MBN	Eto	3-MB		Papilloma		Carcinoma (%)	Endophytic sessile (%)
					(%)	(n)		
Control	–	–	–	100	93	28	0	0
Control	+	–	–	143	100	29	25	7
Control	+	+	–	140	93	29	28	10
Control	+	–	+	128	88	26	20	3
Control	+	+	+	190	96	29	48	28
Zn deficient	–	–	–	220	100	27	0	0
Zn deficient	+	–	–	195	96	26	40	30
Zn deficient	+	+	–	200	100	30	34	50
Zn deficient	+	–	+	212	96	26	43	54
Zn deficient	+	+	+	265	100	24	63	66

MBN, methylbenzylnitrosamine; Eto, ethanol; 3-MB, 3-methylbutanol.
[a] Data from five animals/group 1 week following last of five doses of MBN. Mitotic figures in the control, untreated esophagi were used as the reference point.
[b] Percentage of animals with a given type of neoplasm; the endophytic lesions derived from malignant tumors. Thirty rats per group were started on study.

ethanol, 3-MB) on esophageal tumor incidence and type, as well as basal cell turnover. [³H]Thymidine incorporation yielded a similar pattern (Schrager et al. 1986).

Comparison with Other Species

Spontaneous esophageal cancer is rare in laboratory and other species of animals. It has been reported in sheep (Schutte 1968), dogs (Seibold et al. 1955), and mice (Newberne and McConnell 1982). Esophageal cancer has been induced in rats (Long and Jenner 1963) by dihydrosafrole and in mice by γ-radiation (Gates and Warren 1968). As previously noted, these treatments are less specific for the esophagus than MBN.

The MBN tumor in rats is similar, if not identical, to the squamous cell carcinoma that occurs in humans (Ming 1973). The location of tumors appearing earlier in the rat is more often in the middle third, as in humans, but later tumors occur in the lower third of the esophagus and, to a lesser extent, in the upper third. Uncomplicated MBN-induced tumors in rats are primary squamous papillomas or carcinomas, similar to that seen in humans, where about 95% of primary carcinomas of the esophagus are of the squamous cell type. When dietary zinc deficiency or alcohol is superimposed, there is a shift toward an increased incidence of nonpapillary malignances which appear to develop without going through the papilloma process. As such, the MBN rat model serves as a useful experimental tool to investigate risk factors which may be of consequence to the development of the human counterpart. Histologically, the MBN-induced tumor is more like the human keratinizing invasive squamous cell carcinoma with variable differentiation.

The incidence of esophageal cancer in humans varies widely in different areas of the world (NAS 1982), with 300-fold variation from areas of low to highest incidence. The highest rates are found in the southern Caspian littoral in Iran, the Transkei in South Africa, and in some parts of China. These epidemiologic observations identify environmental factors which need to be verified in appropriate animal models (Doll 1969; Doll and Peto 1981). A number of investigators have found a consistent relationship between mortality from esophageal cancer and ethanol consumption (Chilvers et al. 1979; Lipworth and Rice 1979; Tuyns et al. 1979). These observations have been supported by case control studies (Keller 1980; Pottern et al. 1981; Schmidt and Popham 1981), which have shown strong correlations between alcohol consumption and esophageal cancer. However, there is no consistent association with any specific type of beverage (Williams and Horm 1977; Mettlin et al. 1981). Heavy cigarette use has also been linked to an increased risk of esophageal cancer as well as other neoplasms of the head and neck regions (NAS 1982; Field et al. 1994).

Alcohol consumption cannot account for the pattern of esophageal cancer in Asia and Africa, however (Bradshaw and Schonland 1974; Gatei et al. 1978; Yang 1980), because populations in these areas consume only small quantities of alcoholic beverages, if any. Correlation studies in Iran suggest that the intake of pulses, green vegetables, fresh fruit, animal protein, vitamin A, vitamin C, and riboflavin are lower in areas of high risk (Hormozdiari et al. 1975; Joint Iran-IARC Study Group 1977). Similar studies in China have implicated a low intake of trace elements (molybdenum, zinc), animal products, fruits and vegetables, fat, calcium, and riboflavin (Yang 1980; Lin et al. 1977). A concomitant intake of foods contaminated with N-nitroso compounds and fungal toxins was also reported. Marasas et al. (1979) have also identified mycotoxin contamination of foods in areas of South Africa where there is a high incidence of esophageal cancer compared with patients with other types of cancer or with other diseases, similar to our observations in Chinese (Table 20). Follow-up studies have further supported these earlier observations (Newberne and Schrager 1983).

References

Ahmed SB, Russell RM (1982) The effect of ethanol feeding on zinc balance and tissue zinc levels in rats maintained on zincdeficient diets. J Lab Clin Med 100:211–217

Barch DH, Fox CC, Bennett BT (1987) A simple system of feeding bottles for the study of zinc deficiency and ethanol consumption in the rat. Lab Anim Sci 37:504–506

Blot WJ, Devesa SS, Kneller RW, Fraumeni JF Jr (1991) Rising incidence of adenocarcinoma of the esophagus and gastric cardia. JAMA 265:1287–1289

Bradshaw E, Schonland M (1974) Smoking, drinking and oesophageal cancer in African males of Johannesburg, South Africa. Br J Cancer 30:157–163

Chilvers C, Fraser P, Beral V (1979) Alcohol and esophageal cancer: an assessment of the evidence from routinely collected data. J Epidemiol Commun Health 33:127–133

Doll R (1969) The geographical distribution of cancer. Br J Cancer 23:1–8

Doll R, Peto R (1981) The causes of cancer: quantitative estimates of avoidable risks of cancer in the United States today. J Natl Cancer Inst 66:1191–1308

Field JK, Zoumpourlis V, Spandidos DA, Jonas AS (1994) p53 expression and mutations in squamous cell carcinoma of the head and neck: expression correlates with the patients use of tobacco and alcohol. Cancer Detect Prev 18:197–208

Fong LY, Sivak A, Newberne PM (1978) Zinc deficiency and methylbenzylnitrosamine-induced esophageal cancer in rats. J Natl Cancer Inst 61:145–150

Gabrial GN, Schrager TF, Newberne PM (1982) Zinc deficiency, alcohol, and retinoid: association with esophageal cancer in rats. J Natl Cancer Inst 68:785–789

Gatei DG, Odhiambo PA, Orinda DA, Muruka FJ, Wasunna A (1978) Retrospective study of carcinoma of the esophagus in Kenya. Cancer Res 38:303–307

Gates O, Warren S (1968) Radiation-induced experimental cancer of the esophagus. Am J Pathol 53:667–685

Hormozdiari H, Day NE, Aramesh B, Mahboubi E (1975) Dietary factors and esophageal cancer in the Caspian littoral of Iran. Cancer Res 35:3493–3498

Joint Iran-IARC Study Group (1977) Esophageal cancer studies in the Caspian littoral of Iran: results of population studies – a prodrome. J Natl Cancer Inst 59:1127–1138

Keller AZ (1980) The epidemiology of esophageal cancer in the west. Prev Med 9:607–612

Lin HJ, Chan WC, Fong YY, Newberne PM (1977) Zinc levels in serum, hair and tumors from patients with esophageal cancer. Nutr Rep Int 15:635–643

Lipworth LL, Rice CA (1979) Correlations in mortality data involving cancers of the colorectum and esophagus. Cancer 43:1927–1933

Long EL, Jenner PM (1963) Esophageal tumors produced in rats by the feeding of dihydrosafrole. Fed Proc 22:275

Magee PN, Schoental R (1964) Carcinogenesis by nitroso compounds. Br Med Bull 20:102–106

Marasas WF, van Rensburg SJ, Mirocha CJ (1979) Incidence of fusarium species and the mycotoxins, deoxynival and zearalenone, in corn produced in esophageal cancer areas in Transkei. J Agric Food Chem 27:1108–1112

Mettlin C, Graham S, Priore R, Marshall J, Swanson M (1981) Diet and cancer of the esophagus. Nutr Cancer 2:143–147

Ming SC (1973) Tumors of the esophagus and stomach. In: Firminger HI (ed) Atlas of tumor pathology, 2nd series, fasc 7 AFIP, Washington DC

Mobarhan S, Russell RM, Newberne PM, Ahmed SB (1984) The effect of zinc deficiency and alcohol feeding on esophageal epithelium of rats. Nutr Rep Int 29:639–645

Napalkov NP, Pozharisski KM (1969) Morphogenesis of experimental tumors of the esophagus. J Natl Cancer Inst 42:927–940

National Academy of Sciences (NAS) (1982) Diet, nutrition and cancer. National Academy, Washington DC, chap 17, p 391

Nauss KM, Bueche D, Soule N, Fu P, Yew K, Newberne PM (1986) Effect of dietary selenium levels on methylbenzylnitrosamine-induced esophageal cancer in rats. Cancer Lett 33:107–116

Nauss KM, Bueche D, Newberne PM (1987) Effect of vitamin A nutrition on experimental esophageal carcinogenesis. J Natl Cancer Inst 79:145–147

Newberne PM, McConnell RG (1982) Neoplasms of the digestive system. In: Foster H, Small JD, Fox JG (eds) (1982) The mouse in biomedical research. vol IV. Experimental biology and oncology. Academic, New York, chap 27

Newberne PM, Schrager T (1983) Promotion of gastrointestinal tract tumors in animals: dietary factors. Environ Health Perspect 50:71–83

Newberne PM, Chen T, Bao-Tram N (1994) Effects of zinc deficiency, ethanol, 3-methyl butanol on MBN-induced esophageal cancer in rats. Nutr Cancer (in press)

Pottern LM, Morris LE, Blot WJ, Ziegler RG, Fraument JF Jr (1981) Esophageal cancer among black men in Washington, DC. I. Alcohol, tobacco and other risk factors. J Natl Cancer Inst 67:777–783

Pozharisski KM (1990) Tumours of the esophagus. In: Turusov VS, Mohr U (eds) Pathology of tumours in laboratory animals. IARC Sci publ no 99, Lyon, pp 109–119

Rogers AE, Fox JG, Murphy JC (1981) Ethanol and diet interactions in male rhesus monkeys. Drug Nutr Interact 1:3–14

Russell RM (1980) Vitamin A and zinc metabolism in alcoholism. Am J Clin Nutr 33:2741–2749

Schmidt W, Popham RE (1981) The role of drinking and smoking in mortality from cancer and other causes in male alcoholics. Cancer 47:1031–1041

Schrager TF, Bushy WF Jr, Goldman ME, Newberne PM (1986) Enhancement of methylbenzylnitrosamine-induced esophageal carcinogenesis in zinc-deficient rats: effects on incorporation of [3H]thymidine into DNA of esophageal epithelium and liver. Carcinogenesis 7:1121–1126

Schutte KH (1968) Esophageal tumors in sheep: some ecological observations. J Natl Cancer Inst 41:821–824

Seibold HR, Bailey WS, Hoerlein BF, Jordan EM, Schwabe CW (1955) Observations on the possible relation of malignant esophageal tumors and Spirocerca lupi lesions in the dog. Am J Vet Res 16:514

Tuyns AJ, Pequignot G, Abbatucci JS (1979) Oesophageal cancer and alcohol consumption: importance of type of beverage. Int J Cancer 23:443–447

Williams RR, Horm JW (1977) Association of cancer sites with tobacco and alcohol consumption and socioeconomic status of patients: interview study from the Third National Cancer Survey. J Natl Cancer Inst 58:525–547

Yang CS (1980) Research on esophageal cancer in China: a review. Cancer Res 40:2633–2644

Adenocarcinoma, Esophagus, Rat

Antonio Cardesa, Josep A. Bombi, Manual Pera, and Pedro L. Fernandez

Synonyms. Carcinoma with glandular differentiation, tubular adenocarcinoma, signet ring cell carcinoma

Gross Appearance

Adenocarcinomas of the esophagus are seen with two macroscopic forms (Pera et al. 1989). One form has a diffuse pattern of infiltration with thickening of the esophageal wall, which may measure up to 4 mm in thickness. The second form is seen as a nodular pattern with single or multiple nodules measuring from 1 mm up to 7 mm

in diameter (Fig. 289). In the larger nodular adenocarcinomas, it is common to find an ulcerated surface. Adenocarcinomas are found in the middle and distal thirds of the esophagus.

Microscopic Features

Three microscopic patterns are observed: (1) tubular and glandular structures, (2) signet-ring shaped, diffusely infiltrating cells, and (3) a combination of the first two patterns (Pera et al. 1989). The tubular and glandular pattern is formed by well-developed glands lined with mucous-

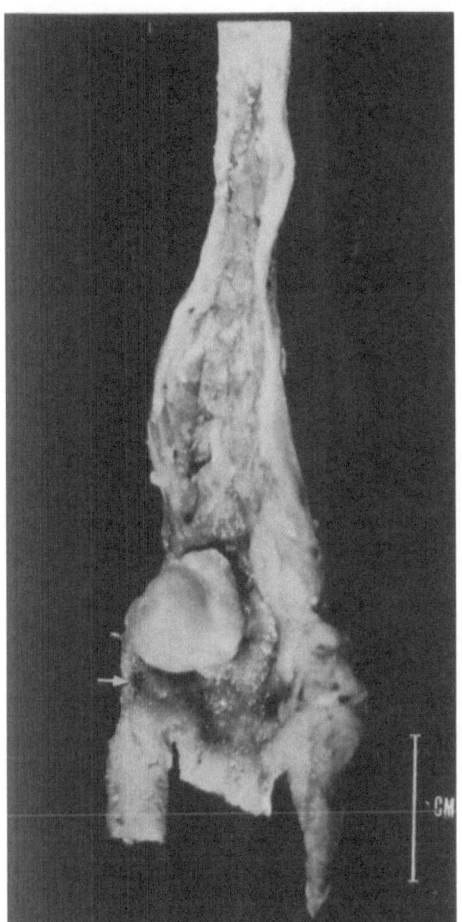

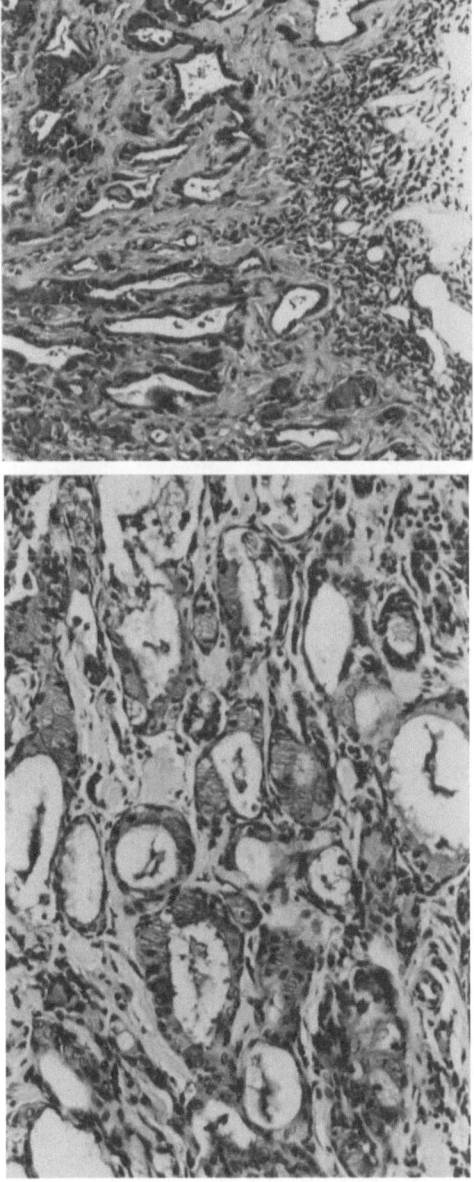

Fig. 289. Nodular adenocarcinoma in the lower third of the esophagus of a Sprague-Dawley rat. Note its origin immediately above one of the suture stitches placed along the line of the esophago-jejunostomy (*arrow*)

Fig. 290. (*above*) Adenocarcinoma, esophagus, rat, with a characteristic tubular and glandular pattern. Alcian blue plus H&E, ×200

Fig. 291. (*below*) Tubular/glandular adenocarcinoma of the esophagus of a rat. Note moderately to well-developed glandular structures with mucin production. H&E, ×250

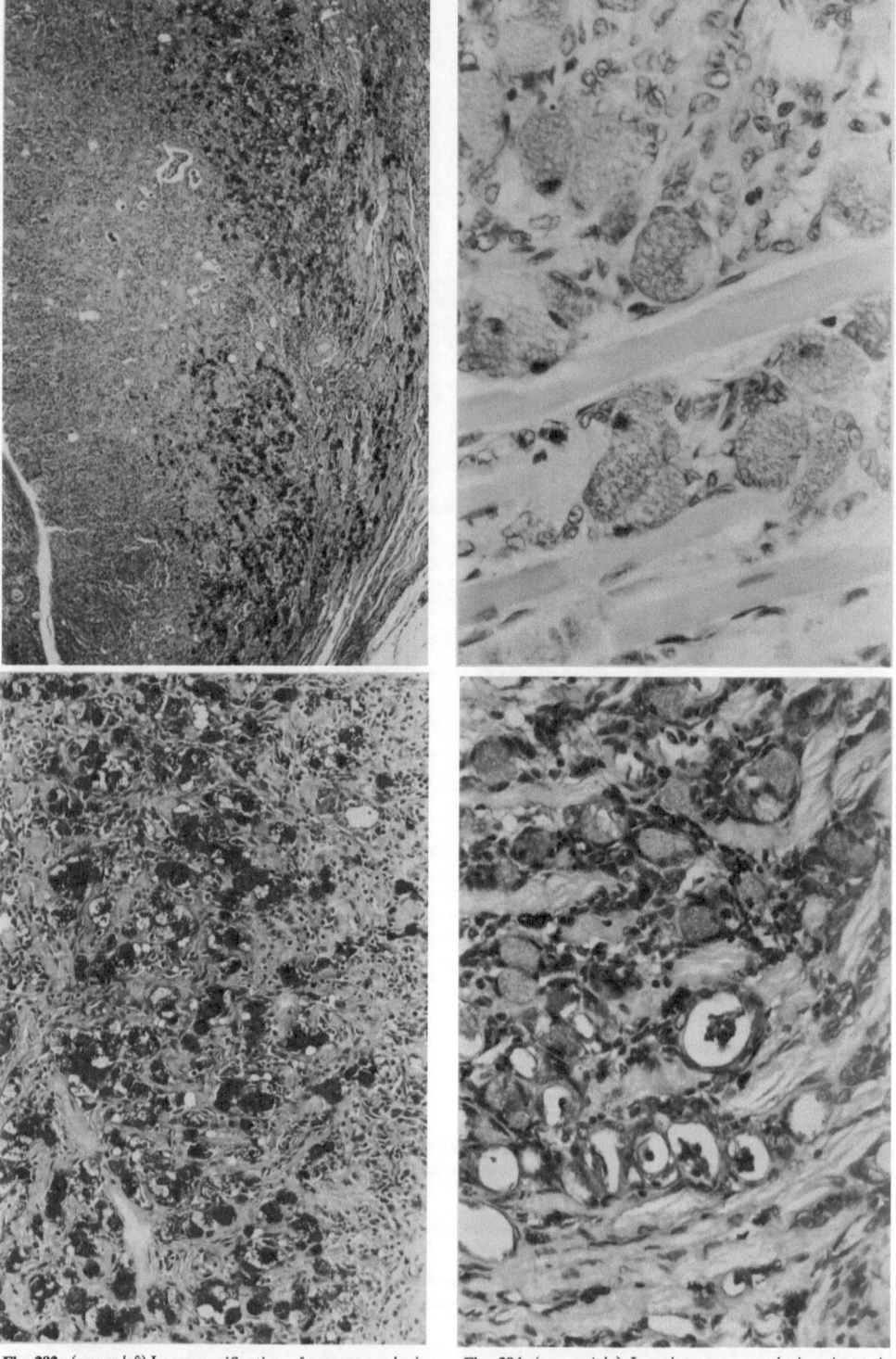

Fig. 292. (*upper left*) Low magnification of a mucus-producing signet-ring cell adenocarcinoma deeply infiltrating the esophageal wall of a rat. Note that the tumor extends into the muscular layer. PAS, ×30

Fig. 293. (*lower left*) Signet-ring cell adenocarcinoma, esophagus, rat. Large, round cells contain mucus. Alcian blue plus H&E, ×200

Fig. 294. (*upper right*) Invasive mucus-producing signet-ring cell adenocarcinoma dissecting the muscle fibers of the esophageal wall, rat. H&E, ×630

Fig. 295. (*lower right*) Adenocarcinoma of the esophagus, rat. Note tubular/glandular areas intermingled with mucus-producing signet-ring cells. Alcian blue, ×200

secreting columnar epithelium (Figs. 290, 291). The nuclei of these columnar cells are often atypical and occasionally contain mitotic figures. The adenocarcinoma with a signet-ring cellular pattern is made up of large, round cells with copious and clear cytoplasm which forms solid cords and stains positively with periodic acid-Schiff (PAS), alcian blue, and mucicarmine (Figs. 292, 293). The neoplastic cells have a definite infiltrative pattern and in some cases penetrate through the muscular coat and reach the periesophageal tissues (Fig. 294). The third pattern is made up of irregular glandular and tubular areas in contact or intermingled with other areas of solid growth consisting of signet-ring cells (Fig. 295). Marked dedifferentiation with solid strings of anaplastic cells can also be seen in these "mixed" adenocarcinomas. They have a strong tendency to invade deeper structures.

Ultrastructure

Electron microscopy study of adenocarcinomas confirms that mucous-secreting cells are the only cell type. When the tumors form tubular or glandular structures, the cells are arranged in a luminal layer with microvilli on the surface and mucinous granules of variable size and electrondensity distributed throughout the cytoplasm (Fig. 296). In addition, the cytoplasm contains scarce rough endoplasmic reticulum, few mitochondria, and occasionally a large Golgi apparatus. The nuclei contain several conspicuous nucleoli. The cells are joined by occasional small desmosomes and are separated from the extracellular matrix by a distinct basement membrane. In the signet-ring cell variety of adenocarcinoma, the cells contain abundant and large mucinous granules in the cytoplasm. These neoplastic cells are in contact with

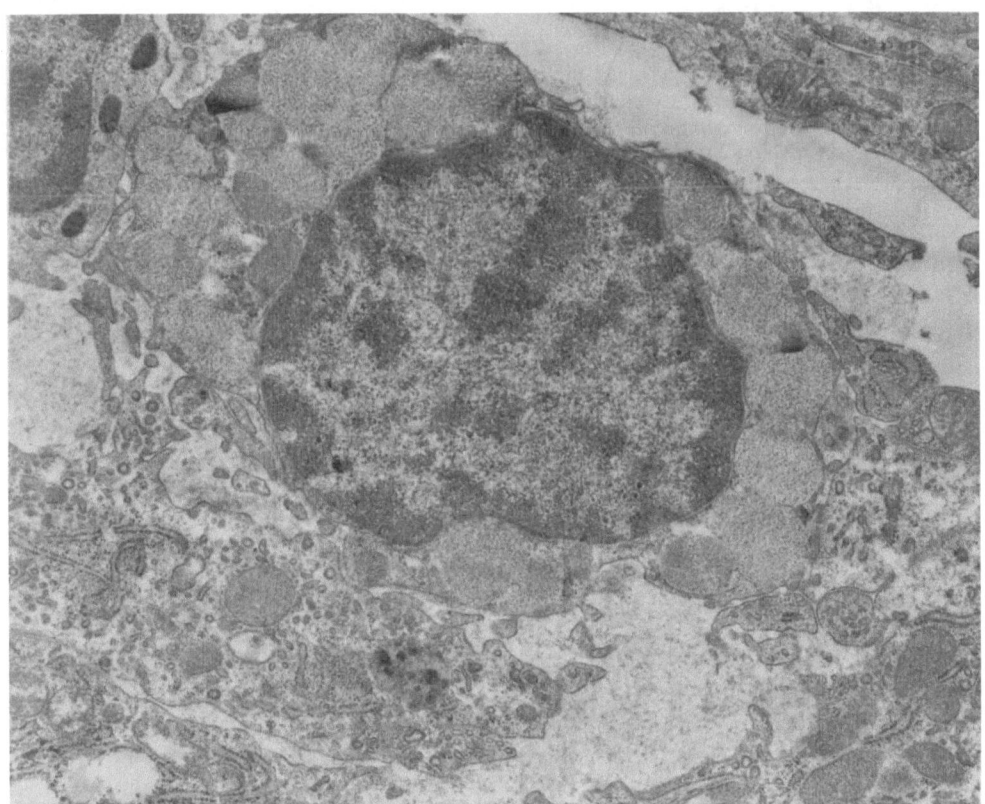

Fig. 296. Mucus-producing signet-ring cell of an adenocarcinoma, esophagus, rat. Note abundant mucinous granules in the cytoplasm and occasional microvilli on the cell surface. TEM, ×3000

the stromal cells without an intervening basement membrane.

Differential Diagnosis

The main differential diagnosis of the adeno-carcinoma of the esophagus in rats is adeno-squamous carcinoma. Within the same ex-perimental model of esophageal carcinogenesis, squamous cell carcinoma, adenocarcinoma, and adenosquamous carcinoma may coexist. There-fore, before typing a given tumor with a predomi-nant malignant glandular differentiation as pure adenocarcinoma, serial sections are needed to exclude the presence of foci with squamous cell differentiation. This latter finding would be indicative of adenosquamous carcinoma (Cardesa et al. 1994).

Biologic Features

The experimental model for inducing the growth of adenocarcinomas in the esophagus was accom-plished in Sprague-Dawley rats under the com-bined influence of chronic reflux esophagitis plus the carcinogenic effect of 2,6-dimethylnitrosamine (2,6-DMNM) (Pera et al. 1989). The induction of a chronic reflux esophagitis was accomplished by the surgical production of an esophago-

jejunostomy with gastric preservation (Fig. 297) according to Levrat's model (Levrat et al. 1962). This procedure diverts the biliary and pancreatic juice into the esophagus and significantly increases the total number of animals with esophageal carci-nomas after the chronic subcutaneous adminis-tration of 2,6-DMNM. In addition and most strikingly, this experimental model resulted in the induction of adenocarcinomas (Pera et al. 1989). The earliest-appearing adenocarcinoma was ob-served in the esophagus of one animal 13 weeks after treatment. The majority of adenocarcinomas were located in the middle and distal thirds of the esophagus, which are the areas where reflux esophagitis lesions were most intense. These findings have been confirmed by others (Attwood et al. 1992; Clark et al. 1994).

Reflux esophagitis of long duration secondary to esophago-jejunostomy without administration of carcinogen induces the development of foci of glandular metaplasia in the esophagus of Sprague-Dawley rats (Pera et al. 1989). This finding suggests that glandular metaplasia may represent a morphological substrate from which the adeno-carcinomas originate, because only squamous cell carcinomas were observed when 2,6-DMNM was given to rats that did not have esophago-jejunostomy (Pera et al. 1989).

In a subsequent study aimed to determine which fraction of the duodenal content reflux, pancreatic

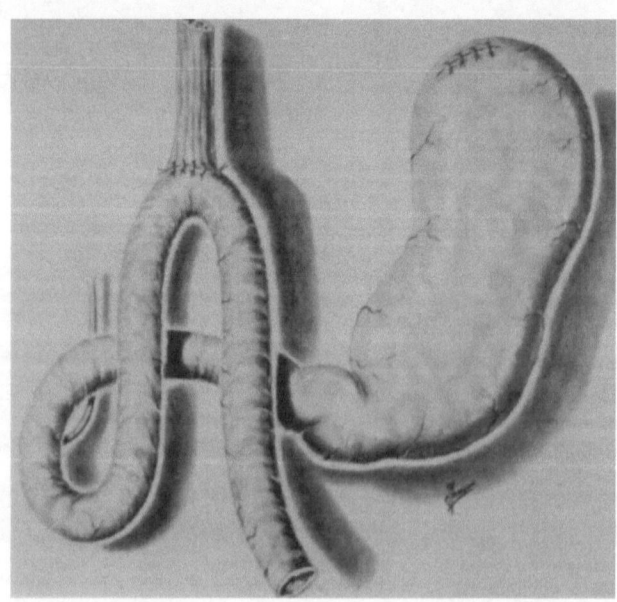

Fig. 297. Esophago-jejunostomy, Sprague-Dawley rat, to produce a chronic reflux esophagitis by diverting the gastric, biliary, and pancreatic juice into the esophagus

or biliary, contributed to the development of esophageal adenocarcinomas in rats treated with 2,6-DMNM at low dosage, it was found that adenocarcinomas developed only in those rats treated with 2,6-DMNM and exposed to reflux of pancreatic secretions, either alone or in combination with bile. Adenocarcinomas were not observed in the group of carcinogen-treated rats exposed to bile reflux alone (Pera et al. 1993a).

Comparison with Other Species

The adenocarcinomas induced under the above-mentioned experimental conditions are morphologically similar to their human counterpart (Webb and Busattil 1978; Gössner 1988). From a comparative standpoint, it seems interesting that the experimental adenocarcinomas were induced only after chronic reflux esophagitis plus subsequent exposure to the carcinogen and that in humans most of the adenocarcinomas of the distal third of the esophagus are associated with longstanding reflux esophagitis, which produces glandular metaplasia (Barrett's esophagus) (Pera et al. 1993b). In humans, the risk of developing glandular metaplasia has been estimated to be between 2% and 11% in patients with reflux esophagitis (Thompson et al. 1983). Numerous authors have noted a strong correlation between the presence of glandular metaplasia and the development of adenocarcinoma of the distal esophagus in humans (Pera et al. 1993b,c; Thompson et al. 1983; Winters et al. 1987).

References

Attwood SE, Smyrk TC, DeMeester TR, Mirvish SS, Stain HJ, Hinder RA (1992) Duodenoesophageal reflux and the development of esophageal adenocarcinoma in rats. Surgery 111:503–510

Cardesa A, Bombi JA, Pera M, Fernandez PL, Campo E, Pera C, Mohr U (1994) Spectrum of glandular differentiation in experimental carcinoma of the esophagus induced by 2,6-dimenthylnitrosomorpholine under the influence of esophago-jejunostomy. Exp Toxicol Pathol 46:41–49

Clark GW, Smyrk TC, Mirvish SS, Anselmino M, Yamashita Y, Hinder RA, DeMeester TR, Birt DF (1994) Effect of gastroduodenal juice and dietary fat on the development of Barrett's esophagus and esophageal neoplasia: an experimental rat model. Ann Surg Oncol 1:252–261

Gössner W (1988) Pathology of adenocarcinoma of the esophagus and the gastroesophageal junction. In: Siewert JR, Hölscher AH (eds) Diseases of the esophagus. Springer, Berlin Heidelberg New York, pp 39–44

Levrat M, Lambert R, Kirshbaum G (1962) Esophagitis produced by reflux of duodenal contents in rats. Am J Dig Dis 7:564–573

Pera M, Cardesa A, Bombi JA, Ernst H, Pera C, Mohr U (1989) Influence of esophago-jejunostomy on the induction of adenocarcinoma of the distal esophagus in Sprague-Dawley rats by subcutaneous injection of 2,6-dimenthylnitrosomorpholine. Cancer Res 49:6803–6808

Pera M, Trastek VF, Carpenter HA, Fernandez PL, Cardesa A, Mohr U, Pairolero PC (1993a) Influence of pancreatic and biliary reflux on the development of esophageal carcinoma. Ann Thorac Surg 55:1386–1293, 1392–1393 (discussion)

Pera M, Trastek VF, Pairolero PC, Cardesa A, Allen MS, Deschamps C (1993b) Barrett's disease: pathophysiology of metaplasia and adenocarcinoma. Ann Thorac Surg 56:1191–1197

Pera M, Cameron AJ, Trastek VF, Carpenter HA, Zinsmeister AR (1993c) Increasing incidence of adenocarcinoma of the esophagus and esophagogastric junction. Gastroenterology 104:510–513

Thompson JJ, Zinsser KR, Enterline HT (1983) Barrett's metaplasia and adenocarcinoma of the esophagus and gastroesophageal junction. Hum Pathol 14:42–61

Webb JN, Busuttil A (1978) Adenocarcinoma of the oesophagus and of the oesophagogastric junction. Br J Surg 65:475–479

Winters C, Spurling TJ, Chobanian SJ, Curtis DJ, Esposito R, Hacker J, Johnson D, Cruess D, Cotelingan J, Gurney S, Cattau E (1987) Barrett's esophagus. A prevalent, occult complication of gastroesophageal reflux disease. Gastroenterology 92:118–124

Adenosquamous Carcinoma, Esophagus, Rat

Antonio Cardesa, Josep A. Bombi, Pedro L. Fernandez, and Manual Pera

Synonyms. Squamous cell carcinoma with glandular differentiation, adenocarcinoma with squamous cell differentiation, mucoepidermoid carcinoma, adenoacanthoma

Gross Appearance

No characteristic macroscopic feature is evident in adenosquamous carcinoma of the esophagus. This

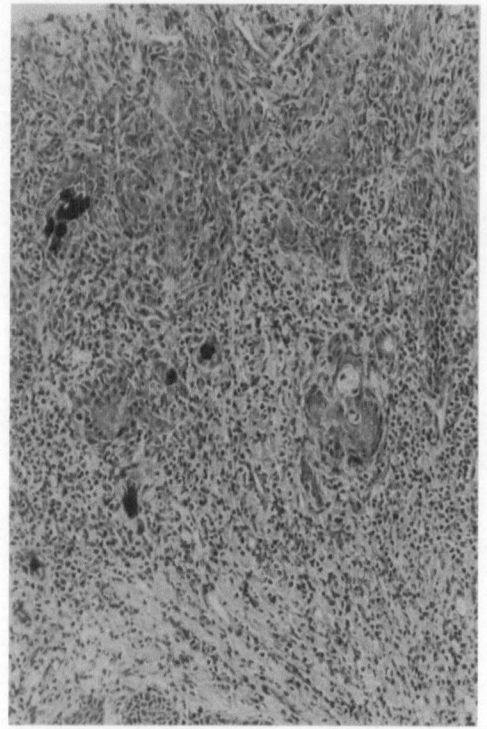

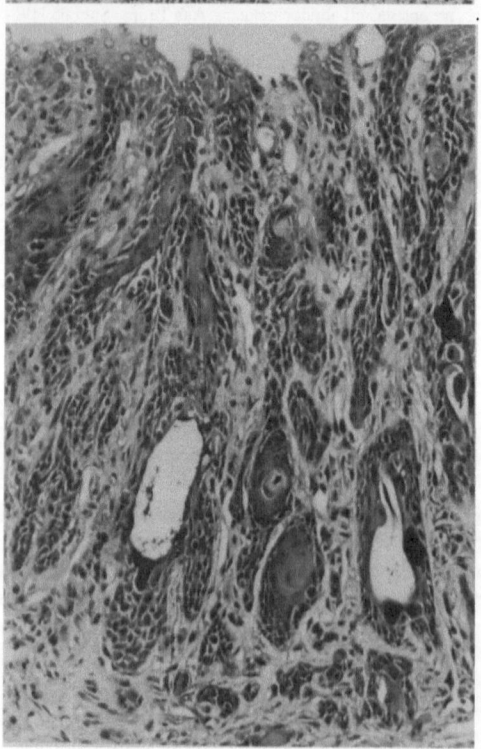

neoplasm may have an appearance similar to that of endophytic squamous cell carcinoma, with irregular ulcerations of the surface of the tumor and a conspicuous thickening of the esophageal wall, due to the inward growth of the tumor. In addition, as seen in adenocarcinomas, it may diffusely infiltrate the esophageal wall or project as a mass into the lumen. Adenosquamous carcinomas are predominantly found in the distal esophagus.

Microscopic Features

Adenosquamous carcinomas are tumors distinguished by the admixture of features of both adenocarcinoma and squamous cell carcinoma. Two main histologic patterns are described: (1) squamous cell carcinoma with focal glandular differentiation and (2) adenocarcinoma with squamous cell differentiation (Cardesa et al. 1994).

The first histologic type is differentiated by the presence, within an otherwise typical squamous cell carcinoma, of isolated foci of mucous epithelial cells with clear and abundant cytoplasm that stains positively with periodic acid-Schiff

◄

Fig. 298. (*above*) Squamous cell carcinoma of esophagus, rat, with focal glandular differentiation. Cords of downward infiltrating squamous cell carcinoma with small groups of periodic acid-Schiff (PAS)-positive mucous cells. PAS plus H&E, ×100

Fig. 299. (*below*) Squamous cell carcinoma of esophagus, rat, with focal glandular differentiation. Cords of downward infiltrating squamous cell carcinoma containing glandular lumini with inspissated mucous secretion. Alcian blue plus H&E, ×200

►

Fig. 300. (*upper left*) Adenocarcinoma, esophagus, rat with squamous cell differentiation. Infiltrating glandular proliferation surrounded by a conspicuous proliferation of neoplastic squamous cells. PAS plus H&E, ×200

Fig. 301. (*lower left*) Adenocarcinoma, esophagus, rat, with squamous cell differentiation. Well-developed atypical glandular formation surrounded by a framework of squamous epithelial cells. Alcian blue plus H&E, ×400

Fig. 302. (*upper right*) Adenocarcinoma, esophagus, rat, with squamous cell differentiation. Numerous signet-ring cells, containing minute mucous droplets, in continuity with cords of squamous epithelium. Mucicarmine ×400

Fig. 303. (*lower right*) Adenocarcinoma, esophagus, rat, with squamous cell differentiation. Large glandular lumini covered by mucus-secreting epithelium. Between the glands there is squamous cell differentiation. Alcian blue plus H&E, ×400

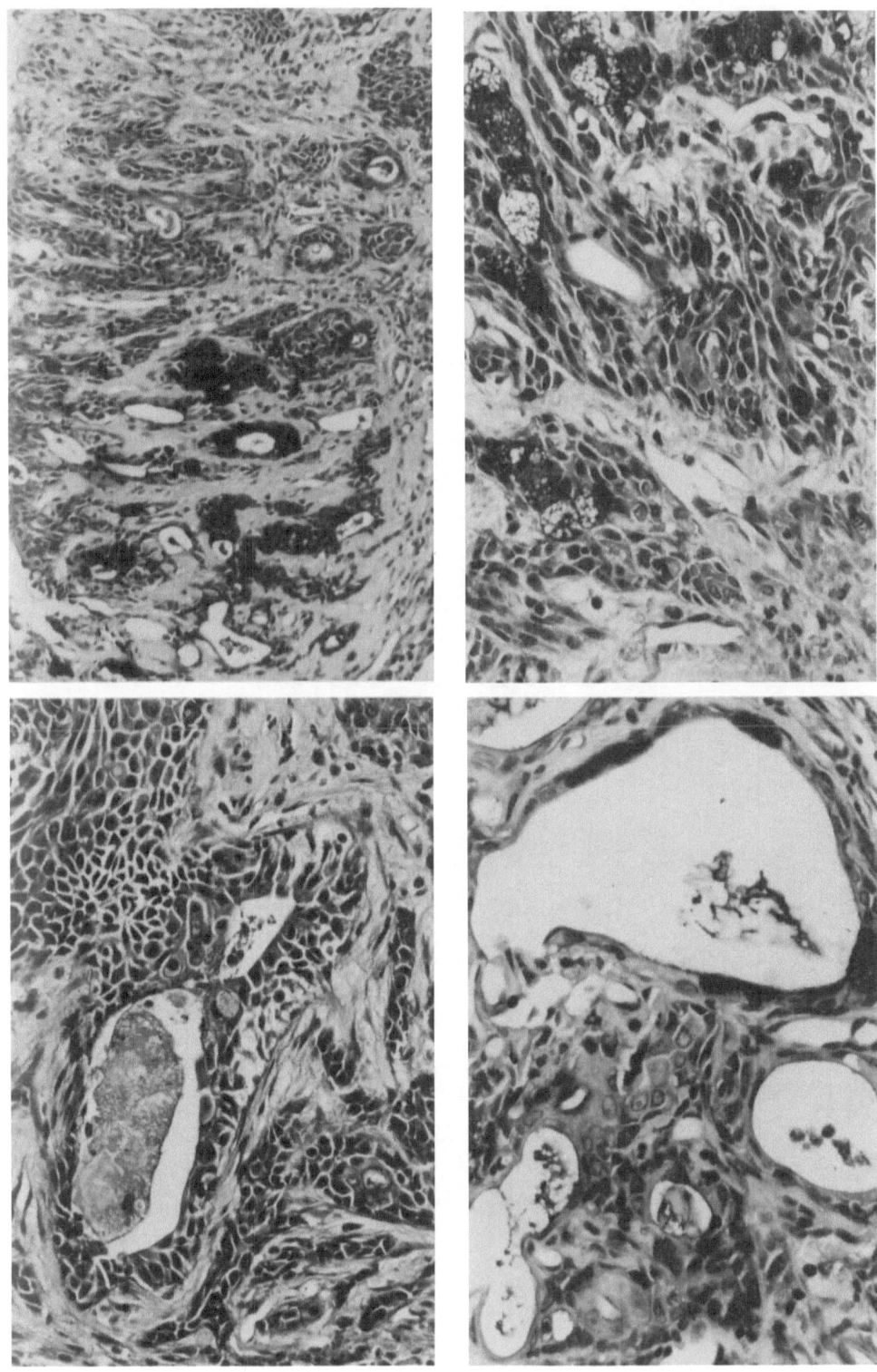

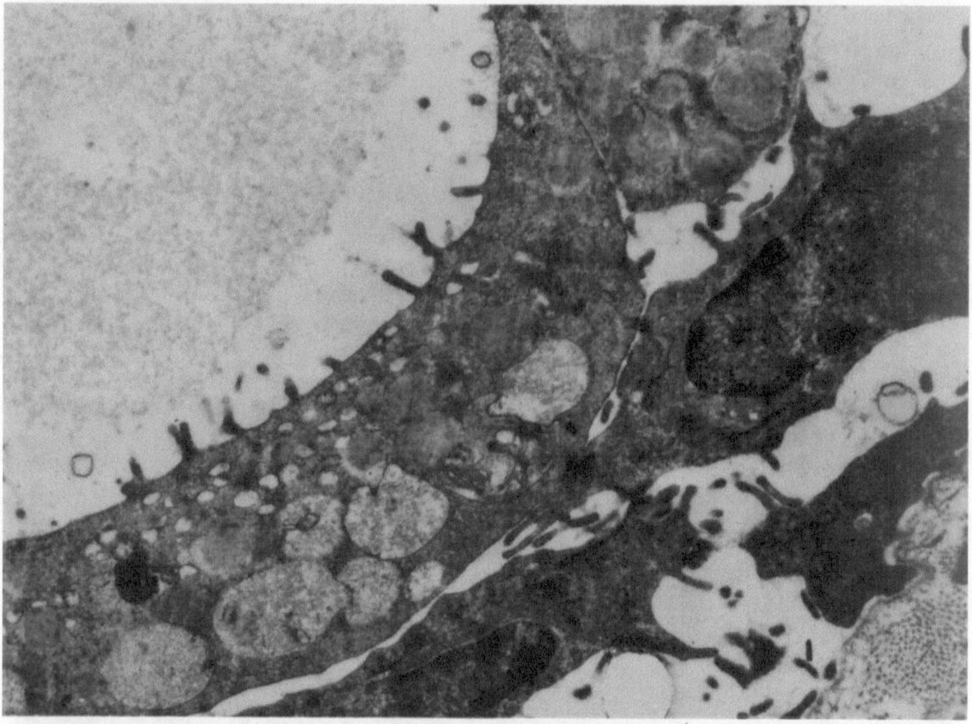

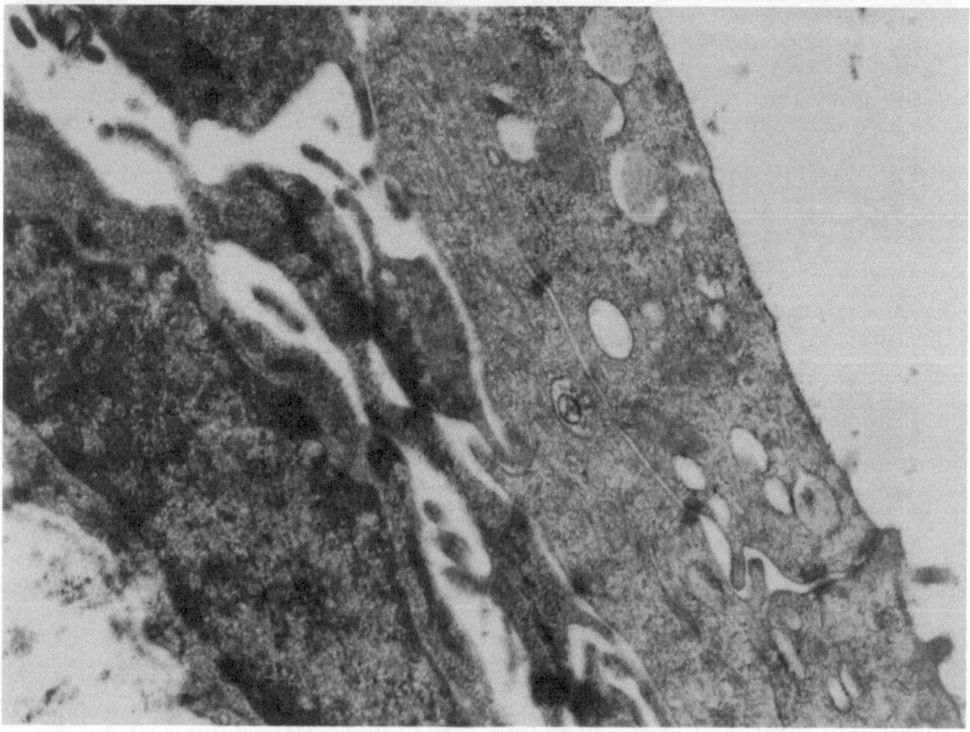

(PAS), alcian blue, or mucicarmine (Figs. 298, 299). In some instances, these mucous epithelial cells give rise to small glandular foci, sharply delineated within predominant areas of squamous cell carcinoma. The squamous cell carcinoma component has an endophytic pattern of growth in the overwhelming majority of cases.

In the second histologic type, the areas with double differentiation, mucinous and squamous, are predominantly found at the deepest invasive parts of the tumors (Fig. 300). At this level they have an abundance of adenocarcinomatous formations, with additional areas where the glands are surrounded by two or more layers of squamous cells (Figs. 301–303). Occasionally, a large framework or even a network of squamous cells is present within an otherwise typical pattern of adenocarcinoma.

Ultrastructure

Electron microscopy study of the glands indicates the cells to be of three different types (glandular, squamous, and intermediate), joined by abundant and characteristic desmosomes. An internal layer of cells on the luminal surface has microvillar structures, and mucinogenous granules of variable size and electron density are distributed throughout the cytoplasm (Fig. 304). Surrounding these cells, others contain typical bundles of tonofilaments in the cytoplasm. Between these two types of cells, there are also some intermediate cells that contain tonofilaments, microvilli, and secretory granules in their cytoplasm. However, these cells only rarely appear as fully mature cells (Figs. 304, 305).

The ultrastructural study of these tumors is sometimes of value to confirm the diagnosis. The obser-

◄

Fig. 304. (above) Adenocarcinoma with of esophagus, rat, with squamous cell differentiation. Lining of the glandular lumen by two cells containing microvilli on the surface and abundant secretory granules within the cytoplasm. Underneath, two squamous epithelial cells have desmosomal attachments to the luminal cells. TEM, ×17500

Fig. 305. (below) Adenocarcinoma, esophagus, rat, with squamous cell differentiation. On the right, two cells with glandular differentiation have pale cytoplasm, a few microvilli, and secretory granules. On the left, cells with squamous differentiation have dark cytoplasm containing bundles of tonofilaments. In the middle, one cell with intermediate differentiation has pale cytoplasm, bundles of tonofilaments, one microvillus on its surface, and desmosomal attachments to surrounding cells. TEM, ×33000

vation of cells with secretory granules, others with tonofilaments, and the finding of abundant desmosomes between mucinous cells further supports the adenosquamous nature of the tumors.

Differential Diagnosis

Serial sectioning of these tumors was essential in order to reach the correct diagnosis. All the cases of squamous cell carcinoma with focal glandular differentiation were encountered after studying serial sections stained with PAS, alcian blue, and mucicarmine of tumors initially considered to be squamous cell carcinoma, but in which foci with unusual vacuolization of the cytoplasm evoked the suspicion of mucous differentiation (Cardesa et al. 1994).

All the cases of adenocarcinoma with squamous cell differentiation were extracted from tumors initially interpreted as adenocarcinoma, but in which later sectioning and staining for mucins revealed glands containing foci of squamous cell differentiation (Cardesa et al. 1994).

Certain areas of adenosquamous carcinoma may pose differential diagnostic problems with mucoepidermoid carcinoma. However, the latter is a tumor which originates from either seromucous or salivary glands, and the esophagus of rats is devoid of these glands.

Biologic Features

The term adenosquamous carcinomas of the esophagus covers a spectrum of differentiation between adenocarcinoma and squamous cell carcinoma of the esophagus. These tumors appear in Sprague-Dawley rats after producing a histologically evident reflux esophagitis by means of an esophago-jejunostomy and by the subsequent administration of the strong esophagotropic carcinogen 2,6-dimethylnitrosomorpholine (Cardesa et al. 1994). No adenosquamous carcinomas are observed when the carcinogen is administered without previous esophago-jejunostomy and subsequent reflux esophagitis.

Comparing the pattern of distribution along the esophagus of pure squamous cell carcinomas and carcinomas with glandular differentiation, it was seen that those carcinomas showing glandular differentiation had a gradual increase in frequency towards the lower half of the esophagus, whereas in the animals with pure squamous cell carcinoma the anatomic distribution of the tumors along the esophagus was more widespread.

A stem cell located at the basal cell layer of the squamous epithelium has also been postulated to give rise to the adenosquamous carcinoma of the upper aerodigestive tract (Hyams et al. 1988). This cell might be the origin of both types of cells, i.e., cells with squamous differentiation and cells with mucinous differentiation. Therefore these two phenotypically different cell proliferations might represent parts of the same tumor and not two distinct collision tumors. The reasons why one or another of these two cell types predominates in one given tumor are not yet clear. In our view, it is likely to be related to the influence of the jejuno-esophageal reflux in the glandular differentiation of the basal cells of the squamous epithelium. Obviously, this influence has to be more prominent in the distal than in the proximal half of the esophagus.

Kuwano et al. (1985) have speculated that the components of glandular differentiation originate from the neighboring submucosal esophageal glands and their ducts. These glands have the same embryologic derivation as the minor salivary glands, where bimorphic tumors such as mucoepidermoid carcinomas can arise. Nevertheless, our experimental model supports the hypothesis that these tumors originate from the basal cell layer of the mucosal squamous epithelium: (a) because we have frequently seen glandular metaplasia with mucus-secreting cells at the base of the squamous epithelium of the esophagus of rats (Pera et al. 1989), which characteristically do not have seromucous glands in their wall, and (b) because, in the tumors reported here, neoplastic glands are frequently seen in the immediate vicinity of an actively proliferating basal cell layer of squamous epithelium of the esophageal mucosa.

Comparison with Other Species

In our view, experimental carcinomas of the esophagus with simultaneous glandular and squamous cell differentiation have many morphological similarities to the variants of adenosquamous carcinoma of the esophagus seen in human pathology (Cardesa et al. 1994). Different authors have reported the presence of glandular or mucus-secreting components in squamous cell carcinoma of the esophagus in humans (Kuwano et al. 1985; Takubo et al. 1989; Bombi et al. 1991). Kuwano et al. (1985) found 41 cases with glandular differentiation in 195 patients with carcinoma of the esophagus. Suzuki and Nagayo (1980), reviewing 256 surgical specimens and 360 autopsy cases of malignant primary esophageal neoplasms other than squamous cell carcinoma, found that about 50% of cases were pure adenocarcinoma or mixed carcinoma containing adenocarcinomatous and squamous components. Smith et al. (1984), in their review of 26 patients with carcinomas arising in Barrett's esophagus, found a spectrum of lesions from well- to poorly differentiated adenocarcinomas. In this study, one adenosquamous carcinoma containing cells with double differentiation and another case of coexisting squamous cell carcinoma and adenocarcinoma were described.

References

Bombi JA, Riverola A, Bordas JM, Cardesa A (1991) Adenosquamous carcinoma of the esophagus. A case report. Pathol Res Pract 187:514–519

Cardesa A, Bombi JA, Pera M, Fernandez PL, Campo E, Pera C, Mohr U (1994) Spectrum of glandular differentiation in experimental carcinoma of the esophagus induced by 2,6-dimethylnitrosomorpholine under the influence of esophagojejunostomy. Exp Toxicol Pathol 46:41–49

Hyams VJ, Batsakis JG, Michaels L (1988) Tumors of the upper respiratory tract and ear. Atlas of tumor pathology, 2nd ser. Fasc 25. AFIP, Washington, pp 104–107

Kuwano H, Ueo H, Sugimachi K, Inokuchi K, Toyoshima S, Enjoi M (1985) Glandular or mucus-secreting components in squamous cell carcinoma of the esophagus. Cancer 56:514–518

Pera M, Cardesa A, Bombi JA, Ernst H, Pera C, Mohr U (1989) Influence of esophagojejunostomy on the induction of adenocarcinoma of the distal esophagus in Sprague-Dawley rats by subcutaneous injection of 2,6-dimethylnitrosomorpholine. Cancer Res 49:6803–6808

Smith R, Hamilton S, Boitnott J, Rogers E (1984) The spectrum of carcinoma arising in Barrett's esophagus. A clinicopathologic study of 26 patients. Am J Surg Pathol 8:563–573

Suzuki H, Nagayo T (1980) Primary tumors of the esophagus other than squamous cell carcinoma-histologic classification and statistics in the surgical and autopsied materials in Japan. Int Adv Surg Oncol 3:73–109

Takubo K, Sasajima K, Yamashita K, Tanaka Y, Fujita K, Mafume K, Wang QH (1989) Morphological heterogeneity of esophageal carcinoma. Acta Pathol Jpn 39:180–189

The Stomach

Anatomy, Histology, Ultrastructure, Stomach, Rat

Norio Matsukura and Goro Asano

Gross Appearance

The stomach is located in close proximity to the abdominal side of the diaphragm. As part of the gastrointestinal tract, on the oral side it is connected to the esophagus at the gastroesophageal junction. This tract is continued into the duodenum at the pyloric ring. Some differences in the terminology of the gross anatomic features of the rat stomach are found in the literature. The terms proposed by Robert (1971) will be included in the following text in parentheses.

The stomach of the rat is divided into the forestomach (pars proventricularis) and glandular stomach (corpus or pars glandularis) (Fig. 306). The forestomach is characteristic of small rodents, pouches outward, and occupies about three fifths of the stomach area. The glandular stomach (corpus) is divided into the fundus and pylorus (or antrum) and communicates with the duodenum at the pyloric ring. In Fig. 307, the features of a stomach which has been opened along the greater curvature can be seen. The mucosal surface of the forestomach is whitish-brown. The borderline between the forestomach and glandular stomach is formed by a distinct elevated structure named the limiting ridge. In the glandular stomach, the fundus is seen as a rugose, reddish mucosa along the greater curvature and the pylorus (antrum) as a relatively whitish mucosa along its lesser curvature. The mucosa is thickest within the fundus.

Blood is supplied to the stomach by the celiac trunk, which is the first branch of the abdominal aorta (Greene 1968). The celiac trunk divides into three branches, i.e., left gastric, splenic, and common hepatic. The common hepatic artery branches into the right gastric, gastroduodenal, and proper hepatic arteries. The left gastric artery anastomoses with the right gastric artery and supplies blood to the lesser curvature. The gastroepiploic artery, which is a branch of gastroduodenal artery, supplies blood along the greater curvature. Veins of the glandular and forestomach are derived from the gastroepiploic and short gastric veins in the greater curvature and from the coronary vein in the lesser curvature. These drain into the splenic vein and finally into the portal vein.

Microscopic Features

The mucosa of the rat forestomach is covered by heavily cornified, stratified squamous epithelium about two to three cells thick. The limiting ridge forms an elevated border as a result of folding at the junction between squamous and columnar epithelium. The mucosal surface of the limiting ridge protrudes into the lumen, and this protrusion is easily recognized from the cut surface (Fig. 308).

The mucosa of the glandular stomach is composed of rows of columnar cells that form numerous gastric pits which are based at the lamina muscularis mucosa and open into the lumen.

The fundic glands are composed of foveolar (surface) epithelium and obtain secretions from chief (zymogenic, peptic), parietal (oxyntic), and mucous neck (accessory) cells (Fig. 309a).

These glands are tubular with small branches. The neck region between the foveolar epithelium and the gland opens into a gastric pit. The foveolar epithelium is a single layer of columnar cells and stains red–purple by Alcian blue periodic acid-Schiff (AB-PAS; Fig. 309b).

Chief cells secrete pepsinogen and are located in relatively deeper parts of the gland. The cytoplasm of these cells stains blue–purple by hematoxylin and eosin (H&E). Nuclei are located in the basal portion of the cells. Parietal cells are eosinophilic and stain red–purple by H&E. These cells are randomly distributed among the gastric glands. Their nuclei are round and located in the centers of the cells. These parietal cells secrete hydrochloric acid. Mucous neck cells are located in the neck region of the gastric pits. These cells stain purple by H&E and light-blue by AB-PAS. Only these cells and not the chief and parietal cells stain brown by paradoxical concanavalin A stain (Fig. 309c; Katsuyama and Spicer 1978).

The pyloric (antrum) mucosa is composed of tubular, branched glands which are covered by foveolar epithelium with the same characteristics as that of fundic glands. The pyloric glands are composed of a single layer of columnar cells (Fig. 310a). Nuclei are located basally and the cytoplasm stains light-blue by AB-PAS and brown by paradoxical concanavalin A (Fig. 310a,b,c). The

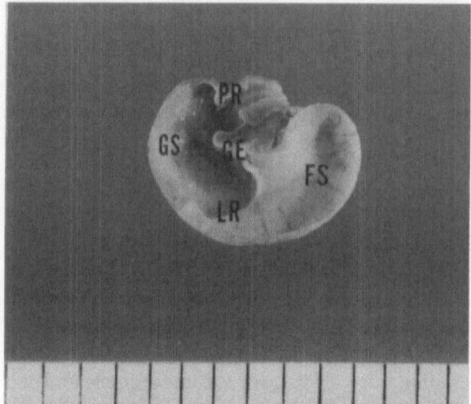

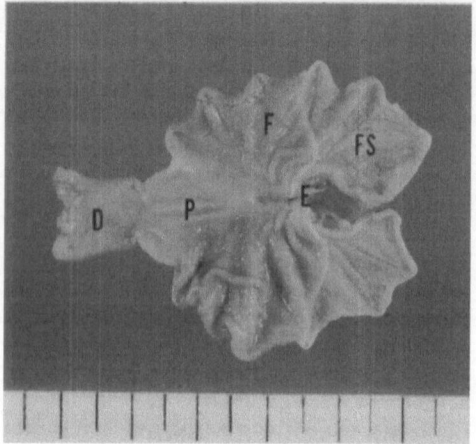

Fig. 306. (*above*) Stomach, Wistar rat. Fixed in distension and cut at the midline along the greater curvature. Forestomach (*FS*) and glandular stomach (*GS*) (corpus) are separated by the limiting ridge (*LR*). *GE*, gastroesophageal junction; *PR*, pyloric ring

Fig. 307. (*below*) Mucosal surface, stomach, rat. Opened along the greater curvature. *F*, fundus; *P*, pylorus (antrum); *D*, duodenum; *E*, esophagus; *FS*, forestomach

generative cell zone of fundic and pyloric glands is located in the neck region between the foveolar epithelium and the glands. The pyloric canal (antrum) joins the duodenum at the pyloric ring. The muscle layer becomes thick in the ring and Brunner's glands are seen on the duodenal side (Fig. 311).

Mucosal blood is supplied from the submucosal arterioles. Terminal branching extends from these arterioles at the base of the mucosa, forming a capillary plexus surrounding the gastric glands. Venules drain from this honeycomb-like network of subsurface capillaries to the submucosal venules (Gannon et al. 1982).

The lamina muscularis mucosae separates the submucosal layer (tunica submucosa) from the mucosal layer (tunica propria mucosae) (Fig. 308). The submucosal layer is composed of loose connective tissue. Blood and lymph vessels are abundant. Muscle layers (tunica muscularis, muscularis propria) are composed of an inner circulatory and an outer longitudinal layer. The muscle layer is thicker in the glandular stomach than in the forestomach, and the inner layer is thicker than the outer layer. Clusters of ganglion cells are seen between the muscle layers. The serosa, which is visceral peritoneum, covers the stomach.

Ultrastructure

Surface Epithelial (Mucous) Cells (Fig. 312). The nucleus is located basally and is round to ovoid in shape. A relatively extensive Golgi apparatus, a rough-surfaced endoplasmic reticulum, and a few mitochondria are seen in the supranuclear cytoplasm. The apical portion of the cytoplasm contains many dark, round mucous granules which are secreted into the lumen in an apocrine fusion. The loss of surface epithelial cells occurs primarily via extrusion of individual interfoveolar cells (Harding and Morris 1977). Tight junctions and desmosomes are seen between cells.

Chief Cells (Fig. 313). The nuclei are round to polygonal and located toward the base of these cells. Rough-surfaced endoplasmic reticulum is well developed and the Golgi apparatus is located in the supranuclear region. Round and large secretory granules occupy a large proportion of the apical portion of the cytoplasm.

Parietal Cells (Fig. 314). The nucleus is round and located near the center of the cell. Many mitochondria are seen in the cytoplasm and appear as pink- to red-staining granules under the light microscope and stained with H&E. Intracellular secretory canaliculi are seen around the nucleus, and the Golgi apparatus is located in the infranuclear region of the cytoplasm.

Mucous Neck Cells. The nucleus occupies a large proportion of the cytoplasm, has an ovoid to angu-

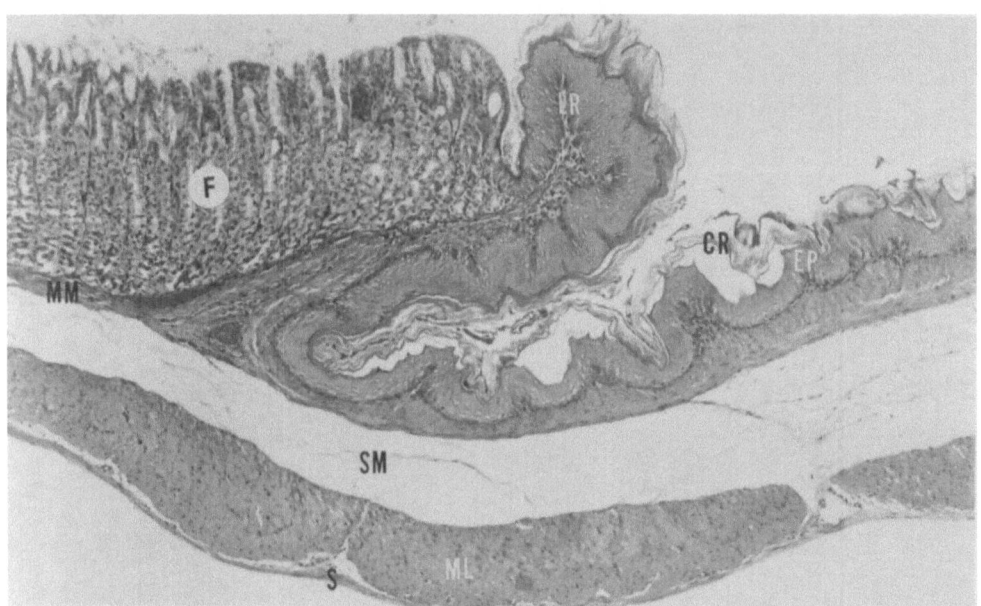

Fig. 308. Stomach, Wistar rat, near the limiting ridge (*LR*); *CR*, cornification; *EP*, stratified squamous epithelium; *F*, fundic epithelium (corpus); *MM*, lamina muscularis mucosa; *SM*, submucosa; *ML*, muscle layer; *S*, serosa. H&E, ×23

lar shape, and is located basally. A few rough-surfaced endoplasmic reticula and many Golgi complexes are seen. Secretory granules occupy the apical portion of the cytoplasm.

Pyloric Glands (Fig. 315). The nucleus is ovoid and positioned crosswise to the length and at the base of the columnar cells. Several Golgi complexes are seen in the supranuclear cytoplasm. Many secretory granules fill the apical portion of the cytoplasm. Intercellular secretory canaliculi with microvilli are present between cells.

Endocrine Cells. In the pyloric glands, gastrin (G), somatostatin (D), enterochromaffin (EC), enterochromaffin-like (ECL), and a few other cells are recognized as endocrine cells. G cells are especially characteristic of the pyloric glands; they secrete gastrin, which acts on fundic glands and strongly promotes hydrochloric acid secretion and weakly promotes pepsinogen. G cells are located mainly at the neck region of the pyloric glands and in the antrum, mostly near the pyloric ring.

The apical portion of gastrin-secreting cells is narrow and has developed microvilli. The basal por-

tion is wider than the apical portion, where the nucleus is located. Well-developed, rough-surfaced endoplasmic reticulum, Golgi apparatus, and many secretory granules with variable density occupy the cytoplasm. D and EC cells are located throughout the gastrointestinal tract, predominantly in the pyloric glands. These cells secrete somatostatin and serotonin, respectively. D cells have long, nonluminal cytoplasmic processes, which terminate on the other cells, such as G cells (Larsson et al. 1979). D cells possess large and round secretary granules, and EC cells have intensely dark secretory granules of irregular shapes and sizes.

In the fundis glands, EC, ECL, A-like, and a few other cells are identified (Forssmann et al. 1969). The ECL and A-like cells are argyrophilic and most numerous. The ECL cells are distinguished by vacuolar granules which occupy most of the cytoplasm, possess APUD (amine precursor uptake decarboxylase) and can synthesize histamine (Rubin and Schwartz 1979). Occupy almost all areas of the cytoplasm. In contrast, the cytoplasm of A-like cells is filled with round, regular, and moderately dense granules. They do not possess APUD and do not secrete histamine.

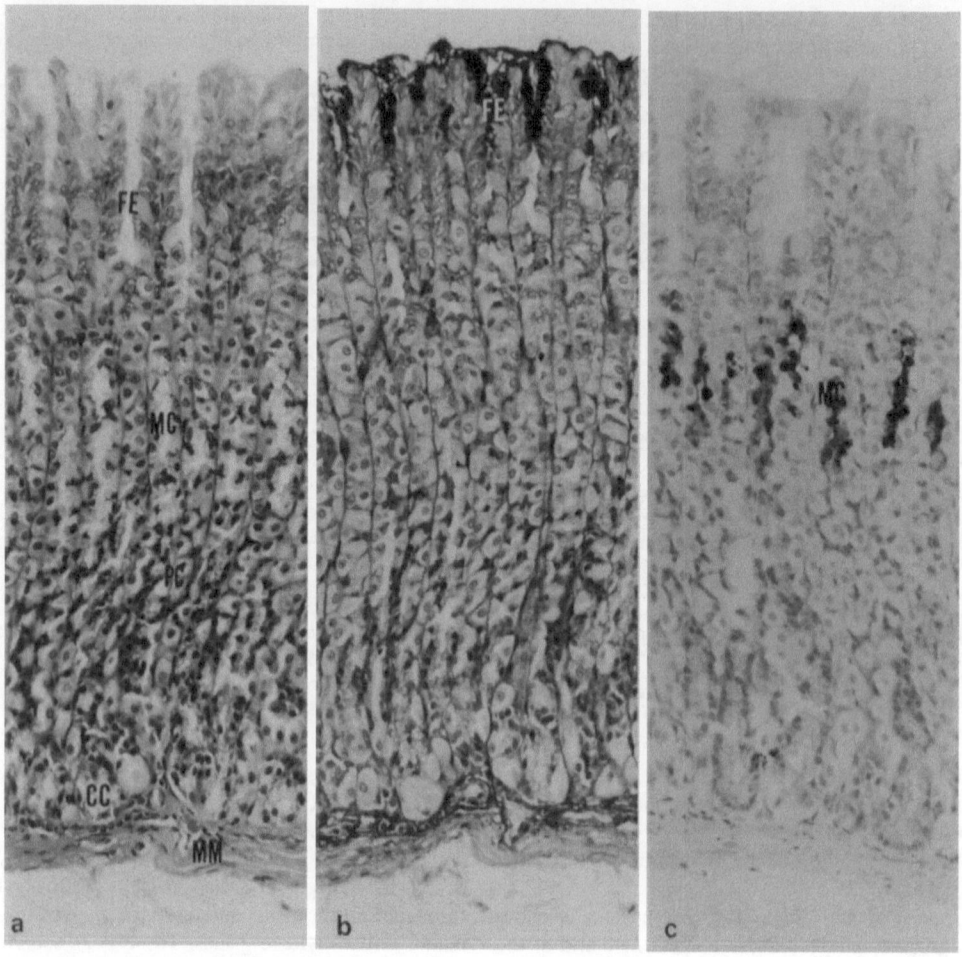

Fig. 309a–c. Fundic (glandular) epithelium, stomach, Wistar rat, serial sections. **a** Foveolar epithelium *(FE)*, chief cells *(CC)*, parietal cells *(PC)*, mucous neck cells *(MC)*, and lamina muscularis mucosa *(MM)*. H&E, ×114. **b** Mucus of foveolar epithelium stains red–purple. AB-PAS, ×114. **c** Mucous neck cells *(MC)* stained brown by paradoxical concanavalin A. ×114

Biologic Features

In newborn rats, inflammatory cell infiltration by cells such as lymph cells, neutrophils, and plasma cells is rarely found, but these cells gradually accumulate with aging.

Intestinal metaplasia, which is frequently found in older human beings, appears less pronounced in older rats and is absent in young rats. This lesion consists of focal changes in gastric epithelial cells into cells which resemble intestinal epithelium, especially by the production of goblet cells. The appearance of intestinal metaplasia is accelerated by gastric carcinogens such as N-methyl-N'-nitro-N-nitrosoguanidine (MNNG) or N-propyl-N'-nitro-N-nitrosoguanidine (Matsukura et al. 1978, 1980; Tatematsu et al. 1983). Metaplastic glands are located in the pyloric region and contain goblet cells and striated borders, but usually no Paneth's cell. Erosion (shallow ulceration) is frequently found in adult rats, but rarely as a deep ulceration. Naturally occurring gastric tumors of either the forestomach or glandular stomach are very rare in rats.

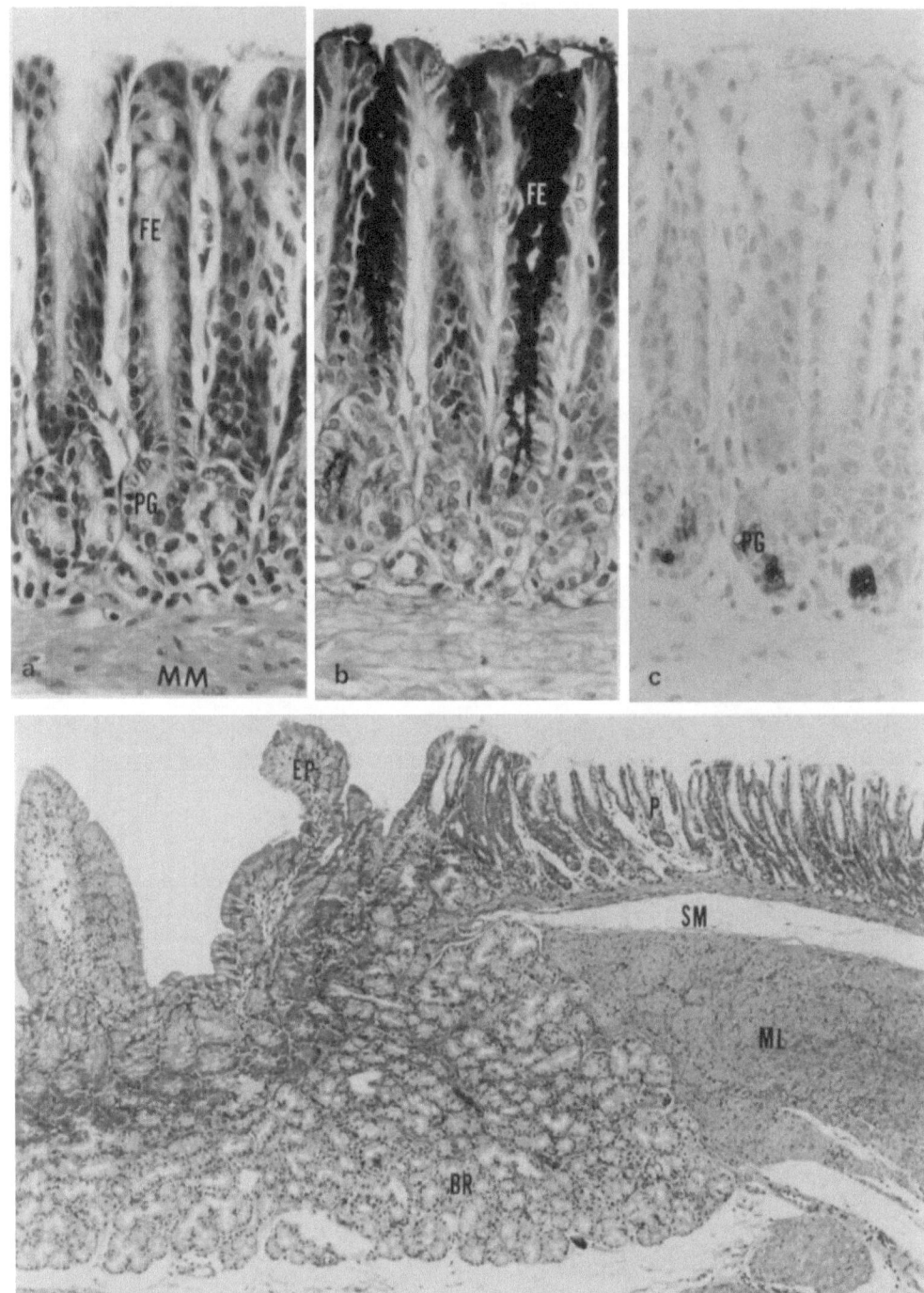

Fig. 310a–c. (*above*) Pyloric epithelium (antrum), stomach, Wistar rat, serial sections. **a** Foveolar epithelium (*FE*) and pyloric glands (*PG*). Lamina muscularis mucosa (*MM*). ×100. **b** Mucus of foveolar epithelium (*FE*) stains red–purple by AB-PAS. ×100. **c** Mucus of pyloric glands (*PG*) stains brown by paradoxical concanavalin A. ×100

Fig. 311. (*below*) Stomach, Wistar rat, section taken near the pyloric ring. *P*, pyloric epithelium; *SM*, submucosa; *ML*, muscle layer; *EP*, duodenal epithelium; *BR*, Brunner's glands. H&E, ×23

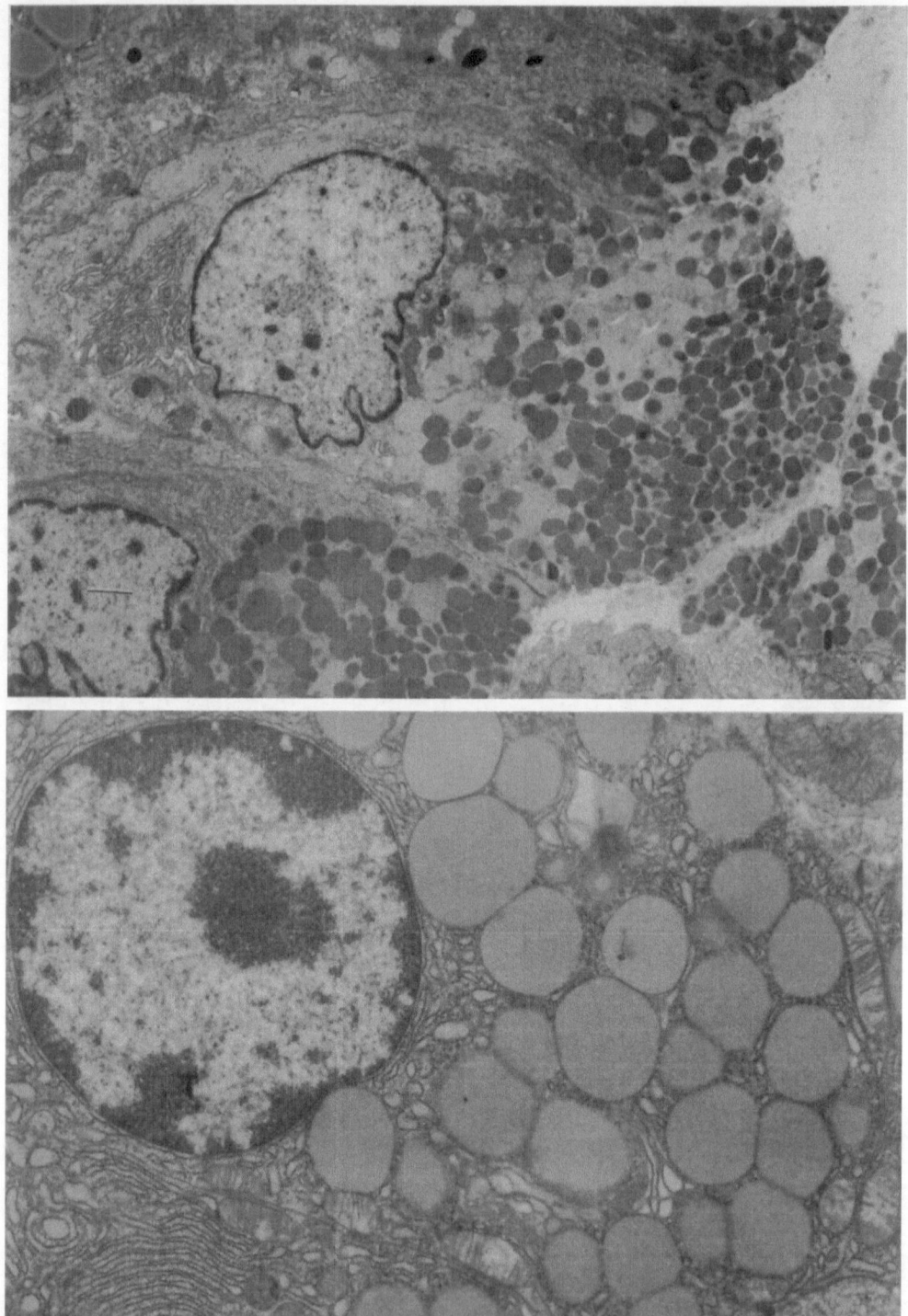

Fig. 312. (*above*) Foveal epithelial cell, stomach, rat. TEM. *Bar*, 1 μm

Fig. 313. (*below*) Chief cell, glandular stomach, rat. TEM. *Bar*, 1 μm

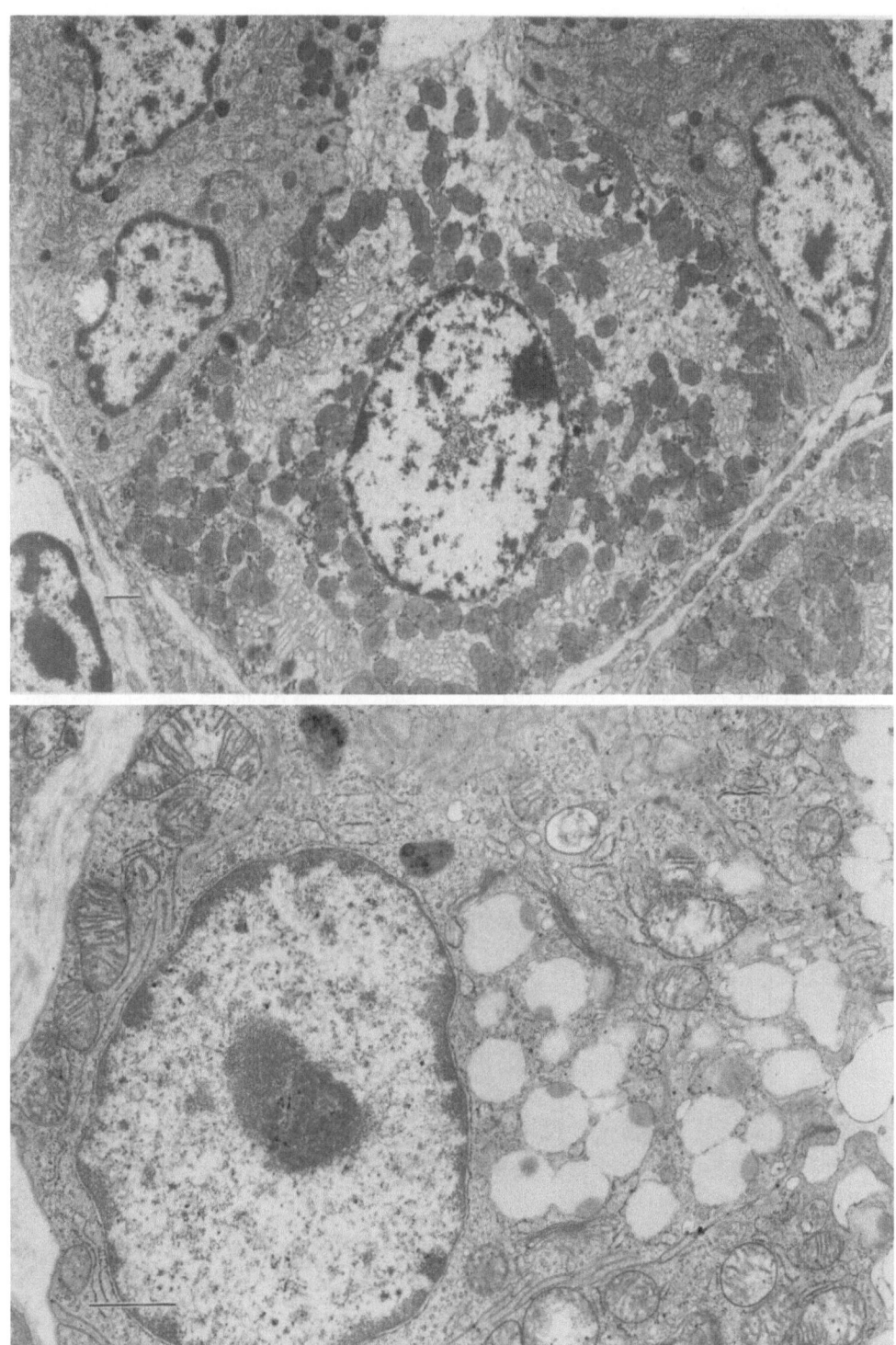

Fig. 314. (*above*) Parietal cell, glandular stomach, rat. TEM. *Bar*, 1 μm

Fig. 315. (*below*) Pyloric gland cell, stomach, rat. TEM. *Bar*, 1 μm

Comparison with Other Species

Small rodents, including rats, have a forestomach which is covered by squamous epithelium, but the forestomach does not exist in the stomach of other experimental animals such as dogs, cats, primates, or humans. Thus the glandular stomach is used as an animal model of the human stomach. The glandular stomach epithelium of rats is composed of fundic and pyloric glands, but without cardiac glands. The boundary line between the fundic and pyloric glands of rats is different from that of humans, monkey, and dog. In the rat, the pyloric glands cover almost the entire lesser curvature, but is humans, monkey, and dog they are restricted to the antral region.

References

Forssmann WG, Orci L, Pictet R, Renold AE, Rouller C (1969) The endocrine cells in the epithelium of the gastrointestinal mucosa of the rat; an electron microscope study. J Cell Biol 40:692–715

Gannon B, Browning J, O'Brien P (1982) The microvascular architecture of the glandular mucosa of rat stomach. J Anat 135:667–683

Greene EC (1968) Anatomy of the rat. Hafner, New York

Harding RK, Morris GP (1977) Cell loss from normal and stressed gastric mucosae of the rat: an ultrastructural analysis. Gastroenterology 72:857–863

Katsuyama T, Spicer SS (1978) Histochemical differentiation of complex carbohydrates with variants of the concanavalin A-horseradish peroxidase method. J Histochem Cytochem 26:233–250

Larsson L-I, Goltermann N, Magistris L, Rehfeld JF, Schwartz TW (1979) Somatostatin cell processes as pathways for paracrine secretion. Science 205:1393–1395

Matsukura N, Kawachi T, Sasajima K, Sano T, Sugimura T, Hirota T (1978) Induction of intestinal metaplasia in the stomachs of rats by N-methyl-N'-nitro-N-nitrosoguanidine. JNCI 61:141–144

Matsukura N, Itabashi M, Kawachi T, Hirota T, Sugimura T (1980) Sequential studies on the histogenesis of gastric carcinoma in rats by a weak gastric carcinogen, N-propyl-N'-nitro-N-nitrosoguanidine. J Cancer Res Clin Oncol 98:153–163

Robert A (1971) Proposed terminology for the anatomy of the rat stomach. Gastroenterology 60:344–345

Rubin W, Schwartz B (1979) Electron microscopic radioautographic identification of the ECL cell as the histamine-synthesizing endocrine cell in the rat stomach. Gastroenterology 77:458–467

Tatematsu M, Furihata C, Katsuyama T, Hasegawa R, Nakanowatari J, Saito D, Takahashi M, Matsushima T, Ito N (1983) Independent induction of intestinal metaplasia and gastric cancer in rats treated with N-methyl-N'-nitro-N-nitrosoguanidine. Cancer Res 43:1335–1341

Papilloma, Forestomach, Rat

Shoji Fukushima, Masao Hirose, and Hideki Wanibuchi

Gross Appearance

Papillomas in the forestomach of the rat project above the mucosa, most often in the form of a polyp with a pedicle, although some are sessile (Fig. 316). Occasionally, they take on cauliflower-shaped papillomatous features. They are more often multiple than single, and white rather than grayish-white. Although tumors may arise anywhere in the forestomach, they are usually located in the region of the limiting ridge between the forestomach and glandular stomach. Tumors in the forestomach sometimes form huge masses on short stalks, much larger than tumors of comparable morphology arising in the oral cavity and esophagus. Occasionally they occupy the entire cavity of the forestomach. When many papillomas are diffusely spread throughout the forestomach, the term papillomatosis is applied.

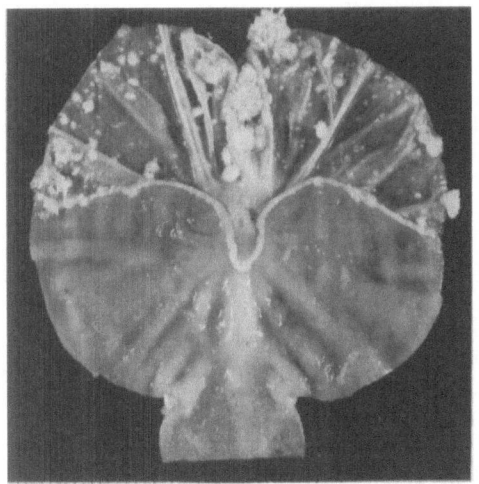

Fig. 316. Papilloma, forestomach, rat. Multiple polyps, arising at random from the epithelium of the forestomach, are evident

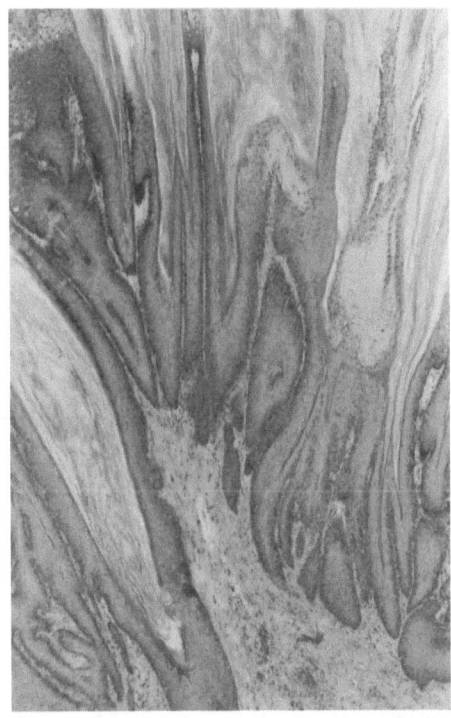

Fig. 317. Papilloma, forestomach, rat. Note the papillary proliferation of squamous epithelium with a stromal core of loose connective tissue. H&E, ×20

Microscopic Features

Papillomas are composed of fronds of hyperplastic squamous epithelium supported by strands of stromal connective tissue (Fig. 317). Varying degrees of epithelial differentiation may be observed, including that of normal squamous epithelium. The superficial layer may become markedly hyperkeratotic and occasionally parakeratotic. Mitotic figures are rare and, when seen, are present in the basal cell layer. Enlarged vessels and occasionally lymphocytic infiltrations

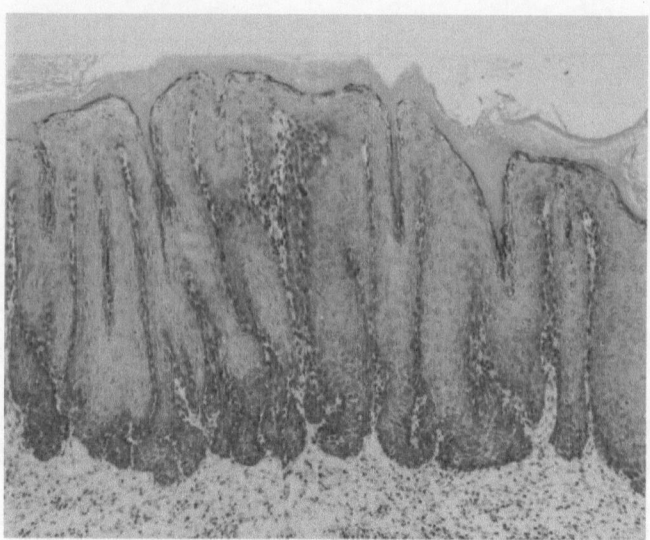

Fig. 318. Sessile papilloma, forestomach, rat. Note the exophytic growth with thickening of the basal cell layer. H&E, ×60

Table 26. Incidence of naturally occurring epithelial tumors in the forestomach of rats

Strain	Sex	Papilloma		Squamous cell carcinoma		Total rats studied	References
		(n)	(%)	(n)	(%)	(n)	
F344	M	2	0.1	3	0.2	1794	Goodman et al. (1979)
	F	1	0.1	1	0.1	1754	
F344	M	0	0	0	0	732	Maekawa et al. (1983)
	F	0	0	0	0	412	
F344	M	0	0	0	0	149	Fukushima et al. (1984)
	F	0	0	0	0	148	
Osborne-Mendel	M	1	0.1	0	0	975	Goodman et al. (1980)
	F	0	0	1	0.1	970	
ACI	M	0	0	0	0	55	Maekawa and Odashima (1975)
	F	0	0	0	0	209	
BN/Bi	M	1	1.4	0	0	74	Burek (1978)
	F	0	0	0	0	236	
WAG/Rij	M	0	0	0	0	124	Burek (1978)
	F	2	2.0	0	0	101	
Wistar CPB	M	0	0	0	0	197	Kroes et al. (1981)
	F	0	0	0	0	182	
Wister SPF Tox	M	0	0	0	0	192	Kroes et al. (1981)
	F	0	0	0	0	192	
Charles River SD	MF	0	0	1	0.2	535	MacKenzie and Garner (1973)
SD	M	1	0.6	1	0.6	179	Prejean et al. (1973)
	F	0	0	0	0	180	
SD	M	0	0	0	0	43	Thompson et al. (1961)
	F	0	0	1	1.2	82	

are present in the stromal stalks. Some papillomas are exophytic with more proliferation of the basal cell layer (Fig. 318). However, tumor cells do not proliferate into the submucosa through the muscularis mucosa.

Ultrastructure

The ultrastructure of these tumors differs little from normal squamous epithelium and is not usually required for diagnosis.

Table 27. Incidence of naturally occurring epithelial tumors in the forestomach of mice and hamsters

Species	Strain	Sex	Papilloma		Squamous cell carcinoma		Total animals studied	References
			(n)	(%)	(n)	(%)	(n)	
Mice	B6C3F₁	M	8	0.3	1	0	2543	Ward et al. (1979)
		F	7	0.3	0	0	2522	
	B6C3F₁	M	1	0.4	0	0	244	Tamano et al. (1988)
		F	2	0.8	2	0.8	246	
	B6C3F₁	M	78	6.9	5	0.4	1130	Hirouchi et al. (1994)
		F	70	6.2	3	0.3	1130	
Hamsters	Syrian, Eppley colony	M	12	8.2	0	0	147	Pour et al. (1976)
		F	7	5.2	0	0	136	
	Syrian, Hannover colony	M	3	4.0	0	0	76	Pour et al. (1976)
		F	3	2.9	0	0	102	
	Syrian	MF	57	5.1	3	0.3	1120	Dontenwill et al. (1973)
	Syrian	M	11	11.7	0	0	94	Fortner (1961)
		F	1	1.1	0	0	87	

Differential Diagnosis

Papillomas may be difficult to distinguish from severe epithelial hyperplasia, especially when they are small. They are benign tumors in which squamous epithelium is arranged in branched finger-like processes, supported by a fibrovascular core. Slight cellular and structural atypia may be seen. Invasion of other tissues or distant metastases serve to distinguish carcinomas from papillomas. However, when squamous cell carcinomas are well differentiated, comparative diagnosis and distinction from papillomas may be difficult. The presence of invasion into the submucosa through the muscularis mucosas is not useful in identifying squamous cell carcinomas. Although tumor tissue of papillomas may also extend into the submucosa, the lack of cellular and structural atypia generally allows a clear distinction to be made.

Biologic Features

Spontaneous appearance of papillomas in the forestomach of rats is rare, even in old animals (Table 26). In spite of this, high incidences of tumors of the forestomach can be induced by chemical carcinogens, including N-nitroso compounds such as 1-methyl-1-nitrosourea, 1-butyl-1-nitrosourea, N-methyl-N-nitrosourethane, N-propyl-N-nitrosourethane, N-butyl-N-nitrosourethane, N-ethyl-N'-nitro-N-nitrosoguanidine, 4-nitroquinoline 1-oxide, 8-nitroquinoline, and antioxidants such as butylated hydroxyanisole (Ito et al. 1983), caffeic acid, sesamol, 4-methoxyphenol, and 4-methoxycatechol.

Comparison with Other Species

The incidence of naturally occurring papillomas and squamous cell carcinomas in the forestomach is quite high in Syrian golden hamsters compared with rats and mice (Table 27). Epithelial hyperplasia of the forestomach is also common in untreated hamsters. They seem to be extremely sensitive to the composition of their diet. The hamster, as opposed to the rat or mouse, may therefore be an advantageous animal model for studying the pathogenesis, underlying mechanisms, biology, and other aspects of tumors of the forestomach.

References

Burek JD (1978) Age-associated pathology. In: Burek JD (ed) Pathology of aging rats. CRC, Boca Raton, chap 4

Dontenwill W, Chevalier HJ, Harke BP, Lafrenz U, Reckzeh G, Schneider B (1973) Spontaneous tumors in Syrian golden hamsters. Z Krebsforsch 80:127–158

Fortner JG (1961) The influence of castration on spontaneous tumorigenesis in the Syrian (golden) hamster. Cancer Res 21:1491–1498

Fukushima, S, Hagiwara A, Ogiso T, Shibata M-A, Kurata Y, Ito N (1984) Spontaneous tumors in F344/DuCrj rats. Nagoya Med J 29:163–172

Goodman DG, Ward JM, Squire RA, Chu KC, Linhart MS (1979) Neoplastic and nonneoplastic lesions in aging F344 rats. Toxicol Appl Pharmacol 48:237–248

Goodman DG, Ward JM, Squire RA, Paxton MB, Reichardt WD, Chu KC, Linhart MS (1980) Neoplastic and nonneoplastic lesions in aging Osborne-Mendel rats. Toxicol Appl Pharmacol 55:433–447

Hirouchi Y, Iwata H, Yamakawa S, Kato M, Kobayashi K, Yamamoto T, Inoue H, Enomoto M, Shiga A, Koike Y (1994) Historical data of neoplastic and nonneoplastic lesions in B6C3F₁ (C57BL/6CrSlc × C3H/HeSlc) mice. J Toxicol Pathol 7:153–177

Ito N, Fukushima S, Hagiwara A, Shibata M, Ogiso T (1983) Carcinogenicity of butylated hydoxyanisole in F344 rats. J Natl Cancer Inst 70:343–352

Kroes R, Garbis-Berkvens JM, de Vries TD, van Nesselrooy HJ (1981) Histopathological profile of a Wistar rat stock including a survey of the literature. J Gerontol 36:259–279

MacKenzie WF, Garner FM (1973) Comparison of neoplasms in six sources of rats. J Natl Cancer Inst 50:1243–1257

Maekawa A, Odashima S (1975) Spontaneous tumors in ACI/N rats. J Natl Cancer Inst 55:1437–1445

Maekawa A, Kurokawa Y, Takahashi M, Kokubo T, Ogiu T, Onodera H, Tanigawa H, Ohno Y, Furukawa F, Hayashi Y (1983) Spontaneous tumors in F-344/DuCrj rats. Gann 74:365–372

Pour P, Kmoch N, Greiser E, Mohr U, Althoff J, Cardesa A (1976) Spontaneous tumors and common diseases in two colonies of Syrian hamsters. I. Incidence and sites. J Natl Cancer Inst 56:931–935

Prejean JD, Peckham JC, Casey AE, Griswold DP, Weisburger EK, Weisburger JH (1973) Spontaneous tumors in Sprague-Dawley rats and Swiss mice. Cancer Res 33:2768–2773

Tamano S, Hagiwara A, Shibata M-A, Kurata Y, Fukushima S, Ito N (1988) Spontaneous tumors in aging (C57BL/6N × C3H/HeN)F₁ (B6C3F₁) mice. Toxicol Pathol 16:321–326

Thompson SW, Huseby RA, Fox MA, Davis CL, Hunt RD (1961) Spontaneous tumors in the Sprague-Dawley rat. J Natl Cancer Inst 27:1037–1057

Ward JM, Goodman DG, Squire RA, Chu KC, Linhart MS (1979) Neoplastic and nonneoplastic lesions in aging (C57BL/6N × C3H/HeN)F₁ (B6C3F₁) mice. J Natl Cancer Inst 63:849–854

Squamous Cell Carcinoma Forestomach, Rat

Shoji Fukushima, Masao Hirose, and Hideki Wanibuchi

Synonyms. Malignant papilloma, epidermoid carcinoma, carcinoma in situ

Gross Appearance

Most squamous cell carcinomas in the forestomach are exophytic (Fig. 319). They are predominantly sessile, have a wide base, and form multiple polypoid excrescences. The surface of the tumors is sometimes ulcerated and necrotic. On the cut surface, prominent intramural growth with necrosis and hemorrhage may sometimes be seen. In some cases, the cancer occupies the entire mucosa of the forestomach, producing large, irregular ulceration. Endophytic carcinomas are occasionally observed which ulcerate and penetrate the muscle layer to various depths (Fig. 320). Large squamous cell carcinomas often invade through to the serosal surface and into the glandular stomach across the limiting ridge (Fig. 321). Metastases to regional lymph nodes are often observed.

Microscopic Features

Carcinomas In Situ. Basal cell carcinoma in situ is almost always observed within areas of hyperplastic epithelium. The tumors are composed of basal cells with atypia. Mitotic figures are common. Extension of the tumor is usually limited to the lamina propria with no invasion into the muscularis mucosa. Squamous cell carcinoma in situ is a relatively rare finding.

Squamous Cell Carcinoma. These carcinomas generally have an exophytic growth pattern and are composed of squamous epithelium with various degrees of atypia. Malignant squamous cells invade the submucosa, muscularis, or serosa (Fig. 322). The cells are irregular in size and shape, and increased numbers of mitotic figures are present. Squamous cell carcinomas may be classified into three types, i.e., well, moderately, and poorly differentiated (Figs. 323–325).

Well- or moderately differentiated squamous cell carcinomas are characterized by the following fea-

Fig. 319. Squamous cell carcinoma, fore-stomach, rat. Large polypoid tumors are seen

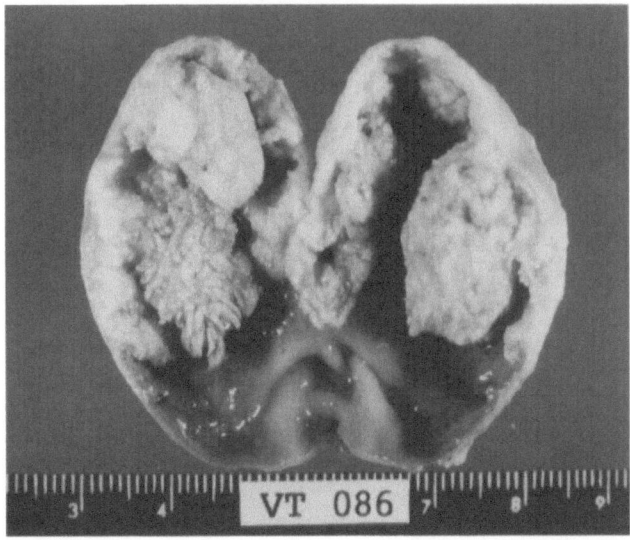

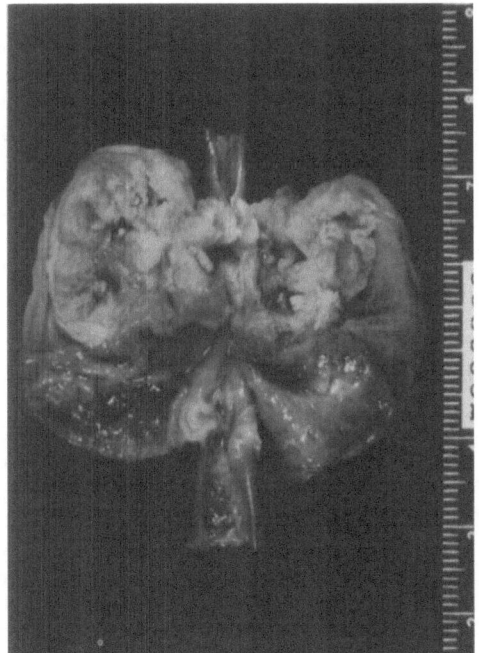

Fig. 320. Squamous cell carcinoma, forestomach, rat. The tumor is necrotic and ulcerated

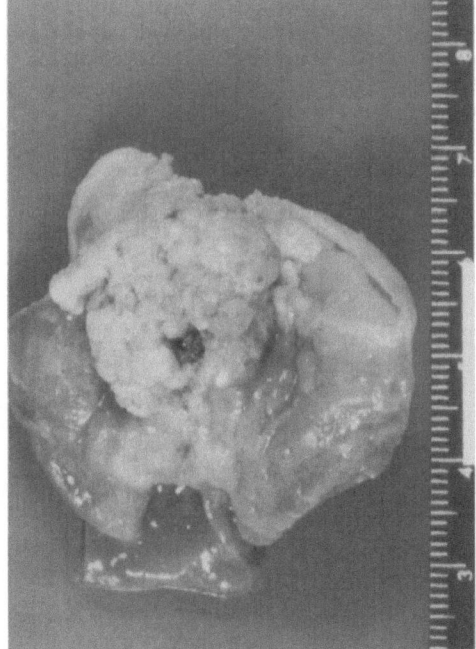

Fig. 321. Squamous cell carcinoma, forestomach, rat. The forestomach tumor has invaded the glandular stomach across the limiting ridge

tures: a distinct tendency for differentiation into squamous epithelium which resembles normal squamous epithelium; varying degrees of keratinization in the center; and keratin pearls (Figs. 323, 324). The malignant tissue consists mainly of cellular masses in which poly-gonal, round, or fusiform cells predominate. The well-differentiated carcinoma has extensive

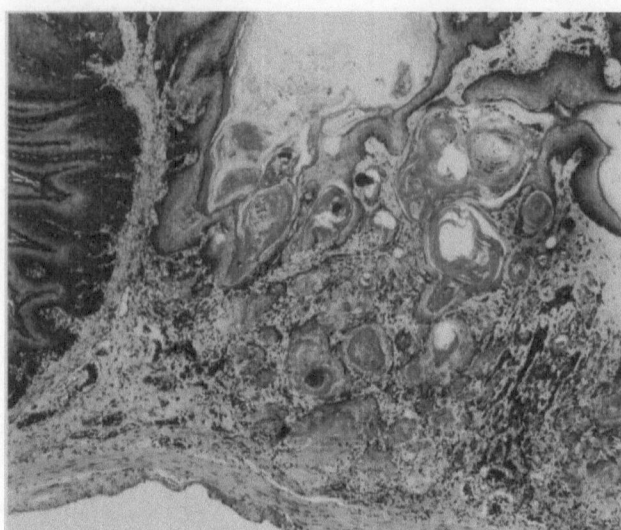

Fig. 322. Squamous cell carcinoma, forestomach, rat. Cancerous tissue is observed invading the submucosa and muscle layer. H&E, ×40

keratinization. The moderately differentiated type is moderately keratinized, and, even in those examples with evidence of decreased differentiation, the presence of keratin allows the tissue of origin to be determined easily.

Tumor cells in poorly differentiated squamous cell carcinomas are arranged in solid sheets separated by a small amount of stromal tissue or as strands or groups of cells intermingled with connective tissue (Fig. 325). In both cases, the typical keratinized cells of squamous epithelium are absent, but intracytoplasmic keratinization is often encountered. Basal cell carcinomas are included in this category and are recognized as endophytic carcinomas. Their superficial epithelial layers remain flat, but atypical basal cells infiltrate the lamina propria and may invade the submucosa or deeper layers.

Ultrastructure

Squamous cell carcinomas have variable features depending on the degree of differentiation. Ultrastructural studies are useful to diagnose poorly differentiated squamous cell carcinomas, since typical desmosomes between neighboring cells are frequently seen in the lesions. Bundles of electron-dense tonofilaments are characteristic, and keratohyalin granules are occasionally present in well-differentiated but not in poorly differentiated tumors. The intercellular spaces between adjacent tumor cells are usually widened.

Differential Diagnosis

Cellular pleomorphism, structural atypism, and, especially, invasion and the presence of distant metastases serve to distinguish carcinomas from hyperplasias and papillomas. However, when the squamous cell carcinoma is well differentiated, distinction from a papilloma is difficult. The presence of tumor tissue invading into the submucosa and through the muscularis mucosa is a useful finding, indicating squamous cell carcinoma.

Biologic Features

Spontaneous squamous cell carcinomas in the rat forestomach are rarely encountered even in aged animals (see Table 26 in the previous chapter; Burek 1978; Dontenwill et al. 1973; Goodman et al. 1979, 1980; MacKenzie and Garner 1973; Ward et al. 1979). However, carcinomas and papillomas can be induced in significant numbers by chemical carcinogens, including genotoxic compounds such as N-methyl-N-nitrosourea, N-butyl-N-nitrosourea, N-methyl-N-nitrosourethane, N-ethyl-N-nitrosourethane, N-propyl-N-nitrosourethane, N-butyl-N-nitrosourethane, N-methyl-N'-nitro-N-nitrosoguanidine, 4-nitroquinoline 1-oxide, 8-nitroquinoline (Takahashi et al. 1978), and the nongenotoxic antioxidants, such as butylated hydroxyanisole (Ito et al. 1983), 4-methoxyphenol (Asakawa et al. 1994), caffeic acid (Hagiwara et al. 1991), sesamol (Tamano et al. 1992), and 4-

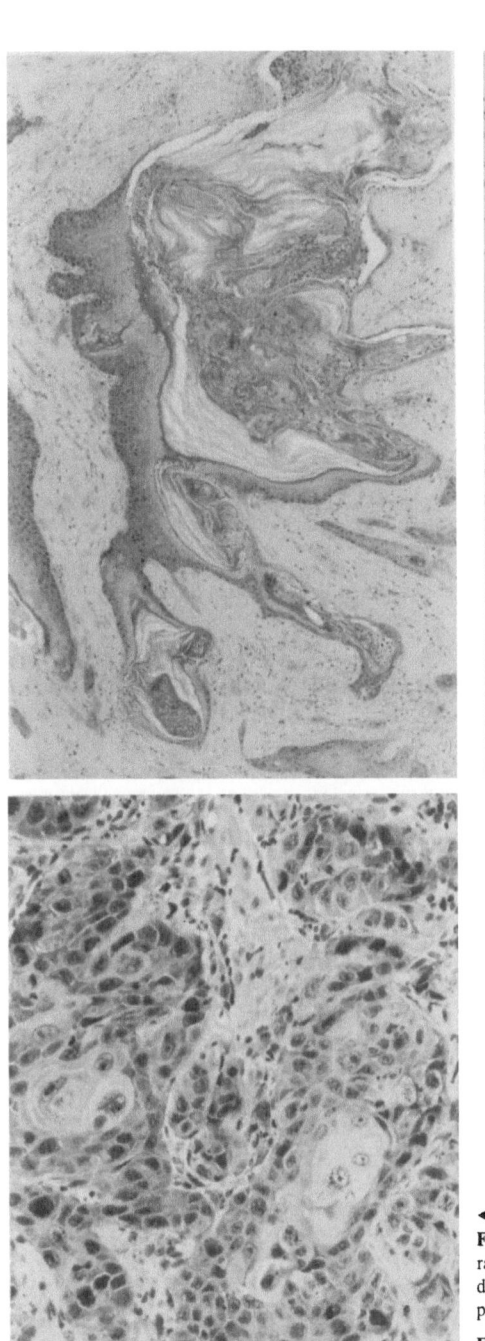

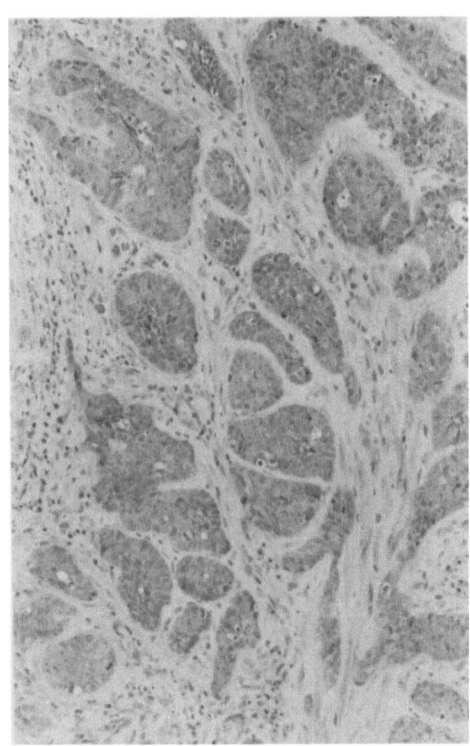

◄

Fig. 323. (*upper left*) Squamous cell carcinoma, forestomach, rat. A well-differentiated type. The tumor displays the typical differentiation of squamous epithelium and keratinization is present. H&E, ×40

Fig. 324. (*below*) Squamous cell carcinoma, forestomach, rat. A moderately differentiated type. Tumor tissue are characterized by moderate atypism and differentiation of epithelium with keratinization. H&E, ×200

Fig. 325. (*upper right*) Squamous cell carcinoma, forestomach, rat. A poorly differentiated lesion. The tumor cells are observed proliferating in small groups without squamous differentiation. H&E, ×100

methylcatechol (Asakawa et al. 1994). Papillomas were also induced in rats treated with sodium nitrite and ascorbic acid or 3-methoxycatechol in combination.

Comparison with Other Species

Spontaneously occurring papillomas and squamous cell carcinomas are found in the forestomach of Syrian golden hamsters more frequently than in rats or mice (see Table 27 in the previous chapter). Epithelial hyperplasia of the forestomach is also common in untreated hamsters, which seem to be sensitive to the composition of the diet. These factors should be borne in mind in experimental investigations of compounds potentially hazardous to humans

References

Asakawa E, Hirose M, Hagiwara A, Takahashi S, Ito N (1994) Carcinogenicity of 4-methoxyphenol and 4-methylcatechol in F344 rats. Int J Cancer 56:146–152

Burek JD (1978) Age-associated pathology. In: Burek JD (ed) Pathology of aging rats. CRC, Boca Raton, chap 4

Dontenwill W, Chevalier HJ, Harke BP, Lafrenz U, Rreckzeh G, Schneider B (1973) Spontaneous tumors in Syrian golden hamsters. Z Krebsforsch 80:127–248

Goodman DG, Ward JM, Squire RA, Chu KC, Linhart MS (1979) Neoplastic and nonneoplastic lesions in aging F344 rats. Toxicol Appl Pharmacol 48:237–248

Goodman DG, Ward JM, Squire RA, Paxton MB, Reichardt WD, Chu KC, Linhart MS (1980) Neoplastic and nonneoplastic lesions in aging Osborne-Mendel rats. Toxicol Appl Pharmacol 55:433–447

Hagiwara A, Hirose M, Takahashi S, Ogawa K, Shirai T, Ito N (1991) Forestomach and kidney carcinogenicity of caffeic acid in F344 rats and C57BL/6N × C3H/HeN F_1 mice. Cancer Res 51:5655–5660

Ito N, Fukushima S, Hagiwara A, Shibata M, Ogiso T (1983) Carcinogenicity of butylated hydroxyanisole in F344 rats. J Natl Cancer Inst 70:343–352

MacKenzie WF, Garner FM (1973) Comparison of neoplasms in six sources of rats. J Natl Cancer Inst 50:1243–1257

Takahashi M, Shirai T, Fukushima S, Hosoda K, Yoshida S, Ito N (1978) Carcinogenicity of 8-nitroquinoline in Sprague-Dawley rats. Cancer Lett 4:265–270

Tamano S, Hirose M, Tanaka H, Asakawa E, Ogawa K, Ito N (1992) Forestomach neoplasm induction in F344/DuCrj rats and B6C3F$_1$ mice exposed to sesamol. Jpn J Cancer Res 83:1279–1285

Ward JM, Goodman DG, Squire RA, Chu KC, Linhart MS (1979) Neoplastic and nonneoplastic lesions in aging (C57BL/6N × C3H/HeN)F$_1$ (B6C3F$_1$) mice. J Natl Cancer Inst 63:849–854

Adenoma, Glandular Stomach, Rat

Michihito Takahashi and Akiyoshi Nishikawa

Synonyms. Adenoma, adenomatous polyp, adenomatous diverticulum, adenomatous cyst

Gross Appearance

Adenomas are grossly observed as plaque-like or nodular lesions covered with glandular epithelium with a smooth and glossy or ulcerated surface. They are often seen as white or translucent nodules of various sizes. In some cases, the lesions appear as well-demarcated thickenings of the mucosa accompanied by polypoid structures on the surface, the thickening mainly being due to hyperplasia of foveolar epithelial cells. Adenomas are predominantly localized at the pyloric antrum, but they also occur in the fundic mucosa when treatment with mucosa-damaging agents such as iodoacetamide (Takahashi et al. 1976) is carried out prior to carcinogen treatment.

Microscopic Features

The majority of adenomas retain a well-organized glandular pattern and show only slight cellular atypia. Although the arrangement of the cells is not particularly irregular, multiple large or small cysts of various shapes, lined by a layer of columnar epithelial cells, are seen in the mucosa (Figs.

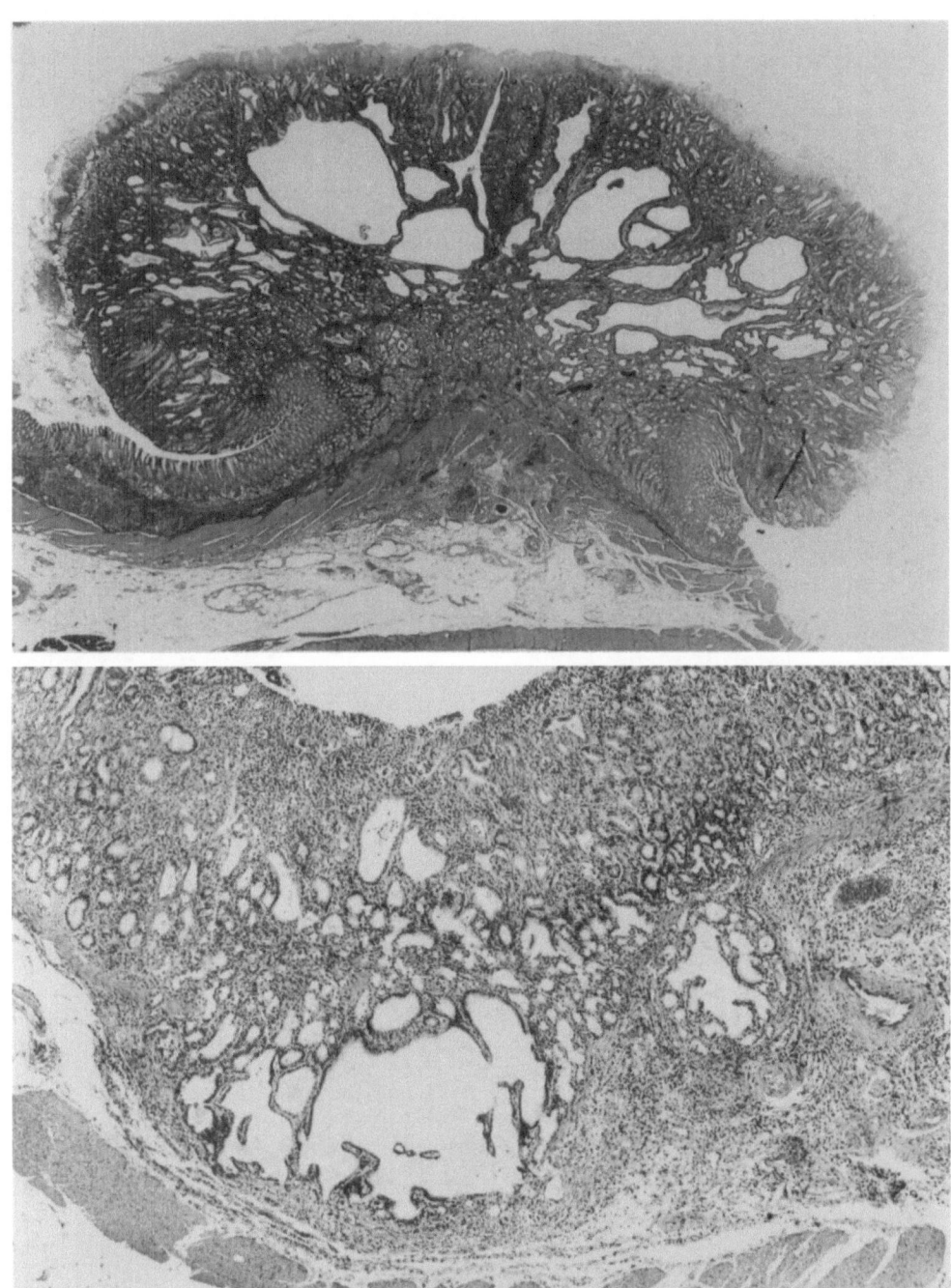

Fig. 326. (*above*) Adenoma, glandular stomach, rat, treated with *N*-methyl-*N'*-nitro-*N*-nitrosoguanidine (MNNG). Pedunculated polyp composed of well-differentiated hyperplastic glands. H&E, ×16

Fig. 327. (*below*) Adenoma, glandular stomach, rat, treated with MNNG. Erosive lesion composed of cystic changes of the proliferated glands. H&E, ×60

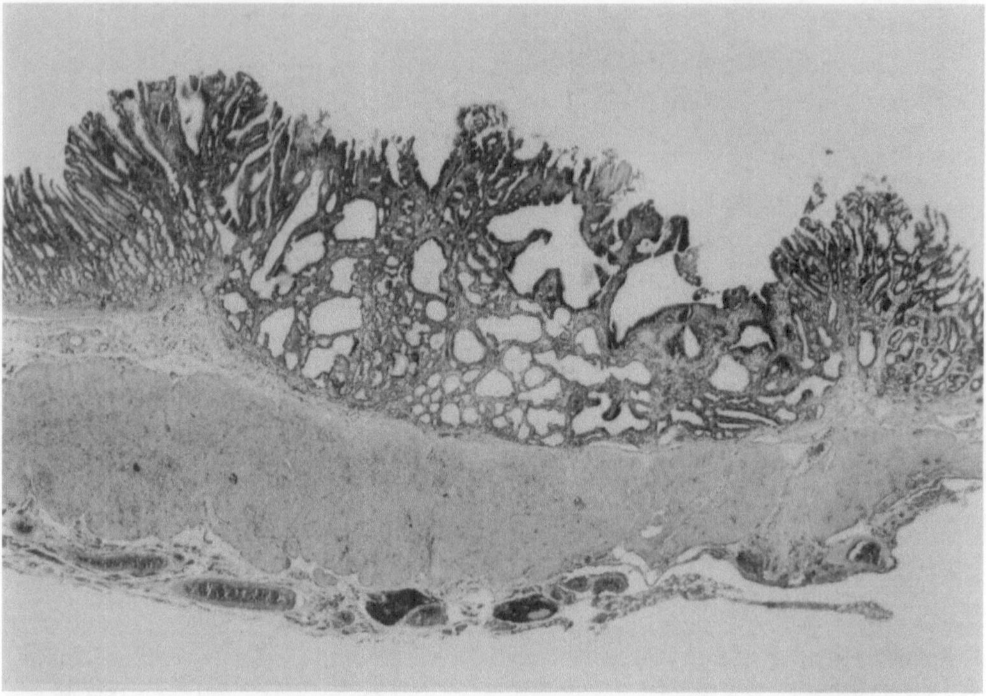

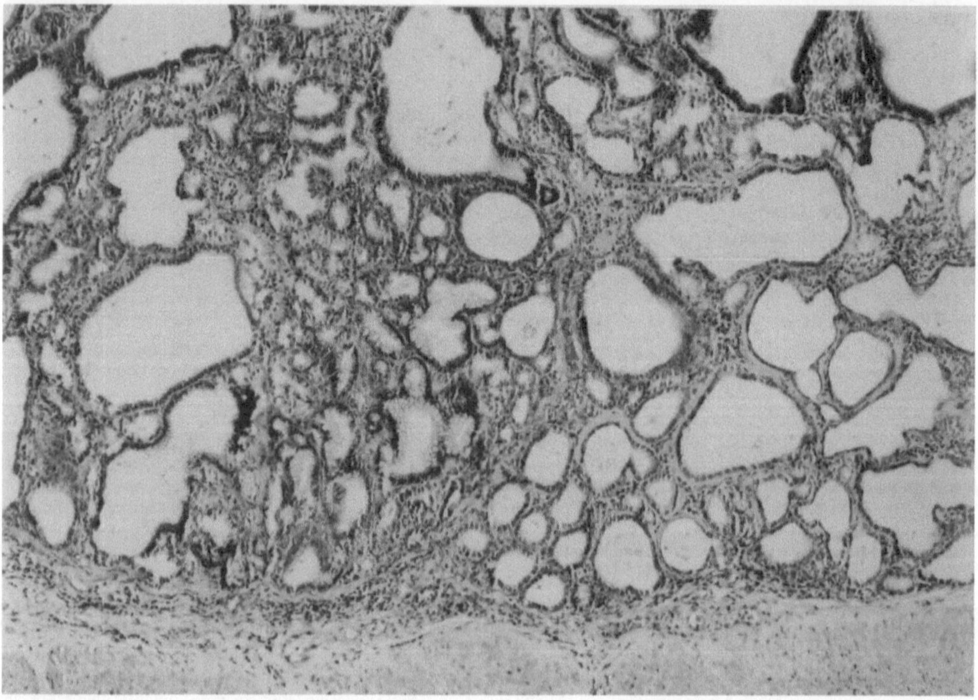

Fig. 328. (*above*) Adenoma, glandular stomach, rat, treated with MNNG. Umbilicated lesions composed of adenomatous tissue. (Courtesy of Dr. O. Kobori) H&E, ×30

Fig. 329. (*below*) Higher magnification of Fig. 328. The adenoma is composed of well-differentiated neoplastic glands with slight cellular or structural atypia. H&E, ×120

326–329). Infiltration into the stroma is not observed, and the boundary between the tumor and surrounding tissues is usually well defined. Adenomas are basically of two different patterns, adenomatous polyp or adenomatous diverticulum. Adenomatous polyps are mainly confined to the mucosa, while in adenomatous diverticula there is prominent downward growth into the stomach wall. The stroma of an adenomatous diverticulum contains fibrous connective tissue and sometimes osteoid tissue.

Glandular cells observed in adenomatous polyps are mainly hyperchromatic, demonstrating cuboidal and/or columnar morphology. The cells are arranged to form acini and branching tubules of various sizes and shapes, independent of whether the lesions are sessile or pedunculated, forming tall polyps (Figs. 326–328). At the bases of these polypoid or nodular lesions, downgrowths of neoplastic cells may form acini, branching tubules, or cysts. The acini, tubules, and cysts are usually lined by a single layer of neoplastic cells, and neither invasion of the lymph or blood vessels nor distant metastasis occurs.

It is sometimes difficult to distinguish between adenoma and well-differentiated tubular or papillary adenocarcinoma. Carcinomas are characterized by atypically branched glands infiltrating the tunica submucosa and the tunica muscularis propria. The glands are lined by cells showing distinctive signs of malignancy, such as hyperchromasia, irregular shape of the nucleus, increase in the nucleus to cytoplasm ratio, and increased mitotic figures.

Ultrastructure

Nuclei of the adenoma cells are irregularly rounded and contain prominent nucleoli and peripherally condensed chromatin. The lateral plasmalemma is highly interdigitated, with desmosomes occasionally observed. The basal plasmalemma is straight and is seen together with an unbroken basement membrane. Stubby microvilli project from the apical surface and some cells demonstrate blebbing into the lumen; these blebs totally lack cytoplasmic organelles

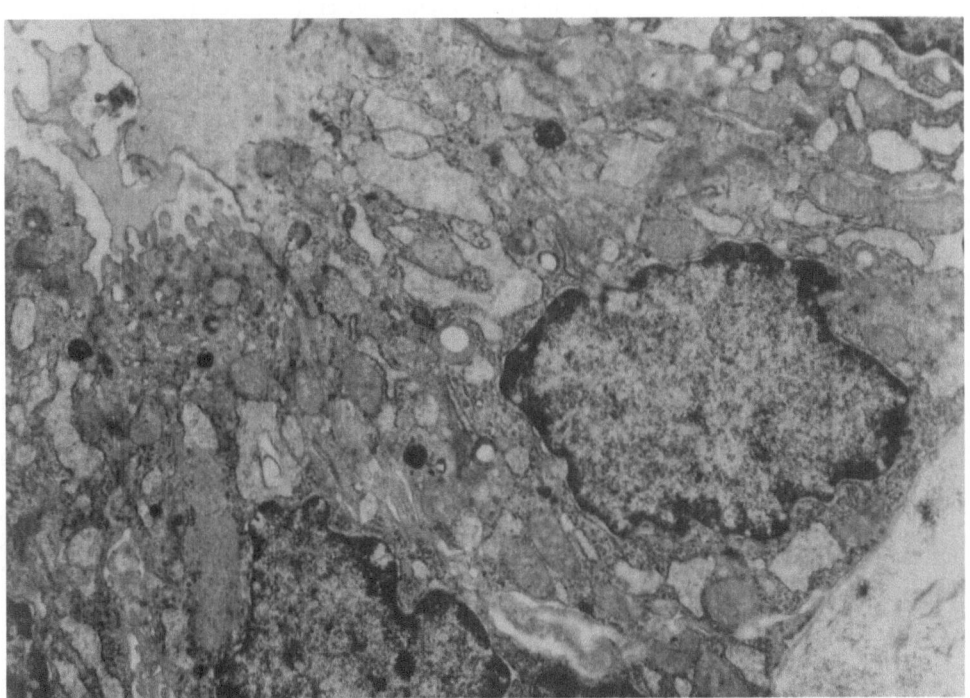

Fig. 330. Neoplastic cells of a gastric adenoma in a rat treated with MNNG. (Courtesy of Dr. O. Kobori.) TEM, ×7600

(Fig. 330). The moderately developed endoplasmic reticulum is predominately of granular type with occasional pronounced dilatation of the cisternae. Numerous free ribosomes and polysomes are distributed throughout the cytoplasm. Components of the Golgi apparatus are well developed in the perinuclear region. Mitochondria are rather few in number, and the cells contain only a few, poorly developed secretory granules, which are usually irregularly scattered. Comparison of adenoma cells with regenerative epithelial cells shows almost complete identity, and characteristically the cells show high differentiation in both papillary and tubular lesions; they frequently cannot be distinguished from the corresponding cells found in normal pyloric mucosa. Goblet cells are only occasionally observed in adenoma tissue.

Differential Diagnosis

Adenomas usually occupy only a relatively small area of pyloric mucosa and involve a small number of pits. They retain a well-organized glandular pattern and show only slight cellular atypia, and no infiltrative growth can be observed. The fact that these features are also characteristic of regenerative hyperplasia presents difficulties in differential diagnosis. Hyperplasias of foveolar epithelium and mucous cells are common in regenerative mucosa of healed ulceration, and it is sometimes difficult, because of transitional features and similarities, to distinguish between hyperplastic and neoplastic changes. However, the distribution of cell types may be helpful in these cases, since the normal separation of foveolar and mucous cells is lost in adenomas. The progression of lesions induced by N-methyl-N'-nitro-N-nitrosoguanidine (MNNG) from adenomas to adenocarcinomas has been well characterized (Kobori et al. 1976, 1977). Even in the early experimental period, mildly atypical adenomatous cells penetrating the muscularis mucosa gave evidence of invasive character. The portion of a tumor actively penetrating the muscularis mucosa has a higher proliferation rate than the tumor portions which are not invading and, in addition, the mucosa may present other features of malignancy. Thus tumors actively penetrating the muscularis mucosae should be diagnosed as adenocarcinomas on the basis of their biologic behavior. On the other hand, although some adenocarcinomas contain irregular glandu-

lar structures lined by atypical, darkly stained, or mucin-producing columnar cells with enlarged nuclei, most cases are relatively well differentiated; therefore, on the grounds of morphological appearance alone, it is sometimes difficult to make a clear diagnosis of malignancy. The degree of invasiveness of adenocarcinomas varies from submucosal penetration to involvement of the deeper layers of the stomach wall.

Pyloric metaplasias are focal abnormalities of the fundic mucosa and represent a type of regenerative lesion observed after carcinogen-induced damage to the fundic mucosa. For example, they are present in the stomach of many MNNG-treated rats (Nishikawa et al. 1992).

Biologic Features

Adenomas can arise spontaneously in any part of the glandular stomach, showing, however, a marked preference for the pyloric areas in some strains of mice. In contrast, spontaneous gastric tumors in rats are rare. Experimentally, adenomas are found at the site of injection of carcinogenic hydrocarbons or in animals given MNNG in drinking water (Nagayo 1973; Sugimura and Kawachi 1978).

When the neoplastic growth is predominantly polypoid in structure, the lesion is referred to as an adenomatous polyp; when cystic downgrowths are apparent, the lesion is referred to as an adenomatous cyst or diverticulum. Proliferative, noninvasive mucosal lesions located predomantly in the pyloric region have been known as either adenomatous hyperplasia or adenoma, but, since these lesions are considered to be the earliest manifestation of carcinogen-induced growth abnormalities in the stomach, the term preneoplastic hyperplasia seems more appropriate. The role of adenomatous hyperplasia in the development of epithelial tumors is, however, difficult to define. Adenomas and adenomatous polyps have benign neoplastic characteristics, and most of them appear to arise from adenomatous hyperplastic lesions. Sometimes, however, the neoplastic cells arise directly from normal cells without a preceding hyperplastic stage. It seems, therefore, that adenomatous hyperplasia, adenomas, adenomatous polyps, and adenocarcinomas may be independent lesions in some cases, whereas in others they are histogenetic stages of a neoplastic process. Concerning the relationship between adenomatous change and carcinoma, the lesions limited to mucosal spread or submucosal invasion can be divided on morphological grounds into the two groups, adenomatous changes and early carcinoma. Moreover, nearly all invasive carcinomas observed in experimental animals show a close positional and chronological relationship to adenomatous lesions. Both are identical with regard to their cytologic properties as revealed by electron microscopy and cytochemistry. It therefore seems appropriate that adenomatous changes induced by MNNG should be hypothesized to be precancerous lesions (Takahashi et al. 1994).

Comparison with Other Species

Spontaneous gastric polyps and adenomas are sometimes present in certain strains of mice and in MNNG-treated dogs and hamsters, but a clear relationship to carcinoma has not been established in any of these species.

References

Kobori O, Gedigk P, Totovic V (1976) Early changes of glandular stomach in Wistar rats ingesting N-methyl-N'-nitro-N-nitrosoguanidine (MNNG): with special reference to light microscopic, electron microscopic, and enzyme histochemical study of the regenerating epithelium induced by MNNG. Z Krebsforsch 87:127–138

Kobori O, Gedigk P, Totovic V (1977) Adenomatous changes and adenocarcinoma of glandular stomach in Wistar rats induced by N-methyl-N'-nitro-N-nitrosoguanidine. An electron microscopic and histochemical study. Virchows Arch (Pathol Anat) 373:37–54

Nagayo T (1973) Tumours of the stomach. In: Turusov VS (ed) Pathology of tumours in laboratory animals, vol I, tumours of the rat, part 1. IARC Sci publ no 5, Lyon, pp 101–118

Nishikawa A, Furukawa F, Mitsui M, Enami T, Kawanishi T, Hasegawa T, Takahashi M (1992) Inhibitory effect of calcium chloride on gastric Carcinogenesis in rats after treatment with N-methyl-N'-nitro-N-nitrosoguanidine and sodium chloride. Carcinogenesis 13:1155–1158

Sugimura T, Kawachi T (1978) Experimental stomach carcinogenesis. In: Lipkin M, Good RA (eds) Gastrointestinal tract cancer. Plenum, New York, chap 13, pp 327–341

Takahashi M, Shirai T, Fukushima S, Hananouchi M, Hirose M, Ito N (1976) Effect of fundic ulcers induced by iodoacetamide on development of gastric tumors in rats treated with N-methyl-N'-nitro-N-nitrosoguanidine. Gann 67:47–54

Takahashi M, Nishikawa A, Furukawa F, Enami T, Hasegawa T, Hayashi Y (1994) Dose-dependent promoting effects of sodium chloride (NaCl) on rat glandular stomach carcinogenesis initiated with N-methyl-N'-nitro-N-nitrosoguanidine. Carcinogenesis 15:1429–1432

Adenocarcinoma, Glandular Stomach, Rat

Zoltán Szentirmay and János Sugar

Synonyms. Tubular adenocarcinoma, adenotubular carcinoma, mucoid-cystic adenocarcinoma, muconodular adenocarcinoma, mucocellular adenocarcinoma, signet-ring cell adenocarcinoma, mucinous adenocarcinoma

Gross Appearance

Most cases of experimentally induced adenocarcinoma arise in the antral region, in the vicinity of the pylorus of the glandular stomach. In early cases, the tumors appear as round or oval, slightly elevated solitary or multiple lesions 2–4 mm in diameter (Fig. 331). Some cases of intramucosal carcinoma with erosion, located in the level of the mucous membrane, have also been observed. The mean diameter of advanced carcinomas is 5–6 mm (Ohgaki et al. 1983), but tumors 20–30 mm in diameter have also been found. Macroscopically, three different types can be distinguished: umbilicated or ulcerated tumors with irregular elevated margins; flowerbed-like lesions with a central area of depression; hemispherical or polypous protruded lesions (Saito et al. 1970). Large tumors sometimes circularly infiltrate the pylorus, penetrate the gastric wall, spread into the serosa, and attach to the adjacent tissue.

Induced rat gastric carcinoma characterized by linitis plastica-like diffuse thickening of the gastric wall has been reported extremely rarely (Takahashi et al. 1973).

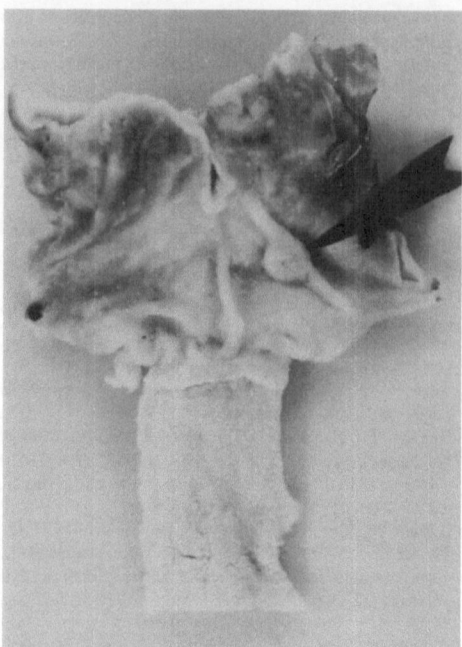

Fig. 331. Adenocarcinoma, protruded type (*arrowhead*) in the glandular stomach of a 283-day-old male WOP rat treated with MNNG. ×4

Table 28. Histologic classification of gastric cancers

Well-differentiated types
  Tubular or papillary
  Mucoid-cystic
  Signet ring cell
    Muconodular
    Mucocellular
  Adenoacanthoma
Poorly differentiated types
  Trabecular
  Scirrhous and/or diffuse
  Mixed forms (signet ring and trabecular)

Microscopic Features

Adenocarcinomas of the glandular stomach can be divided into well and poorly differentiated types (Table 28). In experimentally induced tumors, the well-differentiated type, which has a well-preserved glandular structure with a branching tubular, papillary, or cystic pattern (Fig. 332), is the most frequent. Epithelial cells lining the glands are of two different types; there may be one or more layers of cuboidal or short columnar epithelial cells of varying size and polarity. In this type, the nucleus is round or oval, sometimes dense, hyperchromic, and atypical; in other cases, it is of the vesicular type.

Long columnar cells with slight atypia and pronounced mucus production, accompanied by structural atypia of the glands and by infiltration of the submucosa, the muscular layer, or the serosa, are characteristics of another type of tumor called "mucoid cystic adenocarcinoma." This tumor is sometimes difficult to distinguish from an adenoma. In carcinomas, the irregular glands are lined with markedly atypical cells or with differentiated, slightly atypical columnar epithelial cells closely packed together, which appear in variable numbers in the light microscopy field (Fig. 333).

Tumors with intensive mucus production can form smaller or larger extracellular mucous pools (muconodular carcinoma); the tumor cells become detached from the glands and appear in the mucous pool. The also exhibit abortive glandular formation, bundles, and solid groups (Fig. 334). Signet-ring cell carcinoma infiltrating in single form without extracellular mucus production is rare (Fig. 335).

Squamous cell metaplasia (adenocanthoma) is sometimes encountered in the carcinomatous gland (Fig. 336).

In the stroma of well-differentiated adenocarcinomas, there is pronounced chronic inflammatory infiltration and connective tissue accumulation with hyalinization and myxoid change. Neoplastic cartilage and bone formations may also occur (Fig. 337).

Poorly differentiated adenocarcinomas are characterized by the loss of glandular structure. Tumor cells are highly variable and polymorphic, forming irregularly arborating bundles which may in some focal areas appear as signet-ring cells (Fig. 338).

The scirrhous form, in which the infiltrating single tumor cells are mixed with inflammatory connective tissue stroma, is rare.

Ultrastructure

In well-differentiated adenocarcinoma, the glands are lined by columnar or cuboidal cells with reduced or no polarity. The nuclei are of varying shape and size; irregular, clumped chromatin associated with the nuclear membrane is frequently observed, and nucleoli are significantly enlarged.

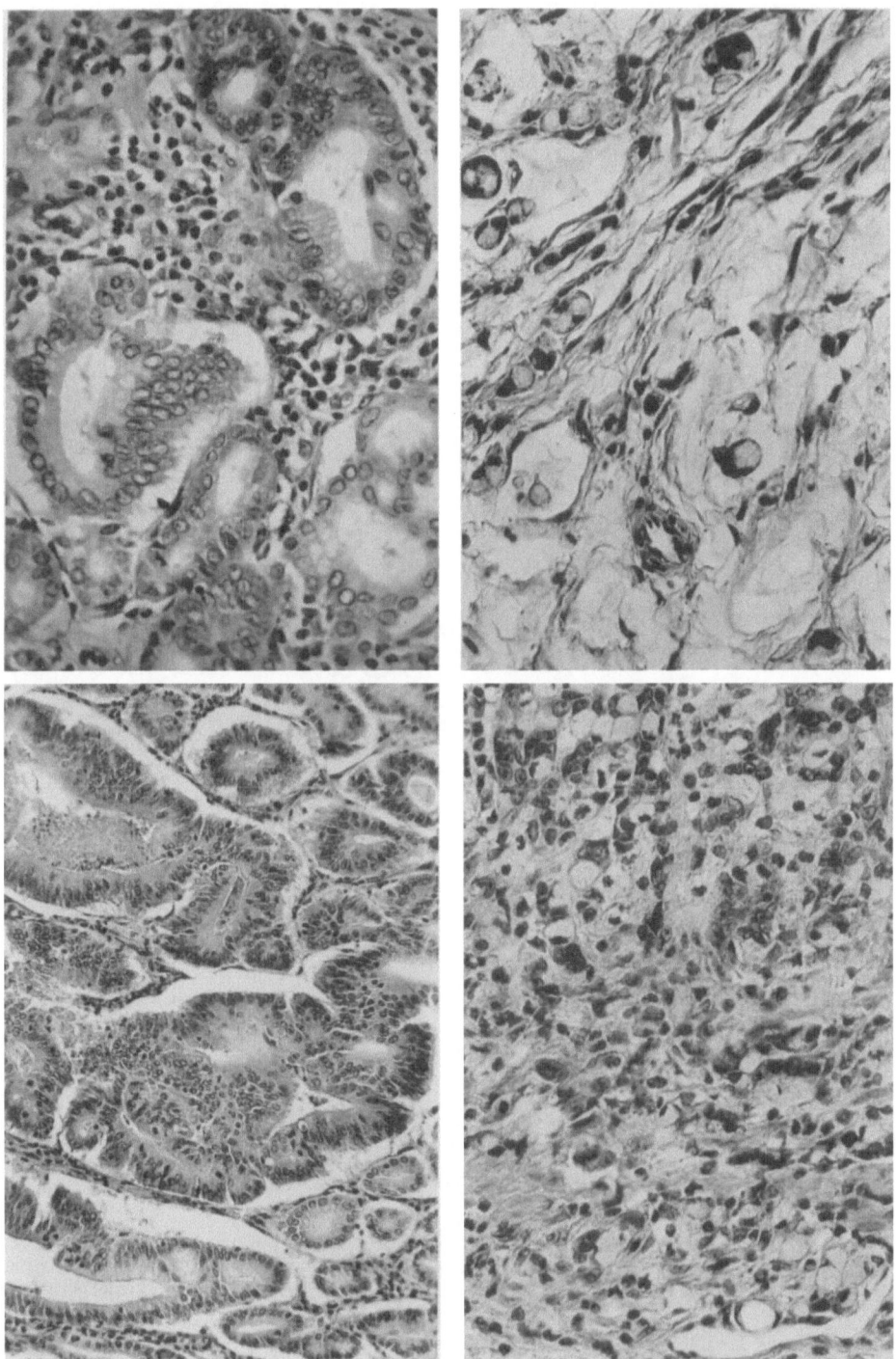

Fig. 332. (*upper left*) Well-differentiated tubular adenocarcinoma with vesicular nuclei. Glandular stomach of a 546-day-old male WOP rat treated with MNNG and SLS. H&E, ×350

Fig. 333. (*lower left*) Tubular adenocarcinoma with well and moderately differentiated glandular pattern. Glandular stomach, 796-day-old male WOP rat treated with MNNG. H&E, ×350

Fig. 334. (*upper right*) Mucus-producing adenocarcinoma (muconodular carcinoma) in the glandular stomach of a 790-day-old male WOP rat treated with MNNG. The tumor cells form small bundles and groups. H&E, ×350

Fig. 335. (*lower right*) Diffusely infiltrating signet-ring cell carcinoma in the glandular stomach of a 795-day-old male WOP rat treated with MNNG. H&E, ×350

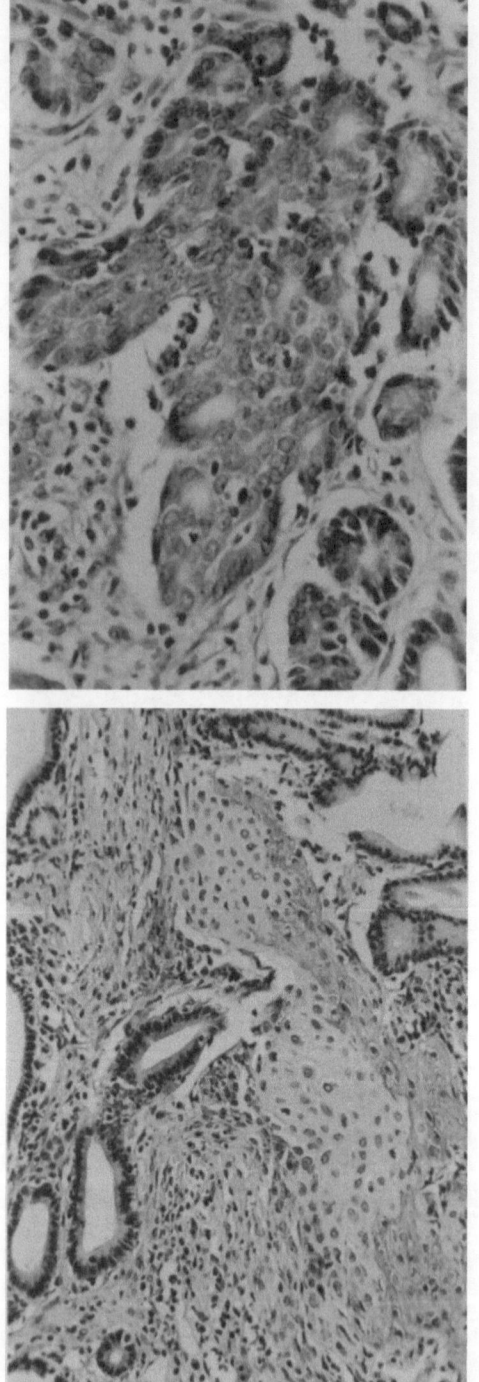

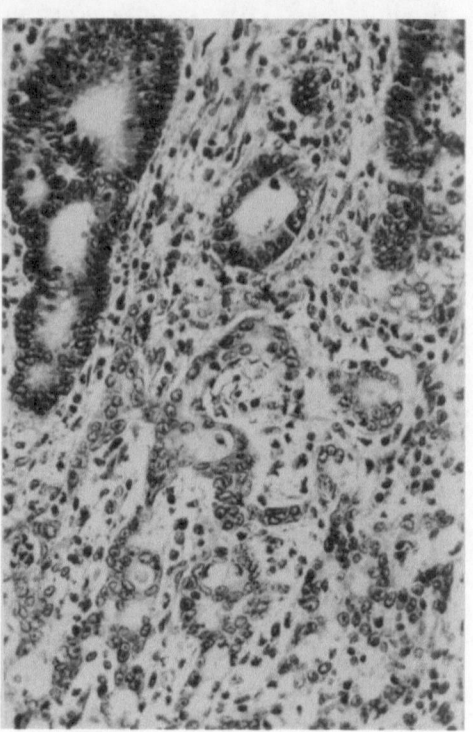

◄
Fig. 336. (*upper left*) Well-differentiated adenocarcinoma with squamous metaplasia (adenoacanthoma). Glandular stomach of a 595-day-old male WOP rat treated with MNNG and Triton X-100. H&E, ×350

Fig. 337. (*below*) Well-differentiated tubular adenocarcinoma with metaplastic cartilage and bone formation within the tumor. Glandular stomach, 595-day-old male WOP rat treated with MNNG. H&E, ×175

Fig. 338. (*upper right*) Poorly differentiated trabecular adenocarcinoma in the glandular stomach of a 780-day-old male WOP rat treated with MNNG and Iriton X-100. H&E, ×250

The atypical cell surface is covered by shorter or longer microvilli of irregular shape and without a central core. Enormous bubble formations are often present on the cell surface facing the glandular lumen. These are formed by a cytoplasmic bulge into the lumen and usually contain free ribosomes and, sometimes, a few secretion granules. Other cells have longer brush border-like microvilli with a central core consisting of microfilaments that lead into the apical cytoplasmic region. These cells are somewhat reminiscent of the intestinal epithelium. The number of mucin granules in the tumor cells is less than those in intact cells, and they are usually found in the apical region of the cell as small, homogeneous formations with reduced electron density. The same type of secretory granules can be found in intact mucous cells of the gastric pit. In other cells, the mucous granules contain electron-dense cores that do not stain by the periodic acid-silver methanamine (PASM) technique. These cells resemble mucus-producing surface cells, but they may also appear farther away from the surface. Mucus-producing cuboid cells, the cytoplasm of which contains abundant amounts of closely packed, often confusing secretion granules of low electron density similar to goblet cells, are also present. In the tumor cells, the usually centrally located Golgi apparatus and mitochondrai are preserved, while the rough surface ergastoplasmic reticulum is scant. Lysosomal heterogeneous dense bodies are present in the cytoplasm, and there are well-developed junctional complexes (i.e., tight junction, desmosome) between adjacent tumor cells.

The poorly differentiated adenocarcinoma is made up of three types of cells: mucin-containing cells, intracellular microcyst cells, and poorly differentiated mucous cells (Hananouchi et al. 1974). At the electron microscopy level, the so-called mucus-containing cells can be divided into two subtypes: those resembling gastric pit mucus cells and goblet cells. The superficial microvilli are poorly formed and irregular, and the number of cellular junctions is reduced. Intracellular microcyst cells are frequently found in those areas of the tumor that display a muconodular structure by light microscopy.

Electron microscopy examination reveals sporadic microvilli on the cellular surface and, in the cytoplasm, smaller or larger cysts with microvilli on their internal surface. These microvilli usually possess a central core with filamentous structure. The cysts contain homogeneous and finely granular PASM-positive material and several myelin-like components. Larger cysts displace the nucleus to one side, giving rise to a signet-ring cell formation. In the cytoplasm, there are mitochondria, endoplasmic reticulum, and abundant microfilaments in the vicinity of the cysts. The poorly differentiated mucous cells have large, irregular nuclei with coarse chromatin. Mitotic figures are frequent. On the surface, there is a small number of poorly developed microvilli.

Mucin granules are scanty in number and are located in the apical cell region.

Cell organelles are reduced in number. The cytoplasm contains free ribosomes and tonofibrillar structures. In some of the tumor cells, intracellular canalicular structures are also observed.

Differential Diagnosis

Well-differentiated adenocarcinomas must be distinguished from adenomas. Macroscopically, adenoma is a polypoid or diverticulum-like, sharply circumscribed alteration. The adenomatous diverticulum, which communicated with the gastric lumen, consists of smaller or larger submucous nodules and the gastric wall is thickened at its site. The tumor displays more or less dilated regular glandulae arranged in a honeycomb pattern. Cells are completely differentiated or slightly atypical. Cystic lumina contain mucus, inflammatory exudate, or necrotic debris. The tumor is frequently isolated from its environment by fibrous inflammatory granular tissue and never grows beyond the stomach.

An adenocarcinoma may also display an expansive growth pattern, but is not accompanied by inflammatory granular tissue in its vicinity. A reliable diagnosis of carcinoma is based on cellular and structural atypia. If cellular atypia is mild, the tumor can be considered carcinoma, provided that the tumorous infiltration has reached the serosa (Nagayo 1973).

Biologic Features

Spontaneous Tumors

Spontaneous adenocarcinoma of the glandular stomach is an extremely rare finding in the rat (Rowlatt 1967). In the course of a rat control program, McCoy (1909) examined 100000 rats and found spontaneous tumors in 103 cases; however, no gastrointestinal tumors were among them. Wolley and Wherry (1911) found 22 spontaneous tumors in 23 000 wild rats (*M. norvegicus*), none of which were gastric carcinoma. Only one gastric adenocarcinoma (with no metastasis) among 33 000 laboratory rats of different ages was found by Bullock and Curtis (1930). Ratcliffe (1940) studied two strains of Wistar rats and found no evidence of gastric carcinoma among the 273 spontaneous tumors observed. Among 184 spontaneous tumors of Osborne-Mendel rats, Saxton et al. (1948) observed a single squamous cell carcinoma in the forestomach. Snell (1966), having examined 488 rats belonging to five inbred strains, observed no gastric carcinoma and only a single case of gastric sarcoma. Prejean et al. (1973) examined 360 male and female Sprague-Dawley rats 18 months of age and could not detect any gastric adenocarcinoma; only a single squamous cell carcinoma was found in the forestomach, although benign and malignant tumors occurred at other sites in 45% of the animals. MacKenzie and Garner (1973) studied a total of 2082 2-year-old rats from six different laboratories; the group contained equal numbers of males and females. No gastric tumors occurred in 258 Sprague-Dawley,

131 Osborne-Mendel, and 676 Oregon rats; one gastric adenocarcinoma was found among 217 SD (Diablo) rats, one squamous cell carcinoma among 535 SD (Charles River) rats, and one gastric fibrosarcoma among 268 SD (Holtzman) rats. In our institute, we have found 47 spontaneous tumors in 164 Wistar rats (H-Riop outbred males and females 24 months of age) and 31 spontaneous tumors in 95 WOP rats (H-Riop inbred males and females of similar age). No gastric adenocarcinoma was found among these tumors.

Induction of Gastric Cancer

Successful experimental induction of gastric carcinoma was accomplished first with aromatic and heterocyclic hydrocarbons and later with aromatic nitrogen compounds (Table 29). In addition to several other tumors, a low percentage of adenocarcinomas was produced by 20-methylcholanthrene (20MC) or 7,12-dimethyl-benz[a]anthracene (DMBA) when applied directly to the stomach (Hare et al. 1952; Grant 1966). These experiments suggested that the glandular stomach is more resistant to carcinogens than the forestomach. Morris et al. (1962) were

Table 29. Incidence of adenocarcinoma of glandular stomach in rat, induced by different carcinogenic agents

Strain	Sex	Types and administration of carcinogens (fotal dose/mg)	Duration of experiments (days)	Effective no. of rats[a]	Incidence of adeno-carcinoma		Reference
					(*n*)	(%)	
Osborne-Mendel Marshall-500 AxC	M/F M/F M/F	20 MC, injected in gastric wall	More than 730	265	8	3	Hare et al. 1952
Sprague-Dowley	?	DMBA, implanted in gastric wall	280	13	3	23	Grant 1966
Buffalo	M	2,7-FAA, oral (583)	317	11	2	18	Morris et al. 1962
	F	2,7-FAA, oral (412)	305	12	1	8	
	M	2,7-FAA, intraperitoneal (511)	250	11	1	9	
	F	2,7-FAA, intraperitoneal (512)	262	8	2	25	
Buffalo	F	4HAQO, instilled into the stomach (30)	610	28	7	25	Mori et al. 1969

20MC, 20-methylcholanthrene; DMBA, 7,12-dimethyl-benz(a)anthracene; 2,7-FAA, 2,7-bis (acetylamino)fluorenylene; 4HAQO, 4-hydroxyaminoquinoline 1-oxide.
[a] The effective number of rats is defined as the number surviving when the first gastric tumor was found in a rat.

Table 30. Incidence of adenocarcinoam of glandular stomach in rat, induced by N-methyl-N'-nitro-N-nitrosoguanidine

Strain	Sex	MNNG + modifying agents in drinking water (mg/l)	Duration of experiments (days)	Effective no. of rats[a]	Incidence of adeno-carcinoma		Reference
					(n)	(%)	
Wistar	M	83167	471	47	21	45	Fujimura et al. 1970
	M/F	5000 intermittent intubation	150	35	7	25	Tabuchi et al. 1974
Wistar	M	83	520	34	21	62	Matsukura et al. 1978
CD CRJ (SD)	?	50	450	16	4	25	Fujii et al. 1980
BDIX	M	83	290–359		6	8	Martin et al. 1974
	F	83	439		2	3	
BN	M	83	325		23	40	
	F	83	336		5	8	
Lewis	M	83	352		8	8	
	F	83	434		5	6	
(Lewis × BN)F$_1$	M	83	342		6	6	
	F	83	295		13	13	
Wistar	M	83	422		6	6	
	F	83	431		8	9	
ACI	M	83	504	15	12	80	Ohgaki et al. 1983
	F	83	504	19	9	47	
Buffalo	M	83	504	17	3	18	
	F	83	504	17	0	0	
(ACI × Buffalo)F$_1$	M	83	504	76	13	17	
	F	83	504	84	7	8	
(ACI × Buffalo)F$_2$	M	83	504	53	19	36	
	F	83	504	59	8	12	
Wistar	M	50	365	32	26	81	Furukawa et al. 1982
	F	50	365	20	0	0	
Castrated	M	50	365	21	6	29	
	F	50	365	20	1	5	
Wistar	M	50 + Surfactants[b]	560	85	62	73	Fukushima et al. 1974
WOP (H-Riop/inbred)	M	30, 100	796	27	12	44	Szentirmay and Sugár (unpublished)
	M	80 + Surfactants[c]	796	22	16	73	
	F	80 + Surfactants[c]	796	21	7	38	
Noninbred albino	M/F	250	540	20	4	20	Chang et al. 1983
	M/F	250 + Aspirin	540	40	21	53	
Wistar	M	100, 150	280	40	6	15	Takahashi et al. 1983
	M	100 + NaCl	280	39	31	79	
	M	NaCl after MNNG 100	280	19	4	21	

MNNG, N-methyl-N'-nitro-N-nitrosoguanine.
[a] The effective number of rats is defined as those surviving when the first gastric tumor was found in a rat.
[b] Tween 20, Tween 40, Tween 60, Tween 80, Span 20 + ethanol, glyceryl monostearate, sucrose monopalmitate, sodium lauryl sulphate, sodium N-lauryl sarcosinate.
[c] Triton X-100, sodium lauryl sulphate.

able to induce gastric carcinoma by the oral or intraperitoneal administration of 2,7-bis(acetylamino)-fluorenylene (2,7-FAA). Mori et al. (1969) induced gastric adenocarcinoma with a relatively high frequency by the intragastric administration of an alcoholic solution of 4-hydroxyaminoquinoline 1-oxide (4HAQO). Results of studies by Butler and Barnes (1966) are also worthy of attention. These investigators observed not only tumors in other sites, but also gastric adenocarcinomas in rats maintained on an aflatoxin-containing diet. However, the most effective stomach carcinogens have proved to be the N-nitroso compounds. Sugimura and Fujimura (1967) induced adenocarcinoma in Wistar rats with N-methyl-N'-nitro-N-nitroso-guanidine (MNNG) dissolved in the drinking water. This experiment has since been repeated by many investigators in the succeeding years, and the effects of the different modifying factors have been subjected to detailed analysis (Table 30).

Modifying Factors

There are several substances that increase the incidence of the induced gastric cancer when administered together with MNNG. Some of them, such as the different surfactants, croton oil, or aspirin, are true promoting agents that induce slight proliferating processes by themselves without causing carcinoma. Experiments performed with such agents have also unequivocally proved the two-step mechanism of carcinogenesis in the formation of gastric tumors (Sugimura and Kawachi 1978). Takahashi et al. (1983) stated that a high-salt diet together with MNNG application also increases the incidence of gastric cancer, while the same diet after MNNG treatment has no significant effect. The promoter effect of a high-salt diet is explained by the damage to gastric mucosa, and sodium chloride is not considered a true promoting agent. Inflammatory and regenerative processes following mucosal damage may thus play some role in tumor formation. This is in accordance with the observation by Dahm and Werner (1973) that, in the rat stomach after Billroth's resection II, MNNG treatment induces carcinoma with high incidence at the anastomotic site.
Tahara and Haizuka (1975) reported that gastrin treatment (20 intraperitoneal injections of 50 mg/kg at 3-day intervals) increases the incidence of MNNG-induced scirrhous gastric cancers.

Serotonin, histamine, glucagon, and insulin, given under similar conditions, were ineffective. The cause of the peculiar effect of gastrin has hitherto not been elucidated.

Strain Specificity

Different strains display varying sensitivity to the gastric cancer-inducing effect of MNNG; the ACI strain is relatively sensitive, and the Buffalo strain relatively resistant (Bralow et al. 1973; Ohgaki et al. 1983). These properties are genetically determined. Relative resistance of the Buffalo strain to MNNG is explained partly by DNA repair and partly by the presence of different protective mechanisms, i.e., more acidic gastric content, a more active enzyme system, and/or more abundant mucus, which protects gastric mucosa from the carcinogenic effect. The incidence of MNNG-induced gastric cancer is often identical or even higher in the inbred than in outbred strains (Martin et al. 1974). The advantage in using inbred strains is that the developed carcinoma is more readily transplantable and immunologic tests are easier to perform.

Natural History

Well-differentiated adenocarcinomas do not give rise to metastases to the lymph nodes and do not grow onto adjacent tissues. With nondifferentiated adenocarcinomas, invasion of blood and lymph vessels and metastases into regional lymph nodes or other organs may take place (Saito et al. 1970; Takahashi et al. 1973).
Kokubo et al. (1981) succeeded in transplanting gastric adenocarcinoma induced by MNNG into newborn outbred Wistar rats. The tumor preserved its glandular structure, ultrastructural properties, and capacity for enzyme production, even during 60 passages over a 4-year period. Well-differentiated adenocarcinoma proliferates slowly. Mathematic analysis of the DNA histogram of tumor cells indicates an approximately 10% average rate of S-phase cells. With regard to DNA ploidy, both hypo- and hyperdiploid aneuploid tumors and even carcinomas consisting of a mixed euploid–aneuploid (heteroploid) cell population are encountered.
In MNNG-induced gastric cancer, the ectopically synthesized placental-type alkaline phosphatase

isoenzyme, considered a marker of malignancy, was also detected (Sugar et al. 1980; Miki et al. 1980).

Comparison with Other Species

Gastric adenocarcinomas can be induced in dogs (Fujita et al. 1974). The mode of induction, course of tumor development, histologic type, and biologic behavior of these tumors are similar to those observed in the rat. The histologic structure of the rat tumor is similar to that of human gastric cancer as well. However, in well-differentiated human tumors, completely differentiated cells do not occur in such high numbers as in the rat tumors. Human tumors also proliferate more readily and usually metastasize with higher frequency.

References

Bralow SP, Gruenstein M, Meranze DR (1973) Host resistance to gastric adenocarcinomatosis in three strains of rats inquesting N-methyl-N'-nitro-N-nitrosoguanidine. Oncology 27:168–180

Bullock FD, Curtis MR (1930) Spontaneous tumors of the rat. J Cancer Res 14:1–115

Butler WH, Barnes JM (1966) Carcinoma of the glandular stomach in rats given diets containing aflatoxin. Nature 209:90

Chang TH, Lee YC, Lee KY, Sun CH, Chano YPC (1983) Cocarcinogenic action of aspirin on gastric tumors induced by N-nitroso-N-methylnitroguanidine in rats. JNCI 70:1067–1075

Dahm K, Werner B (1973) Anastomosenkarzinom in resezierten Magen der Ratte nach Gabe von N-methyl-N'-nitro-N-nitrosoguanidin. Dtsch Med Wochenschr 98:2486–2487

Fujii I, Watanabe H, Terada Y, Naito Y, Naito M, Ito A (1980) Induction of intestinal metaplasia in the glandular stomach of rats by X-irradiation prior to oral administration of N-methyl-N'-nitro-N-nitrosoguanidine. Gann 71:804–810

Fujimura S, Kogure K, Sugimura T, Takayama S (1970) The effect of limited administration of N-methyl-N'-nitro-N-nitrosoguanidine on the induction of stomach cancer in rats. Cancer Res 30:842–848

Fujita M, Taguchi T, Takami M, Usugane M, Takahashi A, Shiba S (1974) Carcinoma and related lesion in dog stomach induced by oral administration of N-methyl-N'-nitro-N-nitrosoguanidine. Gann 65:207–214

Fukushima S, Tatematsu M, Takahashi M (1974) Combined effect of various surfactants on gastric carcinogenesis in rats treated with N-methyl-N'-nitro-N-nitrosoguanidine. Gann 65:371–376

Furukawa H, Iwanaga T, Koyama H, Taniguchi H (1982) Effect of sex hormones on carcinogenesis in the stomach of rats. Cancer Res 42:5181–5182

Grant R (1966) Cancer induction in the glandular stomach of rats at sites of implanted 7,12-diemthylbenz(a)anthracene. JNCI 37:353–364

Hananouchi M, Fukushima S, Takamashi M (1974) Electron microscopic studies on experimental undifferentiated adenocarcinomas of the glandular stomach in rats. Gann 65:323–330

Hare WV, Stewart HL, Bennett JG, Lorenz E (1952) Tumors of the glandular stomach induced in rats by intramural injection of 20-methylcholanthrene. JNCI 12:1019–1055

Kokubo T, Takahashi M, Kurokawa Y, Miyahara M, Furihata C, Matsushima T (1981) Preservation of gastric gland character in transplantable gastric adenocarcinoma (SG2B) of rats. Gann 72:583–589

MacKenzie WF, Garner FM (1973) Comparison of neoplasms in six sources of rats. JNCI 50:1243–1257

Martin MS, Martin F, Justrabo E, Michiels R, Bastien H, Knobel S (1974) Susceptibility of inbred rats to gastric and duodenal carcinomas induced by N-methyl-N'-nitro-N-nitrosoguanidine. JNCI 53:837–840

Matsukura N, Kawachi T, Sasajima R, Sano T, Sugimura T, Hirota T (1978) Induction of intestinal metaplasia in the stomach of rats by N-methyl-N'-nitro-N-nitrosoguanidine. JNCI 61:141–144

McCoy GW (1909) A preliminary report on tumors found in wild rats. J Med Res 21:285–296

Miki K, Oda T, Miyazaki J, Iino S, Niwa H, Oka H, Suzuki S (1980) Alkaline phosphatase isoenzymes in intestinal metaplasia and carcinoma of rat stomach induced by N-methyl-N'-nitro-N-nitrosoguanidine. Oncodev Biol Med 1:313–323

Mori K, Ota A, Murakami T, Tamura M, Knodo M, Ichimura H (1969) Carcinomas of the glandular stomach and other organs of rats induced by 4-hydroxyaminoquinoline 1-oxide hydrochloride. Gann 60:627–630

Morris HP, Wagner BP, Ray FE, Stewart HL, Snell KC (1962) Comparative carcinogenic effects of N,N'-2,7-fluorenylenebisacetamide by intraperitoneal and oral routes of administration to rats, with particular reference to gastric carcinoma. JNCI 29:977–1011

Nagayo T (1973) Tumours of the stomach. In: Turusov VS (ed) Pathology of tumours in laboratory animals, vol I, part I. Tumours of the rat. IARC Sci publ no 5, Lyon, pp 101–118

Ohgaki H, Kawachi T, Matsukura N, Morino K, Miyamoto M, Sugimura T (1983) Genetic control of susceptibility of rats to gastric carcinoma. Cancer Res 43:3663–3667

Prejean JD, Peckham JC, Casey AE, Griswold DP, Weisburger EK, Weisburger JH (1973) Spontaneous tumors in Spraque-Dawley rats and Swiss mice. Cancer Res 33:2768–2773

Ratcliffe ML (1940) Spontaneous tumors in two colonies of rats of the Wistar Institute of Anatomy and Biology. Am J Pathol 16:237–254 + 1 plate

Rowlatt UF (1967) Neoplasms of the alimentary canal of rats and mice. In: Cotchin E, Roe FJC (eds) Pathology of laboratory rats and mice. Blackwell Scientific, Oxford, chap 3

Saito T, Inokuchi K, Takayama S, Sugimura T (1970) Sequential morphological changes in N-methyl-N'-nitro-N-nitrosoguanidinesis in the glandular stomach of rats. JNCI 44:769–783

Saxton JA, Sperling GA, Barnes LL, McCay CM (1948) The influence of nutrition upon the incidence of spontaneous tumors of the albino rat. Acta Unio Int Contra Cancrum 6:423–431

Snell KC (1966) Spontaneous lesions of the rat. In: Ribelin WE, McCoy JR (eds) The pathology of laboratory animals. Thomas, Sprinfield, chap 10

Sugar J, Szentirmay Z, Kralovanszky J (1980) Pathologic features of N-methyl-N'-nitro-N-nitrosoguanidine induced neoplastic and preneoplastic lesions of rat stomach. In: Borzsonyi M (ed) N-Nitroso compounds: analysis, formation, and occurrrence. IARC Sci publ no 31, Lyon, pp 667–675

Sugimura T, Fujimura S (1967) Tumor production in glandular stomach of rat by N-methyl-N'-nitro-N-nitrosoguanidine. Nature 216:943–944

Sugimura T, Kawachi T (1978) Experimental stomach carcinogenesis. In: Lipkin M, Good RA (eds) Gastrointestinal tract cancer. Plenum, New York, chap 13

Tabuchi Y, Ogino T, Mitsuno T, Sugiyama T (1974) Possible role of mucosal damage in stomach carcinogenesis with N-methyl-N'-nitro-N-nitrosoguanidine in the rat. JNCI 52:1589–1594

Tahara E, Haizuka S (1975) Effect of gastro-entero-pancreatic endocrine hormones on the histogenesis of gastric cancer in rats induced by N-methyl-N'-nitro-N-nitrosoguanidine; with special reference to development of scirrhous gastric cancer. Gann 66:421–426

Takahashi M, Fukushima S, Sato H (1973) Carcinogenic effect of N-methyl-N'-nitro-N-nitrosoguanidine with various kinds of surfactant in the glandular stomach of rats. Gann 64:211–218

Takahashi M, Kokubo T, Furukawa F, Kurokawa Y, Tatematsu M, Hayashi Y (1983) Effect of high salt diet on rat gastric carcinogenesis induced by N-methyl-N'-nitro-N-nitrosoguanidine. Gann 74:28–34

Wolley PG, Wherry WB (1911) Notes on twenty-two spontaneous tumors in wild rats (M. norvegicus). J Med Res 25:205–215 + 1 plate

Leiomyoma and Leiomyosarcoma, Stomach, Rat

Michihito Takahashi and Akiyoshi Nishikawa

Synonyms. Leiomyoma, leiomyosarcoma, smooth muscle tumor, leiomyomatous tumor

Gross Appearance

Leiomyomas are well-circumscribed nodular tumors. Leiomyosarcomas are discretely nodular or diffuse gastric masses with varying degress of ulceration and may grow to occupy almost the entire stomach wall; they are usually situated intramurally although they may invade through to either surface. Leiomyosarcomas are a relatively rare finding in the rat stomach after administration of gastric carcinogens. At necropsy they appear as firm, lobulated, nonencapsulated fleshy masses, with varying degrees of hemorrhagic necrosis and cystic changes. This tumor resembles other sarcomas such as fibrosarcoma except that, in contrast to the latter, they have a more grayish color and a less firm consistency. Mucosal ulcerations are commonly present in advanced cases.

Microscopic Features

Microscopically, the leiomyomatous tumor is composed of interlacing bundles of spindle-shaped cells with eosinophilic cytoplasm and fusiform nuclei. Leiomyomas have minimal cellular pleomorphism and mitotic figures.

In leiomyosarcomas, intermingled with the bundles are many giant cells containing one or more large nuclei. The tumore tissue partially consists of sheets of pleomorphic cells or bizarre cells with double to multiple nuclei. Mitotic figures are frequent (Figs. 339–342). In some cases, longitudinal myofibrils are demonstrable in the cytoplasm by Masson's trichrome stain or Mallory's phosphotungstic acid-hematoxylin (PTAH) technique. The tumor stroma is almost entirely composed of blood vessels alone. Azan-Mallory and van Gieson connective tissue stains demonstrate that the tumor cells are of smooth muscle origin. Cross-striation within the cytoplasm of the tumor cells is not seen with PTAH stain. Leiomyosarcomas are rich in reticulin fibrils running parallel to the tumor cells and, in certain areas, individual cells are surrounded by them.

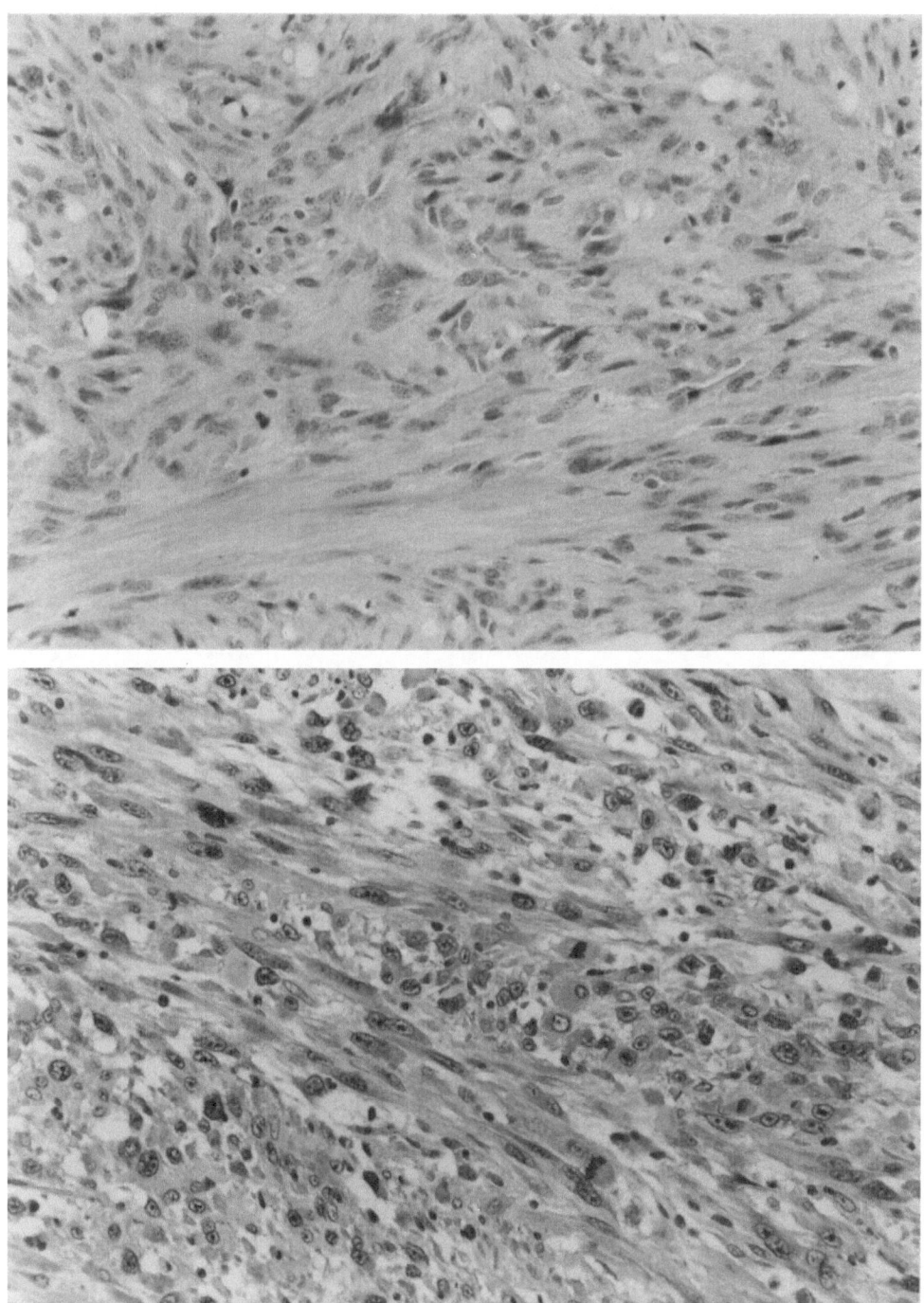

Fig. 339. (*above*) Leiomyosarcoma, stomach, rat, treated with *N*-methyl-*N*'-nitro-*N*-nitrosoguanidine (MNNG). The tumor is composed of elongated fusiform cells with eosinophilic cytoplasm. H&E, ×400

Fig. 340. (*below*) Leiomyosarcoma, stomach, rat, treated with MNNG, composed of spindle cells with blunt-ended nuclei and acidophilic cytoplasm. Note the interlacing pattern with frequent mitotic figures and pleomorphism. H&E, ×600

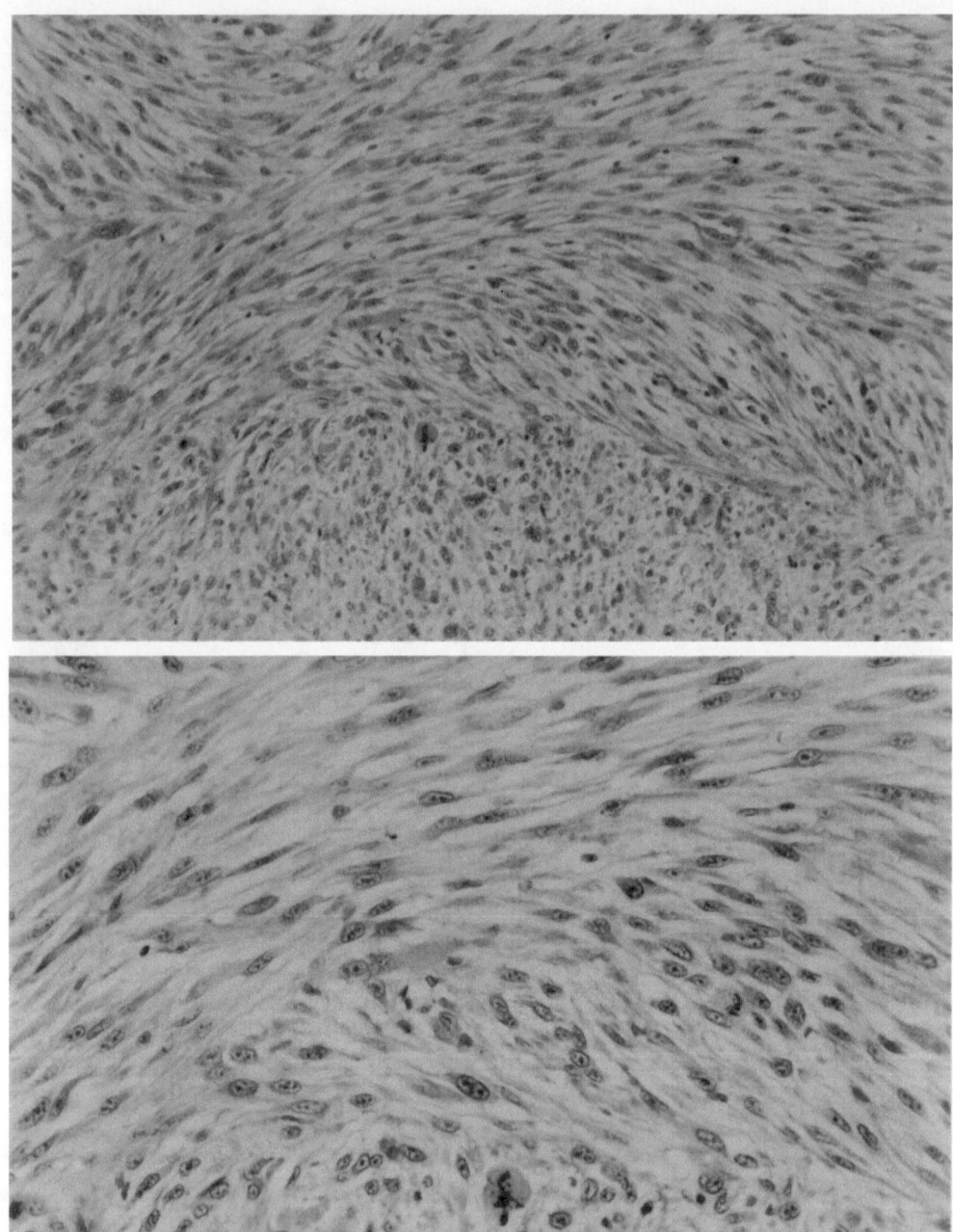

Fig. 341. (*above*) Leiomyosarcoma, MNNG-treated rat. Neo-plastic cells arranged as interlacing bundles within an edematous stroma. H&E, ×300

Fig. 342. (*below*) Higher magnification of Fig. 341. Note many mitotic figures and spaces in the stroma due to edema. H&E, ×600

Collagen formation between tumor cells is absent, but small deposits are found in the stroma around blood vessels. The diagnosis of leiomyosarcoma is established by the presence of highly cellular, nonencapsulated, and infiltrating tumors composed of interlacing bundles of spindle-shaped cells with recognizable myofibrils, varying amounts of eosinophilic cytoplasm, and fusiform nuclei. Infiltration, in most cases, extends into the muscularis propria. The interlacing tumor fascicles tend to be irregular with considerable cellular pleomorphism, uneven nuclear membranes, prominent nucleoli, and frequent mitoses. Well-differentiated leiomyosarcomas resemble benign leiomyomas, but differ in their marked cellularity, hypervascularity, and high mitotic count. Scattered rounded or polygonal eosinophilic cells are observed in a number of tumors, thus showing the features of leiomyoblastomas.

Ultrastructure

Positive ultrastructural criteria for smooth muscle differentiation found in leiomyosarcomas include filament bundles with alternating pale and dense zones, focal anchoring to the plasma membrane, pinocytotic vesicles, undulating nuclear contours, and occasional close cell–cell attachment.

Electron micrographs of leiomyosarcoma cells demonstrate microfilaments, dense bodies, and a basal lamina. The tumors contain myofilaments with densities expected in cells of smooth muscle origin, and desmosome-like structures are frequently found between neighboring tumor cells. The cytoplasm contains rough endoplasmic reticulum cisternae, abundant free ribosomes and polysomes, a few mitochondria, and masses of myofilaments.

Differential Diagnosis

Neither the macroscopic appearance nor the histologic features of nonepithelial tumors induced in the stomach of the rat are specific to that organ. The mucosa of the affected area is raised or remains flattened depending on whether the tumor grows submucosally or intramurally. Ulceration of the mucosa rarely occurs in benign tumors, but is quite frequent in malignant cases. In general, sarcomas grow more rapidly, and at necropsy a greater degree of spread and more irregular ulceration is seen than is epithelial neoplasms. Light microscopy criteria for distinguishing leiomyomas from other sarcomas such as schwannomas are poorly defined and often contradictory. Histologic and histochemical findings should provide sufficient evidence to allow differentiation of leiomyosarcomas from anaplastic carcinoma, fibrosarcoma, pleomorphic rhabdomyosarcoma, or others. On the other hand, in some cases accurate classification of leiomyosarcomas may necessitate ultrastructural analysis, in which positive ultrastructural criteria for smooth muscle differentiation should be demonstrable.

Biologic Features

Benign and malignant mesenchymal tumors of the stomach have been successfully induced in animals given N-methyl-N'-nitro-N-nitrosoguanidine (MNNG) alone (Sugimura et al. 1969) or in combination with other chemical substances (Cohen et al. 1984; Fukushima et al. 1974; Shirai et al. 1982), physical agents (Fukushima et al. 1976), surgical modification of gastric secretory activity (Mori et al. 1981), or after the mechanical production of gastric ulcers (Takahashi et al. 1976, 1981). These tumors have been variously described as leiomyomatous tumor, leiomyoma, leiomyosarcoma, fibrosarcoma, hemangiosarcoma, spindle cell proliferation, spindle cell sarcoma, neurinoma, and sarcoma. In general, the mesenchymal tumors have not received great emphasis, and detailed histologic descriptions of most of these tumors have not appeared in the literature.

Leiomyosarcomas occur in any site of the stomach including not only the lesser and greater curvature and fundic and pyloric areas, but also the forestomach. Since the neoplasms may sometimes be so massive that almost the entire stomach is involved, it may not be possible to determine the point of origin. Sarcomas are more frequently encountered than benign nonepithelial tumors after the oral administration of 4-nitroquinoline 1-oxide or several nitroso compounds (Nagayo 1973). Fibrosarcoma, leiomyosarcoma, and hemangiosarcoma are the sarcoma types most frequently observed, but reticulum cell sarcomas and liposarcomas are also occasionally seen. In some cases it is not possible to make a definitive diagnosis owing to the lack of differentiative character of the tumor cells (undifferentiated sarcoma). Among benign nonepithelial tumors, leiomyoma is the most common. The determination and grading of malignancy in smooth muscle tumors of the stomach is based on evaluation of the degree of cellularity, pleomorphism, presence of mitotic activity, and other cytologic features.

Comparison with Other Species

Leiomyosarcomas in mice and hamsters appear to resemble those in rats. They are rarely reported in mice, but are commonly observed in hamsters after treatment with carcinogens. In humans, leiomyosarcomas are most frequent among nonepithelial tumors, but comprise only about 1% of all stomach tumors. They generally attain large proportions with invasion of the entire stomach wall (Ranchod and Kempson 1977; Shiu et al. 1982).

References

Cohen A, Geller SA, Horowitz I, Toth LS, Werther JL (1984) Experimental models for gastric leiomyosarcoma. The effects of N-methyl-N'-nitro-N-nitrosoguanidine in combination with stress, aspirin, or sodium taurocholate. Cancer 53:1088–1092

Fukushima S, Tatematsu M, Takahashi M (1974) Combined effect of various surfactants on gastric carcinogenesis in rats treated with N-methyl-N'-nitro-N-nitrosoguanidine. Gann 65:371–376

Fukushima S, Hananouchi M, Shirai T, Tatematsu M, Hirose M, Yoshida S, Takahashi M (1976) Effect of plastic bead on gastric carcinogenesis in rats treated with N-methyl-N'-nitro-N-nitrosoguanidine. Gann 67:197–205

Mori H, Domellof L, Weisburger JH, Williams GM (1981) Enhancing effect of vagotomy and pyloroplasty on gastrointestinal carcinogenesis induced by nitrosamide in hamsters. Gann 72:440–445

Nagayo T (1973) Tumours of the stomach. In: Turusov VS (ed) Pathology of tumours in laboratory animals, vol I, part 1. Tumours of the rat. IARC Sci publ no 5, Lyon, pp 101–118

Ranchod M, Kempson RL (1977) Smooth muscle tumors of the gastrointestinal tract and retroperitoneum: a pathologic analysis of 100 cases. Cancer 39:255–262

Shirai T, Imaida K, Fukushima S, Hasegawa R, Tatematsu M, Ito N (1982) Effects of NaCl, Tween 60 and a low dose of N-ethyl-N-nitro-N-nitrosoguanidine on gastric carcinogenesis of rat given a single dose of N-methyl-N'-nitro-N-nitrosoguanidine. Carcinogenesis 3:1419–1422

Shiu MH, Farr GH, Papachristou DN, Hajdu SI (1982) Myosarcomas of the stomach: natural history, prognostic factors and management. Cancer 49:177–187

Sugimura T, Fujimura S, Kogure K, Baba T, Saito T, Nagao M, Hosoi H, Shimosato Y, Yokoshima T (1969) Production of adenocarcinomas in glandular stomach of experimental animals by N-methyl-N'-nitro-N-nitrosoguanidine. In: Yoshida T (ed) Experimental carcinoma of the glandular stomach. Maruzen, Tokyo, pp. 157–196 (Gann monograph 8)

Takahashi M, Shirai T, Fukushima S, Hananouchi M, Hirose M, Ito N (1976) Effect of fundic ulcers induced by iodoacetamide on development of gastric tumors in rats treated with N-methyl-N'-nitro-N-nitrosoguanidine. Gann 67:47–54

Takahashi M, Shirai T, Fukushima S, Ito N, Kokubo T, Furukawa F, Kurata Y (1981) Ulcer formation and associated tumor production in multiple sites within the stomach and duodenum of rats treated with N-methyl-N'-nitro-N-nitrosoguanidine. JNCI 67:473–479.

The Small Intestines

VIRAL INFECTIONS

Mouse Hepatitis Virus Infection, Intestine, Mouse

Stephen W. Barthold

Synonyms. Enterotropic coronavirus infection, mouse, lethal intestinal virus of infant mice (LIVIM)

Gross Appearance

Enterotropic mouse hepatitis virus (MHV) infection is usually subclinical, with no gross lesions. Neonatal mice suffer high mortality when the virus is first introduced to a naive breeding population. They become dehydrated, with soiling of the perineum with yellow, pasty feces. Their stomachs are usually empty, and intestines are thin-walled, flaccid, and contain watery yellow digesta and gas. Juvenile mice are less severely affected, but are often runted with pot bellies and oily-appearing hair. Careful examination of weaning-age or adult mice may reveal dark, sticky feces and opaque, thickened segments of bowel. Livers, if affected, have few to many small, pale or hemorrhagic foci (see "Mouse Hepatic Virus Infection, Liver, Mouse," this volume; Barthold 1986; Barthold and Smith 1984; Barthold et al. 1982, 1993; Broderson et al. 1976; Hierholzer et al. 1979; Ishida and Fujiwara 1979; Ishida et al. 1978; Kraft 1962, 1966). Athymic nude mice infected with enterotropic MHV may have segmental thickening of the cecum and ascending colon, with enlargement of mesenteric lymph nodes (Barthold et al. 1985; Ward et al. 1977).

Microscopic Features

The quality and location of intestinal lesions vary widely, depending on MHV strain and host age. Enterotropic MHV characteristically causes mucosal epithelial necrosis and syncytium formation, while host responses include inflammation and compensatory mucosal hyperplasia. In neonatal mice, viral effects are severe. Syncytia can be pronounced in the small intestine, since they tend to be retained at villus tips or are detached within the lumen. These large, multinucleate cells have been termed "balloon cells" because of their appearance (Fig. 343). Infected enterocytes can have poorly defined eosinophilic intracytoplasmic inclusions, but these are of little diagnostic value. When cytolysis predominates, villi become markedly attenuated (Fig. 344). Syncytia tend to form as large masses in the surface mucosa of the large bowel (Fig. 345), but the mucosal epithelium is often nearly completely effaced (Fig. 346). The ascending colon seems to be a preferential target for enterotropic MHV, regardless of host age or strain. Lesions are most apt to be present in these segments of bowel (Barthold et al. 1993). Surviving mice or weaning-age mice respond to infection with transient mucosal hyperplasia of the involved segment of bowel (Fig. 347). Lesions in adult mice are marginal, consisting of only a few scattered syncytia and modest crypt hyperplasia (Barthold 1986; Barthold et al. 1982, 1993; Biggers et al. 1964; Hierholzer et al. 1979; Ishida and Fujiwara 1979; Ishida et al. 1978; Kraft 1962, 1966). Athymic nude mice can develop chronic mucosal hyperplasia with syncytia, erosion, and inflammation (Barthold et al. 1985; Ward et al. 1977). Alternatively, nude mice infected as adults or with avirulent strains of virus may have minimal enteric lesions, with only a few syncytia present in surface mucosa and minimal hyperplasia or inflammation.

Ultrastructure

Infected enterocytes often have nonspecific degenerative changes or are undergoing necrosis. The endoplasmic reticulum contains 80- to 120-nm

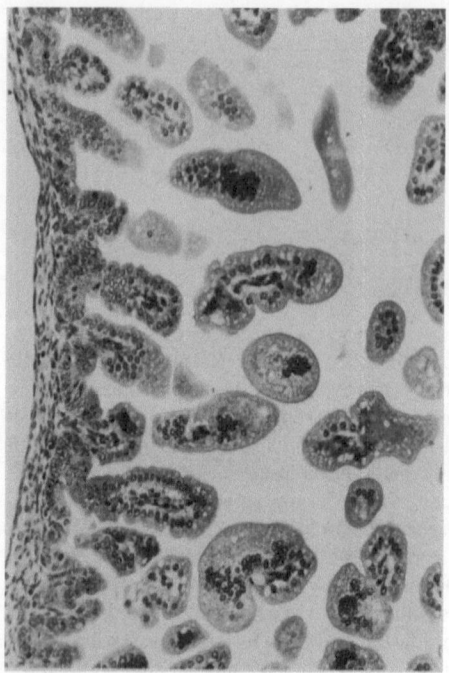

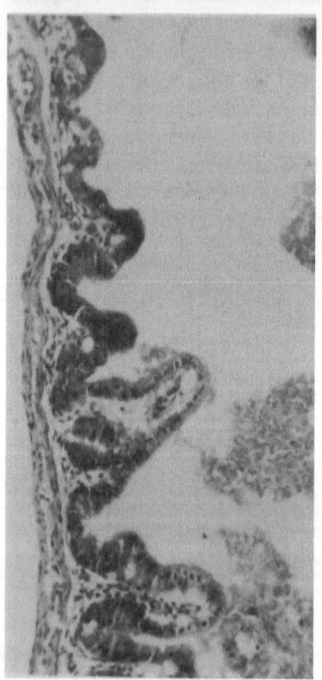

coronaviral particles, and the cytoplasm may possess one or more amorphous electron-dense bodies and complexes of reticular structures (Fig. 348). These changes are seen in other infected tissues as well (see "Mouse Hepatic Virus Infection, Liver, Mouse," this volume). Virus particles can also be found within macrophages of the lamina propria.

Ultrastructural correlates of inclusions seen by light microscopy have not been found (Barthold et al. 1982; Hierholzer et al. 1979; Ishida and Fujiwara 1979; Ishida et al. 1978).

Differential Diagnosis

Enteric disease in mice can be caused by a number of viruses, bacteria, and protozoa, but histologic changes, particularly epithelial syncytia in mice infected with MHV, are diagnostic. Because lesions are often segmental, all levels of bowel should be examined, but lesions are most apt to be present in the ascending colon and cecum. The most sensitive means of detecting MHV exposure in mice is to perform serologic tests for antibody, as most infections are subclinical with minimal microscopic lesions.

Biologic Features

Natural History

Mouse hepatitis virus is a highly mutable coronavirus with many, constantly evolving and changing strains. Host immunity to MHV is virus strain specific, and therefore mice can be repeatedly infected with newly mutated or introduced virus strains (Barthold 1986; Barthold and Smith 1989; Homberger et al. 1992). Enterotropic strains of the virus are highly contagious. When first introduced to a breeding population of mice, epizootics of enteritis with diarrhea can be associated with high morbidity and mortality among in-

◄

Fig. 343. (*above*) Small intestine, neonatal mouse infected with enterotropic mouse hepatic virus (MHV). Villi possess several large multinucleate syncytia ("balloon cells") which are typical for MHV. H&E, ×150

Fig. 344. (*below*) Small intestine, neonatal mouse infected with enterotropic MHV. Villi are markedly attenuated and crypts are hypercellular. (Courtesy of Dr. S. W. Barthold and Hemisphere Publishing Corp.) H&E, ×150

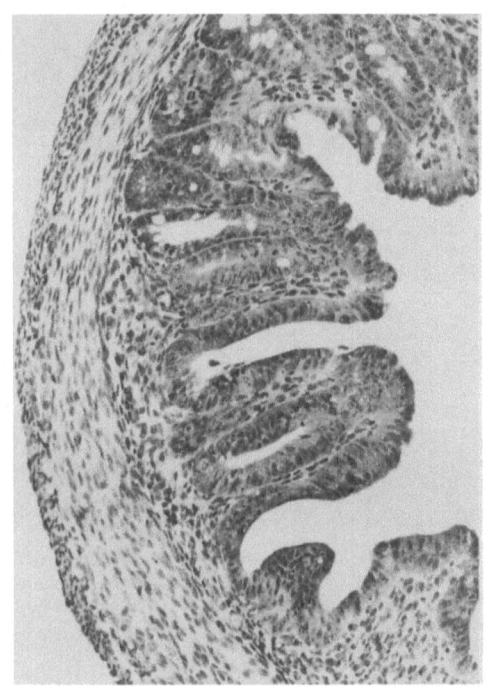

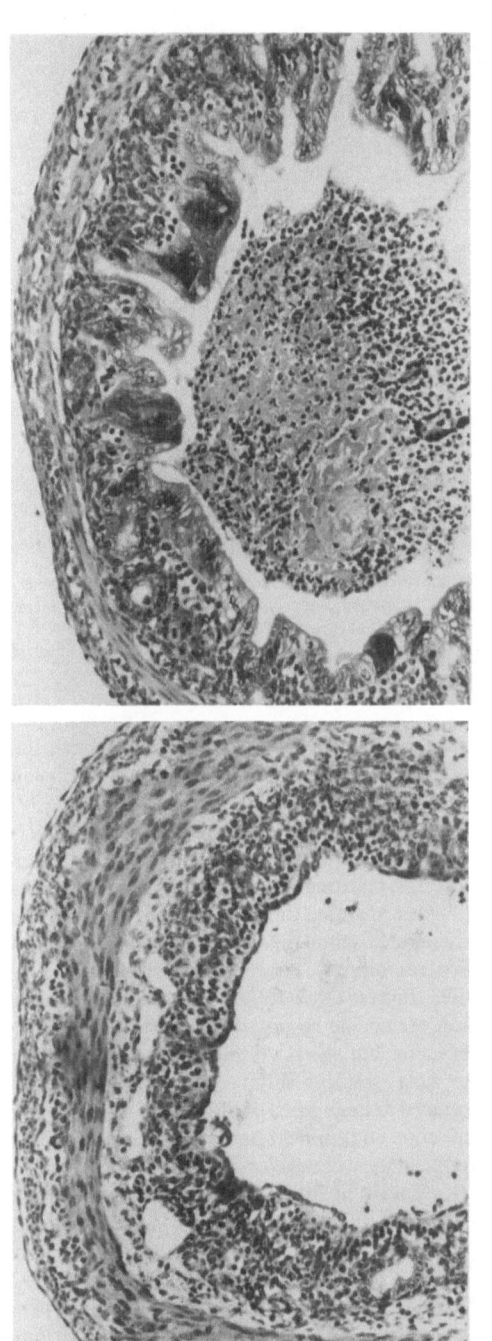

◄

Fig. 345. (*upper left*) Colon neonatal mouse infected with enterotropic MHV. The lumen contains leukocytes and cellular debris. The surface mucosa has multiple prominent epithelial syncytia. H&E, ×150

Fig. 346. (*below*) Cecum, neonatal mouse infected with enterotropic MHV. The mucosal epithelium has been almost completely effaced, leaving only the lamina propria, except for an incomplete layer of surface epithelium. H&E, ×150

Fig. 347. (*upper right*) Cecum, neonatal mouse recovering from enterotropic MHV infection. The mucosa is distorted due to hyperplasia of the crypt epithelium. H&E, ×150

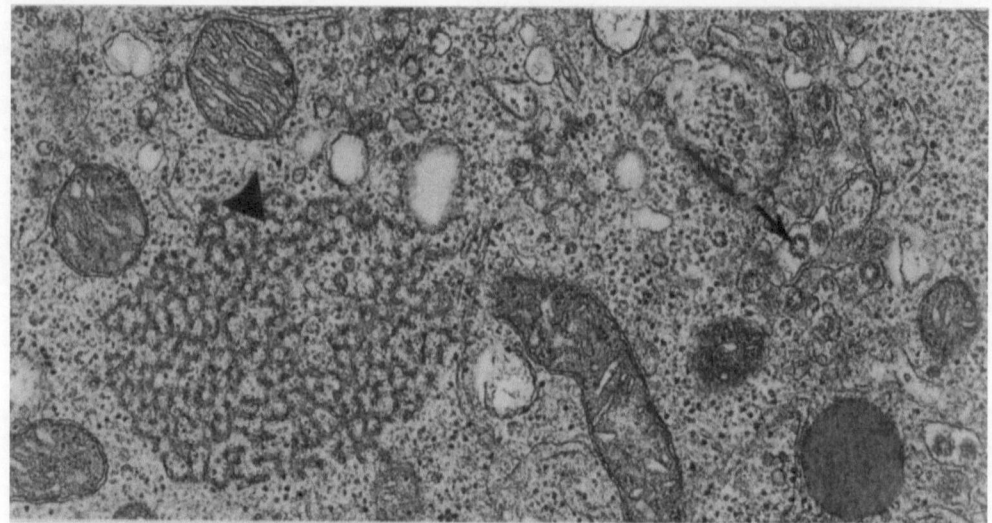

Fig. 348. Electron micrograph of cecal mucosal epithelium of a neonatal mouse infected with MHV. The cytoplasm contains an electron-dense reticular structure. Virus particles are present in cisternae of the endoplasmic reticulum. (Courtesy of Dr. S. W. Barthold and Laboratory Animal Science) Uranyl acetate and lead citrate, ×225000

fant mice. Once enzootic within a population, signs of disease are less obvious, since pups from recovered dams are protected by maternal antibody during the neonatal period (Barthold et al. 1988; Homberger 1992; Homberger and Barthold 1992; Hierholzer et al. 1979; Ishida and Fujiwara 1982; Ishida et al. 1978). Under these circumstances, reduced litter survival and runting occur, but may not be noticed. Mice of all ages and genotypes are susceptible to infection, but infection among adults is usually subclinical. With the exception of athymic nude or other immunodeficient mice which develop persistent infections, mice normally recover within 2 weeks of infection (Barthold 1986; Barthold and Smith 1984; Barthold et al. 1982, 1993; Biggers et al. 1964).

Pathogenesis

Respiratory strains of MHV initially replicate in nasal epithelium and then disseminate to multiple organs. These viruses have been extensively studied and are known to be influenced by host genotype, age, and lymphoreticular function. These viral strains have little or no tropism for bowel (see "Mouse Hepatic Virus Infection, Liver, Mouse," this volume). Enterotropic strains cause infections largely limited to bowel, although limited dissemination to other organs may take place (Barthold and Smith 1984; Barthold et al. 1993; Hierholzer et al. 1979; Ishida and Fujiwara 1979; Ishida et al. 1978). Neonatal mice develop severe enteritis, resulting in malabsorption and diarrhea. As mice mature, intestinal cell turnover rate is accelerated, allowing a compensatory proliferative response to viral damage (Barthold et al. 1993; Biggers et al. 1964). Genotype and maturation of immune response no doubt also influence infection, but age is the most critical determinant of host susceptibility to enterotropic MHV-induced disease. Mice of all ages are susceptible to infection and support nearly equal levels of virus replication, whether or not disease is apparent (Barthold et al. 1993). Colostral immunoglobulin (Ig)A and IgG antibody from immune mothers is protective against fatal infections in neonatal mice (Homberger 1992; Homberger and Barthold 1992; Ishida and Fujiwara 1982). Athymic nude mice develop persistent enteric infections, with chronic mucosal hyperplasia (Barthold et al. 1985). The distribution of lesions within the intestine is probably determined by virus strain. Severe small intestine involvement has been noted with some strains, such as LIVIM, MHV-S/CDC, and MHV-D (Biggers et al. 1964; Hierholzer et al. 1979;

Ishida and Fujiwara 1979; Ishida et al. 1978). Predilection for the large bowel has been noted with other MHV isolates, such as MHV-Y (Barthold et al. 1982, 1993; Barthold and Smith 1984). Intestinal bacteria and protozoa may influence the course of enteric MHV infections, but little work has been done on this aspect. Recovered mice are immune to reinfection with the homologous MHV strain, but can be reinfected with an antigenically heterologous strain (Barthold and Smith 1989; Barthold et al. 1993). Because of the high degree of tissue selectivity, enterotropic strains tend not to disseminate to other tissues and are therefore not likely to be transmitted in utero (Barthold et al. 1988).

Etiology

Many strains of this virus exist which can be partially differentiated antigenically, but all share common antigens with each other and with coronaviruses of rats and humans (Barthold 1986). Antigenic composition does not predict virulence or organotropism (Barthold and Smith 1984).

Frequency

As discussed in "Mouse Hepatic Virus Infection, Liver, Mouse" (this volume), MHV is among the most common viruses of laboratory mice. Enterotropic strains seem to be more prevalent than respiratory strains, but this may be influenced by the relative ease of diagnosing enterotropic infections. With thorough analysis, enteric lesions are consistently found in actively infected mice, but are detected most easily in young mice. Lesions are present for only 1–2 weeks and are absent when seroconversion takes place, unless immune mice have been reinfected with a new strain of the virus.

Comparison with Other Species

Enterotropic MHV infections are analogous to enteric coronavirus infections in many other species of birds and mammals. Enteric coronaviruses have been described in humans, nonhuman primates, turkeys, swine, cattle, sheep, horses, dogs, rats, cats, and rabbits. As in mice, these viruses are generally associated with disease in neonates. Small intestine changes in neonates usually include villus "atrophy," resulting in malabsorption and diarrhea (Barthold 1986).

References

Barthold SW (1986) Mouse hepatitis virus biology and epizootiology. In: Bhatt PN, Jacoby RO, Morse HC III, New AE (eds) Viral and mycoplasmal infections of laboratory rodents: effects on biomedical research, chap 25. Academic, Orlando, pp 571–601

Barthold SW, Smith AL (1984) Mouse hepatitis virus s&m-related patterns of tissue tropism in suckling mice. Arch Virol 81:103–112

Barthold SW, Smith AL (1989) Virus strain specificity of challenge immunity to coronavirus. Arch Virol 104:187–196

Barthold SW, Smith AL, Lord PF, Bhatt PN, Jacoby RO, Main AL (1982) Epizootic coronaviral typhlocolitis in suckling mice. Lab Anim Sci 32:376–383

Barthold SW, Smith AL, Povar ML (1985) Enterotropic mouse hepatitis virus infection in nude mice. Lab Anim Sci 35:613–618

Barthold SW, Beck DS, Smith AL (1988) Mouse hepatitis virus and host determinants of vertical transmission and maternally-derived passive immunity in mice. Arch Virol 100:171–183

Barathold SW, Beck DS, Smith AL (1993) Enterotropic coronavirus (mouse hepatitis virus) in mice: influence of host age and strain on infection and disease. Lab Anim Sci 43:276–284

Biggers DC, Kraft LM, Sprinz H (1964) Lethal intestinal virus infection of mice (LIVIM). An important new model for study of the response of the intestinal mucosa to injury. Am J Pathol 45:413–422

Broderson JR, Murphy FA, Hierholzer JC (1976) Lethal enteritis in infant mice caused by mouse hepatitis virus. Lab Anim Sci 26:824

Hierholzer JC, Broderson JR, Murphy FA (1979) New strain of mouse hepatitis virus as the cause of lethal enteritis in infant mice. Infect Immun 24:508–522

Homberger FR (1992) Maternally-derived passive immunity to enterotropic mouse hepatitis virus. Arch Virol 122:133–141

Homberger FR, Barthold SW (1992) Passively acquired challenge immunity to enterotropic coronavirus in mice. Arch Virol 126:35–43

Homberger FR, Barthold SW, Smith AL (1992) Duration and strain specificity of immunity to enterotropic mouse hepatitis virus. Lab Anim Sci 42:347-351

Ishida T, Fujiwara K (1979) Pathology of diarrhea due to mouse hepatitis virus in the infant mouse. Jpn J Exp Med 49:33–41

Ishida T, Fujiwara K (1982) Maternally derived immune resistance to fatal diarrhea in infant mice due to mouse hepatitis virus. Jpn J Exp Med 52:231–235

Ishida T, Taguchi F, Lee Y-S, Yamada A, Tamura T, Fujiwara K (1978) Isolation of mouse hepatitis virus from infant mice with fatal diarrhea. Lab Anim Sci 28:269–276

Kraft LM (1962) An apparently new lethal virus disease of infant mice. Science 137:282–283

Kraft LM (1966) Epizootic diarrhea of infant mice and lethal intestinal virus infection of infant mice. Natl Cancer Inst Monogr 20:55–61

Ward JM, Collins MJ, Parker JC (1977) Naturally occurring mouse hepatitis virus infection in the nude mouse. Lab Anim Sci 27:372–376

Murine Rotavirus Infection, Intestine, Mouse

Stephen W. Barthold

Synonyms. Epidemic diarrhea of infant mice, epizootic diarrhea of infant mice (EDIM)

Gross Appearance

Mice with murine rotavirus infection less than 18 days of age have mustard-colored to dark sticky feces which adhere to the perineum and give an oily appearance to the hair. Rarely, dried feces obstruct the anus, resulting in obstipation. Affected mice are pot-bellied and runted, but recover their weight by 8 weeks of age. The stomach is full of ingesta, and the intestines are distended with copious mucoid yellow digesta and gas. Older mice do not manifest signs or lesions (Cheever 1956; Cheever and Mueller 1947; Kraft 1957, 1958, 1962; Sheridan et al. 1983).

Microscopic Features

Light microscopic lesions are difficult to recognize and are often absent, particularly in older mice. Tips of small intestinal villi, especially in the jejunum and ileum, appear bulbous due to swelling of epithelial cells, vascular congestion, and dilatation of lymphatics (Fig. 349). Columnar epithelial cytoplasm is swollen due to diffuse, fine vesiculation. Severely affected cells develop large cytoplasmic vacuoles, and 1- to 4-μm acidophilic inclusions, which are not diagnostic, may be randomly distributed within the cytoplasm. Minimal or no inflammation is present (Adams and Kraft 1967; Kraft 1957, 1962; Pappenheimer and Cheever 1948).

Ultrastructure

Enterocytes infected with the virus contain numerous vesicles arising from rough endoplasmic reticulum. These vesicles contain many virus particles, electron-dense granular material, and lipid. The cytoplasm also has aggregates of dense granular material with nascent virus particles that range from 65 to 80 nm in diameter (Fig. 350). The largest particles possess a double set of membranes; the outer membrane is acquired by budding into the lumen of the endoplasmic reticulum. Smaller particles with only a single membrane are also present in vesicles and in the cytoplasmic matrix. Tubular structures with single or double sets of membranes may be present in the cytoplasm, endoplasmic reticulum, and, occasionally, the nucleus. These structures apparently represent products of abnormal virus assembly (Fig. 351). No structures corresponding to the intracytoplasmic inclusions seen by light microscopy can be found ultrastructurally. Virus is present in enterocytes at all levels of the small and large intestine. In the small intestine, there is a gradient of cellular susceptibility toward villus tips. Virus is released into the intestinal lumen by cell disruption and exfoliation (Adams and Kraft 1963, 1967; Banfield et al. 1968; Holmes et al. 1975; Kraft 1962; Osborne et al. 1988; Starkey et al. 1986).

Differential Diagnosis

Agents which produce diarrhea in young mice include enterotropic mouse hepatitis virus (MHV), reovirus, *Salmonella*, and *Spironucleus muris*. Coinfection among these agents is frequent. Murine rotavirus infection is difficult to diagnose, since morphological lesions are often subtle or absent even in the presence of clinical disease. Superb histotechnique is required for appropriate examination, since autolysis and mishandling of tissues are most apt to affect mucosa at the villus tips. Neonatal enterocytes possess apical tubular systems involved in uptake of colostrum, giving

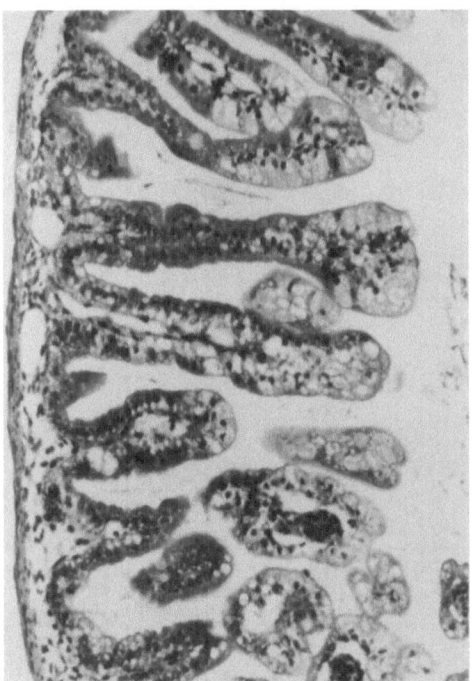

Fig. 349. Small intestine of a young mouse naturally infected with murine rotavirus. Enterocytes at the tips of the villi are swollen and the lymphatic channels are dilated. H&E, ×220

the apical cytoplasm a vacuolated appearance. This must be differentiated from the more diffuse, hydropic swelling with vacuolation that takes place in rotavirus infections. Serology is the most sensitive means of detecting exposure.

Biologic Features

Natural History

Murine rotavirus is moderately contagious by the airborne route, resulting in high morbidity but low mortality in infected populations. Mice of all ages are susceptible to oral infection, but clinical signs and lesions occur only in mice less than 18 days of age. Diarrheal disease lasts for at least 9 days, and virus excretion occurs for approximately 2 weeks, but can persist for up to 21 days in some mice. Dams suckling infected litters are thus asymptomatic, but actively excrete virus. Pups born to previously infected, multiparous females can be equally susceptible to infection, suggesting that

maternally derived passive immunity is ineffective. On the other hand, in some mouse breeding colonies, diarrhea seems to disappear in litters born to recovered dams. This disparity may be due to the type of maternal antibody (see "Pathogenesis" below) and the fact that recovered adult mice can be reinfected (Kraft 1957, 1958, 1961, 1962; Sheridan et al. 1983; Wolf et al. 1981).

Pathogenesis

Although this rotavirus infection occurs in mice of all ages, the course of infection is largely dependent on age. In neonates (less than 1 week of age), virus can be isolated from stomach, small intestine, and large intestine within 3 h of oral inoculation. In both experimental and natural infections, virus replication takes place in the cytoplasm of columnar epithelium from the duodenum to the anus, but not the stomach. Terminally differentiated epithelial cells, which are present on the tips of villi and the surface mucosa, are targeted by the virus. The greatest degree of terminal differentiation is present in the infant bowel, with diminished numbers of these target cells as the mucosal kinetics accelerate with age and acquisition of normal microflora. This is the major determinant of age-related disease susceptibility. Within 22 h, virus can be recovered from blood, liver, spleen, and lungs, and by 30 h virus is also present in kidneys, urinary bladder, urine, and brain. Diarrhea begins as early as 40 h after oral inoculation. Mice remain viremic at 6 days after inoculation and begin to recover from diarrhea by 11 days, despite continued virus excretion in feces. By 3 weeks, virus is no longer detectable in feces. Older mice, including adults, also become viremic, but the distribution of virus changes in intestine, the number of infected enterocytes decreases, and the duration of infection is shorter. As neonatal disease resolves or as the age of the host increases, infection becomes more limited to the distal small intestine rather than the duodenum (Kraft 1958, 1961; Riepenhoff-Talty et al. 1982; Sheridan et al. 1983; Wilsnack et al. 1969; Little and Shadduck 1982). Other studies have found a more restrictive distribution of infection, with only small intestine involvement (Starkey et al. 1986).

Diarrhea is due to a number of pathophysiologic events. Prior to significant virus replication, there is edema of villus tips, constriction of villus bases, villus shortening, and microcirculatory ischemia,

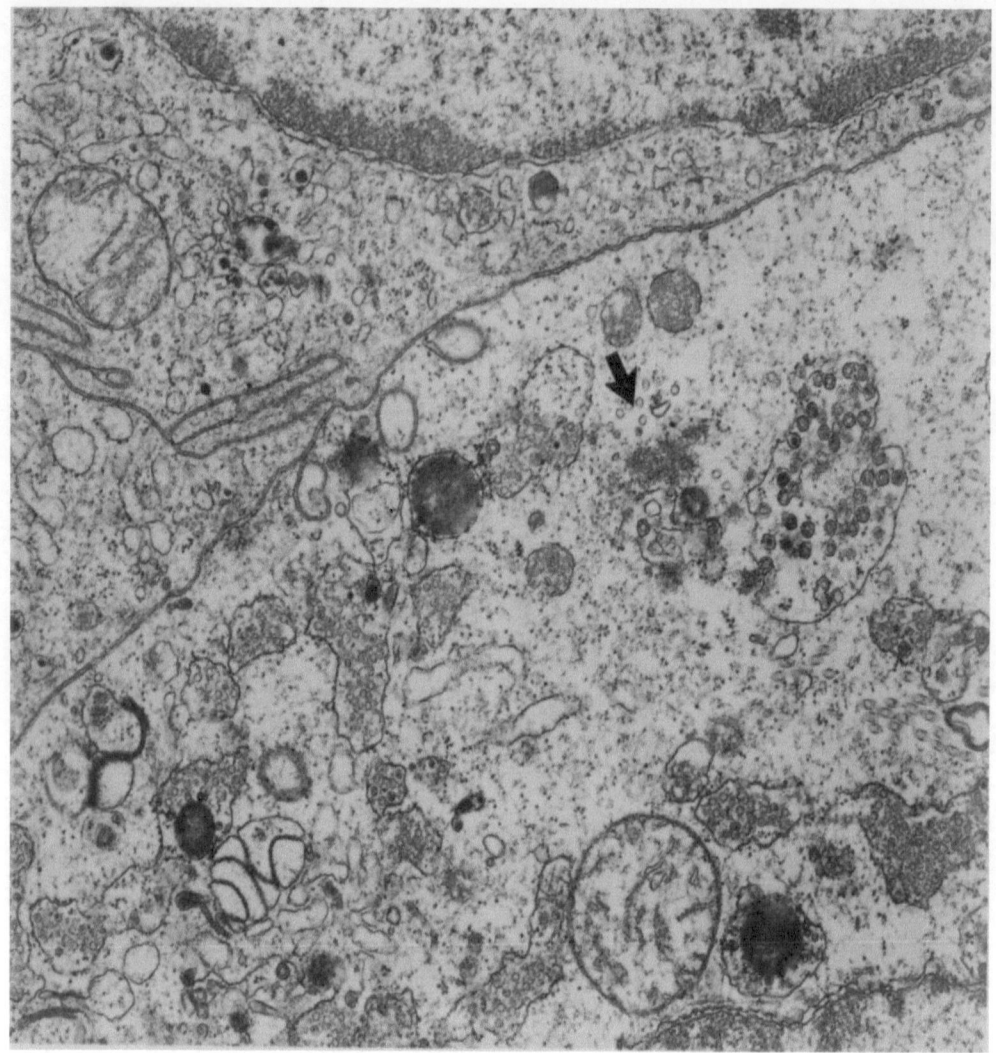

Fig. 350. Degenerating enterocyte infected with murine rotavirus. The cytoplasm contains aggregates of granular material and nascent viral particle (*arrow*). Several dilated vesicles containing single or double membrane-bound virus particles, granular material, and lipid are seen. TEM, ×21 000

which is followed by hyperemia and hyperplasia (Osborne et al. 1991). There are also virus-induced changes in intestinal enzymes and hypersecretion, which affect electrolytes, solutes, and water transport (Collins et al. 1988; Starkey et al. 1990).

Conflicting observations exist on whether maternal antibody is important in the biology of this disease (Kraft 1958, 1961). Protection of neonates from diarrhea (but not infection) occurs with colostrally derived virus-specific immunoglobulin (Ig)G, but not IgA. Thus clinical course in neonates is dependent not only on maternally derived immunity, but more specifically on the class of antibody, whose relative proportions vary from dam to dam. Infection of mice with the virus can take place in the presence of serum and intestinal virus-specific immunoglobulins, and reinfection of adult mice occurs consistently (Sheridan et al. 1983). In other studies, both IgA and IgG were

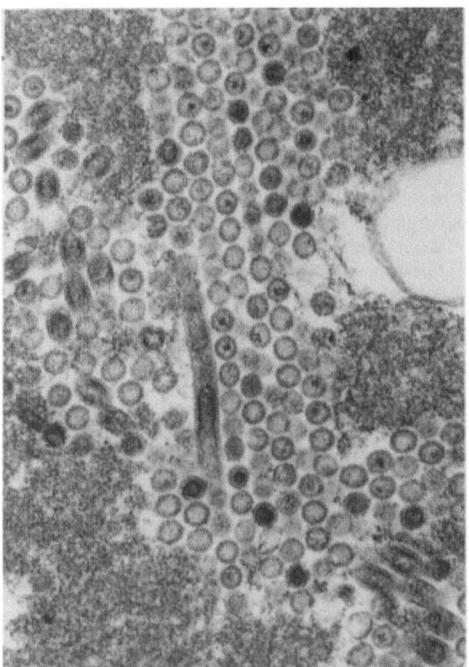

Fig. 351. An aggregate of murine rotavirus particles and tubular structures. TEM, ×72 000

Corticosteroids, which accelerate intestinal maturation, also modulate susceptibility to murine rotavirus (Wolf et al. 1981). Thus infection with this virus can take place at any age, but clinically significant infections only occur in young mice when virus and antigen predominate in the proximal bowel for maximal periods. This site is most apt to be associated with clinical signs of malfunction (malabsorption and diarrhea). Mice with severe combined immunodeficiency, lacking functional T and B cells, cannot recover from rotavirus infection, but disease is age dependent because target cells are restricted to the most terminally differentiated epithelial cells on tips of villi, which are few in the adult intestine (Riepenhoff-Talty et al. 1987; Dharakul et al. 1990). Recovery can be effected by adoptive transfer of T lymphocytes (Dharakul et al. 1990).

Etiology

This infectious agent belongs to the genus *Rotavirus* of the family *Reoviridae*. Rotaviruses are double-stranded RNA viruses which can be separated into groups A–F by serology and relative weights of their nucleic acid segments. The mouse rotavirus is a type-A, or typical rotavirus, whereas groups B–F are often referred to as "atypical" rotaviruses (Huber et al. 1990). Type-A rotaviruses infect a broad range of mammalian and avian species. Rotaviruses are morphologically indistinguishable from one another and are antigenically cross-reactive by complement fixation and indirect immunofluorescence, but serum neutralization can differentiate between strains derived from different species. Mice are susceptible to infection with some, but not all rotaviruses of other species (Kraft 1962; Ramig 1988; Thouless et al. 1977; Woode et al. 1976). Recovered mice resist infection with homotypic virus, but are susceptible to heterotypic rotavirus infection (Ward et al. 1990).

shown to be protective when administered orally to pups, but the greatest protective effect was associated with IgA. No protective effect was afforded by passively transferred serum IgG (Offit and Clarke 1985).

During the course of infection, virus can be isolated from multiple organs, but antigen- or virus-related lesions do not occur in sites other than the bowel (Kraft 1958; Wilsnack et al. 1969). Rotaviruses are believed to cause significant infections in the young because of the presence of specific virus receptors or enzymes in the neonatal bowel. Lactase has been proposed as such a receptor, which may also serve as an uncoating enzyme (Holmes et al. 1976), but presence or absence of lactase in the mouse bowel does not correlate with susceptibility to the virus (Wolf et al. 1981). With a number of rotaviruses, there are striking similarities between the patterns and kinetics of intestinal immunoglobulin-binding sites and virus replication. The receptors for immunoglobulins in the proximal bowel of neonates are Fc receptors, and their age of occurrence correlates with periods of disease susceptibility (Riepenhoff-Talty et al. 1982).

Frequency

Murine rotavirus infection is a common intestinal virus infection of laboratory mice and appears to occur throughout the world. However, recognition of infection is obscured by the frequent absence of clinical disease, difficulty in recognizing lesions, and the lack of a universally applied serologic assay for detection of antibody, even though a sensitive assay exists (Smith et al. 1983).

Comparison with Other Species

This murine disease closely mimics typical rotaviral infections of other species, including humans. Rotaviruses generally produce clinical disease in infants, but with subtle histopathologic lesions (Holmes et al. 1975). Rotaviruses or antibody to rotaviruses has been demonstrated in mice, calves, foals, pigs, sheep, rabbits, deer, monkeys, dogs, guinea pigs, and goats. Cross-infections between species can be induced experimentally and also occur naturally (Holmes 1979; Ramig 1988; Thouless et al. 1977; Woode et al. 1976). Rats are susceptible to infectious diarrhea of infant rats (IDIR), which is caused by an atypical rotavirus (Huber et al. 1990).

References

Adams WR, Kraft LM (1963) Epizootic diarrhea of infant mice identification of the etiologic agent. Science 141:359–360

Adams WR, Kraft LM (1967) Electron-microscopic study of the intestinal epithelium of mice infected with the agent of epizootic diarrhea of infant mice (EDIM virus). Am J Pathol 51:39–60

Banfield WG, Kasnic G, Blackwell JH (1968) Further observations on the virus of epizootic diarrhea of infant mice. An electron microscopic study. Virology 36:411–421

Cheever FS (1956) Epidemic diarrheal disease of suckling mice. Ann N Y Acad Sci 66:196–203

Cheever FS, Mueller JH (1947) Epidemic diarrheal disease of suckling mice. I. Manifestations, epidemiology, and attempts to transmit the disease. J Exp Med 85:405–416

Collins J, Starkey WG, Wallis TS, Clarke GJ, Worton KH, Spencer AJ, Haddon SJ, Osborne MP, Candy DCA, Stephen J (1988) Intestinal enzyme profiles in normal and rotavirus-infected mice. J Pediatr Gastroenterol Nutr 7:264–272

Dharakul T, Rott L, Greenberg HB (1990) Recovery from chronic rotavirus infection in mice with severe combined immunodeficiency: virus clearance mediated by adoptive transfer of immune CD8+ T lymphocytes. J Virol 64:4375–4382

Holmes IH (1979) Viral gastroenteritis. Prog Med Virol 25:1–36

Holmes IH, Ruck BJ, Bishop RF, Davidson GP (1975) Infantile enteritis viruses: morphogenesis and morphology. J Virol 16:937–943

Holmes IH, Rodger SM, Schnagl RD, Ruck BJ, Gust ID, Bishop RF, Barnes GL (1976) Is lactase the receptor and uncoating enzyme for infantile enteritis (rota) viruses? Lancet 1:1387–1388

Huber AC, Yolken RH, Mader LC, Strandberg JD, Vonderfecht SL (1990) Pathology of infectious diarrhea in infant rats (IDIR) induced by an antigenically distinct rotavirus. Vet Pathol 26:376–385

Kraft LM (1957) Studies on the etiology and transmission of epidemic diarrhea of infant mice. J Exp Med 106:743–755

Kraft LM (1958) Observations on the control and natural history of epidemic diarrhea of infant mice (EDIM). Yale J Biol Med 31:121–137

Kraft LM (1961) Responses of the mouse to the virus of epidemic diarrhea of infant mice. Neutralizing antibodies and carrier state. Proc Anim Care Panel 11:125–136

Kraft LM (1962) Two viruses causing diarrhea in infant mice. In: Harris RJC (ed) The problems of laboratory animal disease. Academic, New York, pp 115–127

Little LM, Shadduck JA (1982) Pathogenesis of rotavirus infection in mice. Infect Immun 38:755–763

Offit PA, Clark HF (1985) Protection against rotavirus-induced gastroenteritis in a murine model by passively acquired gastrointestinal but not circulating antibodies. J Virol 54:58–64

Osborne MP, Haddon SJ, Spencer AJ, Collins J, Starkey WG, Wallis TS, Clarke GJ, Worton KJ, Candy DCA, Stephen J (1988) An electron microscopic investigation of time-related changes in the intestine of neonatal mice infected with murine rotavirus. J Pediatr Gastroenterol Nutr 7:236–248

Osborne MP, Haddon SJ, Warton KH, Spencer AJ, Starkey WG, Thornber D, Stephen J (1991) Rotavirus-induced changes in the microcirculation of intestinal villi of neonatal mice in relation to the induction and persistence of diarrhea (see comments). J Pediatr Gastroenterol Nutr 12:111–120

Pappenheimer AM, Cheever FS (1948) Epidemic diarrheal disease of suckling mice. IV. Cytoplasmic inclusion bodies in intestinal epithelium in relation to the disease. J Exp Med 88:317–324

Ramig RF (1988) The effects of host age, virus dose, and virus strain on heterologous rotavirus infection of suckling mice. Microb Pathol 4:189–202

Riepenhoff-Talty M, Lee PC, Carmody PJ, Barrett HJ, Ogra PL (1982) Age-dependent rotavirus-enterocyte interactions. Proc Soc Exp Biol Med 170:146–154

Riepenhoff-Talty M, Dharakul T, Kowalski E, Ogra PL (1987) Persistent rotavirus infection in mice with severe combined immunodeficiency. J Virol 61:3345–3348

Sheridan JF, Eydelloth RS, Vonderfecht SL, Aurelian L (1983) Virus-specific immunity in neonatal and adult mouse rotavirus infection. Infect Immun 39:917–927

Smith AL, Knudson DL, Sheridan JF, Paturzo FX (1983) Detection of antibody to epizootic diarrhea of infant mice (EDIM) virus. Lab Anim Sci 33:442–445

Starkey WG, Collins J, Wallis TS, Clarke GJ, Spencer AJ, Haddon SJ, Osborne MP, Candy DCA, Stephen J (1986) Kinetics, tissue specificity and pathological changes in murine rotavirus infection of mice. J Gen Virol 67:2625–2634

Starkey WG, Collins J, Candy DC, Spencer AJ, Osborne MP, Stephen J (1990) Transport of water and electrolytes by rotavirus-infected mouse intestine: a time course study. J Pediatr Gastroenterol Nutr 11:254–260

Thouless ME, Bryden AS, Flewett TH, Woode GN, Bridger JC, Snodgrass DR, Herring JA (1977) Serological relationships between rotaviruses from different species as studied by complement fixation and neutralization. Arch Virol 53:287–294

Ward RI, McNeal MM, Sheridan JF (1990) Development of an adult mouse model for studies on protection against rotavirus. J Virol 64:5070–5075

Wilsnack RE, Blackwell JH, Parker JC (1969) Identification of an agent of epizootic diarrhea of infant mice by immunofluorescent and complement fixation tests. Am J Vet Res 30:1195–1204

Wolf JL, Cukor G, Blacklow NR, Dambrauskas R, Trier JS (1981) Susceptibility of mice to rotavirus infection: effects of age and administration of corticosteroids. Infect Immun 33:565–574

Woode GN, Bridger JC, Jones JM, Flewett TH, Bryden AS, Davies HA, White GBB (1976) Morphological and antigenic relationships between viruses (rotaviruses) from acute gastroenteritis of children, calves, piglets, mice and foals. Infect Immun 14:804–810

Adenovirus Infection, Intestine, Mouse, Rat

Stephen W. Barthold

Synonyms. None

Gross Appearance

Gross lesions are absent. Affected mice can be runted, but are usually asymptomatic.

Microscopic Features

The only discernible microscopic lesion in infected mice is the presence of characteristically large, basophilic, intranuclear inclusion bodies in the epithelium of the small intestine and cecum (Hashimoto et al. 1970; Luethans and Wagner 1983; Takeuchi and Hashimoto 1976). The nuclei of affected cells are enlarged and are often atypically situated in the apical position of the cell column, readily permitting visualization at low power (Fig. 352). Inclusions are most numerous in infant mice, but can be found in adults in fewer numbers (S.W. Barthold, unpublished observations). Few to numerous inclusions have been observed in the small intestine of clinically normal athymic nude (*nu/nu*) mice (Cohen and de Groot 1976). Other organs are not affected.

Ultrastructure

Nuclei of infected mouse intestinal epithelial cells are enlarged and contain viral masses, often embedded in a homogeneous matrix of chromatin material. Infection is confined to columnar and goblet cells on villi and Paneth and goblet cells in crypts. Viral particles occur in aggregates and crystalline arrays (Fig. 353). They are uniform in size, measuring 75 ± 5 nm in diameter. Most particles have dense cores, but some have electronlucent centers. Free and membrane-bound virions are also present in the cytoplasm (Cohen and de Groot 1976; Luethans and Wagner 1983; Takeuchi and Hashimoto 1976).

Differential Diagnosis

Intestinal lesions associated with mouse adenovirus are nearly pathognomonic and can be confirmed by electron microscopy. K virus, a mouse papovavirus, can also induce intranuclear inclusion bodies in the small intestine of young mice, but unlike mouse adenovirus, they are present in the lamina propria and are not as obvious (see p. 171). Nuclei in the apex of villus epithelium must be differentiated from similarly located nuclei in the process of mitosis and from intraepithelial lymphocytes. Several nuclei may have to be scrutinized before convincing evidence of an inclusion is established.

Biologic Features

Natural History

At least two serologically and biologically distinct types of adenovirus are found in the laboratory

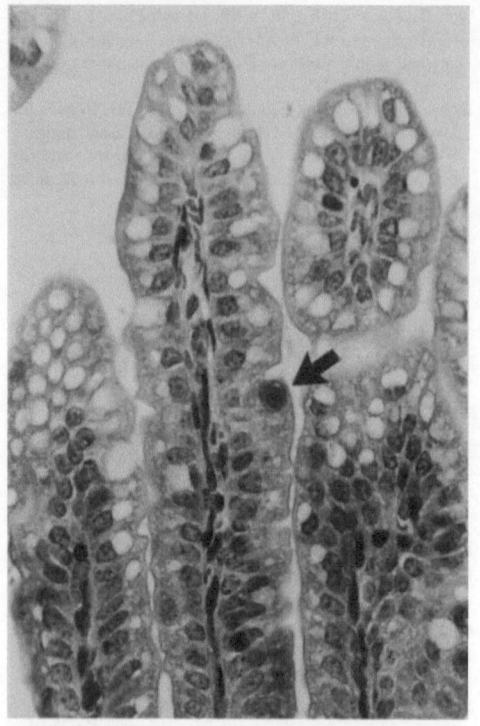

mouse. Adenovirus FL (AdFL) produces a subclinical multisystemic infection with prolonged viruria in adult mice and fatal disease in neonates. The AdFL strain is transmitted by contact and is excreted from urine and probably nasal secretions, but not feces (Hartley and Rowe 1960; van der Veen and Mes 1973). The lack of AdFL enterotropism needs to be verified, as intestinal lesions and inclusions can be seen in severe combined immunodeficient (*scid*) mice experimentally inoculated with AdFL (S.W. Barthold, unpublished observations). Intestinal adenovirus infections are due to a second type of adenovirus, which possesses a high degree of tropism for intestine. The intestinal strain that has been isolated and best characterized is the adenovirus K87 (AdK87) strain of mouse adenovirus. This virus infects only intestine, is excreted in feces, and is inefficiently transmitted (Hashimoto et al. 1966; Smith and Barthold 1987; Sugiyama et al. 1967; Takeuchi and Hashimoto 1976). Rats appear to be infected with a serologically related, but distinct intestinal adenovirus (Smith and Barthold 1987).

Pathogenesis

Intestinal mouse adenovirus infects only intestine, regardless of the route of inoculation, underscoring its high degree of enterotropism. Following oral inoculation of 4-week-old mice, virus can be isolated from feces as early as 3 days and may persist for 3 or more weeks, followed by apparent recovery. Maximum infections occur between 7 and 14 days. The course of infection in younger mice is similar (Hashimoto et al. 1970; Sugiyama et al. 1967). Mice of all ages are usually asymptomatic following experimental and natural infection (Luethans and Wagner 1983; Sugiyama et al. 1967), but transient runting has been observed (S.W. Barthold, unpublished observations). Infection appears to be restricted to the small intestine, particularly distal segments, and cecum. Only

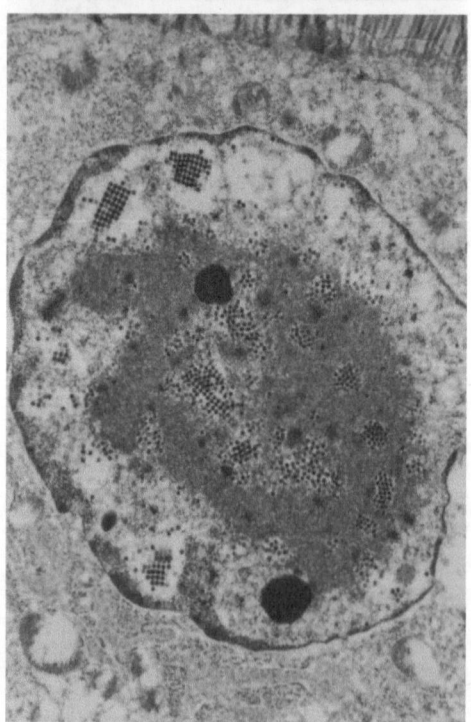

◄

Fig. 352. (*above*) Mucosa, small intestine of a weanling mouse naturally infected with intestinal adenovirus. The nucleus of an atypically aligned enterocyte contains a basophilic inclusion surrounded by a halo (*arrow*). H&E, ×450

Fig. 353. (*below*) Adenovirus, nucleus of enterocyte, small intestine, mouse. Crystalline arrays of virus particles and homogeneous viral matrix. From a naturally infected mouse. (Courtesy of Dr. T.N. Leuthans and Dr. J.E. Wagner and Laboratory Animal Science.) TEM, ×11000

enterocytes are infected with virus (S.W. Barthold, unpublished observations; Cohen and de Groot 1976; Luethans and Wagner 1983; Sugiyama et al. 1967). Recovered mice and adult mice are resistant to AdK87 challenge (Smith and Barthold 1987; Sugiyama et al. 1967). Infected cells bearing intranuclear inclusion bodies possess ultrastructural evidence of degeneration and are often observed in the process of extrusion (Takeuchi and Hashimoto 1976), explaining their atypical position in villar cell columns.

Etiology

Adenoviruses occur in many species, infecting intestine and other organs, but are generally specific (Sugiyama et al. 1967). Intestinal adenovirus infection in mice is due to a specific strain or strains of mouse adenovirus. The enteric AdK87 strain can be differentiated from the nonenteric AdFL strain serologically. Although adenoviruses of many species share common complement-fixing antigens, a lack of complement fixation and serum neutralization reactivity between AdK87 and AdFL strains has been reported (van der Veen and Mes 1974). Others have reported one-way neutralization cross-reactivity, with AdK87 antiserum neutralizing both AdFL and AdK87 strains, but AdFL antiserum only weakly neutralized AdK87 (Wigand et al. 1977). Clinically normal rats can have antibody to AdK87, but rats are not experimentally susceptible to infection with AdK87 (Smith and Barthold 1987). Adenoviral inclusions and virus particles have been observed in intestinal epithelium of immunocompromised rats, but the agent was not isolated (Ward and Young 1976).

Frequency

The frequency of intestinal mouse adenovirus infection is difficult to discern because of the lack of clinical signs, relatively few inclusions in adult mice, and frequent use of the wrong antigen in serologic tests. Antibodies to mouse adenovirus FL strain are rarely identified in contemporary mouse populations. Seroconversion to AdK87 is more common, but prevalence within an infected population is usually low, probably due to inefficient transmission, especially among adult mice (Cohen and de Groot 1976; Luethans and Wagner

1983; Lussier et al. 1987; Smith and Barthold 1987; Smith et al. 1986). The fact that this infection is recognized in the Netherlands, Japan, and the United States suggests that the virus has a widespread geographic distribution. Within an infected population, the chance of encountering lesions is highly dependent on the age of the mice and the stage of infection.

Comparison with Other Species

Intestinal adenoviral infections, with or without clinical signs of illness, occur in several species, including chickens, turkeys, geese, pigs, dogs, horses, rabbits, calves, sheep, rats, mice, nonhuman primates, and humans (Corrier et al. 1982; Ducatelle et al. 1982; Sugiyama et al. 1967; Ward and Young 1976; Wigand et al. 1982).

References

Cohen B, de Groot EG (1976) Adenovirus infection in athymic (nude) mice. Lab Anim Sci 26:955–956

Corrier DE, Montgomery D, Scutchfield WL (1982) Adenovirus in the intestinal epithelium of a foal with prolonged diarrhea. Vet Pathol 19:564–567

Ducatelle R, Coussement W, Hoorens J (1982) Sequential pathological study of experimental porcine adenovirus enteritis. Vet Pathol 19:179–189

Hartley JW, Rowe WP (1960) A new mouse virus apparently related to the adenovirus group. Virology 11:645–647

Hashimoto K, Sugiyama T, Sasaki S (1966) An adenovirus isolated from the feces of mice 1. Isolation and identification. Jpn J Microbial 10:115–125

Hashimoto K, Sugiyama T, Yoshikawa M, Sasaki S (1970) Intestinal resistance in the experimental enteric infection of mice with a mouse adenovirus. I. Growth of the virus and appearance of a neutralizing substance in the intestinal tract Jpn J Microbial 14:381–395

Luethans TN, Wagner JE (1983) A naturally occurring intestinal mouse adenovirus infection associated with negative serologic findings. Lab Anim Sci 33:270–272

Lussier G, Smith AL, Guenette D, Descoteaux LP (1987) Serological relationship between mouse adenovirus strains FL and K87. Lab Anim Sci 37:55–57

Smith AL, Barthold SW (1987) Factors influencing susceptibility of laboratory rodents to infection with mouse adenovirus strains K87 and FL. Brief report. Arch Virol 95:143–148

Smith AL, Winograd DF, Burrage TG (1986) Comparative biological characterization of mouse adenovirus strains FL and K87 and seroprevalence in laboratory rodents. Arch Virol 91:233–246

Sugiyama T, Hashimoto K, Sasaki S (1967) An adenovirus isolated from the feces of mice. II. Experimental infection. Jpn J Microbial 11:33–42

Takeuchi A, Hashimoto K (1976) Electron microscope study of experimental enteric adenovirus infection in mice. Infect Immun 13:569–580

van der Veen J, Mes A (1973) Experimental infection with mouse adenovirus in adult mice. Arch Gesamte Virusforsch 42:235–241

van der Veen J, Mes A (1974) Serological classification of two mouse adenoviruses. Arch Virol 45:386–387

Ward JM, Young DM (1976) Latent adenoviral infection of rats: intranuclear inclusions induced by treatment with a cancer chemotherapeutic agent. J Am Vet Med Assoc 169:952–953

Wigand R, Gelderblom H, Ozel M (1977) Biological and biophysical characteristics of mouse adenovirus, strain FL. Arch Virol 54:131–142

Wigand R, Bartha A, Dreizin RS, Esche H, Ginsberg HS, Green M, Hierholzer JC, Kalter SS, McFerran JB, Pettersson U, Russell WC, Wadell G (1982) Adenoviridae: second report. Intervirology 18:169–176

Infectious Diarrhea of Infant Rats (Rotavirus)

Linden E. Craig and John D. Strandberg

Synonyms. IDIR

Gross Appearance

Affected suckling rats have erythema, cracking, and bleeding of the perianal skin. Stunted growth and generalized dry, flaky skin may also be seen. The colon contains poorly formed fecal pellets along with fluid, gas, and a small amount of mucus. The small intestines contain tan to green fluid and gas. The stomach usually contains milk (Rosenblum and Leach 1985).

Microscopic Features

The most striking histologic features are syncytial cells with numerous intracytoplasmic eosinophilic inclusions in the villous mucosal epithelium of the distal third of the small intestine (Fig. 354). Syncytial cells occur exclusively on the villi and never in the crypts. These striking changes are only seen on the first day following infection. Nonspecific changes such as villous epithelial cell necrosis and villous shortening are also observed early in the infection. By day 3 of infection, no syncytia or inclusions are present; mild crypt hyperplasia is the only finding. By day 5 of infection, the villi appear normal, but the crypts are still slightly elongated (Van Campen and Scaife 1967).

Ultrastructure

The syncytial cells have an intact microvillous border and relatively electronlucent cytoplasm (Fig. 355). Large numbers of viral particles can be found within the cytoplasm of the syncytial cells associated with amorphous or reticular aggregates of electron-dense viral precursor material (Fig. 356). The virions are often found within vesicles of the rough endoplasmic reticulum. The particles are 80 nm in diameter with a thick electron-dense outer coat, a thin inner lucent zone, and an irregular electron-dense core measuring approximately 65 nm in diameter. No syncytia or viral particles are present by day 3 of infection (Van Campen and Scaife 1967).

Differential Diagnosis

There are no other viral diarrheas reported in rats. Bacterial agents causing diarrhea in infant rats include *Escherichia coli* and *Enterococcus faecium-durans-2*. The enterococcal (streptococcal) enteropathy causes high morbidity and mortality (Wolf 1973), in contrast to rotaviral enteropathy, which is rarely, if ever, fatal.

Biologic Features

Natural History

The spontaneous disease is mild and not usually noticed clinically unless the perianal region of the

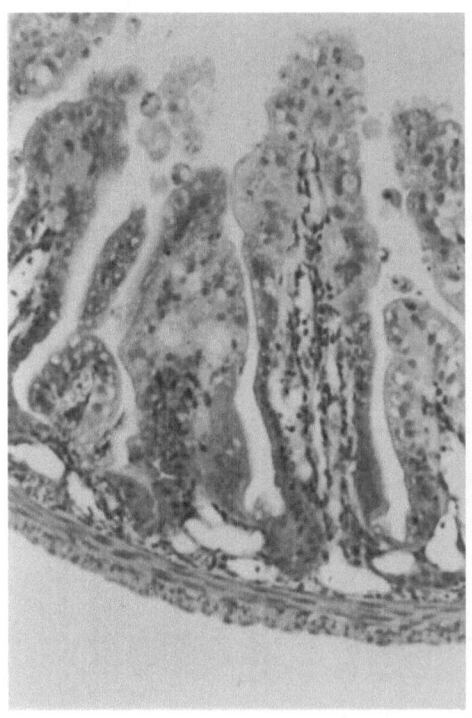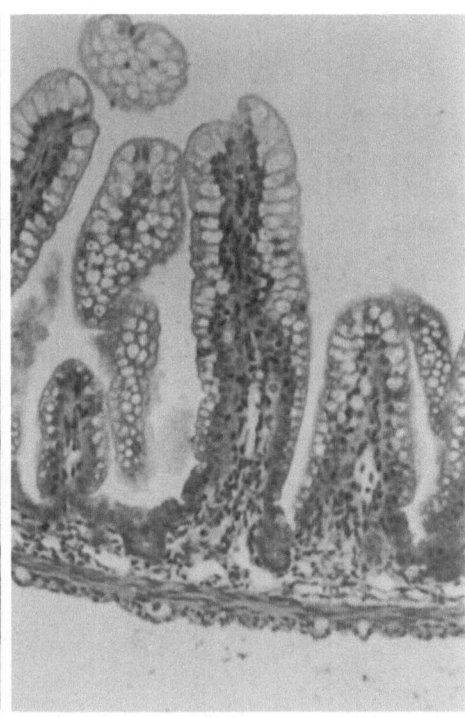

Fig. 354. Jejunal mucosa of a rat 1 day following infection with rotavirus demonstrating syncytial cells containing numerous intracytoplasmic inclusions at the tips of the villi (*left*). Compare with normal jejunal mucosa (*right*). (Courtesy of S. Vonderfecht) H&E, ×250

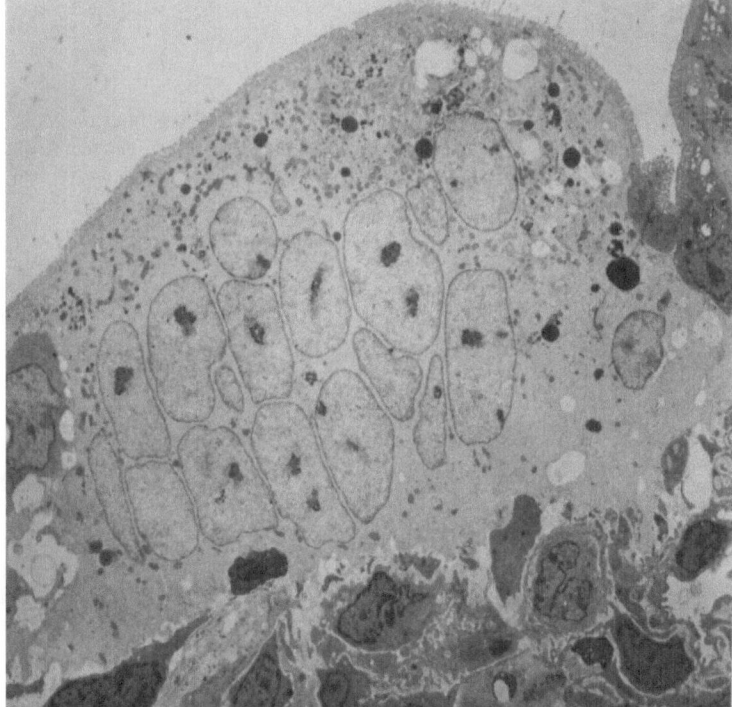

Fig. 355. Rotavirus infection, jejunum, rat. An epithelial syncytial cell demonstrating the intact microvillous border and relatively electronlucent cytoplasm. TEM, ×1000

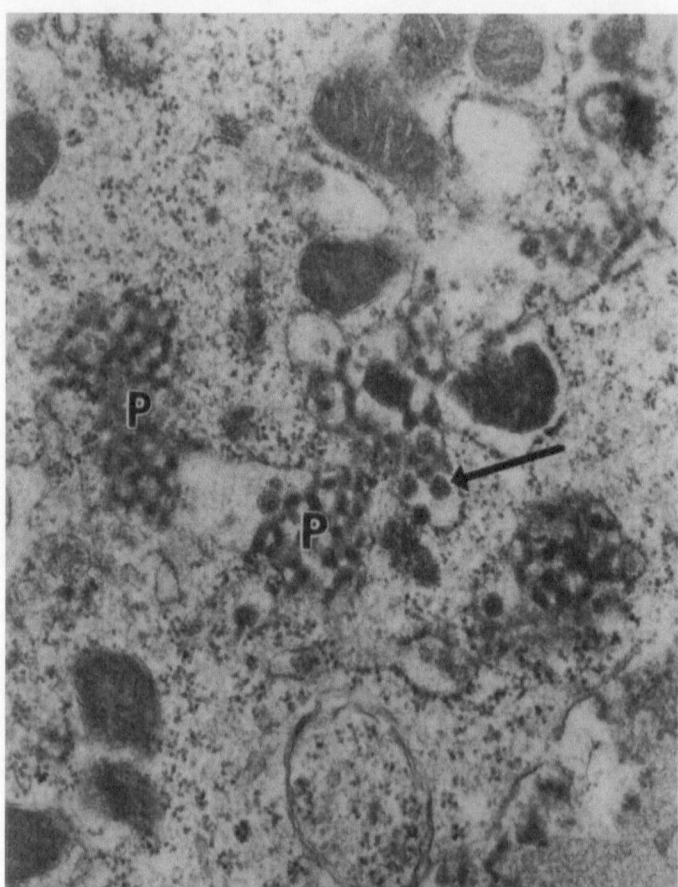

Fig. 356. Viral precursor material (*P*) and virions (*arrow*) within the cytoplasm of an rotavirus-infected epithelial cell. TEM, ×25000

suckling rats is examined. All rats in a litter are usually affected. The diarrhea typically persists for 5–6 days, but no deaths have been reported (Rosenblum and Leach 1985).

Pathogenesis

The route of transmission is fecal–oral. Infection of the terminally differentiated mature enterocytes causes malabsorption of nutrients and watery diarrhea. The affected enterocytes slough and are replaced by the normal migration of unaffected proliferating crypt cells (Peckham 1978).

Etiology

The etiologic agent is a group-B rotavirus. Rotaviruses are a morphologically identical double-stranded RNA viruses with segmented genomes. The non-group-A rotaviruses were previously called atypical rotaviruses, pararotaviruses, rotavirus-like viruses, or novel rotaviruses (Peckham 1978). Rotaviruses are now classified into groups (A–F) based on the migration patterns of their RNA segments (Schmitt et al. 1983). The group-B rotaviruses have been associated with diarrhea in calves, lambs, piglets (Sternlieb 1972), and humans (Sternlieb 1987).

Comparison with Other Species

The rotavirus isolated from rats with diarrhea is antigenically related to other group-B rotaviruses causing diarrhea in calves, lambs, piglets (Sternlieb 1972), and humans (von Arx and

Neher 1963). The human group-B rotaviral diarrhea has been epidemic in China since 1982 (Sternlieb 1987). A group-B rotavirus isolated from diarrheic humans in Baltimore was shown to cause a disease syndrome identical to infectious diarrhea of infant rats (IDIR) when inoculated into infant rats (Laemmli 1970).

References

Laemmli UK (1970) Cleavage of structural proteins during the assembly of the head of bacteriophage T4. Nature 227:685

Peckham MC (1978) Vices and miscellaneous diseases. In: Hofstad MS (ed) Diseases of poultry. Ames, Iowa State University Press, pp 847–893

Rosenblum CI, Leach RM Jr (1985) Biliary copper excretion in the chicken. Biol Trace Elem Res 8:47

Schmitt RC, Darwish HM, Cheney JC, Ettinger MJ (1983) Copper transport kinetics by isolated rat hepatocytes. Am J Physiol 244:G183–G191

Sternlieb I (1972) Evolution of the hepatic lesion in Wilson's disease (hepatolenticular degeneration). In: Popper H, Schaffner F (eds) Progress in liver disease: liver diseases. New York, Grune and Stratton, pp 511–525

Sternlieb I (1987) Hepatic lysosomal copper-thioneine. EXS 52:647–653

Van Campen DR, Scaife PV (1967) Zinc interference with copper absorption in rats. J Nutr 91:473–476

von Arx E, Neher R (1963) Eine multidimentionale Technik zur chromatographischem Identifierung von Aminosauren. J Chromatogr 12:329

Wolf P (1973) Serum ceruloplasmin determination. In: Wolf PL, Williams D (eds) Practical clinical enzymology: techniques and interpretations. Wiley-Interscience, New York, p 26

BACTERIAL INFECTIONS

Clostridial Enteropathies, Hamster*

Jerold E. Rehg

Synonyms. Acute cecitis (typhlitis), acute necrotizing cecitis (typhlitis), hemorrhagic cecitis (typhlitis), enterocolitis, antibiotic-associated cecitis (typhlitis, ileocecitis, enterocolitis, colitis)

Gross Appearance

The cecum is usually distended with gas and a fetid watery or semiliquid content (Figs. 357, 358). Occasionally, the cecum is not distended and contains only a small amount of fluid. The cecal content may be cream, tan, or red (bloody) in appearance. Infrequently, pseudomembranous plaques are adhered to focal areas of the cecal mucosa. The small intestine and colon may be distended with fluid chyme. Often, the distal ileum and proximal colon are also hyperemic or hemorrhagic. If the animal had clinically apparent diarrhea, the distal (terminal) colon and rectum are usually distended with watery stool, but sometimes they are empty.

Microscopic Features

A spectrum of microscopic lesions occurs, ranging from a mild acute typhlitis to a pseudomembranous typhlitis; however, the latter condition is rare in the hamster (Browne et al. 1977; Rehg and Lu 1982). Mild typhlitis is characterized by hyperemia and a few polymorphonuclear cells in the lamina propria, and the surface epithelial cells may be distorted or shortened and may have projecting blebs (Figs. 359, 360). In other instances, the inflammatory reaction also involves the submucosa, and both the lamina propria and submucosa may be edematous (Fig. 361). Occa-

*This review covers *Clostridium* infections other than *Clostridium piliforme*, the etiologic agent of Tyzzer's disease.

sionally, pseudomembranous plaques of cellular debris and exudate are adhered to an ulcerated, hemorrhagic mucosal surface (Fig. 362). In addition, the crypts sometimes may be elongated (Fig. 362; Borriello et al. 1987). Recently, hyperplasia of the cecal mucosa has been reported in association with spontaneous *C. difficile* typhlitis in hamsters (Chang and Rohwer 1991; Ryden et al. 1991). The distal ileum and proximal colon may have microscopic features similar to those in the cecum.

Ultrastructure

As reported by Humphrey et al. (1979), the ultrastructural features of *C. difficile* typhlitis in hamsters are nonspecific. The microvilli of affected surface epithelial cells are distorted, irregular in length, tufted, or absent. Some surface epithelial cells also contain large intracellular vacuoles, indicative of cellular edema. In edematous cells, the endoplasmic reticulum is swollen and the mitochondria are crenated.

Differential Diagnosis

Salmonella (Innes et al. 1956), enteropathogenic *Escherichia coli* (Frisk et al. 1981), and *Bacillus piliformis* (Zook et al. 1977), which is now designated *Clostridium piliforme* (Duncan et al. 1993), each produce a typhlitis in hamsters that is often similar to the disease caused by *C. difficile*. In *Salmonella* and *C. piliforme* infections, other organs are often affected, but intestinal lesions are the only changes reported to be associated with *C. difficile* infections. However, interstitial pneumonia and hepatocellular vacuolization have been described in rats administered *C. difficile* toxin experimentally (Czuprynski et al. 1983). When

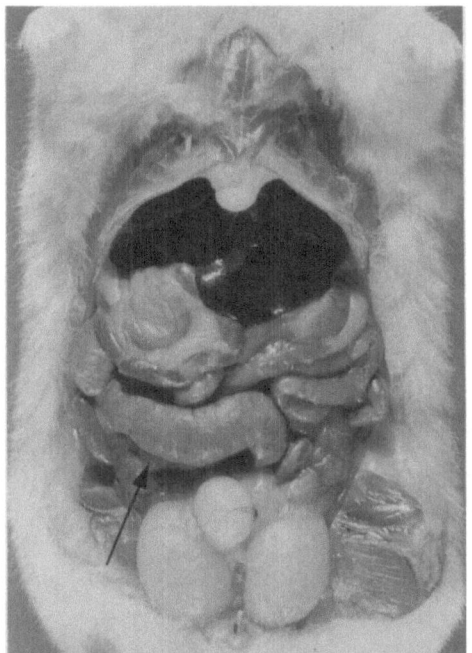

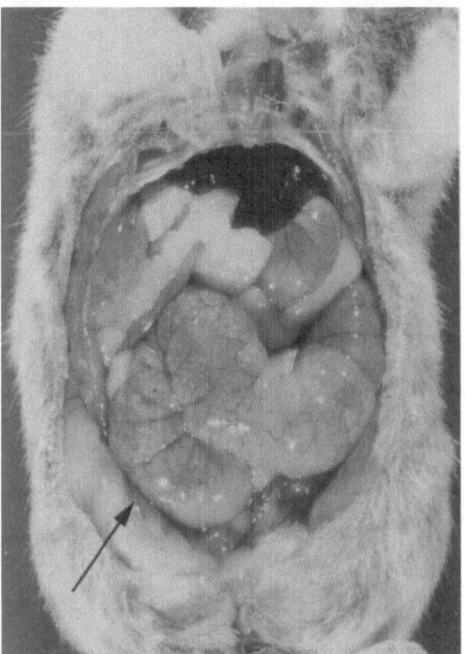

Fig. 357. (*above*) Abdominal viscera of a healthy hamster. Compare the size of the cecum (*arrow*) with the cecum in Fig. 358

Fig. 358. (*below*) Abdominal viscera of a hamster with clindamycin-induced *C. difficile* typhlitis. The cecum (*arrow*) is distended with gas and water. In this case, the cecum is enlarged but not hemorrhagic

culture and toxin assays are inconclusive, special stains and electron microscopy aid in differentiating *C. difficile* typhlitis from that of the other three organisms. *C. difficile* does not invade intestinal epithelial cells, whereas *Salmonella, E. coli* , and *C. piliforme* can be easily observed within the cytoplasm of enterocytes with special stains or electron microscopy.

Cecal mucosal hyperplasia may occur with *C. difficile* infection (Chang and Rohwer 1991; Ryden et al. 1991) and in the hyperplastic enteropathy "transmissible ileal hyperplasia" that has been attributed to a campylobacter-like organism recently designated *Desulfovibrio* sp. (Fox et al. 1994). Generally, *C. difficile* disease can be readily differentiated from *Desulfovibrio* hyperplastic enteropathy. In contrast to *C. difficile* disease, *Desulfovibrio* infection is associated with ileal hyperplasia in addition to cecal hyperplasia and, unlike *C. difficile, Desulfovibrio* is an intracytoplasmic pathogen generally identifiable in infected epithelial cells with silver stains and electron microscopy.

Although not yet reported, concurrent infection by *C. difficile* and one of the other four bacterial pathogens listed above is possible. In these cases, a differential diagnosis would be problematic.

The gross and microscopic lesions of intestinal *C. difficile* infection are similar to those of other *Clostridium* infections (Borriello et al. 1984; Borriello and Carman 1983; Knoop 1979). Although *Clostridium* species other than *C. difficile* and *C. piliforme* have not been established as causes of typhlitis in hamsters, these infections may occur. Not long ago, *C. difficile* itself was considered nonpathogenic. *C. sordellii,* which has been implicated in enteric clostridiosis in calves and foals (Al-Mashat and Taylor 1983; Hibbs et al. 1977), has also been isolated from hamsters with antibiotic-associated colitis. Whether *C. sordellii* is a hamster pathogen remains to be proved. *C. perfringens* type D has also been associated with an enterocolitis in hamsters (Goldman et al. 1972), but there has not been any confirmation that this organism is a cause of disease. Although *C. spiroforme,* a cause of intestinal clostridiosis in the rabbit (Rehg and Pakes 1982), has not been implicated in spontaneous intestinal disease in the hamster, it causes a fatal typhlitis when inoculated experimentally to hamsters administered clindamycin (Borriello and Barclay 1985).

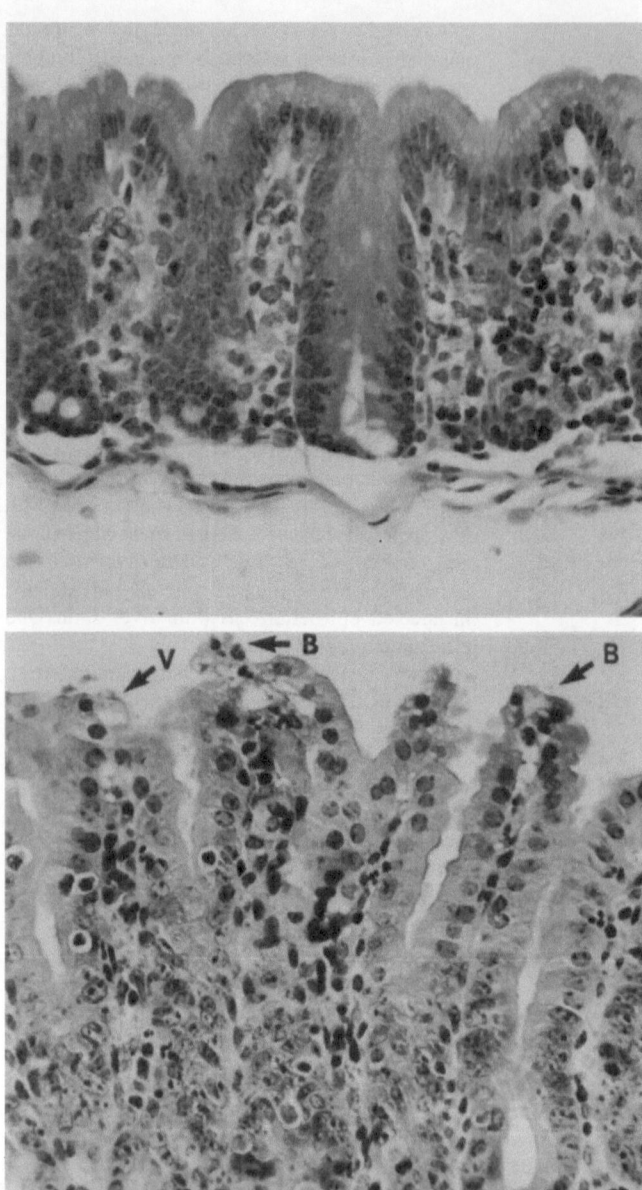

Fig. 359. (*above*) Cecum of a healthy hamster. The mucosal epithelial cells are uniform, the lamina propria has normal cellularity, and the surface contour is smooth. H&E, ×350

Fig. 360. (*below*) Cecum of a clindamycin-treated hamster. The mucosa is thickened, its surface is distorted, and the lamina propria is increased in cellularity. Some degenerative epithelial cells are vacuolated (*V*), others are rounded and are being extruded as blebs (B). H&E, ×350

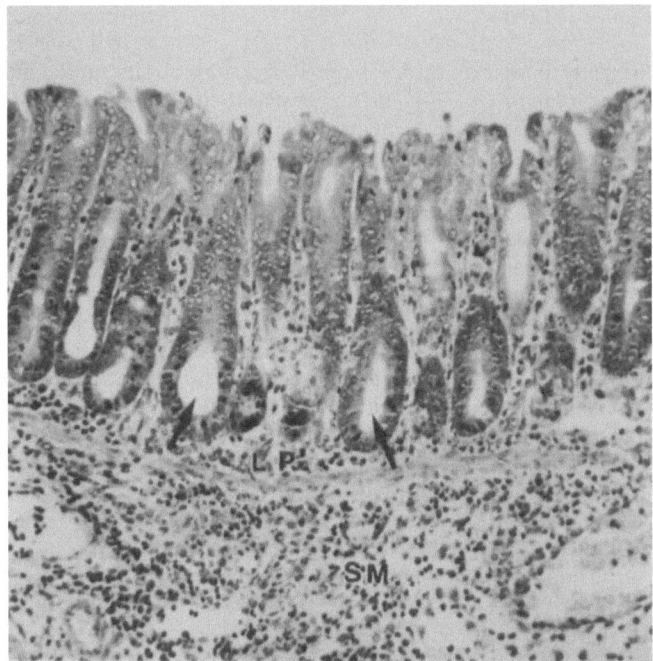

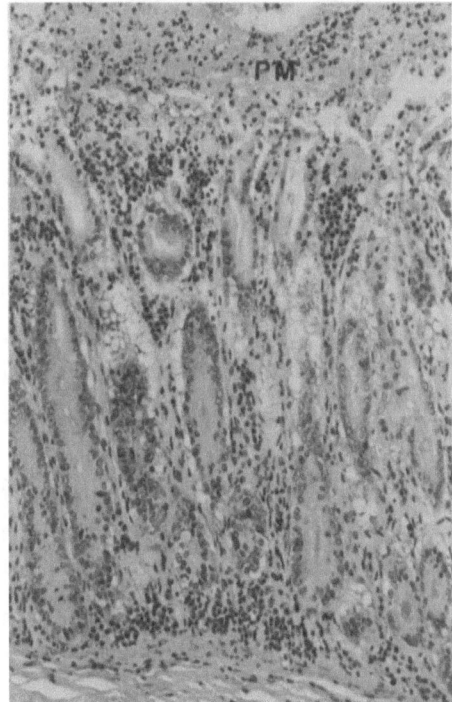

Fig. 361. (*above*) Cecum of a penicillin-treated hamster. The mucosal surface is distorted and the lamina propria (*LP*) and submucosa (*SM*) are hyperemic; both are intensely infiltrated by leukocytes and are edematous. Several crypts (*arrows*) are dilated. H&E, ×300

Fig. 362. (*below*) Typhlitis in the cecum of a hamster that had not received antibiotics. The mucosa is thickened, and the lamina propria is edematous and infiltrated by leukocytes. The surface epithelium is focally eroded. Fibrin, inflammatory cells, and cellular debris form a pseudomembranous plaque (*PM*) that is adherent to the surface. H&E, ×170

Biologic Features

Presently, intestinal clostridiosis in the hamster is caused primarily by *C. difficile*, and treatment with antibiotics usually precipitates the disease. Toxigenic *C. difficile* will colonize the intestines of hamsters that are 4–15 days-old, and high levels of toxin are present in the ceca (Rolfe and Iaconis 1983). However, these animals do not die or develop intestinal disease unless they receive antibiotics (Iaconis and Rolfe 1986). Although *C. difficile* does not colonize the intestinal tract of hamsters less than 4 days of age, these animals also develop disease upon antibiotic treatment. Thus hamsters of any age can develop antibiotic-induced *C. difficile* typhlitis. Further, Griffin et al. (1984) demonstrated that 5-month-old hamsters that received clindamycin are more susceptible to *C. difficile* infection and disease than are similarly treated 25-day-old hamsters. Further, the majority of the hamsters in the reports of spontaneous *C. difficile* in hamsters not receiving antibiotics were 5 months of age or older. These observations suggest that the age of a hamster may influence its susceptibility to *C. difficile* infection.

Although *C. difficile* typhlitis may occur in hamsters that have not been treated with antibiotics (Rehg and Lu 1981; Chang and Rohwer 1991; Ryden et al. 1991), it most commonly develops subsequent to antibiotic use. Erythromycin was the first antibiotic to be implicated with antibiotic-associated lethality in hamsters (Kaipainen and Faine 1954). Since that time, almost all bacterial agents have been implicated in antibiotic-associated typhlitis in the hamster (Bartlett et al. 1978; Fekety et al. 1979). Affected hamsters become lethargic, anorexic, and hypothermic, and approximately 50% of those lethally infected develop diarrhea (Browne et al. 1977). The number of animals in any particular group that die following antibiotic administration varies. In some animal populations no deaths occur, whereas in others the death rate may be as high as 100%. The concentration of *C. difficile* in the environment, the carrier rate in animal colonies, and the type of housing are factors that apparently influence the number of hamsters that develop disease. Onset of the disease usually occurs 5–10 days after initial administration of the precipitating antibiotic, but it may be delayed as much as 2–6 weeks after termination of such treatment. This delayed onset has been noted with ampicillin, vancomycin, chloramphenicol, and other agents (Bartlett et al. 1978; Larson and Borriello 1990).

The precise mechanisms by which antimicrobial agents result in *C. difficile* typhlitis are not entirely understood. However, there is evidence that antimicrobials alter the "inhibitory barrier" (the natural inhibiting effect of intestinal bacteria on other organisms) of the normal intestinal flora. This modulation of the normal flora is thought to enable the intestinal colonization by, and proliferation of, *C. difficile* and the elaboration of its toxin. It is also postulated that β-lactamase-producing bacteria contributes to *C. difficile* colonization of the intestine by inactivating the antibiotics in the intestinal lumen to a level that allows *C. difficile* to multiply and cause disease (Rolfe and Finegold 1983). However, this hypothesis does not account for the development of *C. difficile*-associated typhlitis following administration of antimicrobial agents that do not stimulate β-lactamase production. Further, in some incidences, *C. difficile*-associated intestinal disease occurs after administration of the antimicrobial agent has been terminated (Bartlett et al. 1978). Two mechanisms have been proposed to explain the *C. difficile* intestinal disease that occurs after antimicrobial treatment is discontinued. First, *C. difficile* spores may form during therapy, subsequently multiplying and producing disease upon discontinuation of the antimicrobial agent and its decline to a subinhibitory level (Onderdonk et al. 1980). Second, the intestine might be recolonized by *C. difficile* from the environment after the antimicrobial agent is discontinued and its inhibitory level has declined (Larson and Borriello 1990). In both mechanisms, *C. difficile* would colonize the intestine before the intestine's normal inhibitory bacterial flora reestablished itself. It is possible that different mechanisms may be associated with different classes of antimicrobial agents.

The severity of symptoms in the hamster is closely related to the degree of toxin A production. However, the secondary role of toxin B is probably significant in certain situations, since toxin B acts synergistically with toxin A to cause disease at doses at which the toxins alone are not pathogenic (Lyerly et al. 1985; Mitchell et al. 1986; Borriello et al. 1987). Thus the final outcome of infection depends upon several interacting factors, including the degree of colonization resistance provided by the remaining gut flora, the levels of antibiotic in the gut, the antibiotic sensitivities of *C. difficile,* and the potential virulence of the infecting organisms.

Cecal hyperplasia has been reported in association with *C. difficile* typhlitis (Chang and Rohwer 1991; Ryden et al. 1991). Whether *C. difficile* was associated with the cecal hyperplasia reported by Barthold et al. (1978) is not known. Elongation of the crypts in the cecum (Fig. 362) has been observed in some hamsters with antibiotic-induced *C. difficile* disease (Borriello et al. 1987). However, hyperplasia that is characterized by gland tortuosity with branching, epithelial cell pseudostratification, hyperchromasia, and increased mitosis has not been observed in experimental *C. difficile* disease. It is speculated that the cecal hyperplasia associated with *C. difficile* disease is not due to *C. difficile,* but is a primary entity and an indication that alteration in the intestinal flora has occurred, thereby rendering the host susceptible to *C. difficile* colonization and disease. Toxigenic *C. difficile* isolates vary in virulence (Borriello et al. 1987). Therefore, cecal hyperplasia also may be a manifestation that is associated with specific *C. difficile* strains. Consequently, the pathogenesis of cecal mucosal hyperplasia in conjunction with *C. difficile* typhlitis remains to be elucidated.

Although *C. difficile* typhlitis in the hamster is usually associated with antimicrobial treatment, hypochlorhydria (Gurian et al. 1982), diet (Michelich et al. 1981; Cooperstock et al. 1983; Mahe et al. 1987), surgery (Adler et al. 1981), neuroleptics (Calmat et al. 1982), and antineoplastic drugs (Cudmore et al. 1980; Anand and Glatt 1993) have also been implicated as predisposing factors in this disease. Consequently, *C. difficile* infection is a potential problem in any toxicologic testing program, since any test compound that could alter either gastric acidity or the intestinal bacterial barrier may enable *C. difficile* to colonize the intestinal tract and cause disease. In addition, not every hamster population that receives antibiotics develops typhlitis. In these cases, the hamsters probably were not exposed to spores or vegetative forms of *C. difficile* or the animals were not carriers of the organism.

Some commercial hamster colonies have been reported to have a *C. difficile* carrier rate of 5%–13% (Toshniwal et al. 1981; Hawkins et al. 1984). These carriers are a source of infection for cage mates and of environmental contamination. Special housing systems will not prevent carriers from developing *C. difficile* disease when the inhibitory barrier of their indigenous intestinal flora is altered. However, special caging can protect hamsters that are not carriers from becoming infected with toxigenic *C. difficile* from the environment or from carrier animals (Larson et al. 1980; Hawkins et al. 1984; Borriello et al. 1987).

Thus several effective preventive measures include limiting exposure of susceptible animals to predisposing factors, isolating them from cage mates that are carriers, and placing them in specialized housing systems to prevent their infection by *C. difficile* from the environment. In addition, parenteral and passive immunization are protective. Adult hamsters immunized with either toxoid A or a mixture of toxoids A and B do not develop *C. difficile* disease (Kim et al. 1987; Lyerly et al. 1991).

Treatment with vancomycin, indigenous cecal flora from healthy adult hamsters, and *Saccharomyces boulardii* are all effective in reducing the death rate in *C. difficile* epizootics (Browne et al. 1977; Wilson et al. 1981; Toothaker and Elmer 1984; Larson and Welch 1993). Vancomycin must be administered for months to avoid infection relapse and reinfection when the treatment is discontinued (Boss et al. 1994). Whether long-term vancomycin treatment will prove to be an effective therapeutic regimen for *C. difficile* disease in the hamster remains to be determined.

Comparison with Other Species

For all practical purposes, the gross and microscopic lesions in the hamster are similar to those associated with *C. difficile* typhlitis in the guinea pig (Rehg and Pakes 1981) and rabbit (Rehg and Lu 1981). It is important to note, however, that there are several features of *C. difficile* disease in hamsters that are different from those in humans. First, the hamster rarely develops pseudomembranes, which occur frequently in humans. Second, the disease in hamsters is confined to the terminal ileum, cecum, and proximal colon, and the cecum is the primary intestinal tissue affected, whereas the predominant location of lesions in humans is the distal colon.

References

Adler SP, Chandrika T, Berman WF (1981) Clostridium difficile associated with pseudomembranous colitis. Occurrence in a 12-week-old infant without prior antibiotic therapy. Am J Dis Child 135:820–822

Al-Mashat RR, Taylor DJ (1983) Production of diarrhoea and enteric lesions in calves by the oral inoculation of pure cultures of Clostridium sordellii. Vet Rec 112:141–146

Anand A, Glatt AE (1993) Clostridium difficile infection associated with antineoplastic chemotherapy: a review. Clin Infect Dis 17:109–113

Barthold SW, Jacoby RO, Pucak GJ (1978) An outbreak of cecal mucosal hyperplasia in hamsters. Lab Anim Sci 28:723–727

Bartlett JG, Chang TW, Moon N, Onderdonk AB (1978) Antibiotic-induced lethal enterocolitis in hamsters: studies with eleven agents and evidence to support the pathogenic role of toxin-producing clostridia. Am J Vet Res 39:1525–1530

Borriello SP, Barclay FE (1985) Protection of hamsters against Clostridium difficile ileocaecitis by prior colonization with non-pathogenic strains. J Med Microbiol 19:339–350

Borriello SP, Carman RJ (1983) Association of iota-like toxin and Clostridium spiroforme with both spontaneous and antibiotic-associated diarrhea and colitis in rabbits. J Clin Microbiol 17:414–418

Borriello SP, Larson HE, Welch AR, Barclay F, Stringer MF, Bartholomew BA (1984) Enterotoxigenic Clostridium perfringens: a possible cause of antibiotic-associated diarrhoea. Lancet 1:305–307

Borriello SP, Ketley JM, Mitchell TJ, Barclay FE, Welch AR, Price AB, Stephen J (1987) Clostridium difficile – a spectrum of virulence and analysis of putative virulence determinants in the hamster model of antibiotic-associated colitis. J Med Microbiol 24:53–64

Boss SM, Gries CL, Kirchner BY, Smith GD, Francis PC (1994) Use of vancomycin hydrochloride for treatment of Clostridium difficile enteritis in Syrian hamsters. Lab Anim Sci 44:31–37

Browne RA, Fekety R Jr, Silva J Jr, Boyd DI, Work CO, Abrams GD (1977) The protective effect of vancomycin on clindamycin-induced colitis in hamsters. Johns Hopkins Med J 141:183–192

Calmat A, Faurel JP, Delas N, Bourlioux P (1982) Communication aux actualites de gastroenterologie. Seminar. Hospital St Antonoine, Paris, France

Chang J, Rohwer RG (1991) Clostridium difficile infection in adult hamsters. Lab Anim Sci 41:548–552

Cooperstock M, Riegle L, Woodruff CW, Onderdonk A (1983) Influence of age, sex and diet on asymptomatic colonization of infants with Clostridium difficile. J Clin Microbiol 17:830–833

Cudmore M, Silva J, Fekety R (1980) Clostridial enterocolitis produced by antineoplastic agents in hamsters and humans. In: Nelson JD, Grassi C (eds) Current chemotherapy and infectious disease: proceedings. American Society for Microbiology, Washington DC, p 1460

Czuprynski CJ, Johnson WJ, Balish E, Wilkins T (1983) Pseudomembranous colitis in Clostridium difficile-mono associated rats. Infect Immun 39:1368–1376

Duncan AJ, Carman RJ, Olsen GJ, Wilson KH (1993) The agent of Tyzzer's disease is a Clostridium species. Clin Infect Dis 16:S422

Fekety R, Silva J, Toshniwal R, Allo M, Armstrong J, Browne R, Ebright J, Rifkin G (1979) Antibiotic-associated colitis: effects of antibiotics on Clostridium difficile and the disease in hamsters. Rev Infect Dis 1:386–397

Fox JG, Dewhirst FE, Fraser GJ, Paster BJ, Shames B, Murphy JC (1994) Intracellular Campylobacter-like organism from ferrets and hamsters with proliferative bowel disease is a Desulfovibrio sp. J Clin Microbiol 32:1229–1237

Frisk CS, Wagner JE, Owens DR (1981) Hamster (Mesocricetus auratus) enteritis caused by epithelial cell-invasive Escherichia coli. Infect Immun 31:1232–1238

Goldman PM, Andrews EJ, Lang CM (1972) A preliminary evaluation of Clostridium sp. in the etiology of hamster enteritis. Lab Anim Sci 22:721–724

Griffin GE, McDougall A, Ash SA, Borriello SP (1984) Age related susceptibility of hamsters to clindamycin induced Clostridium difficile caecitis. Microec Ther 14:269–270

Gurian L, Ward TT, Katon RM (1982) Possible foodborne transmission in a case of pseudomembranous colitis due to Clostridium difficile: influence of gastrointestinal secretions on Clostridium difficile infection. Gastroenterology 83:465–469

Hawkins CC, Buggy BP, Fekety R, Schaberg DR (1984) Epidemiology of colitis induced by Clostridium difficile in hamsters: application of a bacteriophage and bacteriocin typing system. J Infect Dis 149:775–780

Hibbs CM, Johnson DR, Reynolds K, Harrington R Jr (1977) Clostridium sordellii isolated from foals. Vet Med Small Anim Clin 72:256–258

Humphrey CD, Lushbaugh WB, Condon CW, Pittman JC, Pittman FE (1979) Light and electron microscopic studies of antibiotic associated colitis in the hamster. Gut 20:6–15

Iaconis JP, Rolfe RD (1986) Clostridium difficile-associated ileocecitis in clindamycin-treated infant hamsters. Curr Microbiol 13:327–332

Innes JR, Wilson C, Ross MA (1956) Epizootic Salmonella enteritidis infection causing septic pulmonary phlebothrombosis in hamsters. J Infect Dis 98:133–141

Kaipainen WJ, Faine S (1954) Toxicity of erythromycin. Nature 174:969–970

Kim PH, Iaconis JP, Rolfe RD (1987) Immunization of adult hamsters against Clostridium difficile-associated ileocecitis and transfer of protection to infant hamsters. Infect Immun 55:2984–2992

Knoop FC (1979) Clindamycin-associated enterocolitis in guinea pigs: evidence for a bacterial toxin. Infect Immun 23:31–33

Larson HE, Borriello SP (1990) Quantitative study of antibiotic-induced susceptibility to Clostridium difficile enterocecitis in hamsters. Antimicrob Agents Chemother 34:1348–1353

Larson HE, Welch A (1993) In-vitro and in-vivo characterization of resistance to colonization with Clostridium difficile. J Med Microbiol 38:103–108

Larson HE, Price AB, Borriello SP (1980) Epidemiology of experimental enterocecitis due to Clostridium difficile. J Infect Dis 142:408–413

Lyerly DM, Saum KE, MacDonald D, Wilkins TD (1985) Effects of Clostridium difficile toxins given intragastrically to animals. Infect Immun 47:349–352

Lyerly DM, Bostwick EF, Binion SB, Wilkins TD (1991) Passive immunization of hamsters against disease caused by Clostridium difficile by use of bovine immunoglobulin G concentrate. Infect Immun 59:2215–2218

Mahe S, Corthier G, Dubos F (1987) Effect of various diets on toxin production by two strains of Clostridium difficile in gnotobiotic mice. Infect Immun 55:1801–1805

Michelich VJ, Nunez-Montiel O, Schuster GS, Thompson F, Dowell VR Jr (1981) Diet as a coadjuvant for development of antibiotic-associated diarrhea in hamsters (Mesocricetus auratus). Lab Anim Sci 31:259–262

Mitchell TJ, Ketley JM, Haslam SC, Stephen J, Burdon DW, Candy DCA, Daniel R (1986) Effect of toxin A and B of Clostridium difficile on rabbit ileum and colon. Gut 27:78–85

Onderdonk AB, Cisneros RL, Bartlett JG (1980) Clostridium difficile in gnotobiotic mice. Infect Immun 28:277–282

Rehg JE, Lu YS (1981) Clostridium difficile colitis in a rabbit following antibiotic therapy for pasteurellosis. J Am Vet Med Assoc 179:1296–1297

Rehg JE, Lu YS (1982) Clostridium difficile typhlitis in hamsters not associated with antibiotic therapy. J Am Vet Med Assoc 181:1422–1423

Rehg JE, Pakes SP (1981) Clostridium difficile antitoxin neutralization of cecal toxin(s) from guinea pigs with penicillin-associated colitis. Lab Anim Sci 31:156–160

Rehg JE, Pakes SP (1982) Implication of Clostridium difficile and Clostridium perfringens iota toxins in experimental lincomycin-associated colitis of rabbits. Lab Anim Sci 32:253–257

Rolfe RD, Finegold SM (1983) Intestinal β-lactamase activity in ampicillin-induced, Clostridium difficile-associated ileocecitis. J Infect Dis 147:227–235

Rolfe RD, Iaconis JP (1983) Intestinal colonization of infant hamsters with Clostridium difficile. Infect Immun 42:480–486

Ryden EB, Lipman NS, Taylor NS, Rose R, Fox JG (1991) Clostridium difficile typhlitis associated with cecal mucosal hyperplasia in Syrian hamsters. Lab Anim Sci 41:553–558

Toothaker RD, Elmer GW (1984) Prevention of clindamycin-induced mortality in hamsters by Saccharomyces boulardii. Antimicrob Chemother 26:552–556

Toshniwal R, Silva J Jr, Fekety R, Kim KH (1981) Studies on the epidemiology of colitis due to Clostridium difficile in hamsters. J Infect Dis 143:51–54

Wilson KH, Silva J, Fekety FR (1981) Suppression of Clostridium difficile by normal hamster cecal flora and prevention of antibiotic-associated cecitis. Infect Immun 34:626–628

Zook BC, Huang K, Rhorer RG (1977) Tyzzer's disease in Syrian hamsters. J Am Vet Med Assoc 171:833–836

Citrobacter freundii Infection, Colon, Mouse

Stephen W. Barthold

Synonyms. Transmissible murine colonic hyperplasia, hyperplastic colitis, catarrhal enterocolitis, colitis cystica

Gross Appearance

Clinical signs and gross lesions in *Citrobacter freundii* in the mouse colon are often difficult to detect. Infected mice can be runted with dark, sticky feces and low mortality. Rectal prolapse is a frequent sign of infection, but prevalence varies within a colony. The descending colon is contracted, is usually devoid of feces, and has an opaque, thickened wall. Other segments of large bowel can also be affected (Barthold et al. 1977, 1978; Bienick and Tober-Meyer 1976; Brennan et al. 1965; Ediger et al. 1974; Silverman et al. 1979).

Microscopic Features

Microscopic changes are consistently present in the distal (descending) colon and, in some mice, are also found in the transverse and ascending colon as well as the cecum. The quality of histologic changes varies with stage of infection. A primary feature is mucosal epithelial hyperplasia, in which crypts are elongated and populated with immature, mitotically active epithelium (Fig. 363). Surface epithelium is covered by myriads of intimately attached coccobacillary bacteria (Fig. 364). This can be accompanied by varying degrees of inflammation, mucosal erosion, and ulceration. Colonies of bacteria may be present in necrotic mucosa. In recovering mice, the mitotic index decreases, the hyperplastic epithelium undergoes maturation, and goblet cells can be abundant (Barthold 1980; Barthold et al. 1977, 1978; Johnson and Barthold 1979).

Ultrastructure

During the early phase of infection, the surface epithelium is carpeted with attached *C. freundii* bacteria which displace and distort the brush border (Fig. 365). As infection progresses, infected cells are extruded and replaced by upwardly migrating hyperplastic cells that remain uninfected. Secondary degenerative changes are present in infected cells, but intracellular bacterial invasion is not a feature of this disease (Johnson and Barthold 1979).

Differential Diagnosis

C. freundii is the most common cause of rectal prolapse in mice, but enteritis by other causes can also be associated with this lesion. Clinical signs of enteritis can be associated with rotavirus, mouse hepatitis virus, reovirus type 3, mouse adenovirus, *Salmonella* sp., *and Bacillus piliformis*, among others. Colonic or cecal hyperplasia and inflammation are also often caused by mouse hepatitis virus and *Bacillus piliformis*. Differentiating fea-

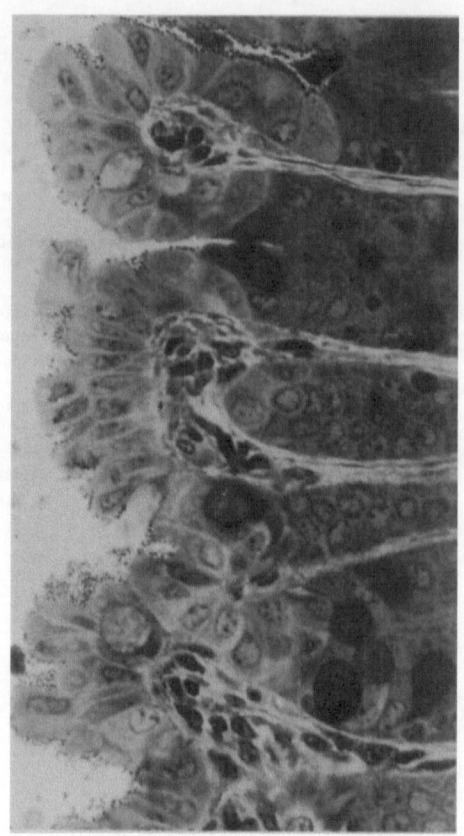

Fig. 364. Descending colon of a mouse, 6 days after inoculation with *C. freundii*. Organisms are attached to the apical cell membrane of the surface mucosal epithelium and crypt neck cells. (Courtesy of Dr. E. Johnson and Dr. S.W. Barthold and *American Journal of Pathology*.) Methylene blue and azure II, ×1100

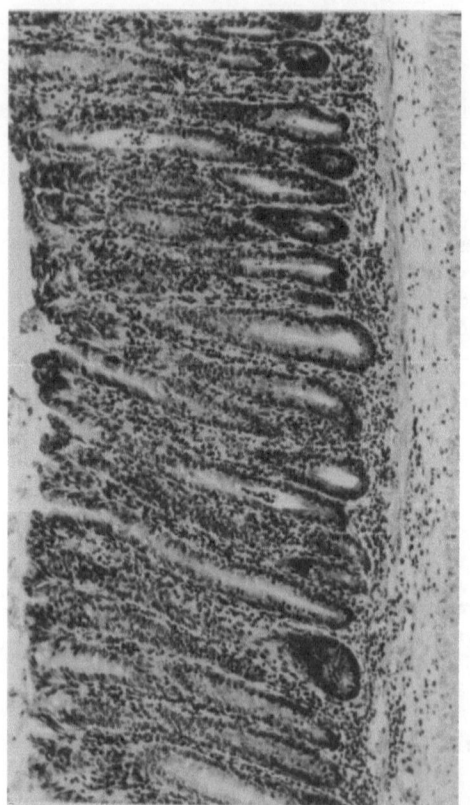

Fig. 363. Descending colon of a mouse experimentally infected with *C. freundii*. Note the crypt hyperplasia and diffuse leukocytic infiltration of the lamina propris. H&E, ×125

tures include the location of the lesion (descending colon) and the presence of *C. freundii*.

Biologic Features

Natural History

C. freundii is transmitted by direct contact and spreads slowly under natural and experimental conditions. Orofecal contamination is the likely means of spread. Husbandry practices can enhance or limit spread between cages. The source of infection in *C. freundii* outbreaks has never been determined. Infection can occur at any age, but clinical signs are most likely to occur around weaning age. The course of infection lasts only 2–

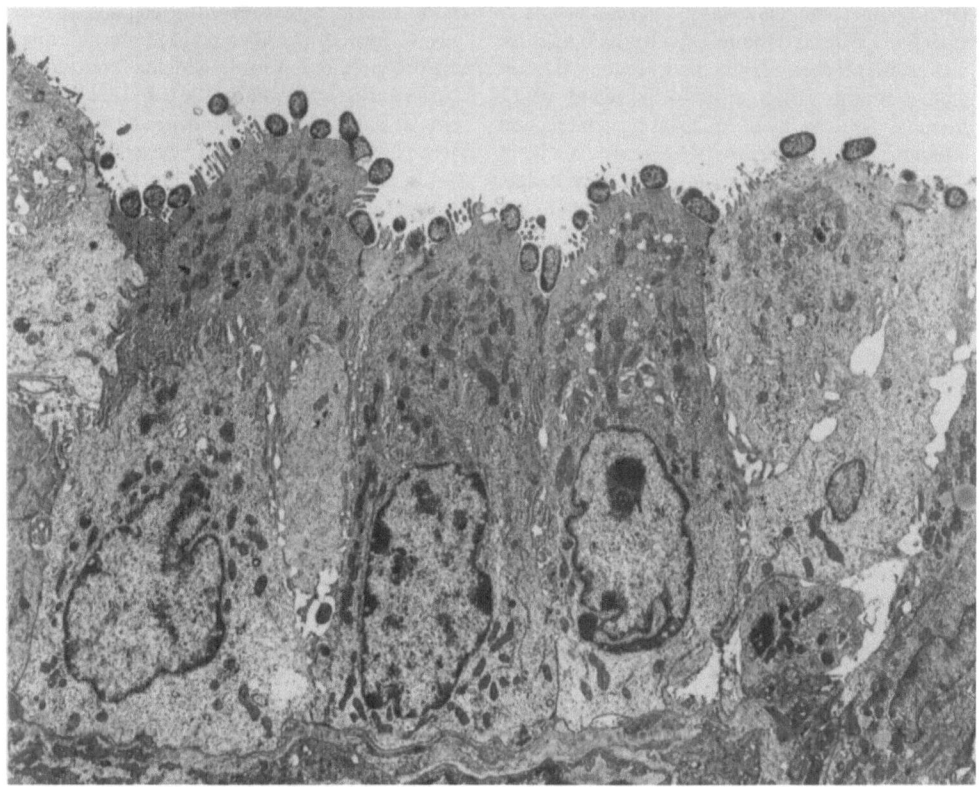

Fig. 365. Descending colon surface epithelial cells, 6 days after inoculation with *C. freundii*. The brush border is disrupted with loss of microvilli due to displacement by attached bacteria. (Courtesy of Dr. E. Johnson and Dr. S.W. Barthold and *American Journal of Pathology*.) Uranyl acetate and lead citrate, TEM, ×15200

3 weeks, with complete recovery and no carrier state (Barthold 1980; Barthold et al. 1978).

Pathogenesis

Following oral inoculation, *C. freundii* is initially isolated from small intestine, but soon colonizes and is restricted to large intestine within 4 days. *C. freundii* bacteria proliferate and cover the surface epithelium of the descending colon. Other aerobic bacteria are usually totally displaced, so that *C. freundii* is the only recoverable aerobe during this phase of infection. Microvilli are effaced, presumably altering colonic resorptive function. *C. freundii* attachment sites are reminiscent of attaching and effacing lesions of enteropathogenic and enterohemorrhagic *Escherichia coli* infections (Schauer and Falkow 1993). Because functional interference is largely restricted to distal colon,

clinical signs are limited to semiformed or sticky feces rather than diarrhea. In response to the altered surface epithelium, crypt cells become mitotically active and the migration rate up the crypt columns is rapid, resulting in a net accumulation of immature epithelial cells with crowding and elongation of crypts and accumulation of epithelium at surface extrusion zones. As hyperplastic cells migrate out of crypt openings, they displace infected cells, which are eventually sloughed. The hyperplastic response reaches its peak at 2–3 weeks after infection. At this time, all infected cells have been extruded and *C. freundii* can no longer be cultivated from the colonic mucosa or lumen. Clinical signs occur on or after the peak of hyperplasia and seem to be more related to degree of inflammation than of hyperplasia. As *C. freundii* is lost from the bowel, crypt mitotic activity undergoes quiescence with a gradual regression of crypt cellularity to normal.

During this phase, there may be excessive differentiation of mucin-producing cells. By 2 months, no residual lesions remain and recovery is complete with no carrier state or shedding of *C. freundii*. The severity of mucosal hyperplasia and inflammation associated with *C. freundii* has been shown to be influenced by mouse genotype, age, and diet. Mice are refractory to reinfection for several weeks after recovery (Barthold 1979, 1980; Barthold et al. 1977, 1978; Johnson and Barthold 1979).

Etiology

Disease in mice is caused by a single or similar variants of *C. freundii*. Most members of the genus *Citrobacter* are motile with peritrichous flagellae. However, pathogenic mouse variants are non-motile and lack both flagellae and fimbriae in vitro and in vivo. *C. freundii ANK, C. freundii* (Ediger), and *C. freundii* (4280) have been isolated from mice with similar enteric disease and appear to be closely related, but minor differences exist. Relatedness of mouse isolates has been tested on a biochemical, but not antigenic or DNA homology basis (Brennan et al. 1965; Barthold et al. 1976; Ediger et al. 1974). *C. freundii* (4280) is the best-characterized strain and seems to be identical to other strains isolated from diseased mice in recent years (Barthold et al. 1976; Silverman et al. 1979).

Frequency

The true frequency of *C. freundii* infections in mice is not known, since clinical disease is often unrecognized. Furthermore, *C. freundii* is often already absent from mice still manifesting clinical signs and lesions. *C. freundii*-associated colitis in mice has been reported in the United States and Europe (Barthold et al. 1976; Bienick and Tober-Meyer 1976; Brennan et al. 1965; Ediger et al. 1974; Silverman et al. 1979). Mice of any age can become infected, but signs of infection are most apt to be seen in weanling-age mice.

Comparison with Other Species

Mice are the only known species susceptible to the enteric effects of *C. freundii*. Among *Citrobacter*

and *C. freundii* isolates from several host species, only *C. freundii* (4280) colonized mouse bowels and produced disease following oral inoculation. *C. freundii* did not colonize the bowel of rats, hamsters, or guinea pigs (Barthold et al. 1977). Colonic hyperplasia associated with *C. freundii* has served as a model system of benign proliferative bowel disease, as it relates to preneoplastic and neoplastic disease in humans. Murine colonic hyperplasia reduces the lag period and increases the yield of early carcinogen-induced neoplastic mucosal changes in mice (Barthold 1979; Barthold and Beck 1980; Barthold and Jonas 1977).

References

Barthold SW (1979) Autoradiographic cytolcinetics of colonic mucosal hyperplasia in mice. Cancer Res 39:24–29

Barthold SW (1980) The microbiology of transmissible murine colonic hyperplasia. Lab Anim Sci 30:167–173

Barthold SW, Jonas AM (1977) Morphogenesis of early 1,2-dimethylhydrazine-induced lesions and latent period reduction of colon carcinogenesis in mice by a variant of Citrobacter freundii. Cancer Res 37:4352–4360

Barthold SW, Beck D (1980) Modification of early dimethylhydrazine carcinogenesis by colonic mucosal hyperplasia. Cancer Res 40:4451–4455

Barthold SW, Coleman GL, Bhatt PN, Osbaldiston GW, Jonas AM (1976) The etiology of transmissible murine colonic hyperplasia. Lab Anim Sci 26:889–894

Barthold SW, Osbaldiston GW, Jonas AM (1977) Dietary, bacterial, and host genetic interactions in the pathogenesis of transmissible murine colonic hyperplasia. Lab Anim Sci 27:938–945

Barthold SW, Coleman GL, Jacoby RO, Livingstone EM, Jonas AM (1978) Transmissible murine colonic hyperplasia. Vet Pathol 15:223–236

Bienick H, Tober-Meyer B (1976) Zur Atiologie der Colitis und des Prelapses recti bei der Maus. Z Versuchstierkd 18:337–348

Brennan PC, Fritz TE, Flynn RJ, Poole CM (1965) Citrobacter freundii associated with diarrhea in laboratory mice. Lab Anim Care 15:266–275

Ediger RD, Kovatch RM, Rabstein NM (1974) Colitis in mice with a high incidence of rectal prelapse. Lab Anim Sci 24:488–494

Johnson E, Barthold SW (1979) The ultrastructure of transmissible murine colonic hyperplasia. Am J Pathol 97:291–313

Schauer DB, Falkow S (1993) Attaching and effacing locus of a Citrobacter freundii biotype that causes transmissible murine colonic hyperplasia. Infect Immun 61:2486–2492

Silverman J, Chavannes JM, Rigotty J, Omaf M (1979) A natural outbreak of transmissible murine colonic hyperplasia in A/J mice. Lab Anim Sci 29:209–213

Proliferative Ileitis, Hamster

Robert O. Jacoby

Synonyms. Transmissible ileal hyperplasia, hamster enteritis, hamster ileitis, regional ileitis, terminal ileitis, enzootic intestinal adenocarcinoma, wet-tail

Gross Appearance

The characteristic lesion of proliferative ileitis is segmental thickening of the small intestine (Jacoby 1978), particularly the distal ileum (Figs. 366, 367), but a similar lesion occasionally develops in the proximal colon. The lesion is usually well demarcated, and the transition from thickened to normal intestine is especially abrupt at the ileocecal junction. The cecum is often flaccid and filled with fetid liquid contents. Affected hamsters are usually dehydrated, and their perianal skin may be wet or matted with liquid feces, a sign for which the term "wet-tail" is commonly used.

Gross lesions are detectable by about 10 days after experimental inoculation and are fully developed by 30 days. For the first 2–3 weeks, thickening of the ileum is due to mucosal hyperplasia. Affected mucosa is raised, roughened, and hyperemic and is often sharply demarcated from adjacent normal mucosa (Fig. 368). The serosa is partially opaque and also may have a reddish hue. As hyperplasia progresses, the ileum becomes firm and rigid and the wall is up to 2.0 mm thick. The mucosal surface can be eroded and discolored brown to red. The lumen may contain blood, but is generally devoid of formed feces. The serosa is pale brown–red to gray, and Peyer's patches are enlarged. Colonic contents vary from normal fecal pellets to liquid feces. Advanced lesions are characterized by severe thickening of the ileum (up to 4.0 mm). The elevated mucosa is partially covered with gray–yellow plaques formed by mucosal necrosis, and the serosa is studded with gray–white nodules that represent pyogranulomatous inflammation and abscesses. If the lumen is stenotic, proximal segments of small intestine may be dilated and filled with yellow–brown liquid. The ileum is often adherent to adjacent mesenteric lymph nodes, which are usually enlarged, and is occasionally adherent to other portions of the intestine and to parietal peritoneum. Colonic intussusception is a

frequent agonal finding. Hamsters that survive acute disease may die after several months from fibrotic stricture of the ileum, which leads to adynamic ileus. Hamsters that succumb during acute disease may have small, gray–white nodules in the liver.

Microscopic Features

The histogenesis of proliferative ileitis occurs in two phases. The first phase is characterized by hyperplasia of mucosal epithelium, and the second phase is marked by pyogranulomatous inflammation (Jacoby et al. 1975; Jacoby 1978). The hyperplastic phase begins by 10 days after experimental inoculation and develops by ascension onto villus walls of immature, pseudostratified, hyperchromatic, crypt-type epithelium, which replaces mature enterocytes (Figs. 369, 370). Affected villi become long and tortuous and fuse. When viewed by scanning electron microscopy, the mucosa consists of individual or fused leaf-like structures in contrast to evenly spaced conical villi (Figs. 371, 372).

Cells in mitosis can be found from apex to crypt on most villus walls. Many cells also contain densely basophilic, intracytoplasmic bodies. Mucosal hyperplasia is often followed by expansion and penetration of hyperplastic crypt epithelium into Peyer's patches and supporting tissue tunics (Fig. 373). This has been interpreted as a neoplastic change by some workers (Jonas et al. 1965), but is more likely due to penetration of proliferating epithelium into ileal walls rendered turgid by preexisting mucosal hyperplasia and reactive hypertrophy of smooth muscle tunics.

The inflammatory phase begins as segments of hyperplastic epithelium covering upper portions of villi and penetrating crypts undergo focal or laminar necrosis, which is frequently accompanied by hemorrhage. In severe cases, transmural necrosis of ileal mucosa occurs. Inflammation is usually observed first in the lamina propria and muscle tunics beneath partially necrotic, dilated crypts (Fig. 374). Infiltrating cells consist primarily of macrophages and neutrophils. The cytoplasm of the macrophages contains periodic acid Schiff

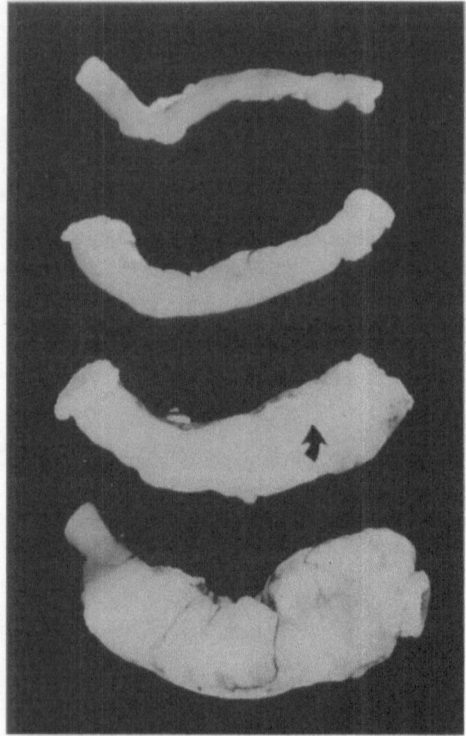

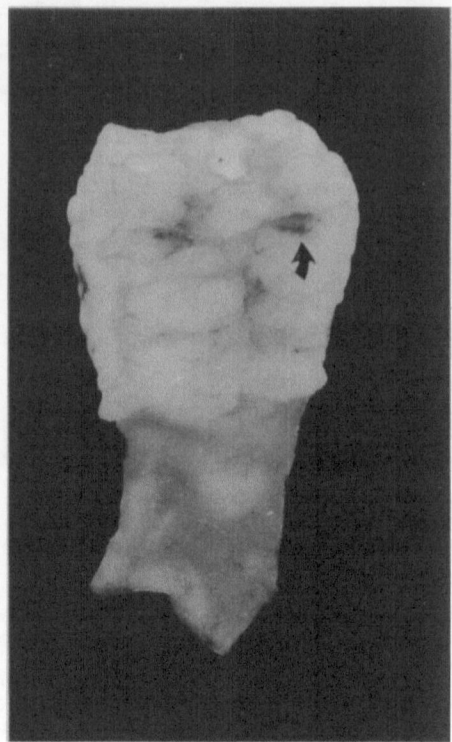

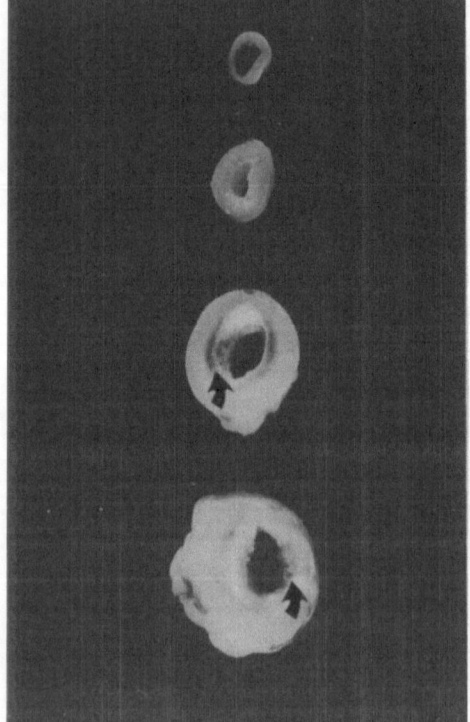

◄

Fig. 366. (*upper left*) Distal segments of ilia from hamsters with experimental proliferative ileitis. The ileocecal junction is at the *right* of each segment. The ileum at the *top* is normal. The three affected segments were collected over a period of 35 days. The two speciments at the *bottom* have small granulomas visible through the serosa (*arrow*). Note the sharp demarcation between normal and affected ileum at the left end of the specimen at the *bottom*. (From Jacoby 1978, with permission of the *American Journal of Pathology*)

Fig. 367. (*upper right*) Transverse sections of the ilea displayed in Fig. 366. The lower two specimens have gray-white mottling of the mucosa from necrosis and hemorrhage (*arrows*) and the specimen at the *bottom* has an irregular serosal contour due to severe pyogranulomatous inflammation. (From Jacoby 1978, with permission of the *American Journal of Pathology*)

Fig. 368. (*below*) Mucosal surface of distal ileum from a hamster with advanced proliferative ileitis. The affected mucosa is thickened and distorted from hyperplasia and fusion of villi. zones of erosion, ulceration, and hemorrhage (*arrow*) are also present. (From Jacoby 1978, with permission of the *American Journal of Pathology*)

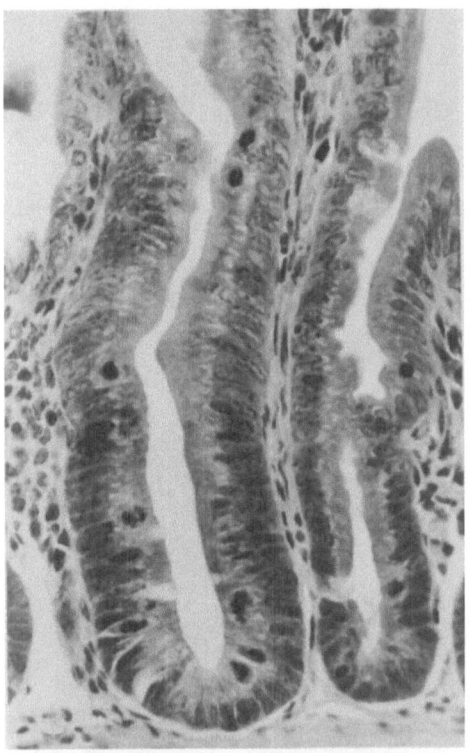

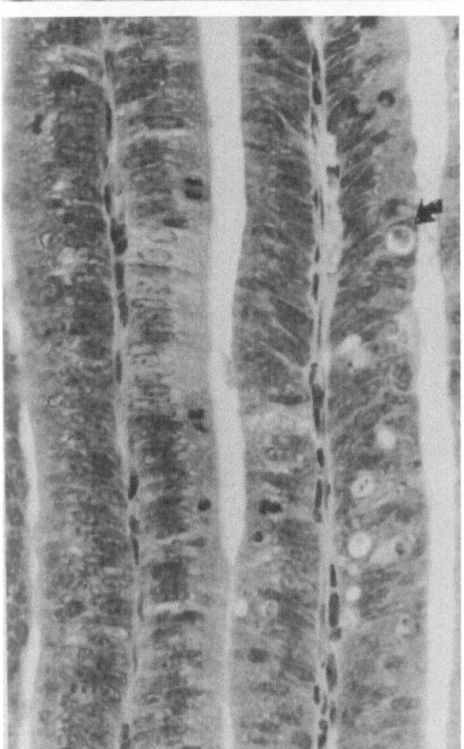

(PAS)-positive granules. Crypt abscesses form, and a thick collar of pyogranulomatous inflammation eventually encircles the hyperplastic mucosa. Fronds of hyperplastic epithelium are dispersed in the chronically inflamed wall (Fig. 375). Inflammation of ileal serosa causes focal peritonitis with fibrinous adhesions between intestine and adjacent mesenteric structures, particularly the mesenteric lymph nodes. The lymph nodes are often hyperplastic and contain clusters of PAS-positive macrophages and polymorphonuclear leukocytes. Small granulomas with PAS-positive macrophages occasionally develop in the liver.

The abrupt transition between affected and normal mucosa evident from gross observation is also apparent histologically. At the ileocecal junction, cecal mucosa, which is usually edematous, lies directly adjacent to massively hyperplastic ileal mucosa.

Mucosal hyperplasia gradually subsides in surviving hamsters. Mucosal epithelium is restored, but villus architecture remains distorted, a finding which suggests that absorptive surfaces are reduced. The new epithelium has an increased proportion of mucin-producing cells. Pyogranulomatous inflammation can persist for several months, but commonly wanes in favor of fibroplasia. Scarring in "healed" lesions can produce fatal stricture of the ileal lumen, as previously mentioned.

Ultrastructure

Hyperplastic epithelial cells in the affected ileum are elongated and, except for cells in mitosis, situated at the basement membrane (Johnson and Jacoby 1978). Cells in mitosis are usually perilumenal and often appear to be separated

◀

Fig. 369. (*above*) Crypt epithelium in early stages (day 10) of proliferative ileitis. The hyperplastic epithelium extends onto the lower portions of the villus walls. Note the abundant mitotic figures. (From Jacoby 1978, with permission of the *American Journal of Pathology*.) H&E, ×470

Fig. 370. (*below*) Upper segments of two villi with typical changes of proliferative ileitis 20 days after inoculation. Mature villus epithelium has been replaced by hyperplastic crypt-type epithelium with numerous dividing cells. Basophilic cytoplasmic inclusions representing phagolysosomes occur in some cells (*arrow*). Note the lack of inflammation in the lamina propria. (From Jacoby 1978, with permission of the *American Journal of Pathology*.) H&E, ×425

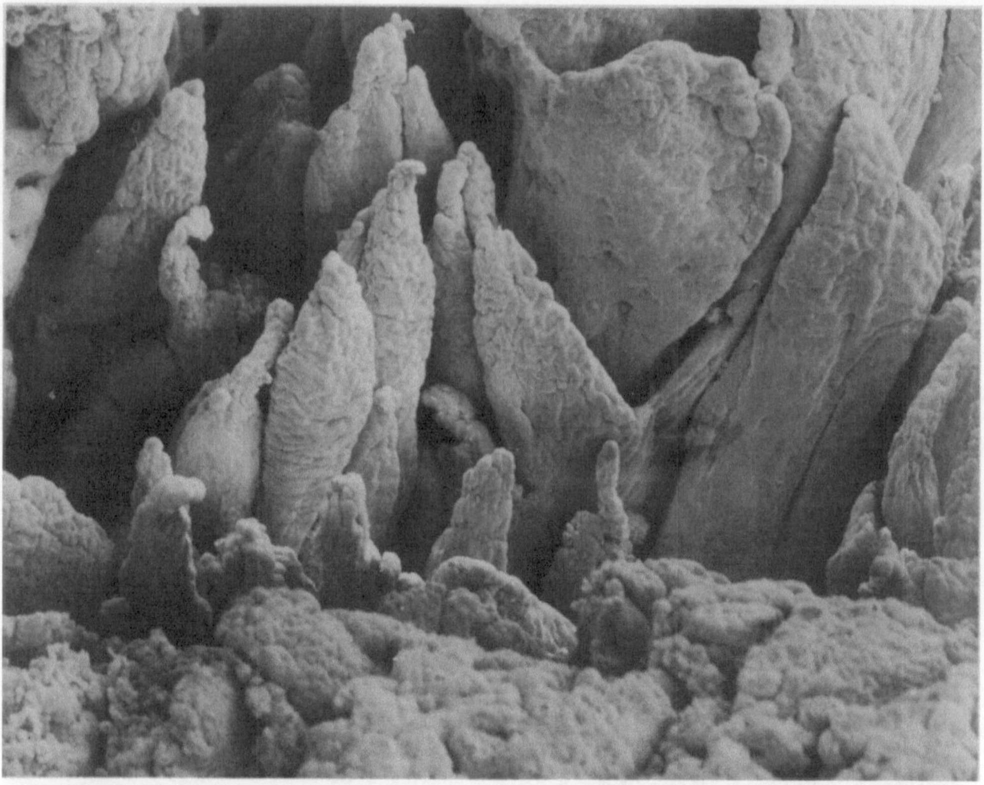

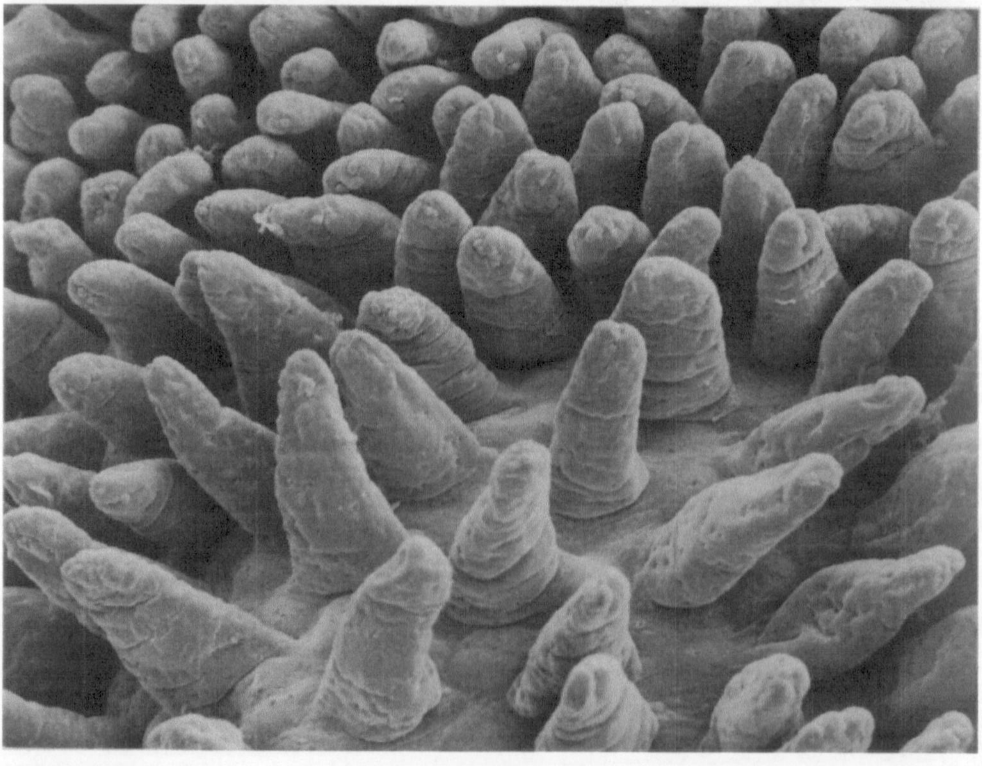

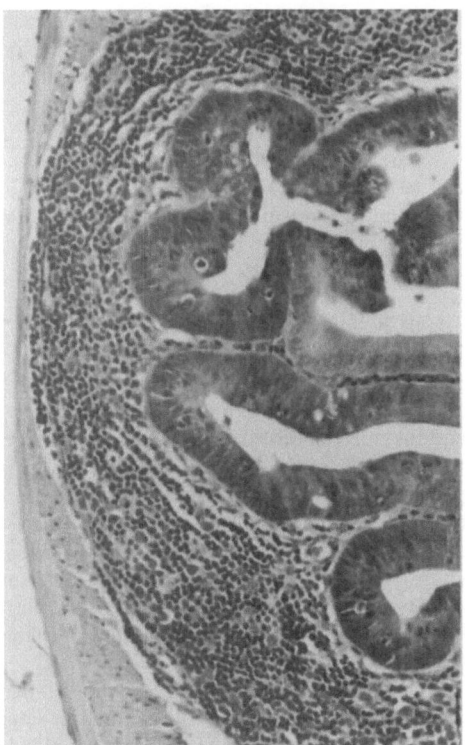

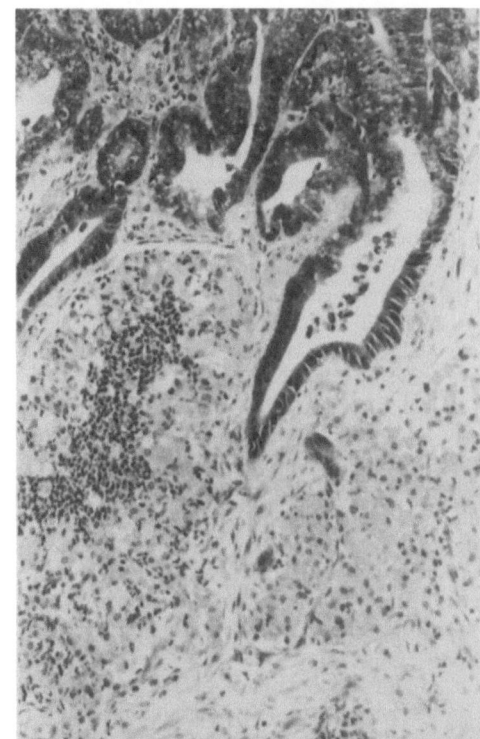

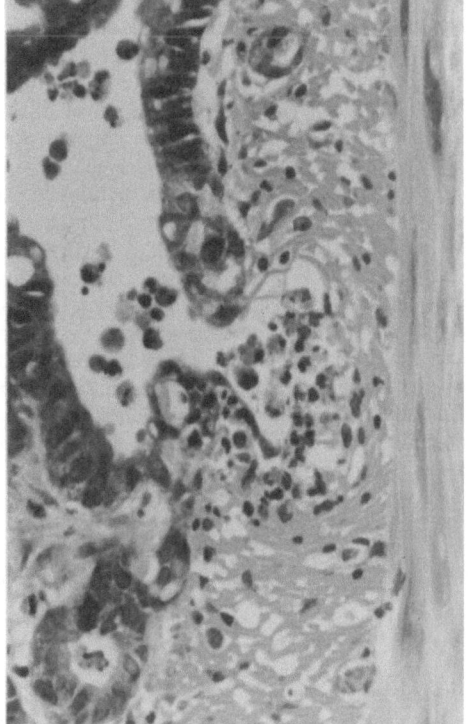

◄

Fig. 371. (*above*) Ileal lesion in the adult hamster. Villus hyperplasia and fusion has formed leaf-like laminae and plateaus. (From Johnson and Jacoby 1978, with permission of the *American Journal of Pathology.*) SEM, ×185

Fig. 372. (*below*) Ileal mucosa from a normal hamster. (From Johnson and Jacoby 1978, with permission of the *American Journal of Pathology.*) SEM, ×185

▶

Fig. 373. (*upper left*) Peyer's patch compressed by expansion of tortuous hyperplastic crypts. Ileum of a hamster with experimentally induced proliferative ileitis. (From Jacoby 1978, with permission of the *American Journal of Pathology.*) H&E, ×185

Fig. 374. (*below*) Early inflammation associated with segmental necrosis of a tortuous dilated crypt that has penetrated the inner muscular tunic of the ileal wall. (From Jacoby 1978, with permission of the *American Journal of Pathology.*) H&E, ×460

Fig. 375. (*upper right*) Ileum, hamster, with typical advanced lesion of proliferative ileitis. Fronds of hyperplastic epithelium are enmeshed in supporting tissues with severe pyogranulomatous inflammation. (From Jacoby 1978, with permission of the *American Journal of Pathology.*) H&E, ×170

from the basement membrane by an underlying cell. Hyperplastic cells resemble undifferentiated crypt epithelium. They have many free ribosomes, a few mitochondria, and an irregular and often rudimentary microvillus border. The intracytoplasmic inclusions mentioned earlier and the PAS-positive granules in macrophages are phagolysosomes consisting of cell debris and rod-shaped bacteria in various stages of degeneration (Fig. 376). The apical cytoplasm of hyperplastic cells contains small, rod-shaped bacteria that are not enclosed by a cytoplasmic membrane. The potential relevance of these organisms to the disease is discussed later.

Although diarrhea ("wet-tail") in the hamster is associated with several diseases, proliferative ileitis is probably the most common underlying cause. Tyzzer's disease can cause diarrhea and high death rates in laboratory rodents, but intestinal lesions are primarily necrotic rather than hyperplastic (Ganaway et al. 1971; see also p. 201, this volume). The causative organism of Tyzzer's disease *(Clostridium piliformis)*, although found intracellularly, has a characteristic morphology that should not be confused with bacteria associated with transmissible ileal hyperplasia. Salmonellosis (Innes et al. 1956), colibacillosis (Thomlinson 1975), or severe enteric parasitism (Frisk and Wagner 1977) are potential causes of diarrhea, but occur only rarely in modern vivaria and can be diagnosed by traditional microbiologic methods. None of these diseases is associated with severe mucosal hyperplasia. Enteritis variably accompanied by necrosis and hemorrhage can occur during natural outbreaks or experimental induction of proliferative ileitis (Jacoby et al. 1975). A hyperplastic component does not develop, and animals often die suddenly. This syndrome may be a variant of proliferative ileitis or a condition that is activated during induction of the disease. Cecal hyperplasia, an acute disease of suckling hamsters (Barthold et al. 1978), can cause diarrhea and high mortality. It is characterized by cecal hyperplasia rather than by ileal hyperplasia. The cause is unknown.

Biologic Features

Proliferative ileitis is a devastating disease. It occurs principally among weanlings and young adults, whereas mature adults are relatively resist-

ant (Jacoby and Johnson 1981). Typical clinical signs include diarrhea, anorexia, rapid weight loss and dehydration, hunched posture, ruffled haircoat, lethargy, and death. The incidence and severity of these signs can vary and are influenced by diet (Jacoby and Johnson 1981). Clinical signs can last from several days to several weeks, but apparently healthy animals occasionally die without premonitory signs. Experimental studies indicate that clinical disease is likely to begin about 3 weeks after infection and that intestinal lesions can be detected by abdominal palpation after 2 weeks (Jacoby 1978).

The disease usually occurs as an explosive epizootic with high morbidity and high mortality and thus is compatible with an infectious etiology. It appears to spread by oral infection, which is probably exacerbated by the hamster's predilection for cannibalizing dead or dying cage mates. Aerosol transmission has not been detected. Hamsters seem to be the only natural host, but rats, mice, and guinea pigs develop specific antibody (see below) after experimental oral infection (R.O. Jacoby, unpublished data).

The pathogenesis and etiology of proliferative ileitis are closely associated with bacterial invasion of ileal mucosal epithelium. Several lines of evidence support this concept. Intracellular bacteria are found in hyperplastic epithelium from natural cases, and morphologically identical organisms are found in experimentally induced disease (Jacoby 1978; Johnson and Jacoby 1978; Wagner et al. 1973). These organisms are slightly curved, rod-shaped, gram-negative bacteria that measure approximately $0.3 \times 2.0\,\mu m$. They can be found in ileal mucosal epithelium by 5 days after oral inoculation of hamsters with suspensions of affected ileum, which is several days before hyperplasia begins. They appear first in crypt epithelium, but are seen later in mature villus epithelium. They do not occur in unaffected segments of intestine. They replicate and accumulate in cells during the development of this lesion, but are not found in intestinal epithelium in surviving animals (Johnson and Jacoby 1978). The disease can be transmitted to hamsters inoculated orally with supernates of ground ileal suspensions passed through filters with a limiting pore size of 650 nm, but not with suspensions passed through filters with a pore size of 220 nm (Jacoby et al. 1975; Jacoby and Johnson 1981). Attempts to cultivate a causative agent in artificial media or tissue culture have been equivocal or unsuccessful (Jacoby and

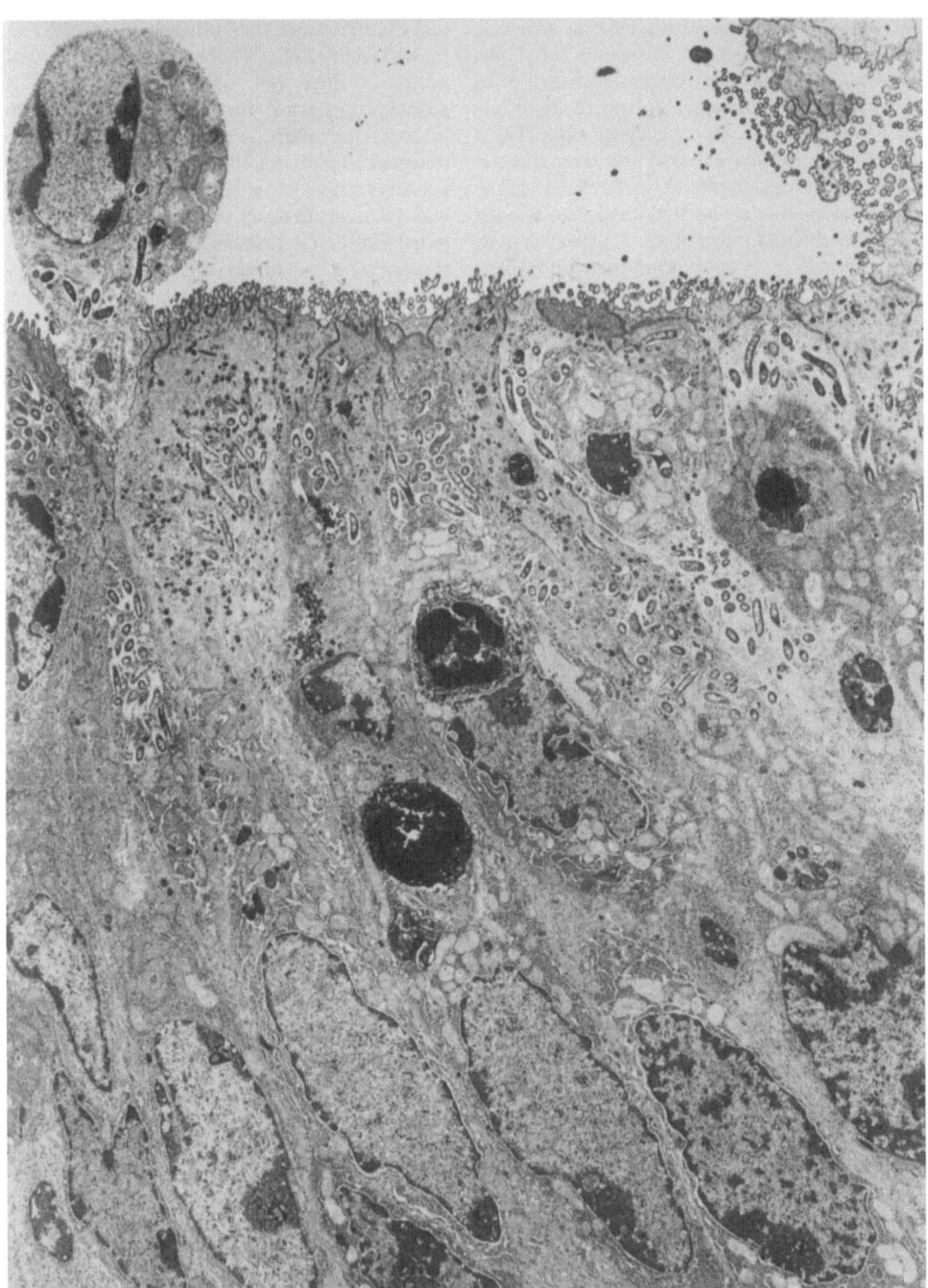

Fig. 376. Ileum, hamster with typical proliferative ileitis. Cells are elongated and pseudostratified. The microvillus border is irregular. Large, dense phagolysosomes that correspond to basophilic inclusions seen by light microscopy are located in the cytoplasm. The cytoplasm contains rod-shaped bacteria and many free ribosomes, which are not easily discerned at this magnification. (From Johnson and Jacoby 1978, with permission of the *American Journal of Pathology*.) Uranyl acetate and lead citrate, ×4625

Johnson 1981). There is speculation that the intracellular organism is a species of *Campylobacter*, but attempts to transmit disease with *Campylobacter* organisms recovered from affected hamsters have not succeeded. Stills (1991) has recently isolated a bacterium from affected hamsters that resembles the intracellular organisms found in tissues and that can induce disease after experimental inoculation. It appears not to be a *Campylobacter*, but has not been firmly identified. Fox and coworkers (1993) have also found evidence of *Chlamydia* infection in affected hamsters, but the role of *Chlamydia* remains speculative.

Another link between proliferative ileitis and intracellular bacteria also provides the basis for a useful diagnostic test. Sera from virtually all infected hamsters contain antibody to the intracellular bacteria. This antibody has been used in an indirect immunofluorescence test to help clarify the epizootiology of experimental disease and to detect intracellular bacteria in affected animals (Jacoby et al. 1975; Jacoby 1978) (Fig. 377). Antibody does not occur in the serum of uninfected hamsters and does not react with the intestinal microflora of normal hamsters. It can be detected as early as 10 days after infection and persists for at least several months. The role of host immunity in the protection of hamsters from lethal infection is unknown.

It is not clear how bacteria infect epithelial cells in this disease or how they may induce hyperplasia. They have not been seen invading cells or in association with the brush border, as reported for other enteric bacterial pathogens (Takeuchi 1971). They also are not enclosed in host cell membranes and are, therefore, in intimate contact with cell organelles. Their predilection for crypt cells provides evidence against a random type of penetration. A more likely possibility is that small numbers of bacteria enter a few crypt cells, perhaps by endocytosis or phagocytosis. Rapid in-

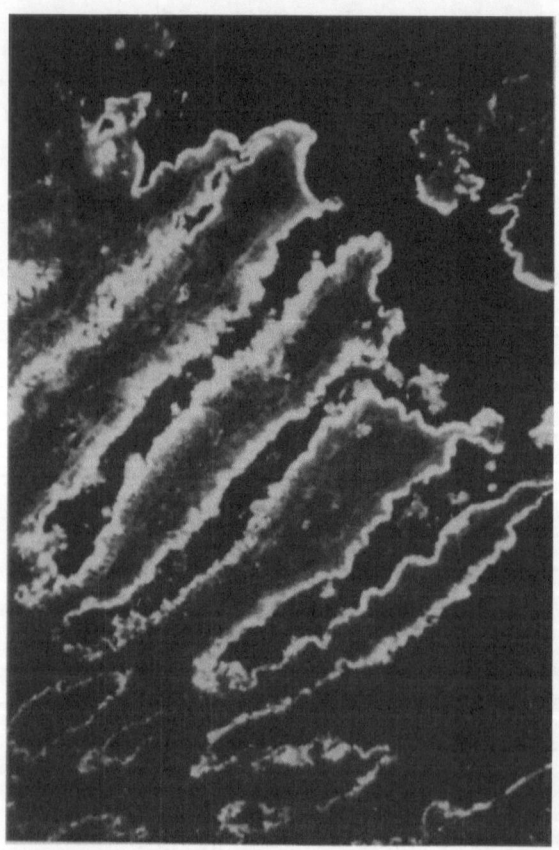

Fig. 377. Ileal mucosa of a hamster stained with specific proliferative ileitis antiserum by indirect immunofluorescence. Fluorescing bacteria in the apical cytoplasm of the hyperplastic epithelium form a typical ribbon pattern which outlines individual villi. (From Jacoby 1978, with permission of the *American Journal of Pathology*.) ×114

tracellular replication may rupture membrane-bound vacuoles. Bacterial replication coincident with crypt cell division and migration might result in the histologic changes described.

Opinion varies as to whether proliferative ileitis is primarily a proliferative or an inflammatory disease (Boothe and Cheville 1967; Jacoby et al. 1975; Jacoby 1978). In experimentally induced disease, mucosal hyperplasia precedes inflammation. Therefore, the term proliferative ileitis, although morphologically sound for mature lesions, does not accurately reflect the histogenesis of the disease. Conversely, the term ileal hyperplasia does not indicate the contribution of inflammation to fully developed lesions.

Comparison with Other Species

Ileal hyperplasia and proliferative ileitis associated with intracellular bacteria have been reported in several species (Cross et al. 1973; Duhamel and Wheeldon 1982; Landsverk 1981; Vandenberghe and Hoorens 1980), but not in other laboratory rodents. Fox and coworkers (Fox et al. 1982) have recently reported a proliferative colitis in ferrets that is associated with intracellular bacteria. A syndrome of swine called intestinal adenomatosis is particularly intriguing from an etiologic aspect. Hyperplastic segments of ileum in swine are morphologically similar to the disease of hamsters and contain intracytoplasmic bacteria which have been identified as *Campylobacter* (Lawson and Rowland 1974; Rowland and Lawson 1974; McOrist et al. 1989).

Proliferative ileitis does not have a true counterpart in humans. Regional hyperplasia of the human small intestine is virtually unknown. Inflammatory bowel disease of the small intestine (i.e., Crohn's disease, regional enteritis) can have a granulomatous component, but mucosal hyperplasia is not characteristic of this syndrome (Kirsner and Shorter 1982; Morson and Dawson 1979; Trier et al. 1965). Lymphoid infiltrates are also common in Crohn's disease, but not in transmissible ileal hyperplasia. Dysplastic changes can occur in the mucosa of patients with Crohn's disease who develop carcinomas of the small bowel (Simpson et al. 1981), but there is no evidence that mucosal hyperplasia in hamsters leads to neoplasia. Finally, there is no firm evidence that inflammatory bowel disease in humans is associated with intracytoplasmic colonization of epithelium by bacteria. Intracellular organisms have been found in Whipple's disease, but they reside in macrophages rather than in mucosal epithelium (Clancy et al. 1975; Kent et al. 1963).

References

Barthold SW, Jacoby RO, Pucak GJ (1978) An outbreak of cecal mucosal hyperplasia in hamsters. Lab Anim Sci 28:723–727

Boothe AD, Cheville NF (1967) The pathology of proliferative ileitis of the golden syrian hamster. Pathol Vet 4:31–44

Clancy RL, Tomkins WA, Muckel TJ, Richardson H, Rawls WE (1975) Isolation and characterization of an aetiological agent in Whipple's disease. Br Med J 3:568–570

Cross RF, Smith CK, Parker CF (1973) Terminal ileitis' in lambs. J Am Vet Med Assoc 162:564–566

Duhamel GE, Wheeldon EB (1982) Intestinal adenomatosis in a foal. Vet Pathol 19:447–450

Fox JG, Murphy JC, Ackerman JL, Prostak KS, Gallagher CA, Rambow VJ (1982) Proliferative colitis in ferrets. Am J Vet Res 43:858–864

Fox JG, Stills HF, Paster BJ, Dewhirst FE, Yan L, Palley L, Prostak K (1993) Antigenic specificity and morphologic characteristics of Chlamydia trachomatis, strain SFPD, isolated from hamsters with proliferative ileitis. Lab Anim Sci 43:405–410

Frisk CS, Wagner JE (1977) Hamster enteritis: a review. Lab Anim 11:79–85

Ganaway JR, Allen AM, Moore TD (1971) Tyzzer's disease. Am J Pathol 64:717–730

Innes JRM, Wilson C, Ross MA (1956) Epizootic Salmonella enteriditis infection causing septic pulmonary phlebo-thrombosis in hamsters. J Infect Dis 98:133–141

Jacoby RO (1978) Transmissible ileal hyperplasia of hamsters. 1. Histogenesis and immunocytochemistry. Am J Pathol 91:433–450

Jacoby RO, Johnson EA (1981) Transmissible ileal hyperplasia. In: Streilein JW, Hart DA, Stein-Streilein J, Ducan WR, Billingham RE (eds) Hamster immune responses in infectious and oncologic diseases. Plenum, New York, pp 267–289

Jacoby RO, Osbaldiston GW, Jonas AM (1975) Experimental transmission of atypical ileal hyperplasia of hamsters. Lab Anim Sci 25:465–473

Johnson EA, Jacoby RO (1978) Transmissible ileal hyperplasia of hamsters. II. Ultrastructure. Am J Pathol 91:451–468

Jonas AM, Tomita Y, Wyand S (1965) Enzootic intestinal adenocarcinoma in hamsters. J Am Vet Med Assoc 147:1102–1108

Kent TH, Layton JM, Clifton JA, Schedl HP (1963) Whipple's disease: light and electron microscopic studies combined with clinical studies suggesting an infective nature. Lab Invest 12:1163–1178

Kirsner JB, Shorter RG (1982) Recent developments in non-specific inflammatory bowel disease. N Engl J Med 306:775–785

Landsverk T (1981) intestinal adenomatosis in a blue fox (Alopex lagopus). Vet Pathol 18:275–278

Lawson GHK, Rowland AC (1974) Intestinal adenomatosis in the pig: a bacteriogical study. Res Vet Sci 17:331–336

McOrist S, Boid R, Lawson GH (1989) Antigenic analysis of Campylobacter species and an intracellular Campylobacter-like organism associated with porcine proliferative enteropathies. Infect Immun 57:957–962

Morson BC, Dawson IM (1979) (eds) Gastrointestinal pathology (with a contribution by A. Spriggs), 2nd edn. Blackwell Scientific, Oxford, pp 293–312

Rowland AC, Lawson GHK (1974) Intestinal adenomatosis in the pig: immunofluorescent and electron microscopic studies. Res Vet Sci 17:323–330

Simpson S, Traube J, Riddell RH (1981) The histologic appearance of dysplasia (precarcinomatous change) in Crohn's disease of the small and large intestine. Gastroenterology 81:492–501

Stills HF (1991) Isolation of an intracellular bacterium from hamsters (Mesocricetus auratus) with proliferative ileitis and reproduction of the disease with a pure culture. Infect Immun 59:3227–3236

Takeuchi A (1971) Penetration of the intestinal epithelium by various micro-organisms. Curr Top Pathol 54:1–27

Thomlinson JR (1975) "Wet-tail" in the Syrian hamster: a form of colibacillosis. Vet Rec 96:42

Trier JS, Phelps PC, Eidelman S, Rubin CE (1965) Whipple's disease: light and electron microscope correlation of jejunal mucosal histology with antibiotic treatment and clinical status. Gastroenterology 48:684–707

Vandenberghe J, Hoorens J (1980) Campylobacter species and regional enteritis in lambs. Res Vet Sci 29:390–391

Wagner JE, Owens DR, Troutt HF (1973) Proliferative ileitis of hamsters: electron microscopy of bacteria in cells. Am J Vet Res 34:249–252

Streptococcal Enteropathy, Intestine, Rat

Stephen W. Barthold

Synonyms. Enterococcal diarrhea

Gross Appearance

Suckling rats develop diarrhea, with dehydration, retarded growth, and retarded hair development. Their abdomens are distended, with fecal staining of the perianal fur. Pasty or liquid feces can be expressed by gentle abdominal pressure. Pups continue to suckle with full stomachs. The distal small intestine and colon are distended with gas and yellow liquid to pasty fecal material. Mortality is variable, but can reach around 50%, particularly among the youngest rats. Surviving pups completely recover, and dams are asymptomatic (Hoover et al. 1985; Etheridge et al. 1988; Etheridge and Vonderfecht 1992).

Microscopic Features

There is striking colonization of villus enterocyte brush borders with large numbers of coccoid bacteria throughout the small intestine, particularly in the duodenum. Colonization does not involve crypt epithelium. Gram stains accentuate the massive bacterial colonization (Fig. 378). This is accompanied by modest crypt hyperplasia, but minimal inflammation (Hoover et al. 1985; Etheridge et al. 1988).

Ultrastructure

Scanning electron microscopy of heavily infected areas reveals bacteria completely covering the surface of villi, whereas in areas where smaller numbers of bacteria are present, they are evenly spread out with filamentous attachments between adjacent bacteria. With transmission electron microscopy, bacteria are covered with microfilamentous strands which interconnect bacteria and the surface of morphologically normal microvilli (Hoover et al. 1985).

Differential Diagnosis

Diarrhea and enteritis are relatively rare in laboratory rats. Other agents of infectious diarrhea in

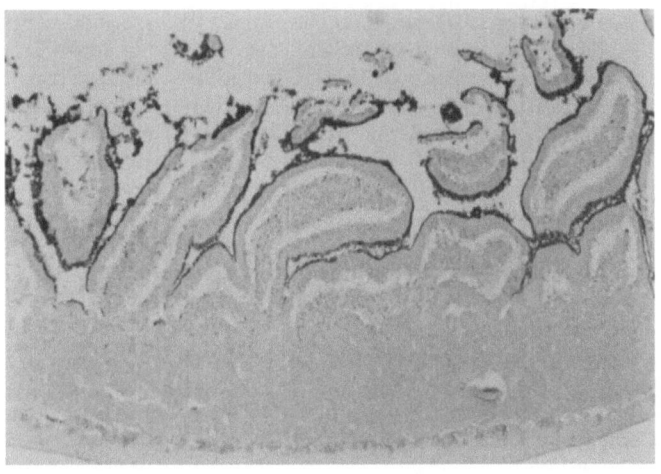

Fig. 378. Small intestine mucosa of a naturally infected infant rat with enterococcal diarrhea. The brush border is heavily colonized by attached coccoid bacteria. Brown and Brenn, ×80

infant rats include *Clostridium piliforme* (Tyzzer's disease), *Salmonella, Campylobacter* and type-B rotavirus. They can be differentiated by culture (*Salmonella, Campylobacter*) or characteristic morphology, such as silver-stained intracellular organisms in enterocytes, hepatocytes, and cardiomyocytes in Tyzzer's disease and villus attenuation with enterocytic syncytia in rotavirus infections.

Biologic Features

Natural History

Enterococci are ubiquitous and part of enteric commensal flora in humans and animals. Thus isolation of these organisms is not diagnostic. Nevertheless, enterococci have been associated with diarrhea and enteritis in a number of other species (Fuller et al. 1981; Tzipori et al. 1984; Collins et al. 1988). The environmental source of outbreaks of infection with pathogenic enterococcus in rat colonies has not been determined.

Pathogenesis

Disease appears to be associated with strains of enterococcus capable of attaching to mucosa, as inoculation of infant rats with nonattaching isolates does not result in disease (Etheridge and Vonderfecht 1992). Experimental inoculation of infant rats with cultured organisms results in clinical signs appearing within 2–6 days, with variable

mortality up to the 12th day, and development of characteristic microscopic lesions (Hoover et al. 1985; Etheridge and Vonderfecht 1992). Inoculation of infant rats beyond 5 days of age resulted in no apparent disease in one study (Etheridge et al. 1988), whereas inoculation of adult rats in another study resulted in soft feces within 3–5 days of inoculation (Hoover et al. 1985). In studies with a similar pathogenic enterococcus in foals, pathogenic enterococci did not produce detectable enterotoxin, but markedly reduced brush border digestive enzyme activity in experimentally infected pigs, suggesting that malabsorption may be an important mechanism in pathogenesis (Tzipori et al. 1984).

Etiology

Isolates associated with disease in infant rats have been identified as nonhemolytic Lancefield group-D enterococci: *Enterococcus hirae* (Etheridge et al. 1988), *E. faecium-durans-2* (Hoover et al. 1985), and *E. fecalis-2* (Etheridge and Vonderfecht 1992). These different isolates can be differentiated antigenically, biochemically, and genetically from one another, but there are no known markers to differentiate pathogenic isolates from commensal isolates (Etheridge and Vonderfecht 1992). Indeed, nonpathogenic *E. durans-2, E. fecalis-3,* and *Streptococcus bovis* were isolated from control rats in a study in which rats were inoculated with pathogenic *E. fecalis-2* (Etheridge and Vonderfecht 1992). Enterococci can be cultured by streaking fecal contents on 5%

sheep blood agar. Pathogenic rat enterococci can be slower growing than other enterococci (Etheridge and Vonderfecht 1992). Colonies can be characterized biochemically with commercially available *Streptococcus* identification kits.

Frequency

Sporadic outbreaks of disease have been noted in recent years in otherwise pathogen-free colonies of rats. Enterococcal diarrhea is a relatively rare, but possibly emerging new entity in rat colonies.

Comparison with Other Species

Enterococci have been implicated as pathogens in a variety of species, including chickens (Fuller et al. 1981; Farrow and Collins 1985), dogs (Collins et al. 1988), horses (Tzipori et al. 1984), and pigs (Hoover et al. 1985). In all of these cases, disease occurred in young animals. As in rats, the attachment of large numbers of coccoid bacteria to brush borders of small intestinal enterocytes is characteristic of pathogenic enterococci in all of these species, and electron microscopy reveals similar microfilamentous structures between bacteria and the brush border (Fuller et al. 1981; Tzipori et al. 1984; Collins et al. 1988). The species

specificity of pathogenic enterococci is not known, but equine isolates were shown to be pathogenic in pigs (Tzipori et al. 1984) and rats developed disease following inoculation with diarrheic material from a human patient (Etheridge et al. 1988).

References

Collins JE, Bergeland ME, Lindeman CJ, Duimstra JR (1988) Enterococcus (Streptococcus) durans adherence in the small intestine of a diarrheic pup. Vet Pathol 25:396–398

Etheridge ME, Vonderfecht SL (1992) Diarrhea caused by a slow-growing Enterococcus-like agent in neonatal rats. Lab Anim Sci 42:548–550

Etheridge ME, Yolken RH, Vonderfecht SL (1988) Enterococcus hirae implicated as a cause of diarrhea in suckling rats. J Clin Microbiol 26:1741–1744

Farrow JA, Collins MD (1985) Enterococcus hirae: a new species that includes amino acid assay strain NCDO 1258 and strains causing growth depression in young chickens. Int J Syst Bacterial 35:73–75

Fuller R, Houghton SB, Brooker BE (1981) Attachment of Streptococcus faecium to the duodenal epithelium of the chicken and its importance in colonization of the small intestine. Appl Environ Microbiol 41:1433–1441

Hoover D, Bendele SA, Wightman SR, Thompson CZ, Hoyt JA (1985) Streptococcal enteropathy in infant rats. Lab Anim Sci 35:635–641

Tzipori S, Hayes J, Sims L, Withers M (1984) Streptococcus durans: an unexpected enteropathogen of foals. J Infect Dis 150:589–593

Spironucleus muris Infection, Intestine, Mouse, Rat, and Hamster

Stephen W. Barthold

Synonyms. Hexamita muris

Gross Appearance

Affected mice are runted and lethargic, with rough hair coats and distended abdomens. Diarrhea or sticky feces may or may not be manifest. The duodenum and occasionally other segments of small intestine are dilated and contain yellow or white watery, foamy digesta. Mesenteric lymph nodes are enlarged, and pancreatic edema and ascites may be present. Mortality up to 50% has been observed (Boorman et al. 1973; Csiza and Abelseth 1973; Flatt et al. 1978; Lussier and Loew 1970; Sebesteny 1969; Wagner et al. 1974). Athymic nude mice develop chronic enteritis with thickening of the small intestine, swollen abdomens, weight loss, and accelerated mortality (Kunstyr et al. 1977a,b).

Microscopic Features

Asymptomatic infections are characterized by the presence of a few *Spironucleus muris* organisms in the intervillar space, crypts, and central lumen, without detectable changes in mucosal morphology. During overt infections, numerous organisms fill crypts and intervillar spaces (Fig. 379). Mucosal changes are variable, including no detectable changes, villus blunting, epithelial damage, mucosal hyperplasia, edema, and leukocytic infiltration (Fig. 380). Crypts can become dilated and cystic in chronic infections (Boorman et al. 1973; Flatt et al. 1978; Lussier and Loew 1970; MacDonald and Ferguson 1978; Sebesteny 1969; Van Kruiningen et al. 1978; Wagner et al. 1974). Trophozoites are found in the upper small intestine and small (5 × 7mm), characteristically banded cysts are found in the remainder of the bowel and feces (Brett and Cox 1982; Kunstyr 1977).

Ultrastructure

Trophozoites are located extracellularly in the lumen or, occasionally, in intercellular spaces of the mucosal epithelium. They are bilaterally symmetric, 2.5–5.2 μm wide, and 12–20 μm long. They possess two anterior nuclei and four pairs of flagellae, three of which emerge laterally from the anterior end. The fourth pair course posteriorly through the body, are encased in flagellar sheaths in the caudal half of the organism, and emerge from the caudal end. Cytoplasm contains extensive endoplasmic reticulum, glycogen particles, and digestive vacuoles (Boorman et al. 1973; Brugerolle et al. 1980; Owen et al. 1979; Van Kruiningen et al. 1978).

Differential Diagnosis

Spironucleus is an opportunistic pathogen under most circumstances. If overgrowth is observed, the underlying cause should be determined (see "Pathogenesis" below). Trophozoites have typical morphology and are most readily found in the duodenum. They can be easily distinguished from the other protozoal intestinal pathogen, *Giardia muris*. Cysts in feces must be differentiated from yeast cells.

Biologic Features

Natural History

S. muris is readily transmitted among susceptible rodents by fecally excreted cysts, which are envi-

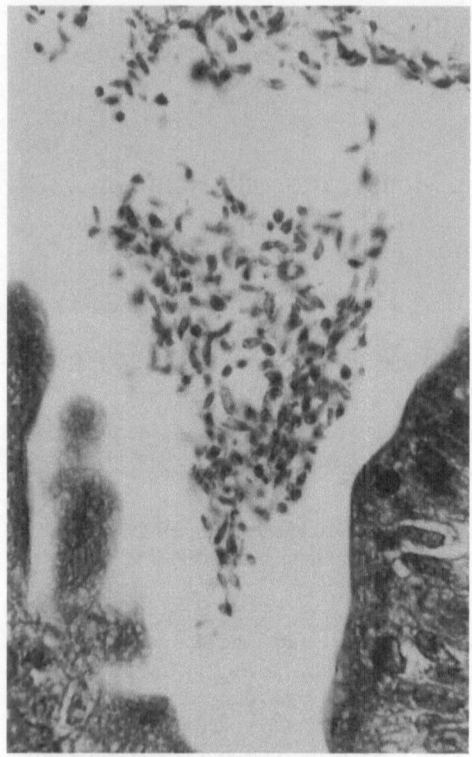

ronmentally resistant (Kunstyr and Ammerpohl 1978). The prepatent period between ingestion and excretion of cysts is 2–3 days, with a maximum of 6 days (Kunstyr 1977; Kunstyr et al. 1977a). Maximal levels of cyst excretion occur at 1–2 weeks, followed by progressive decline. Few parasites are detectable in feces by 9 weeks. Marked differences in duration of infection have been seen between inbred strains of mice (Brett and Cox 1982). During pregnancy and lactation, there is increased excretion of cysts in the feces of mice (Schagemann et al. 1990). Athymic nude mice develop chronic disease. Curiously, outbreaks of clinical disease come and go in mouse colonies that have been and continue to be infected with *S. muris,* suggesting another underlying precipitating factor (Boorman et al. 1973; Csiza and Abelseth 1973; Flatt et al. 1978; Kunstyr et al. 1977a,b; Lussier and Loew 1970; Meshorer 1969; Wagner et al. 1974; Ward 1974).

Pathogenesis

Trophozoites reside in the unstirred layer of intervillar spaces and crypts of the upper small intestine (Owen et al. 1979). They do not generally attach to or invade mucosa, although on occasion they can be found in peripheral blood of hamsters with proliferative enteritis (Wagner et al. 1974). They can also be found in the lamina propria of freshly killed mice, but the significance of this is controversial (Brugerolle et al. 1980; Flatt et al. 1978). Under most circumstances, *S. muris* infection is asymptomatic. It is apparently an opportunistic pathogen, requiring some other factor to allow it to overgrow. Exacerbation of disease seems to be associated with a number of other stressers, including coinfection with *Giardia* or intestinal viruses, X-irradiation, cold, crowding, cadmium toxicity, and diet changes (Barthold 1986; Boorman et al. 1973; Csiza and Abelseth 1973; Exon et al. 1975; MacDonald and Ferguson 1978; Meshorer 1969; Sebesteny 1969).

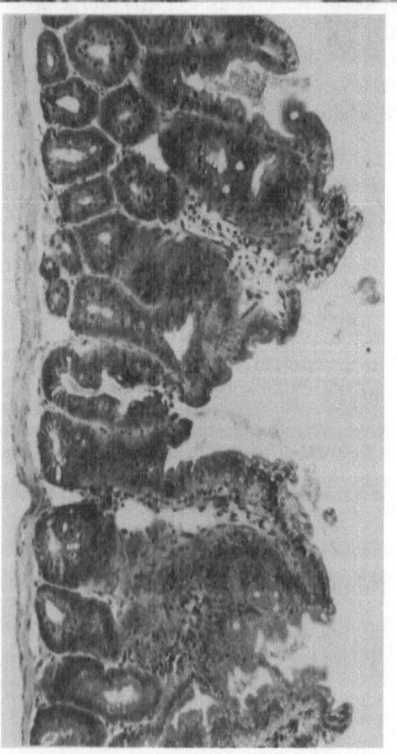

◄

Fig. 379. (*above*) *Spironucleus muris* trophozoites in the duodenal intervillar space of a naturally infected mouse. H&E, ×375

Fig. 380. (*below*) Duodenum of a mouse with spironucleosis. Villi are blunt and crypts are hyperplastic. *Spironucleus* organisms are present in the crypt lumina and intervillar spaces. H&E, ×125

Susceptibility to infection, as measured by fecal cyst excretion, is dependent on mouse genotype, age, and immunologic status. During initial infection, levels of cyst output are similar between mouse genotypes, but rate of elimination varies widely between inbred strains (Brett and Cox 1982). Three-week-old mice are most susceptible, 8-week-old mice are only partially so, and 10-week-old mice are refractory to infection. Susceptibility is enhanced by thymectomy and immunosuppression (Kunstyr et al. 1977a,b). Athymic nude mice are susceptible to infection at any age and develop overt signs of disease. Resistance to reinfection correlates with ability of a genotype to eliminate primary infections (Brett and Cox 1982).

In mice infected with *S. muris,* there is increased crypt cell production, reductionin in villus height, and accelerated migration rate of epithelial cells up the crypt-villus column. Intraepithelial lymphocytes are also increased (Brett and Cox 1982; MacDonald and Ferguson 1978). Aberrations in immune responsiveness and macrophage function have been attributed to *S. muris* in naturally infected mice (Ruitenberg and Kruyt 1975; Keast and Chesterman 1972).

Etiology

S. muris is a flagellated protozoan that has been identified in naturally infected mice, rats, Syrian hamsters, and Chinese hamsters and does not possess species specificity. Disease seems to be restricted to mice and possibly Chinese hamsters (Kunstyr 1977; Wagner et al. 1974). Although a single species, there is variable evidence of host specificity among mouse, hamster, and rat isolates of the organism (Saxe 1954; Sebesteny 1974, 1979; Schagemann et al. 1990).

Frequency

S. muris is common among conventionally housed mice, rats, and hamsters. Disease associated with this organism, on the other hand, is relatively rare, underscoring the importance of some other factor in its pathogenicity. Outbreaks of disease have been described in North America, Europe, and Israel (Boorman et al. 1973; Csiza and Abelseth 1973; Flatt et al. 1978; Kunstyr et al. 1977a,b; Lussier and Loew 1970; MacDonald and Ferguson 1978; Meshorer 1969; Sebesteny 1969; Van Kruiningen et al. 1978; Wagner et al. 1974).

Comparison with Other Species

Lesions are not associated with *S. muris* infections in rats and Syrian hamsters and are poorly characterized in Chinese hamsters. Some members of this genus are free-living, while others are parasitic in invertebrates and all classes of vertebrates, including frogs, birds, and mammals. *Hexamita meleagridis* causes catarrhal enteritis involving the upper small intestine of a variety of commercially important avian species, including turkeys, quail, pheasant, peafowl, and partridge. *Hexamita columbae* is an enteric pathogen of pigeons. Humans are not hosts for *Hexamita* , but *H. pitheci* has been reported in nonhuman primates (Flynn 1973; Kunstyr 1977; Levine 1973; Wagner et al. 1974).

References

Barthold SW (1986) Mouse hepatitis virus biology and epizootiology. In: Bhatt PN, Jacoby RO, Morse HC III, New AE (eds) Viral and mycoplasmal infections of laboratory rodents: effects on biomedical research, chap 25. Academic, Orlando, pp 571–601

Boorman GA, van Hooft JIM, van der Waaij D, van Noord MJ (1973) Synergistic role of intestinal flagellates and normal intestinal bacteria in a post-weaning mortality of mice. Lab Anim Sci 23:187–193

Brett SJ, Cox FE (1982) Immunological aspects of Giardia muris and Spironucleus muris infections in inbred and outbred strains of laboratory mice: a comparative study. Parasitology 85:85–99

Brugerolle G, Kunstyr I, Senaud J, Friedhoff KT (1980) Fine structure of trophozoites and cysts of pathogenic diplomonad Spironucleus muris. Z Parasitenkd 62:47–61

Csiza CK, Abelseth MK (1973) An epizootic of protozoan enteritis in a closed mouse colony. Lab Anim Sci 23:858–861

Exon JH, Patton NM, Koller LD (1975) Hexamitiasis in cadmium-exposed mice. Arch Environ Health 30:463–464

Flatt RE, Halvorsen JA, Kemp RL (1978) Hexamitiasis in a laboratory mouse colony. Lab Anim Sci 28:62–65

Flynn RJ (1973) Parasites of laboratory animals. Iowa State University Press, Ames

Keast D, Chesterman FC (1972) Changes in macrophage metabolism in mice heavily infected with Hexamita muris. Lab Anim 6:33–39

Kunstyr I (1977) Infectious form of Spironucleus (Hexamita) muris: banded cysts. Lab Anim ll:185–188

Kunstyr I, Ammerpohl E (1978) Resistance of faecal cysts of Spironucleus muris to some physical factors and chemical substances. Lab Anim 12:95–97

Kunstyr I, Ammerpohl E, Meyer B (1977a) Experimental spironucleosis (hexamitiasis) in the nude mouse as a model for immunologic and pharmacologic studies. Lab Anim Sci 27:782–788

Kunstyr I, Meyer B, Ammerpohl E (1977b) Spironucleosis in nude mice: an animal model for immuno-parasitologic studies. Proc Int Workshop Nude Mice 2:17–27

Levine ND (1973) Protozoan parasites of domestic animals and of man, 2nd edn. Burgess, Minneapolis

Lussier G, Loew FM (1970) An outbreak of hexamitiasis in laboratory mice. Can J Comp Med Vet 34:350–353

MacDonald TT, Ferguson A (1978) Small intestinal epithelial cell kinetics and protozoal infection in mice. Gastroenterology 74:496–500

Meshorer A (1969) Hexamitiasis in laboratory mice. Lab Anim Care 19:33–37

Owen RL, Nemanic PC, Stevens DP (1979) Ultrastructural observations on giardiasis in a murine model. I. Intestinal distribution, attachment, and relationship to the immune system of Giardia muris. Gastroenterology 76:757–769

Ruitenberg EJ, Kruyt BC (1975) Effect of intestinal flagellates on immune response of mice. Parasitology 71:30 (abstr)

Saxe LH (1954) Transfaunation studies on the host specificity of the enteric protozoa of rodents. J Protozool 1:220–230

Schagemann G, Bohnet W, Kunstyr I, Friedhoff KT (1990) Host specificity of cloned Spironucleus muris in laboratory rodents. Lab Anim 24:234–239

Sebesteny A (1969) Pathogenicity of intestinal flagellates in mice. Lab Anim 3:71–77

Sebesteny A (1974) The transmission of intestinal flagellates between mice and rats. Lab Anim 8:79–81

Sebesteny A (1979) Transmission of Spironucleus and Giardia spp. and some nonpathogenic intestinal protozoa from infested hamsters to mice. Lab Anim 13:189–191

Van Kruiningen HJ, Knibbs DR, Burke CN (1978) Hexamitiasis in laboratory mice. J Am Vet Med Assoc 173:1202–1204

Wagner JE, Doyle RE, Ronald NC, Garrison RG, Schmitz JA (1974) Hexamitiasis in laboratory mice, hamsters, and rats. Lab Anim Sci 24:349–354

Ward JM (1974) Naturally occurring Sendai virus disease of mice. Lab Anim Sci 24:938–942

Giardia muris Infection, Intestine, Mouse, Rat, and Hamster

Stephen W. Barthold

Synonym. Lamblia muris infection

Gross Appearance

Although mice infected with *Giardia* are usually asymptomatic, impairment of weight gain is the most common sign of infection. Severely affected mice are lethargic and have rough hair coats, distended abdomens, and malodorous diarrhea or soft feces. Increased mortality among weanling mice can occur. Gross lesions include a dilated small intestine containing watery fluid and gas. Pancreatic edema, ascites, and mesenteric lymphadenopathy have been described (Boorman et al. 1973; Csiza and Abelseth 1973; Roberts-Thomson et al. 1976a; Sebesteny 1969). As in mice, infections in rats and hamsters are usually asymptomatic, but clinical signs and lesions can occur. Aged hamsters with giardiasis may have chronic diarrhea and weight loss, with thickening of the bowel wall, particularly the cecum.

Microscopic Features

Trophozoites are visualized as 7 to 13 by 5 to 10 μm crescent-shaped structures on the brush border of the small intestine, but free organisms in various planes are also seen in the lumen (Fig. 381). They are found in the anterior fourth of the small intestine; most are concentrated along the periphery of Peyer's patches and the basal third of villi. Ovoid, 8 to 12 by 7 to 10 μm cysts containing four nuclei prevail in the large bowel. Mucosal lesions are often absent, but, if present, include villus blunting, crypt hyperplasia, and mononuclear leukocyte infiltration of the lamina propria and epithelium (Boorman et al. 1973; Brett and Cox 1982; Ferguson et al. 1980; Levine 1973; MacDonald and Ferguson 1978; Owen et al. 1979; Robert-Thomson et al. 1976a,b).

Ultrastructure

Giardia trophozoites are bilaterally symmetrical and piriform with broadly rounded anterior ends

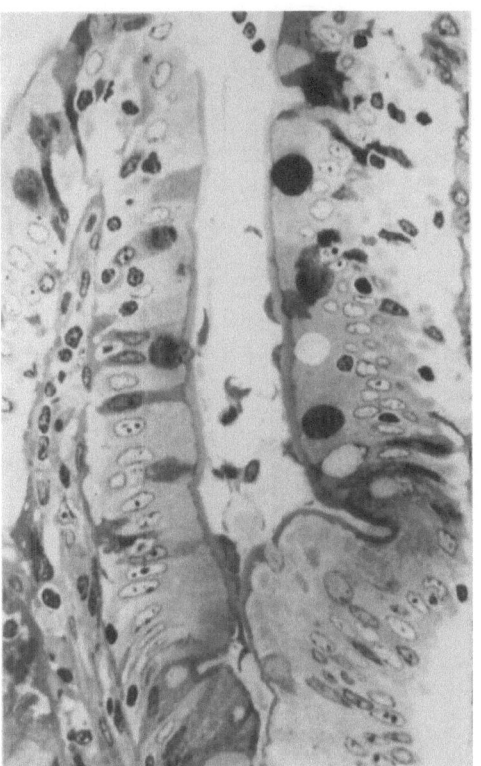

Fig. 381. Giardiasis, small intestine of a rat. *Giardia muris* trophozoites are attached to the brush border of the villi. Note the attachment of a lymphocyte to a detached trophozoite in the intervillar space. Epon, toluidine blue, ×400

trophozoites (Holberton 1973; Levine 1973; Owen et al. 1979).

Differential Diagnosis

Giardia, like *Spironucleus*, is probably an opportunistic pathogen under most circumstances, although further investigation of underlying causes is needed. Trophozoites can be readily identified by their morphology.

Biologic Features

Natural History

Giardia muris is transmitted by fecally excreted cysts. Following oral exposure, cysts appear in feces within 3–5 days, reaching maximal numbers between 7 and 14 days and waning by days 21–28. Marked variation occurs between host genotypes in the time course for elimination of the parasite. Mice infected at birth tend not to clear their infections, but passive immunity to infection is transferred in milk. Resistance to reinfection occurs in recovered mice. Athymic nude mice become persistently infected, without resistance to reinfection (Brett and Cox 1982; MacDonald and Ferguson 1978; Roberts-Thomson et al. 1976a,b; Roberts-Thomson and Mitchell 1978, 1979; Stevens et al. 1978a).

Pathogenesis

Giardia trophozoites reside almost exclusively in the unstirred layer of the anterior small intestine, either attached to mucosal epithelium or within the mucus coat. They adhere by an active suction process achieved with their ventral disk, apparently releasing and reattaching, leaving adhesion marks in the microvillus coat. Although they reside on the surface of the mucosa, they can occasionally be found between epithelial cells. Invasion of the mucosa occurs in heavy infections and, in some cases, organisms can be found in heart, liver, and brain. This invasion is commonplace following irradiation (Lupescu et al. 1970; Owen et al. 1979).

Clinical disease seems to be exacerbated by other underlying causes, including cold and crowding, coinfections with *Spironucleus*, X-irradiation, and

and extended, narrow posterior ends. The outstanding anatomic feature is the large ventral disk, which is supported by a microtubular cytoskeletal element, the striated disk (Fig. 382). The ventral disk is interrupted in its posterior aspect by a median notch. The body of the trophozoite lies above the ventral disk and protrudes over the sides of the disk as cytoplasmic flanges. The cytoplasm contains two axostyles, two anterior nuclei, endoplasmic reticulum, and small and large vesicles as well as endosymbiotic bacteria. They possess four pairs of flagellae that emerge at different locations, but extend posteriorly. Trophozoites attach to the brush border, with most intimate contact at the ventral disk rim, displacing microvilli and leaving temporary depressed rings (adhesive marks) visible by scanning electron microscopy. Lymphocytes are often in the lumen near or in direct contact with

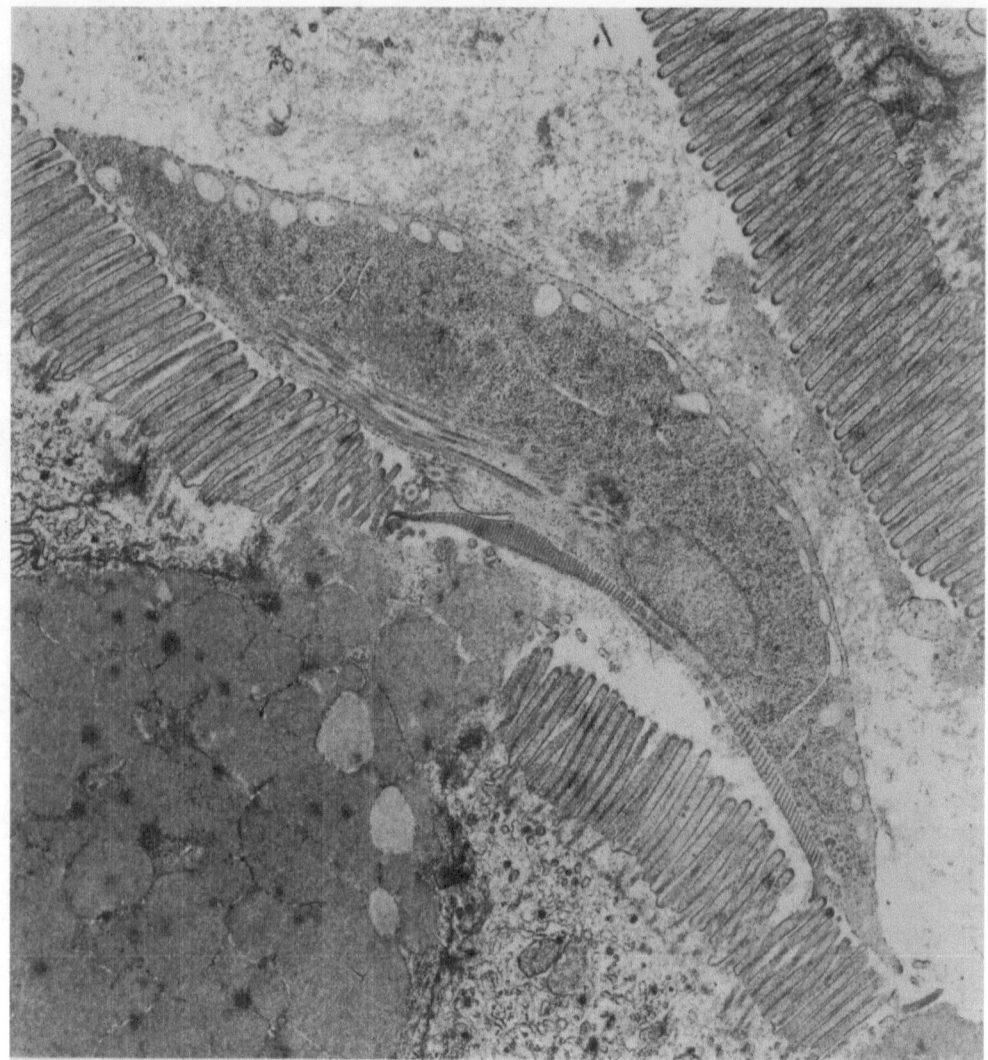

Fig. 382. *Giardia muris* trophozoite attached to the brush border of the intestinal epithelium in a rat. The ventral disk rim is displacing microvilli. Uranyl acetate-lead citrate, TEM, ×16 800

dietary changes (Boorman et al. 1973; Csiza and Abelseth 1973; Sebesteny 1969). Age, host genotype, and immune status are important in determining outcome in *G. muris* infections. Mice infected as neonates develop chronic infections, but protection during the neonatal period is provided by maternal immunoglobulin (Ig)A and IgG antibody (Andrews et al. 1980; MacDonald and Ferguson 1978; Stevens et al. 1978a). Weanling and adult mice usually overcome the infection after several weeks, but rate of recovery varies between genotypes. Among different strains tested, BALB/c mice recovered rapidly, while C57BL/B10, A and C3H/He mice had prolonged infections. Several other strains were intermediate (Brett and Cox 1982; Roberts-Thomson and Mitchell 1978). Recovery from primary infections and resistance to reinfection are due to a specific immune response to the parasite. The mechanisms of this immune response are not well understood. Mice develop circulating IgG and IgA antibody, and intestinal fluid contains low levels of anti-

Giardia secretory IgA (Anders et al. 1982; Brett and Cox 1982; Owen et al. 1979; Roberts-Thomson et al. 1976b; Roberts-Thomson and Mitchell 1979). Thymus-dependent immunity probably plays an important role. Unlike thymus-intact BALB/c mice, athymic BALB/c nude mice do not recover from primary infection and do not develop resistance. Reconstitution of these mice with lymphocytes from thymus-intact mice allows them to recover (Roberts-Thomson and Mitchell 1978; Stevens 1982; Stevens et al. 1978b). During the period of clearance of *G. muris*, intraepithelial and intraluminal lymphocytes increase and can be found in the bowel lumen (Brett and Cox 1982; MacDonald and Ferguson 1978; Owen et al. 1979). Chronic giardiasis is a common finding in aged Syrian hamsters, suggesting that hamsters do not clear the infection, unlike mice and rats (S.W. Barthold, unpublished observations).

Etiology

G. muris is a flagellated protozoan that naturally infects laboratory or wild mice, rats, and hamsters. There are many different named species of *Giardia* in animals and humans, with only minor structural differences. Different names have been given to isolates from different hosts, including *G. lamblia* (humans, nonhuman primates, beavers, swine), *G. bovis* (ox), *G. caprae* (sheep, goat), *G. equi* (horse), *G. canis* (dog), *G. cati* (cat), *G. caviae* (guinea pig), *G. duodenalis* (rabbit), and *G. chinchillas* (chinchilla). *G. muris* and *G. microti* have been recovered from house mice, and *G. muris* and *G. simoni* from Norway rats. Few cross-transmission studies have been performed, but it is clear that host specificity is not absolute. It has been suggested that, on a morphological basis, there are only two major species of *Giardia*, each with a number of races: *G. muris* in mice, rats, and hamsters, and *G. duodenalis* in other species. *G. muris* has small, round median bodies, while *G. duodenalis* has long, transverse, claw-shaped ones. Despite all of this, individual nomenclature continues to prevail (Flynn 1973; Levine 1973; Owen et al. 1979).

Frequency

G. muris is common in conventional laboratory mice and rats and is almost universal in hamsters.

Clinical signs or lesions associated with *G. muris*, on the other hand, are rare.

Comparison with Other Species

In most host species, *Giardia* seems to be only marginally pathogenic. As in rodents, *Giardia* in humans can be paradoxically asymptomatic or cause profound diarrhea and malabsorption. Small-bowel changes also vary from normal mucosa to severe villus attenuation, crypt hyperplasia, and chronic inflammation. Individuals with variable immunodeficiency (hypogammaglobulinemia), but not secretory or cellular immunodeficiency syndromes, appear more susceptible to *Giardia*. People with apparently normal immune function have a variable pattern of illness. Where *Giardia* is endemic with continuous exposure, disease is confined to children and the infection rate among adults is low. In epidemics, all ages are infected and disease is frequent among all ages (Knight 1980; Stevens 1982; Wright 1980). *G. muris* infection in the mouse has served as a highly analogous model for human giardiasis (Roberts-Thomson et al. 1976a).

References

Anders RF, Roberts-Thomson IC, Mitchell SF (1982) Giardiasis in mice: analysis of humoral and cellular immune responses to Giardia muris. Parasite Immunol 4:47–57

Andrews JS Jr, Ellner JJ, Stevens DP (1980) Purification of Giardia muris trophozoites by using nylon fiber columns. Am J Trop Med Hyg 29:12–15

Boorman GA, van Hooft JIM, van der Waaij D, van Noord MJ (1973) Synergistic role of intestinal flagellates and normal intestinal bacteria in post-weaning mortality of mice. Lab Anim Sci 23:187–193

Brett SJ, Cox FE (1982) Immunological aspects of Giardia muris and Spironucleus muris infections in inbred and outbred strains of laboratory mice: a comparative study. Parasitology 85:85–99

Csiza CK, Abelseth MK (1973) An epizootic of protozoan enteritis in a closed mouse colony. Lab Anim Sci 23:858–861

Ferguson A, Gillon J, al Thamery D (1980) Intestinal abnormalities in murine giardiasis. Trans R Soc Trop Med Hyg 74:445–448

Flynn RJ (1973) Parasites of laboratory animals. Iowa State University Press, Ames

Holberton DV (1973) Fine structure of the ventral disk apparatus and the mechanism of attachment in the flagellate Giardia muris. J Cell Sci 13:11–41

Knight R (1980) Epidemiology and transmission of giardiasis. Trans R Soc Trop Med Hyg 74:433–436

Levine ND (1973) Protozoan parasites of domestic animals and of man, 2nd edn. Burgess, Minneapolis

Lupescu GH, Radulescu S, Cernat MJ (1970) The presence of Lamblia muris in the tissues and organs of mice spontaneously infected. J Parasitol 56 [Suppl 2]:444–445

MacDonald TT, Ferguson A (1978) Small intestinal epithelial cell kinetics and protozoal infection in mice. Gastroenterology 74:496–500

Owen RL, Nemanic PC, Stevens DP (1979) Ultrastructural observations on giardiasis in a murine model. I. Intestinal distribution, attachment, and relationship to the immune system of Giardia muris. Gastroenterology 76:757–769

Roberts-Thomson IC, Mitchell GF (1978) Giardiasis in mice. I. Prolonged infections in certain mouse strains and hypothymic (nude) mice. Gastroenterology 75:42–46

Roberts-Thomson IC, Mitchell GF (1979) Protection of mice against Giardia muris infection. Infect Immun 24:971–973

Roberts-Thomson IC, Stevens DP, Mahmoud AA, Warren KS (1976a) Giardiasis in the mouse: an animal model. Gastroenterology 71:57–61

Roberts-Thomson IC, Stevens DP, Mahmoud AA, Warren KS (1976b) Acquired resistance to infection in an animal model of giardiasis. J Immunol 117:2036–2037

Sebesteny A (1969) Pathogenicity of intestinal flagellates in mice. Lab Anim 3:71–77

Stevens DP (1982) Giardiasis: host-pathogen biology. Rev Infect Dis 4:851–858

Stevens DP, Frank DM, Carpenter CC (1978a) Local immunity in murine giardiasis: is milk protective at the expense of maternal gut? Trans Assoc Am Phys 91:268–272

Stevens DP, Frank DM, Mahmoud AA (1978b) Thymus dependency of host resistance to Giardia muris infection: studies in nude mice. J Immunol 120:680–682

Wright SG (1980) Giardiasis and malabsorption. Trans R Soc Trop Med Hyg 74:436–437

The Large Intestine

BACTERIAL INFECTION

Coliform Typhlocolitis, Immunodeficient Mice

Stephen W. Barthold

Synonym. Cecocolitis in immunodeficient mice associated with *Escherichia coli*

Gross Appearance

Infected mice have pasty feces. The colonic mucosa is segmentally thickened and opaque, with occasional blood-tinged contents. The cecum can also be thickened and contracted (Waggie et al. 1988). Chronic body weight loss and rectal prolapse can also be present (S.W. Barthold, unpublished observations).

Microscopic Features

Characteristic histopathology consists of segmental areas of mucosal hyperplasia and inflammation in the cecum and colon. Crypts are elongated and populated with immature, mitotically active enterocytes, and the lamina propria is variably infiltrated with mixed populations of leukocytes (Fig. 383). Areas of mucosal erosion and ulceration can also be present. Typically, the mucosa of both the large and small intestine has segmental, but prominent colonization of the surface epithelium with 3.0 × 0.5-µm, gram-negative coccobacillary bacteria, as well as invasion of degenerating surface enterocytes with bacteria (Waggie et al. 1988).

Ultrastructure

Limited electron microscopy evaluation has been performed, demonstrating intracellular bacteria to be surrounded by electronlucent spaces and not confined to membrane-lined vacuoles. Features of bacterial attachment to the cell surface were not mentioned (Waggie et al. 1988).

Differential Diagnosis

The lesions of coliform typhlocolitis are highly reminiscent of *Citrobacter freundii*-induced colitis, including heavy colonization of surface epithelium with coccobacillary organisms. These organisms can be differentiated by culture and species identification. The *Escherichia coli* responsible for coliform typhlocolitis is atypical in that it does not ferment lactose, thus distinguishing it from more common commensal *E. coli* normally found in mouse intestines and from pathogenic *Citrobacter freundii* (4280). The critical element in coliform typhlocolitis is the immunodeficient status of the mice. The only published report of this syndrome was associated with triple-immune deficient N:NIH(S)III mice, which are deficient in T, B, and NK cells. It could not be reproduced in T cell-deficient nude mice (Waggie et al. 1988). It has also been confirmed in severe combined immunodeficient (SCID) mice on a number of occasions (S.W. Barthold, unpublished observations). Enterotropic mouse hepatitis virus causes similar lesions (without attaching bacteria) in athymic nude mice (see "Mouse Hepatitis Virus Infection, Intestine, Mouse," p. 379, this volume).

A number of typhlocolitides of immunodeficient mice have not been fully characterized, but share similar hyperplastic and inflammatory features. A cluster of simultaneous publications described segmental, hyperplastic typhlocolitis or colitis in mice with gene-targeted disruption of cytokine (interleukin-2, interleukin-10), T cell receptor (TCR-α, TCR-β, TCR-β/γ double mutant), and class II major histocompatibility complex (MHC) genes. No disease was noted in recombination-activating (RAG-1) mutant mice (Kuhn et al. 1993; Mombaerts et al. 1993; Sadlack et al. 1993). In all of these reports, disease was attributed to the immune deficiency, but none properly ruled out an underlying infectious etiology. A similar syndrome involving TCR-α/β mutant mice has

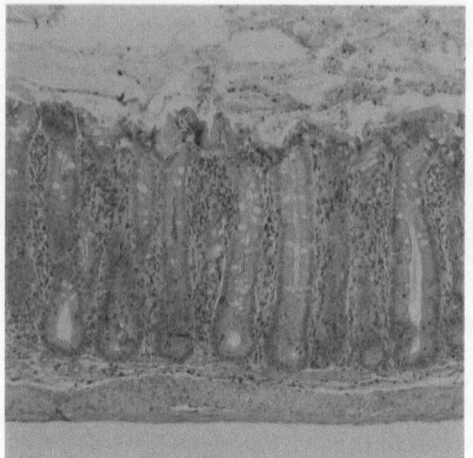

Fig. 383. Colonic mucosa of a severe combined immunodeficient mouse with naturally occurring coliform typhlocolitis. The mucosa is markedly hyperplastic

been evaluated in this laboratory. No known pathogen, including *E. coli*, *Citrobacter*, or viruses could be identified (S.W. Barthold, unpublished observations). However, morphological features of these parallel syndromes strongly suggest an as yet undefined infectious agent that is opportunistic in immunodeficient mice. A recently described, apparently genetically determined syndrome in C3H/HeJ mice is characterized by soft feces, perianal ulceration, and right-sided hyperplastic colitis (Sundberg et al. 1994). Other differential considerations include better-known murine enteropathogens, such as *Salmonella enteritidis* and *Bacillus piliformis* (*Clostridium piliforme*). Older reports of pathogenic *E. coli* in mice must be viewed with caution, as *E. coli* overgrowth is a common feature in underlying viral enteritides of mice.

Biologic Features

Natural History

Although there is only one report of coliform typhlocolitis in the literature (Waggie et al. 1988), it is apparently found but not reported in other immunodeficient mouse colonies (S.W. Barthold, unpublished observations). The source of the *E. coli* is not known.

Pathogenesis

Coliform typhlocolitis shares morphological features with other enteropathogenic coliform bacterial syndromes in mice and other species, but the pathogenesis of this syndrome has not yet been characterized.

Etiology

The *E. coli* associated with this disease is atypical compared to other murine *E. coli* isolates, as it does not ferment lactose but the organism is easily grown on conventional media.

Frequency

This agent is apparently pathogenic only in immunodeficient mice; thus its prevalence in immunocompetent mouse colonies is likely to be unrecognized. Sporadic outbreaks have been observed in colonies of immunodeficient mice (S.W. Barthold, unpublished observations), but its overall frequency is unknown.

Comparison with Other Species

Coliform typhlocolitis is analogous to enteric disease in mice (transmissible murine colonic hyperplasia) and other species caused by attaching and effacing gram-negative enterobacteria (Moon et al. 1983), but is unique in its being restricted to immunodeficient mice. Attaching and effacing lesions have several characteristics, including intimate bacterial adherence to enterocytes with dissolution of the brush border and cytoskeletal rearrangements in the cytoplasm underlying the sites of attachment (Moon et al. 1983). These features are also likely in coliform typhlocolitis, but have not been described.

References

Kuhn R, Lohler J, Rennick D, Rajewsky K, Muller W (1993) Interleukin-10-deficient mice develop chronic enterocolitis. Cell 75:263–274

Mombaerts P, Mizoguchi E, Grusby MJ, Glimcher LH, Bhan AK, Tonegawa S (1993) Spontaneous development of inflammatory bowel disease in T cell receptor mutant mice. Cell 75:274–282

Moon HW, Whipp SC, Argenzio RA, Levine MM, Giannella RA (1983) Attaching and effacing activities of rabbit and human enteropathogenic Escherichia coli in pig and rabbit intestines. Infect Immun 41:1340–1351

Sadlack B, Merz H, Schorle H, Schimpl A, Feller AC, Horak I (1993) Ulcerative colitis-like disease in mice with a disrupted interleukin-2 gene. Cell 75:253–261

Sundberg JP, Elson CO, Bedigian H, Birkenmeier EH (1994) Spontaneous, heritable colitis in a new substrain of C3H/HeJ mice. Gastroenterology 107:1726–1735

Waggie KS, Hansen CT, Moore TD, Bukowski MA, Allen AM (1988) Cecocolitis in immunodeficient mice associated with an enteroinvasive lactose negative E. coli. Lab Anim Sci 38:389–393

NEOPLASMS

Adenocarcinoma, Colon and Rectum, Rat

Paul M. Newberne and Adrianne E. Rogers

Synonyms. Colon carcinoma, rectal carcinoma, cancer of the large bowel

Gross Appearance

Spontaneous tumors of the colon are such rare events in rats and mice that little knowledge about them is available. Several carcinogens, to be described, have been used to induce malignant tumors in the colon and rectum of rats and mice. It is these induced neoplasms that will be considered. Each of these carcinogens produces a similar and predictable series of sequential changes leading to polypoid tumors (Fig. 384), often associated with an intussuspection (Fig. 385), and sessile tumors (Fig. 386), the latter are often mucinous and locally invasive and may metastasize to mesenteric nodes, liver, and lung.

As indicated in Fig. 387, tumors arise in greatest numbers in the distal colon but can occur throughout the colon, particularly if a relatively large dose of carcinogen is administered. In the rat, the invasive mucinous adenocarcinomas tend to be localized in the proximal colon (Nauss et al. 1983; Takemiya et al. 1982), while polypoid tumors are found distally.

Microscopic Features

The earliest lesions appear at or near the mucosal surface as dysplastic glands with abnormal segments of gland continuing in direct apposition with the morphologically normal portion of the gland (Fig. 388). These progress to larger areas with neoplastic glands intermixed with normal-appearing glands (Fig. 389). They are later drawn up into polypoid structures (Fig. 390), some of which exhibit invasion of the stalk (Fig. 391). In rats, the localization of mucinous colon tumors over lymphoid aggregates (Fig. 392), has been rec-

ognized for 10 years (Rogers et al. 1973; Bland and Britton 1984) and recently has been documented in detail. Often, the earliest dysplastic mucosa is found over an aggregate (Nauss et al. 1984). Similar localization has been reported in mice (Wargovich et al. 1983).

The sessile-type adenocarcinomas in rats may produce little mucin; they have well-developed glands and grow exophytically. More often, they develop in the submucosa and are comprised of large mucin-filled glandular structures (Fig. 393). Still others produce a large amount of mucin and are partially or entirely comprised of signet-ring cells with little or no gland formation (Fig. 394). The mucin-producing tumors tend to occur in rats given higher doses of 1,2-dimethylhydrazine (DMH) or azoxymethane (AOM) (Takemiya et al. 1982; Nauss et al. 1983). In both types of tumors and in dysplastic colon epithelium prior to tumor development, abnormalities of mucin synthesis and secretion are prominent.

Biologic Features

Natural History

Signs of developing tumors in rats may go undetected, but close daily observation of freshly voided feces will often reveal bright-red blood. Occult blood tests have not, in our hands, proven to be any better at detection than simple visual observations. The general clinical condition of the rat does not deteriorate until tumor development is well advanced; death may result from massive bleeding or obstruction. Clinical signs are of little help in estimating stage of tumor growth.

The colon epithelium over the lymphoid aggregates in normal rats appears equipped to take up and digest bacteria or other antigens present in the lumen; the antigens may be passed to the adjacent lymphoid cells for processing. Analogies can be

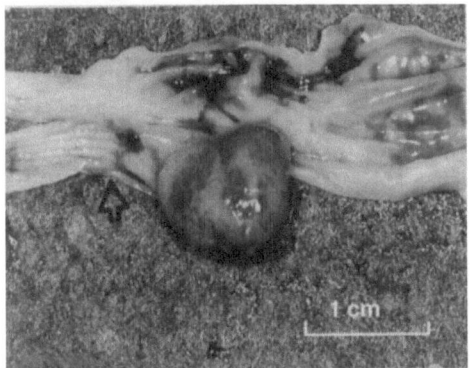

made to the small intestinal epithelium over Peyer's patches (Bland and Britton 1984). It appears possible that this epithelium may differ metabolically from the epithelium in other colon segments. There may be differences in vascular supply to the colon segments that contain lymphoid aggregates, or the lymphoid tissue may modulate the development of intestinal tumors by secretion of lymphokines or by other interactions. An improved prognosis is associated with an active acute and chronic inflammatory response in human colon cancer (Braun and Harris 1981). This inflammatory response is, of course, histologically distinct from lymphoid aggregates and has not found a parallel in the rodent model, but further research may reveal an association.

Etiology

Two types of chemicals have been used to induce adenocarcinomas in the colon of rodents as models for the human disease. One type is the direct-acting (complete) carcinogen, which requires no metabolic activation. The two members of this group used by most investigators are methylnitrosourea (MNU) and methylnitronitrosoguanidine (MNNG). MNU is the more useful of the two; it can be given as a single dose or a small

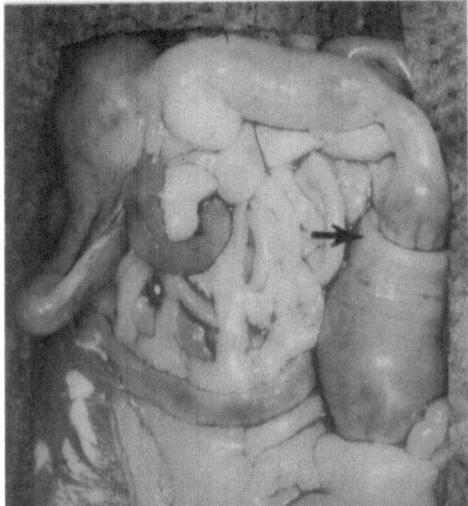

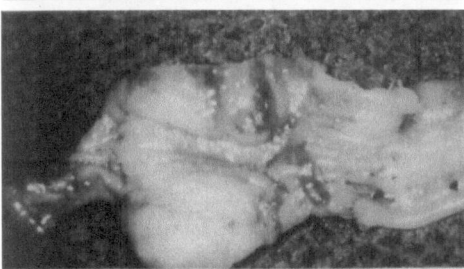

Fig. 384. (*above*) A typical polypoid tumor induced in the rat colon by chemical carcinogens. This one was induced by 1,2-dimethylhydrazine (DMH)

Fig. 385. (*middle*) Intussusception of the colon (*arrow*) due to the presence of a large polypoid tumor, a terminal event in many large neoplasms of this type which develop in the mid- or distal colon

Fig. 386. (*below*) A sessile tumor induced by 1,2-dimethylhydrazine (DMH) in the rat colon. This encompasses most of the circumference of the colon, resulting in a rigid tube with constriction of the lumen

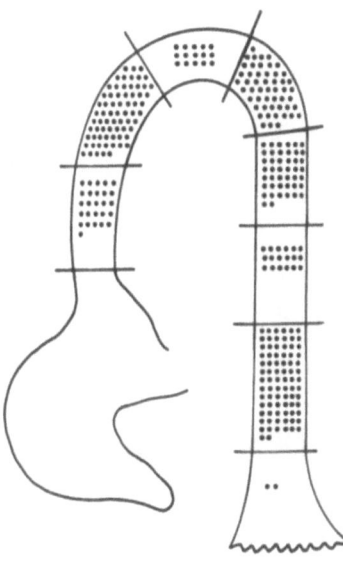

Fig. 387. Distribution of tumors in the colon from one study. Each *closed circle* represents one tumor in one animal in a total of 240 rats

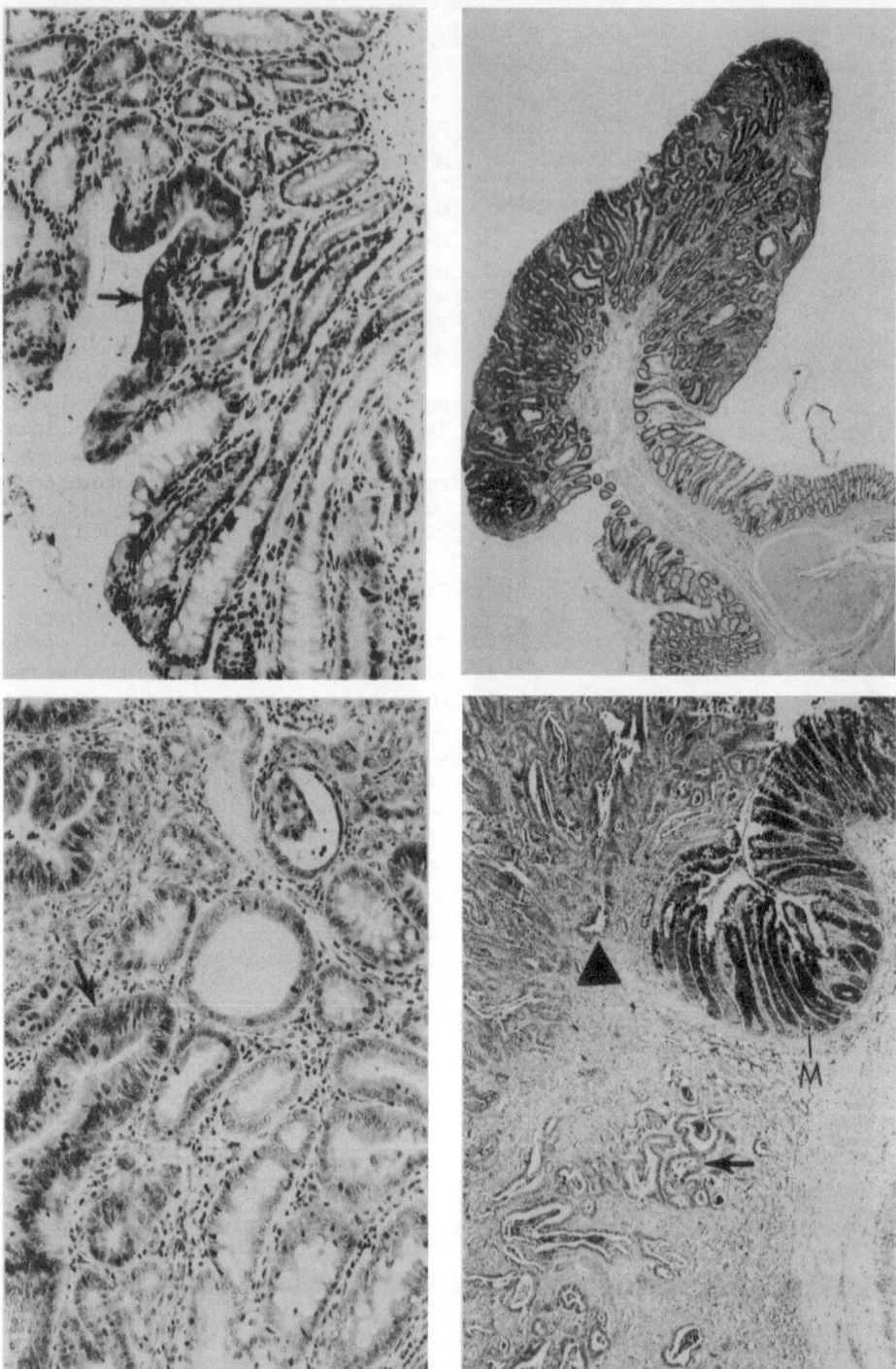

Fig. 388. (*upper left*) A focus of dysplastic epithelium (*arrow*) of the colon adjacent to the normal epithelium. These are situated near the luminal aspect of the mucosa. H&E, ×250

Fig. 389. (*lower left*) Early development of adenocarcinoma of the colon with dysplastic or neoplastic glands (*arrow*) adjacent to normal-appearing mucosa. These foci develop on or near the surface of the mucosa and then appear to be drawn up into polypoid structures, as shown in Figs. 390 and 391. H&E, ×250

Fig. 390. (*upper right*) Polypoid adenocarcinoma (in situ) without detectable invasive properties. H&E, ×25

Fig. 391. (*lower right*) Polypoid adenocarcinoma (*arrow*) invading the stalk of the tumor (*arrowhead*). The more normal-appearing adjacent mucosta (*M*) exhibits characteristic periodic acid-schiff (PAS)-positive staining; the neoplasm does not. PAS, ×100

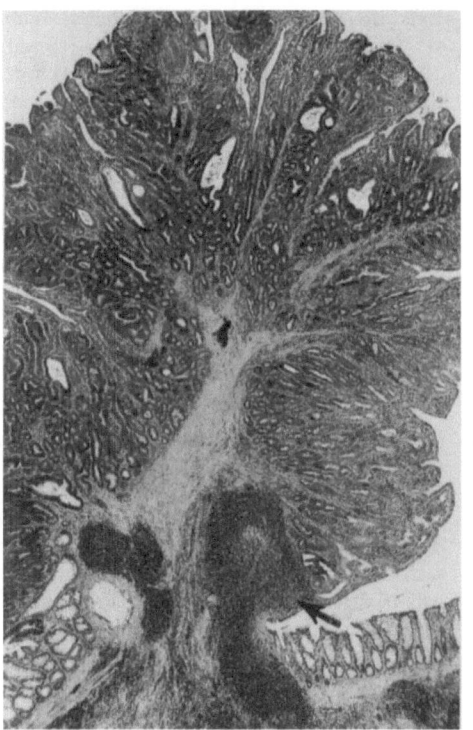

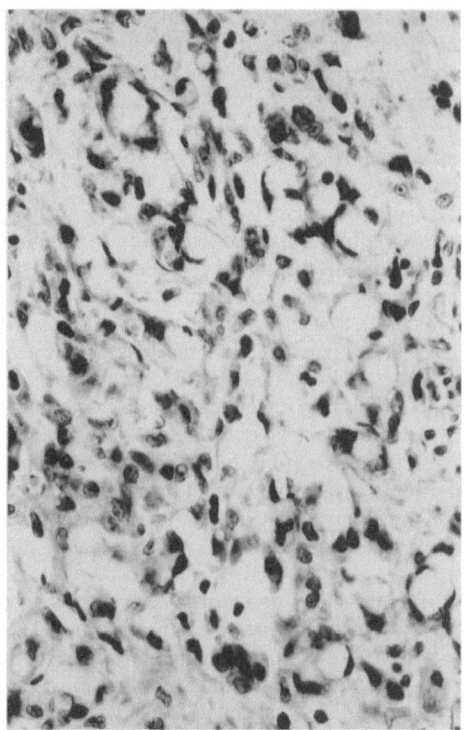

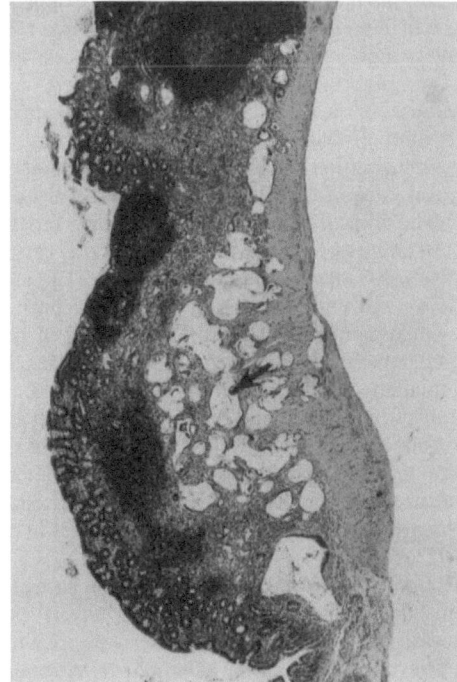

▶

Fig. 392. (*upper left*) A polypoid tumor that has developed over a lymphoid aggregate in the colon, rat. The lymphoid tissue (*arrow*) has moved up from the submucosa to surround partially and infiltrate the stalk of the neoplasm. H&E, ×25

Fig. 393. (*below*) Sessile tumor comprised of mucin-filled glandular structures (*arrow*). Note the association with the lymphoid aggregates. H&E, ×225

Fig. 394. (*upper right*) Mucin in signet-ring cells in a sessile adenocarcinoma, arising in the mucosa of the colon, rat. H&E, ×350

number of doses, which permits accurate studies of factors that influence tumor development before or after exposure to carcinogen. Direct-acting compounds are useful also to separate factors that influence carcinogen metabolism from factors that influence direct interactions between the cell and the active carcinogenic species.

The other group of carcinogens, members of which require metabolic activation, is composed primarily of DMH and its metabolites, AOM and methylazoxymethanol administered as the acetate (MAMA). There are other carcinogens for the large bowel, but they have not been widely used. AOM is useful to study the promotion or inhibition of tumors by dietary or other factors. Administration of 15 mg/kg body weight, two doses at a 2-week interval (total of 30 mg/kg body weight) permits an evaluation of factors superimposed on the carcinogen, because with this regimen the carcinogenic process is slowed, thus facilitating studies on modulating factors.

DMH and its two derivatives are procarcinogens highly specific for intestinal mucosal cells. Their relative potencies are determined by their chemical reactivity and metabolism to the ultimate carcinogen in the target cell. Toxicity is avoided by using small doses. All three of these carcinogens are highly effective in inducing tumors in a relatively short period. Cancer of the large bowel and other gastrointestinal tract neoplasms have been induced in the mouse, rat, and hamster by administration of DMH (Rogers et al. 1973).

Administration by gastric intubation of 10-weekly doses each of DMH (30 mg/kg body weight) dissolved in physiologic saline to weanling rats results in virtually 100% incidence of colon cancer within 5–7 months following exposure. The incidence and rate of appearance can be modulated by treatment schedule and total amount of carcinogen administered. DMH can be given also by subcutaneous administration, while MNU is administered intrarectally directly to the target site with the rat under light anesthesia. Details of the design, including treatment schedule and amount of carcinogen, are available in referenced publications (Nauss et al. 1983; Rogers and Newberne 1973).

Rodent models for colon carcinogenesis have produced important information on factors that retard or enhance tumorigenesis as well as on histology and biochemistry of tumor development. Significant information relative to human colon cancer etiology or factors influencing it has derived from studies using this model. For example, we (Nauss et al. 1987; Newberne and Schrager 1993) have been unable to confirm an effect of dietary fat on the development of colon cancer. On the other hand, we have demonstrated a protective effect of red meat (lean beef) on the induction of colon cancer. These important findings are in contrast to reports of others that dietary fat significantly enhances colon cancer (Reddy et al. 1977) and that red meat increases the risk of colon cancer in humans (Willett 1994). This animal model allows follow-up to be done on important questions raised by epidemiologic observations and mechanisms of carcinogenesis to be explored. Several basic research needs should be resolved in further development of animal models. Major questions remain regarding the role of dietary fat, fiber, minerals, and vitamins in modulating tumor growth as well as the influence of other bowel diseases and of immunologic responses. Investigation of the interactions of environmental chemicals in tumor induction and growth should yield information of value.

Comparison with Other Species

Cancer of the colon and rectum, the second most common visceral cancer in the United States, is a major cause of human morbidity and mortality. About 100000 Americans are diagnosed annualy as having adenocarcinoma of the large bowel, and about half of these die from the disease (Silverberg 1984). This is second in significance only to lung cancer. As with other types of neoplasia, little is known about the etiology of colon carcinoma.

Long-standing ulcerative colitis and other types of inflammatory bowel disease (Yardley et al. 1983), some forms of polyps (Lynch et al. 1979), dietary factors (Newberne 1985), and deranged immunocompetence (Svennevig et al. 1982) have all been linked to colon cancer in one way or another, but the significance of these as etiologic agents or contributing influences is still unclear.

Abnormalities of mucin synthesis and secretion are conspicuous features of induced colonic adenocarcinoma of rats, both in tumors with well-developed glands as well as those that produce large amounts of mucin. Dysplastic colonic epithelium has similar defects of mucin production, even prior to the development of tumor.

Similar abnormalities have been reported in human tumors. The rodent models also show changes in mucosal binding of lectins similar to changes reported in human tumors (Kim and McIntyre 1983).

Tumor location in mice is predominantly in the distal large bowel (Izumi et al. 1979).

1,2-Dimethylhydrazine-induced colon adenocarcinomas in rats are similar grossly and histologically to human colon tumors. A few adenomas have been diagnosed in rats, on the basis of histologic evidence of benign growth and absence of invasion through the muscularis mucosae. Progression from benign to malignant lesions has been described by others (Madara et al. 1983). The rodent adenomas do not exactly mimic tubular or tubulovillous polyps in the human, and the evidence for progression is relatively weak since malignant tumors are usually found earlier than or at the same time as adenomas (Maskens and Dujardin-Loits 1981; Madara et al. 1983; Takemiya et al. 1982). The high dose of carcinogen used in most studies may preclude observation of the sequence of dysplasia, adenoma, and carcinoma that appears to occur in development of at least some colon tumors in humans.

References

Bland PW, Britton DC (1984) Morphological study of antigen-sampling structures in the rat large intestine. Infect Immun 43:693–699

Braun DP, Harris JE (1981) Relationship of leukocyte numbers, immunoregulatory cell function, and phytohemagglutinin responsiveness in cancer patients. J Natl Cancer Inst 67:809–814

Izumi K, Otsuka H, Furuya K, Akagi A (1979) Carcinogenicity of 1,2-dimethylhydrazine dihydrochloride in BALB/c mice. Influence of route of administration and dosage. Virchows Arch (Pathol Anat) 384:263–267

Kim YS, McIntyre LJ (1983) Biochemical changes in experimental colon carcinogenesis. In: Autrup H, Williams GM (eds) Experimental colon carcinogenesis. CRC, Boca Raton, chap 10

Lynch HT, Lynch PM, Follett KL, Harris RE (1979) Familial polyposis coli: heterogeneous polyp expression in two kindreds. J Med Genet 16:1–7

Madara JL, Harte P, Deasy J, Ross D, Lahey S, Steel G Jr (1983) Evidence for an adenoma-carcinoma sequence in dimethylhydrazine-induced neoplasms of rat intestinal epithelium. Am J Pathol 110:230–235

Maskens AP, Dujardin-Loits RM (1981) Experimental adenomas and carcinomas of the large intestine behave as distinct entities: most carcinomas arise de novo in flat mucosa. Cancer 47:81–89

Nauss KM, Locniskar M, Newberne PM (1983) Effect of alterations in the quality and quantity of dietary fat on 1,2-dimethylhydrazine-induced colon tumorigenesis in rats. Cancer Res 43:4083–4090

Nauss KM, Locniskar M, Pavlina T, Newberne PM (1984) Morphology and distribution of 1,2-dimethylhydrazine dihydrochloride-induced colon tumors and their relationship to gut-associated lymphoid tissue in the rat. J Natl Cancer Inst 73:915–924

Nauss KM, Bueche D, Newberne PM (1987) Effect of beef fat on DMH-induced colon tumorigenesis: influence of rat strain and nutrient composition. J Nutr 117:739–747

Newberne PM (1985) Influence of nutrition, immunologic status and other factors on the development of cancer. In: Clayson DB, Krewski D, Munro I (eds) Toxicological risk assessment, vol II. CRC, Boca Raton, chap 3

Newberne PM, Schrager TF (1993) Lipids, lipotropes and malignancy. In: Parke D, Ioannides C, Walker R (eds) Food, nutrition, and chemical toxicity. Smith-Gordon Nishimura, London, pp 227–247

Reddy BS, Watanabe K, Weisburger JH (1977) Effect of high-fat diet on colon carcinogenesis in F344 rats treated with 1,2-dimethylhydrazine, methylazoxymethanol acetate or methyl-nitrosourea. Cancer Res 37:4156–4159

Rogers AE, Newberne PM (1973) Dietary enhancement of intestinal carcinogenesis by dimethylhydrazine in rats. Nature 246:491–492

Rogers AE, Herndon BJ, Newberne PM (1973) Induction by dimethylhydrazine of intestinal carcinoma in normal rats and in rats fed high or low levels of vitamin A. Cancer Res 33:1003–1009

Silverberg E (1984) Cancer statistics, 1984. CA 34:7–23

Svennevig JL, Lunde OC, Holter J (1982) In situ analysis of the inflammatory cell infiltrate in colon carcinomas and in the normal colon wall. Acta Pathol Microbiol Immunol Scand [A] 90:131–137

Takemiya M, Miyayama H, Takeuchi T (1982) Pathogenesis of 1,2-dimethylhydrazine-induced carcinomas in rat intestine. I. The induction of mucin producing carcinomas in the rat intestine. Acta Pathol Jpn 32:257–264

Wargovich MJ, Medline A, Bruce WR (1983) Early histopathologic events to evolution of colon cancer in C57BL/6 and CF1 mice treated with 1,2-dimethylhydrazine. J Natl Cancer Inst 71:125–131

Willett (1994) Diet and health: what should we eat. Science 264:532–537

Yardley JH, Ransohoff DF, Riddell RH, Goldman H (1983) Incidence of inflammatory bowel disease: going up or down? Gastroenterology 85:196–200

Subject Index*

---

*Page numbers in **boldface** indicate the principal discussion; Figures are designated by the letter "f" following the page number; Tables are found on page numbers followed by the letter "t".

Springer
and the
environment

At Springer we firmly believe that an
international science publisher has a
special obligation to the environment,
and our corporate policies consistently
reflect this conviction.
We also expect our business partners –
paper mills, printers, packaging
manufacturers, etc. – to commit
themselves to using materials and
production processes that do not harm
the environment. The paper in this
book is made from low- or no-chlorine
pulp and is acid free, in conformance
with international standards for paper
permanency.

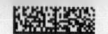